AF905546

Supported Metal Single Atom Catalysis

Supported Metal Single Atom Catalysis

Edited by Philippe Serp and Doan Pham Minh

WILEY-VCH

Editors

Prof. Philippe Serp
Laboratoire de Chimie de Coordination
UPR8241 CNRS
4 Allée Emile Monso
31030 Toulouse Cedex 4
France

Dr. Doan Pham Minh
Université de Toulouse
IMT Mines Albi, UMR CNRS 5302
Campus Jarland
81013 Albi cedex 09
France

Cover image: Courtesy of Dr. Javier Navarro-Ruiz

All books published by **WILEY-VCH** are carefully produced. Nevertheless, authors, editors, and publisher do not warrant the information contained in these books, including this book, to be free of errors. Readers are advised to keep in mind that statements, data, illustrations, procedural details or other items may inadvertently be inaccurate.

Library of Congress Card No.: applied for

British Library Cataloguing-in-Publication Data
A catalogue record for this book is available from the British Library.

Bibliographic information published by the Deutsche Nationalbibliothek
The Deutsche Nationalbibliothek lists this publication in the Deutsche Nationalbibliografie; detailed bibliographic data are available on the Internet at <http://dnb.d-nb.de>.

© 2022 WILEY-VCH GmbH, Boschstr. 12, 69469 Weinheim, Germany

All rights reserved (including those of translation into other languages). No part of this book may be reproduced in any form – by photoprinting, microfilm, or any other means – nor transmitted or translated into a machine language without written permission from the publishers. Registered names, trademarks, etc. used in this book, even when not specifically marked as such, are not to be considered unprotected by law.

Print ISBN: 978-3-527-34844-2
ePDF ISBN: 978-3-527-83015-2
ePub ISBN: 978-3-527-83017-6
oBook ISBN: 978-3-527-83016-9

Typesetting Straive, Chennai, India
Printing and Binding CPI Group (UK) Ltd, Croydon, CR0 4YY

Printed on acid-free paper

C109738_030222

Contents

Foreword

Avelino Corma

Instituto de Tecnología Química, Universitat Politècnica de València - Consejo Superior de Investigaciones Científicas, Avenida de los Naranjos s/n, 46022 Valencia, Spain

It is always very interesting and most rewarding to look back in the history and to see how science and scientific research has developed with time. In a subject such as metal catalysis, research on the nature of the active sites and the mechanism of catalytic process had already started at the beginning of the nineteenth century when L.J. Thénord, Sir Humphry Davy, E. Davy, J.W. Döberciner, and M. Faraday, among others, showed that noble metals as well as finely divided Cu and Fe were able to accelerate the rate of decomposition of H_2, methane, alcohols, and ammonia. This phenomenon was named "catalysis" by J.J. Berzelius, and it was demonstrated that the catalyst does not modify the thermodynamic equilibrium but accelerates the rate toward achieving the equilibrium.

It was at the end of the nineteenth and early twentieth centuries that the work of P. Sabatier on the hydrogenation of unsaturated hydrocarbons, later extended to other molecules, on finely divided metallic nickel as well as on Cu, Fe, Co, and Pt, opened new perspectives. Very soon the importance of the adsorption of reactants on the catalyst surface was recognized as a key issue for the catalysis to occur, as was pointed out already by A. Fusieri, J.W. Döbereinter, M. Faraday, J. Liebig, and P. Sabatier. The latter also claimed that the metal catalyst will be more active when its surface is larger and "in all cases, the catalyst intervened in the reaction by itself or by one of its derivatives giving a transitory and unstable surface product that reacts very fast with the other components of the system to generate the product." The author belongs to the group of scientists that believe that there is a unified mechanism for all the forms of catalysis.

It was in early twentieth century when researchers started to discuss on the nature of the adsorbed species on the surface of the catalyst. In that respect, H. Taylor presented in 1924 the idea that the solid catalysts show heterogeneities on the surface, which can be related with the active centers. Langumir indicated that metallic atoms on the surface have lower coordination than those present in the bulk, and consequently their adsorption and reactivity toward the reactant molecules will be very different.

At this point, two key concepts on metal catalysis were already established. (i) Higher metal surface area gives higher catalytic activity. (ii) Surface metal atoms with lower coordination are responsible for catalysis.

Three prominent catalysis scientists – M. Boudart, G. Ertl, and G. Somorjai – pushed forward the heterogeneous catalysis on metals in the second half of the twentieth century. They established basic principles that are used even today, and sometimes renamed, to explain metal catalysis. Concepts involving reaction sensitivity to crystal size, changes in adsorption and reactivity on different metal crystal facets, and surface dynamics of the metal catalysts and the adsorbed species were already discussed and may need to be revisited today.

It is clear that the synthesis and application of metal and oxide nanomaterials, i.e. materials with well-defined morphologies and with dimensions of a few nanometers, have been present in catalysis by metals and oxides in the last 100 years. This is certainly the case for supported metal catalysts where one of the objectives was to reduce the size of the crystal, following the previously established concept that the higher the number of atoms at the surface, the higher will be the chemical interactions and the reactivity at the solid–gas or solid–liquid interface.

It appears reasonable to think that 50 years ago, researchers were probably preparing supported metal catalysts where some single isolated metal atoms, together with subnanometric metal clusters and nanoparticles, were present. Unfortunately, the characterization techniques available at that time were not able to detect single isolated atoms, if any. The direct visualization probably could not be achieved with nanoclusters with 10 atoms or less either. It is possible that if a fraction of such subnanometric species were present in the catalyst, chemisorption measurements could be sensitive to those species and could explain in more detail why the reactivity for some reactions varies linearly with the number of surface atoms determined by H_2 chemisorption, while for other reactions (those sensitive to the structure of the metal species) the variation was exponential. Boudart's explanation about heterogeneity of metal sites in the supported metal catalysts and the relative variation of catalytic activity with particle size could today be revisited on the basis of considering the co-presence of single isolated metal atoms, subnanometric metal clusters, together with nanoparticles in different catalysts. Furthermore, the contribution of different types of metal species in the solid catalyst may greatly depend on the catalyst preparation method. Taking into account these points, one should logically conclude that not all the metal entities named above should be active, or intrinsically equally active, for all reactions.

Today, we can directly visualize supported single-metal atoms, nanoclusters, and nanoparticles, and it is possible to "see" the size and the shape of the metal entities. Moreover, it is now possible to determine the coordination environment of the metal atoms in supported monometallic or bimetallic catalysts. Techniques such as transmission electron microscopy (TEM), X-ray absorption spectroscopy (XAS), infra-red spectroscopy (IR), X-ray photoelectron spectroscopy (XPS), and Mössbauer spectroscopies can give excellent detailed information at the atomic level. With this information and the advances made on theoretical calculations, it is possible to model the different potential catalytic entities and their chemical reactivity. While this type of work is essential to establish structure–reactivity correlations, we have learned from our predecessors about the dynamic character of the metal catalysts during the catalytic reaction. Then, it becomes essential to figure out the active species under the reaction conditions by means of operando characterization techniques. Fortunately, in situ TEM and near-ambient present XPS, as well as operando XAS techniques are available to elucidate how the supported metal entities may transform during the catalytic process. Based on those structural transformations, the system can be modeled in a more realistic way. With these tools, we are now in a much better position than our predecessor to ask the right questions and look for explanations to the experimental observations. More than that, we are in a good position, starting from the fundamentals, to build the necessary knowledge on surface reactivity to predict the catalysts to be made for a particular reaction, and then to perform the "a priori" synthesis. To achieve that goal, the synthesis of solid catalysts with well-defined single or multiple sites will be the target. This target has been realized in the case of structured micro- and mesoporous materials, including zeolites as a paradigmatic case, where well-defined single metal atoms such as Ti, Sn, Fe, Zn, and Mo have been incorporated into the framework for unique reactivity.

In the case of transition metals (for instance Au, Pt, Pd, Rh, Ir, Ni, and Cu) on other supports (for instance carbon-based materials and oxides), it has been possible to generate single isolated atoms as well as nanoclusters with few atoms on the solid carrier, and the preparation methods have been described in an excellent chapter of this book. In those cases, and from what was said above about the dynamics of the catalyst during the reaction, it is of paramount importance to modulate the support's property to increase the stability of the metal entities. Then, a combination of catalyst preparation, structural characterization, and the theoretical modeling of the adsorption energy of the metal atoms on the particular site of the support versus the cohesive energy of the metal, is key for redesigning and formulating solid catalysts made by single isolated metal sites (SACs) and/or very small clusters. Notice that, though the book is centered in SACs, I have also been continuously referring to metal clusters ranging from 2–10 atoms because, I believe, there is a continuity between the two types of entities, i.e. single sites and subnanometric metal clusters.

When we refer to supported metal catalysts, and going from nanoparticles to single atoms, we can see different chemical entities. In the case of nanoparticles (let's consider nanoparticles with 2–3 nm or more), a considerable fraction of the metal atoms are in the crystal bulk and are not accessible to the reactants. There is another fraction of the metal atoms (whose number depends on the shape of the crystal for a given crystal size) that are coordinated to atoms from the support and to other metal atoms, i.e. at the metal–support interface, can be accessible to the reactants. Finally, there will be a larger fraction of metal atoms located at the external surface of the metal crystals, which are coordinated with other metal atoms. There is no doubt that the supported metal crystallites of a few nanometers in size will present a heterogeneity in terms of the potential active sites. The other extreme is the "supported" isolated single metal atoms that are stabilized by interaction with the support. In that case, the composition, structure, and electronic properties of the surface will determine not only the reactivity of the supported SAC, but also its stability. In a first approximation, we think that all those single isolated atoms on a given support could be equal, and therefore one may have a homogeneous catalyst in a solid form. However, it has to be considered that a given support can have more than one type of anchoring sites for accommodation of the isolated metal atoms, resulting in metal atoms with different adsorption property and catalytic behavior. In fact, one has to consider the catalytic active site as the metal together with the coordinating atoms of the support that involves not only the next but also the next nearest-neighboring atoms of the metal. In fact, the stabilization of the transition states by the dispersion forces will go beyond the specific metallic center. This indicates again the relevance of the support on the catalytic properties and stability of the SACs. The effect can be even more notorious for bimolecular reactions where one of the reactants is adsorbed and activated on the binding atoms of the support, while the other molecule will react on the metal site, as it occurs, for instance, during the catalytic oxidation of CO with Au/TiO_2 catalyst.

In the case of very small supported metal clusters, all the metal atoms could be anchored to the support. Also in this case, the interaction with the support is going to influence the electronic properties of the atoms in the cluster and, consequently, their catalytic activity and stability. There is still another catalytic system that deserves to be systematically studied such as the very small bimetallic supported clusters. One could use in this case the support for fixing a given metal atom whose final catalytic behavior will also be influenced by the second metal. This is a field worth further studying, keeping in mind that when we refer to bimetallic clusters with 2–10 atoms, it may not be appropriate to treat them with the conventional criteria for alloys. In this respect, a very pertinent chapter on supported double- and triple-metal atoms catalyst can be found in the book.

In terms of the catalytic performance of SACs, it appears that the interaction of the single metal atoms with the support will be determinant, and the nearest and next nearest neighbors of the metal atom in the catalytic surface should be carefully considered when modeling SACs and their reactivity.

In the present book, the different authors have considered the key issues for the preparation of SACs, their stability, how to characterize them, and what information can extracted from the different characterization techniques available. I think that the presented considerations about the stability and potential evolution of SACs under reaction conditions are of paramount importance.

The modeling of SACs has also been introduced and should be of interest to inspire readers to better understand their systems and to go into better catalyst design.

There are a series of excellent chapters in the book devoted to catalytic applications of SACs, spanning from activation of small molecules such H_2O, CO, N_2, methane reforming to hydrogenation, oxidation, and coupling reactions for constructing C—C, C—Si, and C—B bonds.

The contribution of the single atoms to the effective separation of the photogenerated charge carriers shows the relevance of the advanced synthesis methods for the preparation of photocatalysis by introducing SACs. Finally, a comprehensive and very pertinent chapter of metal–N–C electrocatalysis has been presented that describes the preparation process, structures identification, and progress of very relevant electrochemical reactions while presenting the challenges and potential improvements.

In my opinion, this is a book of much interest not only for postgraduate students but also for researchers working on metal catalysis in academia and industry. We have read the book knowing that the use of SAC is not a general solution to our catalytic problems, but it is one avenue that should be deeply investigated to improve our knowledge for building a unified theory for catalysis. The recently considered SACs will complement and introduce new properties to the single-site metal catalysts that already exist in structured solids and have shown their benefit for industrial applications. Nevertheless, as we said above, achieving a larger number of single metal atoms on the surface of transition metal oxides is one avenue that also expands to other fields than catalysis, such as for instance optoelectronics, quantum computation and sensing, and medicine.

Valencia
November 2021

Avelino Corma

Preface

Humans have been using natural/synthetic catalysts for thousands of years. The role of these substances is to speed up chemical reactions, without being consumed in the process. A good catalyst must have a high activity and selectivity (produce the right kind of product), and present an acceptable stability/durability, all at an acceptable cost related to the application. Nowadays, catalysis plays a vital role in modern societies, providing pharmaceuticals, commodity and fine chemicals, fuels, and means for protecting the environment. Additionally, catalysis (heterogeneous, homogeneous and enzymatic) is a multidisciplinary science, which has a long-standing importance and impact in pushing forward the frontiers of chemical sciences, posing major fundamental, and conceptual challenges. This has been recognized by more than 20 Nobel prizes in chemistry strongly associated with discoveries related to catalysis. Despite this, still today the precise reasons why catalysts work are often a mystery, even for specialists. However, this does not prevent that at some points, catalysts are involved in more than 80% of the processes of all manufactured products. Consequently, even if most people have no perception of its importance, almost everything in our daily life depends on catalysis.

Catalysis residing in the heart of our modern societies, it is of course also impacted by any paradigm shift. Thus, the relationship between the concepts of resilience and sustainable development represents one of the key challenges of the twenty-first century that also affect catalysis. The ecological and energy transition has an impact on catalysis, orienting research and development toward cheap and green catalysts for clean energy and environmental sustainability. This includes research:

(i) on new types of catalysts that incorporate rational use of strategic metals or their elimination;
(ii) on new activation methods for catalyzed reactions that consume less energy or are compatible with renewable energies and their limitations such as intermittence (photocatalysis, electrocatalysis, magnetic induction, electric internal heating);
(iii) on the discovery of new catalytic transformation and new modes of chemical bond activation and formation for energy and chemistry transition; and

(iv) on the use of new and efficient tools to develop and understand these catalytic transition processes (*in-situ/operando* techniques, high-throughput experimentation/artificial intelligence, multiscale modeling).

Since the beginning of this century, a certain number of researches in catalysis, which integrate the different aspects mentioned above, have developed around the use of isolated supported metal atoms. Conceptually, this approach applies the notion of atom economy to supported catalysis (100% atom utilization of often scarce element), and can be considered as a bridge between conventional supported heterogeneous catalysis, using sets of metal atoms, and homogeneous or enzymatic catalysis, very often using isolated and well characterized active centers. In addition to the economy of atoms, this type of catalysis also makes it possible to envisage new chemical reactivity compared to conventional supported catalysts, because the metal atom undergoes strong electronic and geometric effects from the support. In other words, the same metallic element can have a different reactivity if it appears, when it is deposited on a support, in the state of a metallic particle, of a cluster or of an isolated atom. Consequently, as in Robinson Crusoe adventures where the role of the island has not to be neglected, the role of the support (via the metal-support interaction) plays a pivotal role on the activity, selectivity, and stability of this type of catalysts. In this context, the choice of the support becomes out as important as that of the metal, as for the choice of ligands in homogeneous catalysis. Recent impressive results already obtained with this new type of catalytic materials in fields as broad as energy, the environment or fine-, specialty-, and commodity-chemicals make this type of catalysis one of the most active frontiers in heterogeneous catalysis. If the fact that the active center is an isolated atom can offer interesting perspectives for the understanding of reaction mechanisms by modeling, because of the small size of the systems, the same characteristic also raises important challenges in terms of synthesis of materials and their advanced characterization. We believe that a fundamental understanding of this type of catalysis should lead to a systematic approach for the rational design of innovative catalytic materials with improved performance, which can participate in the necessary industrial transitions of this first half of the twenty-first century to ensure a secure energy/environment future. This is why we have introduced this book.

In the present book, we want to give the reader a comprehensive overview of what is supported metal single atom catalysis, and to rationalize the advantages of these catalytic materials regarding their activity, selectivity, and stability. Thus, each chapter will provide a critical overview of a specific domain through relevant examples of the literature. In this sense, this book is the first to introduce concepts and main achievements, while covering the main aspects of supported metal single atom catalysis. It is our hope that this book will not only prove suitable for self-study and teaching purposes, but will also inspire further research and discovery, thus setting the standard in the field of supported metal single atom catalysis for the years to come.

Finally, we gratefully thank all the authors who contributed to the different chapters of this book. We also thank all the staff members at Wiley for their assistance in editing and publishing this book.

Toulouse, Albi, July 2021

Philippe Serp, LCC,
CNRS-UPR 8241,
ENSIACET

Doan Pham Minh,
Centre RAPSODEE,
CNRS UMR 5302,
IMT Mines Albi

1

Introduction to Supported Metal Single Atom Catalysis

Doan Pham Minh[1] *and Philippe Serp*[2]

[1]*Université de Toulouse, IMT Mines Albi, UMR CNRS 5302, Centre RAPSODEE, Campus Jarlard, F-81013, Albi Cedex 09, France*
[2]*Université de Toulouse, LCC, CNRS-UPR 8241, ENSIACET, 31030 Toulouse, France*

1.1 Introduction

Supported metal catalysts are extensively used in the chemical industry, from bulk and fine chemicals production to petrochemicals. In this family of catalysts, the support contributes to maintain a high dispersion of the active phase, improves catalyst stability (including mechanical) but also determines at a certain extent the physico-chemical properties of the active phase, and thus its activity. When considering most of this support effects, the size of metal particles is an important parameter to consider. For large particles (10–100 nm), the influence of the support will be limited to the interface (interfacial effects), and will involve a limited number of metallic atoms, and the support effect can be diluted by the presence of the numerous metal atoms located far away from the support. The interfacial effects between the support and catalyst could control electron distribution, tune intermediate adsorption, prevent catalysts from dense aggregation, and improve stability. Thus, if we consider charge transfer, which can have a substantial influence on adsorption and, therefore, on reactivity, it can be expected that for large particles the average number of electron transfer per metal atom will be smaller than for small particles [1]. It is however important to note that due to strong charge screening, the excess of charge on metal particle could be expected to accumulate at the direct contact interface (sub-nanometer short-range charge transfer) [2]. Interfacial sites can also contribute to tune intermediate adsorption. For example, in the Sabatier reaction, an enhanced cooperation between nickel and interfacial active sites was reported [3, 4], which leads to rapid dissociative adsorption of H_2 on Ni and hydrogen spillover. The H atoms generated contribute to CO_2 hydrogenation and create oxygen vacancies on the reducible support surface, which contribute to CO_2 adsorption and activation. In that context, it is interesting to decrease the particle size. As particle size decreases (1–10 nm), these interfacial effects will be exacerbated, but other phenomena should be taken into consideration, such as the increase in surface defects (uncoordinated metal atoms) and even quantum size effects in the case of very small

Supported Metal Single Atom Catalysis, First Edition. Edited by Philippe Serp and Doan Pham Minh.
© 2022 WILEY-VCH GmbH. Published 2022 by WILEY-VCH GmbH.

particles (clusters), which can of course dictated the catalytic activity [5, 6]. In the latter case, although tremendous variation in the relationships between size and activity exist depending on the system studied, these relationships are often broken into: positive size-sensitivity reactions, negative size-sensitivity reactions, size-insensitive reactions, and a fourth category composed of reactions for which a local minima or maxima in activity exists at a given particle size [7, 8]. Going back to the Sabatier reaction, structure sensitivity over nickel catalysts was reported in the 1–7 nm range, with a maximum in activity at 2.5 nm [9]. The smallest Ni particles bind CO too weakly, facilitating its easy desorption and modifying the selectivity of the reaction.

Downsizing metal catalyst particles to supported metal single atoms is the ultimate step. In that case, some interfacial effects can be exacerbated like the electronic metal-support interactions and other intrinsic metal effects, such as the electronic quantum size effect and the structure-sensitivity geometrical effect disappear. We thus arrive at a situation where the role of the support, which one has to see as a ligand as in homogeneous or enzymatic catalysis, becomes preponderant. Indeed, it will have a direct influence on the charge state, oxidation state and structure of the frontier orbitals of the active site, which is now limited to an isolated metal single atom (Figure 1.1). Furthermore, in this case, tuning of the support surface chemistry can even be as relevant as the selection of the supported metal itself for regulating the catalytic activity [10].

Of course, such a size reduction raises questions other than chemical reactivity. How to produce such species and how to characterize them reliably? The progress made since the beginning of this century in terms of synthesis, characterization and modeling strategy has made it possible to see the emergence of a new class of catalysts: metal-supported single-atom catalysts (SACs). The catalysis community

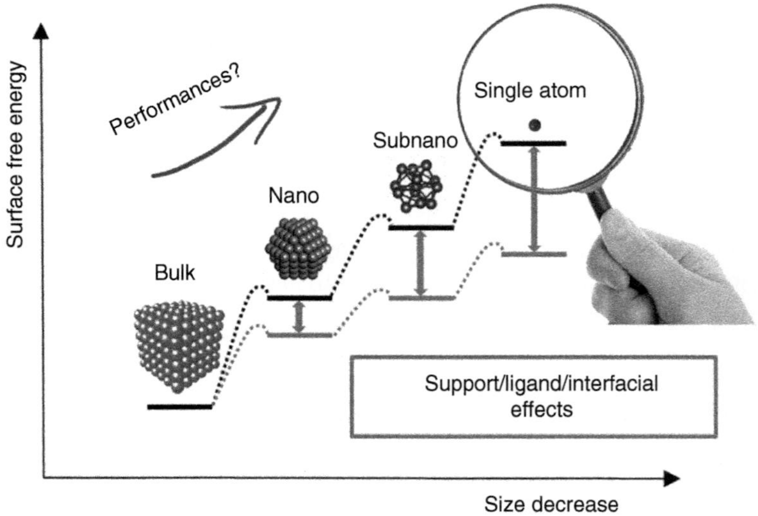

Figure 1.1 Influence of downsizing of catalyst particles on interfacial effects. Source: Philippe Serp.

quickly recognized the importance of metal supported single atom catalysis [11–17], which can be seen as the ultimate consequence of applying the atom economy concept to heterogeneous catalysis. The First International Symposium on Single-Atom Catalysis (ISSAC) was organized in 2016 by T. Zhang, J. Li, and J. Liu at Dalian Institute of Chemical Physics in Dalian, China, in 2016. Since then, an exponential number of publications, including review articles and patents on that subject are published every year.

Supported metal single atom catalysis has often been presented as a bridge between heterogeneous and homogeneous catalysis [17–19]. On one side (where we mainly find the heterogeneous catalysis community), it is true since at the active site, a single metallic atom is surrounded by a ligand (the support) just like in homogeneous system. On the other side (where most molecular chemists stand), a difference exists linked to the presence of a surface that plays the role of a solid, "rigid" ligand with specific redox, acid-base, as well as physical properties (e.g. porosity, hydrophilicity, hydrophobicity, semiconducting properties, etc.). We will also notice that the diversity of direct environment in supported metal SACs, i.e. the first coordination sphere is rather limited compared to homogeneous systems. Indeed, on oxide, it is often limited to a metal-O environment, and on carbon metal-C, metal-N_4 (like in metalloporphyrine) metal-O or metal-S bondings have been regularly reported. Additionally, for oxide supports, beyond the first coordination sphere (O-coordination), metal cations on oxide supports readily serve as the second coordination sphere and are involved in catalytic reactions together with the primary catalytic metal single atoms [20]. The presence of such "macromolecular" ligands can induce specific properties to the metal and *in fine* to the resulting supported catalysts that most of the homogeneous systems do not present [16]. Most, but not all… Thus, Jørgensen introduced the term "non-innocent" ligand in 1966 [21]. He stated "*ligands are innocent when they allow oxidation states of the central atom to be defined*," which correspond to non redox-active ligands [22]. We will see in Section 1.4.3.2 of this chapter that the support can indeed, in some cases, behave as a "non-innocent" ligand and introduce complexity in the system. Additionally, in metal–ligand cooperative catalysis [23], ligands are involved (like in enzyme catalysis [catalytic residues] and in supported metal SACs [24] as we have seen for the Sabatier reaction) in facilitating reaction pathways that would be less favorable to occur solely at the metal center. Also, plastic deformation (a dynamic phenomenon) of the catalytic system can induce difference in reactivity as observed in heterogeneous catalysts [25–27], supported metal SAC [28], and enzymatic catalysis (allosteric sites) [29, 30].

From these various considerations, it appears that a number of factors must be carefully taken into account when trying to make correlation between the structure of supported metal SACs and their catalytic performances (Figure 1.2). Such a level of complexity positions supported metal single atom catalysis closer to enzymatic [31] or supramolecular catalysis [32] than to conventional homogeneous catalysis.

Finally, if metal-support interaction is certainly important for regulating catalytic performance of SACs, other type of cooperativity that are not specific to SACs, such as metal–ligand or heterobimetallic cooperativity, and cooperativity that

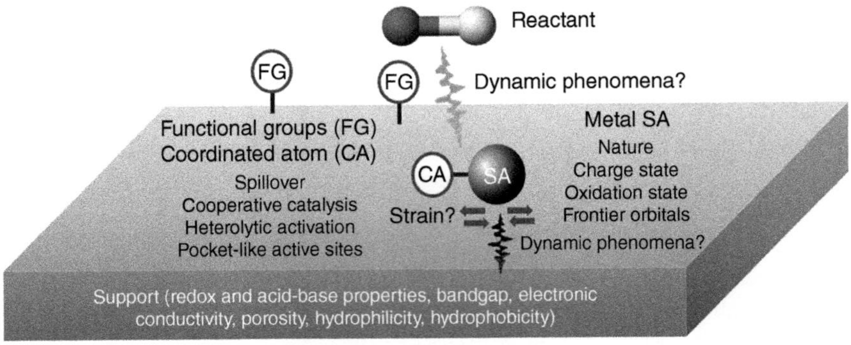

Figure 1.2 Factors affecting catalytic performances of supported metal SAC. Source: Philippe Serp.

are SAC-specific such as nanoparticle-single atom or mixed-valence single atom cooperativity have been reported [33].

1.2 Definition

One discussing about supported metal single atom catalysis, a first question is often asked *how different is tethered homogeneous catalysts or single-site heterogeneous catalysts (SSHCs) to supported metal single atom catalysts (SACs)?* To answer this question, we will rely on some definitions introduced by Thomas et al. [34], Basset and coworkers [16], and Liu [35].

For Thomas et al., an SSHC (catalytically active center) may consist of one (or more in the case of small clusters) atoms spatially isolated from one another. Each single site presents the same energy of interaction between it and a reactant; and is structurally well-characterized as for single site homogeneous molecular catalysts (Figure 1.3a) [34]. SSHCs may be conveniently classified into three main categories: (i) individual isolated ions, atoms, molecular complexes or well-defined clusters anchored to high-area supports; (ii) "ship-in-bottle" structures, in which isolated catalytic molecular entities are entrapped within a zeolite cage; and (iii) solid frameworks that host the catalytically active centers, zeolitic [36], and metal-organic frameworks (MOFs) [37] materials occupying an outstanding position since they are crystalline.

As far as tethered homogeneous catalysts are concerned, Basset distinguishes surface organometallic catalysts (SOMCats) that are directly bound to the surface, and where the surface acts as rigid ligand, and supported homogeneous catalysts (SHCs) where the metal atom is tethered to the support surface via flexible linkers coordinating the metal via covalent or noncovalent interactions (Figure 1.3b) [16]. In SOMCats and SHCs, the catalytically active sites are formed by reacting organometallic or coordination compounds with (functionalized) surfaces (oxides, metal nanoparticles, carbon). For these catalysts, the supported organometallic/coordination species keep at least part of their ligands after grafting. These types of catalysts can be classified as SSHCs if all of the active sites are structurally identical. This of course will

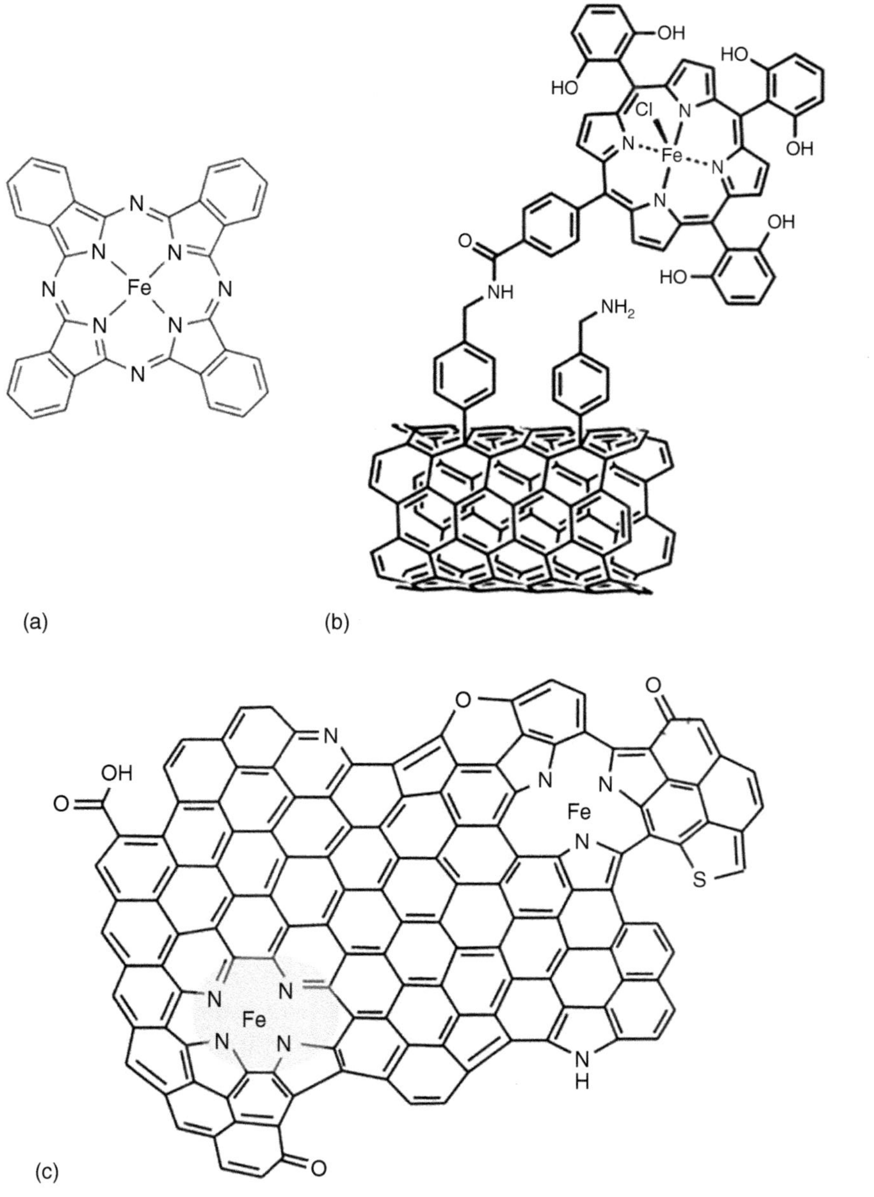

Figure 1.3 (a) A single site homogeneous molecular catalysts (iron(II) phtalocyanine complex); (b) a supported homogeneous catalysts (SHCs); and (c) a supported metal single atom catalyst. Source: Philippe Serp.

depend on the nature of the support. Indeed, if the supporting solid exposes different anchoring situations, the structure of the anchored sites is usually not well known.

In supported metal SACs, only single metal atoms are present, which are dispersed on a solid support (Figure 1.3c) [35]. The interaction between the single atoms (SAs) and the support can be of a different nature involving covalent, coordination, or ionic bonds. There should be no appreciable interactions among the isolated SAs.

The difference between SACs and SOMCats therefore lies in the presence of ligands on the metal generally originating from the precursor used during the synthesis. Of course, if these ligands are removed the SOMCat will transform into SAC. Likewise, if a supported SAC is modified by the addition of a ligand, it can be considered as a SOMCat. The active sites in SACs generally consist of the SAs, but in some cases the immediate neighboring atoms of the support surface can be involved. The catalytic property of the individual active sites can be similar or different depending on the interaction between the SA and its neighboring atoms of the support. When the catalytic behavior of all the metal SAs in a supported SAC is the same, then the SAC can be considered to be an SSHC as well (Scheme 1.1). This should not be the case of the SAC shown in Figure 1.3c, since the two Fe centers do not present the same environment. For supported double and triple metal atom catalysts discussed in *Chapter 15*, the same reasoning can be followed. Supported single, double, triple metal atom catalysts can be considered as a subset of atomically dispersed supported metal catalysts (ADSMCs) [35], together with two-dimensional rafts or very small clusters.

Following these definitions, there is no substantial difference between homogeneous and heterogeneous single-site catalysts since the ligand sphere around the metal, which is accurately engineered, should be the same in both cases. Additionally, supported SACs, SHCs or SOMCats can be considered as SSHCs if all of their active sites are structurally identical (Scheme 1.1), which is not often the case due to support inhomogeneity in term of anchoring sites.

It is worth noting that when considering supported SACs, the rules of molecular chemistry apply, and help to rationalize the structure and the reactivity of these species. This leads us to mention a point that seems important to us in this type of catalysis. If definitions are necessary to clarify the differences between all these types of catalysts, they should not make us forget the similarities. As we have seen, the SSHC family is very large and includes grafted molecular species and crystalline solids, which are generally studied by different communities (molecular chemists and materials chemists). Similarly, the objects involved in catalysis are

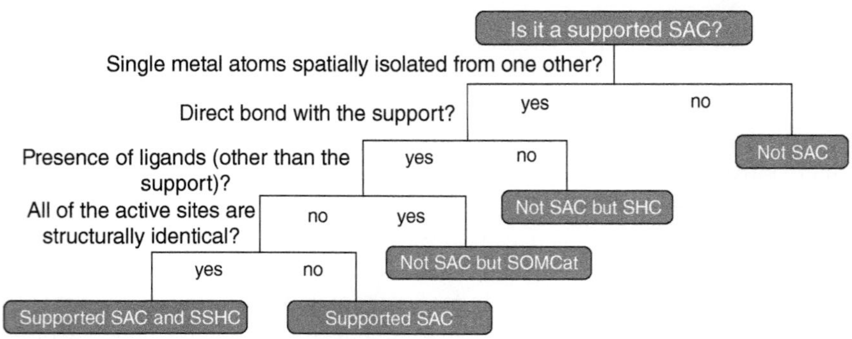

Scheme 1.1 Guide to the definition of supported single metal atom catalysts. Source: Philippe Serp.

often studied in different ways. Studies carried out on the most molecular catalysts (SHCs, SOMCats) are often mechanistic; while those relating to catalytic solids often also integrate traditional problems of heterogeneous catalysis and associated processes (transfer, diffusion, etc.). Interestingly, supported SACs lie in between grafted molecular species and crystalline solids and their study should be performed by multidisciplinary scientists with solid basic knowledge in molecular and material chemistry as well as engineering and technology. In catalysis education, the separation between catalysis science and catalysis technology is a regular issue, we believe that as proposed by Pagliaro new courses in catalysis science and technology should integrate a unified approach [38].

1.3 Origins of Supported Metal Single Atom Catalysts

Most recent reviews dealing with supported metal single atom catalysis underline the innovative character of this type of catalysis:

- *Single-atom catalysis is a recent discipline of heterogeneous catalysis* [16];
- *One of the earliest heterogeneous catalysts with surface metal atoms was reported in 1999* [12];
- *Single-atom catalysis is currently one of the most innovative and fastest growing research areas in the entire field of catalyst science* [14].

While it is true that most of the important work in the field dates from the twenty-first century [39], an analysis of the bibliography associating the words "single atoms" and "catalysis" in web of science (WoS) reveals that research had been initiated long before (Figure 1.4a). We will present below some of the pioneering works (sometimes forgotten) representative of this research (Figure 1.4b) and put them in perspective of certain recent (re)discoveries.

If the failure to obtain a high metal loading can be in certain cases a brake for the industrial development of SACs (for example in electrocatalysis) [40], for certain reactions at very high tonnages involving the use of noble metals, the catalyst must be lightly loaded for economic reasons. This is for example the case of reforming Pt/Al_2O_3 catalysts. A crucial invention was made in 1947 by Haensel at Universal Oil Products, who suggested the use of platinum supported on alumina as a bifunctional catalyst in the refining process [41]. Using such catalyst with an expensive metal for producing a cheap commodity sounded crazy to most of his contemporaries. However, Haensel and coworkers demonstrated that the high activity of a 0.01% Pt/Al_2O_3 catalyst combined with long-term stability and possible *in situ* regeneration justify a high initial price. In fact, this catalyst was economically more efficient in the long run than a "cheap" catalyst with a shorter life. Hydrogen adsorption indicated that 50% or more of the Pt atoms were surface atoms. This was the first industrial bifunctional catalyst, exposing both platinum SA and acid sites, and still today researches are conducted on supported Pt_{SA} as reforming catalysts [42–44]. A combination of nuclear magnetic resonance (NMR) spectroscopy, and high-angle annular dark-field scanning transmission

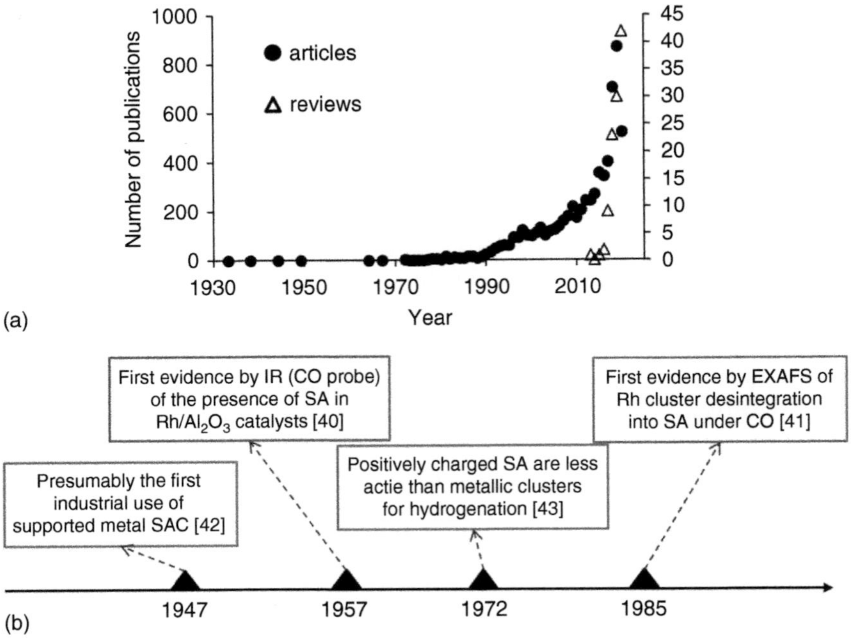

Figure 1.4 (a) Number of publications (research articles or review) per year obtained by the keywords "single atoms" and "catalysis" from Web of Science. Source: Web of Science. (b) Important findings related to supported metal single atom catalysis dating from the last century. Source: Philippe Serp.

electron microscopy (HAADF-STEM) analyses coupled with density functional theory (DFT) calculations was used to reveal the nature of anchoring sites of a catalytically active phase of Pt/γ-Al_2O_3 catalyst [44]. At low ($\leq$1 wt%) Pt loadings, Pt is atomically dispersed on the support surface, and coordinatively unsaturated Al^{3+} centers acts as binding sites for Pt_{SA}. Seventy years after Haensel et al. the group of Hutchings and researchers from Johnson Matthey followed a similar strategy to replace the conventional Hg/C acetylene hydrochlorination catalyst by an Au/C catalyst [45]. They set as a target an Au loading smaller than 0.25 wt%, since higher loadings were considered to be not economically viable. *In situ* X-ray absorption fine structure spectroscopy (EXAFS) experiments have definitively shown that the active catalyst predominantly comprises Au(I) isolated cationic species [46]. Pt_{SA}/C are also promising for this reaction [47]. These two examples show that the concept of SA catalysis is attractive, particularly in the context of sustainable technologies that will make use of critical metals, which are expensive and of limited availability.

Different studies were conducted in the 1970s–1980s on low loading M/Al_2O_3 catalysts (M = Co [48], Cr [49, 50], Re [51] and more particularly Rh) in order to try to elucidate: i) the nature of the metal species (SA [52], highly dispersed 2D phase [δ phase] [53] or clusters [54]/nanoparticles [NP]), ii) the reasons why chemisorption stoichiometries do not prevail for supported Rh catalysts [55], and iii) in some cases the reason of the difficult reduction of some metal species. We will discuss below the case of Rh/Al_2O_3. It was thought that Rh may be present on Al_2O_3 in the form of 3D crystallites, 2D "rafts," and/or isolated SA (ion) sites. This system was investigated by a variety of techniques including chemisorption [53, 55], infrared spectroscopy [56–62], X-ray photoelectron spectroscopy (XPS) [54], electron spin resonance (ESR) [54], EXAFS [52, 54, 63, 64], ^{13}C NMR spectroscopy [65, 66], luminescence spectroscopy [67], and ultra-high resolution electron microscopy [53]. Most of these techniques are still used today for the characterization of supported metal SA (*Chapter 5*). One of the most powerful techniques for the study of the surface chemistry of supported high surface area metal deposits is transmission infrared spectroscopy. The chemisorption of CO by Rh/Al_2O_3 has been well studied using infrared spectroscopy. Preliminary experiments performed by Garland and coworker [56] have shown that three main types of chemisorbed CO are produced on Rh/Al_2O_3 catalysts: (i) a single CO molecule adsorbed on one Rh atom, (ii) a band due to bridged CO, and (iii) two CO molecules adsorbed on one Rh atom (geminal dicarbonyl rhodium species). The latter surface species was particularly present for catalysts with a low Rh loading presenting highly dispersed rhodium species. The other two species are related to CO-covered Rh crystallites. The geminal dicarbonyl rhodium species was associated to a Rh(I) species, which cannot be completely reduced to Rh(0) even at 400 °C [60]. However, infrared (IR) analyses were not conclusive about the exact nature of this species and both monoatomically dispersed Rh(I) species on Al_2O_3 (Rh_{SA}/Al_2O_3), and 2D supported Rh "rafts" [53] were proposed. For the latter case, the geminal dicarbonyl rhodium species formed on the edge atoms of supported Rh "rafts." Later on, an important finding was reported by Primet [68], who discovered that the geminal dicarbonyl species are not formed at low temperature, but develop on warming up to room temperature. Dissociation of CO was assumed on Rh(0), and CO in excess adsorbs on the surface Rh-O species to give the geminal dicarbonyl rhodium(I) species. EXAFS analyses were also conducted on reduced low loading catalysts [52, 54, 63, 64]. In such samples, the measured average coordination number was low (1.5 [52] to 4.9 [64] according to the temperature of reduction). Although the presence of isolated Rh^+ ions in these low loading catalysts was not completely ruled out, it was proposed that they mainly contain small metal clusters, consisting of 6–10 atoms [54]. Another important finding was that adsorption of CO at room temperature changes the system completely. It results in a significant disruption of the Rh clusters, ultimately leading to the isolated Rh(I) geminal dicarbonyl species, in which Rh is surrounded by two CO and three oxygen ions. This study also confirmed that CO dissociation occurs and is followed by CO adsorption on the oxidized rhodium.

These results can be put into perspective with more recent results obtained in studies on supported metal SACs. First, several studies have been published on the role of CO on the stabilization of SA [69], but also on their (dynamic) formation by NP disintegration [70, 71] or reconstruction [72–74]. In particular, it was shown that for CO assisted Ostwald ripening and induced NP disintegration, the strong interaction between CO and the metal is essential [75]. When the formation of the metal–CO complexes becomes exothermic with respect to the supported metal NP of interest, the metal particles will be disintegrated to the individual SA.

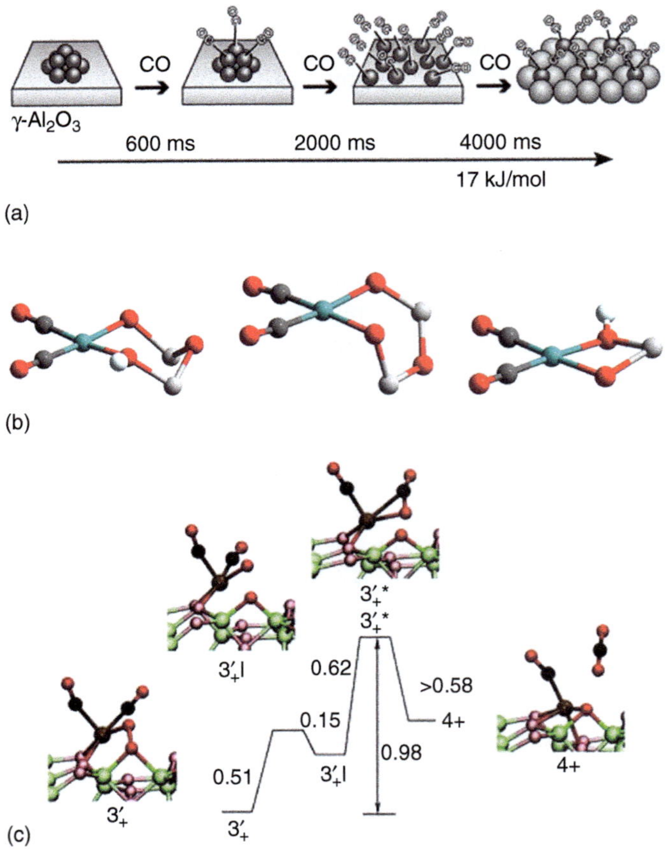

Figure 1.5 (a) An illustrative mechanism, with times given as taken from the beginning of the reaction, of three elementary steps at room temperature for the disintegration of Rh clusters on Al_2O_3 during CO adsorption by time-resolved DXAFS. Source: Suzuki et al. [70]. Reproduced with permission of Wiley-VCH; (b) Three types of connectivity of $Rh^I(CO_2)$ species to the Al_2O_3 observed in the minimum energy geometries obtained by periodic DFT calculations. Source: Roscioni et al. [76]. Reproduced with permission of American Chemical Society; (c) Free energy profile obtained by DFT for CO oxidation on $Rh^I(CO)_2/Al_2O_3$. Source: Ghosh et al. [77]. Reproduced with permission of Wiley-VCH.

Second, some studies on SAs are also dealing with the specific Rh_{SA}/Al_2O_3 system [70, 76, 77]. Thus, time-resolved energy-dispersive EXAFS analyses have revealed three steps for the structural rearrangement of Rh clusters (Figure 1.5a) [70]. Before CO exposure, Rh atoms in the cluster composed of seven Rh atoms in the first layer and three Rh atoms in the second layer interact with the surface oxygen atoms of the support. Each Rh atom in the lower layer interacts with two O atoms of the support. Then CO (0–600 ms) rapidly adsorbs on the cluster without Rh—Rh bond breaking. At the second step after 600 ms, further CO adsorption weakens the Rh—Rh bonds, and the Rh cluster is completely disintegrated at 3000 ms yielding the $[Rh(CO)_2]$ monomer that interacts with three surface O atoms. The slowest step in the cluster disintegration (E_a = 4.1 kcal/mol) is the formation of [$Rh(CO)_2$] monomers, which occurs concertedly with bond rearrangement at the interface, and probably with the oxidation of Rh^0 to Rh^+ by surface OH groups. It is important to place this result in the context of the characterization of supported metal SACs. Indeed, IR spectroscopy using CO as a probe molecule is regularly used to complement more local electron microscopy analyzes (*Chapter 5*) in order to distinguish SA, clusters or NP [78]. The fact that under CO some metallic clusters can disintegrate at room temperature at the second-scale can of course call into question many interpretations, and lead to errors or misunderstanding of the system. First principle plane-wave/DFT calculations were used to determine the exact environment of the [$Rh(CO)_2$] monomer. It was found that Rh_{SA} exhibit a square-planar coordination geometry [76], and is bound to two oxygen atoms of the support (Figure 1.5b) [76, 77]. Finally, the CO oxidation was investigated with this catalyst. CO oxidation over Rh_{NP} is a highly structure sensitive reaction. The observed structure sensitivity involves the formation and interplay of three structurally discrete supported Rh species; Rh_{NP}, a Rh_2O_3-like phase, and monodisperse $Rh^I(CO)_2$ species [79]. Thus, the reactivity of the $Rh^I(CO)_2/Al_2O_3$ was investigated for the CO oxidation reaction by DFT (Figure 1.5c), and compared to that $Rh^0(CO)_2/Al_2O_3$ and Rh_6/Al_2O_3 [77]. The computed activation energy for this reaction were 21.2 ($Rh^I(CO)_2/Al_2O_3$), 9.7 ($Rh^0(CO)_2/Al_2O_3$), and 27.4 kcal/mol (Rh_6/Al_2O_3). The CO oxidation on Rh_6/Al_2O_3 is sluggish compared to SA due to strong Rh—CO bonding on the cluster. For SA, the oxidation of Rh to Rh^I has a negative effect on the rate of CO oxidation, and the first CO oxidation by $Rh^0(CO)_2$ is remarkably faster than that by $Rh^I(CO)_2$, for which the free energy barriers are 9.7 and 21.2 kcal/mol, respectively.

It is worth mentioning the pioneering works of Köpp and coworkers, who investigated in the early 1970s the reactivity of Pd/C catalysts prepared by metal evaporation (Pd_{SA} or Pd_{NP}) for the reverse methane cracking reaction ($C + 2H_2 = CH_4$) [80, 81]. The results of their study on isosteric enthalpies of adsorption of hydrogen on atomically distributed Pd on carbon indicate that the occupation of H_2 molecules per Pd atom in SA catalysts was surprisingly small, and that the Pd_{SA} on the carbon are positively polarized ($Pd^{\delta+}$) [80]. Therefore, they could conclude that the high hydrogenation activity for the investigated reaction was due to Pd clusters and not due to Pd_{SA}. The fact that Pt_{SA} dispersed in sodalite cages cannot chemisorb hydrogen was also evidenced by Imelik and coworkers [82]. Three interpretations were proposed: (i) at least 2 Pt atoms are required to dissociate H_2; (ii) Pt_{SA} have lost their metallic properties including hydrogen chemisorption; and (iii) hydrogen chemisorption is inhibited because a partial electron transfer between Pt atoms and Lewis acid sites occurs. The poor efficiency of electron-deficient supported noble metal SACs for dihydrogen dissociation [83] (and further hydrogenation reactions) [84, 85] has been recently confirmed in different studies. In fact, several studies have shown that in the electron-deficient SA sites, the dissociation of H_2 followed a heterolytic pathway, whose barrier can be higher than that in metallic clusters with homolytic dissociation [86–90]. In that case, the electron depletion of metal SAs induces downshift of the d-band center of

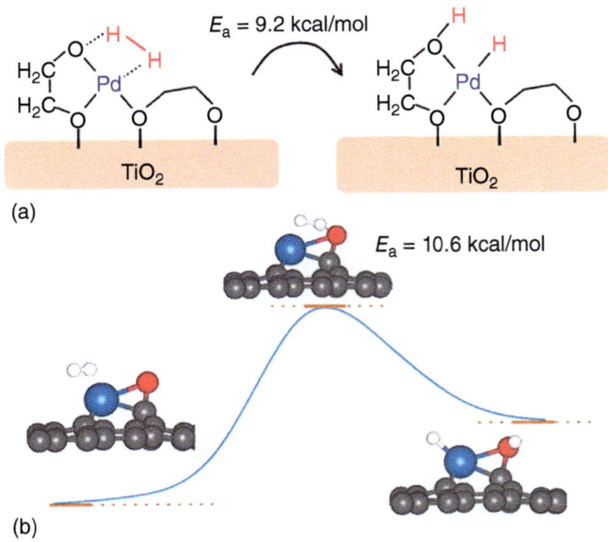

Figure 1.6 (a): Heterolytic dissociation of H_2 on EG-stabilized Pd_{SA}/TiO_2. Source: Philippe Serp. (b): Heterolytic dissociation of H_2 on Pd_{SA}-O/graphene. Source: Yan et al. [92]. Reproduced with permission of Elsevier.

the $M^{\delta+}$ species and consequently reduces, when compared with the metallic clusters/NP, the binding strength of the dissociatively adsorbed hydrogen atoms. This is particularly appealing for selective hydrogenation (see *Chapter 8*), since catalysts allowing activation of H_2 and weak adsorption strength of the dissociated H atoms can exhibit optimal efficiency in balancing activity and selectivity for the targeted product. On other side, if high hydrogenation activity is aimed, it is necessary to modify the local environment of the $M^{\delta+}$ species in order to facilitate the heterolytic dissociation of H_2, or to favor a homolytic cleavage. Such strategies have been followed in some studies. A facilitated heterolytic dissociation or H_2 was obtained on Pd_{SA}/TiO_2 by doping the system with ethylene glycol (EG) [91], or by introducing oxygen functionalities on carbon in Pd_{SA}/C catalysts (Figure 1.6b) [92, 93]. It is generally accepted that upon homolytic dissociation of H_2 on Pd_{NP}, hydrides ($H^{\delta-}$) are formed. When heterolytic activation of H_2 occurs, both $H^{\delta-}$ and $H^{\delta+}$ are produced that should allow better hydrogenation of polar unsaturated bonds. This was demonstrated in a study on EG-stabilized Pd_{SA}/TiO_2 catalysts. First, the presence of EG facilitates the heterolytic dissociation of H_2 (Figure 1.6a). Second, the authors observed a much superior catalytic performance by EG-stabilized Pd_{SA}/TiO_2 than conventional Pd_{NP}/C catalysts in the hydrogenation of benzaldehyde. The activation of H_2 on Pd_{SA}/graphene catalysts was investigated, for which the Pd_{SA} is located in a carbon vacancy [92–94]. If oxygen surface groups are not present in the vicinity of the Pd_{SA}, the hydrogen molecule becomes activated but the H—H bond is not broken. Thus, the electro-deficient Pd_{SA} is not able to dissociate the hydrogen molecule [94]. However, if an oxygen atom is present (Pd_{SA}-O/graphene), the reaction proceeds with an activation energy of 10.6 kcal/mol (Figure 1.6b).

To increase the electronic density on a $Pt^{\delta+}{}_{SA}$ immobilized on ceria, doping with phosphorus, an electron acceptor was proposed [95]. Such electronic interactions between the Pt_{SA} and the P-modulated neighboring oxygen atoms of the support led to a remarkable activity enhancement toward the hydrogenation of styrene, cyclohexene, phenylacetylene, and nitrobenzene. H. Lee and coworkers proposed a more straightforward solution that controls the oxidation state of Pt_{SA} in $Pt_{SA}/CeO_2–Al_2O_3$ catalysts by varying the reduction temperature [96]. Remarkably, the Pt_{SA} were stable up to 500 °C reduction temperature, clusters appearing at 600 °C. EXAFS and XPS analyses have shown that by increasing the reduction temperature from 100 to 500 °C, the oxidation state of Pt_{SA} could be controlled from highly oxidant (Pt^0 16.6%) to highly metallic (Pt^0 83.8%), while maintaining the SA structure.

The low activity of supported metal SAs compared to metal NPs for the reduction of nitric oxide by hydrogen was also experimentally evidenced in the 1970s and 1980s when studying the influence of metal loading in Rh/Al_2O_3 [97] and Pt/Al_2O_3 catalysts [98]. Two explanations were proposed. First, isolated SAs cannot offer an optimum site configuration for a transition complex, since the NO reduction requires at least two nitric oxide molecules and two hydrogen atoms, so that clusters with a minimum number of metal atoms are required for optimal activity. Second, the higher oxidation state of the metal in SAs due to strong interaction with the support is expected to result in a lower turnover frequency, as compared to a lower oxidation state of the metal in the particulate phase [98]. More recently, nitric oxide reduction with H_2 or CO was also investigated on supported metal SA [99–102]. First, in a study on reduction of NO with CO on Rh/SiO_2 catalysts, Tao and coworkers have confirmed that Rh_{SA}/SiO_2 are much less efficient than Rh_{NP}/SiO_2 for this reaction [99]. It was shown from DFT calculations that the significant difference in the reaction mechanisms between Rh_{SA}/SiO_2 and Rh_{NP}/SiO_2 is that NO can directly dissociatively chemisorb on Rh_{NP} due to the multiple adsorption sites; however, for Rh_{SA}/SiO_2, NO hardly dissociates with only one active site. This leads to much higher activity but also selectivity on Rh_{NP} in comparison to Rh_{SA}. Nickel SAs supported on g-C_3N_4 nanosheets shows a comparable reactivity with the Rh_{SA}/SiO_2 catalyst; i.e. complete NO conversion at around 450 °C [100]. An efficient strategy to overcome the limitations of supported single atoms for this reaction is to associate them with a second metal. This can be done either by using single metal alloys (*Chapter 4*) [101], or by diluting noble metal SAs in Co_3O_4 [102]. Thus, Cu_5Pd_{SA}/Al_2O_3 exhibited outstanding catalytic activity and N_2 selectivity in the reduction of NO by CO since complete conversion of NO to N_2 was achieved at 175 °C [101]. Kinetic and DFT studies demonstrated that the rate determining step (N—O bond breaking of the $(NO)_2$ dimer) was promoted by the Pd_{SA} (E_a = 11.4 kcal/mol, to be compared to 25.6 kcal/mol computed for Ni_{SA}/g-C_3N_4) [100], while N_2O decomposition to N_2 smoothly proceeds on the Cu surface, which contributes to the excellent N_2 selectivity. The use of catalysts containing dispersed bimetallic sites Pt_1Co_m and Pd_1Co_n allowed performing the reduction of nitric oxide with hydrogen at temperature as low as 150 °C [102].

1.4 Challenges, Limitations, and Possible Opportunities in Supported Metal Single Atom Catalysis

The unique structural and electronic properties of supported metal SACs that result from multiple factors (unsaturated coordination environment, specific interaction between the SAs and its neighboring atoms) offer great potential in catalysis. All those merits are based on the stability of these catalysts that people are most concerned about. The potential of these catalysts has been explored in diverse thermo- (*Chapters 8–12*), electro-(*Chapter 13*), and photochemical (*Chapter 14*) applications ranging from small-molecule activation (*Chapter 10*) to the production of fine chemicals (*Chapter 12*). If the key factors controlling the activity, selectivity, and stability of this new class of catalysts are mastered (cf. discussion in the following sections of this chapter), thus 100% atom utilization efficiency can be expected. This is particularly appealing for the replacement of noble-metal-based catalysts for energy applications and of homogeneous catalysts in organic synthesis. However, despite 100% atom utilization efficiency, a general question regarding supported metal SA catalysts is whether a SA on a support represents the optimal structure to deliver the highest intrinsic catalytic activity for all reactions. The rate at which an elementary reaction proceeds on the catalyst surface will depend on the strength of reactants/products/intermediates – surface interaction (*electronic effects*), but also on the availability of specific atomic groupings (both in mono- and bimetallic/alloyed systems) that can provide the number of surface atoms necessary for chemisorption, by acting as active sites (atomic *ensemble effects*). In supported metal SA catalysts, if ligand (electronic) effects can be modulated, the so-called ensemble (geometric) effect, which is necessary for some reactions, is not present. Thus, they may be inactive for some complex catalytic reactions where multiple reactants, intermediates, and products are involved, unless the reaction proceeds via a distinct mechanism and/or the support itself can fulfill the required catalytic role. If such reactions are unwanted (undesired side reactions, coking), the isolation of the metallic atoms is a promising strategy [93, 103]. But if atomic ensemble effects are necessary for the aimed reaction to proceed (hydrogenolysis), single metal alloys or double, and triple metal atom catalysts should be preferred [104–106]. Alternatively, the creation of arrays of SAs has been proposed to overcome this limitation of supported metal SA catalysts while maintaining 100% atom utilization [107, 108].

The grand challenge in the development of supported SACs is their low metal-atom loading density, uncontrollable localization and ambiguous interactions with supports, posing difficulty in maximizing their characterization and catalytic performances. The following sections will treat these different aspects of supported single-atom catalysis, highlighting the challenges, but also presenting the limitations and possible opportunities through representative examples.

1.4.1 Metal Loading in Supported Metal Single Atom Catalysts

The low metal loadings generally needed to prevent metal NPs in supported metal SACs is generally presented as a brake on their industrial development. This is a somewhat simplistic vision, which does not integrate the specificities of certain catalytic processes. Thus, we discuss in Section 1.3 the Platforming process developed by Haensel et al. [41], who understood that a Pt-based catalyst presenting a long life and could be regenerated and reused *in situ* would, in fact, be economically efficient. To achieve this goal with an expensive metal as Pt, it was mandatory to minimize the amount of platinum. In 1947, they showed that a catalyst with 0.01% platinum on alumina was both active and stable. Hutchings when looking for a gold catalyst to replace the mercury-based industrial catalyst for hydrochlorination of acetylene to produce vinyl chloride followed the same strategy [45]. Of course, if for some reactions the activities obtained with lightly loaded supported metal SACs are not sufficient, then the low loadings become a major handicap, because their use would imply increasing the size of the reactors, which is not economically sustainable. A current trend in catalysis being to replace noble metals (expensive and for some of them identified as critical elements) by 3d transition metals (Fe, Co, Ni for example), which are generally less active, it also becomes necessary to significantly increase the metal loading. Finally, some difficult reactions, i.e. the oxygen reduction reaction in proton exchange membrane fuel cells, require high loading of an expensive metal as Pt. If the current tendency in fuel cell electrocatalysis is to try to decrease as much as possible the Pt loading [109], significant efforts are still needed to reach the Pt amounts present in catalytic converters [110]. In this context, significant efforts have been made in recent years to increase the loading of single atoms in supported metal SACs (see *Chapters 2 and 3*). To prepare high loading supported metal SACs, three main aspects have to be considered: (i) large-scale synthesis is still challenging and new synthetic strategies should be explored; (ii) suitable metal precursors should be selected to ensure easy decomposition and stabilization; and (iii) the appropriate supports should present high density of potential anchoring sites and allow strong metal-support interactions. Pyrolysis and wet-chemistry methods are the most efficient ones up to now to reach high metal loadings (>5 wt% of metal) [40]. Pyrolysis techniques produce carbon-based catalysts with nitrogen-coordinated metal SA when zeolitic imidazolate frameworks nitrogen-enriched molecular carbon precursors are used as precursors [111–113]. The high temperatures used in these processes usually generate a random distribution of M_{SA}–N–C sites on/in the carbon matrix (even if the M–N_4 environment characteristics of some metalloenzymes is often proposed) [114, 115] making the method relatively unpredictable and difficult to reproduce at the atomic level [113]. Furthermore, it is also worth noting that the exposure and accessibility of the metal SAs is also crucial to fulfill the 100% atomic utilization [116]. Incorporating metal into MOFs will undoubtedly buries some of the SAs deep inside the MOF framework, which would reduce atom utilization [117, 118].

Wet-chemistry routes (e.g. co-precipitation, impregnation) have been widely used to prepare supported metal SACs, due to their simple implementation and potential for large-scale production. To reach high metal loading by these methods, the choice of the support and the type of interaction with the SAs (coordination, electrostatic adsorption or ionic bonding) are crucial. Obviously, supports presenting a high density of well-defined anchoring sites should be preferred. In that context, the use of nitrogen-doped carbon materials such as graphitic carbon nitride is particularly

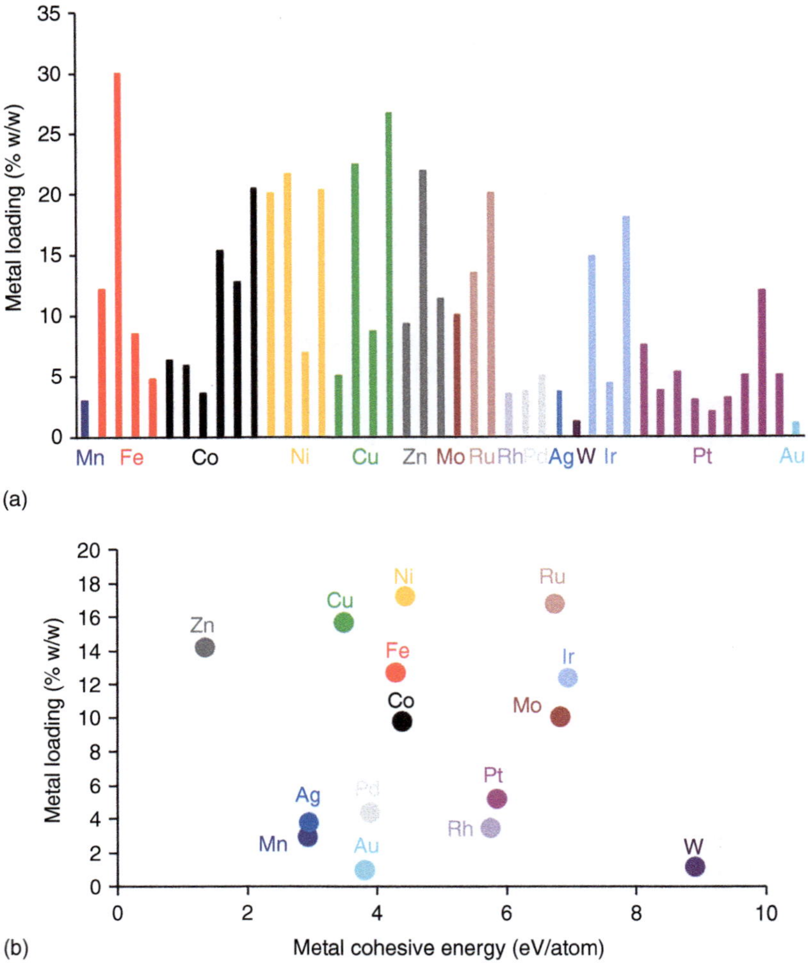

Figure 1.7 (a) Recent advances for supported metal SAC synthesis with high metal loadings. (b) Relation between the metal loading in high loading supported metal SAC and the cohesive energy of the bulk metal. Source: Philippe Serp.

interesting [119]. Other well-organized 2D materials such as hexagonal boron nitride (h-BN) or MoS_2 have also been used.

Figure 1.7a shows recent advances on supported metal SAC synthesis with high metal loadings for Mn [120], Fe [121–125], Co [122, 126–130], Ni [122, 127, 131–133], Cu [122, 134–136], Zn [122, 137, 138], Mo [139], Ru [85, 122], Rh [122, 140], Pd [122, 140, 141], Ag [142], W [143], Ir [122, 140, 144, 145], Pt [140, 146–152], and Au [153]. From this figure, it appears that relatively high metal loadings have been repeatedly obtained by different methods for some metals like Ni, whereas with some others like Pt or Pd, it seems difficult to reach such loadings. In order to explain this tendency, we try to correlate the average value of the (high) metal loading of Figure 1.7a with the cohesive energy of the metals. Indeed, we could expect that metals with a high cohesive energy are difficult to stabilize in the form of isolated atoms, due to their natural tendency to form metal NPs. If high loading were indeed reported for Zn_{SA} catalysts, a metal that present a low cohesive energy; and low loading were reported for W_{SA} catalysts, a metal presenting a high cohesive energy, it is obvious from Figure 1.7b that no correlation really exists between these two parameters. For example, for a metal like ruthenium that presents a relatively high cohesive energy, high metal loadings were obtained by pyrolyzing coordinated polymers, producing Ru_{SA} (13.5 wt%) with a Ru–N environment [122]; or by a bottom up approach, producing Ru_{SA} (20% w/w) with a Ru–C environment [85]. Thus, in addition to the cohesive energy of the metal, the choice of the metal precursor/support couple, which will ultimately dictate the strength of the metal/support interaction must be carefully taken into consideration [154].

1.4.2 Metallic Species Homogeneity in Supported Metal Single Atom Catalysts

Two types of homogeneity can be considered when preparing a supported metal SAC. The first one deals with the selective deposition or not of metal SAs on the support (presence of clusters or NPs). The second one deals with the presence or not of a tailored local environment around the SA (SSHCs or not).

1.4.2.1 Are Clusters or Nanoparticles Present in Supported Metal Single Atom Catalysts?

As far as selective deposition of SAs (or double, triple atoms) is concerned, many synthetic strategies are at the disposal of the chemists (see *Chapters 2, 3, 4, and 15*). It is however important to ensure the homogeneity of the samples, in particular by checking the absence of clusters or nanoparticles. If microscopy offers a clear view, it nevertheless remains local, and analytical techniques offering an overview of the samples must be used in addition (*Chapter 5*). For this purpose, IR spectroscopy (with a probe molecule such as CO), EXAFS or XPS are commonly used. We would like to caution the reader here against the use of these techniques, as they can all present limitations to distinguish supported metal SA from clusters. We have already discussed in Section 1.3, the phenomenon of clusters disintegration under CO at room temperature, which can occur on the scale of the second. IR analyses at sub-ambient temperature is expected to be useful for identifying sample homogeneity, and monitoring possible catalyst reconstruction due to exposure of probe molecules via corollary *in-situ* or *ex-situ* scanning transmission electron microscopy (STEM) and X-ray absorption spectroscopy (XAS) characterization is critical for developing definitive probe molecule IR assignments [78]. For XPS analyses, first supported SACs present some challenges: they are often high surface area insulating powders and the metal loading can be very low (0.5 wt% or lower). In addition, new or inexperienced practitioners should avoid erroneous data collection and interpretation [155]. Finally, to make the distinction between SAs and clusters/NPs, the chemical state and most importantly the electronic state of the metal within a SA-containing material are probed by this technique. Thus, if a significant charge transfer is present, the SA is often found to be electron-deficient ($M^{\delta+}$) [156], while the same metal in an NP will be at zero oxidation state (M^0). It is nevertheless important to note that significant charge transfer can also be measured in the case of clusters of low nuclearity [157–160], making the distinction between SAs and clusters very delicate. Synchrotron radiation-based EXAFS is typically used in combination with HAADF-STEM to identify the SAs. However, it was shown that, in many cases, EXAFS is not sensitive enough to identify SAs, which may easily confuse the contributions from clusters or small nanoparticles due to the polydispersity and disorder effects and similar spectral shape [161]. A very careful measurement by HAADF-STEM and detailed IR/XPS/EXAFS analyses with reasonable comparison to the control samples are thus highly required to thoroughly identify the SAs and distinguish them from clusters or small NPs. Finally, if selective production of supported metal SACs is often the goal to study their specific reactivity, it is important to note that mixtures of SAs and clusters/NPs are often present in commercial catalysts [162]. During a catalytic reaction, all these various "sites" may contribute differently to the observed catalytic performance. If the very large majority of catalytic studies on supported metal SACs are limited to a comparison of the reactivity of SAs with that of metallic NPs; some studies have already demonstrated possible synergy and cooperative catalysis between SAs and NP [163–169]. This is a subject that definitively deserves more study in the future, and in that context, the control of the SA/NP ratio in a given catalyst is also challenging.

1.4.2.2 Control of the Local Environment of Single Atoms in Supported Metal Single Atom Catalysts

Structural regulation of catalytic performances in supported metal SACs involved a control of metal-support interactions (charge state and coordination environment) [24, 170, 171]. These characteristics are playing together to induce steric and electronic effects to determine the catalytic performances of supported metal SACs, as for homogeneous metal catalysts. Of course, and like in homogeneous catalysis, the charge state of the metal can have a pronounced effect on the catalytic activity. The effect of overall charge (particularly of positive charge) on Rh and Ir metal compound reactivity was discussed by Crabtree [172]. He clearly showed that in some cases, overall charge is the major factor determining reactivity, independent of the nature of the ligands. The control of the charge state of a given metal in supported metal SACs can be achieved by changing the nature of the support, if of course the interaction established between the SA and the support has not a predominantly covalent polar nature [173]. Note also that in photo- or electrocatalysis, other characteristics of the support should also be taken in consideration, such as the value of the band gap and its electronic conductivity. Thus, it has been shown that Au_{SA} are positively charged (Au(I)) on TiO_2 and CeO_2 but negatively charged (Au(0)) on ZrO_2, HfO_2, and ThO_2 [174]. This support-dependent oxidation states and charge distribution of Au_{SA} can influence the reactivity toward CO. While CO adsorbs strongly on Au_{SA}/TiO_2, a weaker adsorption occurs for Au_{SA}/ZrO_2. But for a given support the control of the charge state of a given metal (and therefore its reactivity) can also be achieved by changing the nature of the exposed support facet or adsorption site. For example, Au_{SA} on perfect $CeO_2(111)$ are reported to be inactive for CO oxidation due to the formation of negatively charged Au at the oxygen vacancy [175], but Li and coworkers have shown that the CeO_2 support with step sites stabilizes Au_{SA} with a positively charge state of +I, which exhibits much higher reactivity than the Au_{NP} [73]. Similarly, on the (111) and (100) surfaces of CeO_2, a Pt_{SA} exists at an oxidation state of +IV (upon replacement of Ce^{4+}), while on the (110) CeO_2 surface, it exhibits an oxidation state of +II due to the formation of surface peroxide O_2^{2-} species [176]. For CO oxidation, the order of reactivity follows $Pt_{SA}/CeO_2(110) < Pt_{SA}/CeO_2(111)/(100)$; but for methanol oxidation the reactivity order is $Pt_{SA}/CeO_2(100) > Pt_{SA}/CeO_2(110) > Pt_{SA}/CeO_2(111)$. Compared with $Pt_{SA}/CeO_2(110)$ and $Pt_{SA}/CeO_2(111)$, the Pt–O–Ce active interface over $Pt_{SA}/CeO_2(110)$ exhibits the highest intensity of distortion, which was correlated to its high activity. This discussion highlights that the local binding environment plays also a defining role in the catalytic performance, and that for some reactions the charge state alone is insufficient to explain the catalytic activity. Indeed, many different geometries can coexist with similar charge state, and perhaps only one might be catalytically active.

As far as the control of the local environments around the SA is concerned, it is obvious that the ability to tailor the atomic-scale structure of the active site is going to be key going forward in the field of supported metal SACs [177]. Indeed, in the absence of a well-defined coordination sphere around the metal SA, prediction and tuning of catalytic activity with supported metal SAsC will be challenging if the charge state of the metal is not the most important parameter, which is of course system dependent.

Additionally, the applicability of the d-band center and charge states often used to analyze the catalytic activity of transition metal surfaces/clusters can be unsure for supported metal SAC. Indeed, the spatial structure and orientation of frontier orbitals that are closest to the Fermi level of supported metal SACs play also a key role. This has been nicely illustrated by DFT on a study on Au_{SA} deposited on various C_3N supports [178]. In Au_{SA}/C_3N, by modifying the C_3N support, Au_{SA} can exist with different coordination modes and charge states, including Au^+, Au^-, and $Au^{(0)}$. Dioxygen and dihydrogen adsorption was investigated on these different species. Surprisingly, the Au_{SA}^+ is more active for gas adsorption than $Au_{SA}^{(0)}$ or Au_{SA}^-. In fact, the value of adsorption energy increases with the reduction of the number of electrons on Au. This appears to contradict with common believes that Au_{SA}^- are more active to dissociate these molecules, since they can inject electronic density in the unoccupied antibonding molecular orbitals of the adsorbates ($2\pi^*$ for O_2 or σ^* for H_2), which is essential for the dissociation of these molecules. The explanation of the difference of reactivity between the Au_{SA}^- and Au_{SA}^+ species in this specific case is the nature of the frontier orbital. For Au_{SA}^- the d_{z2} orbital is the highest in energy (Figure 1.8a), and for Au_{SA}^+ it is the d_{xy} (Figure 1.8b). The level of hybridization between the frontier orbitals of gold and the lowest unoccupied molecular orbital (LUMO) of the adsorbate dictates the binding energy in these two cases (see Figure 1.8 for the case of O_2). For Au_{SA}^+, the frontier orbital is close to the Fermi level, Au_{SA}–O_2 hybridization is strong and the $2\pi^*$ orbital of O_2 extends

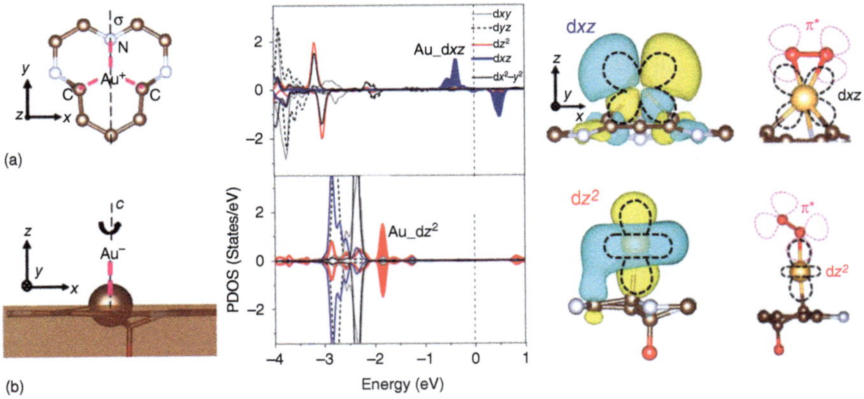

Figure 1.8 Projected density of states (PDOS) of Au d orbitals, calculated Wannier functions of frontier d orbitals of Au_{SA}, and different adsorption configurations of O_2 on Au_{SA} with two different frontier orbitals: (a) Au_{SA}^+; and (b) Au_{SA}^- species. Source: Fu et al. [178]. Reproduced with permission of American Physical Society.

its tail to below the Fermi level and, hence, receives some electrons, even if Au is in a positively charged state. The partial occupation of the antibonding state certainly activates the dissociation of O_2, which is not the case for the Au_{SA}^- species. Thus, the ability of hybridizing with the $2\pi^*$ orbital rather than the negative charge state of gold plays a major role in activating O_2 dissociation on these supported metal SACs.

Finally, beyond the control of the first coordination sphere, the control of the second sphere can also be important [24]. Indeed, atoms/functionalities in this second sphere can participate to the control of the activity [179] or the selectivity [180] of the supported metal SACs.

Various synthetic strategies have been developed for creating supported metal SACs with distinct coordination features. These include among others surface defect engineering, heteroatom doping, functional group grafting, spatial confinement, alloying, galvanic replacement or ionic liquid coating [170, 181, 182]. One exciting avenue is combining metal SA with sosme well-defined 2D materials such as g-C_3N_4 or MoS_2, with the aim to modulate the reactivity of metal atoms through unusual bonding- and electronic-environments. But even on these supports, different adsorption sites are present and both the nature of the adsorption site and energy of adsorption are metal dependent [183, 184]. Thus, the sixfold cavity of g-C_3N_4 is the energetically most favorable site for the deposition of SAs of Ti, Mn, Pd, Ag, Au, and W (site 1 on Figure 1.9a); while for Cr, Fe, Co, Ni, Cu, Mo, Ru, Rh, and Pt, the most favorable anchoring site is located at the corner of the sixfold cavity (site 2 on Figure 1.9a).

For MoS_2, it was found that, except for the Pd, Ag, and Au SAs, which adsorb preferentially on hollow (site 1 on Figure 1.9b), hollow, and topS sites (site 2 on Figure 1.9b), respectively, there is an energetic preference for the threefold top Mo adsorption sites (site 3 on Figure 1.9b). It is obvious that the research of “precise control” of the microenvironments of supported metal SACs is challenging, still at its infancy stage, but highly desired.

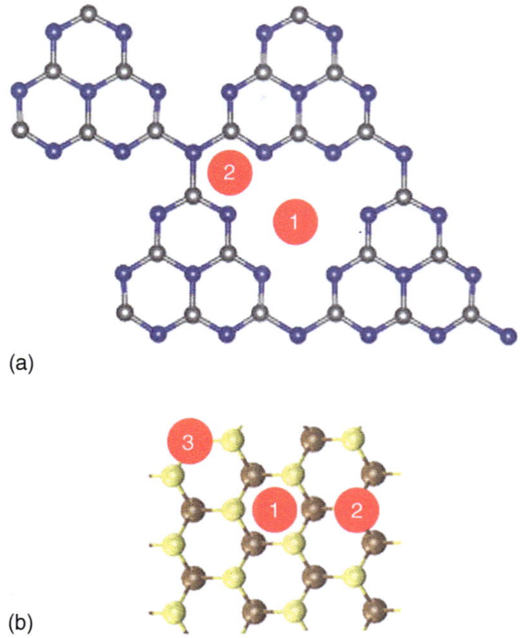

Figure 1.9 Adsorption sites for metal SAs on graphitic carbon nitride (a) and MoS_2 sheets (b). Source: Philippe Serp.

Another option consists in using bottom-up approaches. Thus, it is known that transition metal-metallofullerene can be produced easily by reacting the C_{60} fullerene with metallic precursors providing $C_{60}M_{SA}$ polymers with high metal loading [185–187]. These materials, which contain only metal SAs connected by C_{60} with a well-defined coordination could constitute ideal candidates as model carbon-supported metal SACs.

Unfortunately, precise determination of the active site geometry on a real metal supported SAC is still extremely difficult; and it is thus probably fair to say that in general the primary source of local structural information in SACs is DFT-based calculations. Finally, we should mention that even if a precise location of metal SAs is achieved after the preparation of the catalyst, metal SA stability and dynamic should also be considered, as it will be discussed in Section 1.4.3.

1.4.3 Metal Single Atom Stability and Dynamic in Supported Metal Single Atom Catalysts

1.4.3.1 Thermal and Chemical Stability

The stability of supported metal SACs always comes into question because SAs, if bound only weakly to a support, are susceptible to aggregation either during catalyst pre-activation and/or reaction processes. Developing an active and stable supported metal SA catalyst is challenging due to the high surface free energy of metal atoms. For transition metals commonly used in catalysis, the energy difference (ΔE) between the cohesive energy between metal atoms in the bulk phase and the thermodynamic driving force to form SA-support bonds is generally positive, i.e. thermodynamically it is more favorable to form M—M bonds rather than M-support bonds.

For oxide supports, the ΔE values for Ru, Rh, Pd, Pt, and Au all lay within the range of 100–200 kJ/mol (Figure 1.10a) [188], implying that additional means of stabilization during or after the SA synthesis need to be considered for those metals. The intrinsic stability of SAs arises from the support-assisted lower chemical potential when compared to NPs. When the free-energy change from NPs to SAs is negative, NPs can be dispersed to SAs spontaneously, which leads to thermodynamic stability of SAs (Figure 1.10b, black line). Since the binding strength depends on the p-d coupling between the SA and the support, the stable metal supported SA requests strong p-d coupling and deep states, while the unbound states near Fermi level may weaken the stability. However, if the free-energy change is positive, the SAs will be stable only if the aggregation barrier is high enough to prevent sintering, which corresponds to the kinetic stability of SAs (Figure 1.10b, dotted line) [189].

Even though thermally stable supported metal SAs have been successfully produced using reducible supports prone to generate oxygen vacancies to stabilize SAs [149, 190–194], chemical stability issues due to the presence of oxidative or reducing reactants and high temperature under operating conditions should also be considered [195]. Generally, the more weakly the surface metal atom is attached to the support, the more strongly it binds small adsorbates [196]. Consequently, both thermodynamic and kinetic criteria have to be considered to determine the stability of supported metal SAs [189]. The thermodynamic part includes: (i) energetics of supported metal NPs, which is based on the Gibbs–Thomson relation with considering the adsorbed reactants; and (ii) the chemical potential of *monomers* (both the metal SA and SA-reactant complexes) on the support.

The kinetic part includes: (i) the diffusion barrier of monomers on perfect surfaces and defects; and (ii) the barrier of moving one metal atom from a supported metal NP to a substrate surface with corresponding sintering rate equations [73, 75].

CO stabilize a Pt_{SA} more than a Pt_{NP} because of the stronger adsorption of CO to a Pt_{SA} and also because only a fraction of the Pt atoms in an NP is covered by CO, as shown in Figure 1.10c. We already discussed in Section 1.2, the case of Rh_{NP} that spontaneously disintegrated into Rh_{SA} under CO at room temperature. But at higher temperature, the Rh complexes decomposed, and the Rh_{SA} released started to agglomerate and form metal NPs. Similarly, reactant-assisted ripening/disintegration had also been reported when supported metal NPs were exposed to dioxygen [197–201]. The reason was attributed to the formation of volatile oxygen–metal complexes [202]. Interestingly, Corma and coworkers have shown that by controlling the activation treatment (in that case under H_2 or O_2), it was possible to prepare catalysts in which the nature of the Pt species (Pt_{SA} formed under O_2 or Pt_{NP} formed under H_2) can be tuned reversibly [203]. Note that H_2 activation is not always associated to SA instability. Thus, the addition of H_2 is beneficial for the stabilization of Ir_{SA}/C catalysts by suppressing the formation of volatile Ir complexes during methanol carbonylation to acetic acid [204]. Of course, metal supported SA formation by the disintegration of metallic NPs will be strongly dependent on NP density and the concentration of support defect sites [71, 205]. The energetics of supported NP/SA during Ostwald ripening and reactant assisted (here carbon monoxide-assisted) Ostwald ripening are presented in Figure 1.11a,b [75].

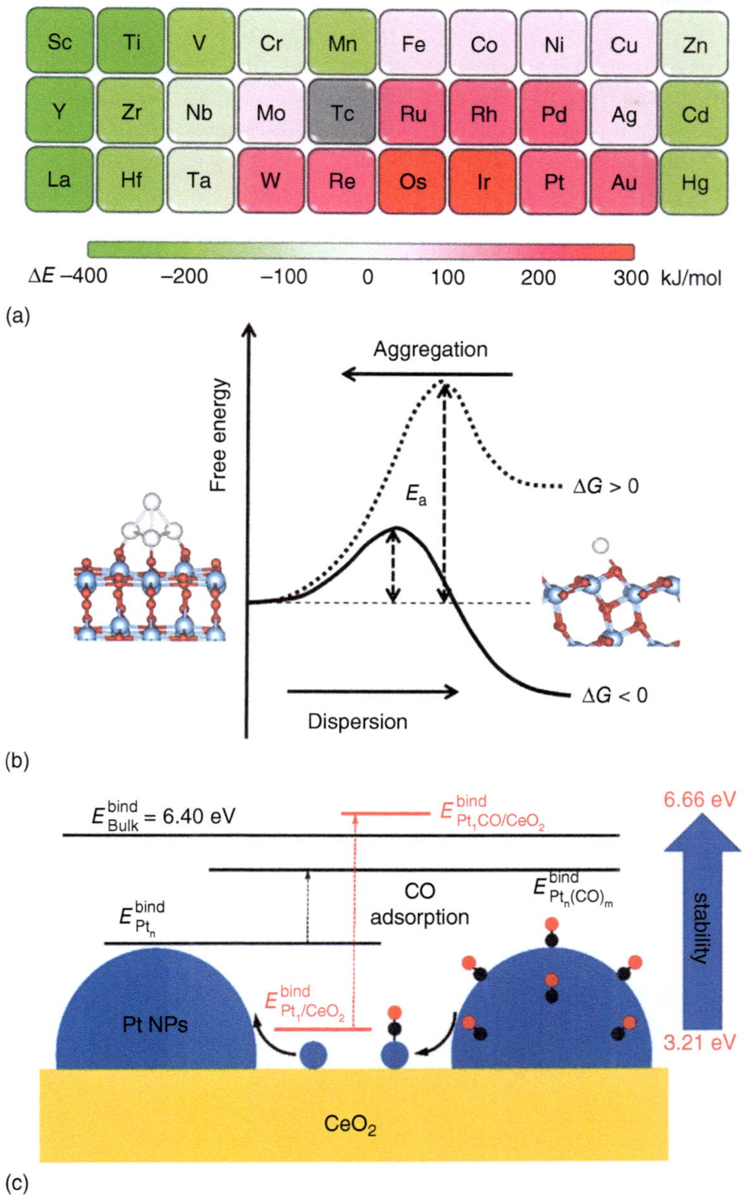

Figure 1.10 (a) Difference (ΔE) between the cohesive energy between metal atoms in the bulk phase and the thermodynamic driving force to form metal–oxygen bonds in kJ/mol. The lowest value indicates the highest stability of M—O bonds and thus the lowest tendency to form a metal NP. Technetium is not considered here. Source: Hülsey et al. [188]. Reproduced with permission of Wiley-VCH. (b) Schematic illustration of free-energy diagram of sintering and dispersion processes between NPs and SAs supported on an oxide. Source: Liu et al. [189]. Oxford University Press. CC BY 4.0. (c) Schematic illustration of the relative stability of Pt_{SA} and Pt_{NP} on CeO_2 in the presence of adsorbed CO molecules. Source: Su et al. [74]. Reproduced with permission of American Chemical Society.

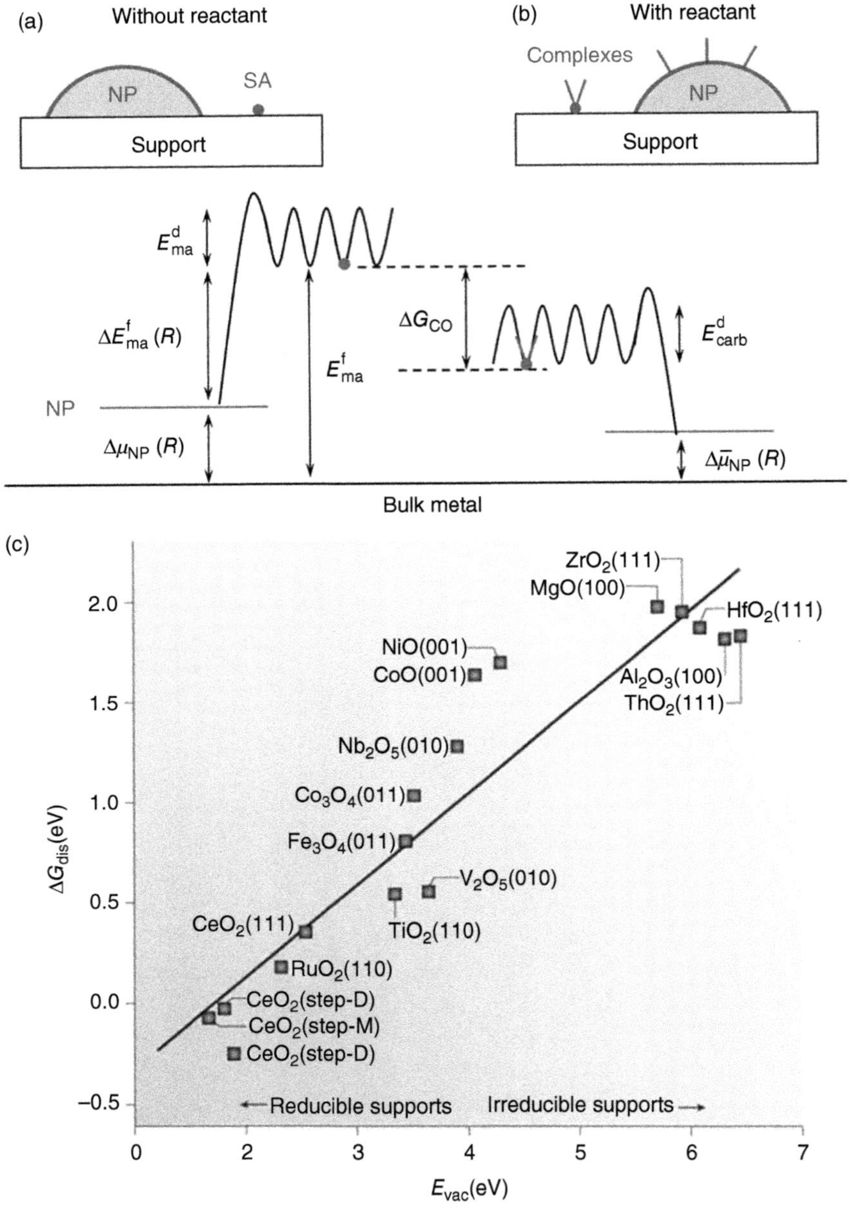

Figure 1.11 Energy diagram of supported NP/SA without (a) and with (b) the presence of CO. Here, $\Delta\mu_{NP}(R)$ and $\Delta\overline{\mu}_{NP}(R)$ are the chemical potentials of supported metal NP, $E^f{}_{ma}$ and $\Delta E^f{}_{ma}$ (R) are the formation energies of monomers (the metal SA) on support with respect to infinite and finite size metal particle, ΔG_{CO} is the Gibbs free energy of adsorption of reactants CO on the metal SA, and $E^d{}_{ma}$ and $E^d{}_{carb}$ are the diffusion barriers of monomers (the metal SA and the SA–reactant complexes) on support. Source: Ouyang et al. [75]. Reproduced with permission of American Chemical Society. (c) The Gibbs free energy change for Au cluster disintegration (ΔG_{dis}) can be plotted against the energy associated with the formation of support vacancies (E_{vac}). The more reducible the support, the more stable the single Au centers are after dissociating from an Au cluster. Source: Wang et al. [12]. Reproduced with permission of Springer Nature.

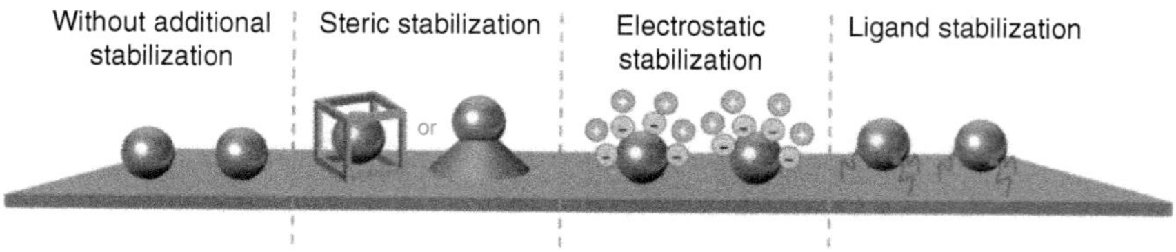

Figure 1.12 Schematic representation of stabilization principles in colloidal chemistry applicable to the synthesis of stable single-atom catalytic systems. Source: Hülsey et al. [188]. Reproduced with permission of Wiley VCH.

The first step of sintering by Ostwald ripening, i.e. the release of metal SAs from NPs is paradoxically the basis for metal redispersion in some specific cases. Ripening occurs when large thermodynamically stable NPs trap the metal SAs released by the smaller NPs after their fragmentation.

But if the support provides strong metal-support interactions such as those typically observed on reducible oxides (CeO_2 or TiO_2), the mobile metal SAs can be intercepted before further aggregation phenomena occur, limiting or even eliminating the problem (Figure 1.11c). In some cases, the support sites necessary for the enthalpic stabilization of SAs, for example O_v, can be formed *in situ* during the catalytic reaction, as shown in the case of Pd/TiO_2 catalysts for the reverse water gas shift reaction [205].

If it appears obvious that the stability control in supported SACs should involve strong electronic metal-support interactions, other strategies have also been proposed, mainly based on knowledge originating from colloidal chemistry (Figure 1.12) [188]. Thus, electrostatic stabilization of supported SA catalysts by ionic liquids (ILs) provides sufficient protection to isolated SAs by increasing the kinetic barrier for the formation of metal–metal bonds on the support [206]. Employing a steric barrier during the reduction of the metal precursor, for example by iced photochemical reduction, led to the formation of SAs even in an aqueous solid matrix that can then be transferred and bind strongly to different supports [207].

Today, a growing body of evidence suggests that supported SAs through strong covalent bonds can be more stable than their NP counterparts and are even resistant to sintering during reactions at high temperatures. If several theoretical models have been developed to correlate the structural stability or catalytic activity with the adsorption strength of reaction intermediate, d-band center of transition metals, and e.g. filling number, and charge transfer ability, the simultaneous correlation of structural stability and catalytic activity with a specific parameter has not been reported so far. Simultaneous enhancement of thermostability and catalytic activity is thus questionable. An attempt was made by Liu and coworkers to obtain a rational description for structural stability and catalytic activity of supported SA catalysts (Figure 1.13) [208]. The parameters associated with competitive distribution of free electrons near the Fermi level should be considered during simultaneously combined optimization of structural stability and catalytic activity. The fully occupied bonding orbitals correspond to the stability between the SAs, and the authors addressed the relationship between the stability of supported SAs and the amount of free electrons near Fermi

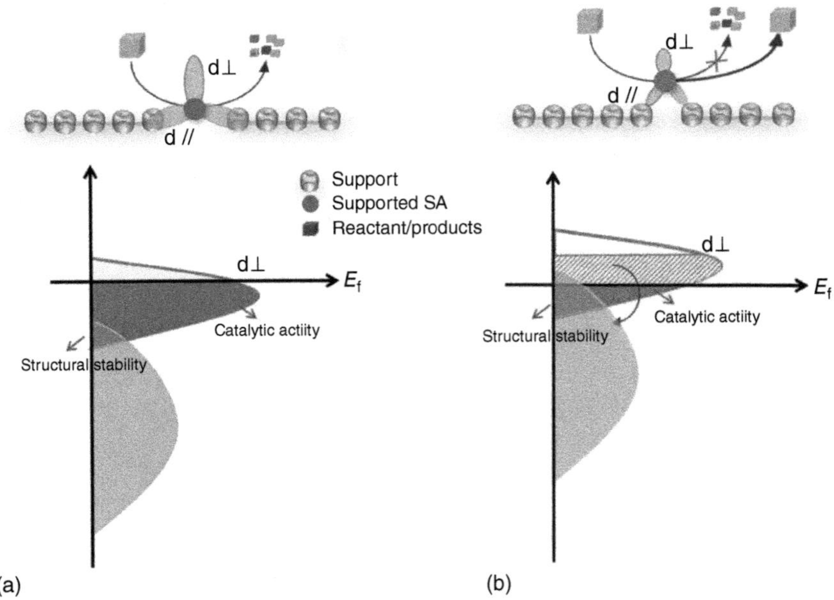

Figure 1.13 Illustration of competitive relationship between (a) catalytic activity and (b) structural stability of supported single atom catalysts. Source: Wang et al. [208]. Reproduced with permission of Elsevier.

level using the parameter of binding strength. The free electrons near Fermi level caused by unsaturated coordination can be responsible for the catalytic activity, and the relationship between catalytic activity and the densities of free electrons was considered to optimize the chemisorption. Compounds without such free electrons may not display excellent catalytic activity. Therefore, the SAs should be strongly coupled to the supports, and the localized structural coordination should be unsaturated to form a dangling bond near the Fermi level [209]. Thus, combining the conventional screening criteria of catalyst stability (formation energy) and catalytic activity (adsorption energy), for example by using high-throughput prediction and machine learning, can provide new insights into understanding and developing new supported SACs combining structural stability and high activity. Thus, combination of DFT calculations and various machine learning methods have already been used to investigate catalyst stability or reactivity [210–214]. Scaling relation between the diffusion activation barrier (E_a) of a metal SA on a support (a most relevant factor to its stability) and $(E_{bind})^2/E_c$ has been reported; where E_{bind} is the binding energy of the SA and E_c the metal cohesive energy [210].

1.4.3.2 Supported Single Atom Dynamics in Chemical Reactions

The structures of supported metal SACs may change during catalysis (dynamic evolution) because of well-known restructuring phenomena induced by chemical reactions and/or high temperatures. Understanding the dynamism of supported metal SACs under reaction conditions is important [215, 216], especially given the possible reversible dynamical conversion of NPs into SAs during catalysis.

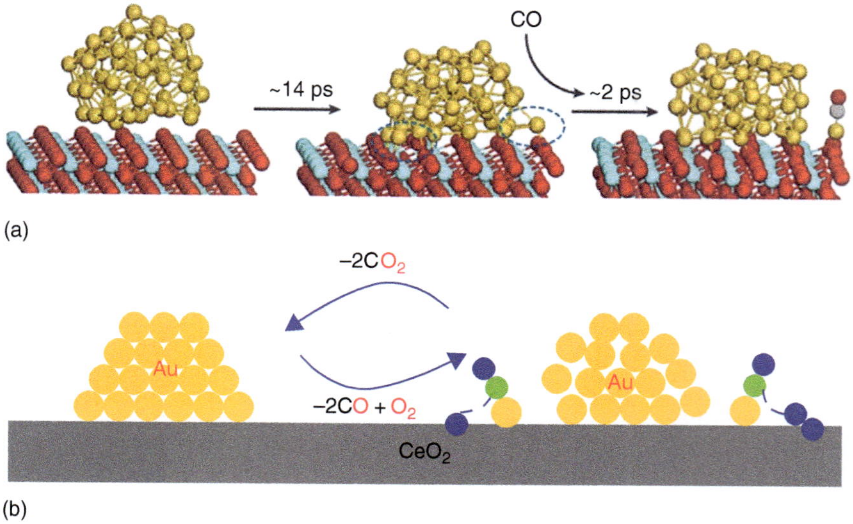

Figure 1.14 (a) Snapshots from a molecular dynamics simulation of a CeO_2-supported Au_{50} cluster show that a Au_{SA} can dissociate from the cluster to bind CO. Source: Wang et al. [12]. Reproduced with permission of Springer Nature. (b) Schematic representation of the dynamic behavior at the interface during CO oxidation on Au/CeO_2 catalysts. Source: Wang et al. [72]. Springer Nature. CC BY 4.0.

Thus, dynamic (co)catalysis of Au_{SA} has been reported in the case of CO oxidation on Au/CeO_2 catalysts [72], and methane pyrolysis for producing carbon and dihydrogen on gold nanoparticles [217]. Rousseau and coworkers used *ab initio* molecular dynamics simulations to reveal dynamic SA catalytic mechanism for the CO oxidation by CeO_2-supported gold clusters [72]. The dynamic SA catalytic mechanism results from the ability of $Au^{\delta+}$ species to strongly couple with the redox properties of CeO_2 in a synergistic manner, thereby lowering the energy of redox reactions. The $Au^{\delta+}$ can break away from the Au_{NP} (Figure 1.14a) to catalyze CO oxidation, adjacent to the metal/oxide interface and subsequently reintegrate back into the NP after the reaction is completed (Figure 1.14b).

Such a dynamic phenomenon is reminiscent of that observed in the case of Pd_{NP} for carbon–carbon coupling reactions [218, 219]. Interestingly, on Pt_{SA}/Al_2O_3 catalysts, a reverse phenomenon was observed [220]. It was first shown that the Pt_{SA}/Al_2O_3 catalysts are poorly active for CO oxidation. However, the Pt_{SA} gradually and irreversibly convert into highly active Pt clusters throughout the heating-cooling reaction cycles, even under highly oxidizing conditions favorable to atomic dispersion. Similarly, the *in situ* transformation of isolated $Cu_{SA}N_4$ into Cu clusters (the real active species) has been evidenced during the electrocatalytic reduction of CO_2 into ethanol [221]. Interestingly, the Cu clusters redispersed into isolated Cu_{SA} following exposure to air. These two latter studies show that care must be taken before concluding on the nature of the active sites in supported metal SACs, even at very low metal loading. Experimental evidence for the co-catalysis between Au_{SA} and Au_{NP} was also evidenced during methane pyrolysis [217]. It was demonstrated that Au_{NP} surface partially disintegrate, releasing Au_{SA}. DFT calculations have shown that the Au_{SA} could co-catalyze the reaction with Au_{NP}. Moreover, the Au_{SA} dynamically aggregate into Au_{NP}, which re-disintegrate back to Au_{SA}.

Another important dynamic phenomenon in supported metal SACs is the change of local coordination environments and/or oxidation states/charge transfer that can occur during the reaction process. A limited number of studies deal with this important point. Dynamic charge and oxidation state of Pt_{SA}/CeO_2 catalysts were examined by combining DFT and first-principles molecular dynamics by N. López and coworkers [222]. The authors also demonstrated how the reactivity for CO oxidation is closely related to this dynamic behavior. Their work raises the question of the common assignment of a fixed oxidation state in supported metal SACs, which is an oversimplification. Indeed, several well-defined charge states that were dynamically interconnected depending on the Pt–CeO_2 combination and thus coexist were identified. Pt^{2+}–4O was the resting state of the material under most conditions; and only after CO (or H_2) treatment a significant pressure, the Pt–4O coordination was reduced to Pt–3O. The Pt–3O system is labile and converts into Pt–2O in <0.5 ps at room temperature. During the CO oxidation cycle on Pt–2O, the electronic structure of the active site needs to be dynamic to allow both CO adsorption (occurring at ionic Pt) and oxidation (occurring on Pt^0) elementary steps in the mechanism. Three Pt oxidation states coexist on Pt–2O (Pt^0 47% total lifetime, Pt^{1+} 49% total lifetime and Pt^{2+} 4% total lifetime). When CO is adsorbed, the new intermediate shows larger lifetimes for the ionic Pt configurations (Pt^{1+} 36% total lifetime and Pt^{2+} 63% total lifetime). Then, O_2 can adsorb, forming a species for which the formal oxidation state of the metal is Pt^{2+}, regardless of its previous state. From this configuration on, CO_2 formation easily pushes the system to the low-lying Pt^0 state (Figure 1.15).

This dynamic evolution of the metal oxidation states reminds the concept of non-innocent ligands in coordination [223–225] or bioinorganic [226] chemistry. This type of ligands is either redox-active ("redox non-innocent") or actively involved in bond-making or -breaking processes ("chemically non-innocent"), and contribute to ligand-mediated oxidation state changes. Structural evolution of atomically dispersed Pt_{SA}/TiO_2 catalysts was also evidenced through a combination of *in situ* atomic-resolution microscopy and spectroscopy based characterizations, and supported by first-principles calculations [227]. It was demonstrated that Pt_{SA} can adopt a range of local coordination environments and oxidation states, which evolve in response to varied environmental conditions, similarly to the dynamic behavior of copper species ions in zeolites [228]. This variation in local coordination and electronic state of Pt species showed a strong influence on the catalytic performance for CO oxidation on systems where the Pt_{SA} active site was adsorbed at well-defined locations on the support. It was proposed, based on EXAFS and STEM analyses and DFT calculations that oxidation of Pt_{SA}/TiO_2 resulted in Pt substituting into the Ti_{6c} position ($Pt^{+4}{}_{Ti6c}$); mild reduction pulled Pt_{SA} out of the lattice to form Pt^{+2}-O_2 species; and harsh reduction induced the formation of Pt^{+1}-OH species that were mobile and adsorbed to both step and terrace sites on the support (Figure 1.16). It was expected that these changes were driven not only by Pt_{SA} energetics, but also by structural transformations of reducible TiO_2

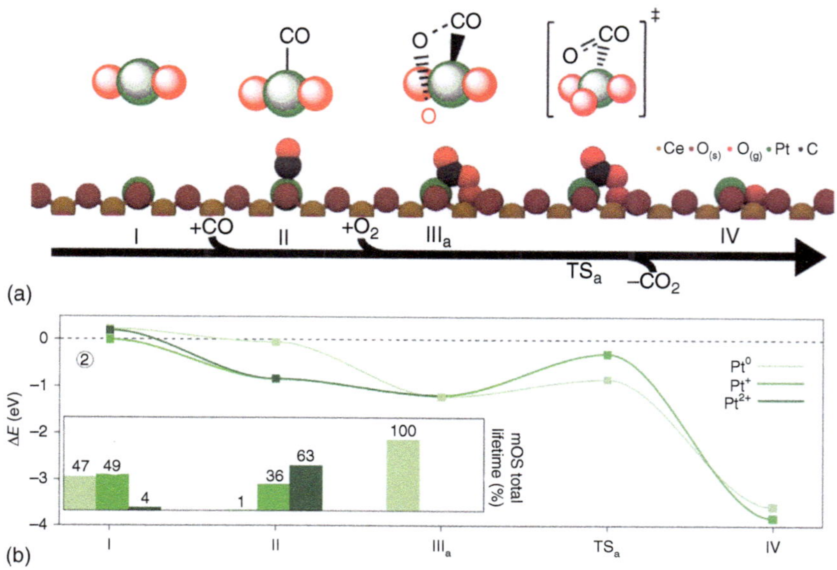

Figure 1.15 (a) Side views of the reaction steps of CO oxidation. Dark red denotes oxygen from the bulk phase (s) and bright red denotes oxygen from the gas phase (g). (b) Reaction profile of CO oxidation for each oxidation state of Pt. Source: Daelman et al. [222]. Reproduced with permission of Springer Nature.

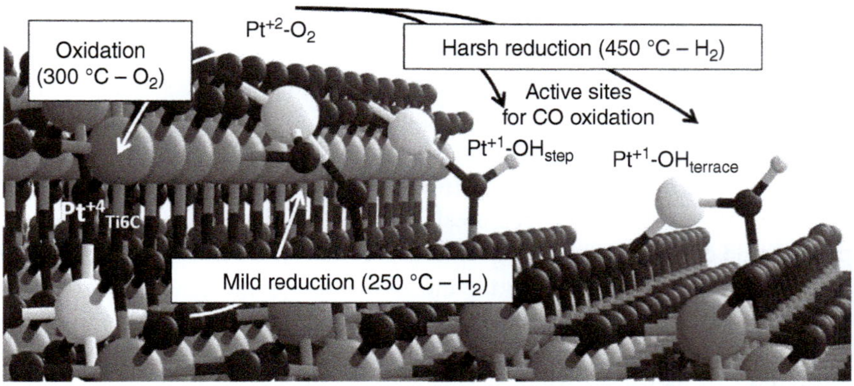

Figure 1.16 Schematic showing the proposed dynamic evolution of Pt_{SA}/TiO_2 catalysts following oxidation, mild reduction, and harsh reduction. Structures were derived from DFT calculations and are consistent with experimental data. Source: DeRita et al. [227]. Reproduced with permission of Springer Nature.

support. Following harsh reduction, Pt_{SA} sites showed an increase in activity for CO oxidation of two to fivefold, depending on temperature, compared to Pt_{SA} sites that have followed oxidation or mild reduction [227].

Direct observation by synchronous illumination XPS of dynamic bond evolution in Pt_{SA}/C_3N_4 catalysts during photocatalytic water splitting was also reported by Bi and coworkers [229].

The graphitic C_3N_4 support containing electron-rich N atoms provides sufficient sites to stabilize $Pt^{2+}{}_{SA}$ through the formation of Pt—N bonds. The authors investigated the charge transfer and chemical bond evolution of such catalyst under light irradiation. The dynamic variations of Pt—N bond cleavage into Pt^0 and C=N bond could be experimentally observed under light irradiation. This transformation implies that in excitation states, the Pt^0 and C_3N_4 should be in separated states without any bonding, and only spatial confinement effects and van der Waals forces may insure the anchoring of the Pt_{SA} atoms on the support under light irradiation. Therefore, in excitation states, the electron-rich $Pt_{SA}{}^0$ atoms and the hole-rich C_3N_4 layers in separated states could participate in water reduction and oxidation, respectively.

These few examples clearly show that an improved understanding and description of dynamic behaviors, especially under operating conditions, will be invaluable.

1.4.4 Obtaining Reliable Information About the Active Sites of Metal SACs

Homogeneous catalysts present the advantage of well-defined single-active site on a molecular level. Supramolecular catalysts have been dominated by enzyme-inspired approaches; i.e. the assembly of catalytic species by harnessing multiple weak intramolecular interactions. In enzymatic catalysis, the simple model often used to describe enzyme activity is known as the lock-and-key model in which enzymes accelerate reactions by providing a tight-fitting area, known as the active site, where substrate molecules can react. For these latter systems, identification of the active site is more complex and relies more and more on computational approaches [230, 231]. In the case of supported metal SACs, obviously some similarities exist with homogeneous or enzymatic catalysis, but also some differences, which make that obtaining reliable information about the active site is complex. The first difficulty arise from the fact that most (if not all) of the supported metal SACs are not SSHCs. Furthermore, there is a possibility that the coordinated atom(s) of the support would be involved in the reactions as a co-catalytic site. Another important aspect to take into consideration is the intrinsic nature of the ligands, a molecular species in one case and a solid for supported metal SACs. Indeed, this solid can present redox properties, be constituted by a second metal, be a platform for reactant diffusion or for spillover (see Figure 1.1); all this richness, this diversity but also complexity are not present, or to a lesser extent in homogeneous catalysis with molecular complexes. In that context, the use of model supports, presenting few and well-defined anchoring possibilities for the metal should be encouraged [156].

We already discussed in this chapter the potentials and limitations of spectroscopic techniques for supported metal SACs, and they are further developed in *Chapter 5*. The advantages of *in situ/operando* techniques for capturing the reaction intermediates, identifying the active sites, and even monitoring the dynamic behaviors of both the geometric structure and electronic environment of catalytic sites in supported metal SACs should of course be considered [216, 232, 233]. They are discussed in *Chapter 6*. Finally, the coupling of spectroscopic studies with modeling works is today a prerequisite to the understanding of the nature of the active site. Combinations of DFT and molecular dynamics and related computational algorithms have already been used with success (see *Chapter 7*) [234]. However, the interaction between the environment and the reaction intermediate species, which plays an important role in the exploration of catalytic mechanisms, is often neglected. Therefore, multiscale computational modeling approach should be developed.

Acknowledgments

This work was supported by the Agence Nationale de la Recherche (project ANR-19-CE07-0030), which is gratefully acknowledged.

References

1 Lykhach, Y., Kozlov, S.M., Skála, T. et al. (2016). Counting electrons on supported nanoparticles. *Nature Materials* 15 (3): 284–288. https://doi.org/10.1038/nmat4500.

2 Binninger, T., Schmidt, T.J., and Kramer, D. (2017). Capacitive electronic metal-support interactions: outer surface charging of supported catalyst particles. *Physical Review B* 96 (16): 165405. https://doi.org/10.1103/PhysRevB.96.165405.

3 Jia, X., Zhang, X., Rui, N. et al. (2019). Structural effect of Ni/ZrO_2 catalyst on CO_2 methanation with enhanced activity. *Applied Catalysis B: Environmental* 244: 159–169. https://doi.org/10.1016/j.apcatb.2018.11.024.

4 Lv, C., Xu, L., Chen, M. et al. (2020). Recent progresses in constructing the highly efficient Ni based catalysts with advanced low-temperature activity toward CO_2 methanation. *Frontiers in Chemistry* 8 (269) https://doi.org/10.3389/fchem.2020.00269.

5 Marzke, R.F. (1979). Quantum size effects in small metallic particles. *Catalysis Reviews* 19 (1): 43–65. https://doi.org/10.1080/03602457908065100.

6 Strizhak, P.E. (2013). Nanosize effects in heterogeneous catalysis. *Theoretical and Experimental Chemistry* 49 (1): 2–21. https://doi.org/10.1007/s11237-013-9297-7.

7 Che, M. and Bennett, C.O. (1989). The influence of particle size on the catalytic properties of supported metals. In: *Advances in Catalysis*, vol. 36 (eds. D.D. Eley, H. Pines and P.B. Weisz), 55–172. Academic Press.

8 Van Santen, R.A. (2009). Complementary structure sensitive and insensitive catalytic relationships. *Accounts of Chemical Research* 42 (1): 57–66. https://doi.org/10.1021/ar800022m.

9 Vogt, C., Groeneveld, E., Kamsma, G. et al. (2018). Unravelling structure sensitivity in CO_2 hydrogenation over nickel. *Nature Catalysis* 1 (2): 127–134. https://doi.org/10.1038/s41929-017-0016-y.

10 Sarma, B.B., Plessow, P.N., Agostini, G. et al. (2020). Metal-specific reactivity in single-atom catalysts: CO oxidation on 4d and 5d transition metals atomically dispersed on MgO. *Journal of the American Chemical Society* 142 (35): 14890–14902. https://doi.org/10.1021/jacs.0c03627.

11 Yang, X.-F., Wang, A., Qiao, B. et al. (2013). Single-atom catalysts: a new frontier in heterogeneous catalysis. *Accounts of Chemical Research* 46 (8): 1740–1748. https://doi.org/10.1021/ar300361m.

12 Wang, A., Li, J., and Zhang, T. (2018). Heterogeneous single-atom catalysis. *Nature Reviews Chemistry* 2 (6): 65–81. https://doi.org/10.1038/s41570-018-0010-1.

13 Liu, L. and Corma, A. (2018). Metal catalysts for heterogeneous catalysis: from single atoms to nanoclusters and nanoparticles. *Chemical Reviews* 118 (10): 4981–5079. https://doi.org/10.1021/acs.chemrev.7b00776.

14 Mitchell, S., Thomas, J.M., and Pérez-Ramírez, J. (2017). Single atom catalysis. *Catalysis Science & Technology* 7 (19): 4248–4249. https://doi.org/10.1039/C7CY90090B.

15 Kaiser, S.K., Chen, Z., Faust, A.D. et al. (2020). Single-atom catalysts across the periodic table. *Chemical Reviews* https://doi.org/10.1021/acs.chemrev.0c00576.

16 Samantaray, M.K., D'Elia, V., Pump, E. et al. (2020). The comparison between single atom catalysis and surface organometallic catalysis. *Chemical Reviews* 120 (2): 734–813. https://doi.org/10.1021/acs.chemrev.9b00238.

17 Cui, X., Li, W., Ryabchuk, P. et al. (2018). Bridging homogeneous and heterogeneous catalysis by heterogeneous single-metal-site catalysts. *Nature Catalysis* 1 (6): 385–397. https://doi.org/10.1038/s41929-018-0090-9.

18 Chen, F., Jiang, X., Zhang, L. et al. (2018). Single-atom catalysis: bridging the homo- and heterogeneous catalysis. *Chinese Journal of Catalysis* 39 (5): 893–898. https://doi.org/10.1016/S1872-2067(18)63047-5.

19 Liu, F., Yang, T., Yang, J. et al. (2019). Bridging the homogeneous-heterogeneous divide: modeling spin for reactivity in single atom catalysis. *Frontiers in Chemistry* 7 (219) https://doi.org/10.3389/fchem.2019.00219.

20 Liu, P. and Zheng, N. (2018). Coordination chemistry of atomically dispersed catalysts. *National Science Review* 5 (5): 636–638. https://doi.org/10.1093/nsr/nwy051.

21 Jørgensen, C.K. (1966). Differences between the four halide ligands, and discussion remarks on trigonal-bipyramidal complexes, on oxidation states, and on diagonal elements of one-electron energy. *Coordination Chemistry Reviews* 1 (1): 164–178. https://doi.org/10.1016/S0010-8545(00)80170-8.

22 van der Vlugt, J.I. (2019). Radical-type reactivity and catalysis by single-electron transfer to or from redox-active ligands. *Chemistry – A European Journal* 25 (11): 2651–2662. https://doi.org/10.1002/chem.201802606.

23 Wodrich, M.D. and Hu, X. (2017). Natural inspirations for metal–ligand cooperative catalysis. *Nature Reviews Chemistry* 2 (1): 0099. https://doi.org/10.1038/s41570-017-0099.

24 Qin, R., Liu, K., Wu, Q. et al. (2020). Surface coordination chemistry of atomically dispersed metal catalysts. *Chemical Reviews* **120** (21): 11810–11899. https://doi.org/10.1021/acs.chemrev.0c00094.

25 Hülsey, M.J., Lim, C.W., and Yan, N. (2020). Promoting heterogeneous catalysis beyond catalyst design. *Chemical Science* 11 (6): 1456–1468. https://doi.org/10.1039/C9SC05947D.

26 Passos, A.R., Rochet, A., Manente, L.M. et al. (2020). Three-dimensional strain dynamics govern the hysteresis in heterogeneous catalysis. *Nature Communications* 11 (1): 4733. https://doi.org/10.1038/s41467-020-18622-2.

27 Khorshidi, A., Violet, J., Hashemi, J. et al. (2018). How strain can break the scaling relations of catalysis. *Nature Catalysis* 1 (4): 263–268. https://doi.org/10.1038/s41929-018-0054-0.

28 Zhao, K., Zhu, Y., Shi, J. et al. (2019). Synergetic effects of strain engineering and substrate defects on generating highly efficient single-atom catalysts for CO oxidation. *Journal of Materials Chemistry A* 7 (15): 9297–9304. https://doi.org/10.1039/C9TA01326A.

29 Kohen, A. (2015). Role of dynamics in enzyme catalysis: substantial versus semantic controversies. *Accounts of Chemical Research* 48 (2): 466–473. https://doi.org/10.1021/ar500322s.

30 Narayanan, C., Bernard, D.N., Bafna, K. et al. (2018). Ligand-induced variations in structural and dynamical properties within an enzyme superfamily. *Frontiers in Molecular Biosciences* 5 (54) https://doi.org/10.3389/fmolb.2018.00054.

31 Kahraman, A. and Thornton, J.M. (2008). Methods to characterize the structure of enzyme binding sites. *Computational Structural Biology*: 189–221.

32 Brown, C.J., Toste, F.D., Bergman, R.G. et al. (2015). Supramolecular catalysis in metal–ligand cluster hosts. *Chemical Reviews* 115 (9): 3012–3035. https://doi.org/10.1021/cr4001226.

33 Serp, P. (2021). Cooperativity in supported metal single atom catalysis. *Nanoscale* 13: 5985–6004. https://doi.org/10.1039/D1NR00465D.

34 Thomas, J.M., Raja, R., and Lewis, D.W. (2005). Single-site heterogeneous catalysts. *Angewandte Chemie International Edition* 44 (40): 6456–6482. https://doi.org/10.1002/anie.200462473.

35 Liu, J. (2017). Catalysis by supported single metal atoms. *ACS Catalysis* 7 (1): 34–59. https://doi.org/10.1021/acscatal.6b01534.

36 Thomas, J.M. (2019). The periodic table, zeolites and single-site heterogeneous catalysts. In: *The Periodic Table II: Catalytic, Materials, Biological and Medical Applications* (ed. D.M.P. Mingos), 37–52. Cham: Springer International Publishing.

37 Rogge, S.M.J., Bavykina, A., Hajek, J. et al. (2017). Metal–organic and covalent organic frameworks as single-site catalysts. *Chemical Society Reviews* 46 (11): 3134–3184. https://doi.org/10.1039/C7CS00033B.

38 Pagliaro, M. (2020). *laquo*Catalysis: a unified approachʊ: a new course in catalysis science and technology. *Journal of Flow Chemistry* 11: 53–58. https://doi.org/10.1007/s41981-020-00100-x.

39 Mitchell, S. and Pérez-Ramírez, J. (2020). Single atom catalysis: a decade of stunning progress and the promise for a bright future. *Nature Communications* 11 (1): 4302. https://doi.org/10.1038/s41467-020-18182-5.

40 Wu, J., Xiong, L., Zhao, B. et al. (2020). Densely populated single atom catalysts. *Small Methods* 4 (2): 1900540. https://doi.org/10.1002/smtd.201900540.

41 National Academy of Engineering (2016). *Memorial Tributes*, vol. 20, 394. Washington, DC: The National Academies Press.

42 Liu, Y., Li, Z., Yu, Q. et al. (2019). A general strategy for fabricating isolated single metal atomic site catalysts in Y zeolite. *Journal of the American Chemical Society* 141 (23): 9305–9311. https://doi.org/10.1021/jacs.9b02936.

43 Zhang, S., Chen, L., Qi, Z. et al. (2020). Insights into the mechanism of n-hexane reforming over a single-site platinum catalyst. *Journal of the American Chemical Society* 142 (39): 16533–16537. https://doi.org/10.1021/jacs.0c07911.

44 Kwak, J.H., Hu, J., Mei, D. et al. (2009). Coordinatively unsaturated Al^{3+} centers as binding sites for active catalyst phases of platinum on γ-Al_2O_3. *Science* 325 (5948): 1670–1673. https://doi.org/10.1126/science.1176745.

45 Johnston, P., Carthey, N., and Hutchings, G.J. (2015). Discovery, development, and commercialization of gold catalysts for acetylene hydrochlorination. *Journal of the American Chemical Society* 137 (46): 14548–14557. https://doi.org/10.1021/jacs.5b07752.

46 Malta, G., Kondrat, S.A., Freakley, S.J. et al. (2017). Identification of single-site gold catalysis in acetylene hydrochlorination. *Science* 355 (6332): 1399–1403. https://doi.org/10.1126/science.aal3439.

47 Kaiser, S.K., Fako, E., Manzocchi, G. et al. (2020). Nanostructuring unlocks high performance of platinum single-atom catalysts for stable vinyl chloride production. *Nature Catalysis* 3 (4): 376–385. https://doi.org/10.1038/s41929-020-0431-3.

48 Yao, H.C. and Bettman, M. (1976). On the existence of two phases of cobalt oxide on a zirconia support. *Journal of Catalysis* 41 (3): 349–358. https://doi.org/10.1016/0021-9517(76)90235-9.

49 Poole, C.P. and MacIver, D.S. (1967). The physical-chemical properties of chromia-alumina catalysts. In: *Advances in Catalysis*, vol. 17 (eds. D.D. Eley, H. Pines and P.B. Weisz), 223–314. Academic Press.

50 O'Reilly, D.E. and MacIver, D.S. (1962). Electron paramagnetic resonance absorption of chromia—alumina catalysts[1]. *The Journal of Physical Chemistry* 66 (2): 276–281. https://doi.org/10.1021/j100808a021.

51 Yao, H.C. and Shelef, M. (1976). Surface interactions in the system Reγ-Al_2O_3. *Journal of Catalysis* 44 (3): 392–403. https://doi.org/10.1016/0021-9517(76)90416-4.

52 Via, G.H., Meitzner, G., Lytle, F.W. et al. (1983). Extended X-ray absorption fine structure (EXAFS) of highly dispersed rhodium catalysts. *The Journal of Chemical Physics* 79 (3): 1527–1529. https://doi.org/10.1063/1.445945.

53 Yates, D.J.C., Murrell, L.L., and Prestridge, E.B. (1979). Ultradispersed rhodium rafts: their existence and topology. *Journal of Catalysis* 57 (1): 41–63. https://doi.org/10.1016/0021-9517(79)90042-3.

54 Van't Blik, H.F.J., Van Zon, J.B.A.D., Huizinga, T. et al. (1985). Structure of rhodium in an ultradispersed rhodium/alumina catalyst as studied by EXAFS and other techniques. *Journal of the American Chemical Society* 107 (11): 3139–3147. https://doi.org/10.1021/ja00297a020.

55 Yao, H.C., Japar, S., and Shelef, M. (1977). Surface interactions in the system $RhAl_2O_3$. *Journal of Catalysis* 50 (3): 407–418. https://doi.org/10.1016/0021-9517(77)90053-7.

56 Yang, C. and Garland, C.W. (1957). Infrared studies of carbon monoxide chemisorbed on rhodium. *The Journal of Physical Chemistry* 61 (11): 1504–1512. https://doi.org/10.1021/j150557a013.

57 Yao, H.C. and Rothschild, W.G. (1978). Infrared spectra of chemisorbed CO on Rh/γ–Al2O3: site distributions and molecular mobility. *The Journal of Chemical Physics* 68 (11): 4774–4780. https://doi.org/10.1063/1.435657.

58 Yates, J.T. Jr., Duncan, T.M., and Vaughan, R.W. (1979). Infrared spectroscopic study of activated surface processes: CO chemisorption on supported Rh. *The Journal of Chemical Physics* 71 (10): 3908–3915. https://doi.org/10.1063/1.438159.

59 Yates, J.T. Jr., Duncan, T.M., Worley, S.D. et al. (1979). Infrared spectra of chemisorbed CO on Rh. *The Journal of Chemical Physics* 70 (3): 1219–1224. https://doi.org/10.1063/1.437603.

60 Rice, C.A., Worley, S.D., Curtis, C.W. et al. (1981). The oxidation state of dispersed Rh on Al_2O_3. *The Journal of Chemical Physics* 74 (11): 6487–6497. https://doi.org/10.1063/1.440987.

61 Cavanagh, R.R. and Jr, J.T.Y. (1981). Site distribution studies of Rh supported on Al_2O_3—An infrared study of chemisorbed CO. *The Journal of Chemical Physics* 74 (7): 4150–4155. https://doi.org/10.1063/1.441544.

62 Yates, J.T. Jr., and Kolasinski, K. (1983). Infrared spectroscopic investigation of the rhodium gem-dicarbonyl surface species. *The Journal of Chemical Physics* 79 (2): 1026–1030. https://doi.org/10.1063/1.445844.

63 Van't Blik, H.F.J., Van Zon, J.B.A.D., Huizinga, T. et al. (1983). An extended X-ray absorption fine structure spectroscopy study of a highly dispersed rhodium/aluminum oxide catalyst: the influence of carbon monoxide chemisorption on the topology of rhodium. *The Journal of Physical Chemistry* 87 (13): 2264–2267. https://doi.org/10.1021/j100236a002.

64 Van Zon, J., Koningsberger, D.C., van't Blik, H.F.J., and Sayers, D.E. (1985). An EXAFS study of the structure of the metal–support interface in highly dispersed Rh/Al_2O_3 catalysts. *The Journal of Chemical Physics* 82 (12): 5742–5754. https://doi.org/10.1063/1.448563.

65 Duncan, T.M. and Root, T.W. (1988). Adsorbed states of carbon monoxide on dispersed metals: quantitative analysis with carbon-13 NMR spectroscopy. *The Journal of Physical Chemistry* 92 (15): 4426–4432. https://doi.org/10.1021/j100326a036.

66 Robbins, J.L. (1986). Rhodium dicarbonyl sites on alumina surfaces. 1. Preparation and characterization of a model system. *The Journal of Physical Chemistry* 90 (15): 3381–3386. https://doi.org/10.1021/j100406a016.

67 Cooper, W.F. and Parker, W.L. (1989). Luminescent detection of Rh^+ in supported metal catalysts. *Chemical Physics Letters* 156 (5): 463–466. https://doi.org/10.1016/S0009-2614(89)87312-9.

68 Primet, M. (1978). Infrared study of CO chemisorption on zeolite and alumina supported rhodium. *Journal of the Chemical Society, Faraday Transactions 1: Physical Chemistry in Condensed Phases* 74: 2570–2580. https://doi.org/10.1039/F19787402570.

69 Kauppinen, M.M., Melander, M.M., and Honkala, K. (2020). First-principles insight into CO hindered agglomeration of Rh and Pt single atoms on m-ZrO_2. *Catalysis Science & Technology* 10 (17): 5847–5855. https://doi.org/10.1039/D0CY00413H.

70 Suzuki, A., Inada, Y., Yamaguchi, A. et al. (2003). Time scale and elementary steps of CO-induced disintegration of surface rhodium clusters. *Angewandte Chemie International Edition* 42 (39): 4795–4799. https://doi.org/10.1002/anie.200352318.

71 Goodman, E.D., Johnston-Peck, A.C., Dietze, E.M. et al. (2019). Catalyst deactivation via decomposition into single atoms and the role of metal loading. *Nature Catalysis* 2 (9): 748–755. https://doi.org/10.1038/s41929-019-0328-1.

72 Wang, Y.-G., Mei, D., Glezakou, V.-A. et al. (2015). Dynamic formation of single-atom catalytic active sites on ceria-supported gold nanoparticles. *Nature Communications* 6 (1): 6511. https://doi.org/10.1038/ncomms7511.

73 Liu, J.-C., Wang, Y.-G., and Li, J. (2017). Toward rational design of oxide-supported single-atom catalysts: atomic dispersion of gold on ceria. *Journal of the American Chemical Society* 139 (17): 6190–6199. https://doi.org/10.1021/jacs.7b01602.

74 Su, Y.-Q., Wang, Y., Liu, J.-X. et al. (2019). Theoretical approach to predict the stability of supported single-atom catalysts. *ACS Catalysis* 9 (4): 3289–3297. https://doi.org/10.1021/acscatal.9b00252.

75 Ouyang, R., Liu, J.-X., and Li, W.-X. (2013). Atomistic theory of ostwald ripening and disintegration of supported metal particles under reaction conditions. *Journal of the American Chemical Society* 135 (5): 1760–1771. https://doi.org/10.1021/ja3087054.

76 Roscioni, O.M., Dyke, J.M., and Evans, J. (2013). Structural characterization of supported $RhI(CO)_2/\gamma\text{-}Al_2O_3$ catalysts by periodic DFT calculations. *The Journal of Physical Chemistry C* 117 (38): 19464–19470. https://doi.org/10.1021/jp405549k.

77 Ghosh, T.K. and Nair, N.N. (2013). $Rh1/\gamma\text{-}Al_2O_3$ single-atom catalysis of O_2 activation and CO oxidation: mechanism, effects of hydration, oxidation state, and cluster size. *ChemCatChem* 5 (7): 1811–1821. https://doi.org/10.1002/cctc.201200799.

78 Asokan, C., DeRita, L., and Christopher, P. (2017). Using probe molecule FTIR spectroscopy to identify and characterize Pt-group metal based single atom catalysts. *Chinese Journal of Catalysis* 38 (9): 1473–1480. https://doi.org/10.1016/S1872-2067(17)62882-1.

79 Newton, M.A., Dent, A.J., Diaz-Moreno, S. et al. (2006). Rapid monitoring of the nature and interconversion of supported catalyst phases and of their influence upon performance: CO oxidation to CO_2 by γ-Al_2O_3 supported Rh catalysts. *Chemistry A European Journal* 12 (7): 1975–1985. https://doi.org/10.1002/chem.200500644.

80 Gentsch, H., Guillen, N., and Köpp, M. (1972). Isostere Adsorptionsenthalpien von Wasserstoff an atomar verteiltem Palladium und Platin auf Kohle. *Zeitschrift für Physikalische Chemie* 82 (1-4): 49. https://doi.org/10.1524/zpch.1972.82.1-4.049.

81 Gentsch, H., Härtel, V., and Köpp, M. (1971). Heterogene Katalyse mit Ni- und Pd-Atomen. Methanbildung aus Kohlenstoff und Wasserstoff. *Berichte der Bunsengesellschaft für Physikalische Chemie* 75 (10): 1086–1092. https://doi.org/10.1002/bbpc.19710751024.

82 Gallezot, P., Alarcon-Diaz, A., Dalmon, J.A. et al. (1975). Location and dispersion of platinum in PtY zeolites. *Journal of Catalysis* 39 (3): 334–349. https://doi.org/10.1016/0021-9517(75)90299-7.

83 Cabria, I., López, M.J., Fraile, S. et al. (2012). Adsorption and dissociation of molecular hydrogen on palladium clusters supported on graphene. *The Journal of Physical Chemistry C* 116 (40): 21179–21189. https://doi.org/10.1021/jp305635w.

84 Rossell, M.D., Caparrós, F.J., Angurell, I. et al. (2016). Magnetite-supported palladium single-atoms do not catalyse the hydrogenation of alkenes but small clusters do. *Catalysis Science & Technology* 6 (12): 4081–4085. https://doi.org/10.1039/C6CY00596A.

85 Rivera-Cárcamo, C., Leng, F., Gerber, I.C. et al. (2020). Catalysis to discriminate single atoms from subnanometric ruthenium particles in ultra-high loading catalysts. *Catalysis Science & Technology* 10 (14): 4673–4683. https://doi.org/10.1039/D0CY00540A.

86 Doudin, N., Yuk, S.F., Marcinkowski, M.D. et al. (2019). Understanding heterolytic H_2 cleavage and water-assisted hydrogen spillover on Fe_3O_4(001)-supported single palladium atoms. *ACS Catalysis* 9 (9): 7876–7887. https://doi.org/10.1021/acscatal.9b01425.

87 Righi, G., Magri, R., and Selloni, A. (2019). H_2 dissociation on noble metal single atom catalysts adsorbed on and doped into CeO_2 (111). *The Journal of Physical Chemistry C* 123 (15): 9875–9883. https://doi.org/10.1021/acs.jpcc.9b00609.

88 Bi, Q., Yuan, X., Lu, Y. et al. (2020). One-step high-temperature-synthesized single-atom platinum catalyst for efficient selective hydrogenation. *Research*: 9140841. https://doi.org/10.34133/2020/9140841.

89 Ye, T.-N., Xiao, Z., Li, J. et al. (2020). Stable single platinum atoms trapped in sub-nanometer cavities in $12CaO \cdot 7Al_2O_3$ for chemoselective hydrogenation of nitroarenes. *Nature Communications* 11: 1020. https://doi.org/10.1038/s41467-019-14216-9.

90 Lou, Y., Wu, H., and Liu, J. (2019). Nanocarbon-edge-anchored high-density Pt atoms for 3-nitrostyrene hydrogenation: strong metal-carbon interaction. *iScience* 13: 190–198. https://doi.org/10.1016/j.isci.2019.02.016.

91 Liu, P., Zhao, Y., Qin, R. et al. (2016). Photochemical route for synthesizing atomically dispersed palladium catalysts. *Science* 352 (6287): 797–800. https://doi.org/10.1126/science.aaf5251.

92 Yan, H., Lv, H., Yi, H. et al. (2018). Understanding the underlying mechanism of improved selectivity in pd_1 single-atom catalyzed hydrogenation reaction. *Journal of Catalysis* 366: 70–79. https://doi.org/10.1016/j.jcat.2018.07.033.

93 Yan, H., Cheng, H., Yi, H. et al. (2015). Single-atom Pd1/graphene catalyst achieved by atomic layer deposition: remarkable performance in selective hydrogenation of 1,3-butadiene. *Journal of the American Chemical Society* 137 (33): 10484–10487. https://doi.org/10.1021/jacs.5b06485.

94 Granja-DelRío, A., Alonso, J.A., and López, M.J. (2017). Competition between palladium clusters and hydrogen to saturate graphene vacancies. *The Journal of Physical Chemistry C* 121 (20): 10843–10850. https://doi.org/10.1021/acs.jpcc.6b12018.

95 Ma, Y., Chi, B., Liu, W. et al. (2019). Tailoring of the proximity of platinum single atoms on CeO_2 using phosphorus boosts the hydrogenation activity. *ACS Catalysis* 9 (9): 8404–8412. https://doi.org/10.1021/acscatal.9b01536.

96 Jeong, H., Shin, D., Kim, B.-S. et al. (2020). Controlling the oxidation state of Pt single atoms for maximizing catalytic activity. *Angewandte Chemie International Edition* 59 (46): 20691–20696. https://doi.org/10.1002/anie.202009776.

97 Yao, H.C., Yao, Y.F.Y., and Otto, K. (1979). Effects of the surface structure of Rhγ-Al_2O_3 on the hydrogenolysis of n-pentane, on the oxidation of n-butane, and on the reduction of nitric oxide. *Journal of Catalysis* 56 (1): 21–31. https://doi.org/10.1016/0021-9517(79)90084-8.

98 Otto, K. and Yao, H.C. (1980). The reduction of nitric oxide by hydrogen over Ptγ-Al_2O_3 as a function of metal loading. *Journal of Catalysis* 66 (1): 229–236. https://doi.org/10.1016/0021-9517(80)90025-1.

99 Zhang, S., Tang, Y., Nguyen, L. et al. (2018). Catalysis on singly dispersed Rh atoms anchored on an inert support. *ACS Catalysis* 8 (1): 110–121. https://doi.org/10.1021/acscatal.7b01788.

100 Zhang, L., Zhang, J., Qi, M. et al. (2020). First-principles investigation of single-atom Ni-g-C_3N_4 as an efficient catalyst for direct reduction of NO with CO. *Energy & Fuels* 34 (10): 12792–12799. https://doi.org/10.1021/acs.energyfuels.0c01908.

101 Xing, F., Jeon, J., Toyao, T. et al. (2019). A Cu–Pd single-atom alloy catalyst for highly efficient NO reduction. *Chemical Science* 10 (36): 8292–8298. https://doi.org/10.1039/C9SC03172C.

102 Nguyen, L., Zhang, S., Wang, L. et al. (2016). Reduction of nitric oxide with hydrogen on catalysts of singly dispersed bimetallic sites Pt_1Co_m and Pd_1Co_n. *ACS Catalysis* 6 (2): 840–850. https://doi.org/10.1021/acscatal.5b00842.

103 Nakaya, Y., Hirayama, J., Yamazoe, S. et al. (2020). Single-atom Pt in intermetallics as an ultrastable and selective catalyst for propane dehydrogenation. *Nature Communications* 11 (1): 2838. https://doi.org/10.1038/s41467-020-16693-9.

104 Pan, Y., Zhang, C., Liu, Z. et al. (2020). Structural regulation with atomic-level precision: from single-atomic site to diatomic and atomic interface catalysis. *Matter* 2 (1): 78–110. https://doi.org/10.1016/j.matt.2019.11.014.

105 Zhao, Y., Yang, K.R., Wang, Z. et al. (2018). Stable iridium dinuclear heterogeneous catalysts supported on metal-oxide substrate for solar water oxidation. *Proceedings of the National Academy of Sciences of the United States of America* 115 (12): 2902–2907. https://doi.org/10.1073/pnas.1722137115.

106 Li, X., Zhong, W., Cui, P. et al. (2016). Design of efficient catalysts with double transition metal atoms on C_2N layer. *The Journal of Physical Chemistry Letters* 7 (9): 1750–1755. https://doi.org/10.1021/acs.jpclett.6b00096.

107 Wang, H., Liu, J.-X., Allard, L.F. et al. (2019). Surpassing the single-atom catalytic activity limit through paired Pt-O-Pt ensemble built from isolated Pt1 atoms. *Nature Communications* 10 (1): 3808. https://doi.org/10.1038/s41467-019-11856-9.

108 Qi, K., Cui, X., Gu, L. et al. (2019). Single-atom cobalt array bound to distorted 1T MoS_2 with ensemble effect for hydrogen evolution catalysis. *Nature Communications* 10 (1): 5231. https://doi.org/10.1038/s41467-019-12997-7.

109 Ganesan, A. and Narayanasamy, M. (2019). Ultra-low loading of platinum in proton exchange membrane-based fuel cells: a brief review. *Materials for Renewable and Sustainable Energy* 8 (4): 18. https://doi.org/10.1007/s40243-019-0156-x.

110 Kongkanand, A., Gu, W., and Mathias, M.F. (2019). Proton-exchange membrane fuel cells with low-Pt content. In: *Fuel Cells and Hydrogen Production: A Volume in the Encyclopedia of Sustainability Science and Technology*, 2e (eds. T.E. Lipman and A.Z. Weber), 323–342. New York: Springer New York.

111 Jiao, L. and Jiang, H.-L. (2019). Metal–organic-framework-based single-atom catalysts for energy applications. *Chem* 5 (4): 786–804. https://doi.org/10.1016/j.chempr.2018.12.011.

112 Han, A., Wang, B., Kumar, A. et al. (2019). Recent advances for MOF-derived carbon-supported single-atom catalysts. *Small Methods* 3 (9): 1800471. https://doi.org/10.1002/smtd.201800471.

113 Hou, C.-C., Wang, H.-F., Li, C. et al. (2020). From metal–organic frameworks to single/dual-atom and cluster metal catalysts for energy applications. *Energy & Environmental Science* 13 (6): 1658–1693. https://doi.org/10.1039/C9EE04040D.

114 Jiao, L., Yan, H., Wu, Y. et al. (2020). When nanozymes meet single-atom catalysis. *Angewandte Chemie International Edition* 59 (7): 2565–2576. https://doi.org/10.1002/anie.201905645.

115 Wu, W., Huang, L., Wang, E. et al. (2020). Atomic engineering of single-atom nanozymes for enzyme-like catalysis. *Chemical Science* 11 (36): 9741–9756. https://doi.org/10.1039/D0SC03522J.

116 Li, M., Wang, H., Luo, W. et al. (2020). Heterogeneous single-atom catalysts for electrochemical CO_2 reduction reaction. *Advanced Materials* 32 (34): 2001848. https://doi.org/10.1002/adma.202001848.

117 Ye, Y., Cai, F., Li, H. et al. (2017). Surface functionalization of ZIF-8 with ammonium ferric citrate toward high exposure of Fe-N active sites for efficient oxygen and carbon dioxide electroreduction. *Nano Energy* 38: 281–289. https://doi.org/10.1016/j.nanoen.2017.05.042.

118 Wan, X., Liu, X., Li, Y. et al. (2019). Fe–N–C electrocatalyst with dense active sites and efficient mass transport for high-performance proton exchange membrane fuel cells. *Nature Catalysis* 2 (3): 259–268. https://doi.org/10.1038/s41929-019-0237-3.

119 Fu, J., Wang, S., Wang, Z. et al. (2020). Graphitic carbon nitride based single-atom photocatalysts. *Frontiers of Physics* 15 (3): 33201. https://doi.org/10.1007/s11467-019-0950-z.

120 Li, J., Chen, M., Cullen, D.A. et al. (2018). Atomically dispersed manganese catalysts for oxygen reduction in proton-exchange membrane fuel cells. *Nature Catalysis* 1 (12): 935–945. https://doi.org/10.1038/s41929-018-0164-8.

121 Zhao, L., Zhang, Y., Huang, L.-B. et al. (2019). Cascade anchoring strategy for general mass production of high-loading single-atomic metal-nitrogen catalysts. *Nature Communications* 10 (1): 1278. https://doi.org/10.1038/s41467-019-09290-y.

122 Xiong, Y., Sun, W., Xin, P. et al. (2020). Gram-scale synthesis of high-loading single-atomic-site Fe catalysts for effective epoxidation of styrene. *Advanced Materials* 32 (34): 2000896. https://doi.org/10.1002/adma.202000896.

123 Lu, C., Chen, Y., Yang, Y. et al. (2020). Single-atom catalytic materials for lean-electrolyte ultrastable lithium–sulfur batteries. *Nano Letters* 20 (7): 5522–5530. https://doi.org/10.1021/acs.nanolett.0c02167.

124 Lei, J., Liu, H., Yin, D. et al. (2020). Boosting the loading of metal single atoms via a bioconcentration strategy. *Small* 16 (10): 1905920. https://doi.org/10.1002/smll.201905920.

125 Yi, J.-D., Xu, R., Wu, Q. et al. (2018). Atomically dispersed iron–nitrogen active sites within porphyrinic triazine-based frameworks for oxygen reduction reaction in both alkaline and acidic media. *ACS Energy Letters* 3 (4): 883–889. https://doi.org/10.1021/acsenergylett.8b00245.

126 Yin, P., Yao, T., Wu, Y. et al. (2016). Single cobalt atoms with precise N-coordination as superior oxygen reduction reaction catalysts. *Angewandte Chemie International Edition* 55 (36): 10800–10805. https://doi.org/10.1002/anie.201604802.

127 Cheng, Y., Zhao, S., Johannessen, B. et al. (2018). Atomically dispersed transition metals on carbon nanotubes with ultrahigh loading for selective electrochemical carbon dioxide reduction. *Advanced Materials* 30 (13): 1706287. https://doi.org/10.1002/adma.201706287.

128 Li, Y., Wu, J., Zhang, B. et al. (2020). Fast conversion and controlled deposition of lithium (poly)sulfides in lithium-sulfur batteries using high-loading cobalt single atoms. *Energy Storage Materials* 30: 250–259. https://doi.org/10.1016/j.ensm.2020.05.022.

129 Li, J., Liu, H., Wang, M. et al. (2019). Boosting oxygen reduction activity with low-temperature derived high-loading atomic cobalt on nitrogen-doped

graphene for efficient Zn–air batteries. *Chemical Communications* 55 (3): 334–337. https://doi.org/10.1039/C8CC08992B.

130 Lin, C., Zhang, H., Song, X. et al. (2020). 2D-organic framework confined metal single atoms with the loading reaching the theoretical limit. *Materials Horizons* 7 (10): 2726–2733. https://doi.org/10.1039/D0MH01061H.

131 Cheng, Y., Zhao, S., Li, H. et al. (2019). Unsaturated edge-anchored Ni single atoms on porous microwave exfoliated graphene oxide for electrochemical CO_2. *Applied Catalysis B: Environmental* 243: 294–303. https://doi.org/10.1016/j.apcatb.2018.10.046.

132 Zhao, S., Cheng, Y., Veder, J.-P. et al. (2018). One-pot pyrolysis method to fabricate carbon nanotube supported Ni single-atom catalysts with ultrahigh loading. *ACS Applied Energy Materials* 1 (10): 5286–5297. https://doi.org/10.1021/acsaem.8b00903.

133 Zhao, S., Wang, T., Zhou, G. et al. (2020). Controlled one-pot synthesis of nickel single atoms embedded in carbon nanotube and graphene supports with high loading. *ChemNanoMat* 6 (7): 1063–1074. https://doi.org/10.1002/cnma.202000223.

134 Guo, T., Tang, N., Lin, F. et al. (2020). High-loading single-atom copper catalyst supported on coordinatively unsaturated Al_2O_3 for selective synthesis of homoallylboronates. *ChemSusChem* 13 (12): 3115–3121. https://doi.org/10.1002/cssc.202000536.

135 Yang, P., Zuo, S., Zhang, F. et al. (2020). Carbon nitride-based single-atom Cu catalysts for highly efficient carboxylation of alkynes with atmospheric CO_2. *Industrial & Engineering Chemistry Research* 59 (16): 7327–7335. https://doi.org/10.1021/acs.iecr.0c00547.

136 Han, G., Zheng, Y., Zhang, X. et al. (2019). High loading single-atom Cu dispersed on graphene for efficient oxygen reduction reaction. *Nano Energy* 66: 104088. https://doi.org/10.1016/j.nanoen.2019.104088.

137 Li, J., Chen, S., Yang, N. et al. (2019). Ultrahigh-loading zinc single-atom catalyst for highly efficient oxygen reduction in both acidic and alkaline media. *Angewandte Chemie International Edition* 58 (21): 7035–7039. https://doi.org/10.1002/anie.201902109.

138 Yang, Q., Yang, C.-C., Lin, C.-H. et al. (2019). Metal–organic-framework-derived hollow N-doped porous carbon with ultrahigh concentrations of single Zn atoms for efficient carbon dioxide conversion. *Angewandte Chemie International Edition* 58 (11): 3511–3515. https://doi.org/10.1002/anie.201813494.

139 Tang, C., Jiao, Y., Shi, B. et al. (2020). Coordination tunes selectivity: two-electron oxygen reduction on high-loading molybdenum single-atom catalysts. *Angewandte Chemie International Edition* 59 (23): 9171–9176. https://doi.org/10.1002/anie.202003842.

140 Wang, L., Chen, M.-X., Yan, Q.-Q. et al. (2019). A sulfur-tethering synthesis strategy toward high-loading atomically dispersed noble metal catalysts. *Science Advances* 5 (10): eaax6322. https://doi.org/10.1126/sciadv.aax6322.

141 Ma, Y., Ren, Y., Zhou, Y. et al. (2020). High-density and thermally stable palladium single-atom catalysts for chemoselective hydrogenations. *Angewandte*

Chemie International Edition 59 (48): 21613–21619. https://doi.org/10.1002/anie.202007707.

142 Jiang, X.-H., Zhang, L.-S., Liu, H.-Y. et al. (2020). Silver single atom in carbon nitride catalyst for highly efficient photocatalytic hydrogen evolution. *Angewandte Chemie International Edition* 59 (51): 23112–23116. https://doi.org/10.1002/anie.202011495.

143 Chen, W., Pei, J., He, C.-T. et al. (2018). Single tungsten atoms supported on MOF-derived N-doped carbon for Robust electrochemical hydrogen evolution. *Advanced Materials* 30 (30): 1800396. https://doi.org/10.1002/adma.201800396.

144 Babucci, M., Sarac Oztuna, F.E., Debefve, L.M. et al. (2019). Atomically dispersed reduced graphene aerogel-supported iridium catalyst with an iridium loading of 14.8 wt %. *ACS Catalysis* 9 (11): 9905–9913. https://doi.org/10.1021/acscatal.9b02231.

145 Wang, Q., Huang, X., Zhao, Z.L. et al. (2020). Ultrahigh-loading of Ir single atoms on NiO matrix to dramatically enhance oxygen evolution reaction. *Journal of the American Chemical Society* 142 (16): 7425–7433. https://doi.org/10.1021/jacs.9b12642.

146 Li, H., Wang, L., Dai, Y. et al. (2018). Synergetic interaction between neighbouring platinum monomers in CO_2 hydrogenation. *Nature Nanotechnology* 13 (5): 411–417. https://doi.org/10.1038/s41565-018-0089-z.

147 Li, T., Liu, J., Song, Y. et al. (2018). Photochemical solid-phase synthesis of platinum single atoms on nitrogen-doped carbon with high loading as bifunctional catalysts for hydrogen evolution and oxygen reduction reactions. *ACS Catalysis* 8 (9): 8450–8458. https://doi.org/10.1021/acscatal.8b02288.

148 Zhu, Y., Cao, T., Cao, C. et al. (2018). One-pot pyrolysis to N-doped graphene with high-density Pt single atomic sites as heterogeneous catalyst for alkene hydrosilylation. *ACS Catalysis* 8 (11): 10004–10011. https://doi.org/10.1021/acscatal.8b02624.

149 Kunwar, D., Zhou, S., DeLaRiva, A. et al. (2019). Stabilizing high metal loadings of thermally stable platinum single atoms on an industrial catalyst support. *ACS Catalysis* 9 (5): 3978–3990. https://doi.org/10.1021/acscatal.8b04885.

150 Cao, S., Zhao, Y., Lee, S. et al. (2020). High-loading single Pt atom sites [$Pt\text{-}O(OH)_x$] catalyze the CO PROX reaction with high activity and selectivity at mild conditions. *Science Advances* 6 (25): eaba3809. https://doi.org/10.1126/sciadv.aba3809.

151 Liu, K., Tang, Y., Yu, Z. et al. (2020). High-loading and thermally stable $Pt_1/MgAl_{1.2}Fe_{0.8}O_4$ single-atom catalysts for high-temperature applications. *Science China Materials* 63 (6): 949–958. https://doi.org/10.1007/s40843-020-1267-2.

152 Zuo, Q., Liu, T., Chen, C. et al. (2019). Ultrathin metal–organic framework nanosheets with ultrahigh loading of single Pt atoms for efficient visible-light-driven photocatalytic H_2 evolution. *Angewandte Chemie International Edition* 58 (30): 10198–10203. https://doi.org/10.1002/anie.201904058.

153 Wang, Z., Gu, L., Song, L. et al. (2018). Facile one-pot synthesis of MOF supported gold pseudo-single-atom catalysts for hydrogenation reactions. *Materials Chemistry Frontiers* 2 (5): 1024–1030. https://doi.org/10.1039/C8QM00081F.

154 Tan, K., Dixit, M., Dean, J. et al. (2019). Predicting metal–support interactions in oxide-supported single-atom catalysts. *Industrial & Engineering Chemistry Research* 58 (44): 20236–20246. https://doi.org/10.1021/acs.iecr.9b04068.

155 Davies, P.R. and Morgan, D.J. (2020). Practical guide for X-ray photoelectron spectroscopy: applications to the study of catalysts. *Journal of Vacuum Science & Technology A* 38 (3): 033204. https://doi.org/10.1116/1.5140747.

156 Hulva, J., Meier, M., Bliem, R. et al. (2021). Unraveling CO adsorption on model single-atom catalysts. *Science* 371 (6527): 375–379. https://doi.org/10.1126/science.abe5757.

157 O'Shea, J.N., Schnadt, J., Andersson, S. et al. (2000). X-ray photoelectron spectroscopy of low surface concentration mass-selected Ag clusters. *The Journal of Chemical Physics* 113 (20): 9233–9238. https://doi.org/10.1063/1.1319700.

158 Vedrine, J.C., Dufaux, M., Naccache, C. et al. (1978). X-ray photoelectron spectroscopy study of Pd and Pt ions in type Y-zeolite. Electron transfer between metal aggregates and the support as evidenced by X-ray photoelectron spectroscopy and electron spin resonance. *Journal of the Chemical Society, Faraday Transactions 1: Physical Chemistry in Condensed Phases* 74: 440–449. https://doi.org/10.1039/F19787400440.

159 Stakheev, A.Y. and Sachtler, W.M.H. (1991). Determination by X-ray photoelectron spectroscopy of the electronic state of Pd clusters in Y zeolite. *Journal of the Chemical Society, Faraday Transactions* 87 (22): 3703–3708. https://doi.org/10.1039/FT9918703703.

160 Bastl, Z. (1986). X-ray photoelectron spectroscopy of supported metal particles. *Vacuum* 36 (7): 447–448. https://doi.org/10.1016/0042-207X(86)90225-3.

161 Feng, K., Zhang, H., Gao, J. et al. (2020). Single atoms or not? The limitation of EXAFS. *Applied Physics Letters* 116 (19): 191903. https://doi.org/10.1063/5.0008748.

162 Petek, U., Ruiz-Zepeda, F., Bele, M. et al. (2019). Nanoparticles and single atoms in commercial carbon-supported platinum-group metal catalysts. *Catalysts* 9 (2): 134.

163 Kuai, L., Chen, Z., Liu, S. et al. (2020). Titania supported synergistic palladium single atoms and nanoparticles for room temperature ketone and aldehydes hydrogenation. *Nature Communications* 11 (1): 48. https://doi.org/10.1038/s41467-019-13941-5.

164 Ma, Y., Yang, T., Zou, H. et al. (2020). Synergizing Mo single atoms and Mo_2C nanoparticles on CNTs synchronizes selectivity and activity of electrocatalytic N_2 reduction to ammonia. *Advanced Materials* 32 (33): 2002177. https://doi.org/10.1002/adma.202002177.

165 Fu, Y., Xu, D., Wang, Y. et al. (2020). Single atoms anchored on cobalt-based catalysts derived from hydrogels containing phthalocyanine toward the oxygen reduction reaction. *ACS Sustainable Chemistry & Engineering* 8 (22): 8338–8347. https://doi.org/10.1021/acssuschemeng.0c02158.

166 Tiwari, J.N., Dang, N.K., Park, H.J. et al. (2020). Remarkably enhanced catalytic activity by the synergistic effect of palladium single atoms and palladium–cobalt phosphide nanoparticles. *Nano Energy* 78: 105166. https://doi.org/10.1016/j.nanoen.2020.105166.

167 Peng, J., Chen, Y., Wang, K. et al. (2020). High-performance Ru-based electrocatalyst composed of Ru nanoparticles and Ru single atoms for hydrogen evolution reaction in alkaline solution. *International Journal of Hydrogen Energy* 45 (38): 18840–18849. https://doi.org/10.1016/j.ijhydene.2020.05.064.

168 Yuan, B., Yao, Z., Qiu, C. et al. (2020). Synergistic effect of size-dependent PtZn nanoparticles and zinc single-atom sites for electrochemical ozone production in neutral media. *Journal of Energy Chemistry* 51: 312–322. https://doi.org/10.1016/j.jechem.2020.03.066.

169 Wang, L., Guan, E., Zhang, J. et al. (2018). Single-site catalyst promoters accelerate metal-catalyzed nitroarene hydrogenation. *Nature Communications* 9 (1): 1362. https://doi.org/10.1038/s41467-018-03810-y.

170 Liu, D., He, Q., Ding, S. et al. (2020). Structural regulation and support coupling effect of single-atom catalysts for heterogeneous catalysis. *Advanced Energy Materials* 10 (32): 2001482. https://doi.org/10.1002/aenm.202001482.

171 Lai, W.-H., Miao, Z., Wang, Y.-X. et al. (2019). Atomic-local environments of single-atom catalysts: synthesis, electronic structure, and activity. *Advanced Energy Materials* 9 (43): 1900722. https://doi.org/10.1002/aenm.201900722.

172 Crabtree, R.H. (1983). Cationic rhodium and iridium complexes in catalysis. In: *Homogeneous Catalysis with Metal Phosphine Complexes* (ed. L.H. Pignolet), 297–316. Boston, MA: Springer US.

173 Tosoni, S. and Pacchioni, G. (2020). Bonding properties of isolated metal atoms on two-dimensional oxides. *The Journal of Physical Chemistry C* 124 (38): 20960–20973. https://doi.org/10.1021/acs.jpcc.0c05958.

174 Tang, Y., Zhao, S., Long, B. et al. (2016). On the nature of support effects of metal dioxides MO_2 (M = Ti, Zr, Hf, Ce, Th) in single-atom gold catalysts: importance of quantum primogenic effect. *The Journal of Physical Chemistry C* 120 (31): 17514–17526. https://doi.org/10.1021/acs.jpcc.6b05338.

175 Camellone, M.F. and Fabris, S. (2009). Reaction mechanisms for the CO oxidation on Au/CeO_2 catalysts: activity of substitutional Au^{3+}/Au^+ cations and deactivation of supported Au^+ adatoms. *Journal of the American Chemical Society* 131 (30): 10473–10483. https://doi.org/10.1021/ja902109k.

176 Tang, Y., Wang, Y.-G., and Li, J. (2017). Theoretical investigations of Pt_1@CeO_2 single-atom catalyst for CO oxidation. *The Journal of Physical Chemistry C* 121 (21): 11281–11289. https://doi.org/10.1021/acs.jpcc.7b00313.

177 Parkinson, G.S. (2019). Single-atom catalysis: how structure influences catalytic performance. *Catalysis Letters* 149 (5): 1137–1146. https://doi.org/10.1007/s10562-019-02709-7.

178 Fu, Z., Yang, B., and Wu, R. (2020). Understanding the activity of single-atom catalysis from frontier orbitals. *Physical Review Letters* 125 (15): 156001. https://doi.org/10.1103/PhysRevLett.125.156001.

179 Chen, P., Zhou, T., Xing, L. et al. (2017). Atomically dispersed iron–nitrogen species as electrocatalysts for bifunctional oxygen evolution and reduction reactions. *Angewandte Chemie International Edition* 56 (2): 610–614. https://doi.org/10.1002/anie.201610119.

180 Lou, Y., Zheng, Y., Li, X. et al. (2019). Pocketlike active site of Rh_1/MoS_2 single-atom catalyst for selective crotonaldehyde hydrogenation. *Journal of the American Chemical Society* 141 (49): 19289–19295. https://doi.org/10.1021/jacs.9b06628.

181 Li, Z., Wang, D., Wu, Y. et al. (2018). Recent advances in the precise control of isolated single-site catalysts by chemical methods. *National Science Review* 5 (5): 673–689. https://doi.org/10.1093/nsr/nwy056.

182 Qi, K., Chhowalla, M., and Voiry, D. (2020). Single atom is not alone: metal–support interactions in single-atom catalysis. *Materials Today* 40: 173–192. https://doi.org/10.1016/j.mattod.2020.07.002.

183 Costa-Amaral, R., Forhat, A., Caturello, N.A.M.S. et al. (2020). Unveiling the adsorption properties of 3*d*, 4*d*, and 5*d* metal adatoms on the MoS_2 monolayer: a DFT-D3 investigation. *Surface Science* 701: 121700. https://doi.org/10.1016/j.susc.2020.121700.

184 Chen, Z., Zhao, J., Cabrera, C.R. et al. (2019). Computational screening of efficient single-atom catalysts based on graphitic carbon nitride (g-C_3N_4) for nitrogen electroreduction. *Small Methods* 3 (6): 1800368. https://doi.org/10.1002/smtd.201800368.

185 Nagashima, H., Nakaoka, A., Saito, Y. et al. (1992). C60Pd: the first organometallic polymer of buckminsterfullerene. *Journal of the Chemical Society, Chemical Communications* 4: 377–379. https://doi.org/10.1039/C39920000377.

186 Lavrentiev, V., Abe, H., Naramoto, H. et al. (2006). Polymeric chains in C_{60} and Co mixture. *Chemical Physics Letters* 424 (1): 101–104. https://doi.org/10.1016/j.cplett.2006.04.046.

187 Leng, F., Gerber, I.C., Lecante, P. et al. (2016). Synthesis and structure of ruthenium-fullerides. *RSC Advances* 6 (73): 69135–69148. https://doi.org/10.1039/C6RA12023G.

188 Hülsey, M.J., Zhang, J., and Yan, N. (2018). Harnessing the wisdom in colloidal chemistry to make stable single-atom catalysts. *Advanced Materials* 30 (47): 1802304. https://doi.org/10.1002/adma.201802304.

189 Liu, J.-C., Tang, Y., Wang, Y.-G. et al. (2018). Theoretical understanding of the stability of single-atom catalysts. *National Science Review* 5 (5): 638–641. https://doi.org/10.1093/nsr/nwy094.

190 Jones, J., Xiong, H., DeLaRiva, A.T. et al. (2016). Thermally stable single-atom platinum-on-ceria catalysts via atom trapping. *Science* 353 (6295): 150–154. https://doi.org/10.1126/science.aaf8800.

191 Qiao, B., Liang, J.-X., Wang, A. et al. (2015). Ultrastable single-atom gold catalysts with strong covalent metal-support interaction (CMSI). *Nano Research* 8 (9): 2913–2924. https://doi.org/10.1007/s12274-015-0796-9.

192 Wei, S., Li, A., Liu, J.-C. et al. (2018). Direct observation of noble metal nanoparticles transforming to thermally stable single atoms. *Nature Nanotechnology* 13 (9): 856–861. https://doi.org/10.1038/s41565-018-0197-9.

193 Lang, R., Xi, W., Liu, J.-C. et al. (2019). Non defect-stabilized thermally stable single-atom catalyst. *Nature Communications* 10 (1): 234. https://doi.org/10.1038/s41467-018-08136-3.

194 O'Connor, N.J., Jonayat, A.S.M., Janik, M.J. et al. (2018). Interaction trends between single metal atoms and oxide supports identified with density functional theory and statistical learning. *Nature Catalysis* 1 (7): 531–539. https://doi.org/10.1038/s41929-018-0094-5.

195 Liu, L. and Corma, A. (2020). Evolution of isolated atoms and clusters in catalysis. *Trends in Chemistry* 2 (4): 383–400. https://doi.org/10.1016/j.trechm.2020.02.003.

196 Campbell, C.T. (2013). The energetics of supported metal nanoparticles: relationships to sintering rates and catalytic activity. *Accounts of Chemical Research* 46 (8): 1712–1719. https://doi.org/10.1021/ar3003514.

197 Rickard, J.M., Genovese, L., Moata, A. et al. (1990). Redispersion of platinum on Pt/Al_2O_3 model catalyst in oxygen studied by transmission electron microscopy. *Journal of Catalysis* 121 (1): 141–152. https://doi.org/10.1016/0021-9517(90)90224-8.

198 Lai, X. and Goodman, D.W. (2000). Structure–reactivity correlations for oxide-supported metal catalysts: new perspectives from STM. *Journal of Molecular Catalysis A: Chemical* 162 (1): 33–50. https://doi.org/10.1016/S1381-1169(00)00320-4.

199 Zhou, J., Kang, Y.C., and Chen, D.A. (2003). Oxygen-induced dissociation of Cu islands supported on TiO_2(110). *The Journal of Physical Chemistry B* 107 (28): 6664–6667. https://doi.org/10.1021/jp0301251.

200 Zhou, J., Kang, Y.C., Ma, S. et al. (2004). Adsorbate-induced dissociation of metal clusters: TiO_2(110)-supported Cu and Ni clusters exposed to oxygen gas. *Surface Science* 562 (1): 113–127. https://doi.org/10.1016/j.susc.2004.05.094.

201 Simonsen, S.B., Chorkendorff, I., Dahl, S. et al. (2010). Direct observations of oxygen-induced platinum nanoparticle ripening studied by in situ TEM. *Journal of the American Chemical Society* 132 (23): 7968–7975. https://doi.org/10.1021/ja910094r.

202 Moliner, M., Gabay, J., Kliewer, C. et al. (2018). Trapping of metal atoms and metal clusters by chabazite under severe redox stress. *ACS Catalysis* 8 (10): 9520–9528. https://doi.org/10.1021/acscatal.8b01717.

203 Moliner, M., Gabay, J.E., Kliewer, C.E. et al. (2016). Reversible transformation of Pt nanoparticles into single atoms inside high-silica chabazite zeolite. *Journal of the American Chemical Society* 138 (48): 15743–15750. https://doi.org/10.1021/jacs.6b10169.

204 Feng, S., Lin, X., Song, X. et al. (2020). The role of H_2 on the stability of the single-metal-site Ir1/AC catalyst for heterogeneous methanol carbonylation. *Journal of Catalysis* 381: 193–203. https://doi.org/10.1016/j.jcat.2019.10.032.

205 Nelson, N.C., Chen, L., Meira, D. et al. (2020). In situ dispersion of palladium on TiO_2 during reverse water–gas shift reaction: formation of atomically dispersed palladium. *Angewandte Chemie International Edition* 59 (40): 17657–17663. https://doi.org/10.1002/anie.202007576.

206 Ding, S., Guo, Y., Hülsey, M.J. et al. (2019). Electrostatic stabilization of single-atom catalysts by ionic liquids. *Chem* 5 (12): 3207–3219. https://doi.org/10.1016/j.chempr.2019.10.007.

207 Wei, H., Huang, K., Wang, D. et al. (2017). Iced photochemical reduction to synthesize atomically dispersed metals by suppressing nanocrystal growth. *Nature Communications* 8 (1): 1490. https://doi.org/10.1038/s41467-017-01521-4.

208 Wang, Y., Song, E., Qiu, W. et al. (2019). Recent progress in theoretical and computational investigations of structural stability and activity of single-atom electrocatalysts. *Progress in Natural Science: Materials International* 29 (3): 256–264. https://doi.org/10.1016/j.pnsc.2019.04.004.

209 Li, J., Guan, Q., Wu, H. et al. (2019). Highly active and stable metal single-atom catalysts achieved by strong electronic metal–support interactions. *Journal of the American Chemical Society* 141 (37): 14515–14519. https://doi.org/10.1021/jacs.9b06482.

210 Su, Y.-Q., Zhang, L., Wang, Y. et al. (2020). Stability of heterogeneous single-atom catalysts: a scaling law mapping thermodynamics to kinetics. *npj Computational Materials* 6 (1): 144. https://doi.org/10.1038/s41524-020-00411-6.

211 Lu, Z., Yadav, S., and Singh, C.V. (2020). Predicting aggregation energy for single atom bimetallic catalysts on clean and O* adsorbed surfaces through machine learning models. *Catalysis Science & Technology* 10 (1): 86–98. https://doi.org/10.1039/C9CY02070E.

212 Lin, S., Xu, H., Wang, Y. et al. (2020). Directly predicting limiting potentials from easily obtainable physical properties of graphene-supported single-atom electrocatalysts by machine learning. *Journal of Materials Chemistry A* 8 (11): 5663–5670. https://doi.org/10.1039/C9TA13404B.

213 Dasgupta, A., Gao, Y., Broderick, S.R. et al. (2020). Machine learning-aided identification of single atom alloy catalysts. *The Journal of Physical Chemistry C* 124 (26): 14158–14166. https://doi.org/10.1021/acs.jpcc.0c01492.

214 Wang, Y., Su, Y.-Q., Hensen, E.J.M. et al. (2020). Finite-temperature structures of supported subnanometer catalysts inferred via statistical learning and genetic algorithm-based optimization. *ACS Nano* 14 (10): 13995–14007. https://doi.org/10.1021/acsnano.0c06472.

215 Boyes, E.D., LaGrow, A.P., Ward, M.R. et al. (2020). Single atom dynamics in chemical reactions. *Accounts of Chemical Research* 53 (2): 390–399. https://doi.org/10.1021/acs.accounts.9b00500.

216 Li, X., Yang, X., Zhang, J. et al. (2019). In situ/operando techniques for characterization of single-atom catalysts. *ACS Catalysis* 9 (3): 2521–2531. https://doi.org/10.1021/acscatal.8b04937.

217 Xi, W., Wang, K., Shen, Y. et al. (2020). Dynamic co-catalysis of Au single atoms and nanoporous Au for methane pyrolysis. *Nature Communications* 11 (1): 1919. https://doi.org/10.1038/s41467-020-15806-8.

218 Gaikwad, A.V., Holuigue, A., Thathagar, M.B. et al. (2007). Ion- and atom-leaching mechanisms from palladium nanoparticles in cross-coupling reactions. *Chemistry A European Journal* 13 (24): 6908–6913. https://doi.org/10.1002/chem.200700105.

219 Trzeciak, A.M. and Augustyniak, A.W. (2019). The role of palladium nanoparticles in catalytic C–C cross-coupling reactions. *Coordination Chemistry Reviews* 384: 1–20. https://doi.org/10.1016/j.ccr.2019.01.008.

220 Dessal, C., Len, T., Morfin, F. et al. (2019). Dynamics of single Pt atoms on alumina during CO oxidation monitored by operando X-ray and infrared spectroscopies. *ACS Catalysis* 9 (6): 5752–5759. https://doi.org/10.1021/acscatal.9b00903.

221 Karapinar, D., Huan, N.T., Ranjbar, S.N. et al. (2019). Electroreduction of CO_2 on single-site copper-nitrogen-doped carbon material: selective formation of ethanol and reversible restructuration of the metal sites. *Angewandte Chemie International Edition* 58 (42): 15098–15103. https://doi.org/10.1002/anie.201907994.

222 Daelman, N., Capdevila-Cortada, M., and López, N. (2019). Dynamic charge and oxidation state of Pt/CeO_2 single-atom catalysts. *Nature Materials* 18 (11): 1215–1221. https://doi.org/10.1038/s41563-019-0444-y.

223 Butin, K.P., Beloglazkina, E.K., and Zyk, N.V. (2005). Metal complexes with non-innocent ligands. *Russian Chemical Reviews* 74 (6): 531–553. https://doi.org/10.1070/rc2005v074n06abeh000977.

224 Ganguly, S. and Ghosh, A. (2019). Seven clues to ligand noninnocence: the metallocorrole paradigm. *Accounts of Chemical Research* 52 (7): 2003–2014. https://doi.org/10.1021/acs.accounts.9b00115.

225 Lyaskovskyy, V. and de Bruin, B. (2012). Redox non-innocent ligands: versatile new tools to control catalytic reactions. *ACS Catalysis* 2 (2): 270–279. https://doi.org/10.1021/cs200660v.

226 Kaim, W. and Schwederski, B. (2010). Non-innocent ligands in bioinorganic chemistry—an overview. *Coordination Chemistry Reviews* 254 (13): 1580–1588. https://doi.org/10.1016/j.ccr.2010.01.009.

227 DeRita, L., Resasco, J., Dai, S. et al. (2019). Structural evolution of atomically dispersed Pt catalysts dictates reactivity. *Nature Materials* 18 (7): 746–751. https://doi.org/10.1038/s41563-019-0349-9.

228 Paolucci, C., Khurana, I., Parekh, A.A. et al. (2017). Dynamic multinuclear sites formed by mobilized copper ions in NO_x selective catalytic reduction. *Science* 357 (6354): 898–903. https://doi.org/10.1126/science.aan5630.

229 Zhang, L., Long, R., Zhang, Y. et al. (2020). Direct observation of dynamic bond evolution in single-atom Pt/C_3N_4 catalysts. *Angewandte Chemie International Edition* 59 (15): 6224–6229. https://doi.org/10.1002/anie.201915774.

230 Qiu, S., Li, Q., Xu, Y. et al. (2020). Learning from nature: understanding hydrogenase enzyme using computational approach. *WIREs Computational Molecular Science* 10 (1): e1422. https://doi.org/10.1002/wcms.1422.

231 Chowdhury, R. and Maranas, C.D. (2020). From directed evolution to computational enzyme engineering—a review. *AIChE Journal* 66 (3): e16847. https://doi.org/10.1002/aic.16847.

232 Fu, L., Tang, Y., and Lin, Y. (2020). Advances in synchrotron radiation-based X-ray absorption spectroscopy to characterize the fine atomic structure of single-atom nanozymes. *Chemistry - An Asian Journal* 15 (14): 2110–2116. https://doi.org/10.1002/asia.202000560.

233 Boyes, E.D., LaGrow, A.P., Ward, M.R. et al. (2020). Visualizing single atom dynamics in heterogeneous catalysis using analytical *in situ* environmental scanning transmission electron microscopy. *Philosophical Transactions of the Royal Society A: Mathematical, Physical and Engineering Sciences* 378 (2186): 20190605. https://doi.org/10.1098/rsta.2019.0605.

234 Li, L., Chang, X., Lin, X. et al. (2020). Theoretical insights into single-atom catalysts. *Chemical Society Reviews* 49: 8156–8178. https://doi.org/10.1039/D0CS00795A.

2

Preparation of Supported Metal Single-Atom Catalysts on Metal Oxides and Hydroxides

Canio Scarfiello[1,2,3], Jeremy Audevard[1], Carole Le Berre[1], Katerina Soulantica[2], Philippe Serp[1], and Doan Pham Minh[3]

[1] *Université de Toulouse, LCC, CNRS-UPR 8241, ENSIACET, 4 allée Emile Monso - CS 44362, 31030 Cedex 4 Toulouse, France*
[2] *Université de Toulouse, LPCNO, CNRS, INSA, UPS, 135 avenue de Rangueil, 31077 Toulouse, France*
[3] *Université de Toulouse, IMT Mines Albi, UMR CNRS 5302, Centre RAPSODEE Campus Jarlard, 81013 Cedex 09, Albi, France*

2.1 Introduction

Catalysis plays a pivotal role in modern chemistry. It takes part in 85% of all chemical processes, with a relevant majority of heterogeneous catalysts (80%) over homogeneous/enzymatic ones (20%) [1]. Metal oxides are largely employed as supports for heterogeneous catalysts in several industrial processes, from oxidation to hydrogenation reactions, including Fischer–Tropsch, steam reforming, and $DeNO_x$ [2]. Indeed, the thermal stability as well as the acid–base properties and the presence of surface defects are only some of the advantages of these porous materials [3]. Therefore, the possibility of boosting the activity of metal oxide–supported catalysts through atomic dispersion of the active phase attracted the interest of several groups, and in the last 10 years, many scientific reports related to this emerging domain have been published. When it comes to the preparation of supported metal single-atom catalysts (SACs), the main classification divides the different techniques into "bottom-up" and "top-down" [4]. The first category consists in the direct deposition of metal single atoms (SAs) onto the support surface. In the second one, SAs result from a two-step procedure, involving breaking metal–metal bonds of nanoparticles (NPs) as precursors into SAs and formation of new strong interactions of these SAs with the support. The latter is an interesting approach, considering that it can be also exploited for the re-dispersion of sintered metallic species of used catalysts. Unfortunately, high-temperature treatments and specific metals or supports are usually necessary, preventing universal utilization of this technique. On the other side, the main problem of "bottom-up" methods is the limited stability of metal SAs, which tend to aggregate under working conditions, especially under reducing atmosphere. Indeed, due to their high surface free energy, SAs are highly mobile and

Supported Metal Single Atom Catalysis, First Edition. Edited by Philippe Serp and Doan Pham Minh.
© 2022 WILEY-VCH GmbH. Published 2022 by WILEY-VCH GmbH.

tend to form aggregates, during either synthesis and/or catalysis [4]. The stability of SAs strongly depends on the metal–support interaction (MSI). Enhanced MSI is necessary to avoid aggregation, but too strong MSI may induce the diffusion into sub-surface layers or bulk phase of the support, decreasing the catalytic activity. Moreover, the catalytic activity is strongly dependent on the SAs local environment, which in turn determines their electronic structure, charge distribution, and oxidation state [5]. Hence, it is clear that there is no perfect method for the preparation of supported metal SAs on metal oxides and hydroxides, but each one has its advantages and drawbacks. In this chapter, we discuss in detail several techniques to give to the reader a complete idea about the different steps of each procedure and about the final stability of the deposited SA. Furthermore, we focus our attention on the "bottom-up" methods, due to their facile utilization and large applicability.

2.2 Gas-Phase Deposition Methods

2.2.1 Mass-Selected Soft-Landing Method

The mass-selected soft landing is a physical vapor deposition technique based on the production of ionic clusters that are subsequently mass selected using a quadrupole before their final deposition onto the support. Abbet et al. [6] used this method for the deposition of Pd_n ($1 \leq n \leq 30$) clusters on a MgO(100) film utilizing the experimental setup shown in Figure 2.1. The Pd_{SA} produced by a laser evaporation source, guided into the quadrupole mass spectrometer by ion optics and deposited [less than 1% of a monolayer (ML)] on the support at −183 °C, were strongly bound to the oxygen vacancies (O_v) (binding energy 3.31 eV vs. 1.16 eV on terrace oxygen sites), remaining stable up to 26 °C [7, 8].

To investigate the CO oxidation on Pd catalysts, Kaden et al. [9] landed Pd_{SA} on a rutile $TiO_2(110)$ single crystal. The laser-vaporized Pd cations were selected employing a quadrupole mass filter and deposited onto the support at room temperature by a 2 mm thickness lens/mask. Such a procedure allowed the deposition of Pd_{SA} and Pd clusters (2–25 atoms). The latter proved to be more active than the isolated species during CO oxidation catalytic tests. More recently, Tong et al. [10] used the same technique to study the influence of water molecules on the adsorption and the stability of Au_{SA} on modified rutile $TiO_2(110)$ containing O_v. They employed a pulsed YAG (yttrium aluminum garnet) laser beam to illuminate a gold rod, producing a gold plasma, from which a beam of Au_{SA}^+ ions was extracted, mass selected, and then focused onto the support in an ultra-high vacuum (UHV) chamber. The outcomes of this work demonstrated the pivotal role of surface hydroxyl groups, resulting from the water molecule dissociation onto the O_v, on the adsorption and stabilization of Au_{SA}. The bonding of isolated Au atoms onto $TiO_2(110)$ surface depended on the temperature, as illustrated in Figure 2.2. At 327 °C, Au_{SA} directly bonded with bridging oxygen rows, while at 27 °C, they directly bonded with fivefold coordinated titanium (5c-Ti) atoms, and this process was reversible.

It has to be noted that due to the high cost of the experimental setup and the low maximum metal loading achievable without aggregation, this technique cannot be

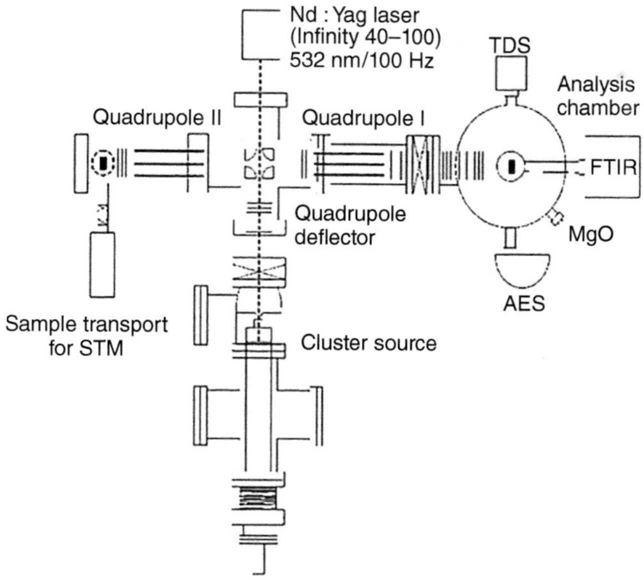

Figure 2.1 Experimental setup for mass-selected soft-landing of Pd_{SA} and Pd clusters deposited onto a MgO film. The analysis chamber, equipped with facilities for thermal desorption spectroscopy (TDS), Fourier transform infrared (FTIR) spectroscopy, and Auger electron spectroscopy (AES), allows following the Pd particle size-dependent reactivity. Source: Abbet et al. [6]. Reproduced with permission of Elsevier.

used for large-scale preparation of commercial catalysts for the moment. However, thanks to the possibility of precisely controlling the number of atoms in the supported clusters, it is extensively used for the preparation of model catalysts that are employed in mechanistic studies [11].

2.2.2 Atomic Layer Deposition (ALD) Method

Atomic layer deposition (ALD) is a chemical vapor deposition (CVD) technique that is attracting great attention owing to the possibility of precisely controlling the deposition of the desired material, spanning from SAs to thin films of specific thickness. During traditional CVD, the volatile precursors are introduced at the same time into the reaction heated chamber, leading to simultaneous homogeneous and heterogeneous reactions, which end up in a poorly controlled film growth or NP deposition. On the contrary, ALD consists of deposition cycles, each cycle being a sequence of self-limited surface reactions between the gaseous precursor and the surface of the support [12]. One cycle usually includes four steps, as shown in Figure 2.3: (i) exposure to the first reagent (metal precursor); (ii) purge of the reaction chamber (to eliminate unreacted precursor and any byproducts); and (iii) exposure to the second reagent, typically oxidants/reductants (to eliminate the remaining ligands of the metal precursor and prepare the surface for the following cycle); and (iv) further purge of the reaction chamber (to eliminate unreacted precursor and any byproducts) [13]. The final metal loading, as well as the particle size and distribution, can be

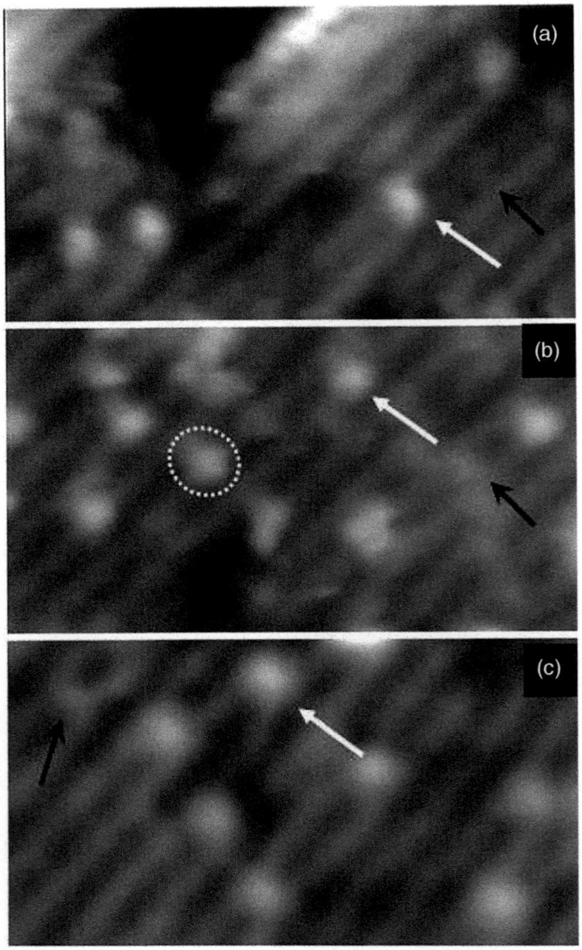

Figure 2.2 Scanning tunneling microscopy (STM) images (70 Å × 40 Å) of a TiO_2(110)-(1 × 1) surface exposed to Au_{SA} at 327 °C and imaged (a) at 327 °C; (b) at 27 °C after cooling from 327 °C; and (c) at 327 °C after heating from 27 °C, respectively; white arrows indicate Au atoms; black arrows in (a) and (c) indicate O_v; black arrow in (b) indicates a hydroxyl group; dashed oval in (b) shows the extent of the contrast of the OH–Au–TiO_2 complex with the brightest spot over the 5c-Ti row and residual contrast over the bridging oxygen row. Source: Tong et al. [10]. Reproduced with permission of American Chemical Society.

accurately tuned by controlling different parameters such as exposure times, number of cycles, deposition temperature, or nature of the support surface. The latter is crucial because the number and the type of the surface functional groups determine the amount and the stability of the SAs, and their eventual aggregation, either during the first or the subsequent steps [14].

Lu's group [15] successfully employed this technique to produce a 0.22 wt% Pt_{SA}/CeO_2 catalyst for the water-mediated CO oxidation. They exposed a commercial CeO_2 powder, previously calcined at 400 °C under an atmosphere of 10% O_2 in Ar

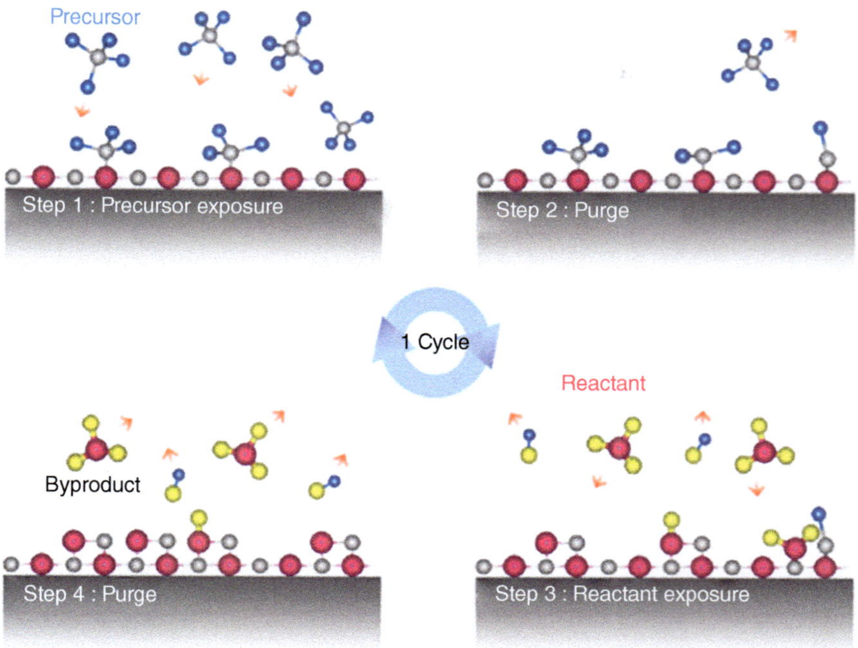

Figure 2.3 Different steps of the ALD process for SAC preparation. Source: Kim et al. [13]. Reproduced with permission of Elsevier.

to remove any carbonaceous species, to a flow of trimethyl(methylcyclopentadienyl)-platinum(IV) [$MeCpPtMe_3$] at 150 °C for 50 seconds, and under nitrogen purge for 120 seconds, respectively.

The sample was then calcined in an oven at 200 °C for 30 minutes under 10% O_2/Ar to eliminate the ligands and obtain the final product. No cluster or NP were formed on the fresh catalysts, and aberration-corrected high-angle annular dark-field scanning transmission electron microscopy (AC-HAADF-STEM) revealed that Pt_{SA} were mainly present on Ce rows of CeO_2 (110) facets. According to DFT calculations, positively charged Pt_{SA} can be stabilized on Ce vacancies by the lattice oxygen via six Pt—O bonds. More recently, the same group used ALD to investigate the effect of defect concentration, metal loading, and high-temperature treatments on the stability of Pt_{SA} on ceria [16]. The preparation was carried out at 150 °C in one cycle, using as supports CeO_2 nanorods with two different densities of Ce^{3+} surface defect sites. The density of the latter was controlled by a thermal treatment of the supports at 60 °C (air, overnight) or 600 °C (10% O_2/Ar for 2 hours). These supports were alternatively exposed to [$MeCpPtMe_3$] and ozone as reagents, and N_2 as purge. By varying the exposure time of the Pt precursor, a set of catalysts with different metal loading, from 0.7 to 4.8 wt%, was obtained. For up to 4 wt% of Pt loading, Pt remained atomically dispersed for all these samples, either on the fresh or post-CO oxidation catalysts. On the other hand, NPs were formed when the loading was further increased to 4.8 wt%. The comparison under the same reduction conditions (10% H_2 in Ar) between two samples with the same amount of metal (c. 1.3 wt%) deposited on the different CeO_2 supports showed a positive effect of the surface defects on SA stability. Indeed, the sample prepared using the defect-rich support exhibited Pt aggregation at 200 °C, while aggregation occurred already at 150 °C on the sample prepared using the support with a lower number of defects. On the latter catalyst, when previously calcined under 10% O_2/Ar at 600 °C, the atomic dispersion was maintained after reduction at 150 and 170 °C, and NPs could be detected only after reduction at 200 °C, revealing a slight improvement of the stability by the high-temperature treatment. Unfortunately, no experimental investigation to thoroughly investigate the reasons for this stabilization was performed.

By using the ALD method, Lu's group also examined Pt_{SA} deposition on Co_3O_4, CeO_2, and ZrO_2 with Pt loadings of 0.5, 1.1, and 0.2 wt%, respectively, to understand how strong electronic metal–support interactions (EMSIs) can affect the activity and stability of these catalysts in the dehydrogenation of ammonia borane [17]. The synthesis was carried out inside a flow reactor in a single ALD cycle, flowing [$MeCpPtMe_3$] at 80, 150, and 120 °C for Pt_{SA}/Co_3O_4, Pt_{SA}/CeO_2, and Pt_{SA}/ZrO_2, respectively, while ozone was used to remove the ligands. Pt was atomically dispersed for all the fresh samples, although after a few catalytic cycles, large NPs and a strong metal loss via leaching were observed on both Pt_{SA}/CeO_2 and Pt_{SA}/ZrO_2 samples. On the contrary, for Pt_{SA}/Co_3O_4, no aggregation or metal loss was detected after several catalytic tests, which was attributed to a strong EMSI, as proposed by DFT calculations and confirmed by extended X-ray absorption fine structure (EXAFS) analysis. Indeed, while a Pt–O coordination was detected on all samples, enhanced interaction between the Pt_{SA} and the support, resulting in the Pt—Co bonds, was evidenced on the Pt_{SA}/Co_3O_4 sample.

As shown in Table 2.1, the ALD can also be used for the preparation of Pt_{SA} on zeolites. Ding et al. [18] employed [$MeCpPtMe_3$] to prepare 0.9, 1.2, and 2.6 wt% Pt_{SA}/HZSM-5, carrying out the depositions at 100 °C with 4 pulses, 100 °C with 10 pulses, and 150 °C with 10 pulses, respectively, to obtain both SAs and NPs on all the samples.

Table 2.1 Selected examples of SACs prepared by ALD.

Sample	Metal loading (wt%)	Presence of NPs[a]	References
Pt_{SA}/CeO_2	0.22	No	[15]
Pt_{SA}/CeO_2	0.7–4	No	[16]
Pt_{SA}/Co_3O_4	0.5	No	[17]
Pt_{SA}/CeO_2	1.1	No	[17]
Pt_{SA}/ZrO_2	0.2	No	[17]
Pt_{SA}/HZSM-5	0.9–2.6	Yes	[18]
Pd_{SA}/Al_2O_3	0.5	Yes	[19]

a) Only on the fresh catalyst.

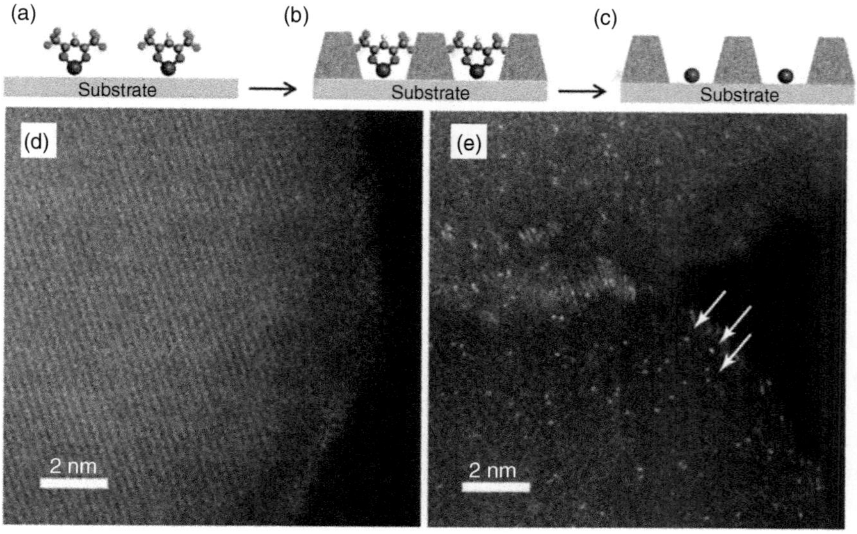

Figure 2.4 Schematics of thermally stable Pd_{SA} catalysts synthesized using ALD. (a) Depositing $[Pd(hfac)_2]$ on spherical Al_2O_3 substrate, (b) creating nanocavities using TiO_2 ALD, and (c) removing hfac ligands using formalin. HAADF-STEM images of: (d) clean spherical Al_2O_3 substrate, and (e) TiO_2 protected Pt_{SA} catalysts on Al_2O_3. Source: Piernavieja-Hermida et al. [19]. Adapted with permission of Royal Society of Chemistry.

An original approach was proposed by Piernavieja-Hermida et al. [19] to deposit Pd_{SA} on Al_2O_3, which were subsequently stabilized by a TiO_2 thin film. Indeed, here, the ALD was exploited twice: the first time to deposit the Pd precursor and, the second, to create a TiO_2 thin film around the precursor, which acted as a barrier against SA aggregation. As summarized in Figure 2.4, the commercial Al_2O_3 support was initially exposed to palladium hexafluoroacetylacetonate $[Pd(hfac)_2]$; subsequently, an alternation of $TiCl_4$ and deionized water led to the formation of protective TiO_2 nanocavities around the Pd precursors, from which ligands were then removed using formalin (an aqueous solution of formaldehyde).

This strategy proved to be effective for the stabilization of Pd_{SA} under both oxidation and reduction conditions. The TiO_2 film thickness is the pivotal parameter to avoid at the same time the aggregation and the complete encapsulation of the metal by overcoating, which would make it inaccessible to the gaseous reactants. The NP formation detected on the 0.55 wt% Pt_{SA}/Al_2O_3 fresh sample was attributed to a reaction between the TiO_2 precursor and the hfac ligands, which enhances the SA mobility. Therefore, the authors believed that this problem could be addressed by the choice of a different TiO_2 precursor, which would not attack the ligands of the $[Pd(hfac)_2]$ precursor.

To date, the ALD technique is mostly employed for the preparation of model catalysts for fundamental studies. Indeed, although recent improvements allowed preparing catalysts with a high metal loading (>1 wt%), the high cost of both ALD reactors and appropriate metallic precursors represents an obstacle to a large-scale industrial application.

2.3 Wet Chemistry Methods

2.3.1 Impregnation Methods

Due to its simplicity and low cost, impregnation is among the most widely employed techniques for the preparation of supported metal catalysts. The support is mixed with a precursor solution. The amount of the latter can be in excess with respect to the pore volume of the support, resulting in wet impregnation (WI) or limited to exactly fill it, resulting in dry impregnation (DI) or incipient wetness impregnation (IWI). In WI, the excess of precursor solution is then removed before the subsequent steps, while in the IWI, this solution completely ends up onto the support. Different processes can take place upon mixing, as well as during the following thermal treatments, which are usually employed to decompose the precursor ligands and determine the dispersion of the metal on the final product [20]. However, employing the traditional preparation, the interaction between the support and the precursor is generally not strong enough to allow the deposition of SAs, and NPs are always present at significant metal loading. To overcome this problem and exploit the advantages of these techniques for the preparation of supported SACs, several groups have been working on the preparation methods, mainly by reducing the amount of metal or by using surface defects for the stabilization of SAs.

2.3.1.1 Wet Impregnation

One of the first successful attempts to prepare supported metal SACs on oxides by WI was carried out by Narula et al. [21, 22]. By adding the support to a solution of H_2PtCl_6 and treating the final powder at 450 °C under air for 4 hours, three $Pt/\theta\text{-}Al_2O_3$ catalysts with a metal loading of 0.18, 1, and 2 wt% were prepared. The low-loading strategy allowed, on the 0.18 wt% sample, the selective deposition of Pt_{SA}, which were, however, mobile on the surface under the beam during aberration-corrected electron microscopy (ACEM) analysis. On the contrary, two different Pt configurations, that is, SAs and 10–20 atoms agglomerates, could be spotted on both 1 and 2 wt% samples (Figure 2.5).

Most of the atoms inside these agglomerates were at more than 3 Å apart from each other, indicating the presence of mainly SAs. Moreover, EXAFS showed that on the 0.18 wt% catalysts, each Pt_{SA} had 6 neighboring oxygens, and no significant change in this configuration was detected after a reduction step at 150 °C under 5% H_2/He. Conversely, at higher metal loading, a strong increase of the Pt–Pt feature, due to sintering, was detected after reduction. Thus, the amount of supported metal is crucial for the stability of Pt_{SA} on Al_2O_3, prepared by WI. Qiao et al. [23] used a similar procedure to synthesize Au_{SA} supported on Co_3O_4 with a low metal loading (0.05 wt%). The desired amount of $HAuCl_4$ was added dropwise to the stirred support suspension, which was then filtered and washed before the drying and calcination steps, which took place for 5 hours at 60 and 450 °C, respectively. As illustrated in Figure 2.6, HAADF-STEM shows only Au_{SA} onto the freshly prepared catalyst. The catalytic results for room temperature CO oxidation showed an increase of the activity during the second cycle of tests. This suggests possible surface structure

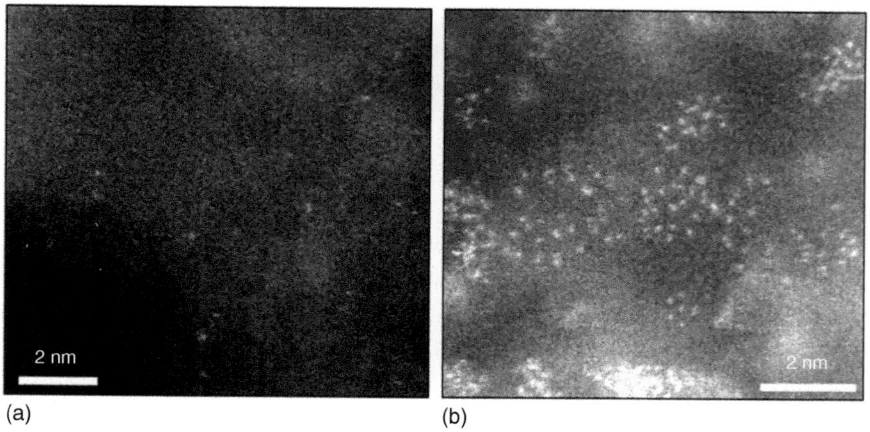

Figure 2.5 ACEM HAADF-STEM images of (a) 0.18% Pt/θ-Al_2O_3; and (b) and 2.0% Pt/θ-Al_2O_3. Source: Moses-DeBusk et al. [22]. Adapted with permission of American Chemical Society.

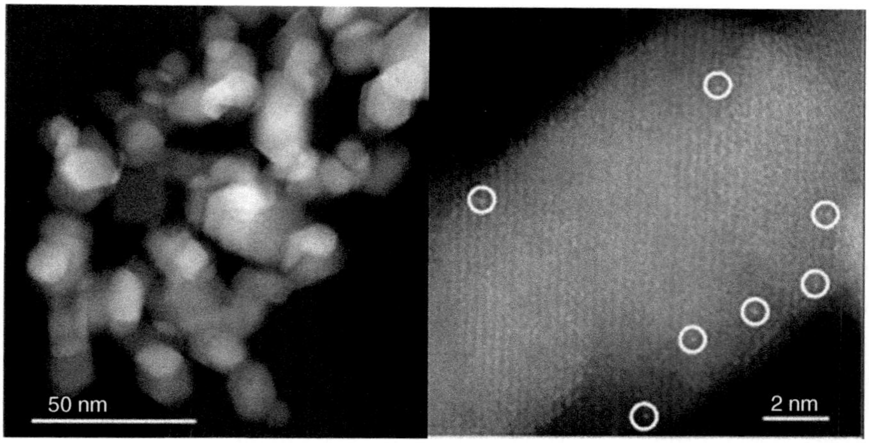

Figure 2.6 STEM-HAADF images of Au_{SA}/Co_3O_4 catalyst at different magnifications. Source: Qiao et al. [23]. Reproduced with permission of Elsevier.

modifications. Also, small changes in the activity were reported when the material went through different pretreatments (e.g. He, 5% O_2/He, 5% CO/He at 200 °C). Unfortunately, the authors did not display further characterization. However, this catalyst showed good stability under mild conditions (20–120 °C) for CO oxidation.

Lang et al. [24] have also employed a simple WI method for the preparation of supported Rh_{SA} on ZnO nanowires. In their synthesis procedure, an aqueous solution of $RhCl_3 \cdot 3H_2O$ with the desired concentration was added dropwise to a water suspension of support and stirred for several hours. The filtered and washed product was then dried at 60 °C overnight and reduced under 20% H_2/He at 200 °C, just before characterization and reaction tests. Three catalysts were prepared following this procedure with 0.006, 0.03, and 0.3 wt% Rh, respectively. Once again, the HAADF-STEM and *in situ* diffuse reflectance infrared Fourier transform spectroscopy (DRIFT) upon CO adsorption confirmed that only Rh_{SA} were present on the two samples containing low metal loading, while Rh clusters with an average diameter of 0.9 nm could be detected on the 0.3 wt% sample. X-ray photoelectron spectroscopy (XPS) analyses showed that the chemical state of Rh species on all samples was different before and after the reduction step. XANES confirmed the presence of almost metallic Rh species on both 0.3 and 0.03 wt% reduced catalysts, as a result of charge transfer from partially reduced ZnO to Rh. Moreover, when the metal loading was maintained below 0.03 wt%, no Rh aggregation took place during the reduction at 200 °C. This indicates that a Rh_{SA} would immediately bind to a Zn atom, which has lost one or more O atoms.

Rh_{SA} were also synthesized on TiO_2 by Matsubu et al. [25] to study how Rh isolated sites control the catalytic stability and selectivity in CO_2 hydrogenation. Thus, catalysts with different metal loadings (0.5, 2, 3, 4, and 6 wt%) and SA/NP ratios were prepared. All samples were synthesized by adding commercial TiO_2-P25 to an aqueous solution of $Rh(NO_3)_3{\cdot}xH_2O$ under stirring. Then, the suspension was dried, ground, and calcined for 4 hours at 450 °C under air. Before catalytic test and characterization, each catalyst was *in situ* reduced under pure H_2 at 350 °C for 1 hour. Only small variation in the average particle size could be detected with the increase of the metal loading. In fact, the Rh average particle size ranged between 1 and 2.5 nm for the catalysts containing 0.5–6 wt% Rh. On the contrary, a large variation in the relative fraction of Rh_{SA} on the surface of TiO_2 was observed: 62% on the 0.5 wt% sample and <20% on the 2 wt% sample. Moreover, to further investigate the activity and selectivity of the different sites toward CO or CH_4, the authors also tested the catalysts with different mixture compositions. Using a feed composition of $CO_2/H_2 = 1/4$, all samples were stable during the first hour of time-on-stream at 200 °C (even if this duration is not long enough to evaluate the stability of a catalyst). On the other hand, upon increasing the CO_2 content of the feed to $CO_2/H_2 = 10/1$, a disintegration of Rh_{NP} into Rh_{SA} was observed.

To deposit a high amount of metal and avoid the formation of any NPs, Chen et al. [26] exploited the surface defects of CeO_2 for the stabilization of Pt_{SA}. By employing ascorbic acid (AA) as reducing agent, they created a controlled amount of surface O_v on the pristine support with the purpose of using these sites for the stabilization of the metal ions during the subsequent impregnation step. The desired amount of AA was added to a suspension of CeO_2 nanorods under stirring for 3 hours before collecting the final product, which was then washed several times using distilled water and dried overnight at 60 °C. This modified support was then re-dispersed in distilled water, mixed with a solution of H_2PtCl_6, and stirred before recovering it by centrifugation. The powder was then washed, dried under vacuum overnight, and finally calcined at 300 °C under static air for 1 hour. Raman and XPS analyses revealed that the modified support had a high amount of O_v and 36.8% of Ce^{3+} on its surface. This support was able to disperse 1 wt% of Pt_{SA}. On the other hand, following the same impregnation procedure onto the pristine support (27.3% of Ce^{3+}), only 0.01 wt% of Pt_{SA} loading could be achieved. The prepared catalyst was active for the CO oxidation (100% of conversion at 60 °C) and showed a good stability during

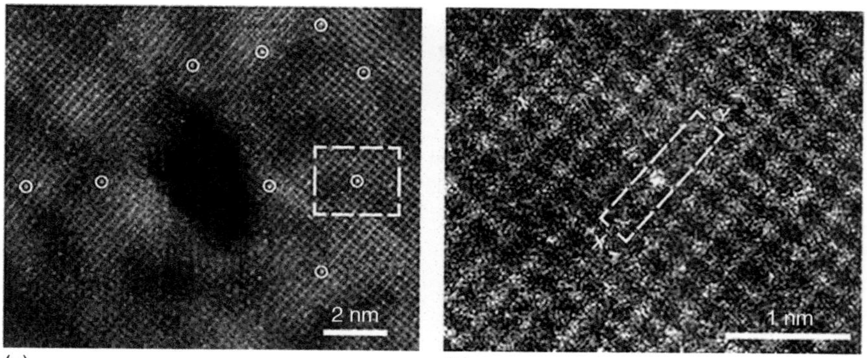

(a) (b)

Figure 2.7 (a) HAADF-STEM image of 0.2% Rh/CoO sample. Rh_{SA} (white circles) are uniformly dispersed on the CoO nanosheets and occupy exactly the positions of Co atoms. (b) Magnified HAADF-STEM image of 0.2% Rh/CoO corresponding to the area of the white square in (a). Source: Wang et al. [27]. Springer Nature/CC BY-SA 4.0.

50 hours of time-on-stream. Nevertheless, as indicated by XPS data, a change in the charge valence of Pt_{SA} from Pt(II) to Pt(0) occurred during the catalytic test.

On the oxide surface, besides the O_v, it is also possible to find their positive counterparts, the cationic vacancies. The latter have been less studied, but can also serve as stabilization sites for metal SAs. In the work of Wang et al. [27], Rh_{SA} were anchored on the position of Co onto CoO nanosheets. Thus, following the addition of a Na_3RhCl_6 to the support suspension, a galvanic replacement reaction tooks place in solution that led to the substitution of Co surface atoms by Rh ones, as highlighted in Figure 2.7.

Only Rh_{SA} were detected by HAADF-STEM on the sample with 0.2 wt% of Rh, while an increase of the metal loading results in the formation of subnanometer clusters of about 10 atoms (at 1 wt% Rh) and NPs of 1.5–2 nm (at 4.8 wt% Rh). XPS results showed that Rh^{3+} was present in the 0.2 wt% sample, which only contained Rh_{SA}, while both metallic Rh and Rh^{3+} were found on the two samples containing higher Rh amounts.

When the same precursor was mixed with VO_2 nanorods, a 0.5 wt% Rh_{SA}/VO_2 catalyst was obtained [28]. Positively charged Rh^+ atoms replaced the V^{4+} ones, which leached into solution as V^{5+} after reducing the initial Rh^{3+} precursor atoms. No particles were detected either before or after NH_3BH_3 hydrolysis catalytic tests between 50 and 85 °C.

Recently, much higher amounts (6.8 wt%) of Rh_{SA} were deposited onto CuO nanowires by Xu et al. [29]. To produce such catalysts onto a copper foam (CF) electrode, they employed a three-step procedure illustrated in Figure 2.8a. First, a nanowire array of $Cu(OH)_2$ with high surface area was synthesized using a wet-chemical oxidation method. This support was then mixed with an aqueous $RhCl_3$ solution for 2 hours at 25 °C, followed by annealing under Ar at 200 °C for 2 hours. AC HAAF-STEM and EXAFS confirmed the complete atomic dispersion

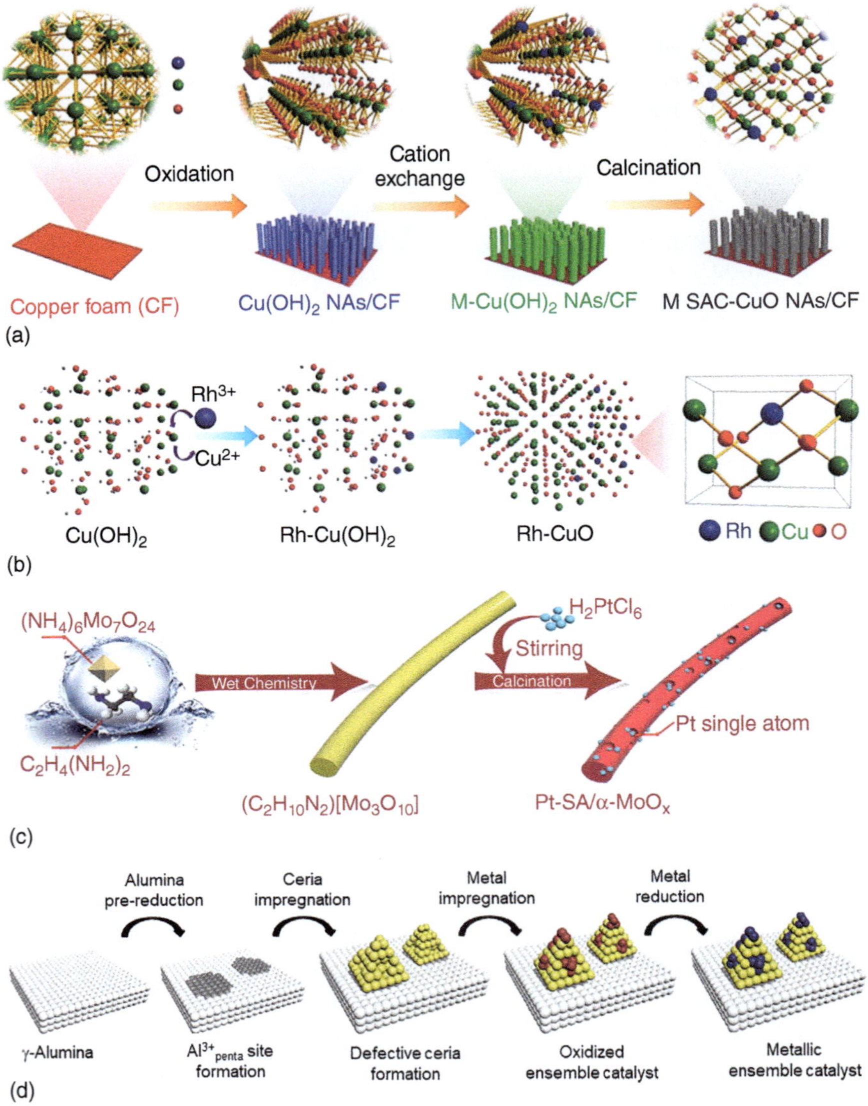

Figure 2.8 Schematic illustration of: (a–b) three-step procedure for 6.8 wt% Rh/CuO nanowires array. Source: Xu et al. [29]. Reproduced with permission of American Chemical Society. (c) Preparation of 0.28 wt% Pt/α-MoO_x. Source: Xu et al. [30]. Adapted with permission of Elsevier. (d) Preparation of 1 wt% Pt, Pd, Rh/CeO_2-Al_2O_3. Source: Jeong et al. [31]. Reproduced with permission of Springer Nature.

of Rh, which carried a trivalent valence state. As summarized in Figure 2.8b, Rh^{3+} species occupied the vacancies of the previously released Cu atoms and were stabilized by surrounding oxygen atoms onto the final CuO nanowires obtained during the calcination step. No NP formation or Rh^{3+} valence modification occurred after the electrocatalytic water-splitting tests, highlighting the robustness of the as-prepared material.

The same type of stabilization was employed by several groups to support different metals onto ZnO nanowires. To succeed in the deposition of only Au_{SA} and Pt_{SA}, Gu et al. [32] decided to employ an extremely low amount of metal (0.0125 wt%), by adding the desired quantity of $HAuCl_4$ or H_2PtCl_4 to the support suspension under stirring, which was then filtered, washed, and dried at 60 °C for 5 hours. The HAADF-STEM analyses showed that Pt_{SA} and Au_{SA} were positioned onto the Zn columns of ZnO nanowires. According to DFT calculations, their stability could be explained by their anchoring onto Zn vacancy positions, and their strong chemical bonding with the coordinated lattice oxygen. Using a similar synthesis procedure, and $Pd(NO_3)_2$ as precursor, Xu et al. [33] succeeded in the selective deposition of Pd_{SA} on the same type of support. Once again, the metal amount was low (0.05 wt%) and the stabilization of the Pd_{SA} was granted by Zn vacancies present on ZnO nanowires.

The stability of metal SAs onto cation vacancies has also been investigated on FeO_x [34] and CeO_2 [35] by Qiao et al. Exploiting the adsorption of $HAuCl_4$ during the mixing step of the impregnation, they succeeded in the preparation of FeO_x-supported catalysts with 0.0015, 0.03, and 0.09 wt% of gold. The low metal loading, along with the presence of surface defects and the absence of any high-temperature calcination steps after drying, allowed the anchoring of Au_{SA} exclusively on the 0.0015 and 0.03 wt% samples, while a few NPs were present on the 0.09 wt% one. The stability of the catalysts, especially the one with 0.03 wt% of Au_{SA} was investigated through CO oxidation tests at 200, 300, and 400 °C, as well as under harsher conditions during the high-temperature reverse water–gas-shift reaction (RWGSR). The material proved to be extremely stable, with no decrease of the activity or NP formation. This was explained by the strong covalent MSI between the positively charged Au_{SA} anchored onto the Fe ion position, and the surrounding oxygen atoms, which could guarantee three (with a stoichiometric surface) or two (with an O_v) covalent Au—O bonds. Employing the same procedure, the authors also managed to support Au_{SA} onto CeO_2 with 0.05 and 0.3 wt% of metal [35]. These catalysts, which showed a great selectivity and stability during the preferential oxidation of CO (PROX) in H_2-rich stream, contained exclusively SAs, even after the catalytic test. Once again, the high sintering resistance of these materials could be attributed to the bond of the positively charged Au_{SA} occupying the surface Ce vacancies with the CeO_2 oxygen. Au_{SA}/CeO_2 catalysts prepared through the same procedure and 0.04 wt% of metal were also very stable and selective for the solvent-free oxidation of alcohols to aldehydes [36].

The stabilization of SAs through cationic vacancies has been employed for the first time on a hydroxide support by Zhang et al. [37]. Working with defect-rich $Ni(OH)_x$ nanoboards, high-loading catalysts (2.3 wt% of Pt) only containing Pt_{SA} were synthesized. HAADF-STEM and XRD showed no sign of cluster or NP, and no Pt–Pt peaks could be detected by EXAFS. During catalyst preparation, the support, which was rich in Ni^{2+} vacancies and Ni^{3+} atoms (naturally present to balance the vacancies), was suspended into EtOH, to which an ethanol H_2PtCl_6 solution was added dropwise. After overnight stirring, the product was collected by centrifugation, dried under vacuum, and reduced under 5% H_2/N_2 to obtain the final

$Pt_{SA}/Ni(OH)_x$. According to EXAFS analyses and DFT calculations, the positively charged Pt_{SA} were adsorbed onto the Ni^{2+} vacancy-rich support with a formation energy of −3.89 eV, much lower than onto the perfect $Ni(OH)_2$ support (−0.72 eV), and stabilized via an enhanced charge-transfer mechanism. When the same preparation procedure was carried out on the support without cationic vacancies ($Ni(OH)_2$), the formation of NPs occurred, even if the metal loading was only 0.9 wt%.

As shown in Table 2.2, the WI method has also been employed for the preparation of Pt_{SA} on an amorphous α-MoO_x support [30]. In a less common synthesis procedure, the impregnation was carried out not directly on the metal oxide, but on one-dimensional (1D) organic–inorganic hybrid nanorods ($(C_2H_{10}N_2)[Mo_3O_{10}]$).

Table 2.2 Selected examples of SACs prepared by WI.

Sample	Metal loading (wt%)	Presence of NPs[a)]	References
Pt_{SA}/θ-Al_2O_3	0.18–2	Yes	[21, 22]
Au_{SA}/Co_3O_4	0.05	No	[23]
Rh_{SA}/ZnO	0.006–0.3	Yes	[24]
Rh_{SA}/TiO_2	0.5–6	Yes	[25]
Pt_{SA}/CeO_2	1	No	[26]
Rh_{SA}/CoO	0.2–4.8	Yes	[27]
Rh_{SA}/VO_2	0.5	No	[28]
Rh_{SA}/CuO	6.8	No	[29]
Pt_{SA}/ZnO	0.0125	No	[32]
Au_{SA}/ZnO	0.0125	No	[32]
Pd_{SA}/ZnO	0.05	No	[33]
Au_{SA}/FeO_x	0.015–0.09	Yes	[34]
Au_{SA}/CeO_2	0.05–0.3	No	[35]
Au_{SA}/CeO_2	0.04	No	[36]
$Pt_{SA}/Ni(OH)_x$	2.3	No	[37]
Pt_{SA}/α-MoO_x	0.28	No	[30]
Pt_{SA}/CeO_2-Al_2O_3	1	No	[31, 38]
Pd_{SA}/CeO_2-Al_2O_3	1	No	[31]
Rh_{SA}/CeO_2-Al_2O_3	1	No	[31]
Ir_{SA}/NiO	2–18	No	[39]
Pt_{SA}/NiO	12	No	[39]
Ru_{SA}/NiO	10	No	[39]
Co_{SA}/NiO	6	No	[39]
Fe_{SA}/NiO	7	No	[39]
Mn_{SA}/NiO	4	No	[39]

a) Only on the fresh catalyst.

H_2PtCl_6 was adsorbed onto the surface functional groups (O^{2-}, OH^-) of this hybrid support, which in turn, during the calcination step, losts its organic part and led to the final 0.28 wt% Pt_{SA}/α-MoO_x catalyst (Figure 2.8c). The characterization results (HAADF-STEM) showed that no NPs were present, and the positively charged Pt_{SA} were uniformly dispersed over the whole support surface. The authors also prepared Ir_{SA}, Au_{SA}, and Pd_{SA} on the same support, proving the universal applicability of this preparation technique, although they did not give any information about the metal loading of these latter samples [30].

Jeong et al. [31] employed a double sequential wet impregnation to obtain Pt_{SA}, Pd_{SA}, and Rh_{SA} on a CeO_2-Al_2O_3 support. Initially, commercial γ-Al_2O_3 was reduced at 350 °C for 1 hour under 10% H_2/N_2 to create Al^{3+}_{penta} defective sites. The latter acted as anchoring sites for the defective ceria NPs, which were synthesized during the first impregnation step using an aqueous $Ce(NO_3)_3 \cdot 6H_2O$ solution. Then, the metal precursor solution was added, and the impregnation was carried out at different temperatures (40, 60, 80, 90 °C) to ensure the desired evaporation rates. Finally, all samples were calcined, then reduced at 500 °C under static air and 10% H_2/N_2 successively. A scheme of the synthesis procedure is shown in Figure 2.8d. A 1 wt% of metal was deposited on the final catalysts, stabilized by the Ce^{3+} defective sites of the CeO_2 NP, which represented 10 wt% of each sample. The characterization results (EXAFS, XANES, HAADF-STEM) showed different metal structures, highlighting the pivotal role of the impregnation temperature. Indeed, depending on this parameter, isolated SAs, small clusters or a mixture of both could be found.

The authors claim that all the samples had a 100% metal dispersion, since the small clusters seemed to be made of only 4 atoms. The latter, defined as ensemble catalysts, showed superior activity for a three-way catalytic reaction (TWC) and great stability during a long-term reaction (420 hours), repeated test, and 900 °C hydrothermal aging (10% H_2O in air for 24 hours). Moreover, the reduction at 500 °C ensured a highly metallic state despite the full dispersion of the metal on all samples. To further clarify this point, the authors investigated also the effect of the reduction temperature on a 1 wt% Pt/CeO_2-Al_2O_3 catalyst prepared following the same procedure [38]. The outcomes of this study were that the variation of the reduction temperature (from 100 to 500 °C) is fundamental to avoid the formation of NPs as well as to control the oxidation state of the Pt_{SA}, from highly oxidized to highly metallic, and consequently the catalytic activity. Finally, it is important to mention that when the preparation of the sample was carried out without the pre-reduction of the Al_2O_3 or the impregnation with CeO_2, it was not possible to achieve a complete dispersion of the metallic species.

Recently, Wang et al. [39] succeeded in the preparation of an ultrahigh-loading catalyst with 18 wt% of Ir deposited onto NiO nanosheets. As illustrated in Figure 2.9, despite the high Ir loading, no NPs were formed, and the uniformly dispersed Ir_{SA} were located at the original nickel positions. Similar atomic structure was obtained when the Ir amount was decreased to 2 and 11 wt%. According to the characterizations and theoretical calculations, Ir_{SA} with almost +4 oxidation state were stabilized via six-coordinated Ir–O on the NiO support. To prepare an electrocatalyst for oxygen evolution reaction (OER), they started from a carbon cloth (CC). The latter was

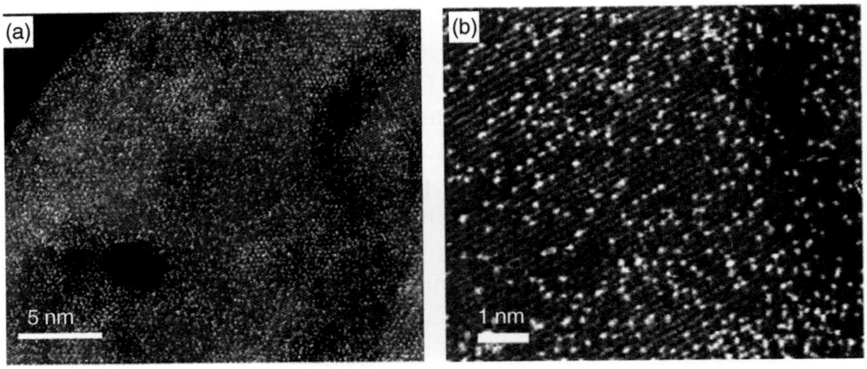

Figure 2.9 HAADF-STEM images of 18 wt% Ir/NiO with different magnifications. Source: Wang et al. [39]. Adapted with permission of American Chemical Society.

immersed into an aqueous solution of $Ni(NO_3)_2{\cdot}6H_2O$ and hexamethylenetetramine (HMT) and kept at 95 °C for 3 hours. After washing and drying, the CC was treated at 350 °C for 2 hours in a muffle furnace, leading to the formation of NiO nanosheets. The NiO/CC was then immersed into a HCl/ethanol solution for 10 minutes, dried at 80 °C, and treated for another 2 hours at 350 °C in a muffle furnace. Finally, after washing and drying, the final catalysts were obtained. Following this procedure, a series of monometallic catalysts with Mn, Fe, Co, Ru, and Pt having 4, 7, 6, 10, and 12 wt%, respectively, were also prepared on the same support.

From the above-mentioned examples, it can be seen that the WI method can be used for the simple preparation of both low and high metal-loading catalysts. Indeed, while the stabilization by O_v is suitable for low loading, the anchoring of the metal atoms onto support's cation positions allows the preparation of highly loaded catalysts with good stability, even at high temperature. Therefore, WI is one of the most promising techniques for an industrial large-scale preparation of metal-supported SACs on oxide supports.

2.3.1.2 Incipient Wetness Impregnation (IWI)

Despite the fact that this preparation technique may lead to very heterogeneous deposition (e.g. egg yolk, egg shell) [20], it was also proved to be effective for the preparation of supported SACs. The Incipient Wetness Impregnation (IWI) of high surface area mesoporous alumina (350 m^2/g) using $Pd(NO_3)_2{\cdot}4NH_3$ allowed Hackett et al. [40] to deposit 0.033 wt% of this metal. The dried sample was then treated for 2 hours under static air and hydrogen at 500 and 400 °C, respectively. EXAFS and HAADF-STEM confirmed the existence of atomically dispersed Pd^{2+}. On the contrary, when Pd loading increased to 0.066 wt% or above, NPs were formed. Moreover, the stability of such atomic sites was proved by *operando* EXAFS during long catalytic tests (alcohol oxidation at 60 °C for several days). A higher Pd loading (0.5 wt%) was achieved by Kwak et al. [41] with the same precursor and a commercial γ-Al_2O_3 (BET surface area = 200 m^2/g) as support. The as-prepared sample was activated by calcination under 6.7% O_2/He for 2 hours before a reduction under 15% H_2/He for 30 minutes, both at 500 °C. Despite the treatments, mainly atomic dispersion of the

Pd onto the fresh material was evidenced by STEM. However, the use of these catalysts in the CO_2 reduction reaction at 25–500 °C induced the formation of metallic clusters.

Using the same activation treatments, the same authors also prepared different Al_2O_3-supported catalysts with 0.1, 0.5, 1, 2, and 5 wt% of Ru [42]. A solution of ruthenium (III) nitrosyl nitrate in diluted nitric acid was chosen for the IWI step. Metal clusters were detected on the highly loaded fresh samples (at and above 1 wt%). Conversely, clusters were formed only after the CO_2 reduction catalytic test in the low-loading samples (<1 wt%).

Finally, a high loading of Pt_{SA} (1–10 wt%) on γ-Al_2O_3 could be obtained by the same research group using IWI method [43]. After impregnating the support with aqueous $Pt(NH_3)_4(NO_3)_2$ solutions, the mixtures were submitted to drying and calcination under dry air at 120 and 300 °C, respectively. The exclusive presence of Pt_{SA} was evidenced by HAADF-STEM on the 1 wt% catalyst, while 2D rafts of PtO made of an average of 20 atoms were spotted on the 10 wt% one. Despite the domination of these agglomerates having sizes around 1 nm, a substantial fraction of metal was still atomically dispersed. Such a result was achieved thanks to a high amount of Al^{3+}_{penta} defective sites, which were able to strongly bind Pt_{SA}. These anchoring points, easily created by dehydration and dehydroxylation, were coordinatively saturated (penta to octahedral conversion) after the formation of oxygen bridges with the Pt atoms. Therefore, the atomic dispersion at 1 wt% of Pt was insured by a $Pt/Al^{3+}_{penta} = 1$, while two-dimensional metal rafts were formed at higher ration. The ability of Al^{3+}_{penta} to inhibit Pt sintering both thermodynamically and kinetically was further rationalized by combining EXAFS experimental results with DFT calculations. Indeed, the strong interaction on Pt–Al^{3+}_{penta} and PtO–Al^{3+}_{penta} (−3.50 and −2.20 eV, respectively) hindered SA migration [44].

The IWI was also successfully employed by Nie et al. [45] for the preparation of Pt_{SA} supported on a polyhedral CeO_2 support for the low-temperature CO oxidation. The support was prepared by heating the $Ce(NO_3)_3{\cdot}H_2O$ precursor under air at 350 °C for 2 hours. Then, the deposition of 1 wt% Pt was achieved by adding dropwise the desired amount of a chloroplatinic acid solution to the support, the latter being constantly ground in a mortar. The final powder was dried (80 °C for 12 hours) and calcined (800 °C for 12 hours). Several characterization techniques (XPS, EXAFS, and DRIFT) showed that at this point, Pt was present exclusively as SAs in an ionic Pt^{2+} form. Furthermore, this material showed good stability under hydrothermal conditions. Indeed, after a treatment under 10% H_2O in Ar at 750 °C for 9 hours, the precious metal kept its isolated and ionic structure, with an enhanced activity in the low-temperature CO oxidation, which was attributed to the activation of the surface oxygen of the support.

A work by Tan et al. [46] pointed out the support particle size as an important parameter in the preparation of stable $Pt_{SA}/CeO_2/Al_2O_3$ catalysts. Indeed, they proved that the dimension of the CeO_2 particles played a pivotal role on the dispersion, activity, and stability of catalysts in both CO and C_3H_6 oxidation reactions. First, two CeO_2/Al_2O_3 supports were prepared by impregnating an aqueous solution of $Ce(NO_3)_3{\cdot}6H_2O$ or an aqueous suspension of commercial CeO_2 NPs on a commercial γ-Al_2O_3, followed by drying (120 °C for 30 minutes) and calcination (550 °C for 2 hours). The mass ratio of CeO_2/Al_2O_3 was fixed at 3/7. The CeO_2 mean particle size of these two supports reached 10 and 20 nm by using a $Ce(NO_3)_3{\cdot}6H_2O$ solution (support denoted as CA-n) and a CeO_2 suspension (support denoted as CA-c), respectively. Then, 1 wt% Pt_{SA}/CA-n and 1 wt% Pt_{SA}/CA-c catalysts were obtained by impregnation of the supports with an aqueous solution of $Pt(NO_3)_2$, followed by calcination at 550 °C for 2 hours. Characterization by several methods revealed that the CA-c support, composed of smaller CeO_2 particles, had a higher surface area and a larger amount of defects than the CA-n support. Thus, the CA-c resulted in the atomic dispersion of the entire precious metal, due to a larger amount of Pt–O–Ce functions. Moreover, upon an activation treatment using 10% H_2 at 400 °C for 1 hour (required before oxidation reactions), both SAs and small clusters were detected. The clusters could be divided into two categories: the first one comprised Pt^0 atoms and the second one was rich in $Pt^{\delta+}$, these latter being more active for CO oxidation. The activated Pt/CA-c catalyst contained mostly small Pt clusters with more active Pt sites ($Pt^{\delta+}$), while the activated Pt/CA-n catalyst contained mostly large Pt clusters with more metallic sites, which were less active for CO oxidation.

Pd_{SA} on γ-Al_2O_3 presenting high activity toward low-temperature CO oxidation were prepared by IWI [47]. In this work, the authors highlighted how the addition of La_2O_3, known for helping the stabilization of the support, could also play an important role in the dispersion of metallic SAs. Therefore, to reach a 0.5 wt% loading of Pd, three consecutive impregnations were carried out on both γ-Al_2O_3 and La-Al_2O_3 (4 wt% La_2O_3) supports. All samples were dried at 110 °C after each impregnation step and eventually calcined at 700 °C for 1 hour under air. The AC-STEM analyses performed on the Pd-La/Al_2O_3 samples revealed the presence of both Pd_{SA} and La_{SA}. Further characterization and catalytic tests proved that the presence of La^{3+} played a key role in the stabilization of atomically dispersed Pd^+_{SA}, which were the active centers for CO oxidation. Indeed, the absence of rare-earth element produced less stable SAs, which could nevertheless be regenerated through a high-temperature (700 °C) calcination step.

Yang et al. [48] worked on a novel stabilization approach of Pt_{SA} on TiO_2, zeolites and mesoporous silica. Indeed, by performing a co-IWI of $[Pt(NH_3)_4](NO_3)_2$ and NaOH (Pt:Na = 1 : 10), 0.5 wt% of precious metal could be loaded onto TiO_2. The resulting material, after a calcination at 400 °C for 4 hours, mostly exhibited isolated Pt species, stabilized through Pt_{SA}–O_x–Na interactions, since no NP could be identified, and X-ray adsorption spectroscopy (XAS) proved that small ensemble of atoms detectable by HAADF-STEM were simply made of isolated atoms without Pt—Pt bonding. On the other hand, when the same synthesis procedure was carried out without NaOH, the final product was full of NPs, thus pointing out the central role of Na in the stabilization of metal SAs. Moreover, the Pt_{SA}–O_x–Na sites were active for the WGSR and showed high stability during cyclic 10 hour tests, with no formation of metallic Pt—Pt bond but only a partial transformation of Pt(IV) to Pt(II). Comparable results were obtained using microporous K-type L-zeolite (KLTL) and mesoporous silica MCM-41 ([Si]MCM41). However, the preparation of these two samples consisted of an initial IWI of the $[Pt(NH_3)_4](NO_3)_2$ alone. Then, the desired amount of NaOH (Pt:Na = 1 : 10) was ground together with the dried powder before calcination at 400 °C for 4 hours [48]. More recently, the same type of stabilization has been employed for the preparation of high-loading Pt SACs using Al_2O_3 and SiO_2 supports [49]. The formation of Pt_{SA}–O_x–K and Pt_{SA}–O_x–Cs bonds on Al_2O_3 and SiO_2 supports, respectively, guaranteed the deposition of significant amounts of metal (up to 2 wt%) by simple IWI. Several samples, from 0.5 to 2 wt% of Pt_{SA}–O_x–K/Al_2O_3 and Pt_{SA}–O_x–Cs/SiO_2 , were prepared by wetting the desired supports with a solution prepared by stirring $H_2Pt(OH)_6$, KOH (K:Pt = 10 : 1) or CsOH (Cs:Pt = 3 : 1) under O_2 purging for 24 hours at 100 °C. The powders, after calcination at 400 °C, were activated under 10% H_2/He at 150 °C before the catalytic test (CO PROX at 110 °C). The atomic dispersion of Pt species on all samples was investigated using DRIFT, EXAFS, AC-HAADF-STEM. Moreover, the catalysts showed high activity and stability during cyclic, multi-temperatures, and 60 hour stability tests, thanks to the considerable amount of accessible Pt_{SA}–O_x active sites, which remained positively charged during the tests. Indeed, only a partial decrease from the initial +4 charge value of the Pt could be detected by EXAFS.

As summarized in Table 2.3, IWI is a suitable method for the preparation of SACs with low metal loading. Indeed, the few examples of SACs with high metal loading and no NPs formation are related to alkaline metal doping for the stabilization of metal SAs.

Table 2.3 Selected examples of SACs prepared by IWI.

Sample	Metal loading (wt%)	Presence of NPs[a)]	References
Pd_{SA}/Al_2O_3	0.033–4.7	Yes	[40]
$Pd_{SA}/\gamma\text{-}Al_2O_3$	0.5	Yes	[41]
$Ru_{SA}/\gamma\text{-}Al_2O_3$	0.1–5	Yes	[42]
$Pt_{SA}/\gamma\text{-}Al_2O_3$	1–10	Yes	[43, 44]
Pt_{SA}/CeO_2	1	No	[45]
$Pt_{SA}/CeO_2/Al_2O_3$	1	Yes	[46]
$Pd_{SA}/La_2O_3\text{-}Al_2O_3$	0.5	No	[47]
Pt_{SA}/TiO_2	0.5	No	[48]
Pt_{SA}/KLTL	0.5	No	[48]
Pt_{SA}/[Si]MCM41	0.5	No	[48]
Pt_{SA}/Al_2O_3	0.5–1	No	[49]
Pt_{SA}/SiO_2	0.5–2	No	[49]

a) Only on the fresh catalyst.

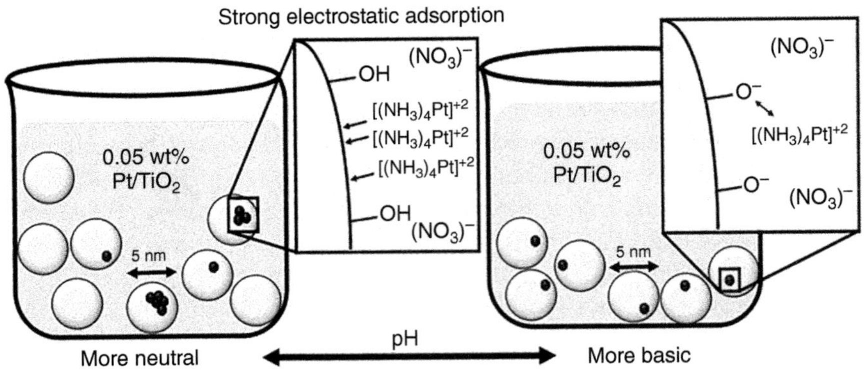

Figure 2.10 SEA mechanism for the deposition of 0.05 wt% Pt/TiO_2 under basic conditions. Source: DeRita et al. [50]. Reproduced with permission of American Chemical Society.

2.3.1.3 Strong Electrostatic Adsorption (SEA)

Strong electrostatic adsorption (SEA) can be considered as a special case of WI in which the pH of the solution is tuned to enhance the electrostatic adsorption of the metal precursors on the oxide surface. Indeed, the hydroxyl (–OH) groups on the oxide surface can be protonated and thus become positively charged or deprotonated and thus negatively charged when the pH of the solution is above or below the oxide PZC (point of zero charge) value, respectively (Figure 2.10). Due to the stronger interactions, the accumulation of metal in the solution during drying is largely reduced, resulting in smaller particle size and size distributions [20]. Therefore, it is not surprising that this technique has been chosen by several scientists for the preparation of SACs with both low and high metal loading.

Fu et al. [51] employed the SEA to deposit ultralow Pt loading (0.04 wt%) on TiO_2 (anatase). Initially, the TiO_2 NPs (5 nm) were suspended in an aqueous solution at pH = 2. Then, the H_2PtCl_6 solution (pH = 2) was slowly added to the first one and aged for 1 hour. The filtrated paste was dried under vacuum overnight and then calcined for the same time at 110 °C under air. Finally, calcination and subsequent reduction were performed at 250 °C under flowing air and 10% H_2/He, respectively. At this metal loading (0.04 wt%), cationic Pt_{SA} were mainly formed, while at higher metal loading, an increasing amount of clusters and NPs was detected by HAADF-STEM and FTIR of CO adsorption.

An even lower amount of Pt was deposited on the same support by DeRita et al. [50]. Indeed, the authors thought that the best way to stabilize metallic SAs without exploiting surface defects was to deposit only one precious metal atom per each support particle. To do so, they decided to employ SEA. Five nanometer diameter TiO_2-anatase with a surface area of 290 m^2/g was dried overnight at 120 °C under vacuum before being added to a NH_4OH solution (pH = 12.2). An aqueous solution of $Pt(NO_3)_2{\cdot}4NH_3$ and NH_4OH (pH = 12.2) was added to the support with an addition rate of 0.025 mg of Pt per hour, followed by heating at 70 °C until complete evaporation. The obtained powder was then dried at 120 °C overnight, calcined under flowing air for 4 hours at 450 °C, and finally reduced *in situ* under H_2 for

1 hour at 240 °C. Despite such harsh treatments, the samples with less than one Pt atom/TiO_2 particle (0.025–0.05 wt%) showed only SAs. On the contrary, when the metal loading was above 1 Pt atom per TiO_2 particle (>0.1 wt%), the formation of large clusters and NPs occurred.

Recently, the same group employed the same procedure to study how, depending on the conditions, the Pt_{SA} supported on TiO_2 anatase can adopt different coordination and oxidation states, with a consequent variation of their catalytic activity [52]. For this, after calcination, 0.025 wt% Pt/TiO_2 underwent three different treatments: oxidation (300 °C under O_2), mild reduction (250 °C under 5% H_2/Ar), and harsh reduction (450 °C under 5% H_2/Ar). Following oxidation and mild reduction, Pt_{SA} kept their cationic state, although with a different coordination environment. Conversely, the harsh reduction led to near-neutral $Pt^{\delta+}$ isolated species. Moreover, *in situ* AC-STEM (Figure 2.11) showed that Pt_{SA} did not move significantly during the first two pretreatments, while they became mobile on the support surface after the high-temperature reduction. These results, combined with EXAFS, XANES, and DFT calculations, allowed the creation of a model explaining the dynamic evolution of the metallic isolated species after the different treatments. The model proposed that after the oxidation, Pt^{4+} species replaced a sixfold coordinated Ti (Ti_{6c}). Upon mild reduction, Pt_{SA} were pulled out from the lattice to form $(PtO_2)_{ads}$ species, where Pt(+2) atoms were bound to two extra O atoms on the surface. Finally, the harsh reduction led to the formation of mobile $(PtOH)_{ads}$ species. These latter, with a +1 oxidation number, were adsorbed on both step and terrace sites of the support. The authors did not figure out the mechanism of these transformations, and how the various species evolved during the CO oxidation reaction. Moreover, such a low loading, while being useful to avoid NP formation, can represent an important drawback for some applications. This work is a perfect example of how the characteristics of the SACs can be specifically tuned by modulating the post-synthesis

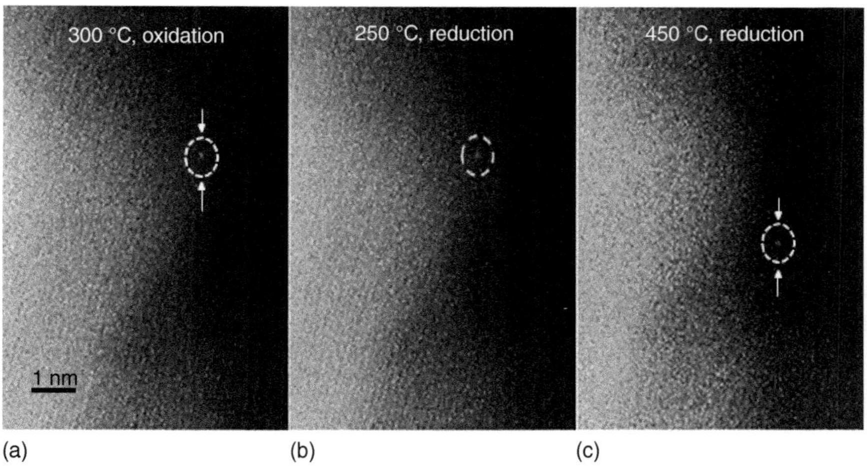

Figure 2.11 *In situ* AC-STEM characterization of Pt_{SA} supported on TiO_2 after (a) 300 °C under O_2, 30 minutes; (b) 250 °C under H_2/Ar, 30 minutes; (c) 450 °C under H_2/Ar, 30 minutes. Source: DeRita [52]. Adapted with permission of Springer Nature.

treatments. Indeed, the evolution of SAs is not only limited to sintering, but can also lead to very active species. The deep investigation of the SA local environment evolution thus attracted the attention of many researchers, and several interesting works were published during the last years.

For example, Tang et al. [53] investigated the ability of Rh_{SA} on TiO_2 to adapt their coordination to the different conditions, pointing out fascinating cooperation between reagent and product during the RWGSR. Combining DFT calculations with experimental studies on a 0.05 wt% Rh/TiO_2 catalyst, they found out that the Rh_{SA} preferred to substitute a cation at the sixfold Ti site on the surface plane under oxidative conditions. These sites were catalytically inactive, and the Rh_{SA} evolved toward a more favorable on-top position on the oxide. This change was insured by the combined action of H_2 and CO. CO was able to stabilize the final on-top Rh_{SA}, but not to interact with the initial substitutional one. Conversely, H_2 could not provide a strong stabilization, but was able to create O_v, which decreased Rh coordination and provided a route for CO adsorption. The catalyst was initially prepared following a SEA procedure using an aqueous solution of rhodium(III) chloride hydrate and a TiO_2 suspension, both at pH = 8.25. The precursor solution was injected in the support suspension and then heated at 70 °C to evaporate the water. The resulting powder was dried overnight at 100 °C and subsequently calcined under 10% O_2 at 350 °C for 4 hours. No metal particles could be detected on both fresh and used catalysts, while clusters were present after a reduction under H_2 at 300 °C. Therefore, for a successful design of SACs, the possible evolution under working conditions and the pivotal role of the support in such processes should also be considered. Indeed, the transformation of Rh_{SA} was driven by the formation of O_v onto the TiO_2 surface. In contrast, the rearrangement of metal clusters under working conditions is mainly driven by thermodynamics and is less influenced by the nature of the support.

The support can also play a direct role on the catalytic performance, as proved by Lou and Liu [54]. To elucidate this role in the CO oxidation reaction, they used SEA to deposit Pt_{SA} on three different oxides: highly reducible Fe_2O_3 crystallites, reducible ZnO nanowires, and irreducible γ-Al_2O_3 powder. After mixing an aqueous solution of $H_2PtCl_6 \cdot 6H_2O$ with each support at controlled pH, the mixtures were aged for 2 hours at room temperature, washed, and filtered. The resulting solids were then dried at 60 °C for 12 hours under air. Then, the catalysts were calcined at 300 °C for 2 hours under air (Pt/γ-Al_2O_3 and Pt/ZnO) or N_2 (Pt/Fe_2O_3) to study the effect of the different gases on the final metal dispersion. No differences could be detected and all catalysts showed the exclusive presence of Pt_{SA}. The final Pt loadings on Fe_2O_3, ZnO, and γ-Al_2O_3 were 0.029, 0.034, and 0.044 wt%, respectively. The catalytic tests showed a great influence of the support resulting from charge transfer from the metal to the support, especially when no –OH/H_2O were present on the surfaces (dry CO oxidation), while no NPs were detected for used catalysts.

All previous examples of SAC preparations involved noble metals. Indeed, the possibility to reduce the high cost by optimizing the catalytic activity of these metals captured the attention of many researchers. Nonetheless, some important reactions can work with less expensive metals, which can as well benefit from an atomic dispersion. Thus, in a recent work, Ni_{SA} showed a high activity and an outstanding coking-resistance during the methane dry reforming reaction [55]. The synthesis of such catalysts was carried out through SEA at room temperature over hydroxyapatite (HAP) and Ce-doped (5 wt%) HAP (HAP-Ce). The desired amount of nickel nitrate was dissolved into deionized water, and the pH of the resulting solution adjusted to 10 using 25 vol% ammonia solution. The support was then added to the as-prepared solution and stirred for 3 hours at room temperature. The resulting powder was filtered, washed, dried overnight at 80 °C, and finally calcined at 500 °C for 4 hours. An additional *in situ* reduction at 500 or 750 °C under 10% H_2/He for 1 hour was performed before the catalytic test. Three different catalysts containing mainly Ni_{SA} having 0.5, 1, and 2 wt% of Ni were prepared following this procedure. After the reduction at 500 °C, the 0.5 wt% Ni/HAP-Ce mostly contained Ni_{SA} and a small fraction *of* c. 1 nm Ni clusters, detected by AC-HAADF-STEM and EXAFS. Ni_{SA} were still predominant in the 2 wt% Ni/HAP-Ce, which contained also Ni clusters and NPs. When the amount of Ni loading increased to 10 wt%, the catalyst mostly contained Ni_{NP} and Ni clusters, without Ni_{SA}. Moreover, XPS and H_2-temperature programmed reduction (TPR) characterization demonstrated that the Ce doping was able to stabilize the atomically dispersed Ni against reduction and sintering. Thus, Ce doping aided to improve the catalytic stability, as later discussed in the Chapter 13 of this book. Finally, increasing the metal loading on the doped support further improved the long-term stability during the catalytic reaction. The authors attributed the enhanced stability of the high-loading catalysts to the high amount of clusters resulting from the reducing pre-treatment. Indeed, the authors suggested that at low Ni loading (0.5 wt%), Ni_{SA}, being initially predominant, could aggregate when they had low coordination (e.g. bonded on the edges of HAP crystals). At higher Ni loading (2 wt%), unstable Ni_{SA} already evolved into Ni clusters, explaining the high catalytic stability.

As shown in Table 2.4, another preparation of high-loading non-noble metal SACs via SEA was performed by Jimenez et al. [56]. A solution of hexamminecobalt(III) chloride was added to a silica suspension, both with a pH value of 11. After stirring and washing to remove the excess of precursor, the powder was dried overnight at 125 °C and then calcined under air at 300 °C for 3 hours. According to the authors, the electrostatic repulsion and the hydration shell of the metal precursor, combined with the high specific surface area of the amorphous silica (300 m^2/g) and the SEA procedure, minimized the formation of adjacent cobalt centers. As a consequence, all the supported metal (1.82 wt%) was atomically dispersed as CoO onto the silica surface. The tetrahedral Co^{2+} SAs showed great stability under reverse-WGSR conditions. Indeed, the cationic nature was completely retained until 600 °C. Above this temperature, a mixture of Co^{2+} and Co^0, due to the formation of Co_{NP}, was detected. A synergistic effect between the metallic and cationic cobalt was proposed by the authors to account for the superior activity of such an ensemble.

To date, the SEA has been proven to be extremely reliable for the preparation of oxide-supported metal SAs. The possibility to support noble and non-noble metals, with both low and high loadings, and without the help of surface defects for the stabilization, makes SEA one of the best candidate methods for future employment in SAC synthesis.

Table 2.4 Selected examples of SACs prepared by SEA.

Sample	Metal loading (wt%)	Presence of NPs[a)]	References
Pt_{SA}/TiO_2	0.04–1	Yes	[51]
Pt_{SA}/TiO_2	0.025–0.5	Yes	[50]
Pt_{SA}/TiO_2	0.025	No	[52]
Rh_{SA}/TiO_2	0.05	No	[53]
Pt_{SA}/Fe_2O_3	0.029	No	[54]
Pt_{SA}/ZnO	0.034	No	[54]
Pt_{SA}/Al_2O_3	0.044	No	[54]
Ni_{SA}/CeO_2-HAP	0.5–2	Yes	[55]
Co_{SA}/SiO_2	1.8	No	[56]

a) Only on the fresh catalyst.

2.3.2 Co-precipitation Method

Co-precipitation is one of the most usually employed techniques for the synthesis of metal SAs on oxides. Its widespread application is mainly due to the fact that all the metal oxides can be simply prepared by precipitation of metallic salts under adequate pH conditions. Therefore, the desired metal atoms can be homogeneously distributed in the structure of the support, which usually also guarantee great stability. However, to avoid the formation of NPs and the embedding of the active species into the bulk support, many parameters must be carefully controlled such as pH, temperature, stirring rate, and precursor addition rate [1, 57].

This technique was employed for the first time with the purpose of preparing SACs by Qiao et al. [58]. In this pioneering work, the authors co-precipitated a solution of $H_2PtCl_6{\cdot}6H_2O$ and $Fe(NO_3)_3$ at 50 °C. Na_2CO_3 was employed as a precipitating agent and the pH was maintained at around 8. The resulting solid was then dried at 60 °C for 5 hours and calcined at 400 °C also for 5 hours. Finally, it was reduced for 0.5 hours at 200 °C under 10% H_2/He. Isolated atoms were exclusively detected on the 0.17 wt% Pt_{SA}/FeO_x sample by HAADF-STEM (Figure 2.12), EXAFS, and CO adsorption. These Pt_{SA} occupied exactly the position of Fe atoms on the surface of the support, anchored onto the threefold hollow sites on the O_3-terminated surface of FeO_x, and were positively charged. On the contrary, when the metal loading was increased to 2.5 wt%, a mixture of Pt_{SA}, Pt rafts, and 3D Pt clusters was identified. Only 1.8 atom% of the total Pt was present as SAs, while a large amount of metallic Pt was present as clusters. The low-loading sample showed great activity and stability during the PROX reaction at 27–80 °C. Indeed, no Pt clusters were detected after catalysis or after two cycles of oxidation–reduction treatments.

After this first development, Zhang's group improved this technique. Indeed, by modifying some parameters of the precipitation or post-treatment procedures, the characteristics of the final material could be specifically tuned. The same catalysts,

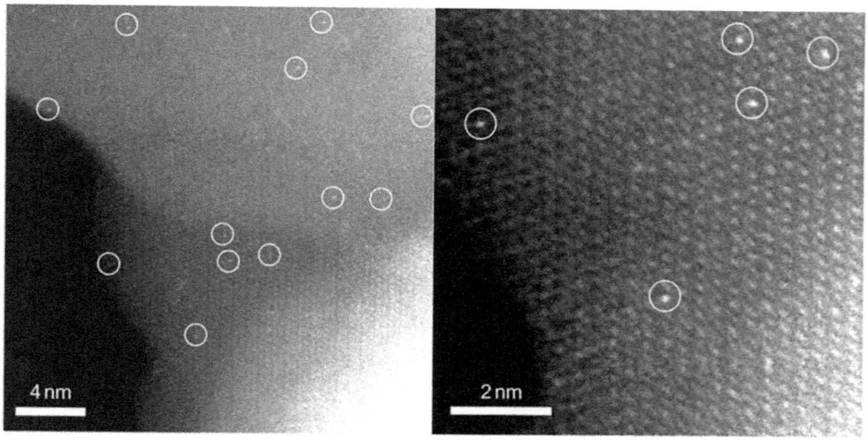

Figure 2.12 HAADF-STEM images of 0.17 wt% Pt/FeO_x sample with different magnification. Source: Qiao et al. [58]. Adapted with permission of Springer Nature.

with a Pt loading between 0.08 and 2.4 wt%, were successfully used as counter electrodes in dye-sensitized solar cells [59]. The analysis of the fresh catalysts showed that the 0.08 wt% sample had only Pt_{SA}, while small clusters were formed at higher metal loadings. When the catalyst was prepared with a lower amount of Pt_{SA} (0.06 wt%) and reduced at 250 °C under 5% H_2/He, it showed superior activity and stability during NO reduction at 200 °C for 6 hours [60]. Moreover, by modulating the amount of Pt and the reduction temperature, the metal dispersion spanned from isolated SAs, to small ensembles, and sub-nanometer clusters to 3D particles, with very different activities for the hydrogenation of nitroarenes [61]. In this case, a 0.08 wt% sample reduced at 200 °C resulted in exclusively Pt_{SA}, while small clusters with a 2D raft-like morphology were formed when the reduction temperature was increased to 250 °C. On the other hand, a mixture of NPs, clusters, and SAs stemmed from increasing the Pt loading (0.31–4.30 wt%) and performing the reduction at 250 °C. Lower Pt loadings not only resulted in higher amount of Pt_{SA} but also, according to XANES characterization, to more electron transfer. Likewise, a lower reduction temperature on the 0.08 wt% sample guaranteed more positively charged Pt species.

This procedure was also used for the preparation of Ir/FeO_x catalysts [62]. Here, to ensure the deposition of the entire amount of H_2IrCl_6, the precipitation temperature was increased to 80 °C. Moreover, the reduction temperature was increased to 300 °C. Low amount of Ir (0.01 wt%) resulted in completely isolated and extremely stable atoms occupying the positions of Fe. Conversely, a 0.22 and 0.32 wt% of Ir led to the formation of metallic clusters. Finally, NPs were detected when the amount of metal was further increased to 2.4 wt%.

A different procedure was employed for the preparation of Pd/FeO_x catalysts for the WGSR [63]. An aqueous solution of $Fe(NO_3)_3 \cdot 9H_2O$ and $PdCl_2$ was added dropwise to an aqueous solution of NaOH, at 80 °C. The resulting mixture was stirred for 3 hours (pH = 8.5) and subsequently aged for 1 hour. Following filtration and washing, the catalyst was dried overnight at 80 °C. Finally, *in situ* reduction at 300 °C for 30 minutes under 10% H_2/He was performed before the catalytic test. An extremely low loading of Pd (0.044 wt%) resulted in complete Pd atomic dispersion, while clusters and NPs were formed at higher loadings (0.22–2.2 wt%).

FeO_x is not the only support employed during co-precipitation. Indeed, Pt_{SA} were deposited on TiO_2 by Xing et al. [64], with a metal loading spanning from 0.2 to 2 wt%. The synthesis was carried out at 50 °C and pH = 8 using a solution of ammonium carbonate and a mixed solution of H_2PtCl_6 and $TiCl_4$. After 3 hours of reaction, the product was washed, dried at 60 °C, and then calcined at 200 °C for 5 hours. Before characterizations and catalytic tests (photocatalytic water-splitting reaction), it was further reduced for 30 minutes at 200 °C under 20% H_2/Ar. The 0.2 wt% Pt/TiO_2 had high surface area (484 m^2/g) and only $Pt_{SA}^{\delta+}$ were identified. No structural changes were detected after catalysis. On the other hand, EXAFS showed a Pt–Pt contribution on the 2 wt% fresh catalyst, highlighting once again the difficulty of preparing high-loading catalysts with only SAs by co-precipitation.

Miller et al. [65] employed a different approach to synthesize several $Ni_xMg_{1-x}O$ catalysts for the RWGSR. The stabilization of Ni_{SA} was achieved by the formation of a solid solution with the substitution of 1–10 atom% of Mg^{2+} by Ni^{2+} inside the MgO lattice. Co-precipitation of Ni and Mg nitrates solution with an ammonia solution at 60 °C allowed the synthesis of $Ni_xMg_{1-x}(OH)_2$ precursors. Then, the solution was aged for 1 hour before cooling. The pH was 8.8 during the whole synthesis and aging processes. The light green precipitate, after filtration and washing, was dried for 15 hours at 80 °C and then calcined for 3 hours at 600 °C under 21% O_2/Ar. Several samples with Ni amounts in the 1–15 atom% range were prepared. Unfortunately, no weight percentage was provided by the authors, but according to our calculations, the maximum amount of Ni (15 atom%) corresponds to 0.24 wt%. Moreover, high metal-loading samples (10 and 15 atom%) resulted in a poor surface Ni concentration. On the contrary, the 1, 3, and 5 atom% samples led to the full metal deposition onto the support surface as SAs, which were active for the RWGSR. The stability against sintering of the Ni_{SA} seemed to be related to a maximum reduction temperature of 300 °C, and a surface reconstruction, which took place at 30 bar under reaction conditions. Indeed, a carbonate-based overlayer was formed at high pressure and its decomposition led to Ni agglomeration.

As shown in Table 2.5, higher loading catalysts have been recently prepared using co-precipitation. Lang et al. [66] succeeded in the preparation of a 1.8 wt% Pt_{SA}/FeO_x catalyst. The synthesis followed a classical procedure, by using $H_2PtCl_6 \cdot 6H_2O$ and $Fe(NO_3)_3 \cdot 9H_2O$ as starting reagents and Na_2CO_3 as precipitating agent at 50 °C. After stirring for 3 hours, followed by static aging for 2 hours, the recovered solid was washed and then dried overnight at 60 °C. A portion of this sample was calcined at 800 °C under air for 5 hours. Both fresh and calcined samples were composed of Pt_{SA} aligned with Fe atomic rows. No clusters or NPs were detected by AC-HAADF-STEM or EXAFS. Moreover, XANES characterization showed that on both samples, the chemical state of Pt resembled that of PtO_2, with tetravalent Pt. The high stability toward high-temperature calcination was attributed to the presence of strong

Table 2.5 Selected examples of SACs prepared by co-precipitation.

Sample	Metal loading (wt%)	Presence of NPs[a)]	References
Pt_{SA}/FeO_x	0.17–2.5	Yes	[58]
Pt_{SA}/FeO_x	0.08–2.4	Yes	[59]
Pt_{SA}/FeO_x	0.06	No	[60]
Pt_{SA}/FeO_x	0.08–4.3	Yes	[61]
Ir_{SA}/FeO_x	0.01–2.4	Yes	[62]
Pd_{SA}/FeO_x	0.04–2.2	Yes	[63]
Pt_{SA}/TiO_2	0.2–2	Yes	[64]
Ni_{SA}/MgO	0.02–0.24	Yes	[65]
Pt_{SA}/FeO_x	1.8	No	[66]

a) Only on the fresh catalyst.

covalent metal–support interactions (CMSIs). Nevertheless, the sample never went through any reduction, which is usually a pivotal step for sintering.

To date, the co-precipitation technique is easy to perform and can be largely employed for the preparation of low-loading SACs, but NPs are formed at high loadings, especially when a reduction step is included.

2.3.3 Deposition–Precipitation Method (DP)

In the deposition–precipitation (DP) method, the support is initially mixed with the metal precursor solution. Then, the metal species are precipitated on the support by addition of a precipitation or reducing agent, or a change in the pH value. To ensure a good metal distribution on the support and avoid the precipitation in solution, it is necessary to fulfill two conditions: (i) a strong interaction between the metal precursors and the surface of the support; and (ii) an accurate concentration of the metal precursor to avoid spontaneous homogeneous precipitation [20]. Therefore, a low metal loading and modification of the oxide support to create more anchoring sites are usually necessary to obtain an atomic distribution of the metal species.

Wang et al. [67] employed the DP method to synthesize Rh_{SA}/Co_3O_4 catalysts for the reduction of NO with H_2. An aqueous solution of $Rh(NO_3)_3$ was mixed with a methanol suspension of Co_3O_4. Then, an aqueous solution of $NaBH_4$ (reducing agent) was added dropwise to this mixture. At this point, Rh^{3+} atoms, weakly bound to OH^- surface groups, were reduced to form Rh—O bonds. The precipitate was then filtered, washed, and dried. On the as-prepared catalysts, EXAFS and XANES analyses revealed the exclusive presence of Rh_{SA}. Moreover, *in situ* EXAFS showed that surface reconstruction occurred at 220 °C with the formation of $RhCo_n$ clusters. Under the atmosphere of $NO:H_2 = 1:1$, the Rh_{SA} immediately reacted with the support Co atoms, which had lost one or more oxygen. Such bonding was thermodynamically and kinetically favored compared to a displacement close to another Rh atom that was 2–3 nm far away on average, due to the extremely low loading. Indeed, even if the authors only provide the atomic ratio of Rh to Co (0.14%), according to our calculations, the Rh metal loading is equal to 0.02 wt%.

Rh_{SA} were also supported on SiO_2 by a modified DP method. First, an aqueous solution of $Rh(NO_3)_3$ was added to a suspension containing dispersed SiO_2 [68]. Then, the pH was gradually adjusted to 9 using a solution of ammonium hydroxide. After 24 hours under stirring, the product was recovered by centrifugation, dried overnight at 50 °C under vacuum, and calcined for 1 hour at 250 °C. Finally, a reduction step at 300 °C for 1 hour under 5% H_2/Ar was performed before characterizations and catalytic tests (NO reduction with CO). This procedure allowed the preparation of catalysts with 0.01–0.03 wt% and 0.05–0.59 wt% metal loadings, which showed different Rh dispersions. Indeed, 0.03 wt% represented the upper limit to obtain only Rh_{SA}, which were also stable during catalysis, while higher metal loadings led to NP formation.

Au/CeO_2 foam catalysts for CO oxidation were prepared using a modified DP method by Wang et al. [69]. The desired amount of CeO_2 foam was added to a solution of $HAuCl_4$ at pH = 9 (NaOH). The suspension was subsequently heated at 65 °C and stirred for 1 hour. The final product was then filtered, washed, dried overnight at 60 °C, and finally calcined at 500 °C for 3 hours. The Au loading spanned from 0.12 to 2.6 wt%. Au SAs were identified via HAADF-STEM only on the 0.12 wt% Au/CeO_2, while a large number of Au_{NP} were detected on all other samples. Moreover, XPS showed the presence, in different ratios, of three types of gold species (Au^0, Au^+, and Au^{3+}) on all catalysts. The increased amount of Au^0 after the calcination at 500 °C was attributed to the electron transfer from the support (probably from O_v) to the Au_{NP}.

Ionic Au_{SA}, with an oxidation state between Au^+ and Au^{3+}, were also prepared on CeO_2 nanorods [70]. In this case, an ammonium carbonate aqueous solution was first added to the support suspension, followed by the dropwise addition of $HAuCl_4{\cdot}3H_2O$. After stirring and aging for 1 hour, the final product was filtered and washed with water at 70 °C, dried and calcined under air for 4 hours at 70 and 400 °C, respectively. Only Au_{SA} were detected by HAADF-STEM and EXAFS, despite the relatively high metal loading of 1.2 wt%. According to the authors, the Au_{SA} stabilization can be attributed to the abundant presence of O_v. Unfortunately, no further investigation was performed to confirm this claim.

Conversely, a deep analysis of the defect contribution in the stabilization of Au_{SA} on TiO_2 was realized by Wan et al. [71]. The synthesis was carried out following the same procedure, but this time defective TiO_2-anatase nanosheets were employed as a support (Figure 2.13a). The latter were rich in surface O_v, created by a reducing treatment under 5% H_2/Ar at 200 °C for 2 hours. EXAFS and HAADF-STEM (Figure 2.13b) confirmed the exclusive existence of Au_{SA}, twofold coordinated by Ti atoms and with two O_2 molecules adsorbed on the opposite sides (Figure 2.13c). DFT calculations highlighted the higher stability of the three-center Ti–Au–Ti structure, with binding energies of −1.28 and −0.69 eV for the twofold (Vo_{2C}) and threefold (Vo_{3C}) O_v, respectively, in comparison with Au_{SA} anchored on the non-defective TiO_2 surface, having a binding energy of 0 eV. Although the metal

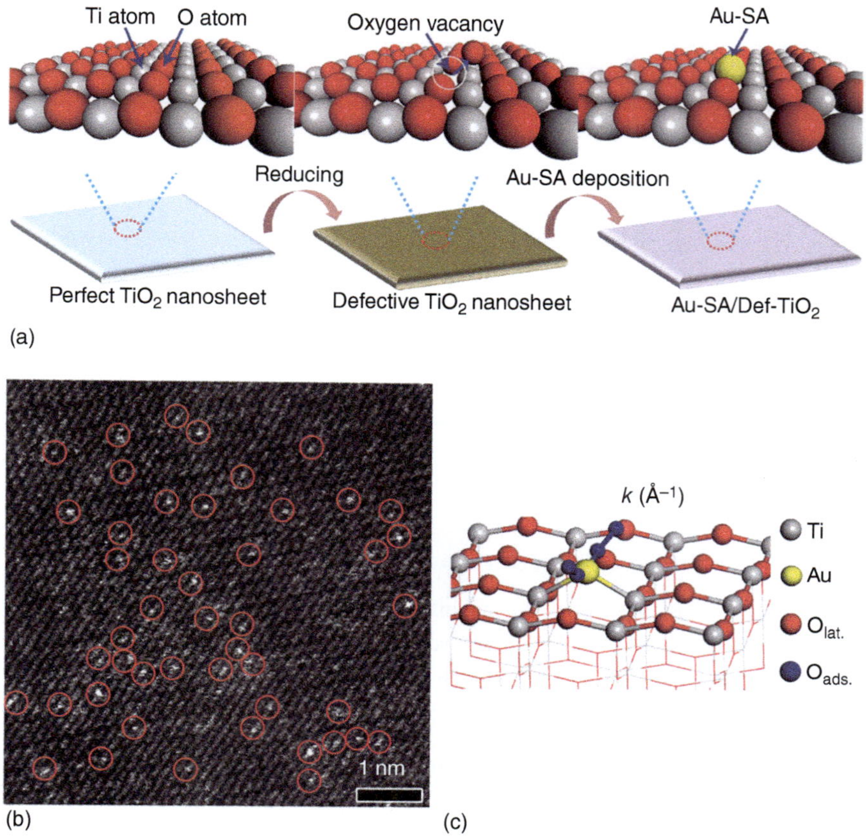

Figure 2.13 (a) Schematic illustration of 0.25 wt% Au/TiO_2 preparation; (b) TEM image of 0.25 wt% Au/TiO_2 sample; (c) local structure of gold SAs in 0.25 wt% Au/TiO_2 sample. Source: Wan et al. [71]. Adapted with permission of WILEY-VCH.

amount was not extremely high (0.25 wt%), the defects played an important role in the stabilization of Au_{SA}, which remained isolated even after a 600 min. stability test (CO oxidation).

As shown in Table 2.6, Au_{SA} were also deposited onto ZrO_2 by adding the support to an aqueous solution of $HAuCl_4$ at room temperature [72]. The pH of the solution was then adjusted to around 9 by addition of a NH_4OH aqueous solution. Before filtration, the solution was stirred for 6 hours and aged for another 2 hours. The recovered powder was washed, dried, and calcined under flowing air at 200 °C. Catalysts with a metal loading of 0.01, 0.05, 0.23, and 0.76 wt% were prepared, and different calcination temperatures were used. Au^{3+} atoms without NPs were detected at low metal loading (0.01 and 0.05 wt%), while upon increasing the metal loading (0.23 and 0.76 wt%), Au_{NP} were formed. Additionally, at 0.76 wt% Au loading, only Au^0 was detected for the sample calcined at 500 °C, while both Au^{3+} and Au^0 were present when the calcination was carried out at 300 °C indicating the impact of the thermal treatment.

Table 2.6 Selected examples of SACs prepared by deposition–precipitation.

Sample	Metal loading (wt%)	Presence of NPs[a]	References
Rh_{SA}/Co_3O_4	0.02	No	[67]
Rh_{SA}/SiO_2	0.01–0.59	Yes	[68]
Au_{SA}/CeO_2	0.12–2.6	Yes	[69]
Au_{SA}/CeO_2	1.8	No	[70]
Au_{SA}/TiO_2	0.25	No	[71]
Au_{SA}/ZrO_2	0.01–0.76	Yes	[72]

a) Only on the fresh catalyst.

All these examples show that this technique can be easily employed for the preparation of low-loading SACs (usually <0.2 wt%), while to achieve higher metal loading, it is necessary to use a support rich in defects to stabilize the SAs.

2.3.4 SAC Synthesis *via* Ion Exchange

Largely employed with zeolites, the ion-exchange technique consists in the replacement of an ion present on the support surface by another ion from the solution. Indeed, the support is usually immersed in an excess of solution containing the desired metal ions, which diffuse inside the pores of the support and gradually substitute the ions on the first layer of the support surface until an equilibrium between the two species is reached [1]. Therefore, it is a suitable preparation procedure for materials like zeolites, where Na^+ and K^+ ions neutralize the negative charge of the AlO_4 tetrahedron. Moreover, the spatial confinement provided by such supports helps the stabilization of the isolated species. Despite its usual application with zeolites, the ion exchange can occur with all types of metal oxides, which carry exchangeable ions, such as HAP ($Ca_{10}(PO_4)_6(OH)_2$).

One of the first reports on the employment of this technique for the preparation of SAC is the work of Lu et al. [73]. The preparation of 1 wt% of gold on NaY zeolite (Si/Al = 2.5) was performed by mixing the previously calcined support with $Au(CH_3)_2(acac)$ (acac - acetylacetonate) in *n*-pentane at 25 °C for one day. Then, the solvent was simply evaporated under vacuum during one day. At this point, the gold precursor was physisorbed on the aluminum sites, interacting with the surface by hydrogen bonding. Upon exposure to CO and O_2 during the catalytic test (CO oxidation), the acetylacetonate ligands were replaced by CO, leading to chemisorbed gold complexes. The Au atoms remained isolated after catalysis, and the change in their oxidation state (from Au^{3+} to Au^{1+}) was associated with the movement between two different anchoring sites. Indeed, initially only 34% of gold complexes was inside the supercages as the first type of anchoring sites, while after catalysis the amount rose to 73%. The remaining gold species were positioned closer to the channel openings connecting adjacent supercages, as the second type of anchoring site. Therefore, the main reason for the stability of the isolated species is the confinement in the zeolite channels.

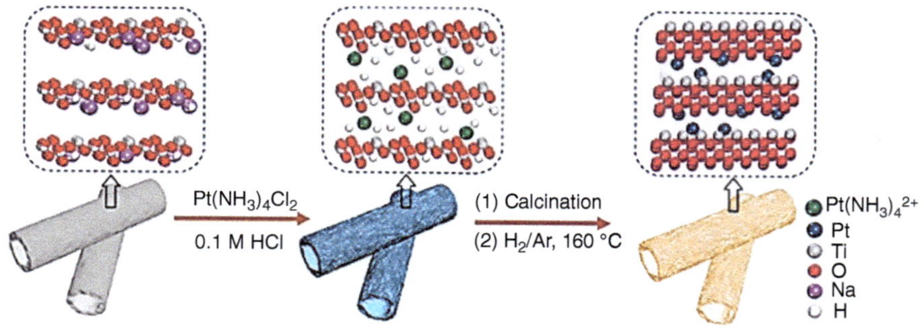

Figure 2.14 Schematic illustration of 0.58 wt% $Pt_{SA}^{\delta+}/TiO_2$ catalyst preparation. Source: Chen et al. [75]. Reproduced with permission of American Chemical Society.

Later, Kistler et al. [74] employed this technique to support 1 wt% of Pt on another zeolite. The KLTL zeolite, after calcination, was mixed for 24 hours with a solution of $Pt(NH_3)_4(NO_3)_2$. After removing water, the $Pt(NH_3)_4^{2+}$/KLTL sample was oxidized under 10% O_2/He at 360 °C for 4 hours. No clusters or NPs were formed either before or after oxidation. On the as-prepared sample, the $Pt(NH_3)_4^{2+}$ species, which replaced the K^+ ions, were mainly present in the largest pores of the zeolite (66%). Upon calcination, only 56% remained in their original position. Moreover, the treatment under O_2 oxidized most of the NH_3 ligands, leading to the formation of platinum oxo species. However, the main reason of the atomic dispersion and stability of the high amount of metal is the structure of the zeolite, which is able to trap the metal atoms inside its channels.

The combined strategy of ion exchange and spatial confinement was also employed for the preparation of partially charged Pt_{SA} on TiO_2-anatase ($Pt_{SA}^{\delta+}/TiO_2$, Figure 2.14) [75]. Titanate nanotubes were added to an aqueous solution of $Pt(NH_3)_4Cl_2$ and HCl. After 30 minutes of sonication, the mixture was stirred for 12 hours at room temperature. The precipitate was recovered, washed, and dried for 12 hours under vacuum at 60 °C. Finally, it was ground, calcined for 1 hour at 400 °C, and then reduced for another hour under 5% H_2/Ar at 160 °C. During this procedure, the cationic metal complexes initially adsorbed on the negatively charged 2D $Ti_3O_7^{2-}$ sheets of titanate, via ion exchange with the interlayered Na^+ ions. Then, the calcination step allowed the phase transformation of the support, and the reduction led to the final $Pt_{SA}^{\delta+}/TiO_2$ catalyst with 0.58 wt% of metal. The 2D spatial confinement was assured by the lamellar structure of this support. Moreover, the choice of the precursor played an important role as well: no cluster or particles were present onto the as-prepared sample, both before and after catalysis (alkene hydrosilylation). Conversely, Pt_{NP} were formed on a control sample prepared using $PtCl_6^{2-}$. Indeed, the choice of a cationic precursor allowed the ion exchange and the isolation of the metallic species.

Therefore, as summarized in Table 2.7, the ion exchange allows the preparation of high-loading and stable SACs. Nevertheless, the necessity of a support with exchangeable ions located in a well-confined structure strongly limits the applications of this technique.

Table 2.7 Selected examples of SACs prepared by ion exchange.

Sample	Metal loading (wt%)	Presence of NPs[a)]	References
Au_{SA}/NaY	1	No	[73]
$Pt_{SA}/KLTL$	1	No	[74]
Pt_{SA}/TiO_2	0.58	No	[75]

a) Only on the fresh catalyst.

Table 2.8 Selected examples of SACs prepared by sol–gel method.

Sample	Metal loading (wt%)	Presence of NPs[a)]	References
Pt_{SA}/Al_2O_3	0.2–2	Yes	[77]
Cu_{SA}/Al_2O_3	8.7	No	[78]

a) Only on the fresh catalyst.

2.3.5 Sol–Gel Solvent Vaporization Self-Assembly Method

The sol–gel preparation involves the initial formation of a stable colloidal solution (sol). The condensation of this "sol" leads to the formation of a 3D solid network of polymeric chains encapsulating the solvent (gel). Solvent removal can occur either by ambient evaporation or under supercritical conditions, leading to "xero-" or "aerogel," respectively [1, 76]. This technique is normally employed for the preparation of porous oxides and by the addition of metal precursors in the starting materials, it can be used for SAC preparation.

As shown in Table 2.8, the sol–gel method can be used for both low- and high-loading catalysts. Zhang et al. [77] prepared different catalysts supported on mesoporous alumina with a Pt loading between 0.2 and 2 wt% (Figure 2.15). Triblock copolymer pluronic P123, $C_9H_{21}AlO_3$, and H_2PtCl_6 were mixed in ethanol and HNO_3. After 48 hours of stirring at room temperature, the ethanol was evaporated at 60 °C for 72 hours. The resulting gels were calcined for 4 hours at 400 °C to obtain the final samples, which in turn were reduced at 400 °C under 5% H_2/N_2 for 1 hour. The assembling of the P123 macromolecules with $C_9H_{21}AlO_3$ during the evaporation of the solvent led to the formation of a gel of hexagonally arranged mesoporous structure. The Pt precursor was encapsulated into this structure. The calcination step allowed the complete removal of the template (P123) and the transformation of aluminum isopropoxide into the final Al_2O_3 framework. The latter had several unsaturated pentahedral Al^{3+} sites available, which played an important role in the stabilization of Pt_{SA}. Indeed, the 0.2 wt% sample showed only $Pt^{\delta+}$ isolated species, extremely stable under both oxidative and reducing conditions. Conversely, NPs were produced at higher metal loading (0.5 and 2 wt%).

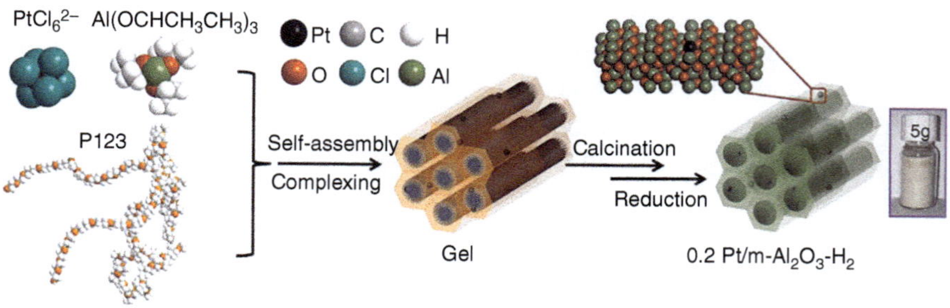

Figure 2.15 Schematic illustration of the sol–gel preparation of Pt_{SA}/Al_2O_3 samples. Source: Zhang et al. [77]. Springer Nature/CC BY-SA 4.0.

The same procedure was used to synthesize a high-loading CuSAC. By using the $Cu(NO_3)_2{\cdot}3H_2O$ precursor and suppressing the reduction step, 8.7 wt% Cu_{SA}/Al_2O_3 was prepared by Guo et al. [78]. Despite the high metal content, all the metal was atomically dispersed as positively charged Cu^{2+}_{SA}, stabilized through Cu–O bridges bonded to coordinatively unsaturated Al^{3+}_{penta}. Indeed, ^{27}Al MAS-NMR of the bare support showed that 40% of the total Al^{3+} ions had pentahedral coordination. This amount decreased to 11% in the Pt/Al_2O_3 sample. No aggregation or metal leaching was detected after the catalytic test (amine-free synthesis of homoallylboronates), proving the high stability provided by Al^{3+}_{penta}-anchoring sites.

To date, this technique has a limited application in the preparation of SACs (Table 2.8). Indeed, compared to other techniques, careful control is necessary during all the different steps, which may be a limitation for large-scale application. Moreover, at high metal loadings, part of the active atoms may be embedded into the bulk support.

2.4 Photochemical Methods

During the conventional photochemical preparation technique, a metal cationic precursor is directly reduced on the surface of the support. Indeed, when a semiconductor support is illuminated at a specific wavelength, a series of photoelectrons can be generated. Therefore, the choice of the specific support, as well as the amount of precursor, the solvent, and the irradiation time are pivotal parameters for the final distribution of the metal [1]. Due to its bandgap, TiO_2 is the most widely employed oxide support in this method, although recently, new materials have been employed such as ZnO and $RuCeO_x$.

High metal-loading SACs have been successfully prepared by tuning the synthesis conditions. Liu et al. [79] prepared a 1.5 wt% Pd_{SA}/TiO_2 starting from two-atom-thick TiO_2 nanosheets protected by ethylene glycolate (EG). The support was mixed with H_2PdCl_4 solution and then irradiated for 10 minutes (365 nm UV). The product was then collected by centrifugation and dried under vacuum. No clusters or NPs were identified on both the fresh and used catalysts recovered after styrene hydrogenation (30 °C, 10 bar), while SA aggregation occurred when the sample was calcined before the test, thus removing the surface EG. Likewise, NPs appeared when the same catalyst was prepared without UV irradiation or starting with TiO_2 without EG. Indeed, the UV light allowed the formation of EG radicals on the TiO_2 nanosheets. The latter played a crucial role in the elimination of Cl^- from the metal precursor and the stabilization of Pd_{SA}. To avoid the aggregation of SAs upon calcination at 350 °C, the authors decreased the metal loading to 1 wt% [80].

Different calcination times allowed the partial or complete removal of EG, producing different amounts of well-defined Ti(III)–O–Pd interfaces. The latter were extremely active for CO oxidation due to the ability to activate O_2 to O_2^- by electron transfer from Ti^{3+} to adsorbed O_2. The O_2^- species strongly interacted with both Ti^{4+} and Pd^{2+}, proving the direct involvement of the oxide support in the metal SA catalysis. Further decrease of the Pd loading (0.1 wt%) was necessary to avoid aggregation on bare TiO_2 [81]. A H_2PdCl_4 aqueous solution was added to the dispersion of either TiO_2-anatase nanocrystals or TiO_2-P25 containing both anatase and rutile phases. After 10 minutes of irradiation (365 nm UV), the recovered product was washed and dried. The Pd/TiO_2-P25 sample was then calcined at 350 °C for 4 hours. Only Pd_{SA} were detected on both supports, before and after catalysis. The elimination of Cl^- atoms, detrimental for single-atoms stabilization, was achieved thanks to the formation of hydroxyl radicals (·OH) during UV treatment.

Following the initial procedure with EG covered two-atom-thick TiO_2 nanosheets, Han et al. [82] prepared a Pt/TiO_2 catalyst. In this case, the mixture of support and H_2PtCl_6 was irradiated for 30 minutes at room temperature. After washing and drying overnight at 60 °C, the product was calcined at 350 °C for 4 hours. Finally, reduction under 10% H_2/He was carried out at different temperatures to investigate the strong metal–support interaction (SMSI). The total amount of supported Pt was equal to 0.77 wt%, either as SAs or NPs (2.1 nm). The characterization results showed that SMSI occurred for both species at very different temperatures: 250 °C for NPs and 600 °C for SAs. Moreover, for NPs, the disappearance of CO adsorption was due to their encapsulation by a TiO_x layer, while for SAs it was due to their coordination saturation by H and Ti^{3+}.

Besides TiO_2, the technique was also employed with other metal oxides. Liu et al. [83] exploited the photoelectron transfer at Mott–Schottky heterojunctions in $RuCeO_x$ to support Pt_{SA} onto RuO_2. Indeed, as shown in Figure 2.16a, during the UV irradiation of the mixture, the photoelectrons generated on CeO_2 moved to the RuO_2 surface promoting the deposition of 0.49 wt% of Pt_{SA}. The resulting Pt–O–Ru moieties showed superior activity and stability for the hydrogen evolution reaction (HER), due to a strong H-spillover effect. For the synthesis of the material, an aqueous solution of H_2PtCl_6 was added to $RuCeO_x$ suspension. After 3 hours of irradiation, the recovered product was washed and dried at 80 °C.

In a different approach, the metal precursor solution was first irradiated without the support to directly reduce the metal atoms, leading to the formation of NPs. To address the problem and maintain the atomic dispersion, Wei et al. [84] froze the precursor solution at −77 °C to prepare an ice with a homogeneous concentration of H_2PtCl_6. The ice was then irradiated for 1 hour at −25 °C and subsequently melted at room temperature (Figure 2.16b). The resulting solution was mixed with the desired

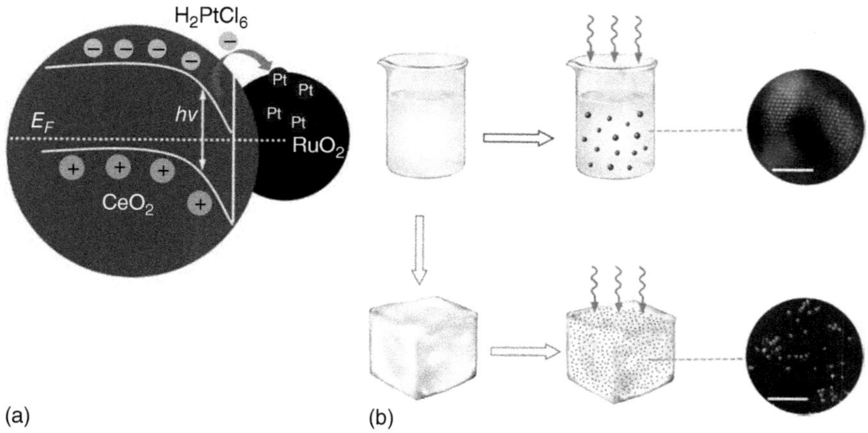

Figure 2.16 (a) Schematic illustration of photoelectron separation at RuO_2/CeO_2 heterojunctions and Pt_{SA} deposition on RuO_2 in the 0.49 wt% $Pt_{SA}/RuCeO_x$ catalyst. Source: Liu et al. [83]. Reproduced with permission of Wiley-VCH GmbH. (b) Schematic illustration of the iced photochemical process. Pt_{NP} are formed in the conventional photochemical reduction (top row). Conversely, SA dispersion is maintained when the H_2PtCl_6 is frozen before UV irradiation (bottom row). Source: Wei et al. [84]. Springer Nature/CC BY-SA 4.0.

support (TiO_2 NPs or ZnO nanowires) and then filtered, washed, and dried at room temperature resulting in the formation of Pt_{SA} catalyst.

No NPs or clusters were detected, proving the ability of ice-photochemical technique to retain the atomic metal dispersion. DFT calculations showed that Pt_{SA} were stabilized on O_v. Unfortunately, no information about the final metal loading of these catalysts was provided by the authors. Therefore, we do not know if such a procedure can be used with these supports, for the preparation of high-loading catalysts.

The photochemical deposition proved to be a route for the preparation of both low- and high-loading catalysts when TiO_2 is used as support for photoelectron generation. The most important limitation toward wider application of this technique is represented by the limited choice of suitable supports. Selected examples of SACs prepared by photochemical methods are shown in Table 2.9.

2.5 Electro-chemical Methods

The electro-chemical methods employ a two- or a three-electrode setup for the cathodic deposition of the metal ions directly on the surface of the support. A slow diffusion rate of the metal ions in the electrolytic solution is a pivotal parameter for the deposition of metal SAs. Hence, the ion concentration must be carefully chosen. The process presents three important advantages: (i) the possibility of regulating the amount and the size of the deposited metal (from SAs to clusters) by varying the plating parameters (e.g. concentration of the plating precursor and plating time); (ii) the efficient utilization of the metal, which is primarily distributed on the surface

Table 2.9 Selected examples of SACs prepared by photochemical methods.

Sample	Metal loading (wt%)	Presence of NPs[a)]	References
Pd_{SA}/TiO_2	1.5	No	[79]
Pd_{SA}/TiO_2	1	No	[80]
Pd_{SA}/TiO_2	0.1	No	[81]
Pt_{SA}/TiO_2	0.77	Yes	[82]
$Pt_{SA}/RuCeO_x$	0.49	No	[83]
Pt_{SA}/TiO_2	Not informed	No	[84]
Pt_{SA}/ZnO	Not informed	No	[84]

a) Only on the fresh catalyst.

of the support; and (iii) the simplicity and scalability of the method, which does not require high-temperature annealing [85]. To investigate the OER activity, Zhang et al. [86] deposited Au_{SA} on a NiFe LDH (layered double hydroxide) support via electrodeposition. An electrolyte solution containing NaCl and $HAuCl_4$ was used with NiFe LDH on Ti mesh as the working electrode. The potential was stepped to −0.6 V vs. saturated calomel electrode (SCE) for 5 seconds and then stepped back to −0.2 V vs. SCE for 5 seconds for five cycles. No clusters or NPs were formed on the 0.4 wt% Au/NiFe LDH catalyst, and HAADF-STEM revealed that only Au_{SA} were present onto the support. To study the stability of this catalyst, a 2000 cycle OER test was performed and no degradation was observed. Moreover, HAADF-STEM images and XANES spectra showed that the atomically dispersed Au was preserved in the 0.4 wt% Au/NiFe LDH after the OER stability test.

Recently, Zhang et al. [87] proposed a universal approach for the electro-chemical deposition of metal SAs on different supports, including $Co(OH)_2$ and MnO_2 nanosheets. Ir, Ru, Rh, Pd, Ag, Pt, and many other metal SAs were successfully supported on both cathode and anode, leading to materials with different electronic states and reactivities. To demonstrate such a procedure, the authors focused their attention on the deposition of Ir_{SA} on $Co(OH)_2$ nanosheets, loaded onto the glassy carbon working electrode in a standard three-electrode system (Figure 2.17). $IrCl_4$ was added to the KOH solution as a metal precursor. During the cathodic deposition, the potential spanned from 0.10 to −0.40 V, driving the $IrCl_3^+$ cations toward the cathode where they coordinated with three O atoms on $Co(OH)_2$. Moreover, a reduction of Ir atoms occurred during $IrCl_3^+$ deposition, leading to an Ir oxidation state lower than +4. During the anodic deposition, $Ir(OH)_6^{2-}$ anions, resulting from the combination of Ir^{4+} of the precursors with the OH^- of the electrolyte, were driven toward the anode. Here, a potential between 1.10 and 1.80 V promoted the oxidation of the anions. Indeed, according to EXAFS, the Ir coordinated with a high number of O atoms (CN of 5.8 from Ir–O contribution), achieving an oxidation state higher than +4. The different oxidation states resulted in a higher activity of the cathodic deposited SACs toward the HER, while those

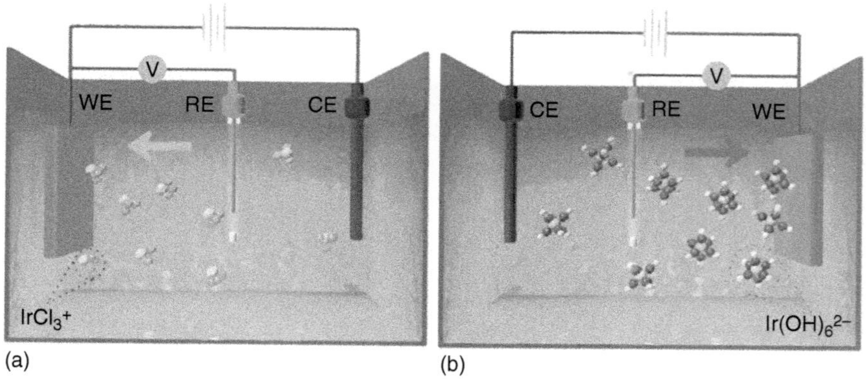

Figure 2.17 Schematic of (a) cathodic and (b) anodic deposition of Ir species, carried out using a standard three-electrode system: $Co(OH)_2$ nanosheets loaded onto a glassy carbon working electrode (WE), carbon rod as the counter electrode (CE), and Ag/AgCl electrode as the reference electrode (RE). Source: Zhang et al. [87]. Springer Nature. CC BY 4.0.

from anodic deposition were more active toward OER. A metal loading of 2 wt% of Ir_{SA} was achieved after 10 scanning cycles for the cathodic deposition. Conversely, three scanning cycles of anodic deposition led to an Ir loading equal to 1.2 wt%. By modifying the Ir precursor concentration, the scanning number, and scanning rate, the atomic dispersion of the supported metal was preserved up to 3.5 wt%, while Ir clusters were formed when the metal loading reached 4.7 wt%. Moreover, the support surface underwent reconstruction during the anodic deposition, generating oxyhydroxides and amorphous phases for $Co(OH)_2$ and MnO_2, respectively.

Nonetheless, the possibility of supporting an extremely wide range of metals with high loadings, different oxidation states and reactivities on several supports makes the electro-chemical deposition method a universal technique for the preparation of single-metal atoms electrocatalysts. Selected examples of SACs prepared by electro-chemical methods are listed in Table 2.10.

2.6 Top-Down Methods

All the above-mentioned techniques involve the direct deposition of mononuclear metal species onto desired supports, avoiding particle formation. Inversely, in top-down techniques, the first step is the deposition of metal NPs. The metal–metal bonds in the NPs are then broken during high-temperature treatments, and the detached metal species are stabilized by enhanced MSI, leading to the final SACs. Datye's group employed the top-down method for the preparation of Pt_{SA} on CeO_2 [88]. These authors found out that when a catalyst containing Pt_{NP} supported on Al_2O_3 was mixed with CeO_2 powder and treated for 10 hours at 800 °C under flowing air, the metal was transferred onto the CeO_2 as SAs. Indeed, under such conditions, Pt was emitted as volatile PtO_2 species, which were in turn trapped on the ceria surface as ionic Pt^{2+} in a sixfold coordination with surrounding oxygens.

Table 2.10 Selected examples of SACs prepared by electro-chemical methods.

Sample	Metal loading (wt%)	Presence of NPs[a)]	References
Au_{SA}/NiFe LDH	0.4	No	[86]
$Ir_{SA}/Co(OH)_2$	1.2–3.5	No	[87]
$Ru_{SA}/Co(OH)_2$	0.9–1.2	No	[87]
$Rh_{SA}/Co(OH)_2$	0.8–1.3	No	[87]
$Pd_{SA}/Co(OH)_2$	0.9–1.0	No	[87]
$Ag_{SA}/Co(OH)_2$	0.8–0.9	No	[87]
$Pt_{SA}/Co(OH)_2$	1.0–1.5	No	[87]
$Au_{SA}/Co(OH)_2$	1.1–1.8	No	[87]
Ir_{SA}/MnO_2	0.8–1.1	No	[87]

a) Only on the fresh catalyst.

XRD, STEM, and DRIFT characterizations confirmed that all the metal (1 wt%) was present as SAs and no NPs were formed on the support. Moreover, polyhedral ceria proved to be more effective than ceria nanorods and cubes for trapping and stabilizing Pt_{SA}. The latter changed their shape during the high-temperature treatment, while the strong interaction between Pt and CeO_2 preserved the surface area of the polyhedral support. The Pt atomic dispersion was also maintained after 3 cycles of CO oxidation test at 300 °C. However, metallic NPs were found on the used catalysts after 3 cycles of propane dehydrogenation at 680 °C [89].

To better understand the mechanism of this "atom-trapping" procedure, the initial Pt_{NP} were directly deposited on the polyhedral ceria [90]. Using $Pt(NH_3)_4(NO_3)_2$ as precursors and IWI technique, samples with 1, 2, 3, and 4 wt% of Pt/CeO_2 were prepared. Then, the samples were dried at 110 °C for 4 hours and calcined under flowing air for 10 hours at 800 °C. AC-STEM and DRIFT measurements confirmed the exclusive presence of Pt_{SA} on the first three samples (1–3 wt%), while a mixture of Pt_{SA} and Pt_{NP} was present on the 4 wt% one, which exceeded the saturation capacity of 1 atom/nm^2. XPS showed that on the 3 wt% sample, the metal was mainly present as Pt^{2+} with a very small fraction of Pt^{4+}. Moreover, the atomic dispersion was preserved after CO oxidation catalytic tests up to 300 °C. Therefore, by combining the experimental results with DFT calculations, the authors concluded that the trapping mechanism and the stability of the as-prepared Pt_{SA} were due to a reaction between the mobile PtO_2 species and undercoordinated Ce^{3+} sites on the support surface. Indeed, (111) facets of polyhedral CeO_2 were rich with step sites, where PtO_2 could strongly bond (adsorption energy of −2.94 eV).

Finally, following the same procedure, two catalysts containing 1 wt% Pt supported on CeO_2 were prepared using $[Pt(NH_3)_4](NO_3)_2$ and H_2PtCl_6 to study the effect of the metal precursors [91]. Both fresh samples presented only Pt_{SA} species, while Pt_{NP} with similar sizes were formed after an activation under CO at 275 °C. Pt_{SA} were also supported onto Fe_2O_3 by Lang et al. [66]. The authors started from a 0.3 wt%

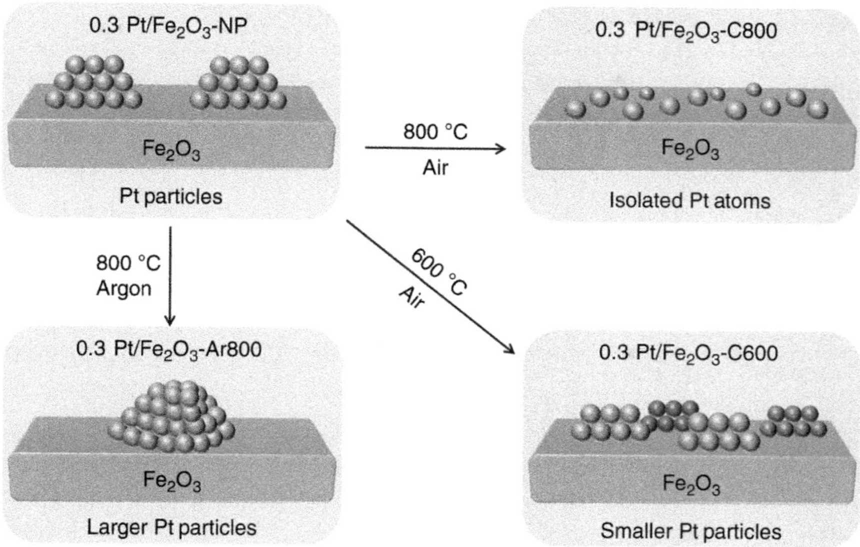

Figure 2.18 Schematic illustration of the structural evolution of Pt_{NP} supported on Fe_2O_3 according to the thermal treatment. Source: Lang et al. [66]. Springer Nature. CC BY 4.0.

Pt/Fe_2O_3 catalyst containing 2–3 nm diameter Pt_{NP}. After 5 hours of calcination at 800 °C, the NPs were completely transformed into Pt_{SA}. Conversely, some NPs were still detectable via STEM when the calcination was carried out at 600 °C and larger particles (3–5 nm) were formed when the sample was treated at 800 °C under Ar (Figure 2.18). Different from CeO_2 support, Pt_{SA} had a high valence (about +4). Furthermore, the dissociative adsorption of PtO_2 volatile species was highly favorable (−2.46 eV) on the oxygen-terminated Fe_2O_3 (0001) surface. Such a strong MSI, responsible for the stability of Pt_{SA}, stemmed from SA coordination with four oxygen atoms in a distorted square geometry. The 800 °C calcination was able to transform all the Pt_{NP} into Pt_{SA} also at 1 wt% loading, while clusters were detected when the Pt amount was further increased to 2 wt%.

As shown in Table 2.11, even higher metal loadings can be obtained with the top-down technique. A catalyst made of 9.7 wt% Ag_{SA} on hollandite-type MnO nanorods (HMO) was synthesized by Huang et al. [92]. The metal was initially deposited onto the HMO nanorods as NPs with an average size of 3.6 nm. The HMO support has one-dimensional square tunnels, and the basic unit of the tunnel structure is built by eight (4 + 4) oxygen atoms to form a tetragonal prism. After 6 hours of calcination in air at 500 °C, the silver atoms moved over the external surfaces to the end of the nanorods, finally reaching the tunnel internal cavities where Ag_{SA} chains were formed. The strong interaction between Ag_{SA} and the internal oxygens made such sites energetically more favorable than the ones on the external surface. The terminal Ag_{SA} exposed at the openings of the HMO tunnels represented the catalytic sites for formaldehyde oxidation.

Top-down methods constitute a good choice to prepare supported metal SAs with specific metal-support systems (e.g. Pt/CeO_2), but the required high-temperature

Table 2.11 Selected examples of SACs prepared by top-down methods.

Sample	Metal loading (wt%)	Presence of NPs[a)]	References
Pt_{SA}/CeO_2	1	No	[88]
Pt_{SA}/CeO_2	1	No	[89]
Pt_{SA}/CeO_2	1–3	No	[90]
Pt_{SA}/CeO_2	1	No	[91]
Pt_{SA}/Fe_2O_3	0.3–1	No	[66]
Ag_{SA}/HMO	9.7	No	[92]

a) Only on the fresh catalyst.

Table 2.12 Selected examples of SACs prepared by other methods.

Sample	Preparation method	Metal loading (wt%)	Presence of NPs[a)]	References
$Fe@SiO_2$	Ball-milling	0.5	No	[93]
Pd@HEFO	Ball-milling	0.5–2	No	[94]
Pd_{SA}/ZnO	Ball-milling	0.25	No	[95]
Rh_{SA}/ZnO	Ball-milling	0.20	No	[95]
Ru_{SA}/ZnO	Ball-milling	0.15	No	[95]
Pd_{SA}/CuO	Ball-milling	0.30	No	[95]
Au_{SA}/CeO_2	Ball-milling	0.10	No	[96]
Ir_{SA}/CeO_2	Ball-milling	0.13	No	[96]
Au_{SA}/NiO	Ball-milling	0.13	No	[96]
Au_{SA}/ZnO	Ball-milling	0.13	No	[96]
Ni_{SA}/TiO_2	Molten salts	0.25–1.41	No	[97]

a) Only on the fresh catalyst.

calcination is detrimental for several other systems, thus preventing its universal application.

2.7 Other Methods

Besides the above-mentioned synthesis strategies, other methods have been developed to generate SACs on oxide supports. Examples of SACs prepared with the most interesting techniques are summarized in Table 2.12.

Ball-milling preparation (mechanochemistry) is an appealing way to support metal SAs on oxides. Due to the absence of solvents and the easy scale-up procedure, it is environmentally benign and industrially interesting. Guo et al. [93] prepared a 0.5 wt% $Fe@SiO_2$ catalyst by ball-milling commercial SiO_2 and Fe_2SiO_4 under high-purity argon at 450 rpm for 15 hours. The mixture was then fused at 1700 °C for 6 hours under air. The resulting catalyst was leached with aqueous HNO_3 solution and finally dried at 80 °C for 12 hours. An activation under 90% CH_4/N_2 at 900 °C allowed the redispersion of the residual 3–4 nm iron oxide particles into Fe_{SA} embedded in the silica matrix. The latter, due to the lattice confinement, proved to be very active and stable during the conversion of methane to ethylene and aromatics. Indeed, after 60 hours on stream at 1020 °C, no NPs were detected by HAADF-STEM and XANES characterizations.

Mechanochemistry was also employed to support Pd_{SA} on a new entropy-stabilized single-phase fluorite oxide (HEFO) $Hf_{0.25}Zr_{0.25}Ce_{0.25}Y_{0.25}O_{2-\delta}$ [94]. Ce, Zr, Hf, La, Ti, and Pd precursors were mixed with fumed silica and treated in a high-speed ball miller twice for 30 minutes each time. The product was then heated at 900 °C for 4 hours under air and after cooling down, it was mixed with 2.5 M NaOH solution for 8 hours, and finally washed with deionized water. This process was repeated four times to remove the silica template before the final drying at 40 °C overnight. Using this procedure, four catalysts with Pd loadings of 0.5, 1, 1.5, and 2.0 wt% were obtained. All the metal was atomically dispersed, most of Pd was incorporated in the bulk phase with a valence state between 0 and +2, while a small part was present on the surface with an electron-deficient state (+4). In fact, the loss of active phase by incorporation in the oxide bulk represents one of the drawbacks of the ball-milling process. Pd_{SA} were also supported on ZnO by He et al. [95]. The authors employed a "precursor-dilution" technique to achieve a 0.25 wt% metal loading by mixing $Pd(acac)_2$ and $Zn(acac)_2$ in a ratio $Pd(acac)_2$:$Zn(acac)_2$ of 1 : 400. The mixture was ball-milled for 10 hours at 400 rpm and then calcined under air at 400 °C for 2 hours. The preparation of 10, 60, 200, and 1000 g batches led to the same final material, made of Pd^{2+} SAs uniformly distributed on the support, proving the large-scale applicability of this procedure. Moreover, following the same procedure, Rh_{SA}/ZnO, Ru_{SA}/ZnO, and Pd_{SA}/CuO with 0.20, 0.15, and 0.30 wt%, respectively, were prepared. Using acetate precursors, the same large-scale technique was employed for the preparation of 0.10 wt% Au_{SA}/CeO_2 [96]. $Au(Ac)_3$ and $Ce(Ac)_3$ were mixed with a ratio of 1 : 1000, ball-milled for 10 hours at 400 rpm, and finally calcined for 5 hours at 500 °C under air. No particles were detected and the presence of metal SAs was proved by HAADF-STEM, both before and after 160 hours on stream during PROX reaction at 120 °C. Moreover, no scaling-up effect was noted for the 10, 50, 100, and 1000 g preparations. The same procedure was successfully employed for the preparation of Ir_{SA}/CeO_2, Au_{SA}/NiO, and Au_{SA}/ZnO with 0.13 wt% of each metal.

Another easily scalable method for the preparation of metal SACs on oxides involves the utilization of molten salts. Xiao et al. [97] used this procedure to prepare a series of catalysts made of Ni_{SA} supported on TiO_2, with a metal loading spanning from 0.25 to 1.41 wt%. To achieve such a result, the authors mixed commercial TiO_2 NPs with $NiCl_2{\cdot}H_2O$, LiCl, and KCl and heated the mixture under N_2 for 2 hours at 500 °C. Above the salts melting point, $NiCl_2$ dissolved in the salts liquid phase reaching atomic dispersion. At the same time, due to the strong polarizing ability of the molten salts, the TiO_2 surface became metastable and the oxygen ions readily reacted with the Ni^{2+} ions to form strong Ni—O bonds. Indeed, during the subsequent washing to remove the salts, the Ni atoms remained onto the TiO_2 surface. The characterizations (EXAFS, AC-STEM) performed on the 0.46 wt% sample showed no Ni_{NP} formation and the exclusive presence of Ni_{SA}, even after an activation under 5% H_2/N_2 at 300 °C for 30 minutes, during which the Ni_{SA} were reduced. Moreover, in the molten salts procedure, all the metal ends up on the surface of the support.

2.8 Conclusions

In this chapter, we have analyzed the main techniques for the preparation of SACs supported on oxides and hydroxides. Each method has advantages and drawbacks. Therefore, a series of parameters must be taken into account for the method choice: (i) the metal/support system, since some techniques can be used only with specific supports or support/metal systems (e.g. photochemical and top-down methods) and the successful stabilization of SAs requires specific binding sites on the different supports (Figure 2.19); (ii) the desired metal loading is fundamental for the choice of the technique, especially when the eventual presence of clusters/NPs is detrimental for the catalytic reaction; (iii) the operating conditions of the catalytic reaction can have a huge impact on the stability and the evolution of SACs; (iv) the SAC local environment deeply influences the reactivity and the stability of the final supported species, and it can be controlled through the different synthesis or post-synthesis treatments; and (v) the amount of catalysts, since most of the analyzed procedures are only suitable for the preparation of small amounts of catalyst. Unfortunately, the development of large-scale preparation methods is fundamental for the industrial application of SACs. Up to now, wet chemistry and ball-milling techniques are the most promising for such a purpose.

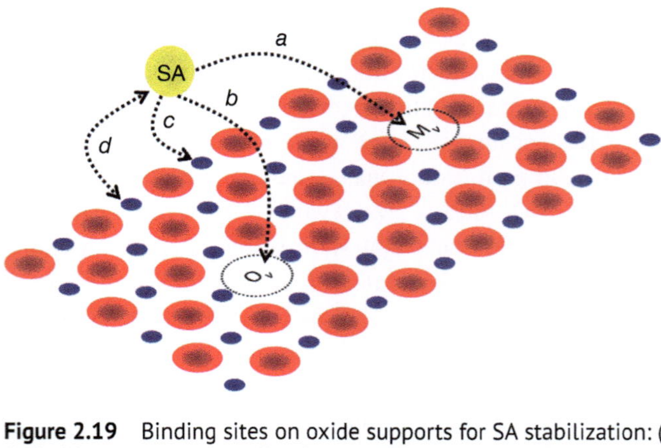

Figure 2.19 Binding sites on oxide supports for SA stabilization: (a) cation (metal) vacancy (M_v); (b) oxygen vacancy (O_v); (c) coordinatively unsaturated centers (i.e. Al^{3+} centers); and (d) sites resulting from ion exchange (for example in zeolites). Source: Philippe Serp.

Acknowledgments

This work was supported by the Agence Nationale de la Recherche (project ANR-19-CE07-0030), which is gratefully acknowledged.

References

1 De Jong, K.P. (ed.) (2009). *Synthesis of Solid Catalysts*. Wiley-VCH Verlag GmbH & Co. KGaA.

2 Védrine, J.C. (2019). Metal oxides in heterogeneous oxidation catalysis: state of the art and challenges for a more sustainable world. *ChemSusChem* 12 (3): 577–588.

3 Védrine, J.C. (2017). Heterogeneous catalysis on metal oxides. *Catalysts* 7 (11): 1–25.

4 Ji, S., Chen, Y., Wang, X. et al. (2020). Chemical synthesis of single atomic site catalysts. *Chemical Reviews* 120 (21): 11900–11955.

5 Lang, R., Du, X., Huang, Y. et al. (2020). Single-atom catalysts based on the metal–oxide interaction. *Chemical Reviews* 120 (21): 11986–12043.

6 Abbet, S., Sanchez, A., Heiz, U. et al. (2001). Tuning the selectivity of acetylene polymerization atom by atom. *Journal of Catalysis* 198 (1): 122–127.

7 Abbet, S., Heiz, U., Häkkinen, H. et al. (2001). CO oxidation on a single Pd atom supported on magnesia. *Physical Review Letters* 86 (26): 5950–5953.

8 Abbet, S., Sanchez, A., Heiz, U. et al. (2000). Acetylene cyclotrimerization on supported size-selected Pd_n clusters ($1 \leq n \leq 30$): one atom is enough! *Journal of the American Chemical Society* 122 (14): 3453–3457.

9 Kaden, W.E., Wu, T., Kunkel, W.A. et al. (2009). Electronic structure controls reactivity of size-selected Pd clusters adsorbed on TiO_2 surfaces. *Science* 326 (5954): 826–829.

10 Tong, X., Benz, L., Chrétien, S. et al. (2010). Direct visualization of water-induced relocation of Au atoms from oxygen vacancies on a TiO_2(110) surface. *The Journal of Physical Chemistry C* 114 (9): 3987–3990.

11 Vajda, S. and White, M.G. (2015). Catalysis applications of size-selected cluster deposition. *ACS Catalysis* 5 (12): 7152–7176.

12 Lu, J., Elam, J.W., and Stair, P.C. (2016). Atomic layer deposition – sequential self-limiting surface reactions for advanced catalyst "bottom-up" synthesis. *Surface Science Reports* 71 (2): 410–472.

13 Kim, H., Lee, H.-B.-R., and Maeng, W.J. (2009). Applications of atomic layer deposition to nanofabrication and emerging nanodevices. *Thin Solid Films* 517 (8): 2563–2580.

14 Cheng, N. and Sun, X. (2017). Single atom catalyst by atomic layer deposition technique. *Chinese Journal of Catalysis* 38 (9): 1508–1514.

15 Wang, C., Gu, X.-K., Yan, H. et al. (2017). Water-mediated Mars–Van Krevelen mechanism for CO oxidation on ceria-supported single-atom Pt_1 catalyst. *ACS Catalysis* 7 (1): 887–891.

16 Ye, X., Wang, H., Lin, Y. et al. (2019). Insight of the stability and activity of platinum single atoms on ceria. *Nano Research* 12 (6): 1401–1409.

17 Li, J., Guan, Q., Wu, H. et al. (2019). Highly active and stable metal single-atom catalysts achieved by strong electronic metal–support interactions. *Journal of the American Chemical Society* 141 (37): 14515–14519.

18 Ding, K., Gulec, A., Johnson, A.M. et al. (2015). Identification of active sites in CO oxidation and water-gas shift over supported Pt catalysts. *Science* 350 (6257): 189–192.

19 Piernavieja-Hermida, M., Lu, Z., White, A. et al. (2016). Towards ALD thin film stabilized single-atom Pd_1 catalysts. *Nanoscale* 8 (33): 15348–15356.

20 Mehrabadi, B.A.T., Eskandari, S., Khan, U. et al. (2017). Chapter 1: A review of preparation methods for supported metal catalysts. In: *Advances in Catalysis* (ed. C. Song), 1–35. Academic Press.

21 Narula, C.K., Allard, L.F., Stocks, G.M. et al. (2014). Remarkable NO oxidation on single supported platinum atoms. *Scientific Reports* 4 (1): 7238.

22 Moses-DeBusk, M., Yoon, M., Allard, L.F. et al. (2013). CO oxidation on supported single Pt atoms: experimental and ab initio density functional studies of CO interaction with Pt atom on θ-Al_2O_3(010) surface. *Journal of the American Chemical Society* 135 (34): 12634–12645.

23 Qiao, B., Lin, J., Wang, A. et al. (2015). Highly active Au_1/Co_3O_4 single-atom catalyst for CO oxidation at room temperature. *Chinese Journal of Catalysis* 36 (9): 1505–1511.

24 Lang, R., Li, T., Matsumura, D. et al. (2016). Hydroformylation of olefins by a rhodium single-atom catalyst with activity comparable to $RhCl(PPh_3)_3$. *Angewandte Chemie International Edition* 55 (52): 16054–16058.

25 Matsubu, J.C., Yang, V.N., and Christopher, P. (2015). Isolated metal active site concentration and stability control catalytic CO_2 reduction selectivity. *Journal of the American Chemical Society* 137 (8): 3076–3084.

26 Chen, J., Wanyan, Y., Zeng, J. et al. (2018). Surface engineering protocol to obtain an atomically dispersed Pt/CeO_2 catalyst with high activity and stability for CO oxidation. *ACS Sustainable Chemistry & Engineering* 6 (11): 14054–14062.

27 Wang, L., Zhang, W., Wang, S. et al. (2016). Atomic-level insights in optimizing reaction paths for hydroformylation reaction over Rh/CoO single-atom catalyst. *Nature Communications* 7 (1): 14036.

28 Wang, L., Li, H., Zhang, W. et al. (2017). Supported rhodium catalysts for ammonia–borane hydrolysis: dependence of the catalytic activity on the highest occupied state of the single rhodium atoms. *Angewandte Chemie International Edition* 56 (17): 4712–4718.

29 Xu, H., Liu, T., Bai, S. et al. (2020). Cation exchange strategy to single-atom noble-metal doped CuO nanowire arrays with ultralow overpotential for H_2O splitting. *Nano Letters* 20 (7): 5482–5489.

30 Xu, J., Zhang, C., Liu, H. et al. (2020). Amorphous MoO_x-stabilized single platinum atoms with ultrahigh mass activity for acidic hydrogen evolution. *Nano Energy* 70: 104529.

31 Jeong, H., Kwon, O., Kim, B.-S. et al. (2020). Highly durable metal ensemble catalysts with full dispersion for automotive applications beyond single-atom catalysts. *Nature Catalysis* 3 (4): 368–375.

32 Gu, X.-K., Qiao, B., Huang, C.-Q. et al. (2014). Supported single Pt_1/Au_1 atoms for methanol steam reforming. *ACS Catalysis* 4 (11): 3886–3890.

33 Xu, J., Song, Y., Wu, H. et al. (2017). Probing the catalytic behavior of ZnO nanowire supported Pd_1 single-atom catalyst for selected reactions. *Chinese Journal of Catalysis* 38 (9): 1549–1557.

34 Qiao, B., Liang, J.-X., Wang, A. et al. (2015). Ultrastable single-atom gold catalysts with strong covalent metal-support interaction (CMSI). *Nano Research* 8 (9): 2913–2924.

35 Qiao, B., Liu, J., Wang, Y.-G. et al. (2015). Highly efficient catalysis of preferential oxidation of CO in H_2-rich stream by gold single-atom catalysts. *ACS Catalysis* 5 (11): 6249–6254.

36 Li, T., Liu, F., Tang, Y. et al. (2018). Maximizing the number of interfacial sites in single-atom catalysts for the highly selective, solvent-free oxidation of primary alcohols. *Angewandte Chemie International Edition* 57 (26): 7795–7799.

37 Zhang, J., Wu, X., Cheong, W.-C. et al. (2018). Cation vacancy stabilization of single-atomic-site $Pt_1/Ni(OH)_x$ catalyst for diboration of alkynes and alkenes. *Nature Communications* 9 (1): 1002.

38 Jeong, H., Shin, D., Kim, B.-S. et al. (2020). Controlling the oxidation state of Pt single atoms for maximizing catalytic activity. *Angewandte Chemie International Edition* 59 (46): 20691–20696.

39 Wang, Q., Huang, X., Zhao, Z.L. et al. (2020). Ultrahigh-loading of Ir single atoms on NiO matrix to dramatically enhance oxygen evolution reaction. *Journal of the American Chemical Society* 142 (16): 7425–7433.

40 Hackett, S.F.J., Brydson, R.M., Gass, M.H. et al. (2007). High-activity, single-site mesoporous Pd/Al_2O_3 catalysts for selective aerobic oxidation of allylic alcohols. *Angewandte Chemie International Edition* 46 (45): 8593–8596.

41 Kwak, J.H., Kovarik, L., and Szanyi, J. (2013). Heterogeneous catalysis on atomically dispersed supported metals: CO_2 reduction on multifunctional Pd catalysts. *ACS Catalysis* 3 (9): 2094–2100.

42 Kwak, J.H., Kovarik, L., and Szanyi, J. (2013). CO_2 reduction on supported Ru/Al_2O_3 catalysts: cluster size dependence of product selectivity. *ACS Catalysis* 3 (11): 2449–2455.

43 Kwak, J.H., Hu, J., Mei, D. et al. (2009). Coordinatively unsaturated Al^{3+} centers as binding sites for active catalyst phases of platinum on γ-Al_2O_3. *Science* 325 (5948): 1670–1673.

44 Mei, D., Kwak, J.H., Hu, J. et al. (2010). Unique role of anchoring penta-coordinated Al^{3+} sites in the sintering of γ-Al_2O_3-supported Pt catalysts. *The Journal of Physical Chemistry Letters* 1 (18): 2688–2691.

45 Nie, L., Mei, D., Xiong, H. et al. (2017). Activation of surface lattice oxygen in single-atom Pt/CeO_2 for low-temperature CO oxidation. *Science* 358 (6369): 1419.

46 Tan, W., Alsenani, H., Xie, S. et al. (2020). Tuning single-atom Pt_1–CeO_2 catalyst for efficient CO and C_3H_6 oxidation: size effect of ceria on Pt structural evolution. *ChemNanoMat* 6 (12): 1797–1805.

47 Peterson, E.J., DeLaRiva, A.T., Lin, S. et al. (2014). Low-temperature carbon monoxide oxidation catalysed by regenerable atomically dispersed palladium on alumina. *Nature Communications* 5 (1): 4885.

48 Yang, M., Liu, J., Lee, S. et al. (2015). A common single-site Pt(II)–$O(OH)_x$ – species stabilized by sodium on "Active" and "Inert" supports catalyzes the water–gas shift reaction. *Journal of the American Chemical Society* 137 (10): 3470–3473.

49 Cao, S., Zhao, Y., Lee, S. et al. (2020). High-loading single Pt atom sites [Pt–$O(OH)_x$] catalyze the CO PROX reaction with high activity and selectivity at mild conditions. *Science Advances* 6 (25): 3809.

50 DeRita, L., Dai, S., Lopez-Zepeda, K. et al. (2017). Catalyst architecture for stable single atom dispersion enables site-specific spectroscopic and reactivity measurements of CO adsorbed to Pt atoms, oxidized Pt clusters, and metallic Pt clusters on TiO_2. *Journal of the American Chemical Society* 139 (40): 14150–14165.

51 Fu, J., Lym, J., Zheng, W. et al. (2020). C—O bond activation using ultralow loading of noble metal catalysts on moderately reducible oxides. *Nature Catalysis* 3 (5): 446–453.

52 DeRita, L., Resasco, J., Dai, S. et al. (2019). Structural evolution of atomically dispersed Pt catalysts dictates reactivity. *Nature Materials* 18 (7): 746–751.

53 Tang, Y., Asokan, C., Xu, M. et al. (2019). Rh single atoms on TiO_2 dynamically respond to reaction conditions by adapting their site. *Nature Communications* 10 (1): 4488.

54 Lou, Y. and Liu, J. (2017). CO oxidation on metal oxide supported single Pt atoms: the role of the support. *Industrial & Engineering Chemistry Research* 56 (24): 6916–6925.

55 Akri, M., Zhao, S., Li, X. et al. (2019). Atomically dispersed nickel as coke-resistant active sites for methane dry reforming. *Nature Communications* 10 (1): 5181.

56 Jimenez, J.D., Wen, C., Royko, M.M. et al. (2020). Influence of coordination environment of anchored single-site cobalt catalyst on CO_2 hydrogenation. *ChemCatChem* 12 (3): 846–854.

57 Schüth, F. and Unger, K. (1999). Precipitation and coprecipitation. In: *Preparation of Solid Catalysts*, 60–84. Wiley.

58 Qiao, B., Wang, A., Yang, X. et al. (2011). Single-atom catalysis of CO oxidation using Pt_1/FeO_x. *Nature Chemistry* 3 (8): 634–641.

59 Shi, Y., Zhao, C., Wei, H. et al. (2014). Single-atom catalysis in mesoporous photovoltaics: the principle of utility maximization. *Advanced Materials* 26 (48): 8147–8153.

60 Lin, J., Qiao, B., Li, N. et al. (2015). Little do more: a highly effective Pt_1/FeO_x single-atom catalyst for the reduction of NO by H_2. *Chemical Communications* 51 (37): 7911–7914.

61 Wei, H., Liu, X., Wang, A. et al. (2014). FeO_x-supported platinum single-atom and pseudo-single-atom catalysts for chemoselective hydrogenation of functionalized nitroarenes. *Nature Communications* 5 (1): 5634.

62 Lin, J., Wang, A., Qiao, B. et al. (2013). Remarkable performance of Ir_1/FeO_x single-atom catalyst in water gas shift reaction. *Journal of the American Chemical Society* 135 (41): 15314–15317.

63 Sun, X., Lin, J., Zhou, Y. et al. (2017). FeO_x supported single-atom Pd bifunctional catalyst for water gas shift reaction. *AIChE Journal* 63 (9): 4022–4031.

64 Xing, J., Chen, J.F., Li, Y.H. et al. (2014). Stable isolated metal atoms as active sites for photocatalytic hydrogen evolution. *Chemistry – A European Journal* 20 (8): 2138–2144.

65 Millet, M.-M., Algara-Siller, G., Wrabetz, S. et al. (2019). Ni single atom catalysts for CO_2 activation. *Journal of the American Chemical Society* 141 (6): 2451–2461.

66 Lang, R., Xi, W., Liu, J.-C. et al. (2019). Non defect-stabilized thermally stable single-atom catalyst. *Nature Communications* 10 (1): 234.

67 Wang, L., Zhang, S., Zhu, Y. et al. (2013). Catalysis and in situ studies of Rh_1/Co_3O_4 nanorods in reduction of NO with H_2. *ACS Catalysis* 3 (5): 1011–1019.

68 Zhang, S., Tang, Y., Nguyen, L. et al. (2018). Catalysis on singly dispersed Rh atoms anchored on an inert support. *ACS Catalysis* 8 (1): 110–121.

69 Wang, H., Shen, J., Huang, J. et al. (2017). Atomically dispersed Au catalysts supported on CeO_2 foam: controllable synthesis and CO oxidation reaction mechanism. *Nanoscale* 9 (43): 16817–16825.

70 Guo, L.-W., Du, P.-P., Fu, X.-P. et al. (2016). Contributions of distinct gold species to catalytic reactivity for carbon monoxide oxidation. *Nature Communications* 7 (1): 13481.

71 Wan, J., Chen, W., Jia, C. et al. (2018). Defect effects on TiO_2 nanosheets: stabilizing single atomic site Au and promoting catalytic properties. *Advanced Materials* 30 (11): 1705369.

72 Zhang, X., Shi, H., and Xu, B.-Q. (2005). Catalysis by gold: isolated surface Au^{3+} ions are active sites for selective hydrogenation of 1,3-butadiene over Au/ZrO_2 catalysts. *Angewandte Chemie International Edition* 44 (43): 7132–7135.

73 Lu, J., Aydin, C., Browning, N.D. et al. (2012). Imaging isolated gold atom catalytic sites in zeolite NaY. *Angewandte Chemie International Edition* 51 (24): 5842–5846.

74 Kistler, J.D., Chotigkrai, N., Xu, P. et al. (2014). A single-site platinum CO oxidation catalyst in zeolite KLTL: microscopic and spectroscopic determination of the locations of the platinum atoms. *Angewandte Chemie International Edition* 53 (34): 8904–8907.

75 Chen, Y., Ji, S., Sun, W. et al. (2018). Discovering partially charged single-atom Pt for enhanced anti-Markovnikov alkene hydrosilylation. *Journal of the American Chemical Society* 140 (24): 7407–7410.

76 Ko, E.I. (1999). Sol–Gel process. In: *Preparation of Solid Catalysts*, 85–98. Wiley-VCH Verlag GmbH.

77 Zhang, Z., Zhu, Y., Asakura, H. et al. (2017). Thermally stable single atom Pt/m-Al_2O_3 for selective hydrogenation and CO oxidation. *Nature Communications* 8 (1): 16100.

78 Guo, T., Tang, N., Lin, F. et al. (2020). High-loading single-atom copper catalyst supported on coordinatively unsaturated Al_2O_3 for selective synthesis of homoallylboronates. *ChemSusChem* 13 (12): 3115–3121.

79 Liu, P., Zhao, Y., Qin, R. et al. (2016). Photochemical route for synthesizing atomically dispersed palladium catalysts. *Science* 352 (6287): 797.

80 Liu, P., Zhao, Y., Qin, R. et al. (2018). A vicinal effect for promoting catalysis of Pd_1/TiO_2: supports of atomically dispersed catalysts play more roles than simply serving as ligands. *Science Bulletin* 63 (11): 675–682.

81 Liu, P., Chen, J., and Zheng, N. (2017). Photochemical route for preparing atomically dispersed Pd_1/TiO_2 catalysts on (001)-exposed anatase nanocrystals and P25. *Chinese Journal of Catalysis* 38 (9): 1574–1580.

82 Han, B., Guo, Y., Huang, Y. et al. (2020). Strong metal–support interactions between Pt single atoms and TiO_2. *Angewandte Chemie International Edition* 59 (29): 11824–11829.

83 Liu, T., Gao, W., Wang, Q. et al. (2020). Selective loading of atomic platinum on a $RuCeO_x$ support enables stable hydrogen evolution at high current densities. *Angewandte Chemie International Edition* 59 (46): 20423–20427.

84 Wei, H., Huang, K., Wang, D. et al. (2017). Iced photochemical reduction to synthesize atomically dispersed metals by suppressing nanocrystal growth. *Nature Communications* 8 (1): 1490.

85 Su, J., Ge, R., Dong, Y. et al. (2018). Recent progress in single-atom electrocatalysts: concept, synthesis, and applications in clean energy conversion. *Journal of Materials Chemistry A* 6 (29): 14025–14042.

86 Zhang, J., Liu, J., Xi, L. et al. (2018). Single-atom Au/NiFe layered double hydroxide electrocatalyst: probing the origin of activity for oxygen evolution reaction. *Journal of the American Chemical Society* 140 (11): 3876–3879.

87 Zhang, Z., Feng, C., Liu, C. et al. (2020). Electrochemical deposition as a universal route for fabricating single-atom catalysts. *Nature Communications* 11 (1): 1215.

88 Jones, J., Xiong, H., DeLaRiva, A.T. et al. (2016). Thermally stable single-atom platinum-on-ceria catalysts via atom trapping. *Science* 353 (6295): 150.

89 Xiong, H., Lin, S., Goetze, J. et al. (2017). Thermally stable and regenerable platinum–tin clusters for propane dehydrogenation prepared by atom trapping on ceria. *Angewandte Chemie International Edition* 56 (31): 8986–8991.

90 Kunwar, D., Zhou, S., DeLaRiva, A. et al. (2019). Stabilizing high metal loadings of thermally stable platinum single atoms on an industrial catalyst support. *ACS Catalysis* 9 (5): 3978–3990.

91 Pereira-Hernández, X.I., DeLaRiva, A., Muravev, V. et al. (2019). Tuning Pt–CeO_2 interactions by high-temperature vapor-phase synthesis for improved reducibility of lattice oxygen. *Nature Communications* 10 (1): 1358.

92 Huang, Z., Gu, X., Cao, Q. et al. (2012). Catalytically active single-atom sites fabricated from silver particles. *Angewandte Chemie International Edition* 51 (17): 4198–4203.

93 Guo, X., Fang, G., Li, G. et al. (2014). Direct, nonoxidative conversion of methane to ethylene, aromatics, and hydrogen. *Science* 344 (6184): 616.

94 Xu, H., Zhang, Z., Liu, J. et al. (2020). Entropy-stabilized single-atom Pd catalysts via high-entropy fluorite oxide supports. *Nature Communications* 11 (1): 3908.

95 He, X., Deng, Y., Zhang, Y. et al. (2020). Mechanochemical kilogram-scale synthesis of noble metal single-atom catalysts. *Cell Reports Physical Science* 1 (1): 100004.

96 Gan, T., He, Q., Zhang, H. et al. (2020). Unveiling the kilogram-scale gold single-atom catalysts via ball milling for preferential oxidation of CO in excess hydrogen. *Chemical Engineering Journal* 389: 124490.

97 Xiao, M., Zhang, L., Luo, B. et al. (2020). Molten-salt-mediated synthesis of an atomic nickel co-catalyst on TiO_2 for improved photocatalytic H_2 evolution. *Angewandte Chemie International Edition* 59 (18): 7230–7234.

3

Preparation of Supported Metal Single-Atom Catalysts on Carbon Supports

Camila Rivera-Cárcamo[1] *and Philippe Serp*[2]

[1] *Institut de Chimie et Procédés pour l'Énergie, l'Environnement et la Santé (ICPEES), CNRS-UMR 7515, 25 Rue Becquerel, 67087, Strasbourg, France*
[2] *Université de Toulouse, Laboratoire de Chimie de Coordination CNRS-UPR 8241, composante ENSIACET, 4 Allée Emile Monso CS 44362, 31030, Toulouse Cedex 4, France*

3.1 Introduction

As presented in *Chapter 2*, a plethora of different preparation protocols (and therefore products) can be found in literature when it comes to the preparation of single-atom catalysts (SACs) over metal oxides supports. This observation is also valid when the goal is to prepare SACs supported on carbon materials. Of course, the selection of the support/metal pair will depend on the application or desired catalytic reaction. For example, a vast number of reports can be found referring to the use of carbon-based SACs for electrochemical, hydrogenation, C–C coupling, hydroalkoxylation, and dehydrogenation reactions. The versatility and readiness of carbon supports to adjust porosity, surface functionalities, surface area, etc., combined with maximal dispersion of the active phase, grant optimal conditions for such reactions, being capable to control accurately the catalytic performance. In contrast, its use under (harsh) thermal catalytic reactions, such as CO oxidation and CO_2 reduction, has not been deeply studied, possibly because of catalyst stability issues.

In the field of carbon-supported SACs [1–5], different classifications can be found trying to group the variety of different synthetic procedures: (i) "top-down" or "bottom-up" if we are referring to the starting nature of the metallic phase [6, 7]; (ii) "low" (<1 w/w%) and "high" (>1 w/w%) metal loading [1]; (iii) depending on the type of support, either graphene, nanostructured carbon (carbon nanotubes [CNT], carbon nanofibers [CNF], etc.), or carbon derived from metalorganic frameworks (MOFs) [8]; or (iv) focusing on the strategy to anchor the active phase (defect/vacancy confinement, spatial confinement, coordination design) [9]. We will not pursue further subdivisions in this chapter, instead, we will center our attention on the description and understanding of key parameters of each method, focusing

Supported Metal Single Atom Catalysis, First Edition. Edited by Philippe Serp and Doan Pham Minh.
© 2022 WILEY-VCH GmbH. Published 2022 by WILEY-VCH GmbH.

on the protocols that have been recently reported on bibliography, including atomic layer deposition (ALD), wet chemistry methods, pyrolysis of metal-containing carbon sources, template sacrificial approach, sputtering, top-down, among others.

3.2 Atomic Layer Deposition (ALD)

After the publication of the first practical preparation of SACs, ALD was raised as a promising synthetic approach due to the possibility of controlling precisely the thickness of the deposit material. ALD has been used to prepare single-atom (SA) materials on various supports, including inorganic nonmetallic materials, metal oxides, and also carbon materials. There are numerous reviews providing interesting perspectives about the utilization of ALD for preparation of atomically dispersed supported catalysts [10–12]. The general principle of this technique is based on the "self-limiting surface reactions." Nevertheless, since the practical aspects of this approach have been already described in *Chapter 2*, we will not elaborate on further details.

On carbon, the first study of ALD was introduced by Sun et al. [13] where they deposited 1.52 wt% Pt_{SA} on reduced graphite oxide (graphene nanosheets, GNS). The support was exposed to intercalated sequences of metallic precursor [$MeCpPtMe_{3,}$] and molecular oxygen to carry out the oxidative decomposition of the complex. The authors highlighted the presence of defects and oxygen functionalities on the support, indicating that Pt_{SA} are preferentially deposited on these defective sites conferring higher stability (Figure 3.1a). The same group proved to be able to deposit likewise 2.1 wt% Pt_{SA} on N-GNS [14]; in that case, instead of oxygen, N-atoms were responsible for the chemical bonding between Pt_{SA} and surface graphene layers ensuring strong metal–support interaction. Some other researchers have used the sixfold cavities from graphitic carbon nitride (g-C_3N_4) to anchor Pd_{SA} (0.5 wt%) (Figure 3.1b). From theoretical calculations, metallic atoms are greatly stabilized on g-C_3N_4 by exchanging electron density [15]. Recently, Qin and coworkers have found a linear correlation between the amount of epoxy-groups on doped graphene and Pt loading [16]. All of this points out to the importance of the existence of anchoring points to tether gaseous metallic species onto the support during ALD operation.

Technical aspects to bear in mind with this technique are, for instance, the nature of the chosen metal precursor. For ALD, organometallic complexes must be liquid at room temperature to better control their evaporation. Cyclopentadienyl complexes have been preferentially used thanks to their low melting points (for instance, 30 °C for [$MePtCpMe_3$]) and because their ligands can be easily removed at reaction temperatures close to 200 °C.

Other combinations of reagents successfully tested are (hexafluoracetate)palladium and formaldehyde [14], bis(cyclopentadienyl)iron with H_2 [17], bis(cyclopentadienyl)cobalt and O_3 [18]. To control the size of the metallic species, there is

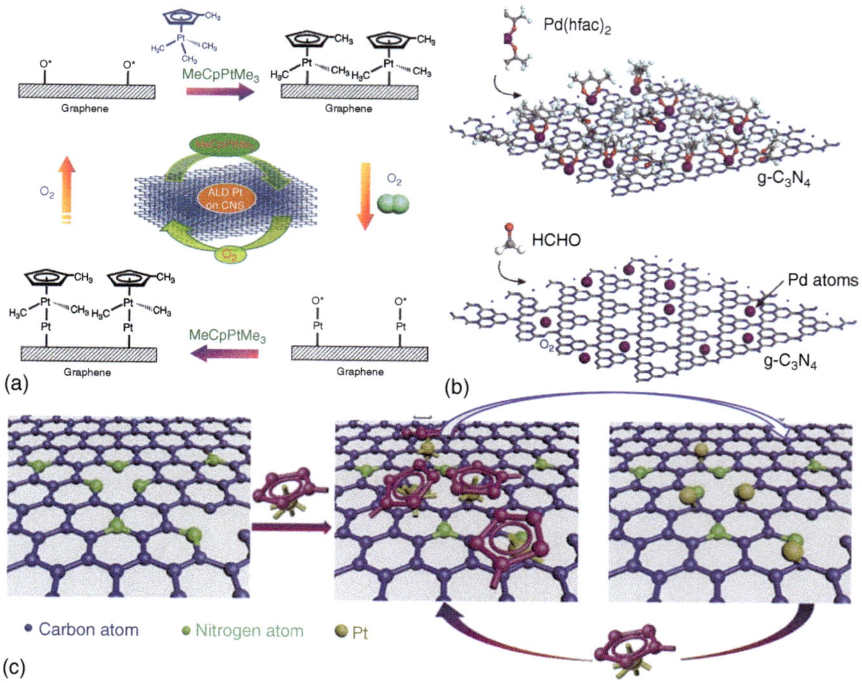

Figure 3.1 Schematic illustrations of SAC preparation through ALD. (a) Pt_{SA}/GNS. Source: Sun et al. [13]. Reproduced with permission of Springer Nature. (b) Pd_{SA}/g-C_3N_4. Source: Huang et al. [15]. Reproduced with permission of Springer Nature. (c) Pt_{SA}/N-GNS. Source: Cheng et al. [14]. Springer Nature, CC BY 4.0.

evidence that three key parameters rule the ALD process on carbon: the number of deposition cycles, the exposure time, and the operating temperature. The trend indicates that only few cycles are enough (<5) to obtain exclusively SAs [15, 17–21]. It is very reasonable to think that a fewer number of cycles and shorter exposure time will turn up into lower metal loading and obviously less willingness to form nanoparticles (NPs). ALD deposition of Ru over graphene show that already after five cycles, there is formation of Ru_{NP} selectively deposited on defects such as wrinkles and grain boundaries [22].

The evolution of metal loading and size of the metal species is strongly impacted on Pd/C_3N_4 photocatalyst going from 1 to 10 cycles, yielding 0.5 wt% of Pd_{SA} and 3.5 wt% Pd_{NP}, respectively [15].

In Figure 3.2a can be seen the linear relationship between Pt loading and number of deposition cycles over an epoxy-rich graphene support. Interestingly, dimers and trimers are observed after three and four cycles, respectively, showing a rigorous deposition control of only one atom at the time [16]. It must be noticed that in the absence of epoxy groups, the linear correlation is not further fulfilled, highlighting the role of surface oxygen groups. On CNT, only Fe_{SA} were obtained on samples with 2, 5, and up to 10 cycles, although the maximal Fe content was 0.36 wt% [17].

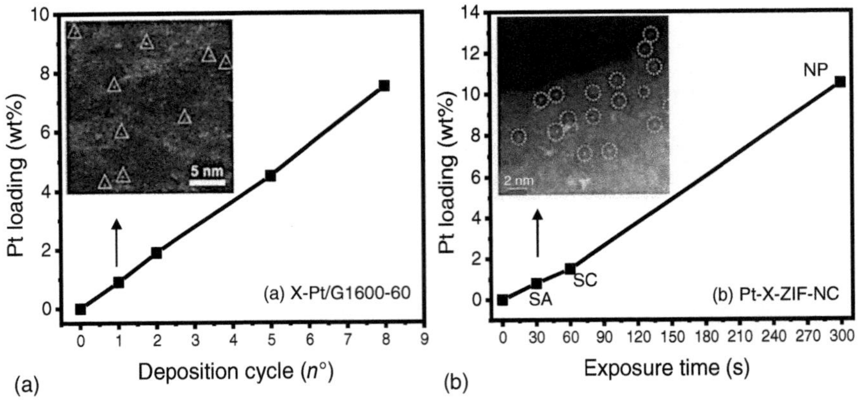

Figure 3.2 Linear relation between (a) Pt loading of *X*-Pt/G1600 changing number of deposition cycles, where *X* is the number of ADL pulses using 10 seconds each. Source: Yang et al. [16]. Adapted with permission of American Chemical Society. (b) Pt loading of Pt-X-ZIF-NC changing exposure time where X is SA, subcluster (SC), or NP, using one cycle on each sample. Source: Song et al. [19]. Adapted with permission of WILEY-VCH.

The second mentioned parameter is the exposition time. By adjusting the time in which the precursor will be exposed during ALD synthesis, it is possible to control the nuclearity of metallic species [19]. Figure 3.2b presents a series of Pt catalysts prepared by deposition in one cycle over N-doped-ZIF-derived carbon (ZIF-NC) changing the exposure time. By adjusting the time to 30, 60, and 300 seconds, it was possible to produce from SAs to subclusters (SCs) to NPs with a linear increase on the metallic charge of 0.8, 1.5, and 10.5 wt% Pt, respectively. The visualization of the atoms is quite easy due to the high contrast of Pt atoms on the carbon support. On Pt–ZIF–NC sample, the bright spots have an average size of 0.85 nm that can be related to groups of less than 5 atoms [23]. After 300 seconds of exposition pulse, Pt(111) crystalline planes can be identified belonging to well-formed Pt_{NP} of 2 nm [19].

As far as temperature is concerned, dispersion tends to increase while decreasing temperature. This behavior has been evidenced on $Co_{SA}/P\text{-}C_3N_4$ by reducing ALD to 150 °C instead of the regular 250 °C, to avoid NP formation after 3 cycles, attaining 1.0 wt% of cobalt [18]. Yan et al. noticed that an important number of Pt_{NP} was created after two successive ALD cycles of Pt at 250 °C. The authors decreased the temperature to 150 °C for the second ALD cycle to avoid metal aggregation. Indeed, Pt_{SA} were only produced however with low loading, reaching a plateau at 0.35 wt% even with a prolonged pulse time [21].

ALD is an efficient technique to control the creation of metallic species at the angstrom level with exceptional uniformity. The total procedure is considerably fast, which makes this approach particularly appealing for the preparation of a great number of samples in a short period of time. On the other hand, although relatively high metallic loading can be reached using ALD, only a small fraction of the metal is actually retained as SAs. Important limitations to its large-scale utilization for catalyst preparation are related to the equipment and precursor's price and availability.

3.3 Solution-Phase Syntheses

3.3.1 Impregnation

In this section, from a very simplified point of view, we are alluding to the syntheses that follow the principle of impregnation. It consits, according to De Jong's definition, to the dissolution of an active phase in a solvent that, when mixed with a solid support, will be absorbed or anchored after removing the solvent (drying) with an eventual reduction [24]. The absorption of the active phase on carbon supports is more likely to occur on reactive surfaces, i.e. over grain defects (imperfections in the graphic network) or adatoms, which are the spots more predisposed to react. Certainly, parameters such as concentration, temperature, and the nature of support and precursor are critical to control high dispersion.

Due to the inert nature of pristine graphitic materials, in comparison with other supports like metal oxides, the generation of strong metal–support interactions with the active phase, at standard impregnation conditions (i.e. room temperature and pressure), is rather challenging. Statistically speaking, reducing the amount of metal supposes a decrease in the possibility of nucleation. Therefore, to avoid NP formation, it should be preferred to work at low metallic loading, with the understanding that impregnation will be an effective protocol to produce modest SA density.

Kochubey et al. have prepared a series of Pd catalysts by wet impregnation with metal loading of 0.04–0.5 wt%. The impregnation was carried out with an aqueous solution of $PdCl_2$ followed by reduction at 250 °C. Stacked-type CNF were used as support in which the basal planes are perpendicular to the axis of the fiber [25]. No NPs are observed when the Pd content is lower than 0.2 wt%. The dispersion was explained in terms of the interaction of Pd_{SA} atoms with the CNF surface. At very low concentration (<0.04 wt%), Pd_{SA} are able to penetrate into CNF interlayers searching for the most energetically favorable location, coordinating with six carbon atoms, three from each graphene sheet as presented in Figure 3.3a.

Samples with a loading between 0.05 and 0.2 wt% started to saturate edges at the surface as evidenced by the change of coordination number to 4.3–4.8 (Figure 3.3b). At concentrations higher than 0.2%, since all possible locations were occupied, Pd_{NP} of 2–3 nm were detected [25]. From this, the need of adding anchoring points became obvious to isolate a higher number of individual atoms. Creating defects on graphitic materials is a good strategy to introduce a higher amount of metal. Theoretical studies have revealed that transition metals (TMs) are thermodynamically favored to be placed on defects such as carbon vacancies [27]. Experimentally, it has been shown that CNTs possessing plenty of vacancies were able to accommodate ~1.0 wt% of ruthenium exclusively as SAs following the procedure described in Figure 3.3c [26]. The preparation consists of a simple impregnation of freshly produced defective-CNTs with a solution of [Ru(COD)(COT)] in pentane [26].

The utilization of such a labile complex allows an easy displacement of the ligands for a fast interaction with very reactive (or unstable) carbon vacancies on the support. Furthermore, the same group claims to be able to modulate the SA/NP ratio

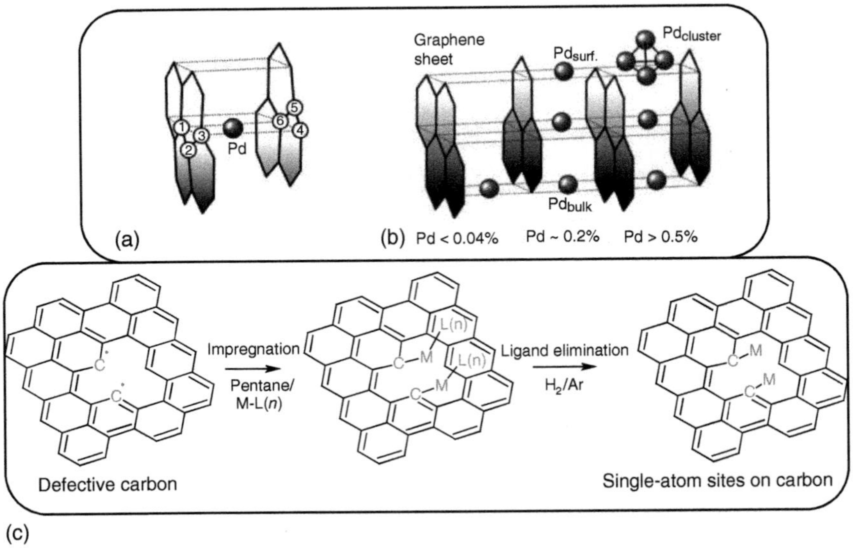

Figure 3.3 (a) Population of the various CNF near-surface sites depending on the Pd concentration; (b) Pd atom position in CNF according to DFT calculation and EXAFS. Source: Kochubey et al. [25]. Reproduced with permission of Elsevier.; and (c) Procedure for the stabilization of SAs on defective CNT. Source: Rivera-Cárcamo et al. [26]. Reproduced with permission of Wiley-VCH.

for a given metallic loading by splitting the required amount of complex in different parts and subsequently adding each portion on consecutive impregnation steps [28]. Figure 3.4 presents the dispersion of four samples containing the same 1.2 wt% of Pd but with different SAs portions. The size of the clusters increases from 0.5, 0.8, 0.9, and 1.1 nm, after one, two, three, and four impregnations, Figure 3.4a–d, respectively. This is undoubtedly a very valuable way to study the influence of the size and nuclearity of metallic species on catalysis, without varying the total metal content and without modifying considerably the textural properties of the catalysts.

Besides, the same group has propose that it is possible to modulate the electronic features of the metallic centers by the addition of dopants, as in the case of sodium [26]. By impregnation of a Ru_{SA}/CNT SAC with an aqueous solution of NaOH, to obtain a ratio 1 : 2 Ru/Na, it was possible to produce Ru_{SA}-Na/CNT. This sample showed an elevated selectivity to CO production under CO_2 reduction reaction. The addition of Na induced a strong charge transfer to generate electron-deficient Ru species, which are known to favor the reverse-water–gas-shift reaction mechanism [26].

From the results presented above, it can be assumed that certain supports or morphologies are more suited for this approach, since edges, vacancies, or any type of reactive emplacement is necessary to adsorb or "trap" metal SAs atoms. In that sense, graphitic carbon nitride (g-C_3N_4) is a good candidate to stabilize metallic SAs from direct impregnation. Thanks to the unique six-folded cavities at the surface of g-C_3N_4, the adsorption of TMs will particularly occur on the sp^2 pyridinic nitrogen atoms surrounding the vacancy, which can tightly retain ions. Multiple transition metals, listed in Table 3.1, have been introduced on its interstitial sites

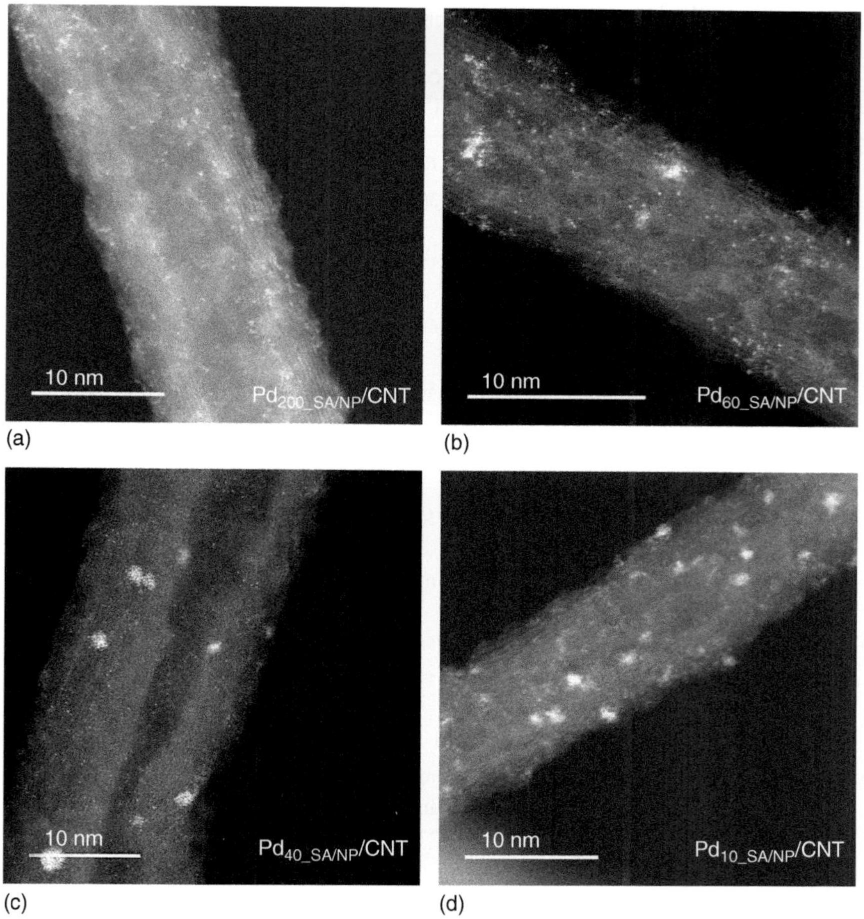

Figure 3.4 STEM-HAADF images of 1.2% Pd/CNT catalysts containing mixtures of SAs and NPs at different ratios prepared by one, two, three, and four impregnation steps. (a) 200 SA/NP, (b) 60 SA/NP, (c) 40 SA/NP, and (d) 10 SA/NP. Source: Rivera-Cárcamo et al. [28]. Adapted with permission of Royal Society of Chemistry.

by impregnation, generating stable single sites due to strong covalent interactions with the support.

Additionally, functional groups of O, N, P, or S can contribute to the stabilization of SAs. Thiolated CNT mixed with H_2PtCl_6 in an aqueous solution yielded surprisingly around 20 wt% Pt_{SA}, which is one of the highest reported in the literature. This high loading can be achieved by means of the known affinity of thiol groups to strongly bond noble metals, avoiding aggregation. In that case and differently than in many other cases, here SAs are mainly in a reduced state. X-ray absorption near edge structure spectroscopy (XANES) analysis indicated indeed that platinum is present in almost zero valence, due to the electronic compensation from S ligands to the Pt_{SA} [40, 41]. A sophisticated nitrogen functionalization was presented by Corma et al. wrapping CNT with polyallylamine hydrochloride and incorporating amino groups

Table 3.1 SACs on carbon prepared by impregnation method.

Entry	Sample	Solution mixture	Drying (°C)	Annealing (°C)	Loading (wt%)	References
1	Fe–CNF	$PdCl_2$/water	120	250/H_2	0.04–0.5	[25]
2	Ru/CNT	Ru(COD)(COT)/ pentane	30	80/Ar/H_2	1.2	[28]
3	Fe–P-C_3N_4	$Fe(NO_3)_3 \cdot 9H_2O$/ water	100	400/inert	0.2–1.0	[29]
4	Pt/C_3N_4	H_2PtCl_6/ water	60	125/Ar	0.075–0.38	[30]
5	Au/g-C_3N_4	$HAuCl_4$/ water	50	100/Ar/H_2	0.1–0.3	[31]
6	Ir/g-C_3N_4	$K_2IrCl_6 \cdot 6H_2O$/ water/$NaBH_4$	50	—	0.5	[32]
7	Pd/g-C_3N_4	$PdCl_2$/water/ $NaBH_4$	50	—	0.5	[33]
8	Ni/N-graphene	$Ni(NO_3)_2 \cdot 6H_2O$/ water	n.i.	300/Ar	0.8	[34]
9	Pd/Graphdiyne	K_2PdCl_4/DMF	50	—	0.42	[35]
10	Au/AC	$HAuCl_4$/acid	110	—	1.0	[36, 37]
11	Au/N-MWCNT	$HAuCl_4$/water/ sodium citrate	−20	—	0.1	[38, 39]
12	Pt/S-MWCNT	$H_2PtCl_6 \cdot 6H_2O$/ water/$NaBH_4$	Not informed	—	20	[40, 41]

at the CNT surface, enhancing the interaction of the support with metal precursor [38, 39].

Incipient wetness impregnation (IWI) was also used to prepare SACs [42, 43]. It is not surprising that in both reported cases, a mixture of SAs and NPs is obtained. IWI usually requires high metal precursor concentration solutions, which as mentioned before, are detrimental to achieve high metal dispersion. Kaiser et al. prepared by IWI a set of 25 bimetallic gold catalysts (1 wt%) using non-noble metals (e.g. Cu, Ni, Co, Ba, La) and precious metals (e.g. Pt, Pd, Ir, Rh) on activated carbon [42]. Multiple parameters were studied, such as pH, the influence of chlorinated precursors, and the concentration of Cl^- ions. Although most of the samples presented both SAs and NPs, surprisingly, the mere addition of platinum as second metal avoids the agglomeration of gold, obtaining samples with 100% dispersion for both metals. Unfortunately, due to limitations in the characterization techniques, the nature of the interactions and/or proximity of Au and Pt sites remains unsolved [42].

Among the greatest advantages of impregnation, the rapidity and simplicity of the method can be counted, without the need for sophisticated equipment. However, despite having good control of the final metal loading, the tendency to form NPs limits the scope, requiring additional procedures to limit agglomeration.

3.3.2 Low-Temperature Techniques

The main goal of reducing temperature or freezing is to tackle the rapid kinetics of nuclei formation in solution. In this sense, a good way to prevent nanocrystal growing is performing syntheses at low temperatures (LT) [44]. In the context of SACs, LT steps have been applied either to produce reduced ions in solution or to improve segregation in solid mixtures that will be posteriorly calcined. For example, in the first scenario, Co^{+2} ions have been reduced by hydrazine hydrate in liquid media (1 : 9 water/alcohol mixtures) at −60 °C. At this ratio, the solution possesses a melting point of −78.5 °C, ensuring reduction at very low temperature over long period of time. After 2 hours of reaction, reduced Co species were absorbed on N-doped mesoporous carbon (N–MC), and let react for 5 additional hours, still at −60 °C. At last, the mixture was annealed at 900 °C to obtain Co_{SA}/N-MC-LT900. The method gives analogous results using iron, silver, and platinum. To prove the role of LT, comparison samples at −30 and +60 °C were prepared, but both generated samples presented clusters and particles [44]. Alternative variations include reducing metallic ions in ice by ultraviolet (UV) irradiation. Specifically, a chloroplatinic acid solution was quickly frozen with liquid nitrogen and exposed to UV-light for 1 hour. With this treatment, Pt^{4+} ions can be reduced in water. Afterward treated ice was melted, and the resulting liquid was used as precursor for the deposition of Pt_{SA} on mesoporous carbon and other substrates (Figure 3.5a) [45]. First-principles molecular dynamic results informed that individual atoms can be stabilized by the formation of H–Pt–OH species due to water dissociation in the reaction media. As other

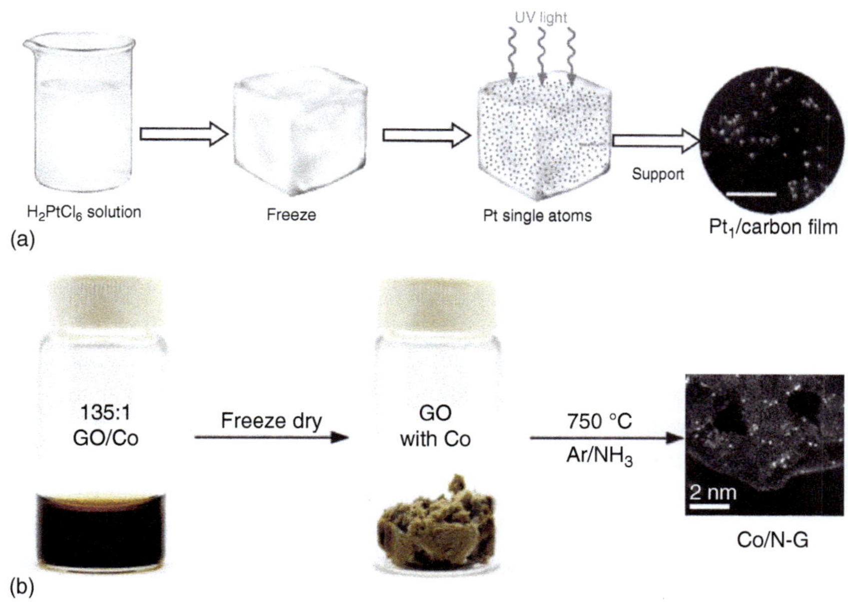

Figure 3.5 (a) Schematic illustration of the iced-photochemical production of Pt_{SA}/carbon catalysts. Source: Wei et al. [45]. Springer Nature / CC BY-SA 4.0. (b) Preparation of atomic cobalt on nitrogen-doped graphene. Source: Fei et al. [46]. Springer Nature / CC BY 4.0.

Table 3.2 SACs prepared with freeze-drying technique.

Sample	Pre-catalysts	Annealing	Loading (wt%)	Active site	References
Ru–NG	$Ru(NH_3)_6Cl_3 + GO + H_2O$ freeze-drying	750 °C NH_3	1.70	Ru–N_x	[47]
Co–NG	$CoCl_2 + GO + H_2O$ freeze-drying	750 °C Ar/NH_3	2.48	Co–N_x	[46]
Ni–NG	$Ni(NO_3)_2 + GO + H_2O$ freeze-drying	750 °C NH_3	0.44[a)]	Ni–N_x	[48]
Sn/NG	$SnCl_2 + GO + H_2O$ freeze-drying	800 °C Ar/NH_3 3 : 1	0.82	N_2–Sn–C_2	[49]

a) Expressed in atomic percentage.

application of LT methods, some authors propose that after homogeneous dispersion of precursor and support in liquid phase, the elimination of the solvent at very low temperature and pressure would be beneficial for the final metallic distribution. The application of lyophilization or cryodesiccation permits the constitution of a tridimensional structure due to solvent freezing, formed by the ice crystals, which can lock and immobilize the dispersed reagents, as a direct way to limit thermal motion of metal ions, avoiding re-stacking and preventing nucleation during posterior treatments. Table 3.2 summarizes some examples exploiting this technique.

Zhang et al. [47] prepared an electrocatalyst for oxygen reduction reaction (ORR) by depositing Ru atoms on graphene oxide (GO). First, they prepared an aqueous suspension of GO and hexaamineruthenium(III) chloride (weight percentage Ru/GO = 0.62), which was frozen and dried for at least 24 hours until obtaining a brown composite. The abundant oxygen groups of GO help to retain metal cations in the matrix. With the lyophilization, a foam-like material is produced ensuring homogeneous distribution of Ru atoms.

This solid was subsequently annealed at 750 °C under a flow of NH_3 with the purpose of adding nitrogen to the matrix. Under such conditions, Ru/N-G sample containing 1.7 wt% of ruthenium was formed. Besides a few isolated 1–2 nm of Ru_{NP}, Ru_{SA} were found in preponderance. It is noteworthy that by doubling the amount of precursor, Ru_{NP} are formed, inferring that the efficiency of lyophilization assistance is limited to a certain concentration. X-ray photoelectron spectrometry (XPS) analyses have revealed the coexistence of Ru^{+3} and Ru^{+4} species. Extended X-ray absorption fine structure spectroscopy (EXAFS) reveals that the first coordination sphere is primordially composed of N atoms; therefore, during annealing there is a transition from Ru–O to Ru–N interactions [47]. Nonetheless, XANES simulations found that the best fitting with experimental data corresponds to a Ru–N_4 moiety having an O_2 molecule adsorbed on the central Ru_{SA}. Ru/NG was assessed in ORR in acidic media. Excellent performance was evidenced having a 7.5 times greater activity than a benchmark Pt/C, normalized by the mass of catalyst. Additionally, impressive stability was monitored after 10 000 cyclic voltammograms, retaining more than

90% of the initial saturated current. Density functional theory (DFT) justify such outstanding catalytic behavior in terms of easier O_2 chemisorption over Ru-oxo-N_4 centers than on Ru–N_4. Analogously, the same group produced atomic cobalt dispersed on nitrogen-doped graphene [46]. In first instance, an aqueous solution of $CoCl_2$ and graphene oxide (ratio GO/Co = 135 : 1 on weight) was well mixed and, secondly, freeze-dried to produce a brownish composite (Figure 3.5b). After treatment at 750 °C under NH_3 atmosphere, the mixture became a black powder containing 8.5 at% N and 0.57 at% Co as single-atom sites. Despite the positive effect on the dispersion thorough freeze-drying, increasing the amount of cobalt (1.23 at%) leads to the formation of Co clusters or nanoparticles.

It must be mentioned that freeze-vacuum drying following identical previous description was used to make the first reported SAC of tin supported on N-doped graphene at *kilogram* scale, generating 828.6 g per batch, loading 0.82 wt% of Sn [49]. Catalytic evaluation in CO_2 electroreduction derives from a new record of turnover frequency (TOF) up to 11 930 h^{-1} for formate formation. These discoveries put into a closer perspective the exploitation of SACs on an industrial scale.

3.4 Sputtering

In the process known as sputtering, a metallic piece is bombarded with energetic particles (either ions or electrons) causing the expulsion of fragments from such material. These ejected species (atoms, dimers, trimers, etc.) can be condensed on a support depositing from atomic species to thin layers. This technique represents one of the most precise ways to design extremely tailored SACs. The landing of the accelerated species can be either on pre-existing anchoring points such as terraces or steps or alternatively over vacancies/defects intentionally produced during the experiment. In the second case, creation of the defect can happen by "knock-on" events, where an accelerated particle hits an atom on the graphitic surface. Upon collision with C atoms, these C atoms are ejected, resulting in a damaged graphitic structure with true holes corresponding to carbon vacancies. Although there are several setups to carry out sputtering, the most interesting is the coupling of the sputtering device with transmission electron microscopy, which allows visualizing the entire procedure *in situ,* and obtaining images in real time of the active sites. Critical values such as emission and adsorption energies and structural rearrangement can be obtained from computational simulations that are often applied in a way to rationalize final products [50]. In theory, both electrons and ions can be used in sputtering machines, but Ar^+, H^+, and Au^+ beams have been popularly used. The characteristics of the final product can be regulated by variating the dose of ions by mm^2, controlling electron beam current density and/or exposure time. Other sensitive parameters to consider during synthesis include beam incidence angle and the location upon collision [50].

Pt atoms were set on freestanding graphene layers using Ar^+ plasma sputtering [51]. There is a strong correlation between the sputtering time and platinum density. SAs are visualized after one second of deposition, as shown in Figure 3.6a.

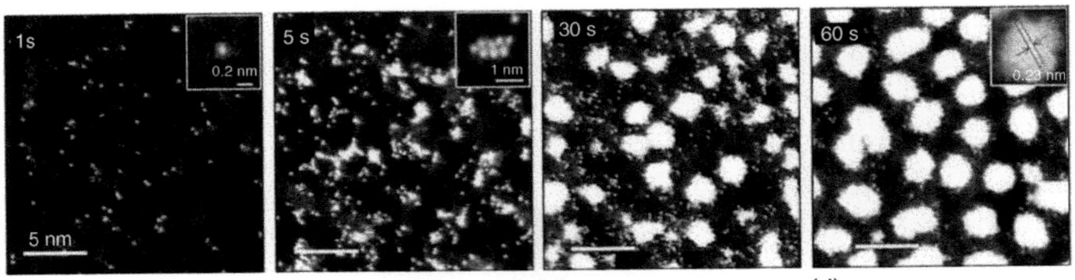

(a) (b) (c) (d)

Figure 3.6 High-angle annular dark field (HAADF) images of Pt/G samples corresponding to sputtering times of (a) 1, (b) 5, (c) 30, and (d) 60 seconds. Source: Yamazaki et al. [51]. Adapted with permission of American Chemical Society.

The prolongation of the deposition time leads to the saturation of Pt on the surface, causing atomic aggregates after 5 seconds (Figure 3.6b). After sputtering, the authors determined that Pt_{SA} tended to be located on the edges (around 90%) and only 10% on terraces of graphene sheets [51]. After 30 or 60 seconds, well-formed Pt_{NP} with a mean diameter of 2 and 4 nm were obtained, respectively (Figure 3.6c,d) [51]. It is worth noting that the sputtered Pt density was estimated around 0.3 atoms/nm^2 after 1 second and 1.5 atoms/nm^2 after 10 seconds, which correspond to Pt monolayer saturation. The knowledge of this value is of interest, since in most of the SAC studies, it is not provided, essentially because it is hard to determine. Its knowledge would accurately allow determining turnover number per atomic unit. Efforts have been devoted to deliver density of sites from direct counting on high resolution transmission electron microscopy (HRTEM) images, but the local scope of that technique delivers just approximate values [52].

A similar process was used to prepare Pt_{SA} on N-doped graphene [53]. The variation with respect to the previous example, focused on the addition of N_2 in the sputtering chamber, generating N reactive species, which were simultaneously embedded in the graphene network. After extensive microscopic and theoretical analysis, it can be confirmed that N and Pt are predominantly absorbed in small graphene islands rather than long-range sheets. The incorporation of nitrogen increases the density of Pt sites, and according to DFT calculations, considerably improves their stability. The same group commented that despite obtaining well-characterized materials, the amount of catalyst that can be prepared on a microscope grid is far away from the hundreds of milligrams that are at least necessary to carry out a single catalytic reaction [53]. Regarding this last point, when the sputtering tool is not coupled to a microscope (therefore, there is no *in situ* observation), it is possible to increase the scale of production. There is an example where a combination of gold species, nanoclusters with a substantial number of isolated atoms have been embedded on activated carbon, loading up to one gram of support in the sputtering chamber [54]. A study dealing with palladium deposition on pre-modified Ar^+ highly oriented pyrolytic graphite using sputtering can be found in the literature [55], although this sample was prepared simply for modeling purposes.

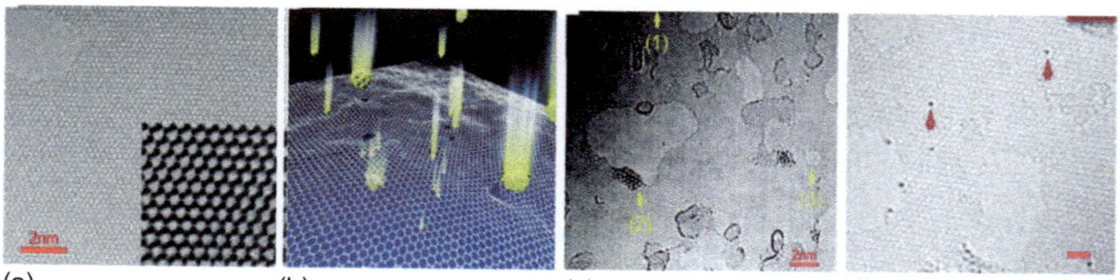

(a) (b) (c) (d)

Figure 3.7 (a) High-resolution micrograph of a graphene monolayer. (b) Schematic illustration of a high-energy particle bombardment process. (c) High-resolution micrograph of graphene layer after bombardment. (d) Pt_{SA} deposited on bombarded graphene. Source: Wang et al. [56]. Reproduced with permission of American Chemical Society.

As mentioned above, sputtering can be applied in a different way. It is possible to carry out an intermediate step during which defects are created, so that later said defects serve as nests to house metal atoms. A study published in 2012 presented how SAs of Pt, Co, and In can be deposited on a monolayer of graphene (Figure 3.7a) by such two-step protocol [56]. During the first step, pristine graphene was irradiated by pieces of gold or boron (either atoms or nanoclusters), ejected from a target employing high-energy pulsed laser deposition of 400 mJ/pulse. These energetic species act as real bullets that pierce the graphitic network as shown in Figure 3.7b,c. Previous studies placed the energy for the generation of monovacancy and divacancy at around 100 eV, which is in the order of what accelerated ions can transfer. The size of the defects varied experimentally from monovacancies to holes of a pair of nanometers. Later improvement of the experimental variables allowed a careful control of vacancy dimensions to produce only single-carbon vacancies [56]. It is important to remark that equal results were obtained with either Au or B projectiles and no remnant atoms stay trapped in graphene, considering a non-pollutant functionalization. During the second step, a conventional sputtering apparatus was used to drag cobalt and indium in vacancies, with a working pressure of 50 mTorr of Ar and plasma power of 10 W. As predicted from DFT calculations, the ensemble Co-vacancy is experimentally magnetic. According to charge analyses, graphene was p-doped with the incrustation of Co and n-doped by In atoms. As comparison and to prove the versatility of created defects, Pt-doping was carried out with a different protocol, using an electron beam to drag Pt atoms from an organic source instead of accelerated atoms. This process was performed immediately after the creation of defects due to the rapid tendency of vacancies to reconstruct (Figure 3.7d) [56]. Indeed, similar results were obtained, pointing to the indifference of using ions or electrons to deposit metal. An identical setup was used to anchor Au_{SA} over a graphene sheet. Metallic species have a prevalence to be placed on the graphene edges. The stability of this location is associated with the higher migration barrier compared to being physisorbed onto the graphene lattice. Also, dangling bonds help to keep Au attached on graphene edges [57].

From a synthetic point of view, the advantage of sputtering application is that no further treatments are required as for liquid-phase techniques (washing, dryings, annealing, calcination, etc.) There is no residual contamination, generating the cleanest catalytic surface. The clear setbacks are the limitations for mass production or scale up constraining the practical application of these highly customized materials, remaining dedicated as an exploratory methodology for producing model catalysts for fundamental research.

3.5 Top-Down Methods

In catalysis, sintering is one of the typical reasons of catalyst deactivation and has been the subject of intense investigations [58]. The two major mechanisms of sintering of metal NPs are atomic migration (Ostwald ripening) and crystallite migration. Specifically, for Oswald ripening, metal atoms detach from crystallites to then agglomerate and reconstruct into larger and more stable particles. The thermodynamic tendency to reconstruct can be interrupted, in the way that the metal–support interaction surpasses the metal–metal attraction (cohesive energy) breaking the sintering cycle, to stabilize mononuclear metallic species over supports by strong metal–support interactions. These observations had been evidenced quite long ago, at very high temperatures and under specific conditions such as oxidation, reduction, or carbonylation [59]. For instance, in 2009, it has been observed that Au_{NP} supported on activated carbon disappeared after methanol carbonylation reaction [60]. Additionally, an induction period was needed to start the reaction. It has been proposed that during that period, Au_{NP} were atomized to form Au dimers or trimers. The use of methyl iodide during the reaction had a role in Au_{NP} redispersion, in some way stabilizing detached Au_{NP}, which apparently were the active sites for catalysis [60].

Nowadays, scientists have taken advantage of this reactivity to synthesize SACs. Based on the previously described finding, Ding's team proposed a method to disperse Ru, Rh, Pd, Ag, Ir, and Pt atoms at loading of 5 wt% over oxygen-containing activated carbon starting from NPs [61]. In a first stage, catalysts with noble metal NPs were prepared by wet impregnation and further reduced at 200 °C for 2 hours, labeled as 5% Rh/AC, taking rhodium as example. Posteriorly, these samples were warmed up at 240 °C in a flow of N_2 for 20 minutes to be changed by a mix of carbon monoxide/CH_3I and kept for 6 hours, to obtain 5% Rh_{SA}/AC. At lower temperatures, Rh NPs did not fully dissolve. The complete transformation of NPs to SAs was confirmed by the total disappearance of Rh–Rh signals in both X-ray diffraction (XRD) and EXAFS analyses [61]. More importantly, from this latter technique, it can be concluded that the atomic coordination environment fits well with $[(CO)_2I_3Rh]$–O–AC sites. For comparison, a fresh catalyst prepared on activated carbon without oxygen groups was treated under the same reaction conditions. In that case, microscopy results indicate that Rh_{NP} became larger. In summary, the combination of adequate temperature, adsorbed carbon monoxide, halogen radicals, and oxygen functionalities on the support surface is primordial to obtain SAs from NPs [61]. The same group has found a promoter effect of lanthanum addition on the disintegration and for the stabilization of Ir_{SA}/AC for methanol carbonylation. Detailed observations of high-resolution micrographs confirm the formation of Ir–La pairs uniformly distributed on the support. The hybrid Ir–La/AC displayed improved stability compared with Ir/AC, for methanol carbonylation to methyl acetate, and the authors proposed a reaction mechanism in which $[Ir^{+1}(CO)_2I_2]$–O–AC and $[La^{+3}I_2]$–O–AC sites are situated next to each other. La^{+3} plays a role of Lewis acid promoting the oxidation of Ir complex to adsorb the reactants [62].

The NP atomization can also be carried out only by the action of temperature. At high enough temperatures, atoms are spontaneously released from the crystal lattice of the nanoparticle. The presence of defects allows caging released atoms, limiting their diffusion and thus avoiding the reconstruction of NPs. The detachment of gold atoms at ~1200 °C from Au_{123} and their diffusion on carbon materials were carefully studied by molecular dynamics simulations [63]. The calculations are conclusive: disjoined Au_{SA} would remain immobile, locked by defect sites producing stable SAs compared with a undefective graphene layer. The need for defects is simply explained in thermodynamic terms. The binding energy of Au_{SA} on carbon vacancies goes from 6 to 18 times higher than for defect-free surfaces [63].

The impact of the presence of nitrogen has also been noted in the atom-trapping technique [64, 65]. A defect-containing N-doped carbon (N-C) obtained from the pyrolysis of ZIF-8 was used to deposit preformed 5 nm Ni_{NP} carefully prepared by reduction of $Ni(acac)_2$. After setting NPs onto the support, the Ni_{NP}/N-C precatalyst was annealed at 900 °C under Ar atmosphere during one hour to produce Ni_{SA}/N-C. DFT studies indicate that the conversion of NPs to Pd-N_4 structure is boosted, due to the superior thermodynamic stability of this structure, when mobile Pd atoms are present on carbon surfaces [65].

Depositing preformed NPs is an easy way to demonstrate the effectiveness of the method. Like this, it can be assumed with certainty that the origin of all atoms, visualized after the heat treatment, comes from the rupture of NPs and not from possible remaining physisorbed precursors. HAADF–STEM analyses confirm Ni atomic species after the total transformation of particles.

Testing the applicability to other supports, samples were prepared identically on activated carbon, showing completely contrasting results [64]. Rather than disappearing, Ni_{NP} grew from 5 up to 50 nm. One might think that the presence of oxygen groups on the activated carbon surface could also serve as a stabilizing agent as observed with nitrogen, but the results indicate that this is not the case. However, it is important to note that the oxygen concentration of that specific AC was quite low (0.6 wt%), which, perhaps, is a negligible concentration to really assist the redispersion. To understand the mechanism of NP-to-SA transformation, the researchers relied on *in situ* environmental transmission electron microscopy (TEM) and energy-dispersive X-ray spectroscopy (EDS) mapping to check the structural evolution of the catalysts after the heat treatment. After heating at 900 °C, there is no trace of Ni_{NP} over the specimen accompanied by signs of erosion throughout the structure (Figure 3.8a,b). The authors call this transformation "surface-enhanced" due to the increase of the surface area and porosity after heating, evolving from 943 m^2/g and 0.22 m^3/g for the pristine NC support, to 1174 m^2/g and 0.44 m^3/g for the final catalyst [64]. This addition of evidence, combined with DFT studies, allows

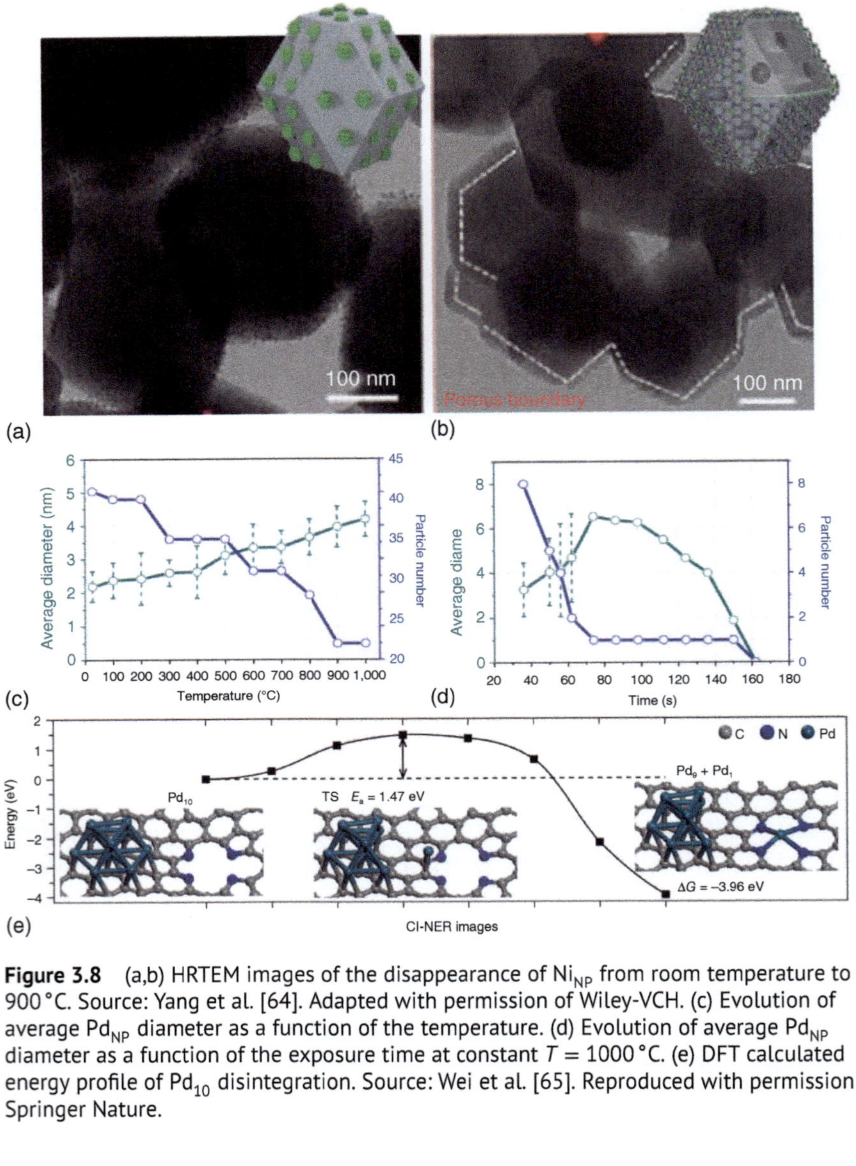

Figure 3.8 (a,b) HRTEM images of the disappearance of Ni_{NP} from room temperature to 900 °C. Source: Yang et al. [64]. Adapted with permission of Wiley-VCH. (c) Evolution of average Pd_{NP} diameter as a function of the temperature. (d) Evolution of average Pd_{NP} diameter as a function of the exposure time at constant $T = 1000\,°C$. (e) DFT calculated energy profile of Pd_{10} disintegration. Source: Wei et al. [65]. Reproduced with permission of Springer Nature.

the researchers to claim that Ni_{NP} move around the graphic matrix in the process, piercing the surface and creating new porosity (Figure 3.8b). During this trajectory, the Ni_{NP} passes over nitrogen sites, and due to strong Ni–N interactions, is capable of rip atoms out from NPs, especially over pyrrolic and pyridinic N. These catalysts were tested in the CO_2 electroreduction reaction to CO, with a good stability, showing a constant faradaic efficiency of 80% even after 60 hours. The authors relate the performance not only to atomically dispersed species, which are well known to be efficient active sites for this reaction, but also to the high porosity that permits propitious mass transfer [64].

In another report, the influence of the temperature over the NPs was carefully studied on Pd_{NP}/N-C catalysts. Contrary to what was expected, a linear increase of NP mean size was observed going from 2 nm to more than 4 nm, with the increase of temperature starting from 100 to 1000 °C, accompanied with a decrease in the total number of particles. This is clearly indicative that sintering also takes place during NP-to-SA transformation, as shown in Figure 3.8c [65]. Furthermore, the general dimension of N-C support shrank 30% approximately, associated with structural collapse. Then, when the temperature is raised up to 900 °C, all the NPs disappeared in less than 3 minutes (Figure 3.8d). Pd_{10} clusters and N_4-C site were used as a platform for DFT studies to calculate the different energies of SA formation mechanism. The anchoring of a Pd_{SA} on a nitrogen defect must overcome an energy barrier of 1.46 eV that leaves a Pd_9 cluster and Pd_{SA}-N_4-C site as represented in Figure 3.8e. The full process is exothermic, releasing 3.96 eV. In comparison, the energy barrier of sintering is lower with only 0.58 eV, which justifies why this process prevailed at lower temperatures (below 900 °C) and atomization at higher (900–1000 °C) [65].

In addition to SAC preparation, the discovery of the atomization phenomenon opens the doors to a possible definitive solution to the great problem of sintering deactivation. The limitations of this method lie mainly in the high energy demand for treating catalysts and the structural modifications that the support suffers during heating. On graphitic supports, it is known that high-temperature treatments under an inert atmosphere can increase graphitization degree or compromise the level of functionalization, which can be detrimental for some applications.

3.6 Pyrolysis Methods

In the Sections 3.2–3.5, we have reviewed protocols in which metal deposition is carried out over previously formed supports, i.e. the addition of the active phase is an additional step. This section discusses "one-pot synthesis" in which metal incorporation is simultaneously done with the formation of the carbon framework. In the majority of cases, this task is achieved by pyrolysis. Thus, the thermal decomposition of a carbon source at elevated temperatures and in a controlled atmosphere will produce a carbon-rich solid residue. When the process is carried out over a source containing metallic species (for example a MOF), such species will end up embedded in the graphitized structure, as SAs, metallic NPs or oxides. Depending on the carbon source and the experimental conditions, it is possible to control the final product to obtain specific morphologies (hierarchical porous carbons, CNT, CNF, carbon nanoflakes, etc.), with different metallic loadings, surface area, and porosity. Different types of materials can be pyrolyzed: MOF, coated templates, polymers, natural products, and synthetic products. Although pyrolysis represents the main technique, one-pot synthesis can also be achieved by polymerization reactions, which in any case also require a high-temperature stage. In the following sections, it will be reviewed the most significant contributions on the topic.

3.6.1 MOF-Derived SACs

Metal–organic frameworks are polymers synthetized via the coordination of transition metal (often Fe, Co, Cu, Zn) ions or clusters with organic ligands (also called "linkers") creating porous networks, tunable by changing the linkers [66]. The use of MOF as a carbon source for the production of catalysts has been broadly studied mainly due to the possibility of preserving some of their inherent features after the pyrolysis, such as high surface area and porosity, but also because it allows the production of doped carbon structures. In 2015, the first report of the production of a MOF-derived carbon SAC was published by Zitolo et al. who reported an Fe-N-C material with 0.5 wt% of Fe_{SA} obtained after pyrolysis at 900 °C of an iron-containing MOF [67].

To date, different types of MOFs have been used for this purpose [68]. Nonetheless, it is interesting to notice that the majority of the studies employs zeolitic imidazole frameworks (ZIFs), a subset of MOFs produced from imidazole ligands and presenting an isomorphic structure with zeolites. Figure 3.9 displays some examples of MOFs used as precursors for SAC preparation. Utilization of ZIF-8 or ZIF-67 is particularly appealing due to their relatively facile synthesis, high surface area, and abundance of nitrogen ligands, which will contribute to the production of N-doped carbons, improving stabilization of SAs through strong metal–nitrogen–support coordinative bonds ($M–N_x–C$).

In some cases, the MOF preparation is relatively simple (coprecipitation or solvothermal methods), justifying its burning [68]. It is common to introduce a second metal for producing bimetallic MOFs (Figure 3.9b) by ion substitution or spatial confinement in the cages. In most studies, one of these metals (with a low boiling point, frequently Zn [73] and even Cd [74]) plays the role of "sacrificial metal" to avoid NP formation during pyrolysis. In that case, both species constitute nodes in the framework. During pyrolysis, an important portion of the sacrificial metal is evaporated, leaving only the target metal with larger interspace between SAs, thus avoiding NP formation [73].

The carbonization can be carried out at different temperatures starting from 500 °C up to 1000 °C, under inert atmospheres (N_2 or Ar). Wang et al. studied the influence of the temperature on final metal loading and coordination environments [75]. They show that Zn/N-C samples prepared at 600 and 900 °C yielded 3.01 and 6.53 wt% of Zn, respectively. In addition, with the increase of temperature, the coordination of the metallic centers changes from tetrahedral to square planar ZnN_4, along with a higher degree of graphitization, which improves the electro-donation from the support to Zn^{+2} centers. Catalytic results have shown that those porphyrin-like ZnN_4 sites present peroxidase-like activity and decompose H_2O_2 at relatively low temperature (40–50 °C) [75].

It is important to mention that after the annealing, acid treatments are often necessary to remove metallic aggregates on the surface; it is very likely that some SAs will be also affected, but in any case, an important portion remains attached to the support. In some cases, NPs can be encapsulated by the carbon shells requiring very

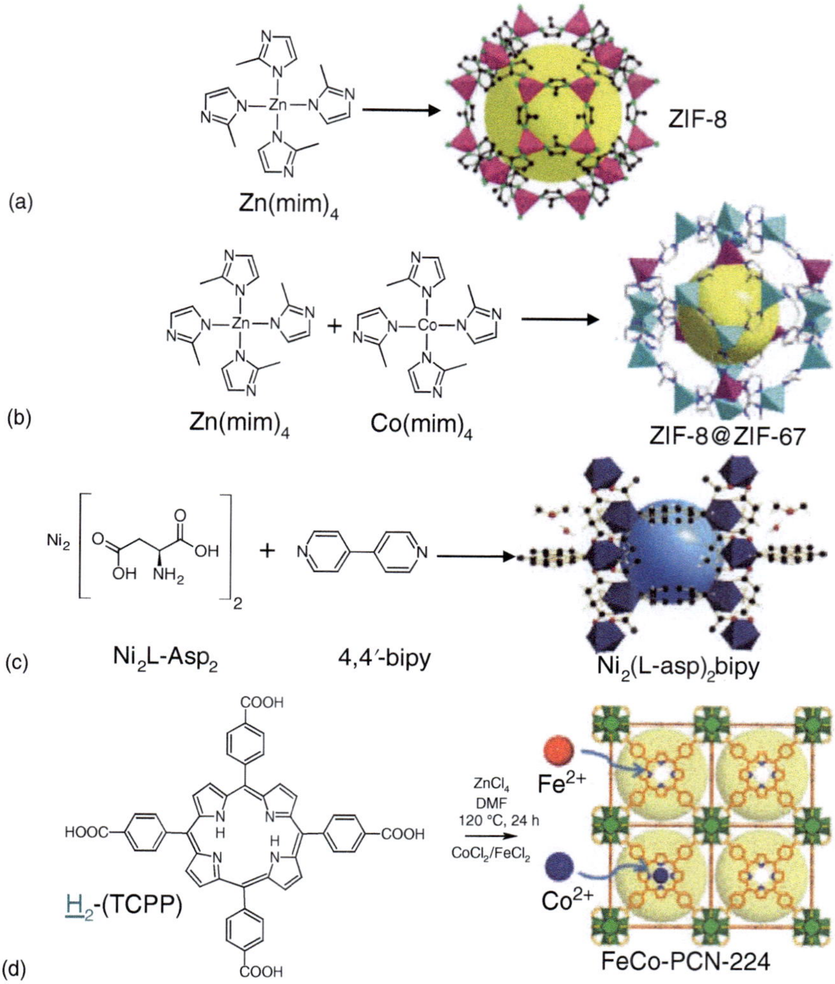

Figure 3.9 MOFs used as substrate for preparation of SACs by pyrolysis. (a) Crystal structure of ZIF-8. Source: Phan et al. [69]. Reproduced with permission of American Chemical Society. (b) Crystal structure of the bimetallic ZIFs. Source: Tang et al. [70]. Springer Nature. CC BY 4.0. (c) Schematic diagram of synthesis of Ni$_2$(L-asp)$_2$bipy. Source: Fan et al. [71]. Springer Nature. CC BY 4.0. (d) Schematic diagram of synthesis of FeCo-PCN-224. Source: Fang et al. [72]. Reproduced with permission of John Wiley and Sons.

harsh conditions to obtain particles-free materials. Table 3.3 summarizes a number of examples of SACs prepared by the pyrolysis of MOFs.

An interesting example is the use of the FeCo-PCN-224 precursor [72]. The authors highlight the use of this MOF due to the porphyrin centers that can be occupied by different single metal ions giving the opportunity to build multi-metal structures. Furthermore, the cages formed by the framework give place to mesopores and high specific surface area that remains after annealing (Figure 3.9d). PCN-224 was prepared starting from tetrakis(4-carboxyphenyl)porphyrin (H_2TCPP) and $ZrCl_4$. Once

Table 3.3 Summary of MOF-derived SACs.

Catalyst	SAs or SAs + NPs	Precursor	Conditions	Loading (wt%)	References
Ni_{SA}/onion-like carbon	SAs + NPs after acid leaching	Ni_2(L-asp)$_2$bipy	700 °C/5 h/N_2	1.5	[71]
$FeCo_{SA}$/N-doped porous carbon	SAs (after acid leaching)	$FeCo_x$-PCN-224	900 °C/2 h/N_2	0.43 Fe	[72]
				0.37 Co	
Fe_{SA}/N-C	SAs after acid leaching	Fe-MOF-545	800 °C/2 h/N_2	1.76	[76]
	NPs at higher Fe content				
Ni_{SA+NP}/N-CNT	SAs + NPs after acid leaching	Ni-MOF-74	1000 °C/2 h/Ar	2.36	[77]
Zn_{SA}/N-HOPCP	SAs after basic etching	Zn-ZIF-8	800 °C/2 h/N_2	5.8	[78]
Co_{SA}/N-C	SAs (no leaching)	Co-ZIF-67	600 °C/2 h/vacuum	No informed	[79]
	SAs + NPs at higher pyrolysis T°				
Co_{SA+NP}/CNT	SAs + NPs	Co-ZIF-67	750 °C/2 h/Ar	0.14[a]	[80]
H-PtCo/Pt_{SA}N-C	SAs (after acid leaching)	Pt-Co-ZIF-67	600 °C/3 h/H_2/Ar	0.47 Pt[a]	[81]
				0.16 Co [a]	
Ru_{SA}/N-C	SAs (no leaching)	Ru-Zn-ZIF-8	900 °C/3 h/N_2	0.18	[82]
Ru_{SA}/N-C	SAs	Ru-Zn-ZIF-8	900 °C/3 h/N_2	0.50–2.48	[83]
	NPs + SAs at higher wt%			6.82–8.50	
Mn_{SA}/N-C	SAs (after acid leaching)	Mn-Zn-ZIF-8	1100 °C/1 h/N_2	0.68	[84]

the MOF is formed, the solid is treated with a solution of $FeCl_2$ and $CoCl_2$ to substitute porphyrin centers to produce FeCo-PCN-224. Different catalysts, $FeCo_2$-NPC-T (T = 800, 900, 1000 °C), were synthetized by the pyrolysis of FeCo-PCN-224. After the thermal treatment and HF etching, a 3D hollow N-doped porous carbon was formed, having 1240 m^2/g of Brunauer-Emmett-Teller (BET) surface area 0.43 and 0.37 wt% of Fe and Co content, respectively. The crystalline cubic structure prevents the agglomeration of the active phase and allows homogeneous dispersion of CoN_x and FeN_x sites. $FeCo_2$-NPC-900 performs an outstanding oxygen reduction reaction activity under both acidic and alkaline conditions [72].

Despite being a method in which non-noble metals prevail, ruthenium ZIF were also synthesized and used for the preparation of Ru_{SA}/N-doped carbon catalysts [82, 83]. Remarkably, the specific activity of Ru_{SA}/N-C was 2.1-fold higher compared with supported Ru_{NP} for the electrochemical N_2 reduction to NH_3. It must be mentioned that the yield rate of 120.9 $\mu g_{NH3}/mg_{catalysts} \cdot h$ was significantly higher than the highest value ever reported. TPD-N_2 analysis indicates that the Ru_{SA}^{+3} binds strongly with N_2, explaining the good performance [82].

3.6.2 Template Sacrificial Approach

Continuing with the concept of carbonization of carbon sources, unlike MOF-based SACs where the MOF (or ZIF) structure itself induces SAC morphology and porosity, the template sacrificial method relies on the use of an inorganic template, in most of the cases an oxide, to achieve a specific tridimensional carbon shape (spheres, tubes, sheets, etc.), with the goal of modulating the porosity. The experimental methodology includes three steps: (i) coating the template with a metal-containing nitrogen-rich precursor, (ii) carbonization of the composites at high temperatures (500–950 °C), and (iii) removal of the template by etching.

Along the pyrolysis, carbon-containing molecules decompose and graphitize on the surface of the rigid solid adopting the shape as an inverse casting. Then, a leaching step (acid or basic) is required to remove the mold, to finally obtain atomically dispersed metallic species anchored on the porous carbon. As a consequence of the leaching, part of the initial amount of metal is wasted by the dissolution of metallic and/or oxide NPs. This is why this approach is not applied to expensive precious metals. Nonetheless, the production of total NP-free catalyst is appealing to test the performance of samples containing exclusively SAs. An important advantage that the template sacrificial approach has over MOF-derived SACs is that despite the initial very high surface area and porosity of MOFs, during pyrolysis, an important aggregation takes place, and therefore their textural properties are negatively affected. Contrarily, with the template sacrificial approach, thanks to the presence of the mold, the aggregation of the carbon framework is avoided and very high surface area can be obtained, even above 1000 m^2/g [85].

In an effort to substitute Pt/C catalyst that is commonly used for various reactions (HER, ORR, CO_2 electroreduction), this approach has been followed for the preparation of non-noble catalysts such as Fe [86–91], Co [92–94], and Ni [95, 96]. EXAFS, and XANES allow confirming that the fixation of the atomic species in the

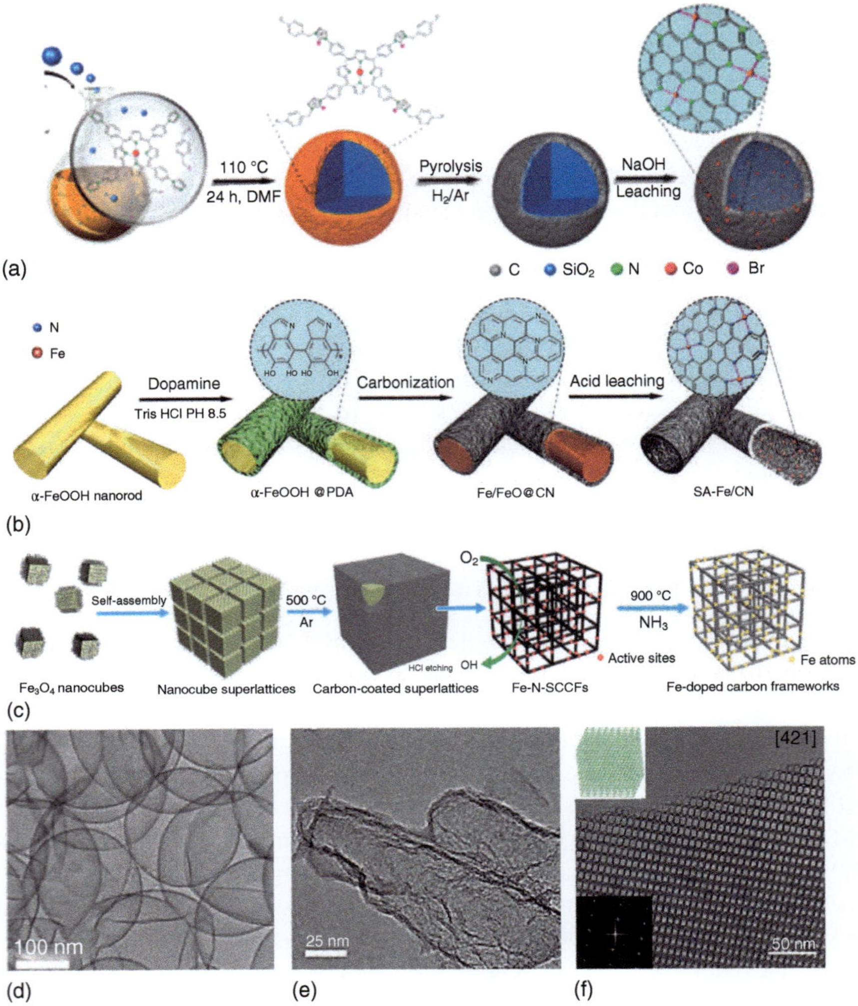

Figure 3.10 (a) Schematic illustration of the synthesis of ISAS-Co/HNCS to obtain hollow nanospheres. Source: Han et al. [93]. Reproduced with permission of American Chemical Society. (b) Schematic illustration of the synthesis of Fe_{SA}/CN to obtain hollow pipes. Source: Zhang et al. [90]. Reproduced with permission of American Chemical Society. (c) Schematic illustration of the fabrication of Fe_{SA}/N-SCCF. Source: Wang et al. [101]. Reproduced with permission of American Chemical Society. (d–f) correspond to TEM images of the products of (a), (b), and (c), respectively. Source: (d) Han et al. [93]. Reproduced with permission from American Chemical Society. (e) Zhang et al. [90]. Reproduced with permission of American Chemical Society. (f) Wang et al. [101]. Reproduced with permission of American Chemical Society.

graphitic framework is achieved through the formation of tetradentate N coordinative sites that can strongly stabilize metal atoms. Thanks to the combination of these techniques with DFT calculations [97], it has been possible to unequivocally identify the reactive sites; and porphyrin-like [98] or phthalocyanine-like [85] structures are the most recurrent centers. The presence of these M-N_4-C moieties has been historically related to improvements of the catalytic performances, even since the early 1960s, particularly on ORR [99]. It is important to mention that other type of coordination has been proposed, such as Mo_1-N_1-C_2 sites in the case of molybdenum [100].

Figure 3.10 displays three examples of template sacrificial approach, and as can be seen, different morphologies were obtained depending on the template that was used. For instance, isolated single atomic sites anchored on hollow N-doped carbon spheres (ISAS-Co/HNCS) with diameters around 150 nm and thickness of 5 nm were obtained, when SiO_2 spheres were coated with metalloporphyrins (M-TIPP, M = Co, Fe, Cu, Fe), which polymerize to generate SiO_2@M-TIPP-polymer (Figure 3.10a). Afterward, this composite was treated at 800 °C for 3 hours under a mixture of H_2/Ar to finally remove the template by etching using 6 M NaOH during 24 hours, withholding 2.2 wt% of cobalt [93]. According to XPS analysis, graphitic, pyridinic, and pyrrolic nitrogen species coexist, suggesting that part of the tetrapyrrolic ring from porphyrin evolves to other configurations to form Co-N_4-C sites. Besides, XANES situates cobalt valence between Co^0 and Co^{2+}. This catalyst showed outstanding activity for ORR in acidic media and good stability after 10 000 cycles. The good catalytic performance was attributed to the hollow structure and mostly to Co_{SA} that facilitates proton and charge exchange to adsorbed *OH species to produce H_2O. A slightly different protocol was followed by Zhang et al. [90] (Figure 3.10b), using iron (III) oxide-hydroxide (α-FeOOH) nanorods of 0.5–1.0 μm length as template and iron source. The coating of the rods is fulfilled using dopamine as monomer, due to its strong chelating capability with metal ions. The α-FeOOH@polydopamine composite was converted to hollow pipe-like N-doped carbon at 700 °C under an Ar atmosphere, reaching 891 m^2/g of surface area. Under these conditions, α-FeOOH is transformed into Fe_{SA} that remains attached onto the internal walls of the carbon framework even after leaching, yielding 0.9 wt% of Fe. Additionally, comparative samples were prepared using polyaniline and polypyrrole as carbon sources, giving identical results. The performance of these catalysts was evaluated in the reaction of hydroxylation of benzene to phenol, showing 45% benzene conversion after 23 hours, which is one of the best results found in literature taking into account the low Fe content.

In that case, where the template is the source of the metal, the active sites can end up on the internal faces of the carbon structure, which can limit their accessibility during catalysis. In an effort to overcome this issue, structures that are more open can be designed such as 3D frameworks (simple-cubic carbon frameworks [SCCF]) (Figure 3.10c) [101] from a self-assembled organization of Fe_2O_4 nanocubes, to benefit utterly accessible active sites enhancing mass transfer process. In that case, oleic acid acts as stabilizer and carbon source. After carbonization and acid etching, it was necessary to perform a NH_3 activation at 900 °C to obtain Fe_{SA}/N-SCCF with a surface of 1180 m^2/g.

It is important to keep in mind that the inclusion of the dopant is not only to stabilize SAs but also to modulate the electronic properties of active sites. Although most studies use essentially nitrogen, the addition of a second heteroatom has also been tested, such as sulfur [102, 103] or phosphorus [104], by using monomers with double functionality such as 1-allyl-2-thiourea or mixtures as phytic acid/dopamine.

By incorporating less electronegative atoms such as S and P compared with N, the coordination and thus the electronic density on the metal can be modulated, tuning the catalytic performance. For instance, different coordination spheres are observed when Fe_{SA}, Co_{SA}, or Ni_{SA} are supported on N,S-codoped porous carbon. For cobalt and nickel, S_1-Co-N_3 and S_1-Ni-N_3 sites were detected compared to Fe-N_4-S_2 for iron. DFT calculations show that when S atoms are directly bonded with the metal, a lower density of charge remains on the metallic center, throwing values of 0.09, 0.03, and 0.11 for Co, Ni, and Fe, respectively [102].

3.6.3 Other Sources

In comparison with previous pyrolysis methods (Sections 3.6.1 and 3.6.2), Section 3.6.3 presents examples where in an attemp to simplify SAC production, cheap and abundant feedstock, such as algae [105], blood [106], silk [107], and even guano [108], have been used, eliminating tedious steps such as MOF preparation or template leaching. Table 3.4 groups some significant examples of pyrolysis of different sources, besides MOFs and coated templates.

Chlorella is an easily cultivable nitrogen-rich tiny alga characterized by its great absorbent power, including for heavy metal ions. A solution of cobalt was used for its cultivation to obtain cobalt–chlorella bio-precursor that gives rise to Co_{SA}/N-C with a surface area of 2907 m^2/g by pyrolysis. The pyrolysis was assisted by an eutectic mixture of NaCl/KCl, which at 800 °C is a liquid that prevents the agglomeration during graphitization [105].

A SAC with at least 0.3 atom% iron was obtained from the direct pyrolysis of dried pig blood without any additives, with an acceptable surface area of 215 m^2/g. After the thermal treatment at 900 °C, the heme groups from the hemoglobin in blood have evolved toward a more stable structure such as Fe phthalocyanine [106]. Silk, a nitrogen-rich natural fiber, is very abundant and can absorb and retain transition metal ions owing to its hydrophobic and hydrophilic fibroin blocks. In addition, thanks to its lamellar structure, it is the perfect candidate to generate 2D carbon nanosheets [107]. Zhang and coworkers introduced Fe^{+3} ions into the silk framework to produce highly porous nanosheets of 2105 m^2/g. Inspired by pyrolysis of Zn-M-ZIF, the addition of Zn allowed ensuring a large distance between Fe atoms [107]. Lignin, an abundant organic polymer, is a potential candidate for the large-scale production of catalysts. In addition, being a product of wood production, its cost is very low. There are reports mentioning the intrinsic properties of lignin to coordinate transition metals through its phenolic groups [109]. Nitrogen-doped carbon SACs of Fe, Co, Ni, and Cu were prepared by mixing lignin and dicyandiamide [109]. Profiting of strong interactions, metallic atoms became a node in which multiple phenolic ends coordinate, forming a supramolecular network. Further pyrolysis results in catalysts with atomic dispersion, for which unlike other previous examples, M-N_3-C species were identified for all four metals.

Table 3.4 SACs supported on carbon prepared by pyrolysis of different sources.

Sample	SAs/SAs + NPs	Carbon source/precursor	Annealing	Loading (wt%)	Metal valence	References
Co_{SA}/N-C layers	SAs	Chlorella NaCl + KCl/$Co(NO_3)_2$	800 °C 2 h N_2	3.1	$Co^{\delta+}$ ($2 < \delta < 3$)	[105]
Fe_{SA}/N-C	SAs	Dried pig blood	900 °C 1 h Ar	0.12[a]	Fe^{+2}	[106]
Fe_{SA}/carbon nanosheets	SAs (after acid leaching)	Silk + $FeCl_3/ZnCl_2$	900 °C 2 h Ar/H_2	0.12	Fe^{δ}	[107]
M_{SA}/RGO	SAs + NPs	Bird droppings + GO	800 °C N_2	0.16 Fe 0.0072 Mn 0.0013 Ni traces Co	Not informed	[108]
Co_{SA}/N-C	SAs	Lignin + dicyandiamide + $Co(NO_3)_2$ + $Zn(NO_3)_2$	550 °C 1 h Ar 1000 °C 1 h Ar	1.54	Co^{+2} to Co^{+3}	[109]
Ni_{SA}/N-CNT	SAs	Dicyandiamide + $Ni(acac)_2$	350 °C 3 h/Ar 650 °C 3 h/Ar 700 °C 1 h/Ar	20.3	Not informed	[110]
Ni_{SA}/fluor-nitrogen-nanosheets	SAs	Melamine + glucose + PTFE $Ni(NO_3)_2$	600 °C 1 h/Ar 900 °C 1 h/Ar	5.92 (sheets) 6.35 (CNT)	Ni^0 to Ni^{+2}	[111]
Pd_{SA}/N-G	SAs	Dicyandiamide + glucose Na_2PdCl_4	800 °C 2 h Ar	2.3	Not informed	[112]
Fe_{SA}/carbon aerogel	SAs (after acid leaching)	Chitosan + SiO_2 + Fe-phenanthrolene + $Zn(OAc)_2$	900 °C 3 h Ar/H_2	0.72	Fe^{+2}	[113]
Co_{SA}/CB	SAs + NPs	B_{12} vitamin + CB	700 °C 2 h/N_2	0.95	$Co^{\delta+}$ ($2 < \delta < 3$)	[114]
Co_{SA}/RGO	SAs (after acid leaching)	B_{12} vitamin + GO	800 °C	n.i.	$Co^{\delta+}$ ($2 < \delta < 3$)	[115]

a) Expressed in atomic percentage.

Moving on to the use of synthetic molecules, melamine or dicyandiamide has been broadly used in the industry of resins, and lately for the manufacture of carbon nitrides. Due to their high nitrogen content and very low price, they are potential candidates to produce catalysts on a large scale. For example, the synthesis of sheets or nanotubes decorated with Ni_{SA} at an ultrahigh loading (20.3 wt%) was achieved thanks to pyrolysis of dicyandiamide and Ni acetylacetonate [110]. The obtained morphology changed depending on the pyrolysis temperature. According to the proposed mechanism, at low temperatures, between 350 and 650 °C, the nitrogen-containing precursor begins to polymerize, forming melem (2,5,8-triamino-heptazine) fragments and further laminar C_3N_4, conferring the flat structure to the product. If the temperature continues to increase above 800 °C, graphitization takes place, giving way to a "rolling-process," shaping the nanotubes. The bamboo-like CNT had a tubular diameter of 20–50 nm, and no NPs were spotted [110]. In another report, melamine in combination with glucose, a nickel precursor, and polytetrafluoroethylene (PTFE), allowed building highly dispersed nickel catalysts on fluor-nitrogen-doped carbons (Ni_{SA}/F-N-C) [111]. Interestingly, the addition of PTFE plays as much in the dispersion as in the morphology of the carbon support, changing the tendency previously observed. After pyrolysis of the melamine + glucose + PTFE composite at 900 °C, and without the need of leaching, the formation of F-N-doped carbon nanosheets decorated with Ni_{SA} was observed, totally free of Ni_{NP}. Samples prepared without PTFE yielded a completely different form, adopting a bamboo-like tubular structure and with numerous Ni_{NP} embedded inside the tubes. Apparently, the presence of F-containing gases during pyrolysis assists the segregation of Ni atoms and avoids the transformation of sheets into tubes. XPS shows that the presence of fluorine has a direct impact on the electronic environment of the metallic centers [111]. Consequently, F-doped Ni SACs present superior performance in electro-reduction of CO_2 to CO. Mechanistic studies of the reaction assert that the presence of F near the Ni-N_4 active sites reduces the CO_2 activation barrier, making the formation of the *COOH intermediate more favorable.

Some other works have replaced the carbonization of commercially available metallic sources by the use of macrostructures with fine molecular control. Specifically, spatial distancing was identified as the key parameter by He et al. who developed a "precursor-dilution strategy" synthetizing poly-tetra(4′-vinylphenyl) porphyrin (p-TVPP) matrix allowing the modulation of the proximity between two metallic atoms. This was achieved by copolymerizing tetra(4′-vinylphenyl)porphyrin iron, FeTVPP, and iron-free tetra(4′-vinylphenyl)porphyrin, TVPP moieties. By increasing the proportion of iron-free fragments (FeTVPP : TVPP = 1 : 20), the physical spacing of Fe atoms was extended. The effectiveness of this input was evidenced after pyrolysis, without formation of Fe clusters [116].

In some cases, substrates can be loaded on a preformed support to be further annealed. The support is used to avoid agglomeration, just as the template does in the template sacrificial approach. For example, cyanocobalamin, one type of B_{12} vitamin, has been loaded on carbon black [114] or GO [115] to produce Co SACs. Chang et al. proposed the synthesis of a py-B12/C catalyst by the pyrolysis of vitamin B_{12} over Vulcan-72X carbon black [114]. Only 9.0% of total weight loss was measured after pyrolysis, meaning that CB resists well to the pyrolysis conditions. These catalysts were tested on ORR in proton-exchange membrane exhibiting high activity and selectivity at 80 °C and 2 atm of pressure. As mentioned before, four-coordinated compounds have been identified to be the best active sites for ORR. Thus, EXAFS indicates that the coordination of the cobalt center evolved from sixfold octahedral geometry to a four-coordinated cobalt (Co-N_4), with both axial positions vacant. Additionally, the oxidation state of vitamin B_{12} changes from Co^{+3} to Co^{+2}, an important aspect to consider since ORR activity is ruled by the redox potential of M^{+2} to M^{+3} transition [114].

3.7 Polymerization

In polymer chemistry, small molecules combine through covalent bonds to form supramolecular assemblies. In the context of SACs, monomer assembly will produce a carbon support hosting atomic metallic species. The term co-polymerization has also been used, referring to the fact of having both metal and carbon in the same reaction medium. In general, polymerization for SAC formation follows a polymerization by condensation, where polyfunctional N-containing monomers such as urea [117], melamine [118, 119], phthalocyanine [120], porphyrin [116, 121], phenanthroline [122] have been used. The metal can come from the monomer itself, thus being a metallic monomer, or from the addition of a metallic precursor during the polymerization process that will coordinate in some of the available N-positions in the structure. The utilization of polymerization techniques to obtain SACs has attracted attention due to the possibility to finely design coordination and electronic environments.

One of the pioneering works in the field, published by Chen et al. [123, 124], proposes the construction of silver carbon nitride from the condensation of monomeric silver tricyanomethanide (AgTCM) and cyanamide, as depicted in Figure 3.11a. The variation of the ratio between cyanamide/AgTCM concentration during the polymerization was changed from 1/500 to 1/50 targeting from 1% to 10% of silver content. It was observed that not only there is an impact on the amount of silver, but also on the density of negative charges introduced into the polymeric network. Catalysts containing 0.74–6.44 wt% Ag_{SA} were obtained, without any Ag_{NP} formation. The resultant structure fits well with mesoporous graphitic carbon nitride (mpg-C_3N_4), according to XRD analysis, although with a slight distortion in the structure, due to the presence of Ag ions. XPS highlights the effect of the addition of AgTCM: at higher concentrations, C^- and N^- signals are detected, and negative charges prevail over N, which is rationalized simply by its higher electronegativity. Besides, XPS Ag signals show contribution of both oxidized and reduced states; the presence of the latter one was rationalized by electron back-donation from the electron-rich support. Doubtless, the presence of negatively charge species had an influence on the electronic properties of Ag_{SA}/mpg-C_3N_4. Optical spectroscopy, electrochemical impedance, and Motty–Schottky analyses confirm modifications on its semiconductor features, for instance by an increase

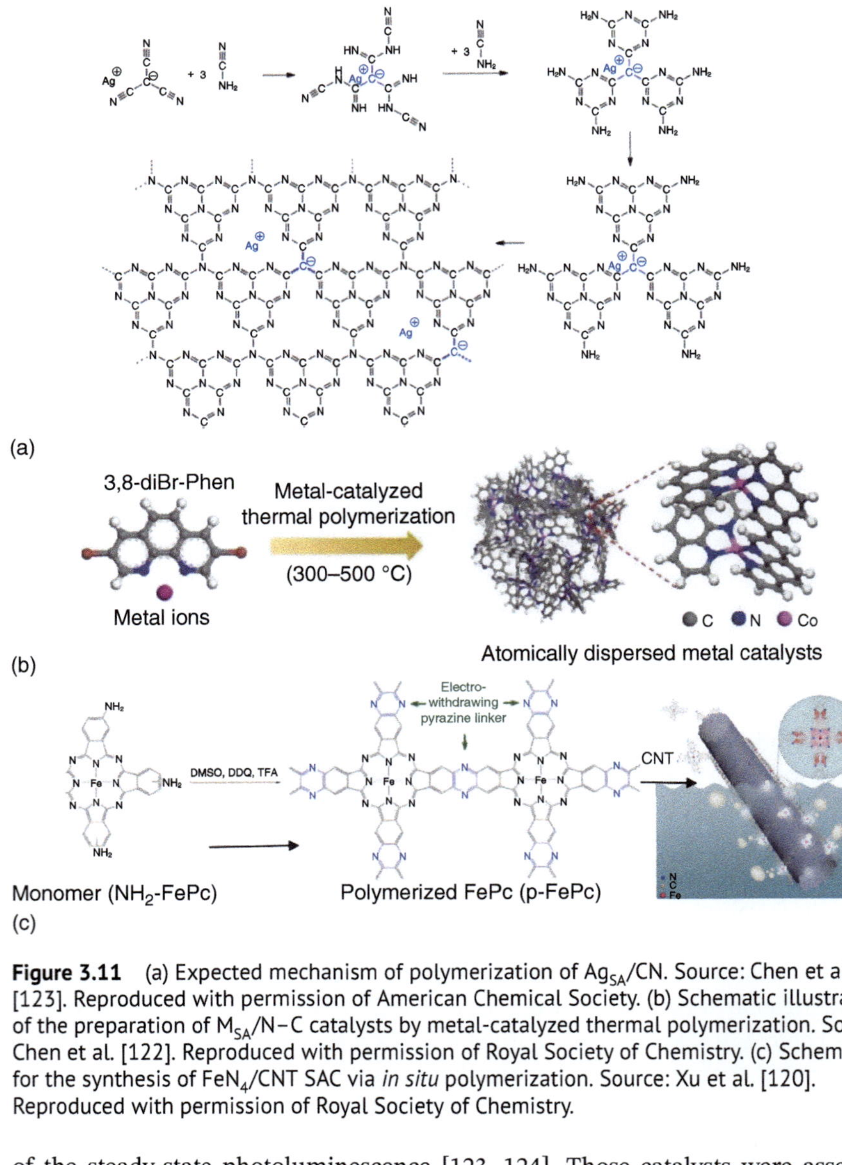

Figure 3.11 (a) Expected mechanism of polymerization of Ag_{SA}/CN. Source: Chen et al. [123]. Reproduced with permission of American Chemical Society. (b) Schematic illustration of the preparation of M_{SA}/N–C catalysts by metal-catalyzed thermal polymerization. Source: Chen et al. [122]. Reproduced with permission of Royal Society of Chemistry. (c) Schemes for the synthesis of FeN_4/CNT SAC via *in situ* polymerization. Source: Xu et al. [120]. Reproduced with permission of Royal Society of Chemistry.

of the steady-state photoluminescence [123, 124]. Those catalysts were assessed for the semihydrogenation of 1-hexyne to 1-hexene, having better performances compared to NPs. The authors direct the attention to the advantages of polymeric synthesis over impregnation-chemical reduced procedures, mentioning that at similar metallic loading, formation of Ag_{NP} is always observed.

Other strategies have been reported for the synthesis of metal-doped C_3N_4 materials, such as Cu [118, 125], W [119], and Eu [126]. For instance, preassembled melamine–Cu complexes and cyanuric acid were combined [118] to fabricate porous hollow g-C_3N_4 spheres (HCNS) with 0.85 wt% of embedded Cu_{SA} through thermal polymerization at 550 °C. X-ray absorption fine structure (XAFS)/XANES experiments reveal an oxidized state of Cu_{SA} with an average valence of +1.6 and a Cu–N_3 coordination. Another example with copper involved the preparation of Cu_{SA}/C_3N_4 by polymerization of $[H_2\ mela]_2[CuCl_5]Cl$ with a final Cu atomic fraction of almost 11% [125]. The process is assisted by molten salts to promote photogenic electron transfer while increasing crystallinity. Lanthanides, such as europium ions, have been inserted as single atoms on C_3N_4 due to their particular luminescent properties that originate in their 4f–4f transitions [126]. The polymerization at 550 °C of dicyandiamide and $EuCl_3{\cdot}6H_2O$ leads to the formation of Eu_{SA}/g-C_3N_4 up to 4 wt% of Eu. The authors claim that the addition of NH_4Cl in the reaction media helps as a dynamic gas template, giving shape of nanosheets to the final product. EXAFS analysis evidences that Eu^{+3} ions are embedded on top of the six-membered cavities of g-C_3N_4 [126].

If polymerization is an efficient way to obtain a well-defined atomic structure of the atomic center, low surface area and stacking issues can limit the accessibility to active sites. This problem can be solved by the addition of an agent that provides porosity such as a hard template [122] or a porous support [120]. Chen et al. reported a general method to prepare atomically dispersed Fe, Ni, and Co catalysts by dehalogen polymerization of the bidentate monomer, 3,8-dibromo-1,10-phenanthroline (3,8-diBr-Phen) and metallic salts with the aid of SiO_2 particles as promoter of porosity [122] (Figure 3.11b). They proposed that metallic species catalyze the 500 °C thermal polymerization (debromination and C–C coupling). The Co-PPhen-500 sample holds more than 2.0 wt% of metal in the form of SAs after template etching. Higher temperature treatment leads to pyrolysis and no polymerization, leading to the formation of Co_{NP}. Similar results were obtained when Ni or Fe nitrites were used. ^{13}C-^{1}H nuclear magnetic resonance (NMR) analyses reveal the prevalence of phenanthroline units. The catalysts were tested for ethylbenzene oxidation to acetophenone, obtaining comparable performance with the most active catalysts found in literature [122]. On the other hand, high surface area CNTs were used, as a porosity promoter, to avoid thickening of polymeric Fe-phthalocyanine (p-FePc) structures, as well as keeping good conductivity properties. Amino iron phthalocyanine monomers (NH_2-FePc) are adsorbed on CNT surface via π–π non-covalent interactions. Oxidative polymerization takes place due to the action of an oxidative agent (DDQ [2,3-dichloro-5,6-dicyanobenzoquinone]/TFA [trifluoroacetic acid]) (Figure 3.11c). As a result, CNT walls were coated with a polymer-like structure of ~1.5 nm thickness and an iron content of 0.89 wt%. XPS and Fourier transform infrared (FTIR) analyses convey that pyrazine rings are formed as the linker between FePc units. With respect to Fe-active centers, they present oxidation states between +2 and +3 due to the electron density–withdrawing effect of the pyrazine linkers. It is interesting to note that after polymerization, the Fe-N_4 centers remain practically unchanged, and DFT calculations have shown that the most stable configuration is associated with the adsorption of an O_2 molecule to form O–O–Fe-N_4 species [120].

Although the aim of this section was to discuss catalysts prepared through polymerization, it is also worth mentioning SACs prepared on polymers. Polymers have rarely been used to house single atoms, but there is an interesting example using polythiophene (PTh) as a support. Due to its excellent electrical conductivity and its heterocyclic structure containing sulfur (less electronegative than N), this polymer can generate different electronic response. The introduction of cobalt was done by a liquid–liquid interfacial process mixing an ethyl acetate solution of PTh with an aqueous solution of $CoCl_2$/NaCl to produce 0.2 wt% Co/PTh [127]. After evaporation of the organic fraction, a reddish suspension appears on the surface that is recovered by centrifugation; finally the powder is vigorously washed with ethanol to remove all physically adsorbed cobalt. Transition electron microscopy reveals that Co/PTh retains PTh sheet-like film morphology with wrinkles and folding. No cobalt-related crystallographic patterns were observed by XRD either for Co^0, CoO, or CoS. HAADF-STEM and electron energy loss spectroscopy (EELS) mapping confirm the absence of agglomerates in the sample. Fourier-transformed EXAFS spectra point to a coordination structure $Co_xO_yS_4$, where O atoms derived from oxygen were adsorbed at Co/PTh surface. Regarding electronic analysis, XPS spectra show C–S binding energy to lower values compared with pristine PTh as a result of the coordination of Co with the thiophene ring. Furthermore, Co peak deconvolution presents a significant contribution of Co^{+3}, which is another sign of the strong metal–support interaction [127].

0.2 wt% Co/PTh catalyst was used for the oxygen evolution reaction, showing superior performance compared with benchmark IrO_2 and RuO_2 catalysts [127] . Besides, acceptable stability for a non-nobel metal was measured, with no more than 5% of current density decreasing after 15 hours of test. The good catalytic performance was ascribed to the rapid charge transfer on Co/PTh, due to an extended electron distribution thanks to cobalt sites that act as linkages forming a single interconnected structure.

3.8 Other Methods

Besides the approaches presented above, other methods for synthesizing SACs on carbon have been reported. Some of them propose different energetic activations such as temperature shockwaves [128], laser pulses [129], or microwaves [130]. Table 3.5 summarizes some particular methods employed for the preparation of SACs.

In the shockwave synthesis, a thermal shock is applied to a material by the application of electrical joule heating on intermittent periods. The temperature can vary in the range 850 °C–1200 °C, and the exposure time is limited to a couple of milliseconds. The parameters that can be controlled are on–off durations and number of cycles. CO_2-activated CNFs were impregnated with H_2PtCl_6 (0.24 wt% Pt) and submitted to this treatment. After one pulse at 1773 °C, Pt_{NP} were partially transformed into SAs, and clusters still can be visualized on the surface. After ten shocks, no Pt_{NP} were observed, assuming that the totality of the metal was disintegrated into SAs on the support. Subsequent EXAFS analyses confirm this fact (absence of the signal corresponding to the Pt—Pt bond). It is important to highlight the importance of CO_2-support activation, because it introduces a high amount

Table 3.5 Other methods of synthesis of SACs supported on carbon materials.

Synthetic method	Metal	Support	Loading (wt%)	Catalytic reaction	References
High-temperature shockwave	Pt	CNF	0.24	CO oxidation	[128]
Laser pulses	Pt	C_3N_4	1.4	Water splitting	[129]
Microwave	Co	C_3N_4	~2.0	CO_2 photo reduction	[130]
Dealloying	Ni	Graphene	4–8[a)]	Hydrogen evolution reaction	[131]
Ball milling	Fe	C_3N_4	1.5–4.0	Benzene oxidation	[132]
Arc discharge	Nb	Nano-onions	13.7	ORR	[133]
Electrochemical deposition	TM	N-carbon	Up to 4.7	HER and ORR	[134]
Decomposition of organometallic complex	Ru	Fullerene	10–20	Nitrobenzene hydrogenation	[135]

a) Expressed in atomic percentage.

of defects and micropores, which are critical to accommodate single atoms [128]. Testing the scope of the technique, samples with a content of Pt five and ten times higher were treated, but under the same operating conditions, NPs remained on the support at the end of the procedure. This suggests that there exists a saturation level of the sites that received the metal atoms. Comparison of samples produced with different heating strategies, low-temperature but prolonged time (300 °C, 1 hour) or high-temperature annealing (1230 °C, 10 minutes) leads to dramatic particle agglomeration. The authors proposed that the shockwave pulses should be short enough to avoid structural modifications of the support, such as graphitization, but energetic enough to detach atoms off from the nanoparticles. These atomic species present one of the best thermal stabilities reported so far. Indeed, *in situ* STEM observations reveal static and unaffected Pt atoms after one hour of heating at 1000 °C [128].

Dealloying is a peculiar way to synthetize SACs [131]. A $Ni_{30}Mn_{70}$ plaque was attacked with $(NH_4)_2SO_4$ aqueous solution at 50 °C to partially dissolve the alloy. This treatment leaves a 3D hollowed structure. Then, this plate was used to grow carbon by chemical vapor deposition. The decomposition of CH_4 on this surface gives place to the formation of few-layer graphene. After total dissolution of the remaining alloy, nickel-doped nanoporous graphene (Ni/np-G) was formed. In that process, the alloy acted both as template and Ni source, leaving a very distinct porosity with internal channels and pores of 100–300 nm long (Figure 3.12). From EDS analysis, it is seen this catalyst possessed between 4 and 8 atom% of Ni and a negligible amount of Mn [131]. The amount of nickel varied with the time of etching, but after 7 hours, the concentration remained constant with an approximate density of 10^{18}–10^{19} atoms/m^2. For etching time <6 hours, several NPs were still observed

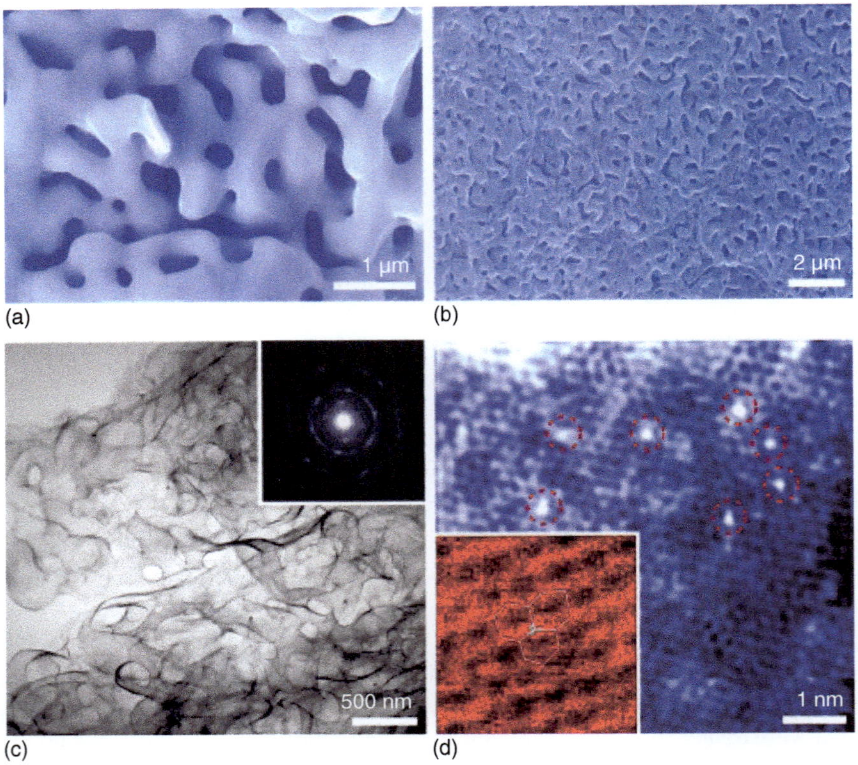

Figure 3.12 (a) SEM micrograph of the as-prepared Ni/graphene composite. (b) SEM micrograph of Ni-doped np-G after Ni dissolution for six hours. (c) TEM micrograph of Ni/np-G. (d) HAADF-STEM image of Ni/np-G. Source: Qiu et al. [131]. Reproduced with permission of John Wiley and Sons.

on the surface. The authors claimed that Ni atoms were in an intermediate electronic state, neither zero valence nor completely oxidized, as a consequence of the formation of a strong interaction with the support. Ni/np-G shows remarkable performance for hydrogen evolution reaction (HER), but more notably, an excellent stability in acidic media, which is always a problem due to the dissolution of the active phase [131].

Solvent-free reactions constitute an interesting option within conventional chemistry, and studies using ball milling have been applied with success for the preparation of carbon-supported SACs [132, 136]. In ball milling, chemical reactions are activated by the action of mechanical energy. Deng et al. reported a one-step protocol to produce graphene-confined coordinatively unsaturated iron sites using high-energy ball milling. The catalysts (FeN_4/GN) were prepared inside a glove box under a high-purity Ar atmosphere, mixing graphene nanosheets and iron phthalocyanine (FePc) for 20 hours with a frequency of 450 rpm. Thanks to the energy transferred by the steel balls, the structure of the phthalocyanine can be broken leaving residual Fe–N_4 fragments that will interact with the damaged graphene nanosheets. Carbon atoms near the defect are supposed to be able to restructure to merge with the iron centers in the graphite matrix. Raman and

XRD analyses did not display FePc signals, confirming that the ball-milling treatment is sufficiently energetic to break the phthalocyanine molecules. Scanning tunneling microscopy (STM) shows intense bright spots corresponding to Fe, but interestingly, atoms that are close to the metal appear brighter than atoms that are distant. Further STM simulations confirm that this response corresponds to C and N atoms richer in electron density due to a transfer effect from Fe to the support.

In another work, the preparation of precious metal catalysts (Pt, Ru, Rh) on N-doped carbon black was reported. The embedding of the metal was achieved with a ball mill machine agitated at 400 rpm for 20 minutes containing a mixture of N-Vulcan XC72 and K_2PtCl_4, using zirconia beads of 3 mm diameter. With this treatment, a sample with almost 2 wt% Pt was obtained but inevitably after grinding, few crystals of the precursor salt were still distinguished. For comparison, samples with the same amounts of support and precursor were prepared in liquid phase with $NaBH_4$ as reducing agent, yielding exclusively Pt_{NP} [136]. If the former catalyst was treated with $NaBH_4$, despite Pt_{NP} formation, a considerable number of Pt_{SA} remain on the surface.

Fullerenes is one of the least used carbon materials in the field of carbon-supported SACs. However, its strong electron acceptor and great tendency to polymerize can confer interesting reactivity when mixed with metallic precursors [135]. In fact, the preparation of metal fullerides is not new, and 3D arrangement from SAs to NPs with fullerene can be produced by adjusting the experimental conditions. For instance, $Ru_{SA}@C_{60}$ was synthetized by the decomposition under H_2 of the Ru complex [Ru(COD)(COT)] dissolved in toluene in presence of fullerene, leading to the formation of $-(M_1@C_{60})_n-$ chains [135]. The selection of the solvent was crucial to obtain high density of Ru_{SA} (up to 20% Ru w/w) without Ru_{NP} formation. A combination of solvents, toluene–methanol (T–M), toluene dichloromethane (T–D), led to the formation of different Ru nuclearity. Nevertheless, the visualization of the different metallic species was difficult, and wide angle X-ray scattering (WAXS) or EXAFS could not discriminate among them. In front of this complication, catalysis was used as a characterization tool to assess the presence of Ru_{SA} or clusters. In theory, electron-deficient Ru_{SA} should be less active than Ru_{NP}, particularly for hydrogenation reaction where the dissociation of H_2 usually takes place on an electron-rich center. Despite the high amount of metal on $Ru_{SA}@C_{60}$, this catalyst was poorly active for nitrobenzene hydrogenation, behavior that may confirm electron-poor Ru sites. Bader charge analysis backs up this hypothesis, showing a charge loss of −0.61electron on each Ru_{SA} on $Ru_{SA}@C_{60}$ structures. Contrarily, catalysts produced in solvent mixtures showed an extremely high activity related to the presence of Ru clusters/nanoparticles [135].

3.9 Conclusion

The reasons why carbon supports are used in catalysis have been already discussed (high surface area and porosity, functionalization versatility, etc.) [137], but their use in SACs is extremely appealing because it comes to exacerbate their own catalytic properties. Thus, the preparation of carbon-supported SACs is an exciting research topic within the catalysis community. As discussed in this chapter, dozens of preparation methods can be used for SAC preparation on carbon materials, and the final choice for a given catalytic application will depend from a very general point of view on three variables: (i) support, (ii) aimed reaction, and (iii) available equipment.

The technological applications of SACs should not be limited by the problem of large-scale production, and the use of biosourced precursors constitutes a green chemistry approach. This also makes us reflect on the real necessity of using highly sophisticated preparation techniques (ALD, sputtering, low-temperature techniques, etc.) when it is possible to resort to much simpler alternatives. A controversial article that uses guano to produce TM-doped graphene electrocatalysts illustrates this dilemma.

In terms of identification of the active sites, undoubtedly many advances have been achieved in the last two decades. The improvements of HAADF-STEM techniques that allow easier visualization of metallic atomic species, due to the lower density of carbon compared to transition metal atoms, facilitate greatly their study. The accurate identification of the active sites is one of the crucial issues for the correct correlation between the structural and catalytic properties. *In situ* STEM visualization during sputtering deposition has allowed to directly demonstrate the crucial role that defects and vacancies play in the accommodation of individual atoms. Besides, techniques such as EXAFS, being pushed to the limit, allow us to distinguish among first, second, and even third coordination spheres very accurately [127]. XPS allows to establish that in the majority of SACs, atomic species resides on the support in an electron-deficient state (δ^+) due to electron transfer from the metal to the graphitic network. The modulation of such phenomena can be engineered with the addition of dopants such as F [111], S [102], or Na [26]. Despite having reviewed examples with oxygen-, phosphorus-, or sulfur-doping, without any doubt the most studied are dealing with nitrogen-doping [4]. Its introduction can be accomplished in a straightforward manner by pyrolysis techniques. Again, EXAFS has unambiguously shown that upon the addition of N, the metallic centers will be stabilized, forming mainly species of the M-N_4-C type. In addition, it is in the case of carbons as for the other supports: a fine control of the direct environment of the SAs (first sphere of coordination) remains a primary objective.

Among the factors that can be improved in the synthesis of carbon-supported SACs are metal loading and stability. It is true that results are already found with up to 20% metal in the exclusive form of single atoms, but some catalysts present a large part of the SAs that are buried and inaccessible. Therefore, in addition to having a high density of sites, it is also important that ideally, all sites came out exposed to truly have 100% utilization of the active phase. In this sense, template sacrificial approach is a good solution to solve this challenge, with specific areas that goes beyond 1000 m^2/g. This method has mastered the creation of porosity of specific size and shape, allowing reducing possible diffusion limitations. On the other hand, there is still a long way to go in terms of stability. Top-down method is a good option to recover catalysts that have sintered, but it is not feasible as a primary method of preparation due to the low metal loading that it can handle, and the great energy demand that its application implies.

Acknowledgments

This work was supported by the Agence Nationale de la Recherche (project ANR-19-CE07-0030), which is gratefully acknowledged.

References

1 Rivera-Cárcamo, C. and Serp, P. (2018). Single atom catalysts on carbon-based materials. *ChemCatChem* 10 (22): 5058–5091.
2 Li, H., Zhang, H.-x., Yan, X.-l. et al. (2018). Carbon-supported metal single atom catalysts. *New Carbon Materials* 33 (1): 1–11.
3 Peng, Y., Lu, B., and Chen, S. (2018). Carbon-supported single atom catalysts for electrochemical energy conversion and storage. *Advanced Materials* 30 (48): 1801995.
4 Sun, J.-F., Xu, Q.-Q., Qi, J.-L. et al. (2020). Isolated single atoms anchored on N-doped carbon materials as a highly efficient catalyst for electrochemical and organic reactions. *ACS Sustainable Chemistry & Engineering* 8 (39): 14630–14656.
5 Gawande, M.B., Fornasiero, P., and Zbořil, R. (2020). Carbon-based single-atom catalysts for advanced applications. *ACS Catalysis* 10 (3): 2231–2259.
6 Ji, S., Chen, Y., Wang, X. et al. (2020). Chemical synthesis of single atomic site catalysts. *Chemical Reviews* 120 (21): 11900–11955.
7 Zhang, Q., Zhang, X., Wang, J. et al. (2020). Graphene-supported single-atom catalysts and applications in electrocatalysis. *Nanotechnology* 32 (3): 032001.
8 Khalid, M., Bhardwaj, P.A., Honorato, A.M.B. et al. (2020). Metallic single-atoms confined in carbon nanomaterials for the electrocatalysis of oxygen reduction, oxygen evolution, and hydrogen evolution reactions. *Catalysis Science & Technology* 10 (19): 6420–6448.
9 Zhao, D., Zhuang, Z., Cao, X. et al. (2020). Atomic site electrocatalysts for water splitting, oxygen reduction and selective oxidation. *Chemical Society Reviews* 49 (7): 2215–2264.
10 Lu, J. (2020). A perspective on new opportunities in atom-by-atom synthesis of heterogeneous catalysts using atomic layer deposition. *Catalysis Letters* 151: 1535–1545.
11 Wang, H. and Lu, J. (2018). Atomic layer deposition: a gas phase route to bottom-up precise synthesis of heterogeneous catalyst. *Acta Physico-Chimica Sinica* 34 (12): 1334–1357.
12 Cheng, N. and Sun, X. (2017). Single atom catalyst by atomic layer deposition technique. *Chinese Journal of Catalysis* 38 (9): 1508–1514.

13 Sun, S., Zhang, G., Gauquelin, N. et al. (2013). Single-atom catalysis using Pt/graphene achieved through atomic layer deposition. *Scientific Reports* 3 (1): 1775.

14 Cheng, N., Stambula, S., Wang, D. et al. (2016). Platinum single-atom and cluster catalysis of the hydrogen evolution reaction. *Nature Communications* 7 (1): 13638.

15 Huang, X., Xia, Y., Cao, Y. et al. (2017). Enhancing both selectivity and coking-resistance of a single-atom Pd_1/C_3N_4 catalyst for acetylene hydrogenation. *Nano Research* 10 (4): 1302–1312.

16 Yang, J., Fu, W., Chen, C. et al. (2021). Atomic design and fine-tuning of subnanometric Pt catalysts to tame hydrogen generation. *ACS Catalysis* 11 (7): 4146–4156.

17 Wang, X., Jin, B., Jin, Y. et al. (2020). Supported single Fe atoms prepared via atomic layer deposition for catalytic reactions. *ACS Applied Nano Materials* 3 (3): 2867–2874.

18 Cao, Y., Chen, S., Luo, Q. et al. (2017). Atomic-level insight into optimizing the hydrogen evolution pathway over a Co_1-N_4 single-site photocatalyst. *Angewandte Chemie International Edition* 56 (40): 12191–12196.

19 Song, Z., Zhu, Y.-N., Liu, H. et al. (2020). Engineering the low coordinated Pt single atom to achieve the superior electrocatalytic performance toward oxygen reduction. *Small* 16 (43): 2003096.

20 Yan, H., Cheng, H., Yi, H. et al. (2015). Single-atom Pd_1/graphene catalyst achieved by atomic layer deposition: remarkable performance in selective hydrogenation of 1,3-butadiene. *Journal of the American Chemical Society* 137 (33): 10484–10487.

21 Yan, H., Lin, Y., Wu, H. et al. (2017). Bottom-up precise synthesis of stable platinum dimers on graphene. *Nature Communications* 8 (1): 1070.

22 Kim, M., Kim, K.-J., Lee, S.-J. et al. (2017). Highly stable and effective doping of graphene by selective atomic layer deposition of ruthenium. *ACS Applied Materials & Interfaces* 9 (1): 701–709.

23 Imaoka, T., Akanuma, Y., Haruta, N. et al. (2017). Platinum clusters with precise numbers of atoms for preparative-scale catalysis. *Nature Communications* 8 (1): 688.

24 Marceau, E., Carrier, X., and Che, M. (2009). Impregnation and drying. In: *Synthesis of Solid Catalysts* (ed. K.P. de Jong), 59–82. Wiley-VCH.

25 Kochubey, D.I., Chesnokov, V.V., and Malykhin, S.E. (2012). Evidence for atomically dispersed Pd in catalysts supported on carbon nanofibers. *Carbon* 50 (8): 2782–2787.

26 Rivera-Cárcamo, C., Scarfiello, C., García, A.B. et al. Stabilization of metal single atoms on carbon and TiO_2 supports for CO_2 hydrogenation: the importance of regulating charge transfer. *Advanced Materials Interfaces*: 2001777. https://doi.org/10.1002/admi.202001777.

27 Bhattacharya, A., Bhattacharya, S., Majumder, C. et al. (2010). Transition-metal decoration enhanced room-temperature hydrogen storage in a defect-modulated graphene sheet. *The Journal of Physical Chemistry C* 114 (22): 10297–10301.

28 Rivera-Cárcamo, C., Gerber, I.C., del Rosal, I. et al. (2021). Control of the single atom/nanoparticle ratio in Pd/C catalysts to optimize the cooperative hydrogenation of alkenes. *Catalysis Science & Technology* 11: 984–999.

29 Hu, S., Ma, L., You, J. et al. (2014). Enhanced visible light photocatalytic performance of g-C_3N_4 photocatalysts co-doped with iron and phosphorus. *Applied Surface Science* 311: 164–171.

30 Li, X., Bi, W., Zhang, L. et al. (2016). Single-atom Pt as Co-catalyst for enhanced photocatalytic H_2 evolution. *Advanced Materials* 28 (12): 2427–2431.

31 Chen, Z., Zhang, Q., Chen, W. et al. (2018). Single-site AuI catalyst for silane oxidation with water. *Advanced Materials* 30 (5): 1704720.

32 Chen, Z., Mitchell, S., Vorobyeva, E. et al. (2017). Stabilization of single metal atoms on graphitic carbon nitride. *Advanced Functional Materials* 27 (8): 1605785.

33 Vilé, G., Albani, D., Nachtegaal, M. et al. (2015). A stable single-site palladium catalyst for hydrogenations. *Angewandte Chemie International Edition* 54 (38): 11265–11269.

34 Bi, W., Li, X., You, R. et al. (2018). Surface immobilization of transition metal ions on nitrogen-doped graphene realizing high-efficient and selective CO_2 reduction. *Advanced Materials* 30 (18): 1706617.

35 Yin, X.-P., Tang, S.-F., Zhang, C. et al. (2020). Graphdiyne-based Pd single-atom catalyst for semihydrogenation of alkynes to alkenes with high selectivity and conversion under mild conditions. *Journal of Materials Chemistry A* 8 (40): 20925–20930.

36 Liu, X., Conte, M., Elias, D. et al. (2016). Investigation of the active species in the carbon-supported gold catalyst for acetylene hydrochlorination. *Catalysis Science & Technology* 6 (13): 5144–5153.

37 Malta, G., Kondrat, S.A., Freakley, S.J. et al. (2017). Identification of single-site gold catalysis in acetylene hydrochlorination. *Science* 355 (6332): 1399–1403.

38 Corma, A., Concepción, P., Boronat, M. et al. (2013). Exceptional oxidation activity with size-controlled supported gold clusters of low atomicity. *Nature Chemistry* 5 (9): 775–781.

39 Corma, A., Salnikov, O.G., Barskiy, D.A. et al. (2015). Single-atom gold catalysis in the context of developments in parahydrogen-induced polarization. *Chemistry European Journal* 21 (19): 7012–7015.

40 Kim, Y.-T., Ohshima, K., Higashimine, K. et al. (2006). Fine size control of platinum on carbon nanotubes: from single atoms to clusters. *Angewandte Chemie International Edition* 45 (3): 407–411.

41 Woo, H., Lee, E.-K., Yun, S.-W. et al. (2017). Platinum single atoms on carbon nanotubes as efficient catalyst for hydroalkoxylation. *Bulletin of the Korean Chemical Society* 38 (10): 1221–1225.

42 Kaiser, S.K., Clark, A.H., Cartocci, L. et al. Sustainable synthesis of bimetallic single atom gold-based catalysts with enhanced durability in acetylene hydrochlorination. *Small*: 2004599. https://doi.org/10.1002/smll.202004599.

43 Bulushev, D.A., Zacharska, M., Lisitsyn, A.S. et al. (2016). Single atoms of Pt-group metals stabilized by N-doped carbon nanofibers for efficient hydrogen production from formic acid. *ACS Catalysis* 6 (6): 3442–3451.

44 Huang, K., Zhang, L., Xu, T. et al. (2019). −60 °C solution synthesis of atomically dispersed cobalt electrocatalyst with superior performance. *Nature Communications* 10 (1): 606.

45 Wei, H., Huang, K., Wang, D. et al. (2017). Iced photochemical reduction to synthesize atomically dispersed metals by suppressing nanocrystal growth. *Nature Communications* 8 (1): 1490.

46 Fei, H., Dong, J., Arellano-Jiménez, M.J. et al. (2015). Atomic cobalt on nitrogen-doped graphene for hydrogen generation. *Nature Communications* 6 (1): 8668.

47 Zhang, C., Sha, J., Fei, H. et al. (2017). Single-atomic ruthenium catalytic sites on nitrogen-doped graphene for oxygen reduction reaction in acidic medium. *ACS Nano* 11 (7): 6930–6941.

48 Jiang, K., Siahrostami, S., Zheng, T. et al. (2018). Isolated Ni single atoms in graphene nanosheets for high-performance CO_2 reduction. *Energy & Environmental Science* 11 (4): 893–903.

49 Zu, X., Li, X., Liu, W. et al. (2019). Efficient and robust carbon dioxide electroreduction enabled by atomically dispersed $Sn^{\delta+}$ sites. *Advanced Materials* 31 (15): 1808135.

50 Krasheninnikov, A.V. and Nordlund, K. (2010). Ion and electron irradiation-induced effects in nanostructured materials. *Journal of Applied Physics* 107 (7): 071301.

51 Yamazaki, K., Maehara, Y., Lee, C.-C. et al. (2018). Atomic structure and local electronic states of single Pt atoms dispersed on graphene. *The Journal of Physical Chemistry C* 122 (48): 27292–27300.

52 Kuo, C.-T., Lu, Y., Kovarik, L. et al. (2019). Structure sensitivity of acetylene semi-hydrogenation on Pt single atoms and subnanometer clusters. *ACS Catalysis* 9 (12): 11030–11041.

53 Sugimoto, R., Segawa, Y., Suzuta, A. et al. (2021). Single Pt atoms on N-doped graphene: atomic structure and local electronic states. *The Journal of Physical Chemistry C* 125 (5): 2900–2906.

54 Veith, G.M., Lupini, A.R., Pennycook, S.J. et al. (2007). Magnetron sputtering of gold nanoparticles onto WO_3 and activated carbon. *Catalysis Today* 122 (3): 248–253.

55 Kettner, M., Maisel, S., Stumm, C. et al. (2019). Pd-Ga model SCALMS: characterization and stability of Pd single atom sites. *Journal of Catalysis* 369: 33–46.

56 Wang, H., Wang, Q., Cheng, Y. et al. (2012). Doping monolayer graphene with single atom substitutions. *Nano Letters* 12 (1): 141–144.

57 Wang, H., Li, K., Cheng, Y. et al. (2012). Interaction between single gold atom and the graphene edge: a study via aberration-corrected transmission electron microscopy. *Nanoscale* 4 (9): 2920–2925.

58 Hansen, T.W., DeLaRiva, A.T., Challa, S.R. et al. (2013). Sintering of catalytic nanoparticles: particle migration or ostwald ripening? *Accounts of Chemical Research* 46 (8): 1720–1730.

59 Ouyang, R., Liu, J.-X., and Li, W.-X. (2013). Atomistic theory of ostwald ripening and disintegration of supported metal particles under reaction conditions. *Journal of the American Chemical Society* 135 (5): 1760–1771.

60 Goguet, A., Hardacre, C., Harvey, I. et al. (2009). Increased dispersion of supported gold during methanol carbonylation conditions. *Journal of the American Chemical Society* 131 (20): 6973–6975.

61 Feng, S., Song, X., Liu, Y. et al. (2019). In situ formation of mononuclear complexes by reaction-induced atomic dispersion of supported noble metal nanoparticles. *Nature Communications* 10 (1): 5281.

62 Feng, S., Song, X., Ren, Z. et al. (2019). La-stabilized, single-atom Ir/AC catalyst for heterogeneous methanol carbonylation to methyl acetate. *Industrial and Engineering Chemistry Research* 58 (12): 4755–4763.

63 Wu, L., Hu, S., Yu, W. et al. (2020). Stabilizing mechanism of single-atom catalysts on a defective carbon surface. *npj Computational Materials* 6 (1): 23.

64 Yang, J., Qiu, Z., Zhao, C. et al. (2018). In situ thermal atomization to convert supported nickel nanoparticles into surface-bound nickel single-atom catalysts. *Angewandte Chemie International Edition* 57 (43): 14095–14100.

65 Wei, S., Li, A., Liu, J.-C. et al. (2018). Direct observation of noble metal nanoparticles transforming to thermally stable single atoms. *Nature Nanotechnology* 13 (9): 856–861.

66 Batten, S.R., Champness, N.R., Chen, X.-M. et al. (2013). Terminology of metal–organic frameworks and coordination polymers (IUPAC Recommendations 2013). *Pure and Applied Chemistry* 85 (8): 1715–1724.

67 Zitolo, A., Goellner, V., Armel, V. et al. (2015). Identification of catalytic sites for oxygen reduction in iron- and nitrogen-doped graphene materials. *Nature Materials* 14 (9): 937–942.

68 Huang, H., Shen, K., Chen, F. et al. (2020). Metal–organic frameworks as a good platform for the fabrication of single-atom catalysts. *ACS Catalysis* 10 (12): 6579–6586.

69 Phan, A., Doonan, C.J., Uribe-Romo, F.J. et al. (2010). Synthesis, structure, and carbon dioxide capture properties of zeolitic imidazolate frameworks. *Accounts of Chemical Research* 43 (1): 58–67.

70 Tang, J., Salunkhe, R.R., Zhang, H. et al. (2016). Bimetallic metal-organic frameworks for controlled catalytic graphitization of nanoporous carbons. *Scientific Reports* 6 (1): 30295.

71 Fan, L., Liu, P.F., Yan, X. et al. (2016). Atomically isolated nickel species anchored on graphitized carbon for efficient hydrogen evolution electrocatalysis. *Nature Communications* 7 (1): 10667.

72 Fang, X., Jiao, L., Yu, S.-H. et al. (2017). Metal–organic framework-derived FeCo-N-doped hollow porous carbon nanocubes for electrocatalysis in acidic and alkaline media. *ChemSusChem* 10 (15): 3019–3024.

73 Yin, P., Yao, T., Wu, Y. et al. (2016). Single cobalt atoms with precise N-coordination as superior oxygen reduction reaction catalysts. *Angewandte Chemie International Edition* 55 (36): 10800–10805.

74 Al-Zoubi, T., Zhou, Y., Yin, X. et al. (2020). Preparation of nonprecious metal electrocatalysts for the reduction of oxygen using a low-temperature sacrificial metal. *Journal of the American Chemical Society* 142 (12): 5477–5481.

75 Wang, Q., Ina, T., Chen, W.-T. et al. (2020). Evolution of Zn(II) single atom catalyst sites during the pyrolysis-induced transformation of ZIF-8 to N-doped carbons. *Science Bulletin* 65 (20): 1743–1751.

76 Jiao, L., Wan, G., Zhang, R. et al. (2018). From metal–organic frameworks to single-atom Fe implanted N-doped porous carbons: efficient oxygen reduction in both alkaline and acidic media. *Angewandte Chemie International Edition* 57 (28): 8525–8529.

77 Yang, S., Zhang, J., Peng, L. et al. (2020). A metal–organic framework/polymer derived catalyst containing single-atom nickel species for electrocatalysis. *Chemical Science* 11 (40): 10991–10997.

78 Guo, Y., Feng, L., Wu, C. et al. (2020). Confined pyrolysis transformation of ZIF-8 to hierarchically ordered porous Zn-N-C nanoreactor for efficient CO_2 photoconversion under mild conditions. *Journal of Catalysis* 390: 213–223.

79 Sun, X., Sun, S., Gu, S. et al. (2019). High-performance single atom bifunctional oxygen catalysts derived from ZIF-67 superstructures. *Nano Energy* 61: 245–250.

80 Dilpazir, S., He, H., Li, Z. et al. (2018). Cobalt single atoms immobilized N-doped carbon nanotubes for enhanced bifunctional catalysis toward oxygen reduction and oxygen evolution reactions. *ACS Applied Energy Materials* 1 (7): 3283–3291.

81 Lai, W.-H., Zhang, B.-W., Hu, Z. et al. (2019). The quasi-Pt-allotrope catalyst: hollow PtCo@single-atom Pt1 on nitrogen-doped carbon toward superior oxygen reduction. *Advanced Functional Materials* 29 (13): 1807340.

82 Geng, Z., Liu, Y., Kong, X. et al. (2018). Achieving a record-high yield rate of 120.9 for N_2 electrochemical reduction over Ru single-atom catalysts. *Advanced Materials* 30 (40): 1803498.

83 Hu, X., Luo, G., Zhao, Q. et al. (2020). Ru single atoms on N-doped carbon by spatial confinement and ionic substitution strategies for high-performance Li–O_2 batteries. *Journal of the American Chemical Society* 142 (39): 16776–16786.

84 Li, J., Chen, M., Cullen, D.A. et al. (2018). Atomically dispersed manganese catalysts for oxygen reduction in proton-exchange membrane fuel cells. *Nature Catalysis* 1 (12): 935–945.

85 Gu, W., Wu, M., Sun, J. et al. (2019). Atomically dispersed Fe–Nx active sites within hierarchical mesoporous carbon as efficient electrocatalysts for the oxygen reduction reaction. *Journal of Materials Chemistry A* 7 (35): 20132–20138.

86 Xie, J., Yin, K., Serov, A. et al. (2017). Selective aerobic oxidation of alcohols over atomically-dispersed non-precious metal catalysts. *ChemSusChem* 10 (2): 359–362.

87 Liang, H.-W., Wei, W., Wu, Z.-S. et al. (2013). Mesoporous metal–nitrogen-doped carbon electrocatalysts for highly efficient oxygen reduction reaction. *Journal of the American Chemical Society* 135 (43): 16002–16005.

88 Serov, A., Artyushkova, K., Niangar, E. et al. (2015). Nano-structured non-platinum catalysts for automotive fuel cell application. *Nano Energy* 16: 293–300.

89 Wang, Y., Kong, A., Chen, X. et al. (2015). Efficient oxygen electroreduction: hierarchical porous Fe–N-doped hollow carbon nanoshells. *ACS Catalysis* 5 (6): 3887–3893.

90 Zhang, M., Wang, Y.-G., Chen, W. et al. (2017). Metal (Hydr)oxides@polymer core–shell strategy to metal single-atom materials. *Journal of the American Chemical Society* 139 (32): 10976–10979.

91 Liu, W., Zhang, L., Liu, X. et al. (2017). Discriminating catalytically active FeNx species of atomically dispersed Fe–N–C catalyst for selective oxidation of the C—H bond. *Journal of the American Chemical Society* 139 (31): 10790–10798.

92 Jie, Y., Wenzhao, F., Chaoqiu, C. et al. (2021). Research Square. https://doi.org/10.21203/rs.3.rs-84814/v1. 11, 7, 4146–4156

93 Han, Y., Wang, Y.-G., Chen, W. et al. (2017). Hollow N-doped carbon spheres with isolated cobalt single atomic sites: superior electrocatalysts for oxygen reduction. *Journal of the American Chemical Society* 139 (48): 17269–17272.

94 Liu, W., Zhang, L., Yan, W. et al. (2016). Single-atom dispersed Co–N–C catalyst: structure identification and performance for hydrogenative coupling of nitroarenes. *Chemical Science* 7 (9): 5758–5764.

95 Yuan, C.-Z., Zhan, L.-Y., Liu, S.-J. et al. (2020). Semi-sacrificial template synthesis of single-atom Ni sites supported on hollow carbon nanospheres for efficient and stable electrochemical CO_2 reduction. *Inorganic Chemistry Frontiers* 7 (8): 1719–1725.

96 Hou, Y., Liang, Y.-L., Shi, P.-C. et al. (2020). Atomically dispersed Ni species on N-doped carbon nanotubes for electroreduction of CO_2 with nearly 100% CO selectivity. *Applied Catalysis B: Environmental* 271: 118929.

97 Xu, H., Wang, D., Yang, P. et al. (2020). A theoretical study of atomically dispersed MN_4/C (M = Fe or Mn) as a high-activity catalyst for the oxygen reduction reaction. *Physical Chemistry Chemical Physics* 22 (48): 28297–28303.

98 Wu, F., Pan, C., He, C.-T. et al. (2020). Single-atom Co–N_4 electrocatalyst enabling four-electron oxygen reduction with enhanced hydrogen peroxide tolerance for selective sensing. *Journal of the American Chemical Society* 142 (39): 16861–16867.

99 Guo, J., Yan, X., Liu, Q. et al. (2018). The synthesis and synergistic catalysis of iron phthalocyanine and its graphene-based axial complex for enhanced oxygen reduction. *Nano Energy* 46: 347–355.

100 Chen, W., Pei, J., He, C.-T. et al. (2017). Rational design of single molybdenum atoms anchored on N-doped carbon for effective hydrogen evolution reaction. *Angewandte Chemie International Edition* 56 (50): 16086–16090.

101 Wang, B., Wang, X., Zou, J. et al. (2017). Simple-cubic carbon frameworks with atomically dispersed iron dopants toward high-efficiency oxygen reduction. *Nano Letters* 17 (3): 2003–2009.

102 Zhang, J., Zhao, Y., Chen, C. et al. (2019). Tuning the coordination environment in single-atom catalysts to achieve highly efficient oxygen reduction reactions. *Journal of the American Chemical Society* 141 (51): 20118–20126.

103 Wu, K., Chen, X., Liu, S. et al. (2018). Porphyrin-like Fe-N_4 sites with sulfur adjustment on hierarchical porous carbon for different rate-determining steps in oxygen reduction reaction. *Nano Research* 11 (12): 6260–6269.

104 Zhu, X., Tan, X., Wu, K.-H. et al. (2019). N,P co-coordinated Fe species embedded in carbon hollow spheres for oxygen electrocatalysis. *Journal of Materials Chemistry A* 7 (24): 14732–14742.

105 Wu, D., Hu, J., Zhu, C. et al. (2021). Salt melt synthesis of Chlorella-derived nitrogen-doped porous carbon with atomically dispersed CoN_4 sites for efficient oxygen reduction reaction. *Journal of Colloid and Interface Science* 586: 498–504.

106 Jiang, W.-J., Hu, W.-L., Zhang, Q.-H. et al. (2018). From biological enzyme to single atomic Fe–N–C electrocatalyst for efficient oxygen reduction. *Chemical Communications* 54 (11): 1307–1310.

107 Wang, C., Chen, W., Xia, K. et al. (2019). Silk-derived 2D porous carbon nanosheets with atomically-dispersed Fe-Nx-C sites for highly efficient oxygen reaction catalysts. *Small* 15 (7): 1804966.

108 Wang, L., Sofer, Z., and Pumera, M. (2020). Will any crap we put into graphene increase its electrocatalytic effect? *ACS Nano* 14 (1): 21–25.

109 Zhou, H., Hong, S., Zhang, H. et al. (2019). Toward biomass-based single-atom catalysts and plastics: Highly active single-atom Co on N-doped carbon for oxidative esterification of primary alcohols. *Applied Catalysis B: Environmental* 256: 117767.

110 Zhao, S., Cheng, Y., Veder, J.-P. et al. (2018). One-pot pyrolysis method to fabricate carbon nanotube supported Ni single-atom catalysts with ultrahigh loading. *ACS Applied Energy Materials* 1 (10): 5286–5297.

111 Han, S.-G., Ma, D.-D., Zhou, S.-H. et al. (2021). Fluorine-tuned single-atom catalysts with dense surface Ni-N4 sites on ultrathin carbon nanosheets for efficient CO_2 electroreduction. *Applied Catalysis B: Environmental* 283: 119591.

112 Zhou, S., Shang, L., Zhao, Y. et al. (2019). Pd single-atom catalysts on nitrogen-doped graphene for the highly selective photothermal hydrogenation of acetylene to ethylene. *Advanced Materials* 31 (18): 1900509.

113 He, T., Lu, B., Chen, Y. et al. (2019). Nanowrinkled carbon aerogels embedded with FeN_x sites as effective oxygen electrodes for rechargeable zinc-air battery. *Research* 2019: 6813585.

114 Chang, S.-T., Wang, C.-H., Du, H.-Y. et al. (2012). Vitalizing fuel cells with vitamins: pyrolyzed vitamin B12 as a non-precious catalyst for enhanced oxygen reduction reaction of polymer electrolyte fuel cells. *Energy & Environmental Science* 5 (1): 5305–5314.

115 Sabhapathy, P., Liao, C.-C., Chen, W.-F. et al. (2019). Highly efficient nitrogen and carbon coordinated N–Co–C electrocatalysts on reduced graphene oxide derived from vitamin-B12 for the hydrogen evolution reaction. *Journal of Materials Chemistry A* 7 (12): 7179–7185.

116 He, Q., Meng, Y., Zhang, H. et al. (2019). Facile synthesis of impurity-free iron single atom catalysts for highly efficient oxygen reduction reaction and active-site identification. *Catalysis Science & Technology* 9 (23): 6556–6560.

117 Yang, Y., Zeng, G., Huang, D. et al. (2020). In situ grown single-atom cobalt on polymeric carbon nitride with bidentate ligand for efficient photocatalytic degradation of refractory antibiotics. *Small* 16 (29): 2001634.

118 Zhang, T., Zhang, D., Han, X. et al. (2018). Preassembly strategy to fabricate porous hollow carbonitride spheres inlaid with single Cu–N_3 sites for selective oxidation of benzene to phenol. *Journal of the American Chemical Society* 140 (49): 16936–16940.

119 Kumru, B., Cruz, D., Heil, T. et al. (2020). In situ formation of arrays of tungsten single atoms within carbon nitride frameworks fabricated by one-step synthesis through monomer complexation. *Chemistry of Materials* 32 (21): 9435–9443.

120 Xu, H., Xi, S., Li, J. et al. (2020). Chemical design and synthesis of superior single-atom electrocatalysts via in situ polymerization. *Journal of Materials Chemistry A* 8 (34): 17683–17690.

121 He, X., He, Q., Deng, Y. et al. (2019). A versatile route to fabricate single atom catalysts with high chemoselectivity and regioselectivity in hydrogenation. *Nature Communications* 10 (1): 3663.

122 Chen, L.-W., Wu, Z.-Y., Nan, H. et al. (2019). A metal-catalyzed thermal polymerization strategy toward atomically dispersed catalysts. *Chemical Communications* 55 (77): 11579–11582.

123 Chen, Z., Pronkin, S., Fellinger, T.-P. et al. (2016). Merging single-atom-dispersed silver and carbon nitride to a joint electronic system via copolymerization with silver tricyanomethanide. *ACS Nano* 10 (3): 3166–3175.

124 Wang, Y., Zhao, X., Cao, D. et al. (2017). Peroxymonosulfate enhanced visible light photocatalytic degradation bisphenol A by single-atom dispersed Ag mesoporous g-C_3N_4 hybrid. *Applied Catalysis B: Environmental* 211: 79–88.

125 Li, Y., Li, B., Zhang, D. et al. (2020). Crystalline carbon nitride supported copper single atoms for photocatalytic CO_2 reduction with nearly 100% CO selectivity. *ACS Nano* 14 (8): 10552–10561.

126 Mori, K., Murakami, T., and Yamashita, H. (2020). Luminescent single-atom Eu-coordinated graphitic carbon nitride nanosheets for selective sensing of acetone and cyclohexane. *ACS Applied Nano Materials* 3 (10): 10209–10217.

127 Xing, D., Zhou, P., Liu, Y. et al. (2021). Atomically dispersed cobalt-based species anchored on polythiophene as an efficient electrocatalyst for oxygen evolution reaction. *Applied Surface Science* 545: 148943.

128 Yao, Y., Huang, Z., Xie, P. et al. (2019). High temperature shockwave stabilized single atoms. *Nature Nanotechnology* 14 (9): 851–857.

129 Wu, C., Xue, S., Qin, Z. et al. (2021). Making g-C_3N_4 ultra-thin nanosheets active for photocatalytic overall water splitting. *Applied Catalysis B: Environmental* 282: 119557.

130 Huang, P., Huang, J., Pantovich, S.A. et al. (2018). Selective CO_2 reduction catalyzed by single cobalt sites on carbon nitride under visible-light irradiation. *Journal of the American Chemical Society* 140 (47): 16042–16047.

131 Qiu, H.-J., Ito, Y., Cong, W. et al. (2015). Nanoporous graphene with single-atom nickel dopants: an efficient and stable catalyst for electrochemical hydrogen production. *Angewandte Chemie International Edition* 54 (47): 14031–14035.

132 Deng, D., Chen, X., Yu, L. et al. (2015). A single iron site confined in a graphene matrix for the catalytic oxidation of benzene at room temperature. *Science Advances* 1 (11): e1500462.

133 Zhang, X., Guo, J., Guan, P. et al. (2013). Catalytically active single-atom niobium in graphitic layers. *Nature Communications* 4 (1): 1924.

134 Zhang, Z., Feng, C., Liu, C. et al. (2020). Electrochemical deposition as a universal route for fabricating single-atom catalysts. *Nature Communications* 11 (1): 1215.

135 Rivera-Cárcamo, C., Leng, F., Gerber, I.C. et al. (2020). Catalysis to discriminate single atoms from subnanometric ruthenium particles in ultra-high loading catalysts. *Catalysis Science & Technology* 10 (14): 4673–4683.

136 Jin, H., Sultan, S., Ha, M. et al. (2020). Simple and scalable mechanochemical synthesis of noble metal catalysts with single atoms toward highly efficient hydrogen evolution. *Advanced Functional Materials* 30 (25): 2000531.

137 Gerber, I.C. and Serp, P. (2020). A theory/experience description of support effects in carbon-supported catalysts. *Chemical Reviews* 120 (2): 1250–1349.

4

Single-Metal Alloys

Jianyu Han[1,2], Junju Mu[1], and Feng Wang[1,]*

[1] State Key Laboratory of Catalysis, Dalian National Laboratory for Clean Energy, Dalian Institute of Chemical Physics, Chinese Academy of Sciences, Dalian, China
[2] University of Chinese Academy of Sciences, Beijing, China

4.1 Introduction

Supported metal catalysts exhibit unique catalytic properties for various reactions [1–5]. The activation step in heterogeneous catalysts usually occurs at surface sites, and it is accepted that only surface and subsurface atoms on the catalyst are capable of affecting the catalytic performance [6, 7]. Also, abundant coordinate unsaturated sites on metal nanoparticles (NPs) are benign to catalyst's performance. These catalyst design criteria lead to the ultimate dispersion of metal atoms, namely single-atom catalysts (SACs). Therefore, it can maximize the atom efficiency, which brings better catalytic performance and reduces the cost of the catalyst. Although SACs are proved to be suitable for many reactions, methods for optimizing their property are rather limited. Alloying single atoms with another kind of metal offers an effective way of tuning the catalytic performance of a single atom. It is well known that there are two key factors in alloy catalyst that contributes to the overall catalytic performance, i.e. ligand effect [8] and coordination effect [9]. The ligand effect originates from the redistribution of electrons between guest metal and host metal. The coordination effect arises from the lattice distortion and strain due to the substitution of host-metal atoms by guest-metal atoms. Additionally, organic substrates and intermediates are usually adsorbed on catalyst surfaces in bidentate or more complex conformations. Single-atom active site could not provide a suitable adsorption site for complex conformation unless it is densely distributed on the surface of the support. Thus, synthesizing single-atom alloy (SAA) catalyst is considered as a promising way to further improve the catalytic performance compared to SACs.

Pioneering works have outlined the blueprint of developing SAA catalyst as shown in Figure 4.1. In our opinion, corresponding researches in literature could be categorized into three kinds of SAAs. The first kind is atomically dispersing several

* Corresponding author email: wangfeng@dicp.ac.cn

Supported Metal Single Atom Catalysis, First Edition. Edited by Philippe Serp and Doan Pham Minh.
© 2022 WILEY-VCH GmbH. Published 2022 by WILEY-VCH GmbH.

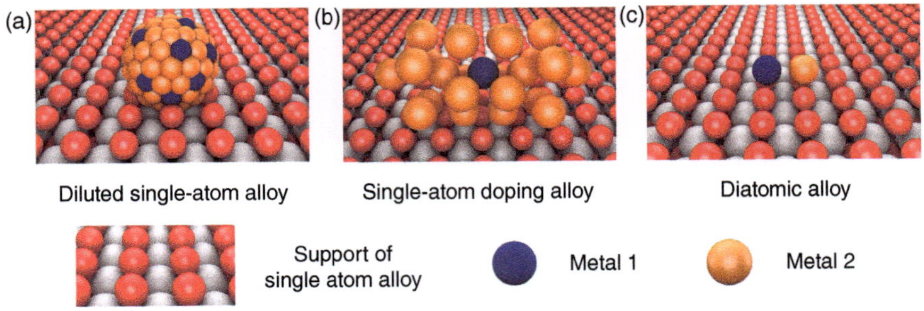

Figure 4.1 Schematic illustration of three kinds of single-atom alloy: (a) dilute single-atom alloy, (b) single-atom doping alloy, and (c) diatomic alloy. Source: Jianyu Han, Junju Mu, and Feng Wang.

guest-metal atoms onto the host-metal nanocrystals, which is called diluted SAAs in this chapter (Figure 4.1a). The second kind of SAAs is doping a single guest atom onto an Au or Ag cluster with a magic number, which is called single-atom doping alloy in this chapter (Figure 4.1b). The third kind of SAAs is the heteroatom pair, which is called diatomic alloy in this chapter (Figure 4.1c).

Diluted SAA catalysts are widely applied in catalysis due to their stability compared to the other two kinds of SAA catalysts. Diluted SAA structures provide proper binding sites for a broad scope of organic substrates and intermediates [10]. The catalyst model is usually depicted as depositing guest-metal single atoms on the surface of host-metal or dispersing guest-metal single atoms in host-metal NPs. Overall, the amount of guest-metal atoms is dramatically lower than the amount of host-metal atoms. Single-atom doping alloy catalysts are mainly confined in Au or Ag nanoclusters due to the structural stability. Redistribution of the electrons, which is caused by a single heteroatom, could be demonstrated in single-atom doping alloy. Synthesis processes and dopant elements mainly affect substitution sites in Au/Ag nanoclusters. The well-defined structure of such kind of catalyst relies on accurate theoretical study with first-principle calculations and multiple characterizations. Diatomic alloy is an emerging scientific frontier, which represents the highest level of manipulating catalyst surface. The active sites are composed of two heteroatoms such as Pt–Ru, Ni–Fe, or Sn–Zn pairs [11–13]. The potential of activating inert molecules such as O_2, CO_2, and N_2 under mild conditions through diatomic alloys also attracts the attention of scientists. However, both synthesis and characterization of diatomic alloys face great challenges, resulting in scarce reports in the corresponding field.

4.2 Diluted Single-Atom Alloy Catalysts

4.2.1 Synthesis of Diluted Single-Atom Alloy Catalysts

Dispersing single atoms of guest metal into host-metal's nanocrystals is an effective way to prepare SAAs. The first report of SAAs was published by Sykes and coworkers in 2012. In this work, a Pd–Cu SAA was used to catalyze alkyne selective hydrogenation [10]. Sykes and coworkers adopted the surface chemistry method to

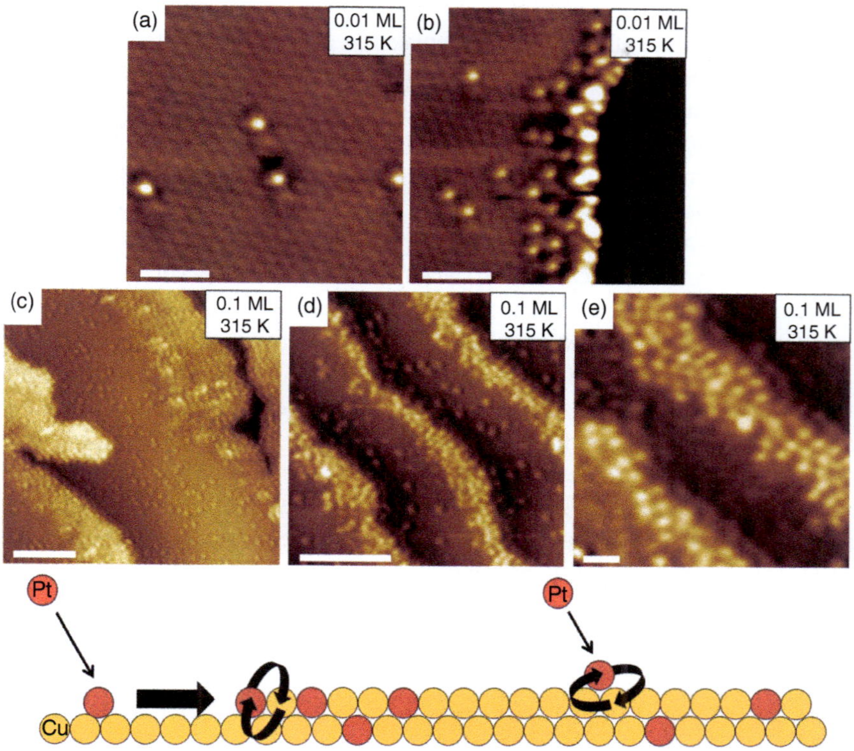

Figure 4.2 STM images of Pt–Cu alloys. 0.01 ML Pt was deposited on Cu (111) (a, b). 0.1 ML Pt was deposited on Cu (111) (c, d, e). Source: Lucci et al. [18]. Reproduced with permission of American Chemical Society.

deposit isolated Pd atoms on the surface of Cu(111), and the Pd–Cu alloy formed as the temperature of the Cu(111) substrate was increased. Pd atoms were trapped on the edge of the Cu(111) surface, and subsequently alloyed with Cu, where Pd was dispersed atomically [14–16]. This catalyst conforms to the definition of diluted SAA catalyst: (i) guest metal was dispersed atomically and (ii) guest metal could sustain its dispersion thermodynamically.

However, when applied under realistic conditions, Pd_{SA} atoms tended to aggregate as NPs on the Cu(111) surface. Researchers proposed several methods to solve this problem, including changing substrate metal, dispersing guest-metal atoms into lattices, preparation methods, etc. [17]. For example, physical vapor deposition was used to prepare Pt–Cu diluted SAAs by heating a Cu(111) substrate from 315 to 550 K [18] as shown in Figure 4.2.

Pt was deposited on the Cu(111) surface by a competing mechanism: Pt could be both directly incorporated into Cu terrace and trapped at Cu edge. This was attributed to the surface free energy difference between Cu (1.566 J/m^2) and Pt (2.204 J/m^2) [19]. Both Pt and Cu have an fcc crystal structure, which leads to the occupation of fcc vacancy sites on Cu(111) by Pt atoms. Sykes and coworkers discovered that Cu(111) substrate temperature is the main factor for regulating the

Pt-exchange mechanism competition. At relatively low temperature (315 K) of the Cu(111) substrate, the alloy formation between Pt and ascending step edges Cu is the dominant process. At a relatively high temperature (450 K) of the Cu(111) substrate, Pt atoms can diffuse into subsurface layers, generating a more homogeneous alloy. There is ~70% Pt on Cu(111) surface at 450 K, while only ~20% at 550 K.

Sykes and coworkers and other researchers reported many studies focusing on diluted SAA model catalysts such as Ni–Au [20], Pt–Cu [21–25], Ni–Cu [26], Pd–Cu [27, 28], and Pd–Au [29, 30]. These studies presented typical platforms for investigating diluted SAA surface chemistry. Density functional theory (DFT) calculations also illustrated that Pd–Cu SAAs would reduce H_2 dissociation energy [31], which supported previous experimental works [10, 18].

Applying diluted SAA catalysts into realistic reaction conditions is an important issue. To achieve this goal, scientists developed several supports such as carbon materials, metal oxides, nitride materials, perovskites, and metal–organic frameworks (MOFs). Metal oxides are the most extensively studied support due to their stability and low prices. After discovering the advantages of Pt–Cu diluted SAA catalysts, Sykes and coworkers immobilized Pt–Cu diluted SAA nanoparticles on γ-Al_2O_3 through galvanic-exchange method [32], which was applied in butadiene hydrogenation [33]. The catalyst was prepared by a modified galvanic-exchange method from Bouche's work [34]. The galvanic replacement reaction consists of a redox process between a metal (employed as a sacrificial template), and metal ions in solution, which is driven by the difference in electrochemical potential between two metals. Once this reaction takes place on the crystal surface, it provides the ability to control nanostructure. Cu NPs were firstly *in situ* synthesized on the γ-Al_2O_3 surface and calcined in air at 350 °C. Pt precursor (H_2PtCl_6) was solubilized in the Cu/γ-Al_2O_3 suspension with the protection of N_2 atmosphere and refluxed at temperature ~100 °C.

The impregnation method can also be adopted to prepare diluted SAA catalysts, as Meyer and coworkers reported [35]. They synthesized Pd–Ag diluted SAA catalysts and evaluated their performances in acrolein hydrogenation. In this work, sequential impregnation was found more favorable for the preparation of diluted SAA catalysts than co-impregnation. The typical synthesis process dissolves the Ag precursor and Pd precursor separately, and Pd was first impregnated on a silica support. Then Ag was impregnated on Pd/SiO_2, and the catalyst powder was reduced at 325 °C for 2 hours.

4.2.2 Characterizations of Diluted Single-Atom Alloy Catalysts

Temperature programmed desorption (TPD) and DFT study was adopted for Pd–Cu diluted SAA catalyst in Skyes and coworker's work. They mutually verified that Pd–Cu diluted SAAs could facilitate H_2 dissociation [10]. Meanwhile, TPD and DFT studies proved the facile removal of absorbed H species from Pd–Cu diluted SAAs. H_2 dissociation on Pd(111) surface was barrierless, readily activating the hydrogen molecules. In comparison, the dissociation of H_2 was impeded on Cu(111) since the barrier was higher (0.4 eV). The use of Pd–Cu diluted SAAs reduced the energy barrier of H_2 dissociation. In contrast, the trend was reversed in the situation of H atoms' binding energy.

The Pt–Cu/γ-Al_2O_3 dilute SAA prepared by Skyes and coworkers was characterized by high angle annular dark field image (HADDF)-spherical aberration corrected transmission electron microscope (STEM) and extended X-ray absorption fine structure (EXAFS) [33]. Pt was proved to be atomically dispersed, and the absence of Pt—Pt bond in EXAFS data consolidated the formation of Pt–Cu diluted SAA.

Meyer et al. also scrutinized diluted SAAs by EXAFS and the fitting results. The as-prepared Pd–Ag alloy was proved as diluted SAA since the Pd—M bond distance matched neither Pd—Pd bond distance nor Ag—Ag bond distance. The coordination number of Pd in diluted single atom was similar to the coordination number of Ag, which also reflected the atomic dispersion of Pd.

4.2.3 Catalytic Performances of Diluted Single-Atom Alloy Catalysts

Diluted SAA catalysts were mainly applied in hydrogenation reactions [35–39], and there are also several reports of diluted SAA-catalyzed dehydrogenation [23, 40] and oxidation [41] reactions. Sykes and coworkers applied Pd–Cu diluted SAA alloy catalyst in selective hydrogenation of styrene and acetylene [10]. It showed that only a little Pd_{SA}, which are dispersed on the Cu(111) surface, could lower the onset temperature of hydrogenation. In comparison, a full layer of Pd on Cu(111) surface would drop the selectivity for ethylbenzene to ~50%. Acetylene hydrogenation also showed the same trend on 0.01 and 1 Pd layer-covered Cu(111) surface.

Pt was considered unsuitable for butadiene hydrogenation to butene due to over hydrogenation, while the activity of butadiene hydrogenation on Cu catalyst was low, which limits its usage in semihydrogenation. Pt–Cu/Al_2O_3 SAA improved butadiene hydrogenation rate about 10 times compared to Cu without the loss of selectivity [33]. Besides, Pt–Cu SAA catalyst possesses superior stability than Cu catalyst. Alkadiene hydrogenation in the presence of excess propylene is an important industrial reaction, and Pt–Cu SAA catalysts exhibited attractive performance in this reaction. Butadiene was selectively hydrogenated by Pt–Cu SAA catalysts without side reaction (propylene hydrogenation). The robustness of Pt–Cu SAA was also proved by its lifetime over 46 hours under relatively harsh reaction conditions (145 and 160 °C).

Meyer and coworkers discovered that atomically dispersed Pd would improve both allyl alcohol selectivity and turnover frequency (TOF) of Pd–Ag alloy catalyst in the selective hydrogenation of acrolein. The better performance of Pd–Ag diluted SAA was described by DFT calculations. The binding energy of H species on Pd–Ag diluted SAA surface is ~0.9 eV lower than that on pure Pd surface, which facilitates H_2 spillover on SAA surface. Additionally, the activation barrier of H_2 dissociation on the Pd–Ag diluted SAA surface is ~0.84 eV lower than that on the surface of Ag. Conclusions from DFT calculations were also consistent with previous reports [10, 42–44]. The high reaction temperature employed in the reaction (200 °C) demonstrates that the coimpregnated Pd–Ag diluted SAA catalyst was very stable. The application of diluted SAA/MO_x catalysts in various realistic reaction conditions was boosted after Meyer and coworker's work [34, 35, 38, 45].

4.3 Single-Atom Doping Alloy Catalysts

4.3.1 Synthesis of Single-Atom Doping Alloy Catalysts

The single-atom doping alloy catalysts represent more precise control on heterogeneous catalyst structure than diluted SAAs since there is only one guest atom in every single-atom doping alloy crystal. This is a challenging object, and it has only been synthesized with few nanocrystals, namely Au and Ag nanoclusters [46, 47] with specific magic number (e.g. 25). Gold is the most famous host metal in single-atom doping alloys. Murray and coworkers reported the first example of single-atom doping alloy, $Au_{24}Pd(SC_2Ph)_{18}$, in 2009 [48]. The synthesis of $Au_{24}Pd$ alloy was based on the report about Brust method [49]. Precursors of Au and Pd ($AuCl_4^-$ and $PdCl_4^{2-}$) were dissolved in a toluene solution of Oct_4NBr and phenylethanethiol ($HSCH_2CH_2Ph$). Then, Au and Pd precursors were reduced by $NaBH_4$ at 0 °C in the solution. The excess ligands were removed by rotovapping and ethanol extraction. The $Au_{24}Pd$ alloy nanocrystals were precipitated by methanol. The number of Pd atoms in Au nanoclusters could be controlled by the Au/Pd precursor molar ratio. $Au_{24}Pd(SC_2Ph)_{18}$ was proved to be generated at two Au/Pd precursor molar ratios, i.e. 9 : 1 and 13 : 12.

Although Murray et al.'s work was impressive, they found it challenging to isolate the $Au_{24}Pd(SC_2Ph)_{18}$ single-atom doping alloy NPs from $Au_{25}(SC_2Ph)_{18}$. Negishi et al. reported a purification method for single-atom doping catalyst. They employed solvent fractionation and high-performance liquid chromatography (HPLC) method and changed dodecanethiolate as the ligand of AuPd single-atom doping alloy [50]. HPLC results showed two metal nanocluster compositions. The peak with a shorter retention time was identified as $Au_{25}(SC_{12}H_{25})_{18}$, and the peak with longer retention time was identified as $Au_{24}Pd(SC_{12}H_{25})_{18}$.

Based on these two studies, many metals such as Cu, Pt, Cd, Hg, and Ag were used as single-atom doping into Au_{25} nanoclusters [51–54]. Au nanoclusters with more atoms could also be doped by single Pd or Ag atoms [55–65]. Later, Jin and coworkers reported a new method of synthesizing $Pt_{SA}Au_{24}$ NPs. They produced a mixture of $Pt_{SA}Au_{24}(SR)_{18}$ and $Au_{25}(SR)_{18}$ via a size-focusing process, and then $Pt_{SA}Au_{24}(SR)_{18}$ NPs were obtained by selective decomposition of $Au_{25}(SR)_{18}$ in the mixture with concentrated H_2O_2 followed by purification via size-exclusion chromatography (Figure 4.3a) [54]. According to this report, the Pt_{SA}-doped Au nanocluster was proved to be more stable than the $Au_{25}(SR)_{18}$ nanocluster.

Yang and coworkers reported the third single-atom dopant element, Hg, for Au_{25} nanocrystals [66]. They firstly synthesized $[Au_{25}(\text{phenylethanethiolate})_{18}]^-$ according to an earlier report [67]. Then an aqueous solution of $Hg(NO_3)_2$ and phenylethanethiolate/CH_3CN solution were added into the $[Au_{25}(\text{phenylethanethiolate})_{18}]^-$ solution. The formed precipitate was washed by MeOH and redispersed in toluene/acetonitrile (1 : 4, v/v) mixture. There is no separation step in synthesis, suggesting that this method had high selectivity for Hg single-atom-doped Au nanoclusters.

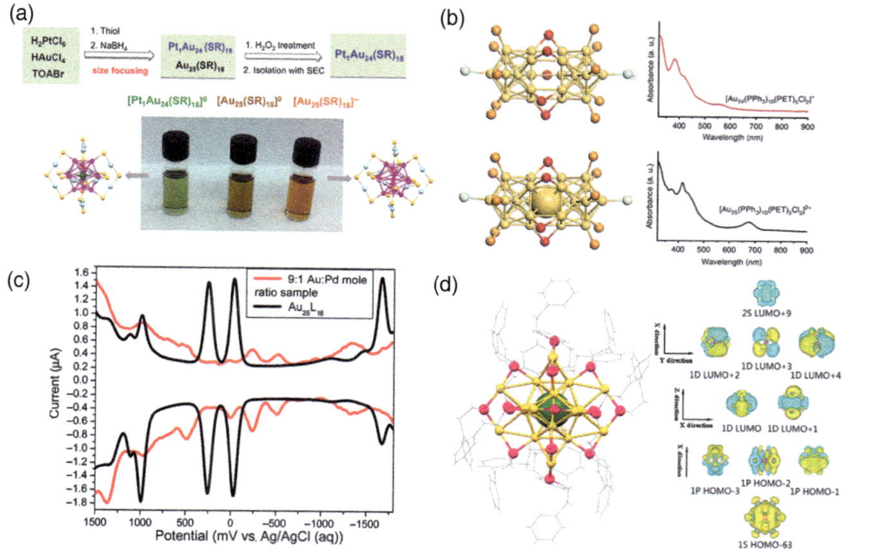

Figure 4.3 (a) Procedure for $Pt_{SA}Au_{24}(SR)_{18}$ synthesis. Source: Qian et al. [54]. Reproduced with permission of American Chemical Society. (b) X-ray structures and UV–vis spectra of Au_{24} and Au_{25} nanoclusters. Source: Wang et al. [46]. Springer Nature. CC BY 4.0. (c) Voltammograms of monometal $Au_{25}(SC_2Ph)_{18}$ (red curve) and of an extensively purified NP sample prepared with a 9 : 1 Au:Pd mole ratio (black curve). Source: Fields-Zinna et al. [48]. Reproduced with permission of American Chemical Society. (d) Left: total structure of $Cd_{SA}Au_{24}(PhC_2H_4S)_{18}^0$ NCs. Right: superatomic orbital of $[Cd_{SA}Au_{12}]^{6+}$. Source: Wang et al. [52]. Reproduced with permission of American Chemical Society.

Zhu and coworkers reported the fourth single-atom dopant element, Cd, for Au_{25} nanocrystals [52]. The $Cd_{SA}Au_{24}$ nanoclusters were prepared by a metal-exchange process. Cd thiolate complex was pre-synthesized and reacted with $Au_{25}(SR)_{18}^-$ at room temperature. Meanwhile, Pt, Pd, and Ni elements could not be doped into the Au_{25} nanocrystal. Additionally, the number of Hg, Ag, and dopant atoms could not be controlled by the metal-exchange process. Only Cd atoms were doped into Au nanoclusters with a determined number (only one atom dopant).

Cu and Ag are more challenging dopants compared to Pd, Pt, Hg, and Cd because of their similarity to Au atoms [68–72]. Recently, Jin and coworkers reported an atom shuttle process for synthesizing $Cu_{SA}Au_{24}$ and $Ag_{SA}Au_{24}$ single-atom doping catalysts. This method took advantage of the stable central hollow $[Au_{24}(PPh_3)_{10}(SC_2H_4Ph)_5Cl_2]^+$ nanocrystal displayed in Figure 4.3b [73, 74]. The aqueous solution of Au was added to tetraoctylammonium bromide/toluene solution. After the aqueous layer was removed, the PPh_3 and $NaBH_4$ solution was rapidly added into the Au precursor suspension. Excess amounts of phenylethylthiol were introduced into the Au nanocluster suspension as the second ligand. Figure 4.3b illustrates that Au atom could readily fill the vacancy in Au_{24} nanocrystal [74]. The Au vacancy in $[Au_{24}(PPh_3)_{10}(SC_2H_4Ph)_5Cl_2]^+$ nanoparticle could also be filled by other elements such as Cu and Ag.

4.3.2 Characterizations of Single-Atom Doping Catalysts

Murray and coworkers [48] reported that the electrochemical and optical properties of $Au_{25}(SC_2Ph)_{18}$ could be altered by Pd atom doping. The highest occupied molecular orbital (HOMO)–lowest unoccupied molecular orbital (LUMO) energy gap of $Au_{24}Pd(SC_2Ph)_{18}$ was measured by voltammetric reaction pattern (~0.5 eV), which is lower than that of the Au_{25} nanocluster as shown in Figure 4.3c.

Negishi et al. reported that HPLC could be used to separate $Pd_{SA}Au_{24}$ nanoclusters from Au_{25} nanoclusters [50]. Negative-ion matrix assisted laser desorption ionization (MALDI) mass spectra could clearly discriminate $Au_{25}(SC_{12}H_{25})_{18}$ clusters from $Pd_{SA}Au_{24}(SC_{12}H_{25})_{18}$ clusters. The $Pd_{SA}Au_{24}(SC_{12}H_{25})_{18}$ cluster was proved to be neutral since it has a shorter retention time. Negishi et al. also simulated the structure of $Pd_{SA}Au_{24}(SC_{12}H_{25})_{18}$ by DFT calculations and pinpointed the location of Pd single atom. It showed that the Pd atom occupied the central site in $Pd_{SA}Au_{24}$ nanocluster.

Jin and coworkers adopted high-precision electrospray ionization (ESI) MS to investigate $Pt_{SA}Au_{24}$ nanoclusters prepared by the H_2O_2 decomposition method [54]. They found out that both types of ligands bonded to Au atoms instead of Pt atom, proving that the Pt_{SA} dopant is located at the center of the $Pt_{SA}Au_{24}$ nanocluster. DFT calculations also supported the conclusion of ESI-MS. Although both Pt and Pd single-atom dopants tended to occupy the center of the Au nanocluster [50, 70, 75], $Pt_{SA}Au_{24}$ and $Pd_{SA}Au_{24}$ exhibited a different peak in the NIR region [76, 77]. This suggests that the electronic structure of a single-atom doping alloy could be altered dramatically by even one single-atom dopant [71].

Yang and coworkers further studied UV–vis–NIR spectra of $[Hg_{SA}Au_{24}(PET)_{18}]^-$, $[Au_{25}(PET)_{18}]^-$ and $[Au_{25}(PET)_{18}]^0$ nanoclusters [66]. $[Hg_{SA}Au_{24}(PET)_{18}]^-$ exhibited two bands at ~700 and 405 nm, and $[Au_{25}(PET)_{18}]^-$ and $[Au_{25}(PET)_{18}]^0$ nanoclusters possessed another band located at ~450 nm. The charge state of $Hg_{SA}Au_{24}(PET)_{18}$ nanoclusters was determined as neutral according to X-ray photoelectron spectroscopy (XPS) characterization. DFT calculations pointed out that the Hg_{SA} was located at the outer shell of the $Hg_{SA}Au_{24}(PET)_{18}$ nanocluster, which was different from the location of Pt and Pd [48, 50, 68, 78–80]. Since there was no reducing agent involved in the preparation, Hg atom was reduced through the galvanic exchange. They found out that Au_{25} nanocluster has stronger reducing capability than bulk Au, in accordance with other reports [81–84]. The authors also studied $Hg_{SA}Au_{24}(PET)_{18}$ nanocluster by differential pulse voltammetry (DPV) [85–89]. HOMO–LUMO energy gaps of $Au_{25}(SC_6H_{13})_{18}$, $Au_{25}(PET)_{18}$, $Au_{25}(SC_6H_5)_{18}$, and $Hg_{SA}Au_{24}(PET)_{18}$ could be calculated from the gaps between oxidation and reduction peaks.

Zhu and coworkers studied $Cd_{SA}Au_{24}$ nanocrystal by X-ray powder diffraction (XRD), and the results are displayed in Figure 4.3d [52]. The Cd_{SA} was located at the center of $Cd_{SA}Au_{24}$ nanocrystal, whose location was similar to Pt and Pd. Bader charge analysis based on DFT calculations showed that the electrons on the Cd_{SA} dopant are transferred to Au atoms, making Cd_{SA} positively charged. Walter et al. proposed a rule for counting the charge state of nanocrystals in their work [90].

The authors hypothesized that d^{10}–d^{10} interactions between Cd and Au atoms maintained the structure of the $Cd_{SA}Au_{24}$ nanocrystal. This report validated the feasibility of the metal ion–exchange method, which bypassed the limitation of metal reduction activity in the galvanic-exchange method [48, 50, 52].

UV–vis spectrometry was also applied for $Cd_{SA}Au_{24}(SC_2H_4Ph)_{18}$ nanocluster characterization [53]. Three peaks (400, 480, and 680 nm) were observed in the spectrum of $Au_{25}(SC_2H_4Ph)_{18}$. The peak at 680 nm shifted to 650 nm in the spectrum of $Cd_{SA}Au_{24}(SC_2H_4Ph)_{18}$ nanocluster, suggesting that Cd dopant changed the HOMO–LUMO band gap. Deng et al. studied nanocrystal stability by photon energy spectra. $Cd_{SA}Au_{24}(SC_2H_4Ph)_{18}$ did not decompose in 98 hours under oxygen atmosphere. In comparison, $Au_{25}(SC_2H_4Ph)_{18}$ was thoroughly decomposed within 98 hours.

4.3.3 Catalytic Applications of Single-Atom Doping Alloys

Tsukuda and coworkers reported the first example of a single-atom doping alloy catalyst. The catalytic performance of $Pd_{SA}Au_{24}$/carbon nanotubes (CNTs) was evaluated for benzyl alcohol aerobic oxidation [51]. Benzyl alcohol oxidation conversion on $Pd_{SA}Au_{24}$ nanocrystals was about three times higher than the value on Au_{25} nanocrystals. This was ascribed to electronic change from Pd_{SA} dopant since Pd is located at the center of $Pd_{SA}Au_{24}$ nanocrystal [50]. Pd could not interact with substrate directly, it can only impose an influence on the electronic structure of the $Pd_{SA}Au_{24}$ nanocrystals.

Single-atom doping alloy catalysts could also be immobilized on metal oxide supports. Jin and coworkers reported a $Pt_{SA}Au_{24}/TiO_2$ catalyst for styrene oxidation [54]. $Pt_{SA}Au_{24}$ single-atom doping alloy catalyst had about twice higher conversion than the Au_{25} nanocrystal catalyst. The selectivity for benzaldehyde was also improved on the $Pt_{SA}Au_{24}/TiO_2$ catalyst. These experiments demonstrated that even Pt_{SA} dopant could dramatically change the catalytic performance of Au nanocrystal.

4.4 Diatomic Alloy Catalysts

4.4.1 Synthesis of Diatomic Alloy Catalysts

There is only one kind of metal that is atomically dispersed in both diluted SAA and single-atom doping alloys. In comparison, the development of diatomic alloys is dedicated to disperse two kinds of metal as single atoms, which may lead to exotic synergistic effects. Diatomic alloys represent the highest level of manipulating atoms at the microscopic scale; thus, corresponding research also faced great challenges from synthesis to characterization. In this section, we will briefly introduce the synthesis method of diatomic alloys. There are two main problems in the synthesis process of diatomic alloys: (i) preventing the formation of mono metal dimers and (ii) ensuring atomic dispersion of both metals. To overcome these issues, researchers realize that support with proper and continuous binding sites for single-metal atoms plays the most important role in diatomic alloy synthesis. Nitrogen-doped carbon supports such as CNTs, g-C_3N_4, and MOFs are the most prevailing supports for diatomic alloys [11, 12, 91–94]. N dopants and neighboring defects on carbon could stabilize single atom of different metals. Zhou et al. proved that Pt–Ru diatomic sites surrounded by C and N atoms are more stable than isolated Pt–Ru diatomic alloy [11]. There are several methods for depositing metal atoms on the surface of the support. It is also noteworthy that some diatomic alloy catalysts have bridge atoms, N or O, in their structures, which verifies again that C, N, and O atoms contribute to the formation of diatomic alloy catalysts [13, 95–97]. Li and coworkers reported a host–guest strategy to synthesize Fe–Co diatomic alloy catalysts as shown in Figure 4.4 [94]. Fe precursors were embedded in a Zn/Co bimetallic MOF (ZIF-8). Then, the MOF was pyrolyzed at ~1173 K. During this process, Fe^{3+} and Co^{2+} were reduced by ZIF-8 support to form Fe–Co diatomic alloy, and Zn was vaporized due to its low boiling point. Yao and coworkers reported a two-step synthesis method for rod-like Co–Pt diatomic alloy fabrication [98]. The first step is simultaneous nitrogen doping and carbonization of Co-MOF. The Co-MOF was mixed with dicyandiamide

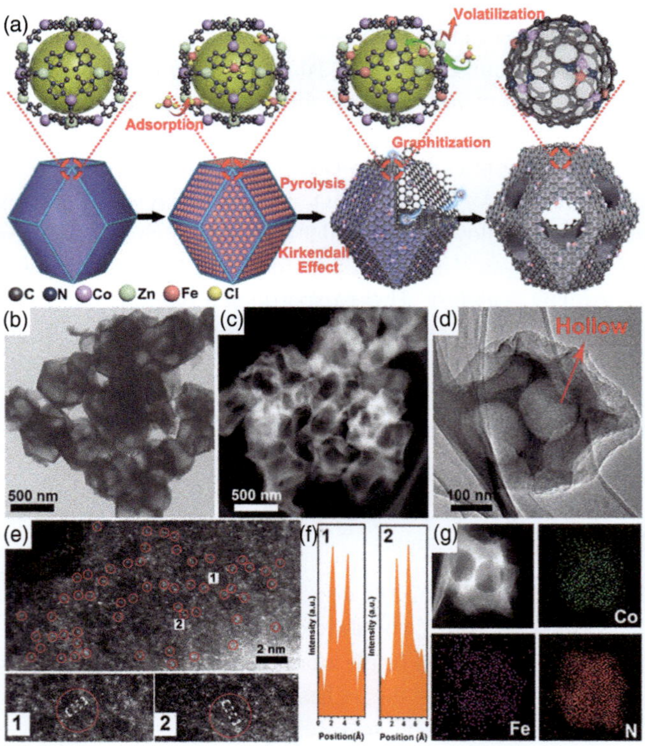

Figure 4.4 (a) Preparation of Fe–Co diatomic alloy. (b, c, d, e) TEM images of Fe–Co diatomic alloy. (f) Corresponding intensity profiles obtained on the zoomed areas in (e). (g) Electron energy loss spectroscopy mapping of Co, Fe, and N. Source: Wang et al. [94]. Reproduced with permission of American Chemical Society.

and annealed at 850 °C under N_2. Then, the Co-MOF was treated with an electrochemical cyclic potential generated between a glassy carbon electrode and a Pt wire counter electrode. Yu and coworkers proposed a precise single-atom to single-atom grafting method to synthesize Pt–Fe diatomic alloy [95]. The method is based on incipient wetness impregnation of the H_2PtCl_6 Pt precursor. Pt atoms were immobilized onto Fe–N–C support by stirring a diluted H_2PtCl_6 water solution at 70 °C. Then the powder was calcinated at 450 °C for 2 hours in Ar. This method generates Pt_1–O_2–Fe_1–N_4 moiety where the bridging O atoms and anchoring N atoms stabilizing this moiety. Hu and coworkers also synthesized an O bridging Co–Fe diatomic alloy [96]. They used phenanthroline as the ligand of Co ions and MgO as the support for Co-phenanthroline. The MgO was then leached by HNO_3. Fe was incorporated onto the Co–N–C support by an electrochemical activation process. Hu and coworkers also employed pyrolysis method to produce a Co–Ni diatomic alloy [99]. They firstly synthesized a Co–Ni bimetallic MOF and coated it with polydopamine. Then, the nanocrystals were annealed at 500 °C in NH_3 to form the Co–Ni diatomic alloy. Zhao and coworkers developed an ion-exchange strategy to generate a Ni–Fe diatomic alloy on nitrogen-doped carbon support [12]. Fe was chemically bonded on ZIF-8 by the ion-exchange method. Then, a Ni precursor was encapsulated and reduced within Fe-ZIF-8 cavities to form the Ni–Fe diatomic alloy. Sun and coworkers adopted atomic layer deposition (ALD) to immobilize Pt and Ru sequentially on N-doped CNTs from organometallic precursors [91]. The ALD method provides precise control of atom positions in diatomic alloy. They performed two-step ALD to successively deposit Pt atoms and Ru atoms on N-doped CNTs. The Ru atoms were selectively deposited on the Pt atoms due to the closer affinity between Pt and Ru than the affinity between Ru- and N-doped CNTs. Guo and coworkers reported a light irradiation synthesis method for Pt–Ru diatomic alloy on N-vacancy-riched g-C_3N_4 [11]. Pt and Ru precursors (H_2PtCl_6 and $RuCl_3$) were mixed in a suspension of N-vacancy-riched g-C_3N_4 and then irradiated by a 300 W Xe lamp with a light filter of 420 nm. Pt and Ru were reduced by the support and formed the Pt–Ru diatomic alloy. Yang et al. developed a controllable self-assembly and nitridation process for W–Mo diatomic alloy [97]. W and Mo precursors ($Na_2WO_4 \cdot 2H_2O$ and $(NH_4)_6Mo_7O_{24} \cdot 4H_2O$) were mixed with the suspension of graphene oxide by sonicating. Then the mixture was treated by a hydrothermal process during which the W–Mo pairs formed on graphene oxide. Finally, the catalyst was produced by a chemical vapor deposition (CVD) treatment in NH_3/Ar gas at 800 °C.

4.4.2 Characterizations of Diatomic Alloy Catalysts

The ultimate goal of diatomic alloy characterization is to prove the atomic dispersion of two metals as well as the formation of dimers. There are three major characterization techniques, i.e. transmission electron microscopy (TEM), X-ray absorption spectroscopy (XAS), and DFT calculations. As shown in Figure 4.4, high-resolution TEM could tell us the distribution of metal atoms on the support surface. Li and coworkers shown that bright dots (the metal atoms) were homogeneously dispersed, and metal atoms formed diatomic pairs on the support surface. Yao and coworkers discovered that metal atoms were mainly dispersed at defect sites of the support. The metal atom distance in the Co_{SA}–Pt_{SA} diatomic alloy was about 0.34 nm [92]. However, it is very difficult to discriminate heteroatoms through TEM. It can only prove the formation of dimers. Electron energy-loss spectroscopy (EELS) may provide an auxiliary proof of the coexistence of dual metals. But the resolution of EELS could not support the observation on a sub-angstrom scale. Overall, TEM itself could not provide solid evidence of diatomic alloy formation.

The EXAFS is a versatile characterization method for diatomic alloy. It can figure out coordination number, neighboring atom, and bond distances in the alloy, making this technique very complementary to TEM and EELS. Lin and coworkers studied novel electrocatalyst with Fe–Co dual sites embedded in N-doped CNTs, and found out that the Fe K-edge X-ray absorption near edge structure (XANES) signal is more similar to FePc compared to Fe_2O_3 and Fe foil [100]. The Fe K-edge EXAFS showed that neighboring atoms of Fe are mainly N and Co, suggesting the formation of Fe–Co diatomic alloy. The fitting result of Fe K-edge EXAFS demonstrated that Fe—Co bond distance was 2.22 Å. Furthermore, the absence of Fe—Fe bond and Co—Co bond proved that there were no Fe and Co clusters on the catalyst surface. Thus, the formation of Fe–Co diatomic alloy was confirmed. Yu and coworkers studied the anchoring of Fe and Pt atoms on the surface of a nitrogen-doped carbon support, which stabilized diatomic alloy [95]. Fourier-transformed EXAFS spectra showed the existence of Fe–N_4 moieties in the catalyst. Besides, EXAFS verified the existence of bridging -O–O- structure in the Pt–Fe diatomic alloy, which enlightened the investigation on bridging atoms in diatomic alloy and their function. Zhao and coworkers also applied EXAFS to analyze the Ni–Fe diatomic alloy [12]. The Ni—Fe bond distance was determined as 0.206 nm, which is a distinct signal in Ni–Fe diatomic alloy sample. There was no such peak in pure Ni and Fe samples, suggesting the formation of Ni–Fe diatomic alloy.

Since the diatomic alloy represents a smaller scale, i.e. atomic level, compared to SAAs, the great difficulty in diatomic alloy characterization rendered DFT an important tool. Guo and coworkers performed a series of DFT calculations that are related to Pt–Ru diatomic alloys [11]. They studied the stability of the Pt–Ru diatomic alloy with various coordination environments, and single-atom Pt, Ru with the same coordination environments. C–Pt–Ru–N exhibited the highest stability among all structures that are investigated, which pointed out that C, N coordination is necessary to the formation of diatomic alloys. Partial density of state (PDOS) analysis also showed that Pt was the electron-rich site owing to the electron transfer within the Pt–Ru diatomic alloy.

4.4.3 Catalytic Applications of Diatomic Alloys

The first example of diatomic alloy catalyst use was for the oxygen reduction reaction (ORR) [94]. An Fe_{SA}–Co_{SA}/N–C diatomic alloy catalyst possessed good ORR activity with $E_{1/2}$ = 0.863 V and E_{onset} = 1.06 V, while Fe_{SA}/N–C and Co_{SA}/N–C had ~0.1 V lower $E_{1/2}$ potential. The ORR activity of Fe_{SA}–Co_{SA}/N–C diatomic alloy catalyst was not only better than most transition metal catalysts, but also comparable to the Pt/C catalyst. Zhang and coworkers further reported Pt_{SA}–Co_{SA}/N–C catalyzed ORR with $E_{1/2}$ = 0.96 V, which is slightly higher than the Fe_{SA}–Co_{SA}/N–C catalyst [98]. There are also various diatomic alloy catalysts that were reported as good catalysts for the ORR. For example, Fe_{SA}–Co_{SA}/CNT with $E_{1/2}$ = 0.945 V [100], Ni_{SA}–Co_{SA}/N–C with $E_{1/2}$ = 0.76 V [99].

Diatomic alloy catalysts were also applied for the water-splitting reaction recently. Zhang and coworkers reported that a Ru_{SA}–Pt_{SA}/N-CNT catalyst could achieve a mass activity of 23.1 A/mg at 0.05 V overpotential [91], while that of Pt_{SA}/N-CNT was 10 A/mg and Pt/C was 0.43 A/mg. In 2019, Zhang and coworkers reported that Pt_{SA}–Co_{SA}/N–C showed better catalytic performance in hydrogen evolution reaction (HER) compared to Pt/C and Co/C catalysts. Pt_{SA}–Co_{SA}/N–C exhibited better catalytic performance in a wide pH range. In acidic solution, Pt_{SA}–Co_{SA}/N–C had an overpotential of 27 mV at 10 mA/cm^2, which was ~30 mV lower than the overpotential for Pt/C. In alkaline solution, Pt_{SA}–Co_{SA}/N–C had an overpotential of ~57 mV at 10 mA/cm^2, which was ~10 mV lower than the overpotential for Pt/C.

4.5 Machine Learning-Guided Single-Atom Alloy Catalyst Design

Recently, theoretical chemists have been paying more attention to SAAs as they have become a hot topic in heterogeneous catalysts. It has been a long tradition that theoretical chemists study SAAs with first-principles calculations such as DFT, as we introduced above. However, first-principles calculations are time consuming and computationally expensive, making high-throughput screening of SAAs a challenging task. In 2020, Rajan and coworkers firstly proposed that this problem could be overcome by employing machine-learning algorithms [101]. The basic assumption of machine-learning-assisted first-principles investigation on SAAs is that there is a relationship between SAA structures and the activation barriers of substrates, which absorbed on their surfaces. Once the relationship is pinpointed, chemists are able to predict characteristics of a large variety of SAAs without experiments and first-principles calculations. This relationship is hidden behind high-dimensional data, which happens to be the specialty of machine-learning algorithms. So far, machine-learning studies on SAAs are focused on two aspects, i.e. SAA stability [102–105] and SAA catalytic performances on certain reactions [101, 104, 106]. As shown in Figure 4.5a, Chen and coworkers outlined a typical process of machine-learning-assisted first-principles investigation on SAAs [106]. Briefly, it can be divided into four parts: constructing dataset, descriptor engineering, machine-learning model fitting and generalization, and predicting SAA characteristics. The first step of machine learning is constructing the dataset for machine learning by DFT calculations. For example, Singh and coworkers defined and calculated the aggregation energy of SAA in presence and absence of O* adatoms [102]. Gao and coworkers calculated H-binding energy, H_2 dissociation energy barrier on SAA surfaces, and aggregation energy of SAAs [104]. The second step of machine learning is descriptor engineering. Researchers proposed several

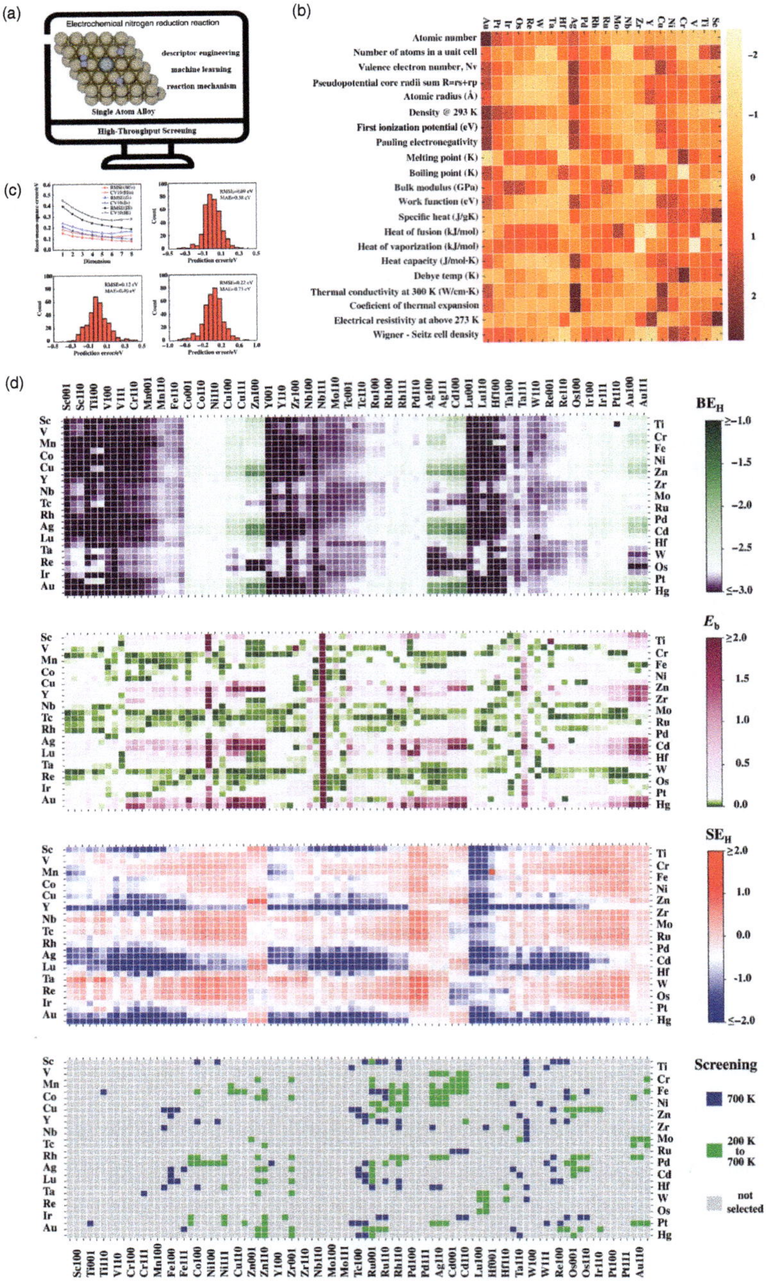

Figure 4.5 (a) Schematic illustration of machine-learning guided SAA design and development. Source: Zheng et al. [106]. Reproduced with permission of American Chemical Society. (b) An example of SAA surfaces descriptors. Source: Dasgupta et al. [101]. Reproduced with permission of American Chemical Society. (c) Top-left: Root mean square error (RMSE) and averaged RMSE of the 10-fold cross-validation. Top-right, bottom-left, bottom-right: Prediction errors vs. DFT calculations for binding energies of H (BE_H), the dissociation barriers of H_2 (E_b), and segregation energies (SE). (d) Results of high-throughput screening of SAAs. Source: Han et al. [104]. Springer Nature. CC BY 4.0.

main factors to describe the structures of SAAs since raw data of first-principles investigation on SAAs could not be directly interpreted by machine-learning algorithms, as shown in Figure 4.5b. There are many kinds of descriptor engineering methodology, for example, Gao and coworkers employed a recently developed sure independence screening and sparsifying operator (SISSO) to extract descriptors from SAA structures [104]. The third step of machine learning is model fitting and generalization. Theoretical chemists tend to apply existing machine-learning models to SAAs such as the Gaussian process regression [101, 102], the support vector machine [102, 103], decision tree [103], random forest [106], and neural network [103]. In this step, researchers should carefully avoid underfitting and overfitting of machine-learning model. The dataset is split into training sets and test sets, and usually the cross-validation is performed to test the fitting status of machine-learning model. Root mean square error of training set and test set is the most important criterion to judge the quality of fitting. Once the machine-learning model exhibits similar root mean square error in training set and test set as shown in Figure 4.5c, it is generalized to a broader scope as shown in Figure 4.5d. The final step of machine learning is predicting SAA characteristics with well-trained machine learning model. For example, Rajan and coworkers predicted that VCr and VZr are efficient catalysts for activating H_2, VW and AgAu are efficient catalysts for activating CH_4, Sc–Ti and V–Au are efficient catalysts for activating CH_3OH, and CuAg and NiAg are efficient catalysts for activating CO_2 [101]. Gao and coworkers reported that the presence of H adatoms on Pd/Ag(111) and Pt/Ag(111) SAAs could reverse the segregation direction of Pd and Pt guest atoms [104].

4.6 Perspectives

Diluted SAAs are considered as the bridge between model catalysts and catalysts applied under realistic reaction conditions. They often exhibit better catalytic performance than the prediction of Sabatier principle. Current research in the field of a diluted single atom is also limited, and breakthroughs can be expected in the following aspects: (i) expanding the applications of diluted SAAs. Diluted SAA catalysts are mainly studied for hydrogenation probe reactions, while there is a report suggesting that they may be suitable catalysts for dehydrogenation [23]. (ii) Discovering more characterization methods for diluted SAAs. The most convincing evidence of diluted SAA is that the guest-metal atoms only have host-metal atoms in its coordination environment. However, this conclusion could only be obtained from EXAFS fitting. This characterization is not available for most research teams, and it could not satisfy all diluted SAA compositions. For example, it cannot discriminate neighboring elements of the periodic table of elements. (iii) Developing a scale-up synthesis method for diluted SAAs. For example, decreasing the loading amount of guest metal and employing specific support with strong interaction with metal atoms could make impregnation a suitable method for synthesizing diluted alloys.

Single-atom doping alloy is an ideal platform for studying the ligand effect and coordination effect of single atoms. There are several unsolved problems in the research of single-atom doping alloys. (i) The galvanic-exchange method in single-atom doping alloys is not fully demonstrated. The exchange sequence of metal does not conform to the classical metal activity sequence. The reason for this anti-galvanic-exchange phenomenon was not studied and explained. (ii) Many potential dopant elements cannot be incorporated into Au and Ag nanoclusters. (iii) The location of single-atom dopant varies from center site to outer sphere site. It is desirable to figure out the key factors for controlling the location of dopant, which may require *in situ* characterization of the single-atom doping alloy formation. (iv) Current examples of SAAs are limited to Au and Ag clusters. Expanding to other kinds of host-metal nanocrystals for single-atom doping would generate more knowledge of the ligand effect and coordination effect of single-atom dopants.

Diatomic alloy catalysts constitute a recently emerging family of heterogeneous catalysts. Both host- and guest-metals are dispersed as single atoms in diatomic alloy catalysts, which could further improve the catalytic performances. Since diatomic alloy catalysts are in their infancy of development, corresponding knowledge is still limited. In our opinion, further studies may focus on the following aspects: (i) Developing suitable supports for the dimer formation. Such kind of support should possess strong adsorption sites for single metal atoms to prevent their agglomeration. Besides, these adsorption sites should exist as pairs to favor the formation of alloy dimers. (ii) Exploiting diatomic alloy catalyst and corresponding support through machine-learning algorithms. Developing diatomic alloy is a time-consuming and resource-consuming process if mainly based on experience, since there is no accumulation of experience for this kind of recently designed catalysts. Machine learning would accelerate this process by predicting strong adsorption sites on various supports, and highly active diatomic alloy sites. (iii) Broadening the application of diatomic alloy catalysts. Diatomic alloy catalysts are mainly applied for electrocatalytic water-splitting and oxygen reduction reactions. The applications of single-atom doping alloy catalysts and diluted SAA catalysts suggest that diatomic alloy catalysts may also exhibit good performances in hydrogenation/dehydrogenation reactions and many other reactions.

Combining machine-learning algorithms with first-principles investigation on SAAs is a promising and effective way to improve their catalytic performances and stability. However, there are still several questions to be addressed in future study: (i) Validating machine-learning prediction with experiments. Current publications about machine-learning-assisted first-principles investigation on SAAs lack solid experimental evidence to support their predictions. We expect that there will be closer collaboration between theoretical chemists and experimental chemists in SAA development in the future. (ii) Extending dataset for machine learning to experimentally collected data. Present dataset is completely based on DFT calculations, since DFT calculations are performed in uniform conditions and they are easier to be obtained than experimental data. However, DFT calculations could not simulate the entire reaction environment, which results in a systematic deviation of machine-learning prediction based on DFT calculations dataset. Meanwhile, recent breakthrough of robotic chemists [107] makes it more realistic to construct machine-learning dataset with experimentally collected data, which may prevent the systematic deviation from DFT calculation. (iii) Exploiting SAAs with interpretable machine-learning models. Although scientists are concerned about fitting and generalization performance of machine learning, they are more interested in the scientific law lying behind machine-learning models. However, some machine-learning models used in publications are black-box models. It is unknown to what extent the predictions based on black-box machine learning models are reliable in various SAA compositions and catalytic reactions combinations. Additionally, even well-trained black-box machine learning models could not throw a hint about the mechanism under observations. On the other hand, interpretable machine-learning models may provide us suggestions about the hidden laws of nature. Thus, we look forward to more applications of interpretable machine-learning models in SAA catalysts development in the future.

References

1 Liu, C., Yang, B., Tyo, E. et al. (2015). Carbon dioxide conversion to methanol over size-selected Cu_4 clusters at low pressures. *Journal of the American Chemical Society* 137 (27): 8676–8679.

2 Liu, Y., Deng, D., and Bao, X. (2020). Catalysis for selected C_1 chemistry. *Chem* 6 (10): 2497–2514.

3 Zhong, M., Tran, K., Min, Y. et al. (2020). Accelerated discovery of CO_2 electrocatalysts using active machine learning. *Nature* 581 (7807): 178–183.

4 Yao, Y.G., Huang, Z.N., Xie, P.F. et al. (2018). Carbothermal shock synthesis of high-entropy-alloy nanoparticles. *Science* 359 (6383): 1489–1494.

5 Xin, Y., Li, S.H., Qian, Y.Y. et al. (2020). High-entropy alloys as a platform for catalysis: progress, challenges, and opportunities. *ACS Catalysis* 10 (19): 11280–11306.

6 Wu, Z., Bukowski, B.C., Li, Z. et al. (2018). Changes in catalytic and adsorptive properties of 2 nm Pt_3Mn nanoparticles by subsurface atoms. *Journal of the American Chemical Society* 140 (44): 14870–14877.

7 Cai, W.T., Mu, R.T., Zha, S.J. et al. (2018). Subsurface catalysis-mediated selectivity of dehydrogenation reaction. *Science Advances* 4 (8): eaar5418–eaar5426.

8 Bligaard, T. and Nørskov, J.K. (2007). Ligand effects in heterogeneous catalysis and electrochemistry. *Electrochimica Acta* 52 (18): 5512–5516.

9 Liu, P., Qin, R., Fu, G. et al. (2017). Surface coordination chemistry of metal nanomaterials. *Journal of the American Chemical Society* 139 (6): 2122–2131.

10 Kyriakou, G., Boucher, M.B., Jewell, A.D. et al. (2012). Isolated metal atom geometries as a strategy for selective heterogeneous hydrogenations. *Science* 335 (6073): 1209–1212.

11 Zhou, P., Hou, X., Chao, Y. et al. (2019). Synergetic interaction between neighboring platinum and ruthenium monomers boosts CO oxidation. *Chemical Science* 10 (23): 5898–5905.

12 Ren, W., Tan, X., Yang, W. et al. (2019). Isolated diatomic Ni-Fe metal–nitrogen sites for synergistic electroreduction of CO_2. *Angewandte Chemie International Edition* 58 (21): 6972–6976.

13 Shi, Q., Ji, Y.J., Chen, W.X. et al. (2020). Single-atom Sn-Zn pairs in CuO catalyst promote dimethyldichlorosilane synthesis. *National Science Review* 7 (3): 600–608.

14 Tierney, H.L., Baber, A.E., Kitchin, J.R. et al. (2009). Hydrogen dissociation and spillover on individual isolated palladium atoms. *Physical Review Letters* 103 (24): 246102–246106.

15 Bellisario, D.O., Han, J.W., Tierney, H.L. et al. (2009). Importance of kinetics in surface alloying: a comparison of the diffusion pathways of Pd and Ag atoms on Cu(111). *Journal of Physical Chemistry C* 113 (29): 12863–12869.

16 Tierney, H.L., Baber, A.E., and Sykes, E.C.H. (2009). Atomic-scale imaging and electronic structure determination of catalytic sites on Pd/Cu near surface alloys. *Journal of Physical Chemistry C* 113 (17): 7246–7250.

17 Zhang, H.J., Watanabe, T., Okumura, M. et al. (2012). Catalytically highly active top gold atom on palladium nanocluster. *Nature Materials* 11 (1): 49–52.

18 Lucci, F.R., Lawton, T.J., Pronschinske, A. et al. (2014). Atomic scale surface structure of Pt/Cu(111) surface alloys. *Journal of Physical Chemistry C* 118 (6): 3015–3022.

19 Tyson, W.R. and Miller, W.A. (1977). Surface free-energies of solid metals - estimation from liquid surface-tension measurements. *Surface Science* 62 (1): 267–276.

20 Wang, Z.T., Darby, M.T., Therrien, A.J. et al. (2016). Preparation, structure, and surface chemistry of Ni-Au single atom alloys. *Journal of Physical Chemistry C* 120 (25): 13574–13580.

21 Marcinkowski, M.D., Liu, J.L., Murphy, C.J. et al. (2017). Selective formic acid dehydrogenation on Pt-Cu single-atom alloys. *ACS Catalysis* 7 (1): 413–420.

22 Liu, J.L., Lucci, F.R., Yang, M. et al. (2016). Tackling CO poisoning with single-atom alloy catalysts. *Journal of the American Chemical Society* 138 (20): 6396–6399.

23 Marcinkowski, M.D., Darby, M.T., Liu, J.L. et al. (2018). Pt/Cu single-atom alloys as coke-resistant catalysts for efficient C-H activation. *Nature Chemistry* 10 (3): 325–332.

24 Shan, J.J., Lucci, F.R., Liu, J.L. et al. (2016). Water co-catalyzed selective dehydrogenation of methanol to formaldehyde and hydrogen. *Surface Science* 650 (650): 121–129.

25 Wang, Z.T., Hoyt, R.A., El-Soda, M. et al. (2018). Dry dehydrogenation of ethanol on Pt-Cu single atom alloys. *Topics in Catalysis* 61 (5–6): 328–335.

26 Shan, J.J., Liu, J.L., Li, M.W. et al. (2018). NiCu single atom alloys catalyze the C—H bond activation in the selective non- oxidative ethanol dehydrogenation reaction. *Applied Catalysis B: Environmental* 226 (226): 534–543.

27 Boucher, M.B., Marcinkowski, M.D., Liriano, M.L. et al. (2013). Molecular-scale perspective of water-catalyzed methanol dehydrogenation to formaldehyde. *ACS Nano* 7 (7): 6181–6187.

28 Kruppe, C.M., Krooswyk, J.D., and Trenary, M. (2017). Selective hydrogenation of acetylene to ethylene in the presence of a carbonaceous surface layer on a Pd/Cu(111) single-atom alloy. *ACS Catalysis* 7 (12): 8042–8049.

29 Lucci, F.R., Darby, M.T., Mattera, M.F.G. et al. (2016). Controlling hydrogen activation, spillover, and desorption with Pd-Au single-atom alloys. *Journal of Physical Chemistry Letters* 7 (3): 480–485.

30 Han, S. and Mullins, C.B. (2018). Surface alloy composition controlled O_2 activation on Pd–Au bimetallic model catalysts. *ACS Catalysis* 8 (4): 3641–3649.

31 Cao, X.R., Fu, Q., and Luo, Y. (2014). Catalytic activity of Pd-doped Cu nanoparticles for hydrogenation as a single-atom-alloy catalyst. *Physical Chemistry Chemical Physics* 16 (18): 8367–8375.

32 Lu, X., Chen, J., Skrabalak, S.E. et al. (2007). Galvanic replacement reaction: a simple and powerful route to hollow and porous metal nanostructures. *Proceedings of the Institution of Mechanical Engineers, Part N: Journal of Nanoengineering and Nanosystems* 221 (1): 1–16.

33 Lucci, F.R., Liu, J.L., Marcinkowski, M.D. et al. (2015). Selective hydrogenation of 1,3-butadiene on platinum-copper alloys at the single-atom limit. *Nature Communications* 6 (6): 8550–8558.

34 Boucher, M.B., Zugic, B., Cladaras, G. et al. (2013). Single atom alloy surface analogs in $Pd_{0.18}Cu_{15}$ nanoparticles for selective hydrogenation reactions. *Physical Chemistry Chemical Physics* 15 (29): 12187–12196.

35 Aich, P., Wei, H.J., Basan, B. et al. (2015). Single-atom alloy Pd-Ag catalyst for selective hydrogenation of acrolein. *Journal of Physical Chemistry C* 119 (32): 18140–18148.

36 Giannakakis, G., Trimpalis, A., Shan, J.J. et al. (2018). NiAu single atom alloys for the non-oxidative dehydrogenation of ethanol to acetaldehyde and hydrogen. *Topics in Catalysis* 61 (5–6): 475–486.

37 Pei, G.X., Liu, X.Y., Yang, X.F. et al. (2017). Performance of Cu-alloyed Pd single-atom catalyst for semihydrogenation of acetylene under simulated front-end conditions. *ACS Catalysis* 7 (2): 1491–1500.

38 Pei, G.X., Liu, X.Y., Wang, A.Q. et al. (2015). Ag alloyed Pd single-atom catalysts for efficient selective hydrogenation of acetylene to ethylene in excess ethylene. *ACS Catalysis* 5 (6): 3717–3725.

39 Long, R., Li, Y., Liu, Y. et al. (2017). Isolation of Cu atoms in Pd lattice: forming highly selective sites for photocatalytic conversion of CO_2 to CH_4. *Journal of the American Chemical Society* 139 (12): 4486–4492.

40 Shan, J.J., Janvelyan, N., Li, H. et al. (2017). Selective non-oxidative dehydrogenation of ethanol to acetaldehyde and hydrogen on highly dilute NiCu alloys. *Applied Catalysis B: Environmental* 205 (205): 541–550.

41 Wrasman, C.J., Boubnov, A., Riscoe, A.R. et al. (2018). Synthesis of colloidal Pd/Au dilute alloy nanocrystals and their potential for selective catalytic oxidations. *Journal of the American Chemical Society* 140 (40): 12930–12939.

42 Mohammad, A.B., Lim, K.H., Yudanov, I.V. et al. (2007). A computational study of H_2 dissociation on silver surfaces: the effect of oxygen in the added row structure of Ag(110). *Physical Chemistry Chemical Physics* 9 (10): 1247–1254.

43 Lim, K.H., Mohammad, A.B., Yudanov, I.V. et al. (2009). Mechanism of selective hydrogenation of α,β-unsaturated aldehydes on silver catalysts: a density functional study. *The Journal of Physical Chemistry C* 113 (30): 13231–13240.

44 Bligaard, T., Nørskov, J.K., Dahl, S. et al. (2004). The Brønsted–Evans–Polanyi relation and the volcano curve in heterogeneous catalysis. *Journal of Catalysis* 224 (1): 206–217.

45 Zhang, L., Wang, A., Miller, J.T. et al. (2014). Efficient and durable Au alloyed Pd single-atom catalyst for the ullmann reaction of aryl chlorides in water. *ACS Catalysis* 4 (5): 1546–1553.

46 Wang, S.X., Abroshan, H., Liu, C. et al. (2017). Shuttling single metal atom into and out of a metal nanoparticle. *Nature Communications* 8 (8): 848–854.

47 Bootharaju, M.S., Sinatra, L., and Bakr, O.M. (2016). Distinct metal-exchange pathways of doped Ag_{25} nanoclusters. *Nanoscale* 8 (39): 17333–17339.

48 Fields-Zinna, C.A., Crowe, M.C., Dass, A. et al. (2009). Mass spectrometry of small bimetal monolayer-protected clusters. *Langmuir* 25 (13): 7704–7710.

49 Brust, M., Walker, M., Bethell, D. et al. (1994). Synthesis of thiol-derivatised gold nanoparticles in a two-phase Liquid–Liquid system. *Chemical Communications* 7 (7): 801–802.

50 Negishi, Y., Kurashige, W., Niihori, Y. et al. (2010). Isolation, structure, and stability of a dodecanethiolate-protected Pd_1Au_{24} cluster. *Physical Chemistry Chemical Physics* 12 (23): 6219–6225.

51 Xie, S.H., Tsunoyama, H., Kurashige, W. et al. (2012). Enhancement in aerobic alcohol oxidation catalysis of Au_{25} clusters by single Pd atom doping. *ACS Catalysis* 2 (7): 1519–1523.

52 Wang, S.X., Song, Y.B., Jin, S. et al. (2015). Metal exchange method using Au_{25} nanoclusters as templates for alloy nanoclusters with atomic precision. *Journal of the American Chemical Society* 137 (12): 4018–4021.

53 Deng, H.J., Wang, S.X., Jin, S. et al. (2015). Active metal (cadmium) doping enhanced the stability of inert metal (gold) nanocluster under O_2 atmosphere and the catalysis activity of benzyl alcohol oxidation. *Gold Bulletin* 48 (3–4): 161–167.

54 Qian, H.F., Jiang, D.E., Li, G. et al. (2012). Monoplatinum doping of gold nanoclusters and catalytic application. *Journal of the American Chemical Society* 134 (39): 16159–16162.

55 Negishi, Y., Chaki, N.K., Shichibu, Y. et al. (2007). Origin of magic stability of thiolated gold clusters: a case study on $Au_{25}(SC_6H_{13})_{18}$. *Journal of the American Chemical Society* 129 (37): 11322–11323.

56 Qian, H.F., Ellen, B., Zhu, Y. et al. (2011). Doping 25-atom and 38-atom gold nanoclusters with palladium. *Acta Physico-Chimica Sinica* 27 (3): 513–519.

57 Negishi, Y., Igarashi, K., Munakata, K. et al. (2012). Palladium doping of magic gold cluster $Au_{38}(SC_2H_4Ph)_{24}$: formation of $Pd_2Au_{36}(SC_2H_4Ph)_{24}$ with higher stability than $Au_{38}(SC_2H_4Ph)_{24}$. *Chemical Communications* 48 (5): 660–662.

58 Kumara, C. and Dass, A. (2011). $(AuAg)_{144}(SR)_{60}$ alloy nanomolecules. *Nanoscale* 3 (8): 3064–3067.

59 Tracy, J.B., Kalyuzhny, G., Crowe, M.C. et al. (2007). Poly(ethylene glycol) ligands for high-resolution nanoparticle mass spectrometry. *Journal of the American Chemical Society* 129 (21): 6706–6707.

60 Dass, A., Stevenson, A., Dubay, G.R. et al. (2008). Nanoparticle MALDI-TOF mass spectrometry without fragmentation: $Au_{25}(SCH_2CH_2Ph)_{18}$ and mixed monolayer $Au_{25}(SCH_2CH_2Ph)_{18-x}L_x$. *Journal of the American Chemical Society* 130 (18): 5940–5946.

61 Chaki, N.K., Negishi, Y., Tsunoyama, H. et al. (2008). Ubiquitous 8 and 29 kDa gold: alkanethiolate cluster compounds: mass-spectrometric determination of molecular formulas and structural implications. *Journal of the American Chemical Society* 130 (27): 8608–8610.

62 Jadzinsky, P.D., Calero, G., Ackerson, C.J. et al. (2007). Structure of a thiol monolayer-protected gold nanoparticle at 1.1 angstrom resolution. *Science* 318 (5849): 430–433.

63 Dass, A. (2009). Mass spectrometric identification of $Au_{68}(SR)_{34}$ molecular gold nanoclusters with 34-electron shell closing. *Journal of the American Chemical Society* 131 (33): 11666–11667.

64 Heaven, M.W., Dass, A., White, P.S. et al. (2008). Crystal structure of the gold nanoparticle $[N(C_8H_{17})_4][Au_{25}(SCH_2CH_2Ph)_{18}]$. *Journal of the American Chemical Society* 130 (12): 3754–3755.

65 Akola, J., Walter, M., Whetten, R.L. et al. (2008). On the structure of thiolate-protected Au_{25}. *Journal of the American Chemical Society* 130 (12): 3756–3757.

66 Liao, L.W., Zhou, S.M., Dai, Y.F. et al. (2015). Mono-mercury doping of Au_{25} and the HOMO/LUMO energies evaluation employing differential pulse voltammetry. *Journal of the American Chemical Society* 137 (30): 9511–9514.

67 Yao, C.H., Chen, J.S., Li, M.B. et al. (2015). Adding two active silver atoms on Au_{25} nanoparticle. *Nano Letters* 15 (2): 1281–1287.

68 Negishi, Y., Iwai, T., and Ide, M. (2010). Continuous modulation of electronic structure of stable thiolate-protected Au_{25} cluster by Ag doping. *Chemical Communications* 46 (26): 4713–4715.

69 Negishi, Y., Munakata, K., Ohgake, W. et al. (2012). Effect of copper doping on electronic structure, geometric structure, and stability of thiolate-protected Au_{25} nanoclusters. *Journal of Physical Chemistry Letters* 3 (16): 2209–2214.

70 Walter, M. and Moseler, M. (2009). Ligand-protected gold alloy clusters: doping the superatom. *Journal of Physical Chemistry C* 113 (36): 15834–15837.

71 Jiang, D.-e. and Dai, S. (2009). From superatomic $Au_{25}(SR)_{18}^-$ to superatomic $M@Au_{24}(SR)_{18}^q$ core–shell clusters. *Inorganic Chemistry* 48 (7): 2720–2722.

72 Gottlieb, E., Qian, H.F., and Jin, R.C. (2013). Atomic-level alloying and de-alloying in doped gold nanoparticles. *Chemistry European Journal* 19 (13): 4238–4243.

73 Das, A., Li, T., Nobusada, K. et al. (2012). Total structure and optical properties of a phosphine/thiolate-protected Au_{24} nanocluster. *Journal of the American Chemical Society* 134 (50): 20286–20289.

74 Shichibu, Y., Negishi, Y., Watanabe, T. et al. (2007). Biicosahedral gold clusters $Au_{25}(PPh_3)_{10}(SC_nH_{2n+1})_5Cl_2^{2+}$ (n = 2–18): a stepping stone to cluster-assembled materials. *Journal of Physical Chemistry C* 111 (22): 7845–7847.

75 Kacprzak, K.A., Lehtovaara, L., Akola, J. et al. (2009). A density functional investigation of thiolate-protected bimetal $PdAu_{24}(SR)_{18}^z$ clusters: doping the superatom complex. *Physical Chemistry Chemical Physics* 11 (33): 7123–7129.

76 Zhu, M.Z., Eckenhoff, W.T., Pintauer, T. et al. (2008). Conversion of anionic $Au_{25}(SCH_2CH_2Ph)_{18}^-$ cluster to charge neutral cluster via air oxidation. *Journal of Physical Chemistry C* 112 (37): 14221–14224.

77 Zhu, M., Aikens, C.M., Hollander, F.J. et al. (2008). Correlating the crystal structure of A thiol-protected Au_{25} cluster and optical properties. *Journal of the American Chemical Society* 130 (18): 5883–5885.

78 Kumara, C., Aikens, C.M., and Dass, A. (2014). X-ray crystal structure and theoretical analysis of $Au_{25-x}Ag_x(SCH_2CH_2Ph)_{18}^-$ alloy. *Journal of Physical Chemistry Letters* 5 (3): 461–466.

79 Wang, S.X., Meng, X.M., Das, A. et al. (2014). A 200-fold quantum yield boost in the photoluminescence of silverdoped Ag_xAu_{25-x} nanoclusters: the 13 th silver atom matters. *Angewandte Chemie International Edition* 53 (9): 2376–2380.

80 Kauffman, D.R., Alfonso, D., Matranga, C. et al. (2013). A quantum alloy: the ligand-protected $Au_{25-x}Ag_x(SR)_{18}$ cluster. *Journal of Physical Chemistry C* 117 (15): 7914–7923.

81 Wu, Z.K. (2012). Anti-galvanic reduction of thiolate-protected gold and silver nanoparticles. *Angewandte Chemie International Edition* 51 (12): 2934–2938.

82 Li, H.W., Yue, Y., Liu, T.Y. et al. (2013). Fluorescence-enhanced sensing mechanism of BSA-protected small gold-nanoclusters to silver(I) ions in aqueous solutions. *Journal of Physical Chemistry C* 117 (31): 16159–16165.

83 Sun, J., Wu, H.X., and Jin, Y.D. (2014). Synthesis of thiolated Ag/Au bimetallic nanoclusters exhibiting an anti-galvanic reduction mechanism and composition-dependent fluorescence. *Nanoscale* 6 (10): 5449–5457.

84 Wang, S.X., Meng, X.M., Feng, Y. et al. (2014). An anti-galvanic reduction single-molecule fluorescent probe for detection of Cu(II). *RSC Advances* 4 (19): 9680–9683.

85 Lee, D., Donkers, R.L., Wang, G.L. et al. (2004). Electrochemistry and optical absorbance and luminescence of molecule-like Au_{38} nanoparticles. *Journal of the American Chemical Society* 126 (19): 6193–6199.

86 Aikens, C.M. (2010). Geometric and electronic structure of $Au_{25}(SPhX)_{18}^-$ (X = H, F, Cl, Br, CH_3, and OCH_3). *The Journal of Physical Chemistry Letters* 1 (17): 2594–2599.

87 Parker, J.F., Kacprzak, K.A., Lopez-Acevedo, O. et al. (2010). Experimental and density functional theory analysis of serial introductions of electron-withdrawing ligands into the ligand shell of a thiolate-protected Au_{25} nanoparticle. *Journal of Physical Chemistry C* 114 (18): 8276–8281.

88 Jung, J., Kang, S., and Han, Y.K. (2012). Ligand effects on the stability of thiol-stabilized gold nanoclusters: $Au_{25}(SR)_{18}^-$, $Au_{38}(SR)_{24}$, and $Au_{102}(SR)_{44}$. *Nanoscale* 4 (14): 4206–4210.

89 Tlahuice-Flores, A., Whetten, R.L., and Jose-Yacaman, M. (2013). Ligand effects on the structure and the electronic optical properties of anionic $Au_{25}(SR)_{18}$ clusters. *The Journal of Physical Chemistry C* 117 (40): 20867–20875.

90 Walter, M., Akola, J., Lopez-Acevedo, O. et al. (2008). A unified view of ligand-protected gold clusters as superatom complexes. *Proceedings of the National Academy of Sciences of the United States of America* 105 (27): 9157–9162.

91 Zhang, L., Si, R., Liu, H. et al. (2019). Atomic layer deposited Pt-Ru dual-metal dimers and identifying their active sites for hydrogen evolution reaction. *Nature Communications* 10 (1): 4936–4946.

92 Zhang, L.Z., Jia, Y., Liu, H.L. et al. (2019). Charge polarization from atomic metals on adjacent graphitic layers for enhancing the hydrogen evolution reaction. *Angewandte Chemie International Edition* 58 (28): 9404–9408.

93 Zhao, R., Liang, Z.B., Gao, S. et al. (2019). Puffing up energetic metal-organic frameworks to large carbon networks with hierarchical porosity and atomically dispersed metal sites. *Angewandte Chemie International Edition* 58 (7): 1975–1979.

94 Wang, J., Huang, Z.Q., Liu, W. et al. (2017). Design of N-coordinated dual-metal sites: a stable and active Pt-free catalyst for acidic oxygen reduction reaction. *Journal of the American Chemical Society* 139 (48): 17281–17284.

95 Zeng, X., Shui, J., Liu, X. et al. (2018). Single-atom to single-atom grafting of Pt_1 onto $Fe\text{-}N_4$ Center: Pt_1@Fe-N-C multifunctional electrocatalyst with significantly enhanced properties. *Advanced Energy Materials* 8 (1): 1701345–1701353.

96 Bai, L., Hsu, C.-S., Alexander, D.T.L. et al. (2019). A cobalt–iron double-atom catalyst for the oxygen evolution reaction. *Journal of the American Chemical Society* 141 (36): 14190–14199.

97 Yang, Y., Qian, Y.M., Li, H.J. et al. (2020). O-coordinated W-Mo dual-atom catalyst for pH-universal electrocatalytic hydrogen evolution. *Science Advances* 6 (23): eaba6586–eaba6599.

98 Zhang, L., Fischer, J.M.T.A., Jia, Y. et al. (2018). Coordination of atomic Co–Pt coupling species at carbon defects as active sites for oxygen reduction reaction. *Journal of the American Chemical Society* 140 (34): 10757–10763.

99 Han, X., Ling, X., Yu, D. et al. (2019). Atomically dispersed binary Co-Ni sites in nitrogen-doped hollow carbon nanocubes for reversible oxygen reduction and evolution. *Advanced Materials* 31 (49): 1905622–1905631.

100 Wang, J., Liu, W., Luo, G. et al. (2018). Synergistic effect of well-defined dual sites boosting the oxygen reduction reaction. *Energy & Environmental Science* 11 (12): 3375–3379.

101 Dasgupta, A., Gao, Y., Broderick, S.R. et al. (2020). Machine learning-aided identification of single atom alloy catalysts. *The Journal of Physical Chemistry C* 124 (26): 14158–14166.

102 Lu, Z., Yadav, S., and Singh, C.V. (2020). Predicting aggregation energy for single atom bimetallic catalysts on clean and O* adsorbed surfaces through machine learning models. *Catalysis Science and Technology* 10 (1): 86–98.

103 Rao, K.K., Do, Q.K., Pham, K. et al. (2020). Extendable machine learning model for the stability of single atom alloys. *Topics in Catalysis* 63 (7): 728–741.

104 Han, Z.-K., Sarker, D., Ouyang, R. et al. (2021). Single-atom alloy catalysts designed by first-principles calculations and artificial intelligence. *Nature Communications* 12 (1): 1833–1842.

105 Liu, M., Yang, Y., and Kitchin, J.R. (2021). Semi-grand canonical Monte Carlo simulation of the acrolein induced surface segregation and aggregation of AgPd with machine learning surrogate models. *The Journal of Chemical Physics* 154 (13): 134701–134711.

106 Zheng, G., Li, Y., Qian, X. et al. (2021). High-throughput screening of a single-atom alloy for electroreduction of dinitrogen to ammonia. *ACS Applied Materials & Interfaces* 13 (14): 16336–16344.

107 Burger, B., Maffettone, P.M., Gusev, V.V. et al. (2020). A mobile robotic chemist. *Nature* 583 (7815): 237–241.

5

Characterization of Supported Metal Single-Atom Catalysts

Lei Zhang[1], *Kieran Doyle-Davis*[2], *and Xueliang Sun*[2]

[1] *Shenzhen University, College of Chemistry and Environmental Engineering, 3688 Nanhai Avenue, Shenzhen 518060, P. R. China*
[2] *The University of Western Ontario, Department of Mechanical and Materials Engineering, 1151 Richmond Street, London, ON N6A 5B9, Canada*

5.1 Introduction

The development of highly efficient catalysts is one of the most important research directions in catalysis. Metal catalysts are widely used in many industrial catalytic reactions due to their good activity and high selectivity. For supported catalysts, the number of active sites on the metal catalyst is closely related to the particle size [1, 2]. This is because, during the catalytic process, the reactants only react with the active sites of the metal surface, and the unexposed metal atoms inside the catalyst do not participate in the reaction. Therefore, to make full use of each metal atom on the metal catalysts, researchers attempted to reduce the size of the metal catalysts to achieve higher atom utilization. Studies have shown that sub-nanoclusters exhibited better catalytic activity or selectivity than nanoscale particles [3–5]. Theoretically, 100% dispersed supported metal catalysts results in uniformly dispersed single atoms. On a supported single-atom catalyst (SAC), every atom of the metal catalyst is fully utilized and acts as an active site, which can significantly improve the catalytic efficiency [6–18].

Single-atom catalysis is different from nanocatalysis and sub-nanocatalysis because when the particle size reaches single-atom size, its energy level structure and electronic structure will undergo fundamental changes, such as a sharp increase in surface free energy, unsaturated coordination environment, and the strong interaction between metal and support, etc. [19, 20]. Because of these unique structural features, SACs often exhibit distinct activity, selectivity, and stability compared with traditional nanocatalysts. In addition, SACs can not only improve the utilization efficiency of metal atoms but also change the adsorption and desorption energy of active components on the catalyst, thereby affecting the reaction kinetics.

The study of single-atom catalysts has an important relationship with advanced characterization techniques. With the development of transmission electron microscopy (TEM) and scanning tunneling microscopy (STM), we can directly

Supported Metal Single Atom Catalysis, First Edition. Edited by Philippe Serp and Doan Pham Minh.
© 2022 WILEY-VCH GmbH. Published 2022 by WILEY-VCH GmbH.

observe the dispersion of single atoms on the support. X-ray absorption near-edge spectroscopy (XANES) and extended edge X-ray absorption fine structure spectroscopy (EXAFS) can provide information on the oxidation state and coordination of single atoms. Diffuse reflectance Fourier transform infrared spectroscopy (DRIFTS) can analyze the adsorption structure of gas on the SAC. In this chapter, recent progress on the characterization techniques of SAC is described. First, the fundamentals of TEM and STM are briefly introduced, which can directly observe the SAC. Then the techniques for the investigation of electronic structures of SAC are systematically summarized, such as X-ray absorption spectroscopy (XAS), Fourier transform infrared spectroscopy (FTIR), X-ray photoelectron spectroscopy (XPS), and Mössbauer spectroscopy. Furthermore, characterization techniques to determine the mass loading of SACs are briefly introduced. The systematic summary of characterization techniques for SACs will provide useful experience for the investigation of newly developed SACs.

5.2 Morphology Characterization

5.2.1 Transmission Electron Microscopy (TEM)

5.2.1.1 Introduction of TEM

The operating principles of electron microscopy and optical microscopy are the same; however, while an optical microscope uses visible light and a glass lens, electron microscopy uses an electron beam as the light source with an electromagnetic field as the lens. TEM transmits the accelerated and concentrated electron beam to a very thin sample. The electrons collide with the atoms in the sample to change direction, thereby generating solid angle scattering. The size of the scattering angle is related to the density and thickness of the sample, so images with different brightness and darkness can be formed.

Due to the existence of aberrations (spherical aberration, astigmatism, coma, and chromatic aberration), the electromagnetic lens system cannot be perfect. The electrical resolution of regular TEM is only 0.8 nm, and it is very difficult to observe single-atom catalysts. Spherical aberration is one of the main contributors to this resolution limit. In an optical lens, the combination of convex lens and concave lens can be used to more narrowly focus the incoming light, more than possible through the use of just a convex lens. For electromagnetic lenses, we do not have concave lenses. Spherical aberration has become the most important and difficult issue affecting TEM resolution. A transmission electron microscope can use a spherical aberration correction device to act as a concave lens to correct the spherical aberration, creating a new technique: aberration-corrected transmission electron microscope (AC-TEM) [21–23]. Since TEM is divided into ordinary TEM and STEM for fine structure imaging, the spherical aberration electron microscope can also be divided into AC-TEM (the spherical aberration corrector is installed at the objective lens position) and AC-STEM (the spherical aberration correction device is installed at the condenser position). In addition, there are two rectifiers installed on a TEM at the same time to correct the double spherical aberration correction TEM of the probe and the image.

Compared with traditional TEM, AC-TEM effectively reduces aberrations, and the resolution of AC-TEM is significantly improved. The resolution of traditional TEM and STEM is at the nanometer and sub-nanometer level, while the resolution of AC-TEM and aberration-corrected scanning tunneling microscopy (AC-STEM) can reach the angstrom and sub-angstrom levels. The increase in resolution means that SACs can be characterized accurately by AC-TEM.

5.2.1.2 TEM Characterization Technique

Aberration-corrected transmission electron microscopy (AC-TEM). AC-TEM has a higher resolution than ordinary electron microscopes. Indeed, ordinary TEM can only be used to see the appearance, it is difficult to see the internal structure, such as interplanar spacing, atomic arrangement, and other information. Therefore, AC-TEM is used to observe the internal structure of crystals, atomic arrangement, and identifying the SAC.

Scanning transmission electron microscope (STEM). While TEM and high resolution TEM (HRTEM) irradiate the entire surface when collecting their image, AC-STEM is obtained by scanning individual sections, and then collecting the entire image. Obviously, STEM is more detailed in characterizing the structure. STEM and conventional TEM also distinguish bright field and dark field, but STEM is often used in conjunction with HAADF (a high-angle ring dark field detector) to obtain information on the micro-area structure and elemental distribution of the material.

Energy-dispersive X-ray spectroscopy (EDX) mapping. EDX is an analytical method used for elemental analysis and chemical characterization. First, incident electrons (or protons, photons) excite the inner electrons of the ground state atoms. The inner electrons leave holes after they leave the atoms. When the outer electrons at higher energy levels fill these lower energy holes, the excess energy may be released in the form of X-rays. Energy scattering X-ray spectrometers collect and measure the energy and intensity of these X-rays. Since the energy distribution of these X-rays can reflect the atomic characteristics of specific elements, EDX can be used to obtain the elemental distribution in alloys, nanotubes, shell materials, etc. to assist in phase identification or structural analysis.

Electron energy loss spectroscopy (EELS). The use of incident electrons to cause surface electron ionization, valence band electron excitation, oscillation, etc. can cause inelastic scattering, and the lost energy can be used to obtain physical and chemical information of surface atoms. Compared with EDX, the electron energy loss spectrum has a better resolution for light elements. The energy resolution is 1–2 orders of magnitude better than EDX. Due to the resolution of the electron energy loss spectrum in electron volts or even sub-electron volts, it can be used for elemental valence analysis, which EDX is not best suited for.

5.2.1.3 Characterization of Typical SACs

Isolated Single Atoms Electron microscopy technology can be used to directly observe the isolated atoms on the support. In the past few years, the aberration-corrected high-angle-annular-dark-field scanning transmission electron microscopy (AC-HAADF-STEM), as an indispensable characterization method, has provided

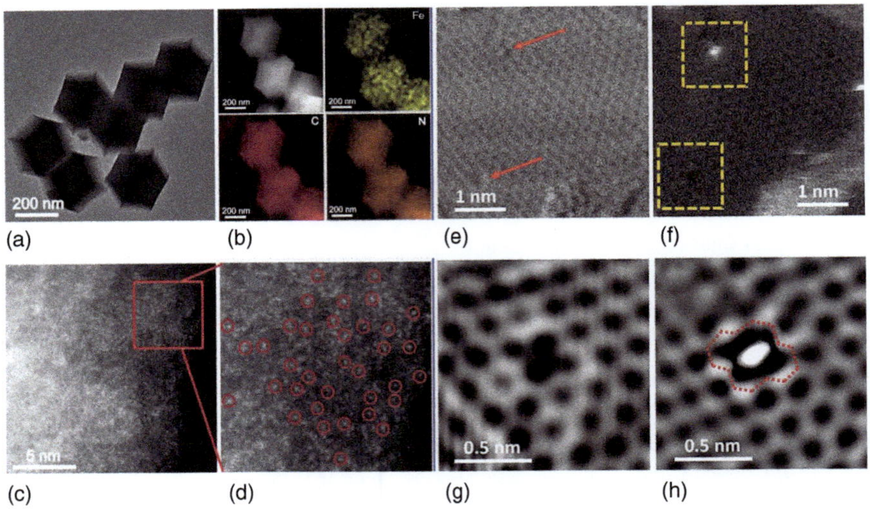

Figure 5.1 (a) TEM, (b) HAADF-STEM image, corresponding element maps showing the distribution of Fe, C, and N, (c, d) HAADF-STEM images and enlarged images of the Fe single atoms. Source: Chen et al. [24]. Adapted with permission of Wiley-VCH. (e) Bright-field STEM image of A–Ni@DG with high resolution. (f) The corresponding HADDF-STEM image of Ni_{SA}@DG of (e). (g) The zoomed-in image of the defective area (vacancy). (h) The zoomed-in image of the defective area (with atomic Ni trapped). Source: Zhang et al. [25] Adapted with permission of Elsevier.

valuable information for the characterization of supported SACs. In addition, atomic resolution EDX and EELS can analyze the composition of individual metal atoms. Taking meta-organic framework (MOF)-derived SACs as an example, the morphology, composition, and dispersion of SACs was characterized by TEM, EDX, and AC-HADDF-STEM images, respectively [24]. Figure 5.1a shows that the Fe SAC on N-doped carbon support retained its initial dodecahedral shape after pyrolysis, while its surface became much rougher. The EDX mapping images showed that Fe, C, and N were well-dispersed on the whole particle (Figure 5.1b). Importantly, the AC-HADDF-STEM image clearly shows the formation of isolated Fe single atoms on the support (Figure 5.1c,d).

In addition to the observation of isolated single atoms, their coordination environments can also be analyzed by the atomic resolution characterization of the substrates. Zhang et al. report a graphene defect trapped Ni_{SA} (Ni_{SA}@DG) as an active site for unique electrocatalytic reactions [25]. EDX mapping and the spectrum of Ni_{SA}@DG show that the Ni element exists uniformly on the defective graphene, which indicates some trivial Ni is invisible in these relatively low magnitude observations. The HAADF-STEM image (Figure 5.1f) confirms the abundance of Ni_{SA} (white dots) present in the Ni_{SA}@DG catalyst and the average dot size is 0.24 nm. The defective graphene provides sufficient defects as trapping sites for the deposition of Ni single atoms. The defective areas are further confirmed by the high-resolution HADDF-STEM image (Figure 5.1f). The sags and crests are attributed to the di-vacancy (Figure 5.1g) and the di-vacancy with trapped Ni atom (Figure 5.1h), respectively.

Single-Atom Alloy Single-atom alloy (single-atom alloy (SAA), see Chapter 4) catalysts are based on the deposition of isolated reactive metal adatoms onto host metal surfaces. Pt-based SAAs have generated significant interest due to their application in several selective hydrogenation/dehydrogenation reactions [26–28]. Note that the contrast of the HAADF-STEM image is proportional to the atomic number of the elements in the material. In this case, the heavier Pt atoms show a brighter contrast compared to the lighter atoms in the substrates, such as Cu or Pd.

For example, Gong and coworkers synthesized the Pt_{SA}/Cu SAA supported on γ-alumina through atomic dilution. The structure of Pt_{SA}/Cu SAA was addressed by AC-HAADF-STEM images. Single Pt atoms can clearly be distinguished from Cu atoms due to differences in the Z-contrast. The lattice spacing of Pt_{SA}/Cu SAA is confirmed to be 0.21 nm, which is in good agreement with the lattice spacing of Cu(111), indicating dilute dispersion of Pt atoms on Cu nanoparticles (NPs) [29]. In addition to the Cu substrates, Pt_{SA} was also deposited onto a Pd surface for electrochemical applications [30]. Figure 5.2a,b display HAADF-STEM images of Pt_{SA}/Pd SAA catalysts, which was obtained by atomic layer deposition (ALD) of Pt atoms on octahedral Pd particles. In this case, the heavier Pt atoms show a brighter contrast compared to the lighter Pd atoms. As seen in Figure 5.2a, the Pd octahedral particles were dispersed on nitrogen-doped carbon nanotubes (N-CNTs). The majority of the Pd particles maintained their original morphology after Pt-ALD. The atomic-resolution STEM image (Figure 5.2b) clearly illustrates the presence of a few brighter atoms at the surface of octahedral Pd particle, suggesting the presence of Pt clusters on the Pd octahedral particle seed. In addition, some individual bright spots on the surface of octahedral Pd particles can be identified, indicating the formation of Pt isolated atoms on Pd surfaces (Figure 5.2c).

Dimers Recently, dimer catalysts were found to exhibit greatly enhanced activities (Chapter 15), which were attributable to the synergistic effects between the electronic structures of the two elements [31–40]. The intensity of the HAADF-STEM image is proportional to the square of the atomic number (Z), which means that the brighter part of the image indicates the greater atomic number. This feature can be used to detect the formation of heteronuclear dimer catalysts. In addition, atomic resolution energy distribution X-ray spectroscopy (EDX) and EELS can also provide useful information of the composition of individual metal atoms.

Recently, Fe–Co dual sites embedded in N-doped CNTs were constructed by precisely controlling the bonding between Fe^{3+} precursors and Co nodes of Zn/Co bimetallic metal-organic frameworks (BMOFs) [41]. Regular TEM and HAADF-STEM images indicate the absence of metal NPs in the carbon nanotube (CNT) (Figure 5.3a). To further investigate the Fe and Co atoms, a magnified AC-HAADF-STEM was carried out (Figure 5.3b). The atomic coordination between Fe, Co, and N can be directly observed using HAADF-STEM images coupled

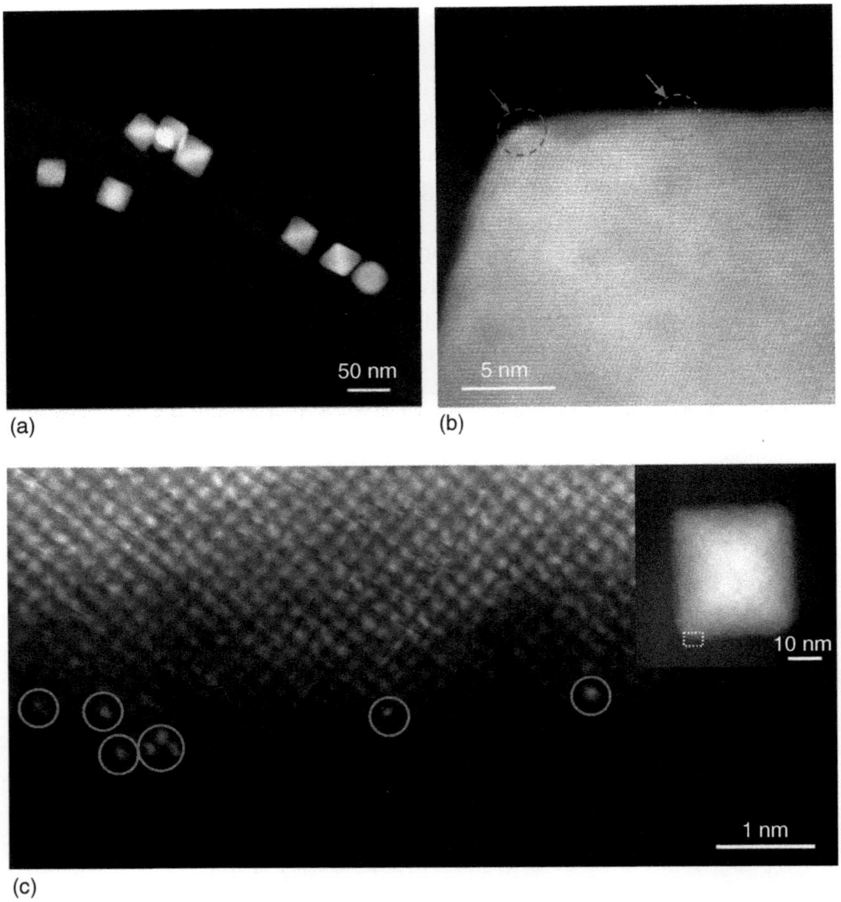

Figure 5.2 (a) Low-magnification STEM image of the octahedral Pt_{SA}/Pd SAA catalysts on N-CNT. (b) Atomic-resolution STEM image of an individual octahedral Pt_{SA}/Pd SAA particle. (c) High-resolution STEM image showing the surface of one individual octahedral Pt_{SA}/Pd SAA (inset), indicating the formation of Pt isolated atoms on Pd particles. Source: Zhang et al. [30]. Adapted with permission of American Chemical Society.

with EELS at the atom level. The homogeneously distributed bright dual dots confirmed the existence of Fe–Co dual sites because of the heavier Fe and Co atoms as compared to C and N atoms. This hypothesis was further verified using EELS. As shown in Figure 5.3c,d, when the electron beam was placed on the red rectangle, both Fe and Co were present in the EELS, suggesting the coexistence of Fe and Co in Fe–Co dual sites. This observation is supported by the homogeneous elemental distribution of Co, Fe, C, and N on the (Fe, Co)/CNT via EELS (Figure 5.3d). The atomic resolution HAADF-STEM images can also illustrate a Pt–Ru dimer-like structure, which was successfully synthesized by a designed ALD process (Figure 5.3e–f) [38]. As shown in Figure 5.3h, the two atoms in the dimer show different contrast, which indicates the dimer-like structure is composed of two different elements (in this case are Pt and Ru atoms). Although there are some

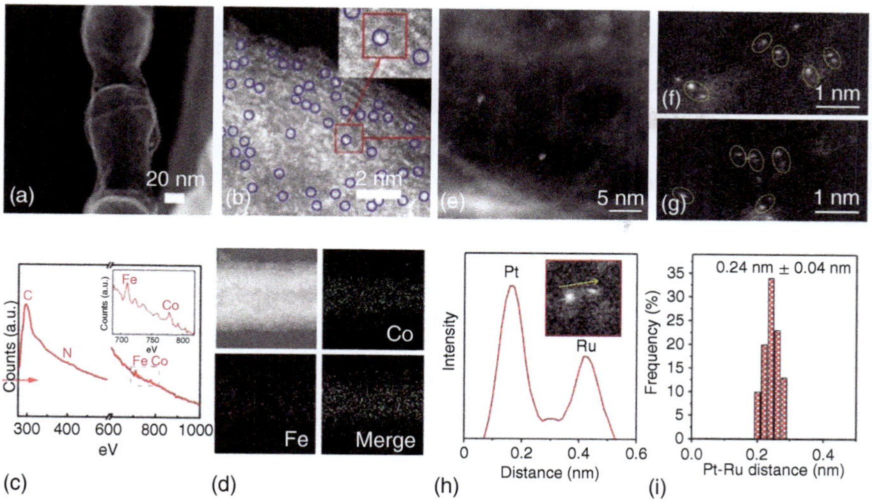

Figure 5.3 (a) HAADF-STEM image of the (Fe, Co)/CNT. (b) Magnified HAADF-STEM images of the (Fe, Co)/CNT. (c) The EEL spectrum shows that Fe, Co is coordinated with N at the atomic scale. (d) Corresponding EELS mapping images of Co, Fe, and N. Source: Wang et al. [41]. Adapted with permission of Royal Society of Chemistry. (e–g) Aberration-corrected HAADF-STEM images of Pt–Ru dimers/N-CNTs. (h) The intensity profile obtained on one individual Pt–Ru dimer. (i) Pt–Ru distance in the observed Pt–Ru dimers. Source: Zhang et al. [38]. Springer Nature/CC BY 4.0.

Pt_{SA} existing on the substrates, the ratio of dimer structures is around 70% of the total Pt content, indicating the significant amount of these types of structures in the material prepared. The Pt–Ru bonding of the dimers was statistically analyzed and showed a distance of 0.24 ± 0.04 nm (Figure 5.3i). This bond distance information can be used as a reference for the creation of density functional theory (DFT) models and the XAS fitting.

There are also some limits and challenges when using TEM for these systems. For example, TEM and STEM images can only provide the atomic structure of SACs in a limited region. Although several examples of the identification of single atoms by AC-STEM, EELS, and EDX spectroscopy have been reported, it is not always clear whether the atoms were located at a surface or inside the support [42]. The detection of isolated atoms inside pores could be also delicate. In addition, for transition-metal and rare-earth atoms, EELS may be more effective for a sufficiently thin substrate. To avoid sample damage, it should be noted that measurement time must be reduced due to the occurrence of atomic displacement.

5.2.2 Scanning Tunneling Microscopy (STM)

5.2.2.1 Introduction of STM

STM enables us to observe the arrangement of individual atoms on the surface of a substance and the physical and chemical properties related to surface electronic behavior [43]. The basic principle of STM is to use the tunneling effect in quantum theory. An atomic-scale ultrafine probe and the surface of the substance to be studied are used as two electrodes. When the distance between the sample and the needle tip is very close (usually less than 1 nm), the electrons will overcome the barrier between the two electrodes and flow to the opposite electrode under the action of an external electric field. This phenomenon is the tunnel effect [44]. The tunnel current intensity is very sensitive to the distance between the tip and the sample surface. If the distance is reduced by 0.1 nm, the tunnel current will increase by an order of magnitude. When operating, an electronic feedback circuit is used to control the constant tunnel current, and piezoelectric ceramic materials are used to control the tip movement on the sample. The height changes of the probe in the direction perpendicular to the sample reflect the undulations of the sample surface. The trajectory of the needle tip when scanning the sample surface can be directly displayed on the screen or recording paper, showing the distribution of the density of states or the atomic arrangement of the sample surface. For a sample surface with little undulation (surface fluctuations less than 1 nm), such as SACs, the height of the needle tip can be controlled for constant scanning, and the distribution of the surface state density can be obtained by recording the change of the tunnel current. This scanning method is characterized by fast scanning speed, which can reduce the influence of noise and thermal drift on the signal. The STM performs nondestructive detection on the surface of the sample, avoiding changes to the sample, and without subjecting the sample to destructive high-energy radiation. In addition, any microscope that uses lenses to focus light or other radiation is inevitably subject to a fundamental limitation: the phenomenon of light diffraction. Due to the diffraction of light, details smaller than half of the wavelength of the incident light will become blurred under the microscope. The STM can easily overcome this limitation and achieve atomic-level resolution.

5.2.2.2 STM Characterization of SACs

The atomic-resolution STM has proved to be an excellent technique for imaging supported individual metal atoms [26–28, 45, 46]. For example, Lucci and coworkers prepared Pt_{SA}/Cu SAA using physical vapor deposition of Pt onto a clean Cu(111) surface at 107 °C [27]. STM images show that at low Pt coverages (0.02 monolayer), Pt atoms exist as individual, isolated species substituted in the Cu surface lattice (Figure 5.4a). The Pt_{SA}/Cu SAA catalysts exhibited high performance for the selective hydrogenation of 1,3-butadiene to butenes under mild conditions. Liu and coworkers identified isolated Pt atoms on the surface of Cu(111) and investigated the fundamentals of CO adsorption on Pt_{SA}/Cu SAA using STM. They found that CO binds more weakly to single Pt atoms on Cu compared to larger Pt ensembles or monometallic Pt. Thus, when CO is present in the gas phase, more CO-free Pt sites are available on the SAA than the monometallic Pt, yielding higher H_2 activation reactivity and sequential hydrogenation activity under realistic conditions. Lucci et al. observed the Pd_{SA}/Au(111) SAA with isolated Pd atoms present on the surface layer by STM (Figure 5.4b). Using the atomically resolved model systems of dilute Pd–Au surface alloys, they determined the key aspects of H_2 activation, diffusion, and desorption on Au–Pd surface [46].

To improve the resolution of STM for examining single atoms, STM usually operates at low temperatures under ultrahigh vacuum (UHV). Krenner used low-temperature STM to investigate the organization and bonding of Co atoms on the self-assembly of *N,N*$_0$-diphenyl oxalic amide molecules on the Ag(111) surface (Figure 5.4c) [45]. The STM images show that Co atoms prefer to bind on top of the phenyl rings, and thus the realization of Co-phenyl complexes is preferred over metal cluster growth on the bare Ag(111) surface (Figure 5.4d,e). In addition, the optimal temperatures for the fraction of monomeric Co species on the template are from −93 to −73 °C. Around 80% of the total amount of Co can be deposited on the substrates.

5.3 Structure Characterization

5.3.1 Synchrotron Radiation X-ray

5.3.1.1 Fundamentals of Synchrotron Radiation X-ray

Although AC-HAADF-STEM images can provide direct evidence for the formation of single atoms on the support surface, the detected area is limited to several nanometers in scale, which means that AC-HAADF-STEM can only provide local information on the catalyst. Synchrotron radiation technology is an important characterization method to provide information from the whole sample, in complement to TEM techniques. To conclude the formation of high quality SACs, the metal–metal bond should be absent in the EXAFS spectra. Furthermore, the XAS data can also identify the information of SAC from the oxidation state, electronic structure, coordination environment, and detailed coordination number [47, 48].

Thanks to the high-quality incident light, XAS has a very sensitive detection limit to test the ultralow loading of SACs on the supports. According to the energy level, X-rays can be divided into soft X-rays and hard X-rays. The latter has an energy of 5+ keV and wavelengths below 0.2–0.1 nm, which have been widely applied in the analysis of the SACs [49–51]. When the X-ray energy is able to excite an inner electron to a higher orbital, resonance absorption occurs. The X-ray absorption coefficient is abrupt, and this jump is called the absorption edge. The absorption edges of electrons with different main quantum numbers in atoms are named as *K, L,* etc. absorption edges. As each element has its characteristic absorption edges, XAS can be used for qualitative analysis of elements. XANES and EXAFS are two of the most widely used synchrotron-based techniques for SACs, which can provide information about the dispersibility of single atoms and their oxidation state.

5.3.1.2 XANES

In XAS, the spectrum in the low energy region within 60 eV above the threshold has strong absorption characteristics, which is called the XANES. XANES generally has a sufficiently high signal quality, and a lot of fingerprint information, which can give reliable information. The most common one is the valence analysis. The positions of the absorption edges can vary according to the valence state of the element of interest. The higher the valence state, the more the absorption edge moves toward

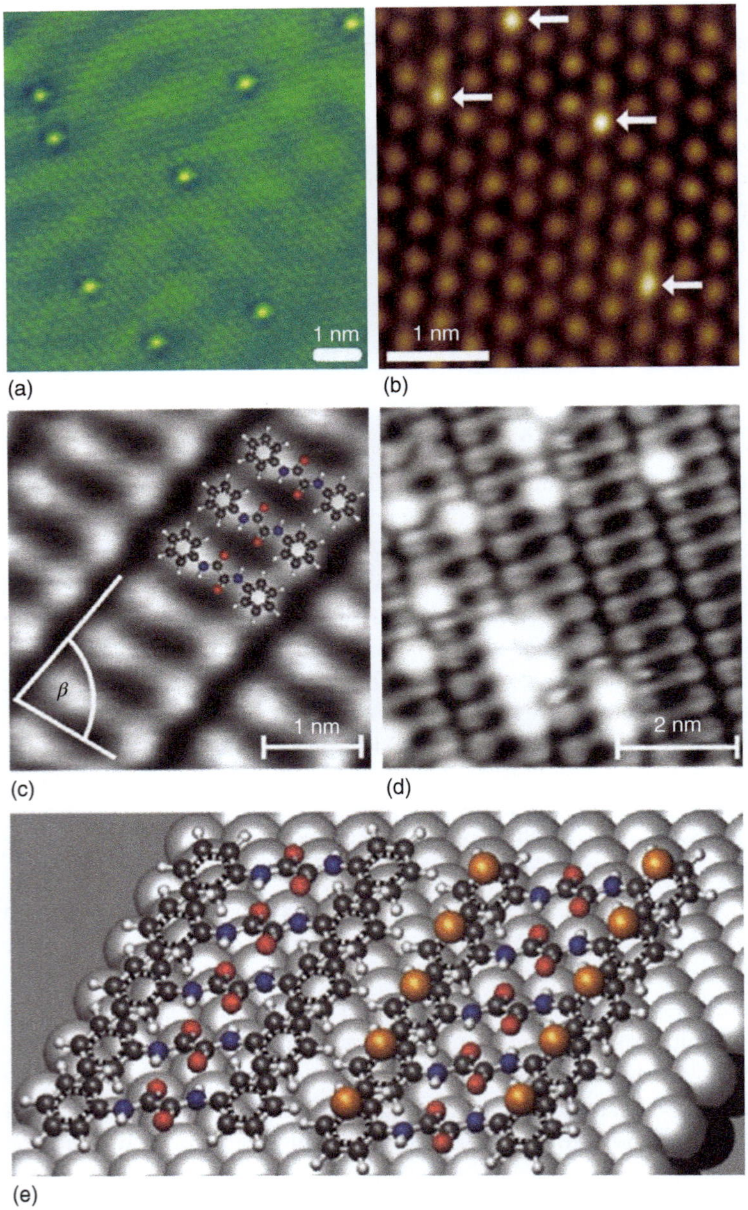

Figure 5.4 (a) STM image of 0.02 monolayer Pt_{SA}/Cu(111) SAA surface. Source: Lucci et al. [27]. Springer Nature/CC BY 4.0. (b) STM image of PdSA–Au(111) SAA. Source: Lucci et al. [46]. Adapted with permission of American Chemical Society. (c) High-resolution STM image with a superimposed molecular model of a dense-packed monolayer of *N,N′*-diphenyloxalic amides. (d, e) STM image and 3D visualization show that Co atoms (bright protrusions) adsorb monomerically on top of the phenyl rings of the molecules. Source: Krenner et al. [45]. Reproduced with permission of American Chemical Society.

higher energy. In addition, absorption features can be used for determining the valence state of the element. For a SAC sample, by comparing the position of the corresponding absorption edge obtained from the standard metal foil and metal oxide, the average valence state of the corresponding element can be obtained through the linear combination of the two spectra.

For example, when Fe single atoms are anchored on MOF-derived N–C support, their XANES curves show near-edge absorption energy between those of metal oxide and metal foil, implying the single atoms carried positive charges [24]. In addition, the positive shift of near-edge absorption energy, in general, comes along with an increase in the white line peak (the first major peak at the edge) intensity. As shown in Figure 5.5a, the threshold energy of Ni_{SA}/N–C and the intensity of the line in XANES spectra, both located between those for the Ni foil and NiO clearly showed the partly charged electronic structure of $Ni^{\delta+}$ ($0 < \delta < 2$) [52]. When the single atoms are deposited on transition metals, the investigation of their bonding environment is also very important. The electronic state determined by XANES spectra can provide strong evidence to show the formation of metal–metal bonds or metal–O bonds. For instance, single Pt atoms can be uniformly dispersed on FeO_x support of high surface area by coprecipitation method [6]. The white line (WL) intensities in the Pt/FeO_x spectra is between the intensities of Pt foil and PtO_2 (Figure 5.5b), suggesting that the Pt single atoms carry positive charges. Wang and Li found that Pt single atoms are positively charged on anatase TiO_2 support [54]. Their results indicated that when the single atoms are deposited onto metal oxides, they are bonded with the O sites of the metal oxide, causing the partially positively charged electronic state.

In addition to the position of the absorption edge, the height of the absorption edge (WL) or the area of the corresponding interval can also be normalized. The XANES can be quantitatively analyzed for determining the electronic structure of metals. Sun and coworkers first used quantitative analysis of WL intensity to explore the implication of the unoccupied densities of 5d states in Pt single atoms [53]. The Pt $L_{3,2}$-edge WL intensity can be analyzed using the Au metal $L_{3,2}$-edge XANES as the background. Pt single atoms on graphene (50ALDPt/GNS) were prepared by the ALD method and their electronic structure of Pt was investigated based on this quantitative analysis method. The 50ALDPt/GNS sample showed the highest total unoccupied density of Pt 5d character states (Figure 5.5c) [8]. In addition to carbon-based supports, the Pt isolated atoms were also deposited on metal surfaces (octahedral Pd particles) to form an SAA structure [30]. The density of Pt d states on Pd can also be analyzed by this method. According to the quantitative analysis results of the WL intensity, Pt on Pt-SAA had an unoccupied density of Pt 5d state character of 0.8176, which is much higher than that of Pt foil sample (0.6754). This result indicates the strong interaction between Pt single atoms and Pd surfaces, which increases the durability of Pt in oxygen reduction reaction (ORR) and hydrogen evolution reaction (HER) tests.

5.3.1.3 EXAFS

The EXAFS is generated from the scattering of absorbing atoms and other atoms around them, starting at roughly 50 eV and extending to about 1000 eV above the

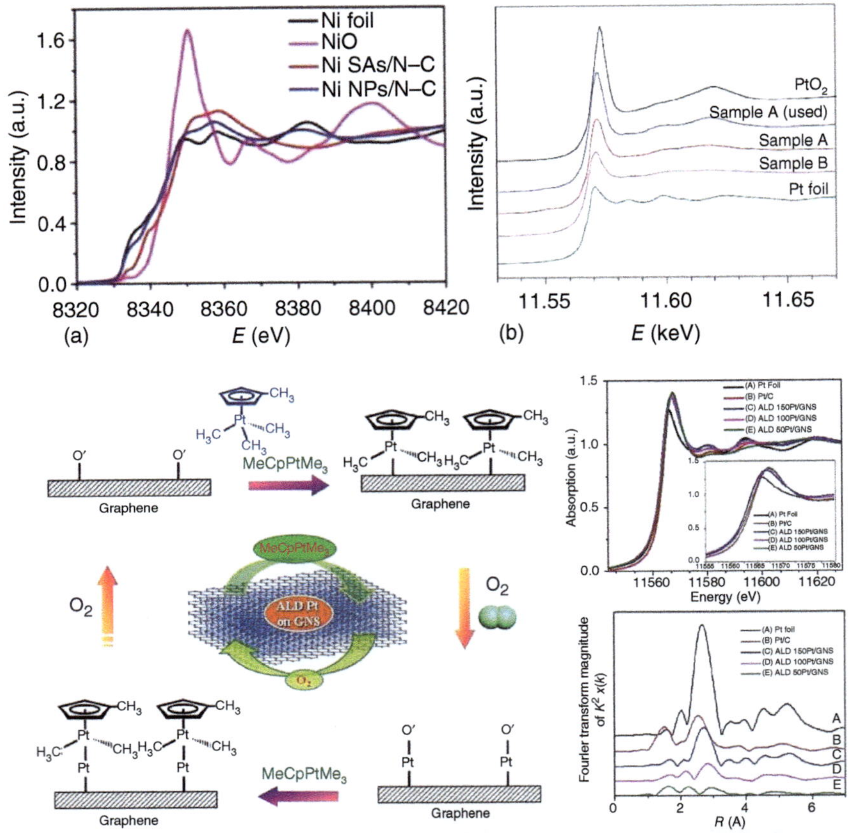

Table 1 | Pt $L_{3,2}$-edge threshold and whiteline parameters

	Pt L_3-edge WL				Pt L_2-edge WL					
Sample	E_0(eV)[a]	E(eV)[b]	Γ(eV)[c]	ΔA_3[d]	E_0(eV)[a]	E(eV)[b]	Γ(eV)[c]	ΔA_2[d]	$h_{5/2}$	$h_{3/2}$
Pt foil	11563.8	11565.7	4.67	6.78	13269.5	13272.5	7.48	2.95	0.53	0.12
ALD 50 Pt/GNS	11563.6	11567.1	8.43	7.47	13269.7	13272.9	7.67	6.15	1.20	0.25
ALD100 Pt/GNS	11563.8	11567.5	8.76	6.88	13269.9	13273.1	8.24	5.31	1.13	0.21
ALD150 Pt/GNS	11563.6	11567.0	8.66	6.98	13269.5	13273.5	8.55	5.47	0.93	0.22

[a] Position of the point of inflection of the rising edge; [b] Peak position; [c] Linewidth at Half maximum of the WL, see inset of Fig. 4 a; [d] Area under the difference curve (see Fig. S10, marked by the vertical bar) for unity edge jump, the unity edge jump for the Pt L_3 and L_2 edge corresponds to a value of 2.5×10^3 cm^{-1} and 1.16×10^3 cm^{-1}, respectively.

(c)

Figure 5.5 The normalized XANES spectra of (a) Ni single atoms at Ni K-edge and (b) Pt single atoms on FeO_x at the Pt L_3 edge. Source: Zhao et al. [52]. Reproduced with permission of American Chemical Society; and Qiao et al. [6]. Reproduced with permission of Nature Publishing Group. (c) The normalized XANES and EXAFS spectra at Pt L_3 edge of Pt single atoms on graphene obtained by the ALD method. Source: Sun et al. [53]. Reproduced with permission of Nature Publishing Group.

edge in the XAS spectra. Therefore, EXAFS can reflect the neighboring structure around the absorbing atoms and detailed parameters such as the interatomic distance and coordination number. EXAFS can also identify the neighbor structure of the specified element atoms and distinguish the types of neighbor atoms. To determine the peak attribution in the spectra, the EXAFS spectra of standard materials are

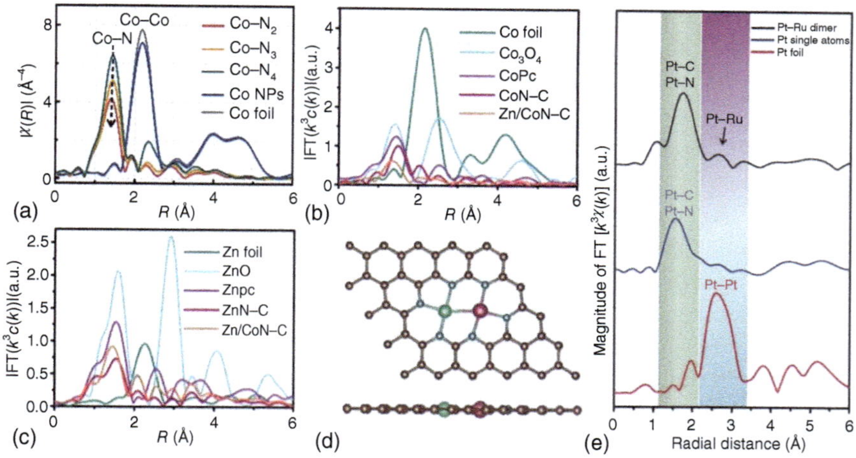

Figure 5.6 (a) EXAFS spectra confirm the atomic dispersion of Co atoms in Co–N_2, Co–N_3, and Co–N_4. Source: Wang et al. [55]. Reproduced with permission of Wiley-VCH. (b) Co K-edge and (c) Zn K-edge EXAFS spectra of Zn/CoN–C. (d) The relative position of Zn, Co, and N for the Zn/CoN–C considered in the calculations. Source: Lu et al. [32]. Reproduced with permission of Wiley-VCH. (e) EXAFS of the Pt–Ru dimers, Pt single atoms, and Pt foil at Pt L_3 edge. Source: Zhang et al. [38]. Springer Nature/CC BY 4.0

required. Then, the peaks can be defined as specific bonding, such as metal–metal, metal–N, and metal–O bonds. By fitting the EXAFS curves, the specific coordinated atoms with a single atom and the coordination distance can be obtained.

Taking atomically dispersed Co atoms as an example, Li and coworkers constructed Co SAC on a nitrogen-doped porous carbon support by *in situ* pyrolysis of Zn/Co BMOFs precursors. EXAFS can effectively confirm the coordination environment of Co sites, which were obtained at different pyrolysis temperatures in the range of 800–1000 °C. A prominent peak at 1.4 Å can be observed from all the obtained Co SACs, which can be attributed to the Co–N coordination. More importantly, the intensity of the Co–N peak implies the surrounding *N* number of Co center. According to the fitting results, the Co–N coordination numbers within these three samples are 4.1, 3.1, and 2.2, respectively, which agrees well with the Co–N_4, Co–N_3, and Co–N_2 structures (Figure 5.6a) [55].

Recently, dual-metal sites (Chapter 15) attracted more and more attention due to their greatly enhanced activities compared with SACs. EXAFS analysis can be applied for identifying the formation of a dual metal bond. For example, Zn–Co atomic pairs coordinated on N-doped carbon support were prepared by using chitosan as C and N sources and zinc chloride and cobalt acetate as metal precursors [32]. The EXAFS fitting results reveal that the coordination number of Co–N is 3.5 and Co–M is 0.5 ± 0.1 (Figure 5.6b). Similarly, for Zn of the Zn/CoN–C, it is dominated by the Zn–N coordination ($N = 3.5$) in the first shell, together with a weak second shell due to Zn–M (Zn–M coordination number = 0.5 ± 0.1), confirming the existence of $ZnCoN_6$ structure in Zn/CoN–C (Figure 5.6c,d). Sun and coworkers prepared Pt-Ru dimers on N-CNTs and investigated the coordination bond of Pt and Ru by EXAFS [38]. As a comparison, Pt single atoms exhibit only the peak at 1.6 Å, indicating that no Pt—Pt bond was detected. In addition to the Pt–N/Pt–C peak, a relatively weak feature is resolved at around 2.6 Å for the Pt–Ru dimer structure (Figure 5.6e). The EXAFS R-space curve fitting result indicated the coordination number 1 for Pt–Ru. Furthermore, from the Ru R space fitting results, the Ru atoms have a coordination number 0.7 for Ru–Pt, also suggesting the formation of Pt–Ru dimer structure.

5.3.1.4 In situ XAS Study on Structural Evolution During Catalytic Reaction

During the catalytic reaction, the catalyst might restructure due to the influence of the reaction conditions, such as the temperature, applied potentials, adsorbed species, and other factors. Therefore, the characterization of prepared or post-testing catalysts might not clearly show the active sites during the reactions. As such, the development of corresponding *in situ* experimental techniques is very important to identify the structural information of the active centers and key intermediates in the catalytic reaction process (see also Chapter 6). Until now, real-time XAS characterization can be carried out in several electrocatalytic reactions, such as ORR, oxygen evolution reaction (OER), CO_2 reduction, and HER. During testing, a homemade organic glass electrochemical cell was required for real-time XAFS experiments [56–58]. The measurements are usually conducted simultaneously on a computer-controlled electrochemical analyzer and the spectra are recorded on the as-prepared catalysts at different potentials. The *in situ* XANES and EXAFS information of the catalysts at different applied potentials can clearly show the structure evolution of the catalysts during the catalytic process.

For example, Li et al. prepared Cu SACs on carbon support with excellent ORR activity [58]. The structural evolution of Cu atoms at the atomic interface was investigated under electrocatalytic conditions by using potential-dependent *in situ* XAFS measurements. The *in situ* XANES spectra were recorded at 1.0, 0.893, and 0.7 V. The results show that the absorption edge was gradually shifted to lower energy, and the white line peak intensity reduced as the potential dropped from 1.0 to 0.7 V. The average oxidation states of Cu decreased from 1.59 to 1.09 under the working conditions, which implied that low valence Cu (+I) species work as the catalytic sites during the ORR process. The real-time EXAFS at the Cu K-edge presents that the peak position of the Cu–N displays an obvious low-R shift from 1.44 to 1.35 Å. According to the fitting results, under the catalytic conditions (1.0, 0.893, and 0.7 V), the $Cu–N_4–C_8S_2$ moiety were adsorbed with OOH*, O*, or OH* intermediates. These results indicated that the formation of bond-shrinking low-valence $HOO–Cu_1–N_4–C_8S_2$, $O–Cu_1–N_4–C_8S_2$, and $HO–Cu_1–N_4–C_8S_2$ sites under working conditions is responsible for the high ORR activity of Cu-SA/SNC. Besides ORR, transition metals (Fe, Co, and Ni) SAC can effectively convert CO_2 into CO during electrochemical CO_2 reduction [9, 57, 59, 60]. Deng et al. systematically investigated the active sites of a series of metal phthalocyanines (MePcs with Me = Mn, Fe, Co, Ni, and Cu) catalysts by *in situ* XAFS [60]. They found that no significant changes were detected in both XANES profiles and EXAFS spectra at the Co K edge of the CoPc catalyst when exerting −0.6 V to reduce CO_2, indicating that the valence state and the coordination structure of Co^{2+} remained at the reduced potential.

In addition to these electrocatalytic reactions, *in situ* XAS tests were also used for detecting the electronic structure of catalysts during preferential oxidation of CO in hydrogen (PROX) reaction. Lu and coworkers used *in situ* XAS to investigate the electronic structure evolution of isolated $Fe(OH)_x$ sites on Pt/SiO_2 during the PROX process [12]. They first pressed the sample into a pellet and then loaded it into a homemade quartz reaction cell, which can be heated to 500 °C with external heating. Kapton foil was used as the X-ray window material. A K-type thermocouple, protected by a closed-end quartz tube, was located near the sample pellet to measure the sample temperature. The XAFS spectra of the sample were recorded according to the various reaction temperatures and different atmosphere treatments. Thanks to the *in situ* XAFS spectra, the coordination information and even the reaction intermediate can be concluded for the specific catalysts. Yan and coworkers studied the structural change of well-defined atomically dispersed Rh on phosphotungstic acid by a suite of *operando/in situ* spectroscopic experiments during CO oxidation [61]. The SACs were treated in CO flow with temperature stepwise increased from 27 °C to 200 °C. They found two intense peaks at 23 237 and 23 255 eV at 50 °C, indicating the structural change of the catalyst. Additionally, the formation of a pre-edge feature at 23 220 eV can also be observed at a temperature as low as 50 °C. The strong mixture of 5p and 4d orbitals caused the appearance of pre-edge features, and thus enhanced the intensity of electric dipole forbidden transitions from 1s to 4d orbitals.

Although XAS analysis has been widely used for identifying SACs, it also has some limitations in its application, which could provide the wrong message [62]. For example, EXAFS struggles to identify mixed materials with different particle sizes, as this technique is insensitive to polydispersity. EXAFS reflects an average signal of the compositions and cannot clearly show the features for each component. Also, for materials with disordered atomic structures, EXAFS can show a clear first-shell peak, while the second-shell peak may be significantly suppressed. The EXAFS peak intensity may also sharply decrease with decreased particle size. Specifically, when the disordered compositions are small to sub-nano size, the second shell of M–M peaks cannot be observed due to the ultralow intensity, which could be very similar to that for single atoms. Therefore, if the SAC is only characterized by EXAFS, it is hard to avoid the coexistence of other compositions in the preparation process, especially when synthesized by chemical methods. Small disordered clusters or NPs may also be prepared in the same materials. In addition, atomic resolution STEM images can only show the structure in very local regions, which may easily miss broader area trends. Therefore, detailed XAS analysis must be used in conjunction with morphology characterization methods, such as TEM and STM, to thoroughly identify isolated atoms.

5.3.2 Infrared (IR) Spectroscopy

Infrared spectroscopy is the branch of spectroscopy that studies the infrared part of the electromagnetic spectrum. It includes many techniques, the most commonly used so far is absorption spectroscopy. Like all spectroscopic techniques, it can be used to identify a compound and study the composition of a sample. When the vibration frequency or rotation frequency of a chemical group in the substance molecule is the same as the frequency of infrared light, the molecule absorbs energy and transitions from the original ground state vibration (rotation) kinetic energy level and jump to a higher vibrational (rotational) kinetic energy level. As the molecules absorb infrared radiation and undergo a transition between vibrational and rotational energy levels, the light of this wavelength is absorbed by the substance. Therefore, infrared spectroscopy is essentially an analytical method for determining the molecular structure of substances and identifying compounds, which is based on the information of the relative vibrations and molecular rotations between the atoms within the molecules. The absorption of infrared light by the molecule is recorded with an instrument to obtain an infrared spectrum. Infrared spectrograms usually use wavelength (λ) or wavenumber (σ) as the abscissa to indicate the position of the absorption peak and use transmittance ($T\%$) or absorbance (A) as the ordinate to indicate the absorption intensity. When measuring a sample, a beam of infrared light passes through the sample, and the energy absorption at each wavelength is recorded. This can be achieved by continuously changing the monochromatic wavelength used, or by using Fourier transform (FT) to measure all wavelengths at once. In this case, the transmission spectrum or the absorption spectrum may be recorded to show the wavelength absorbed by the sample infrared, so that the chemical bonds contained in the sample can be analyzed. FT-IR is an extremely effective measurement method for recording infrared spectral signals.

IR spectroscopy was also demonstrated as a fast and convenient characterization method to directly distinguish and quantify Pt single atoms from NPs [16, 63–68]. For example, Ding et al. investigated the IR spectra of adsorbed CO on four Pt/H-form of Zeolite Socony Mobil-5 (HZSM-5) samples with different loadings [16]. The IR bands centered at 2115 cm^{-1} to CO molecules adsorbed on Pt single atoms can be observed, which is significantly different from the IR peak at 2070–2090 cm^{-1} to CO molecules linearly adsorbed on Pt atoms (Figure 5.7a). DeRita used CO probe molecule IR spectroscopy to identify unique vibrational signatures of CO adsorbed to Pt single atoms (Pt_{iso}) and preoxidized (Pt_{ox}) and prereduced (Pt_{metal}) Pt clusters [63]. It can be seen that when only Pt_{SA} are deposited on TiO_2 particles, a single and sharp CO band with a frequency at 2112 cm^{-1} was observed. When the Pt clusters form, a second broad CO band formed with a frequency of 2040–2090 cm^{-1}, signifying CO adsorbed to Pt_{metal} sites. In addition, by using the infrared spectroscopic signatures, they found that the order of CO adsorption energy on these sites was identified to be $Pt_{iso} \ll Pt_{metal} < Pt_{ox}$. For Rh single atom and clusters, the peaks at ~2097 and ~2028 cm^{-1} in the CO probe molecule IR spectra are associated with the symmetric and asymmetric stretches of the $Rh(CO)_2$ gem-dicarbonyl species that exist uniquely at Rh_{SA} sites (Figure 5.7b) [65, 66]. The peaks at 2068 and 1860 cm^{-1} are associated with CO adsorbed in linear and bridge-bound geometries at Rh_{NP} sites. In addition, the absolute intensity of the bridge peak had a minimal impact on the quantified Rh_{NP} and Rh_{SA} site fractions, which can be used for determining the ratio of each site type (Rh_{SA} and Rh_{NP}).

IR spectroscopy was also used to characterize the formation of dimer structures on the support. Yang and coworkers confirmed the adsorption of the precursors of $Rh_2(OAc)_4$ on MgO support by IR spectra [67]. Zhao et al. prepared dinuclear

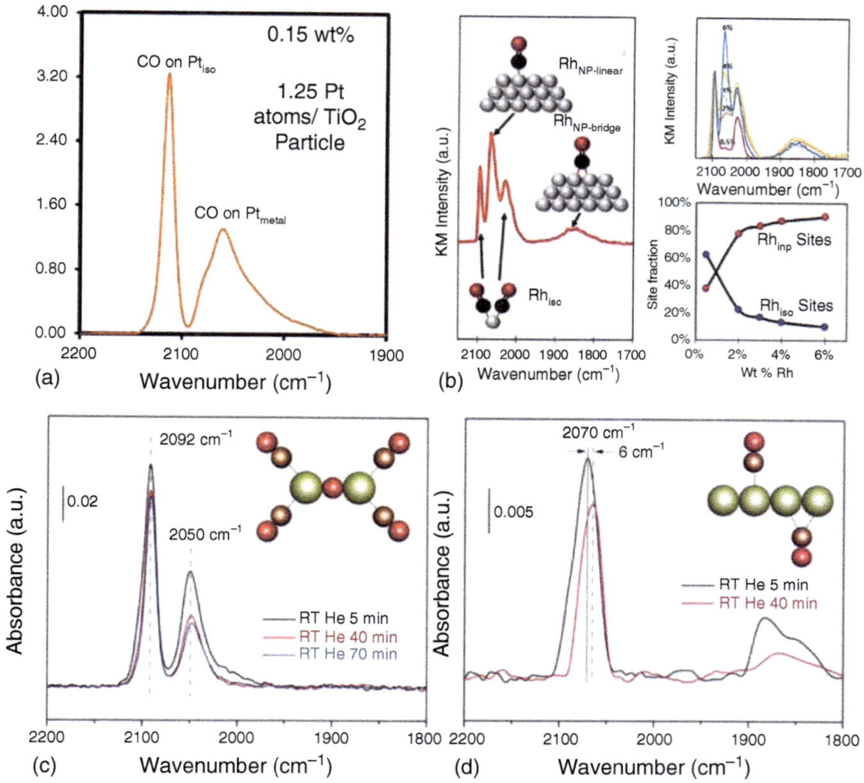

Figure 5.7 (a) IR spectra of CO adsorbed at saturation coverage and room temperature onto 0.15 wt% Pt/TiO_2 catalysts. Source: DeRita et al. [63]. Reproduced with permission of American Chemical Society. (b) DRIFT spectrum obtained from a saturated layer of CO adsorbed at 27 °C on Rh/TiO_2 with different Rh loading. Source: Matsubu et al. [65]. Reproduced with permission of American Chemical Society. (c, d) *In situ* DRIFT spectra of Ir DHCs and NPs in He flow after CO adsorption at room temperature. Source: Zhao et al. [68]. Reproduced with permission of Nature Publishing Group.

iridium heterogeneous catalysts (DHCs) via a facile photochemical method. To confirm the Ir–O–Ir arrangement within the Ir DHCs, they used *in situ* CO diffuse reflectance infrared Fourier transform (DRIFT) spectroscopic characterization (Figure 5.7c,d) [68]. As shown in the *in situ* CO DRIFT experiment, the arrangements of Ir–Ir and Ir–O–Ir are expected to be readily distinguished by the characteristic absorption features at ~1850 cm^{-1}, which is indicative of the bridge adsorption of CO at the Ir–Ir site. The broad peak at ~1850 cm^{-1} was observed for Ir_{NP}, while this peak is absent from Ir DHCs, indicating the formation of Ir–O–Ir.

Although IR spectroscopy has been used for characterizing SACs and dimers, there are still some challenges with using probe molecule FTIR to characterize isolated atoms species. The first challenge is the homogeneity of the SAC local environments on the support [69]. Only when the isolated atom species exist in a single location, the insights into the local environment from IR spectroscopy is useful. Furthermore, the SACs on the support may move or reposition themselves during the calcination, reduction, and exposure of probe molecules to catalysts, which causes additional challenges for characterization. Therefore, the characterization of the samples during and after various treatments via corollary *in situ* or *ex situ* STEM, XAS characterization, and DFT calculation is critical for developing definitive probe molecule IR assignments to different adsorption sites.

5.3.3 Mössbauer Spectroscopy

The Mössbauer spectrum is a γ-ray absorption spectrum measured by the Mössbauer spectrometer based on the Mössbauer effect [70]. The Mössbauer effect is the phenomenon in which atomic nuclei emit or resonantly absorb γ-ray without recoil. When γ-rays pass through a material, if the energy of the incident γ photon is equal to the energy level transition energy of some nuclei in the material, the γ photons of this energy will be resonantly absorbed by the atom. The corresponding relationship between the number of γ photons and energy measured after absorption is the Mössbauer spectrum. Mössbauer spectroscopy is similar to infrared absorption spectroscopy, but the electromagnetic wave source for the excitation is γ rays with extremely short wavelengths. Since the Mössbauer effect involves the properties of the nucleus, including the energy level structure of the nucleus and the chemical environment in which the nucleus is located, the Mössbauer spectrum can be used to determine the valence state of the atom, the ionicity of the chemical bond, and the coordination number.

Liu et al. prepared three different types of Fe–N–C catalysts (Fe–N–C-600, Fe–N–C-700, and Fe–N–C-600,) by pyrolyzing the precursors at 600, 700, and 800 °C under N_2 atmosphere, respectively [71]. Mössbauer spectra provided solid evidence that the relative concentration of each FeN_x species is critically dependent on the pyrolysis temperature [71]. For both Fe–N–C-600 and Fe–N–C-700 samples, the spectra can be well-fitted with three doublets. The absence of a sextet and singlet indicates the absence of Fe^0 species. By contrast, in the Fe–N–C-800 sample, besides two doublets, one singlet (S1), and one sextet (Sext1) corresponding to γ-Fe and Fe_xC species appear in the spectrum. In comparison, the doublets for high-spin X–$Fe^{III}N_4$–Y exists in all the three samples, and even become the predominant Fe species in both Fe–N–C-600 and Fe–N–C-800 samples, while it appears to be the least populated species in Fe–N–C-700. On the other hand, the concentration of doublets for low-spin N–($Fe^{III}N_4$)–N increases with the pyrolysis temperature, until it completely disappears in Fe–N–C-800. This result seems to suggest that the doublets for low-spin N–($Fe^{III}N_4$)–N structure is destroyed and then the aggregation of Fe^0 occurs as a consequence of pyrolysis at higher temperatures (800 °C or above). Note that the concentration of doublets for medium-spin N–($Fe^{III}N_4$) gets maximized in Fe–N–C-700, which might bring about activity enhancement. By combining the Mössbauer spectra analysis with performance and DFT simulations, the medium-spin $Fe^{III}N_5$ was confirmed to possess the most active sites for selective oxidation of the C—H bond.

Besides SACs, Mössbauer spectra can also determine the formation of bimetallic bonds. Wang and coworkers develop a host–guest strategy to construct an electrocatalyst with Fe–Co dual sites embedded on N-doped porous carbon [34]. To better understand the coordination environment of Fe–Co dual sites, Mössbauer spectroscopic analysis was conducted. The Mössbauer spectrum of Fe_{SA}/N–C can be fitted with three doublets. The doublets D1, D2, and D3 can be assigned to square-planar $Fe^{II}N_4$ coordination with Fe(II) in a high-, low-, and intermediate-spin state, respectively. The (Fe, Co)/N–C also consists of three FeN_4-centers (D1–D3), which are similar to those of the Fe_{SA}/N–C. More specifically, the occurrence of the minor amount of iron species (singlet component) in (Fe, Co)/N–C positively demonstrates the presence of Fe—Co bonding.

5.3.4 X-ray Photoelectron Spectroscopy (XPS)

XPS is a technique for analyzing the chemical properties of the surface of a substance. It is a quantitative energy spectroscopy technique used to determine the composition and experimental formula of elements in materials, as well as the chemical and electronic states of the elements contained therein. This technique uses X-rays to irradiate the material to be analyzed, and at the same time measures the kinetic energy and number of electrons that escape from the surface of the material in the range of 1–10 nm, thereby obtaining an X-ray photoelectron spectrum. XPS has been widely used in characterizing the elemental composition, chemical state of the support, and electronic state of the SAC.

Carbon-based supports are commonly used for the preparation of SACs. XPS can distinguish between a detailed carbon and doped N structure according to the peak positions. For instance, Zhang et al. [72] employed three different types of N-doped CNT as supports (denoted hereafter as CNT, N-CNT, and HN-CNT) for the deposition of Pt_{SA}. XPS results indicate that the N contents in the CNT, N-CNT, and HN-CNT are 1.7%, 3.6%, and 9.4%, respectively. The fitted peaks of the C 1s spectral components at bond energies of 284.5, 285, and 286.8 eV correspond to the C—C bonds, C—C structural defects, and C=O chemical bonds, respectively. With an increase in N-doping, the ratio of the C—C defect peak increased gradually, indicating the creation of defects by N-doping. The asymmetric N 1s spectra can be fit into six peaks at 398.2 eV (N1), 399.8 eV (N2), 401.0 (N3), 402.0 (N4), 403.9 eV (N5), and 405.0 eV (N6). The peaks at 398.2, 399.8, and 401.0 eV are attributed to pyridine-like nitrogen, pyrrole-like nitrogen, and quaternary N, respectively. The peaks N3, N4, and N5 can be assigned to the triply coordinated configuration, with oxidized nitrogen and molecular nitrogen inside the cavities of the nanotubes. By comparing the area of different peaks, it can be found that the ratio of pyrrole-like nitrogen to pyridine-like nitrogen is close to 1 : 1 in HN-CNT, while the ratio in N-CNT is only around 1 : 3 (Figure 5.8a,b) [72]. This result indicated that the pyrrole-like nitrogen exists widely in the HN-CNT structure, which might be the main reason for the well-dispersed Pt SAC and effectively enhanced HER activity.

XPS is also widely used to characterize the valence state and composition evolution during the synthesis. For example, Zhao and coworkers employed MOFs to assist the preparation of a catalyst containing Ni single atoms (Ni_{SA}/N–C) for efficient electroreduction of CO_2 [52]. In the XPS spectra, the binding energy of the Ni $2p_{3/2}$ peak was 855.1 eV for Ni_{SA}/N–C, which is higher than that reported for Ni^0

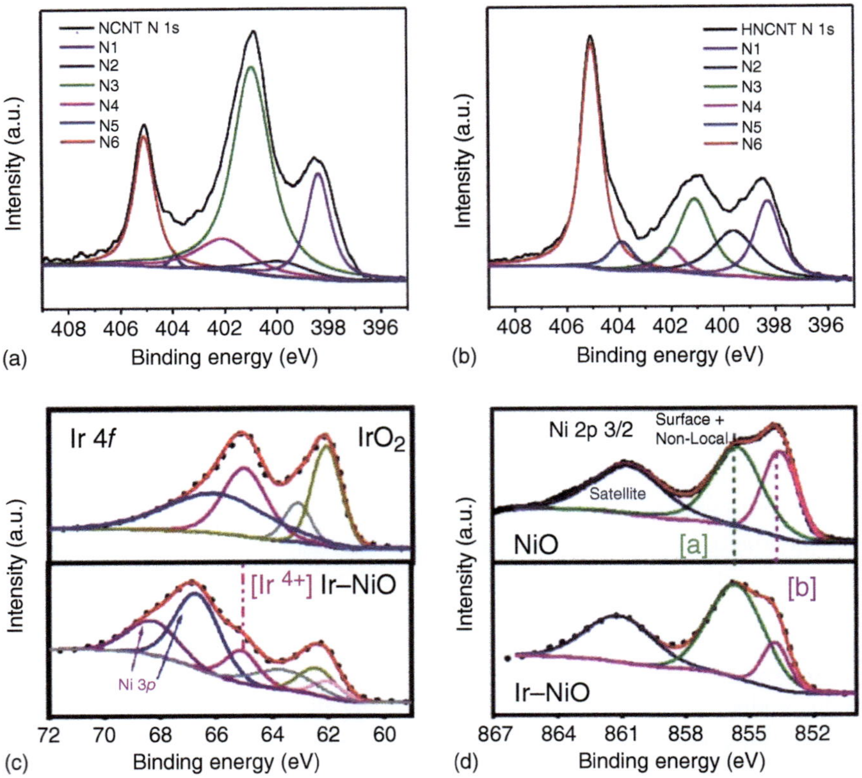

Figure 5.8 (a, b) XPS spectra of N 1s for N-CNT and HN-CNT. Source: Zhang et al. [72]. Reproduced with permission of Wiley-VCH. (c) Curve-resolved XPS of the Ir 4f region for IrO_2 and Ir–NiO. (d) Curve-resolved XPS of the Ni 2p region for NiO and Ir–NiO. Source: Wang et al. [73]. Reproduced with permission of American Chemical Society.

(852.5–853.0 eV) and lower than that for Ni^{2+} (855.7 eV), indicating the ionic $Ni^{\delta+}$ ($0 < \delta < 2$) nature of Ni in Ni_{SA}/N–C. Wang et al. [73] obtained an unprecedented high-loading of single Ir atoms, up to ~18 wt%, on a nickel oxide (NiO) matrix as the electrocatalyst for OER. They used XPS to study the electronic structure of Ir and Ni. As shown in Figure 5.8c, a typical IrO_2 XPS spectrum shows two sets of doublets centered at 62.1 and 65.0 eV, 63.1 and 66.0, which can be attributed to mainly Ir^{4+}. While for the Ir–NiO, the Ir 4f spectrum can be deconvoluted into two sets of doublets centered at 62.3 and 65.1 eV, 61.9 and 63.8 eV, indicating the presence of Ir^{4+} and Ir^{3+}. In addition, the deposition of Ir atoms onto the NiO surface would affect the electronic structure of Ni. As shown in Figure 5.8d, the main peaks of Ni $2p_{2/3}$ at 853.7 eV for Ir–NiO are much higher than that of NiO, indicating an increased valance state of Ni atoms on the surface. The chemical state and most importantly the electronic state of the metal within an SA-containing material is probed by this technique. In most supported SACs, a significant charge transfer between the isolated atom and support is present. The SA is often found to be electron-deficient, while the same metal in a NP will be at zero oxidation state [74]. It is nevertheless important to note that significant charge transfer can also be measured in the case of low atom count clusters, making the distinction between SA and clusters very delicate.

Besides the detection of the electronic state, XPS can determine the metal loading of the SAC by calculating the integral area of the elemental peaks. Because of the limited detection depth (less than 10 nm), XPS is usually used for semiquantitative analysis of the mass loading of single-atom materials, which provides complementary information to inductively coupled plasma or thermogravimetric analysis (TGA) results. For example, Liu et al. [75] prepared Co–N–C catalysts wherein cobalt is dispersed exclusively as single atoms by using a support-sacrificial approach. They found that the atomic concentration of Co determined by XPS was 0.7 at%, corresponding to a weight percentage of 3.5 wt%, which was very close to the inductively coupled plasma atomic emission spectrometry (ICP-AES) analysis (3.6 wt%).

5.3.5 Solid-State Nuclear Magnetic Resonance

Solid-state nuclear magnetic resonance (SSNMR) is an analysis technique that uses solid samples as the research object. In a liquid sample, the rapid movement of molecules will cause the nuclear magnetic resonance (NMR) spectrum to average out various interactions (such as chemical shift anisotropy and dipole–dipole interactions, etc.), thereby obtaining high-resolution liquid NMR spectra. For solid samples, the rapid movement of molecules is restricted, and the existence of various effects such as chemical shift anisotropy makes the spectral line broaden seriously, so the resolution of solid nuclear magnetic resonance technology is lower than that of liquid. Magic-angle spinning (MAS) makes the sample tube (rotor) rotate rapidly in the direction of 54.7° to the static magnetic field, achieving a result similar to the rapid movement of molecules in the liquid, and thus improving the spectral resolution. Usually, the speed range is 0–80 kHz. MAS can effectively average various anisotropic effects and achieve high resolution for solid-state NMR spectroscopy.

SSNMR can be applied for studying the different local environments around various nuclei, that is, short- and medium-range interactions, which can provide rich and detailed structural information. It can be used for structural analysis of solid materials with high crystallinity, low crystallinity, and amorphous materials. The spectra can reflect the bond length, bond angle, hydrogen bond formation, intra- and intermolecular interference in the molecular structure, and can be used with X-ray diffraction to study the overall structure of the materials. It has also been applied for the characterization of supported single atoms [76]. For example, SSNMR can directly distinguish alkyl, alkylidene, and alkylidyne by ^{13}C NMR5 or amido, imido, and nitrido groups by ^{15}N NMR in surface complexes. In addition, ^{27}Al with different coordination numbers and ^{29}Si in different environments have been widely investigated by their NMR signatures [77]. Recently, Wu et al. report amorphous phosphorus nitride imide nanotubes (PNs) as a support to stabilize Ru single metal sites materials by its abundant dangling unsaturated P vacancies [78]. The P L-edge EXAFS spectra of Ru SACs on PNs and PNs have little changes in the peak position and shape, which means that the P skeleton is stable after the loading of single Ru atoms and the N coordination is decisive to the stabilization. The

Solid-state MAS NMR spectrum further demonstrates that there is little chemical ^{31}P difference between Ru SACs on PNs and PNs.

5.3.6 Electron Paramagnetic Resonance (EPR)

Electron paramagnetic resonance (EPR) is a magnetic resonance technique, which is very similar to NMR (nuclear magnetic resonance). However, this technique does not detect the nuclear transition in the sample but detects the transition of unpaired electrons in an external magnetic field. EPR can be used to qualitatively and quantitatively detect the unpaired electrons contained in the atoms or molecules of substances and explore the structural characteristics of the surrounding environment. For free radicals, the orbital magnetic moment has almost no effect. The majority of the total magnetic moment (more than 99%) contributes to the electron spin, so EPR is also called electron spin resonance (ESR).

There are many types of materials containing unpaired electrons, which contain free radicals, multiple transition metal ions, or defects. The lifetime of a free electron is usually very short, but it can still play a vital role in many processes, such as photosynthesis, oxidation, catalysis, polymerization, and so on. Therefore, EPR is a technique that spans multiple disciplines, including chemistry, physics, biology, materials science, medicine, etc. EPR technique can clearly detect unpaired electrons. Other techniques such as fluorescence detection can only provide indirect evidence about free radicals, and only EPR can accurately prove the existence of free radicals. In addition, EPR also has the unique ability to detect the local environment of supports [79]. For example, Baumann and Paul combined the high-energy resolution of conventional spin resonance with STM to measure the EPR of individual iron (Fe) atoms placed on a magnesium oxide film. They drove the spin resonance with an oscillating electric field (20–30 GHz) between tip and sample. In addition, they found that the spin resonance signals of different Fe atoms differ much more than their resonance linewidth [80]. Liu and coworkers reported the successful synthesis of a Co–N–C catalyst with single-atom dispersion (Figure 5.9a). EPR further proved the existence of Co^{2+} on the supports (Figure 5.9b) [75].

5.3.7 Photoluminescence

Photoluminescence (PL) is a type of luminescence, which refers to the process of absorbing photons (or electromagnetic waves) and then reradiating photons (or electromagnetic waves). From the theory of quantum mechanics, matter, under radiation, absorbs a photon and transitions to a higher-energy excited state, and then returns to an initial state, while emitting photons. The energy of photoluminescence radiation is related to the different energy level differences between two electronic states, which involves the transition between the excited state and the equilibrium state. The amount of excitation light is related to the contribution of the radiation process. Photoluminescence allows detecting the electronic structure of a material and is a nondestructive technique. It can be applied to band gap detection,

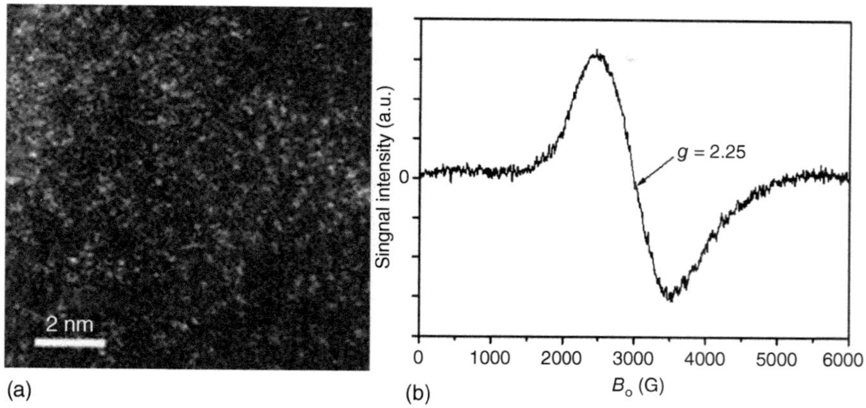

Figure 5.9 (a) Atomic resolution STEM and (b) EPR spectra obtained at room temperature on the Co–N–C catalyst. Source: Liu et al. [75]. Adapted with permission of Royal Society of Chemistry.

impurity level and defect detection, composite mechanisms, and material quality identification.

Mori et al. investigated the luminescent property of a single-atom Eu^{3+} complex on a framework of exfoliated g-C_3N_4 nanosheets. The photoluminescence (PL) spectra of the samples are investigated at room temperature. The Eu SACs exhibited characteristic Eu^{3+} emissions at around 579, 592, 614, 650, and 700 nm, which are attributed to the $5D_0 \rightarrow 7F_2$ transitions [81]. Ishikawa investigated the luminescent centers of individual Ce dopants in wurtzite-type aluminum nitride (w-AlN). The homogeneous, pink-colored luminescence shows that the single Ce atoms are uniformly doped throughout the single crystals. A single broadband peak at 600 nm can be assigned to the 4f–5d electron transition of Ce^{3+} dopants in a bulk material [82].

5.4 Loading Amount Characterization

5.4.1 Inductively Coupled Plasma Atomic Emission Spectrometry

Inductively coupled plasma atomic emission spectrometry (ICP-AES) and inductively coupled plasma optical emission spectrometry (ICP-OES) are a type of atomic emission spectrometry analysis method that uses plasma as the excitation light source, which allows simultaneous qualitative and quantitative analysis of multiple elements of a liquid sample. The analysis of a solid sample needs a mineralization step, usually by an acid solution. First, the liquid sample enters the central channel of the plasma in the form of aerosol, and is fully evaporated, atomized, ionized, and excited in a high temperature and inert atmosphere, such that the contained elements can emit their respective characteristic spectral lines. According to the presence or absence of the characteristic spectrum of each element, we can identify whether the sample contains a certain element (qualitative analysis). In addition, the content of the corresponding element in the sample can be determined by the intensity of the characteristic spectrum (quantitative analysis).

To obtain the loading content of the supported SAC, the samples are usually dissolved in hot fresh aqua regia first. Because the carbon support cannot be dissolved, the solution should be filtered for further testing. For example, Zhao et al. confirmed that the actual loading of Ni in Ni_{SA}/N–C was 1.53% by ICP-AES [52]. Cheng and coworkers confirmed the Pt loading after ALD was 2.1 and 7.6 wt% for the 50 and 100 ALD cycled samples using ICP-AES [8].

5.4.2 Thermogravimetric Analysis

TGA is a method of measuring the relationship between the mass of a substance and temperature or time under program control temperature (and atmosphere). The instrument for TGA, called thermogravimeter, is mainly composed of three parts: temperature control system, detection system, and recording system. TGA is an analysis method of changing the physical and chemical properties of materials with the increase of temperature or time. TGA can provide information about physical phenomena, including evaporation, sublimation, absorption, adsorption, and desorption. In addition, TGA can also provide information about chemical phenomena, including chemical adsorption, desolvation, decomposition, and solid–gas phase reactions. Oxidation loss is the most commonly observed loss in TGA under air, which can be used in determining the mass loading of SAC.

For example, Zhao et al. develop a general cascade anchoring strategy for the mass production of a Fe SAC, where the amount of Fe in Fe–NC SAC is determined by TGA [83]. The initial weight loss below 100 °C should be attributed to the evaporation of adsorbed H_2O. The weight loss between 300 and 550 °C comes from carbon combustion. By heating the sample to 700 °C, the only product is Fe_2O_3 and its corresponding weight is 12.71 wt%. Therefore, the loading content of Fe (8.9%) can be calculated according to this equation: m% (Fe) = residual mass *M (Fe)/M (Fe_2O_3). In addition to Fe, the metal loading of other M–NC SAC (M = Mn, Co, Ni, Cu, Mo, Pt, etc.) was determined from 12.1 to 4.5 wt% by TGA.

5.5 Summaries and Outlook

In this chapter, we summarized the characterization methods for the supported single atoms. With the atomic resolution of TEM and STM, the morphology of isolated single atoms and detailed structure of supports can be clearly observed. The metal loading of the supported SAC can be determined by ICP-AES, TGA, and XPS analyses. More importantly, the catalytic performance of SAC is highly dependent on the coordination environment of the atoms. XAS, XPS, FTIR, and Mössbauer spectroscopy provide several helpful routes to identify the electronic structure, coordination number, and oxidation state of the SAC, which can systematically investigate the enhanced mechanism of the SAC for specific catalytic reactions.

Although the morphology and structure of the supported SAC can be characterized, the characterization methods of SACs are not yet satisfactory. With the development of SACs, more complicated catalysts will be developed. For example, dimer catalysts show distinct catalytic activities compared with SACs, due to the strong perturbation of the second metal single atom. It can be predicted that trimer catalysts will be prepared in the future (see Chapter 15), which will require more precise characterization methods to identify their morphology and electronic structure.

Additionally, it is of great importance to develop advanced reaction devices for further *in situ* analysis, such as *in situ* TEM and *in situ* XAS (see Chapter 6). As described in the chapter, *in situ* measurements can systematically analyze the morphology and coordination environment change of single atoms during the catalytic reactions. Although several studies have reported *in situ* XAS analysis for electrocatalytic reactions, the detailed morphology change cannot be observed by *in situ* TEM testing. In addition, few studies reported *in situ* XAS study for SACs in flow cell systems and membrane electrode assembly for CO_2 reduction and fuel cells. Therefore, the development of advanced reaction devices for direct investigation of the catalyst evolution is still a great challenge.

Another major challenge, which hinders the broad application of SACs, is their stabilization under realistic working conditions. During the catalytic reactions, the atoms tend to move on the surface of the support and aggregate to clusters with lower energy, which would cause a significant decrease in activity. Atomic-scale information on the diffusion of single atoms on supports and the evolution of SAC structures during reactions is extremely useful to establish sintering mechanisms of SACs. In addition, further characterization on the correlation between single atom coordination structures, quantitative analysis of the degree of metal-support interactions is also important to the stabilization of SACs.

References

1 Zhang, L., Doyle-Davis, K., and Sun, X. (2019). Pt-Based electrocatalysts with high atom utilization efficiency: from nanostructures to single atoms. *Energy & Environmental Science* 12 (2): 492–517.

2 Zhang, W., Liu, Y., Zhang, L. et al. (2019). Recent advances in isolated single-atom catalysts for zinc air batteries: a focus review. *Nanomaterials* 9 (10): 1402.

3 Yao, Y., Huang, Z., Xie, P. et al. (2019). Ultrafast, controllable synthesis of sub-nano metallic clusters through defect engineering. *ACS Applied Materials & Interfaces* 11 (33): 29773–29779.

4 Liu, Q. and Wang, X. (2020). Polyoxometalate clusters: sub-nanometer building blocks for construction of advanced materials. *Matter* 2 (4): 816–841.

5 Zhang, L., Roling, L.T., Wang, X. et al. (2015). Platinum-based nanocages with subnanometer-thick walls and well-defined, controllable facets. *Science* 349 (6246): 412.

6 Qiao, B., Wang, A., Yang, X. et al. (2011). Single-atom catalysis of CO oxidation using Pt1/FeO_x. *Nature Chemistry* 3 (8): 634–641.

7 Chen, Y., Ji, S., Zhao, S. et al. (2018). Enhanced oxygen reduction with single-atomic-site iron catalysts for a zinc-air battery and hydrogen-air fuel cell. *Nature Communication* 9: 5422.

8 Cheng, N., Stambula, S., Wang, D. et al. (2016). Platinum single-atom and cluster catalysis of the hydrogen evolution reaction. *Nature Communication* 7: 13638.

9 Genovese, C., Schuster, M.E., Gibson, E.K. et al. (2018). Operando spectroscopy study of the carbon dioxide electro-reduction by iron species on nitrogen-doped carbon. *Nature Communication* 9: 935.

10 Li, H., Wang, L., Dai, Y. et al. (2018). Synergetic interaction between neighbouring platinum monomers in CO_2 hydrogenation. *Nature Nanotechnology* 13 (5): 411–417.

11 Wei, S., Li, A., Liu, J.C. et al. (2018). Direct observation of noble metal nanoparticles transforming to thermally stable single atoms. *Nature Nanotechnology* 13 (9): 856–861.

12 Cao, L., Liu, W., Luo, Q. et al. (2019). Atomically dispersed iron hydroxide anchored on Pt for preferential oxidation of CO in H_2. *Nature* 565 (7741): 631–635.

13 Lin, L., Zhou, W., Gao, R. et al. (2017). Low-temperature hydrogen production from water and methanol using Pt/alpha-MoC catalysts. *Nature* 544 (7648): 80–83.

14 Li, J., Chen, M., Cullen, D.A. et al. (2018). Atomically dispersed manganese catalysts for oxygen reduction in proton-exchange membrane fuel cells. *Nature Catalysis* 1 (12): 935–945.

15 Yao, Y., Hu, S., Chen, W. et al. (2019). Engineering the electronic structure of single atom Ru sites via compressive strain boosts acidic water oxidation electrocatalysis. *Nature Catalysis* 2 (4): 304–313.

16 Ding, K., Gulec, A., Johnson, A.M. et al. (2015). Identification of active sites in CO oxidation and water-gas shift over supported Pt catalysts. *Science* 350 (6257): 189–192.

17 Liu, P., Zhao, Y., Qin, R. et al. (2016). Photochemical route for synthesizing atomically dispersed palladium catalysts. *Science* 352 (6287): 797–801.

18 Malta, G., Kondrat, S.A., Freakley, S.J. et al. (2017). Identification of single-site gold catalysis in acetylene hydrochlorination. *Science* 355 (6332): 1399.

19 Cai, Z.X., Wang, Z.L., Kim, J. et al. (2019). Hollow functional materials derived from metal-organic frameworks: synthetic strategies, conversion mechanisms, and electrochemical applications. *Advanced Materials* 31 (11): e1804903.

20 Kim, J., Kim, H.E., and Lee, H. (2018). Single-atom catalysts of precious metals for electrochemical reactions. *ChemSusChem* 11 (1): 104–113.

21 Pennycook, S.J., Rafferty, B., and Nellist, P.D. (2000). Z-contrast imaging in an aberration-corrected scanning transmission electron microscope. *Microscopy and Microanalysis* 6 (4): 343–352.

22 Dahmen, U., Erni, R., Radmilovic, V. et al. (2009). Background, status and future of the transmission electron aberration-corrected microscope project. *Philosophical Transactions of the Royal Society A* 367 (1903): 795–3808.

23 Lentzen, M., Jahnen, B., Jia, C.L. et al. (2002). High-resolution imaging with an aberration-corrected transmission electron microscope. *Ultramicroscopy* 92 (3–4): 233–242.

24 Chen, Y., Ji, S., Wang, Y. et al. (2017). Isolated single iron atoms anchored on N-doped porous carbon as an efficient electrocatalyst for the oxygen reduction reaction. *Angewandte Chemie International Edition* 56 (24): 6937–6941.

25 Zhang, L., Jia, Y., Gao, G. et al. (2018). Graphene defects trap atomic Ni species for hydrogen and oxygen evolution reactions. *Chem* 4 (2): 285–297.

26 Liu, J., Lucci, F.R., Yang, M. et al. (2016). Tackling CO poisoning with single-atom alloy catalysts. *Journal of the American Chemical Society* 138 (20): 6396–6399.

27 Lucci, F.R., Liu, J., Marcinkowski, M.D. et al. (2015). Selective hydrogenation of 1,3-butadiene on platinum-copper alloys at the single-atom limit. *Nature Communication* 6: 8550.

28 Zhou, X., Yang, W., Chen, Q. et al. (2016). Stable Pt single atoms and nanoclusters on ultrathin CuO film and their performances in CO oxidation. *The Journal of Physical Chemistry C* 120 (3): 1709–1715.

29 Sun, G., Zhao, Z.J., Mu, R. et al. (2018). Breaking the scaling relationship via thermally stable Pt/Cu single atom alloys for catalytic dehydrogenation. *Nature Communication* 9 (1): 4454.

30 Zhang, L., Liu, H., Liu, S. et al. (2019). Pt/Pd single-atom alloys as highly active electrochemical catalysts and the origin of enhanced activity. *ACS Catalysis* 9 (10): 9350–9358.

31 Li, Z., He, H., Cao, H. et al. (2019). Atomic Co/Ni dual sites and Co/Ni alloy nanoparticles in N-doped porous Janus-like carbon frameworks for bifunctional oxygen electrocatalysis. *Applied Catalysis B: Environmental* 240: 112–121.

32 Lu, Z., Wang, B., Hu, Y. et al. (2019). An isolated zinc-cobalt atomic pair for highly active and durable oxygen reduction. *Angewandte Chemie International Edition* 58 (9): 2622–2626.

33 Ren, W., Tan, X., Yang, W. et al. (2019). Isolated diatomic Ni-Fe metal-nitrogen sites for synergistic electroreduction of CO_2. *Angewandte Chemie International Edition* 58 (21): 6972–6976.

34 Wang, J., Huang, Z., Liu, W. et al. (2017). Design of N-coordinated dual-metal sites: a stable and active Pt-free catalyst for acidic oxygen reduction reaction. *Journal of the American Chemical Society* 139 (48): 17281–17284.

35 Xiao, M., Zhang, H., Chen, Y. et al. (2018). Identification of binuclear Co_2N_5 active sites for oxygen reduction reaction with more than one magnitude higher activity than single atom CoN_4 site. *Nano Energy* 46: 396–403.

36 Ye, W., Chen, S., Lin, Y. et al. (2019). Precisely tuning the number of Fe atoms in clusters on N-doped carbon toward acidic oxygen reduction reaction. *Chem* 5 (11): 2865–2878.

37 Zhang, L., Fischer, J., Jia, Y. et al. (2018). Coordination of atomic Co-Pt coupling species at carbon defects as active sites for oxygen reduction reaction. *Journal of the American Chemical Society* 140 (34): 10757–10763.

38 Zhang, L., Si, R., Liu, H. et al. (2019). Atomic layer deposited Pt-Ru dual-metal dimers and identifying their active sites for hydrogen evolution reaction. *Nature Communication* 10 (1): 4936.

39 Zhao, J., Zhao, J., Li, F. et al. (2018). Copper dimer supported on a C_2N layer as an efficient electrocatalyst for CO_2 reduction reaction: a computational study. *The Journal of Physical Chemistry C* 122 (34): 19712–19721.

40 Zhu, W., Zhang, L., Liu, S. et al. (2020). Enhanced CO_2 electroreduction on neighboring Zn/Co monomers by electronic effect. *Angewandte Chemie International Edition* 59 (31): 12664–12668.

41 Wang, J., Liu, W., Luo, G. et al. (2018). Synergistic effect of well-defined dual sites boosting the oxygen reduction reaction. *Energy & Environmental Science* 11 (12): 3375–3379.

42 Egerton, R.F. and Watanabe, M. (2018). Characterization of single-atom catalysts by EELS and EDX spectroscopy. *Ultramicroscopy* 193: 111–117.

43 Binnig, G. and Rohrer, H. (1983). Scanning tunneling microscopy. *Surface Science* 126 (1): 236–244.

44 Binnig, G. and Rohrer, H. (1987). Scanning tunneling microscopy – from birth to adolescence. *Reviews of Modern Physics* 59 (3): 615–625.

45 Krenner, W., Klappenberger, F., Kühne, D. et al. (2011). Positioning of single Co atoms steered by a self-assembled organic molecular template. *The Journal of Physical Chemistry Letters* 2 (13): 1639–1645.

46 Lucci, F.R., Darby, M.T., Mattera, M.F. et al. (2016). Controlling hydrogen activation, spillover, and desorption with Pd–Au single-atom alloys. *The Journal of Physical Chemistry Letters* 7 (3): 480–485.

47 Catlow, C.R.A. and Greaves, G.N. (1990). *Applications of Synchrotron Radiation.* New York: Springer Netherlands.

48 Iwasawa, Y. (1996). *X-ray Absorption Fine Structure for Catalysts and Surfaces.* River Edge, N.J.: World Scientific.

49 Koch, E.-E., Brown, G.S., and Moncton, D.E. (1983). *Handbook on Synchrotron Radiation.* New York: Elsevier Science Pub. Co.

50 Rullhusen, P., Artru, X., and Dhez, P. (1998). *Novel Radiation Sources Using Relativistic Electrons : from Infrared to X-Rays.* River Edge, N.J.: World Scientific.

51 Saisho, H. and Gohshi, Y. (1996). *Applications of Synchrotron Radiation to Materials Analysis.* New York: Elsevier.

52 Zhao, C., Dai, X., Yao, T. et al. (2017). Ionic exchange of metal-organic frameworks to access single nickel sites for efficient electroreduction of CO_2. *Journal of the American Chemical Society* 139 (24): 8078–8081.

53 Sun, S., Zhang, G., Gauquelin, N. et al. (2013). Single-atom catalysis using Pt/graphene achieved through atomic layer deposition. *Scientific Reports* 3: 1775.

54 Chen, Y., Ji, S., Sun, W. et al. (2018). Discovering partially charged single-atom Pt for enhanced anti-markovnikov alkene hydrosilylation. *Journal of the American Chemical Society* 140 (24): 7407–7410.

55 Wang, X., Chen, Z., Zhao, X. et al. (2018). Regulation of coordination number over single Co sites: triggering the efficient electroreduction of CO_2. *Angewandte Chemie International Edition* 57 (7): 1944–1948.

56 Fang, S., Zhu, X., Liu, X. et al. (2020). Uncovering near-free platinum single-atom dynamics during electrochemical hydrogen evolution reaction. *Nature Communication* 11: 1029.

57 Jiang, K., Siahrostami, S., Zheng, T. et al. (2018). Isolated Ni single atoms in graphene nanosheets for high-performance CO_2 reduction. *Energy & Environmental Science* 11 (4): 893–903.

58 Jiang, Z., Sun, W., Shang, H. et al. (2019). Atomic interface effect of a single atom copper catalyst for enhanced oxygen reduction reactions. *Energy & Environmental Science* 12 (12): 3508–3514.

59 Yang, H.B., Hung, S.-F., Liu, S. et al. (2018). Atomically dispersed Ni(i) as the active site for electrochemical CO_2 reduction. *Nature Energy* 3 (2): 140–147.

60 Zhang, Z., Xiao, J., Chen, X.J. et al. (2018). Reaction mechanisms of well-defined metal-N_4 sites in electrocatalytic CO_2 reduction. *Angewandte Chemie International Edition* 57 (50): 16339–16342.

61 Hulsey, M.J., Zhang, B., Ma, Z. et al. (2019). In situ spectroscopy-guided engineering of rhodium single-atom catalysts for CO oxidation. *Nature Communication* 10: 1330.

62 Feng, K., Zhang, H., Gao, J. et al. (2020). Single atoms or not? The limitation of EXAFS. *Applied Physics Letters* 116 (19): 191903.

63 DeRita, L., Dai, S., Lopez-Zepeda, K. et al. (2017). Catalyst architecture for stable single atom dispersion enables site-specific spectroscopic and reactivity measurements of CO adsorbed to Pt atoms, oxidized Pt clusters, and metallic Pt clusters on TiO_2. *Journal of the American Chemical Society* 139 (40): 14150–14165.

64 Hoffman, A.S., Fang, C.Y., and Gates, B.C. (2016). Homogeneity of surface sites in supported single-site metal catalysts: assessment with band widths of metal carbonyl infrared spectra. *The Journal of Physical Chemistry Letter* 7 (19): 3854–3860.

65 Matsubu, J.C., Yang, V.N., and Christopher, P. (2015). Isolated metal active site concentration and stability control catalytic CO_2 reduction selectivity. *Journal of the American Chemical Society* 137 (8): 3076–3084.

66 Yang, D., Xu, P., Browning, N.D. et al. (2016). Tracking Rh atoms in zeolite HY: first steps of metal cluster formation and influence of metal nuclearity on catalysis of ethylene hydrogenation and ethylene dimerization. *The Journal of Physical Chemistry Letter* 7 (13): 2537–2543.

67 Yang, D., Xu, P., Guan, E. et al. (2016). Rhodium pair-sites on magnesium oxide: synthesis, characterization, and catalysis of ethylene hydrogenation. *Journal of Catalysis* 338: 12–20.

68 Zhao, Y., Yang, K.R., Wang, Z. et al. (2018). Stable iridium dinuclear heterogeneous catalysts supported on metal-oxide substrate for solar water oxidation. *Proceedings of the National Academy of Sciences of the United States of America* 115 (12): 2902–2907.

69 Asokan, C., DeRita, L., and Christopher, P. (2017). Using probe molecule FTIR spectroscopy to identify and characterize Pt-group metal based single atom catalysts. *Chinese Journal of Catalysis* 38 (9): 1473–1480.

70 Stankov, S., Ślęzak, T., Zając, M. et al. (2013). In situ Mössbauer spectroscopy with synchrotron radiation on thin films. In: *Mössbauer Spectroscopy*, 1–42. Wiley.

71 Liu, W., Zhang, L., Liu, X. et al. (2017). Discriminating catalytically active FeN_x species of atomically dispersed Fe–N–C catalyst for selective oxidation of the C-H bond. *Journal of the American Chemical Society* 139 (31): 10790–10798.

72 Zhang, L., Wang, Q., Si, R. et al. (2021). New insight of pyrrole-like nitrogen for boosting hydrogen evolution activity and stability of Pt single atoms. *Small* 17 (16): 2004453. https://doi.org/10.1002/smll.202004453.

73 Wang, Q., Huang, X., Zhao, Z.L. et al. (2020). Ultrahigh-loading of Ir single atoms on NiO matrix to dramatically enhance oxygen evolution reaction. *Journal of the American Chemical Society* 142 (16): 7425–7433.

74 Davies, P.R. and Morgan, D.J. (2020). Practical guide for x-ray photoelectron spectroscopy: applications to the study of catalysts. *Journal of Vacuum Science & Technology A* 38 (3): 033204.

75 Liu, W., Zhang, L., Yan, W. et al. (2016). Single-atom dispersed Co–N–C catalyst: structure identification and performance for hydrogenative coupling of nitroarenes. *Chemical Science* 7 (9): 5758–5764.

76 Coperet, C., Liao, W.C., Gordon, C.P. et al. (2017). Active sites in supported single-site catalysts: an NMR perspective. *Journal of the American Chemical Society* 139 (31): 10588–10596.

77 Dědeček, J., Sobalík, Z., and Wichterlová, B. (2012). Siting and distribution of framework aluminium atoms in silicon-rich zeolites and impact on catalysis. *Catalysis Reviews* 54 (2): 135–223.

78 Yang, J., Chen, B., Liu, X. et al. (2018). Efficient and robust hydrogen evolution: phosphorus nitride imide nanotubes as supports for anchoring single ruthenium sites. *Angewandte Chemie International Edition* 57 (30): 9495–9500.

79 Seifert, T.S., Kovarik, S., Nistor, C. et al. (2020). Single-atom electron paramagnetic resonance in a scanning tunneling microscope driven by a radio-frequency antenna at 4 K. *Physical Review Research* 2 (013032).

80 Baumann, S., Paul, W., Choi, T. et al. (2015). Electron paramagnetic resonance of individual atoms on a surface. *Science* 350 (6259): 417.

81 Mori, K., Murakami, T., and Yamashita, H. (2020). Luminescent single-atom Eu-coordinated graphitic carbon nitride nanosheets for selective sensing of acetone and cyclohexane. *ACS Applied Nano Materials* 3 (10): 10209–10217.

82 Ishikawa, R., Lupini, A.R., Oba, F. et al. (2014). Atomic structure of luminescent centers in high-efficiency Ce-doped w-AlN single crystal. *Scientific Reports* 4: 3778.

83 Zhao, L., Zhang, Y., Huang, L.B. et al. (2019). Cascade anchoring strategy for general mass production of high-loading single-atomic metal-nitrogen catalysts. *Nature Communication* 10: 1278.

6

In situ/Operando Techniques for Characterization of Supported Metal Single-Atom Catalysts

Alberto Casu[1], *Samy Ould-Chikh*[2], *Gavin Mountjoy*[3], *Anna Corrias*[3], *and Andrea Falqui*[4]

[1]*King Abdullah University of Science and Technology (KAUST), Biological and Environmental Sciences and Engineering (BESE) Division, 23955-6900 Thuwal, Saudi Arabia*
[2]*King Abdullah University of Science and Technology, KAUST Catalysis Center (KCC), Physical Sciences and Engineering (PSE) Division, 23955-6900 Thuwal, Saudi Arabia*
[3]*University of Kent, School of Physical Sciences, Ingram Building, Canterbury CT2 7NH, UK*
[4]*University of Milan, Department of Physics "Aldo Pontremoli", Via Celoria 16, 20133 Milan, Italy*

6.1 Introduction

In situ and *operando* characterization techniques are of paramount importance in the field of catalysis for two reasons. The first is because catalytic conditions involve applications of gas, temperature, or voltage, and these conditions can be achieved during *in situ* measurements. The second is because catalysis involves cycles, such as oxidation and reduction, and the different stages of these cycles can be studied during *operando* measurements. In this chapter, we will review progress of *in situ/operando* characterization techniques for studying supported metal single-atom catalysts (SACs). In particular, this chapter will cover spectroscopic techniques, such as X-ray absorption spectroscopy (XAS), X-ray photoelectron spectroscopy (XPS), IR spectroscopy, and UV–vis spectroscopy. Spectroscopy techniques are able to provide key information about the bonding and local atomic configurations around SACs. This chapter will also cover electron microscopy techniques of transmission electron microscopy (TEM) and scanning transmission electron microscopy (STEM). Due to the focus on single atoms (SAs), emphasis is placed on electron microscopy techniques that provide atomic-resolution imaging, which are high-resolution transmission electron microscopy (HRTEM), and high-angle annular dark field (HAADF) imaging in STEM. Such techniques are key in providing direct confirmation of atomically dispersed sites, and a wealth of details about the structural relationship between SAs and their supports.

Supported Metal Single Atom Catalysis, First Edition. Edited by Philippe Serp and Doan Pham Minh.
© 2022 WILEY-VCH GmbH. Published 2022 by WILEY-VCH GmbH.

6.2 *In situ/Operando* XAS

6.2.1 Method

XAS is an element-specific technique, which is sensitive to the local atomic structure around the excited atom. There are many references available that introduce XAS (e.g. [1]). XAS is ideal for studying SACs because it targets the element of the SAC even when this element is present in very low concentration, and it does not require the element to be present in a crystalline, or even solid, phase.

In an XAS experiment, the absorption of X-rays is measured as a function of the energy of the X-rays (E), and when E corresponds to the binding energy of core electrons in the target element, they are excited to unoccupied states (i.e. become photoelectrons). This causes a sudden, large, jump in the absorption, which is called an absorption edge. Absorption edges occur for each shell of core electrons of each element present in the sample. In particular, K-edge refers to excitations from the $1s_{1/2}$ ($n = 1$) shell, and L_3-edge refers to excitations from the $2s_{3/2}$ ($n = 2$) shell.

XAS experiments involve scanning the energy of high-intensity X-rays, and therefore they are conducted at synchrotrons. All major synchrotrons have XAS beamlines, and these are usually equipped to allow *in situ* and *operando* measurements. XAS is ideal for studying SACs because there are many XAS beamlines available to study metal atoms with typical absorption edge energies of 5–25 keV, and also because X-rays with such energies can easily pass through X-ray windows into *in situ* and *operando* apparatus. In the following sections, we review *in situ/operando* XAS studies of the K-edges of transition metals in period 3 (Fe, Ni, Co, Cu in Section 6.2.4) and precious metals in period 4 (Ru, Rh, Pd in Section 6.2.5), and of the L_3-edges of precious metals in period 5 (Pt in Sections 6.2.6 and 6.2.7, and Ir and Au in Section 6.2.8).

In situ and *operando* measurements rely on the availability of suitable reactors in an XAS beamline. Various reactive cells have been developed and can be divided into two general categories: (i) packed-bed reactors using pellets optimized for XAS at the detriment of the catalytic performance; and (ii) plug-flow reactors based on powdered samples ideal for mimicking catalytic reactions. XAS measurements are usually made by detecting transmitted X-rays in ion chamber detectors or fluorescent X-rays in semiconductor detectors. It is more convenient to make XAS measurements on reactors in the fluorescent detection mode, and this mode is also more sensitive to target elements at low concentrations.

The measurement time depends on the need to scan X-ray energy by rotating monochromator crystals. *In situ* and *operando* XAS studies benefit from short measurement times, which can be achieved either using the "quick EXAFS" (QEXAFS), or the energy-dispersive EXAFS (EDE) techniques. QEXAFS uses special oscillating monochromator crystals, which rapidly increase and decrease the X-ray energy within one oscillation, and thus the measurement rate is twice the oscillation frequency. EDE uses special bent polychromator crystals to diffract polychromatic X-rays to a focus on the sample that then diverges toward a position-sensitive detector with position correlated to energy; thus, the whole XAS measurement is collected without moving the polychromator crystal.

6.2.2 X-ray Absorption Near-Edge Structure (XANES)

The measurement of X-ray absorption vs. X-ray energy over an absorption edge is referred to as the XAS spectra. In the pre-edge region, the absorption smoothly decreases as E increases, because of the X-rays becoming more penetrating. The region around the edge is called X-ray absorption near-edge structure (XANES), or sometimes near-edge X-ray absorption fine structure (NEXAFS). There are many references about the XANES technique available (e.g. [2]).

The studies reviewed in this chapter routinely analyze the XANES to extract information about the oxidation states of metal atoms in SACs. In K-edge XANES, the energy of the absorption edge increases with the oxidation state (see Figure 6.1), and this is used to study SACs of transition metals and precious metals in period 4. In L_3-edge XANES, the intensity of prominent main absorption edge peak (called the white line) increases with oxidation state, and this is used to study SACs of precious metals in period 5. In both cases, quantitative information can be extracted by careful comparison of the XANES with reference compounds having known oxidation states.

The next most common method for analyzing *in situ/operando* XANES of SACs is known as the "fingerprint" method and compares the peak positions and intensities in the XANES spectra with those from reference compounds with known local atomic environment of the target metal atom (see Figure 6.1 for a Fe SAC). This can be made quantitative in two ways. One is to interpret an observed XANES spectrum as a linear combination of two (or more) specific environments of the target metal atom. The other is to compare the experimental XANES spectrum with a quantum mechanics simulation, for example, using density functional theory (DFT), for a hypothetical environment of the target metal atom (see Figure 6.2 for a Co SAC).

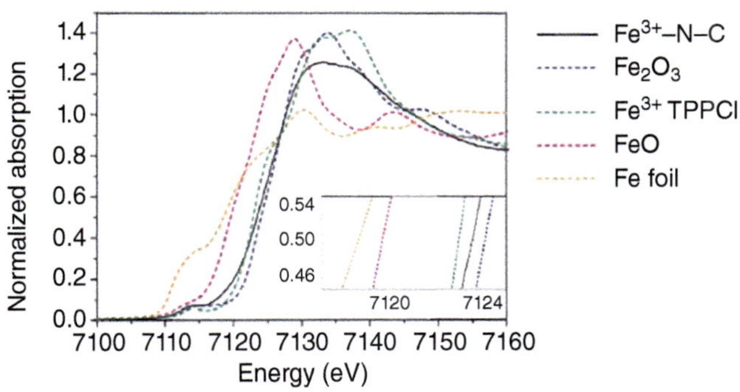

Figure 6.1 Fe K-edge XANES spectra for Fe SAC (Fe^{3+}–N–C), which illustrates sensitivity to oxidation state (see comparison of Fe-foil, FeO, and Fe_2O_3 in inset) and local atomic environment (see comparison of Fe_2O_3 and Fe^{3+} TPPCl). Source: Gu et al. [3]. Reproduced with permission of American Association for the Advancement of Science.

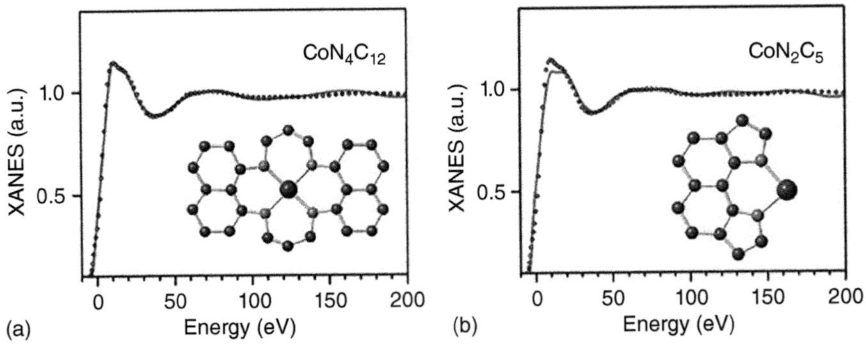

Figure 6.2 (a) and (b) Co K-edge XANES for Co single-atom catalyst from experiment (symbols) and simulations (lines) for hypothetical local atomic environments. Source: Zitolo et al. [4]. Springer Nature CC BY 4.0.

In the case of K-edge XANES of transition metals, a useful approach to interpretation concerns small peaks, which appear at energies slightly before the main absorption edge, which are known as pre-edge peaks. These represent $s \rightarrow p$ transitions of electrons to non-bonding states with p character, which depend on the site symmetry of the target metal atom. In particular, the pre-edge peak is small for centrosymmetric sites (e.g. octahedral) and larger for non-centrosymmetric sites (e.g. tetrahedral).

There is an advanced method of making XAS measurements, which increases the information available in XANES spectra. It involves detecting fluorescent X-rays with a crystal analyzer to greatly increase the energy resolution, and the measurements obtained are referred to as high-energy resolution fluorescence detection (or HERFD) XANES. This method has a lower data acquisition rate compared to normal XANES, which can make it more challenging for target atoms with low concentrations and for short measurements under *operando* conditions.

6.2.3 Extended X-ray Absorption Fine Structure (EXAFS)

At energies well beyond the absorption edge, there is seen to be oscillations in absorption, on top of the general trend, which is to decrease as the energy increases. The oscillations occur when the excited-core electron emitted from the absorbing atom (a photoelectron) is scattered by the neighboring atoms. This region is named EXAFS, which stands for extended X-ray absorption fine structure (or sometimes XAFS). There are many references about the EXAFS technique available (e.g. [5]).

There is a highly quantitative method for analyzing EXAFS spectra (or EXAFS oscillations) by taking the Fourier transform (FT) and fitting the so-called EXAFS equation. This enables determination of the details of neighboring atoms such as atom type, coordination numbers, and interatomic distance (i.e. bond length), and this method does not require reference samples (see Figure 6.3). This method has been routinely applied for *in situ/operando* EXAFS studies of SACs reviewed in this chapter. The application of the method varies, from a simpler approach of only fitting

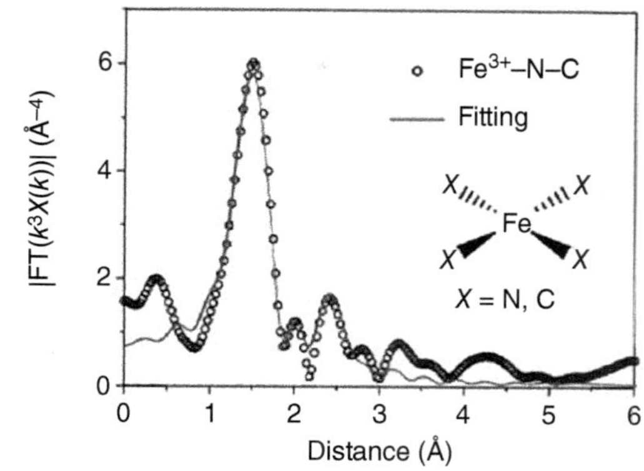

Figure 6.3 Fe K-edge EXAFS Fourier transform (FT) for iron SAC (Fe^{3+}–N–C), which has been fitted using the so-called EXAFS equation to determine the type, number, and bond lengths of nearest-neighbor atoms. Source: Gu et al. [3]. Reproduced with permission of American Association for the Advancement of Science.

one shell of nearest-neighbor atoms, to a more detailed approach of fitting multiple shells corresponding to the geometry of the hypothetical local atomic environments. However, it should be noted that the sensitivity of the EXAFS technique might limit the ability to differentiate SAs from small clusters, and nanoparticles (NPs) [6].

There is a more advanced method of interpreting EXAFS spectra, which has been little used to date. Recently, it has been shown that by using the wavelet transform instead of the Fourier transform it becomes easier to distinguish the type of neighboring atom, including distinguishing neighboring atoms at similar distances and having similar atomic number, e.g. N and O [7].

6.2.4 *In situ/Operando* XAS of SACs Consisting of Transition Metals in Period 3

In situ XAS was used by Jia et al. [8] to study the oxygen reduction reaction (ORR) involving Fe–N_x–C catalysts. The results show that it is the dynamic structure accompanying the $Fe^{2+/3+}$ redox transition, which governs the activities, and not the static structure of the bare sites. This is one of a few studies, which compare experimental XANES spectra with simulations based on hypothetical metal atom sites, in this case the lengths of Fe—N bonds in FeN_4 sites.

The electroreduction of CO_2 (CO_2RR) to CO using an iron catalyst supported on N-doped carbon was studied using *operando* XAS by Gu et al. [3]. It was found that discrete Fe^{3+} ions coordinated to pyrrolic N atoms are active sites, which maintain the +3 oxidation state throughout the reaction.

Wang et al. [9] used *operando* XAS to study a Fe_{SA}/N–C catalyst for high-efficiency alkaline hydrogen evolution reaction (HER). The results showed that the Fe–N_x moiety works as the active site, and the Fe–N coordination number and Fe oxidation state decreased to cause weakening of the Fe–support interaction under working states. Hence, the HER performance is boosted because a more favorable structure is provided by the evolved Fe_{SA} site with more d electrons. This is one of two studies reported in this chapter, which also measured Fe L_3-edge XANES to get more details about Fe 2p to 3d transitions, in this case to confirm the presence of Fe—N bonds and absence of Fe—O bonds.

Operando XAS was used by Genovese et al. [10] to study iron species on N-doped carbon as catalyst for CO_2RR. The results showed that the interface between iron oxyhydroxide and the N-doped carbon provided a location for the formation of nitrogen-coordinated iron (II) sites as single-atoms or polyatomic species. This is one of two studies in this chapter, which also measured Fe L_3-edge XANES to get more details about Fe 2*p* to 3*d* transitions, in this case to scrutinize the contributions from iron (II) in octahedral and square planar sites.

Nakatsuka et al. [11] used *in situ* XAS together with a combination of other techniques to study carbon-supported Co catalysts. These had been prepared by the structural transformation of the deposited Co(salen) complex precursor under heat treatment. The results showed that at first, isolated Co_{SA} are obtained by decomposition of the Co(salen) complex up to 400 °C. The single-site catalyst is highly dispersed and electron deficient due to interaction with the carbon support, and is active and selective for the oxidation of ethylbenzene. Further decomposition when the heat treatment temperature exceeds 450 °C leads to the formation of Co oxide clusters. At higher heat treatment temperature, there is growth of Co_{NP}.

Zitolo et al. [4] synthesized a Co/N–C material pyrolytically to comprise only atomically dispersed cobalt ions. *Operando* XAS in acidic medium was used to compare Co/N–C with an Fe/N–C catalyst prepared similarly. At voltages from 0.0 to 1.0 V vs. a reversible hydrogen electrode (RHE), the Co moieties were found to be unmodified. Instead, there were changes to the structure and electronic state of Fe-based moieties. This is one of a few studies, which compare experimental XANES spectra with simulations based on hypothetical metal atom sites, in this case the number and arrangement of Co—N and Co—O_2 bonds.

The CO_2RR using catalysts with well-defined Co–N_4 sites (cobalt phthalocyanine as model catalyst) was studied by Zhang et al. [12]. The valence state and coordination structure of Co sites was studied using *in situ* XAS, and neither XANES nor EXAFS spectra revealed any significant changes.

Single Ni atoms dispersed onto graphene nanosheets were studied by Jiang et al. [13] using *in situ* XAS. The goal was to achieve active sites for the CO_2RR to CO. It was found that graphene vacancies provide different Ni sites with or without neighboring N coordination.

Atomically dispersed nickel on nitrogen-doped graphene was studied by Yang et al. [14] with the aim of achieving a durable and efficient electrocatalyst for CO_2 reduction. *Operando* XAS was used and the results showed that the catalytically active site was a monovalent Ni atomic center with a d^9 electronic configuration.

A Cu SAC, which is stable and highly active for the oxidation of CO at temperatures up to 350 °C, and for CO removal in H_2, was studied by Abdel-Mageed et al. [15] using time-resolved *operando* XAS. The catalyst was prepared using the metal–organic framework (MOF) UiO-66 by attaching Cu atoms covalently to defect sites at the zirconium oxide clusters. The results indicate that atomically dispersed, positively charged, ionic Cu species are responsible for the catalytic activity. This is one of the few studies reported in this chapter, which analyzed XANES spectra in terms of linear combinations of two states. This was done using reference compounds to estimate the proportions of Cu^{1+} and Cu^{2+} as a function of reaction time.

6.2.5 *In situ/Operando* XAS of SACs Consisting of Precious Metals in Period 4

In situ XAS was used by Aitbekova et al. [16] to study Ru_{NP} supported on ceria. The results show that low temperature (210 °C) redispersion of the Ru_{NP} under oxygen led to formation of stable single-site RuO_x/CeO_2 species. In the reverse water–gas shift reaction, these showed remarkable selectivity for CO production. This is the only study reported in this chapter, which analyzes the pre-edge peak of the K-edge XANES in terms of the degree of centrosymmetry of the metal atom site. The results are interpreted as showing Ru sites, which are initially octahedral and change to square pyramidal or square planar with increasing reaction temperature.

Li et al. [17] showed by using *operando* XAS that single-atomic ruthenium catalyst anchors on the surface of cobalt iron–layered double hydroxides. These catalysts showed exceptional activity for oxygen evolution reaction (OER). Due to synergetic electron coupling between ruthenium and layered double hydroxides, the ruthenium was kept in the oxidation states of +4 even at high overpotential.

Cao et al. [18] used *operando* XAS combined with synchrotron radiation infrared spectroscopy to study a ruthenium SAC. The catalyst has an atomically dispersed Ru_1-N_4 site, which is anchored on nitrogen-carbon support (Ru_{SA}/N–C); and for acidic OER it is found to be an efficient and durable electrocatalyst. The study with combined techniques enabled the identification of the dynamic adsorption of single oxygen atoms on the Ru site under working potentials.

In a study of isolated rhodium catalysts by Shan et al. [19], XAS data were collected *ex situ* and on catalysts suspended in water, using oxygen and carbon monoxide under mild conditions to mimic the reaction conditions for the direct conversion of methane to methanol and acetic acid. The isolated rhodium catalysts were anchored on either a zeolite or TiO_2 support. The results indicated the single-site distribution of isolated Rh^+ cations with no noticeable difference between supports.

Hülsey et al. [20] used *operando* XAS to investigate rhodium SACs during carbon monoxide oxidation. The catalysts consisted of atomically dispersed Rh on phosphotungstic acid, and *operando* measurements were made between 300 and 473 K, under air and 5% CO/H_2 atmospheres. It was found that Rh undergoes a cycle between Rh^{3+} and Rh^+ in association with creation of oxygen vacancies (O_v) on the support, and reoxidation was the rate-controlling step. This is one of a few studies that compare experimental XANES spectra with simulations based on hypothetical metal atom sites; in this case the effect of O_v and the number of CO ligands.

Operando XAS was used by Peterson et al. [21] to investigate isolated palladium atoms during carbon monoxide oxidation. These were prepared on industrially relevant γ-alumina supports with added lanthanum. The latter was found to help in the ionic Pd_{SA} stabilization, because palladium and lanthanum became mixed on the γ-alumina surface. The atomically dispersed Pd^{1+} is the active centre for CO oxidation. The added lanthanum also improved alumina stability. This is one of the few studies reported in this chapter, which analyzed XANES spectra in terms of linear combinations of two states, in this case Pd^{0+} and Pd^{2+}, as a function of reaction time.

Piernavieja-Hermida et al. [22] studied Pd SACs using combined *in situ* Pd K-edge XAS and infrared spectroscopy. They were thin-film-stabilized Pd_{SA} catalysts that had been prepared using atomic layer deposition (ALD) on alumina, and creating a titania nanocavity thin-film structure significantly enhances the thermal stability of the Pd_{SA}. The results showed that Pd_{SA} are anchored on the surface through chlorine sites, with the chlorine coming from the $TiCl_4$ used as titania precursor in the ALD process. It was also found that these catalysts are thermally stable in both oxidation and reduction, and that in the methanol decomposition reaction their catalytic performance depends on the thickness of the protecting TiO_2layers.

Graphene-supported Pd_{SA} catalysts synthesized by ALD were studied by Yan et al. [23] for selective hydrogenation of 1,3-butadiene. To determine how the Pd_{SA} bond to the graphene support, they used *in situ* XAS, and it was found that three Pd—C and one Pd—O—C bonds are likely to be formed.

A nitrogen-doped carbon-supported palladium SAC was investigated using *in situ* XAS by He et al. [24]. This catalyst can catalyze the reduction of CO_2 into CO, and the results showed that the most likely active center for CO production is the Pd–N_4 site, without the formation of palladium hydride (Pd-H) due to the isolated nature of the Pd, which is essential for typical Pd_{NP} catalysts. In addition, the CO_2RR capability at low overpotentials is enhanced because the stabilization of the adsorbed CO_2 intermediate is facilitated by the well-dispersed Pd–N_4 sites. This is the first of two studies reported in this chapter, which use the advanced wavelet transform method to analyze EXAFS spectra, but this was done to study Pd—Pd bonds in a comparison Pd_{NP} catalyst rather than to study the Pd–N_4 site.

In situ XANES was used by Bugaev et al. [25] to study how Pt and Pd species supported on functionalized UiO-67 MOF evolve. It was found that Pd active sites grafted on the MOF framework are not very stable under reaction conditions (activation procedure by heating in inert and H_2-containing atmospheres). This is one of a few studies, which compare experimental XANES spectra with simulations based on hypothetical metal atom sites, in this case the number and arrangement of Pd—N and Pd—Cl bonds. This is also one of the few studies reported in this chapter, which analyzed XANES spectra in terms of linear combinations of initial and final states, in this case by using principal component analysis (PCA), and the proportions of each state were determined as a function of reaction time and temperature.

Nguyen et al. [26] studied the reduction of nitric oxide with hydrogen by singly dispersed bimetallic species, namely Pd/Co. They used *in situ* XPS and EXAFS, and the formation of bimetallic sites was confirmed by EXAFS, by showing bonding of Pd_{SA} to cobalt. These bimetallic sites were found to be stable during catalysis up to 300 °C.

6.2.6 *In situ/Operando* XAS of SACs Consisting of Platinum on Oxide Supports

Yang et al. [27] used *in situ* XAS to study Pt_{SA} stabilized by sodium through O ligands. These were studied in realistic water–gas shift reaction mixtures from 120 to

400 °C. The properties were found to be unaffected by the choice of support (TiO_2, L-zeolites, and mesoporous silica MCM-41). The atomic dispersion of Pt is preserved upon 0.5 wt% Pt loading on all of these supports.

In situ XAS was used by Dessal et al. [28] to study Pt/γ-Al_2O_3 catalyst. It was observed that in an O_2-rich atmosphere the Pt_{SA} are stable, immobile, and strongly oxidized. Instead, under a reducing atmosphere of H_2, the Pt_{SA} aggregate into reduced, mobile, sub-nanometer size clusters. This indicates that the single-atom nature might not be retained in all catalytic applications. This is the second of two studies reported in this chapter, which use the advanced wavelet transform method to analyze EXAFS spectra, and this was done to search for Pt—Pt bonds, the absence of which confirmed the presence of Pt_{SA}.

Dessal et al. [29] used XAS combined with diffuse reflectance infrared spectroscopy to monitor the dispersion, oxidation state, and activity of Pt during CO oxidation on Pt/γ-Al_2O_3. It was shown that although single Pt^{m+} atoms ($m \geq 2$) are poorly active, they irreversibly and gradually convert into highly active ~1 nm sized Pt^{1+} clusters. This happens even under conditions favorable to atomic dispersion, i.e. highly oxidizing conditions during heating and cooling cycles.

Isolated Pt catalysts were studied by DeRita et al. [30] using a combination of *in situ* atomic-resolution microscopy, *in situ* IR spectroscopy, and *in situ* XAS. These are sinter-resistant catalysts where Pt_{SA} have been uniformly deposited on well-defined locations on anatase TiO_2 nanoparticle supports. The results showed that a range of oxidation states and local coordination environments can be adopted by isolated Pt species, and these evolve when environmental conditions are varied.

Yoo et al. [31] studied CO-tolerant Pt_{SA} using *operando* XAS. These catalysts are highly reactive for the CO oxidation reaction. It was found that Pt_{SA} are stabilized by strong electronic interactions thanks to the formation of a CeO_x-TiO_2 interface when 1 wt% of Ce is added to TiO_2 support particles. This is one of the few studies reported in this chapter, which analyzed XANES spectra in terms of linear combinations of two or more metal oxidation states, in this case Pt^0 and Pt^{2+}.

Dispersed Pt species on CeO_2 supports were studied using *in situ* XAS by Resasco et al. [32], and the samples show different resistance to reduction and sintering, and interaction with CO, depending on Pt loadings.

Wei et al. [33] studied high-loading Pt/FeO_x catalyst using *in situ* XAS. This was for chemoselective hydrogenation of substituted nitroarenes to form the corresponding functionalized anilines. The results showed that the catalyst can be made highly chemoselective by including alkali metals (Li^+, Na^+, K^+, etc.). It was proposed that the active site is a Pt–O–Na–O–Fe-like species, and this ensures that the Pt centers are isolated and positively charged, so assists the preferential adsorption of the $-NO_2$ group.

In situ XAS was used together with ambient pressure XPS by Zhang et al. [34] to study singly dispersed Pt atoms anchored on Co_3O_4 nanorods as catalysts in the water gas shift reaction (WGSR). The aim was to correlate surface chemistry with the catalytic performance. It was found that singly dispersed $Pt_{SA}Co_n$ anchored on Co_3O_4 nanorods makes up the active surface of this catalyst. Nanoclusters of $Pt_mCo_{m'}$ are formed in the temperature range of 200–300 °C as the Co_3O_4 support is reduced to CoO_{1-x}. The synergy between these entities might be the reason for the high activity. A combination of in situ XAS and in situ IR was used by Liu et al. [35] to study Pt catalysts containing selected types of Pt entities. These varied from single atoms to clusters and nanoparticles, and the study aimed to determine the active sites under working conditions. Results from a variety of heterogeneous catalysis reactions, including selective hydrogenation reactions, CO oxidation, dehydrogenation of propane, and photocatalytic H_2 evolution reaction, gave evidence that clusters or nanoparticles can form during the reaction due to sintering of single-atom sites. The way in which these evolve depends on the metal–support and metal–reactant interactions. Hence, the combination of suitable ex situ and in situ characterization techniques is very important to characterize the coordination environment of the metal entities.

6.2.7 *In situ/Operando* XAS of SACs Consisting of Platinum on Non-oxide Supports

In situ XAS was used by Li et al. [36] to study the formation of Pt_3Ti nanoparticles on a two-dimensional transition metal carbide. These were used as catalysts for HER. The results showed that they form Pt_{SA} due to a strong binding of the precursor on the support.

Nguyen et al. [26] studied the reduction of nitric oxide with hydrogen by singly dispersed bimetallic species, namely Pt/Co. They used *in situ* XPS and EXAFS, and the formation of bimetallic sites was confirmed by EXAFS, by showing bonding of Pt_{SA} to cobalt. These were found to be stable during catalysis up to 250 °C.

Operando XAS was used by Fang et al. [37] to study the details of atomic structure associated with Pt_{SA}/N-C. It was found that Pt_{SA} evolve into a near-free state under HER conditions. The results indicated that during the reaction, there is dynamic release from the nitrogen-carbon support, in which the geometric structure is less coordinated to the support, and the electronic properties are closer to valence of zero.

Single-atoms and few-atom clusters of platinum were studied by Kim et al. [38]. These had been placed on the zirconia nodes of a NU-1000 MOF. These catalysts are resistant to sintering up to 200 °C and showed high activity for ethylene hydrogenation. *In situ* IR spectroscopy showed the presence of both SAs and few-atom clusters depending upon synthesis conditions. *Operando* XAS and X-ray pair distribution function measurements revealed changes in chemical bonding environment and stability of cluster size stability under reaction conditions.

In situ XANES was used by Bugaev et al. [25] to study how Pt and Pd species on functionalized UiO-67 MOF evolve. In the temperature range from 200 to 300 °C, under the conditions of inert flow, they found isolated Pt^{2+} active sites grafted on the MOF framework. This is one of a few studies, which compare experimental XANES spectra with simulations based on hypothetical metal atom sites, in this case the number and arrangement of Pt—N and Pt—Cl bonds. This is also one of the few studies reported in this chapter, which analyzed XANES spectra in terms of linear combinations of initial and final states, in this case using PCA, and the proportions of each state were determined as a function of reaction time and temperature.

6.2.8 *In situ/Operando* XAS of SACs Consisting of Precious Metals in Period 5 Other than Platinum

Jiang et al. [39] used *operando* XAS to study isolated iridium atoms anchored on oxyhydroxides. Due to the effect of the support, the isolated iridium sites remain dispersed during OER. These catalysts have good OER performance, and the results showed that they form multiple active sites during OER due to a deprotonation process, promoting O–O coupling.

In situ IR and *operando* XAS were used by Lu et al. [40] to study Ir/$MgAl_2O_4$ catalysts during low-temperature CO oxidation. The goals were to study the resting state of the catalysts, as well as to identify the active single-atom complex. While Ir_{NP} were poisoned by CO, the results showed that Ir_{SA} could form Ir(CO). This results in a different reaction mechanism whereby the ability of SA to coordinate with multiple ligands enables Ir(CO) to provide an interfacial site for facile O_2 activation between Ir and Al. This is the first of two studies reported in this chapter, which used high-energy resolution (or HERFD) XANES to obtain more detail in the experimental Ir L_3-edge XANES spectrum to aid in comparison with DFT simulations of XANES. This is also one of a few studies, which compare experimental XANES spectra with simulations based on hypothetical metal atom sites, in this case the number and arrangement of Ir—O and Ir—CO bonds.

Atomically dispersed iridium sites on indium tin oxide were studied by Lebedev et al. [41] using *in situ* XAS. These catalysts were prepared via surface organometallic chemistry, and they show excellent catalytic activity in OER. The results showed the formation of Ir^{5+}, which is then oxidized to Ir^{6+} to promote fast water nucleophilic attack and oxygen evolution.

Hoffman et al. [42] used *in situ* XAS to study isolated iridium centers adsorbed on HY zeolite and on MgO supports. The results gave information on the structures of Ir sites in terms of the roles of ligands such as ethylene and CO. This is the second of two studies reported in this chapter, which used high-energy resolution (or HERFD) XANES to obtain more detail in the experimental Ir L_3-edge XANES spectrum to aid in comparison with DFT simulations of XANES. This is also one of a few studies, which compare experimental XANES spectra with simulations based on hypothetical metal atom sites, in this case the number and arrangement of Ir—O, Ir—CO, and Ir—C_2H_4 bonds.

A combination of *in situ* X-ray absorption fine structure analysis and *in situ* IR spectroscopy was used by Guo et al. [43] to study the contribution of distinct gold species (single atoms, clusters, and particles) supported on ceria to CO oxidation. The results showed that cationic Au_{SA} as the active sites were less critical to the activity, and that metallic clusters or particles were more critical.

6.2.9 *In situ/Operando* XAS of Other Atoms in SACs and Further Studies

This chapter has discussed on a number of recent *in situ* and *operando* XAS studies of SACs of metal atoms in periods 3, 4, and 5. The focus of these studies, and of this chapter, is the role in the catalytic process of the SA centers. However, a number of the studies discussed also present concurrent XAS results for target atoms, which are ligands, such as N (see Refs Wang et al. [9], Jiang et al. [13], Cao et al. [18], and Fang et al. [37]), or which are on the support, such as transition metal atoms in transition-metal-oxide supports (see Ref. Yoo et al. [31]).

In principle, ion-exchanged metals supported on porous materials such as MOF and zeolites, as well as organometallic complexes anchored to supports, could also be viewed as SACs. Examples of *in situ* XAS studies on these materials can be found in a review by Li et al. [44].

6.3 Other *In situ/Operando* Spectroscopies: IR, UV–vis and Mössbauer Spectroscopies, and XPS

6.3.1 Methods

The other *in situ/operando* spectroscopies, which are used for the study of metal SACs are IR, UV–vis and Mössbauer spectroscopies, and XPS. Among these, IR has been applied more extensively, especially in the diffuse reflectance configuration, which is generally named as diffuse reflectance Fourier transform infrared spectroscopy (DRIFTS). Note that several of the references reviewed in this chapter feature a combination of both *in situ/operando* XAS and IR spectroscopy. IR spectroscopy is an extremely well-known, laboratory-based characterization technique, and there are many references available that describe the use of IR to study catalysis (for example [45]). IR spectra show well-defined peaks whose frequencies correspond with specific chemical groups, and it is possible to observe subtle shifts in frequency due to changes in the local atomic environment of the chemical group. The *in situ* and *operando* IR spectroscopy studies referred to in this chapter largely focus on peaks around 2000 cm^{-1} due to CO adsorbed on metal atom sites (see Figure 6.4 and Table 6.1).

UV–vis spectroscopy is also an extremely well-known, laboratory-based characterization technique. For a review of the use of UV–vis spectroscopy to study catalysis, see [54]. The UV–vis spectrum shows absorption bands due to electronic transitions such as transitions among *d* and *f* electrons in metal atoms, charge transfer transitions between metal atoms and ligands, and π–π^* transitions in ligands. The energies of such transitions can depend on the oxidation state and bonding of the metal atom (see Figure 6.5). The use of UV–vis spectroscopy in *in situ* and *operando* measurements of SACs is less common than the other spectroscopy methods reviewed in this chapter.

Mössbauer spectroscopy of ^{57}Fe is based on the resonant absorption of γ-radiation by the ^{57}Fe nuclei in a sample, and it is sensitive to changes in the chemical environment of the iron atoms, including oxidation state, coordination number, and spin

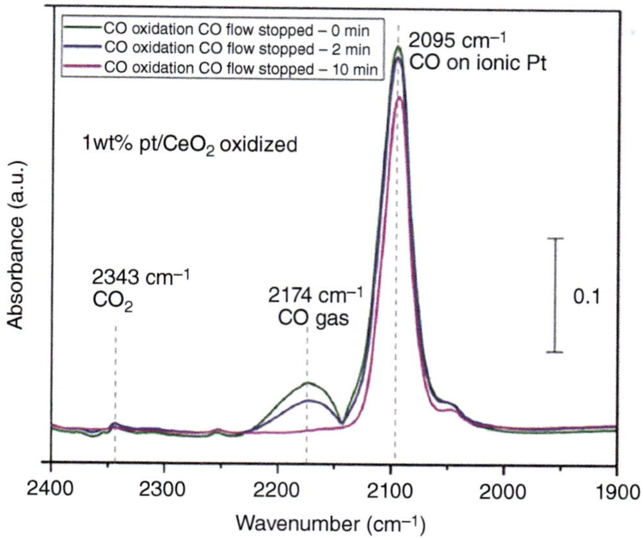

Figure 6.4 DRIFTS of an 1 wt% Pt/CeO_2 catalyst during CO oxidation. The symmetrical feature at 2095 cm^{-1} is assigned to isolated ionic Pt sites on the ceria. Source: Jones et al. [46]. Reproduced with permission of American Association for the Advancement of Science.

Table 6.1 CO frequencies obtained on SAs of different metals and on different supports.

Metal atom	Support	Frequency (cm^{-1})	References
Pt	Al_2O_3	2150-2130 and 2095	[35]
Pt/Pd	Al_2O_3	2065	[47]
Pt	Mesoporous Fe_2O_3	2083	[48]
Pt	Mg(Sn)(Al)O/Al_2O_3	2086 and 2066	[49]
Pt	TiO_2	2135, 2112, 2077, and 2050	[50]
Pt	CeO_2	2095	[30]
Pt	CeO_2	2095	[46]
Pt	PDMS-PEG/SiO_2	2084	[51]
Cu	UiO-66	2138, 2132, and 2113	[15]
Ru	Zeolite	2174, 2120, 2060, and 2000	[52]
Rh	Phosphotungstic acid	2110, 2085, and 2039	[20]
Pd	Al_2O_3-TiO_2	2160, 2116, 2094, 1975, 1944, and 1882	[22]
Pd/Ag	SiO_2	2041 and 2028	[53]
Ir	$MgAl_2O_4$	2070 and 1989	[40]

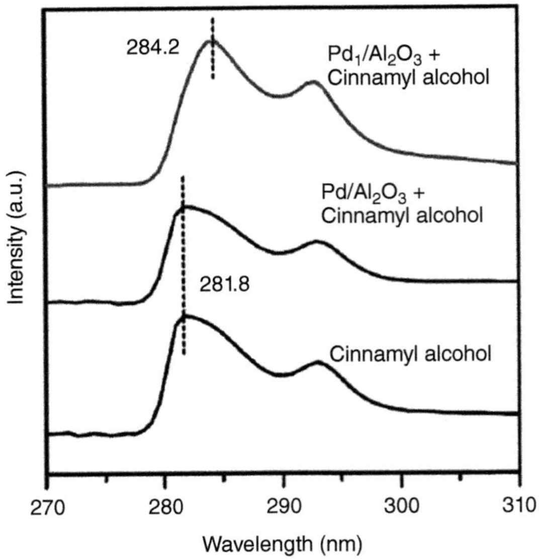

Figure 6.5 UV–vis spectra of cinnamyl alcohol in toluene, which shows an absorption peak at 281.8 nm assigned to the π–π^* transition. An increase in wavelength to 284.2 nm occurred on contact with Pd single atoms on Pd_{SA}/Al_2O_3 catalysts. Source: Shang et al. [55]. Reproduced with permission of Elsevier.

state. The use in *in situ* and *operando* measurements of SACs has been so far very limited, possibly because it can only be used to study Fe SACs.

XPS requires a dedicated instrument or access to an XPS beamline at an X-ray synchrotron. XPS, in common with XAS, is able to target specific elements and uses X-rays to penetrate sample cells. XPS is based on detecting the emission of photoelectrons, and the energies of these are both element specific and sensitive to the oxidation state and type of bonding of the target element(s) (see Figure 6.6). There are many references available that describe the use of XPS to study catalysis (for example [57]). However, XPS is highly surface sensitive and normally requires a vacuum, so the use of XPS for *in situ* and *operando* measurements has benefited from the recent development of ambient pressure-X-ray photoelectron spectroscopy (AP-XPS, see review in [58]).

6.3.2 *In situ/Operando* IR Spectroscopy of SACs Consisting of Platinum

Dessal et al. [29] studied CO oxidation on $Pt/\gamma\text{-}Al_2O_3$ by monitoring it using *operando* DRIFTS. The changes in the IR absorption peaks indicate some Pt clustering occurs, but at the end of the CO oxidation cycles, a large proportion of Pt_{SA} remain.

Moses-DeBusk et al. [59] used *in situ* DRIFTS to study CO oxidation on Pt_{SA} on $\theta\text{-}Al_2O_3(010)$ surface. Their results were used to support a proposed reaction mechanism where Pt_{SA} bond preferentially to O_2 over CO, with CO then bonding with the oxygenated Pt_{SA} and forming a carbonate, which dissociates to liberate CO_2, leaving an oxygen atom on Pt. Subsequent reaction with another CO molecule regenerates the SAC.

DeRita et al. [30] used CO probe molecule Fourier-transform infrared spectroscopy (FTIR) following pretreatment of the catalyst (consisting of Pt_{SA} deposited

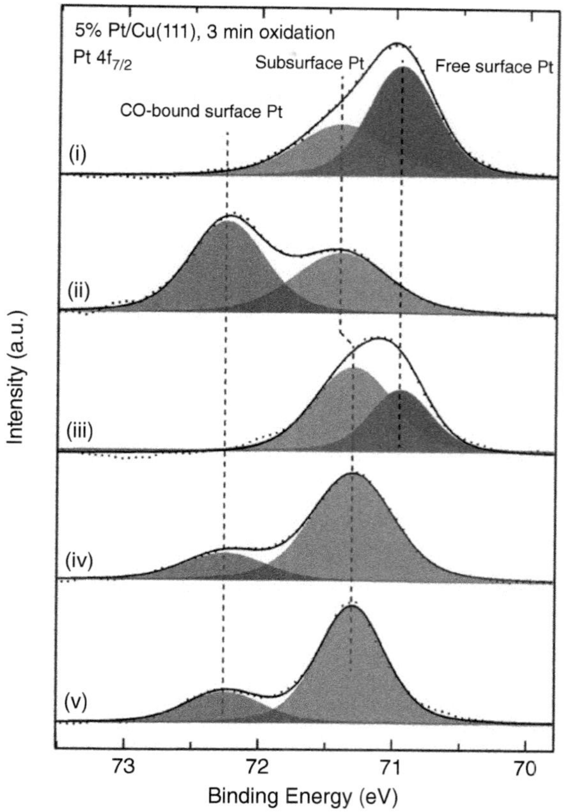

Figure 6.6 Pt $4f_{7/2}$ XPS spectra of 5% PtCu(111) SAC (i) as deposited, and under an atmosphere (not greater than 10^{-4} Torr) of (ii) CO, (iii) O_2, (iv) CO, and (v) H_2. Source: Schilling et al. [56]. Reproduced with permission of American Chemical Society.

on TiO_2 anatase NPs) in oxidative and reductive conditions. Under oxidizing and mildly reducing conditions, the observed bands (at 2135 and 2112 cm^{-1} , respectively) were consistent with CO bound to cationic Pt, whereas under harshly reducing conditions the observed bands (2077 and 2050 cm^{-1}) were consistent with near-neutral $Pt^{\delta+}$ sites. The results from *in situ* IR spectroscopy and XAS (indicating the mildly reduced Pt_{SA} species to be step edge) were combined with DFT simulations to develop a detailed scheme for the different states of Pt_{SA} species (see Figure 6.7).

Yang et al. [48] used *in situ* DRIFTS to identify phenolate (band at 1572 cm^{-1}) and benzoquinone as well as cyclohexanone (band at 1720 cm^{-1}) and maleate (band at 1532 cm^{-1}) as the main intermediates in the oxidation of benzene catalyzed by Pt_{SA} deposited on three-dimensionally ordered mesoporous iron oxide. The results are explained in terms of the strong reaction between Pt_{SA} and Fe_2O_3 causing reduction of Fe^{3+} to Fe^{2+}, creating surface oxygen vacancies, which then interact with gaseous oxygen to form active oxygen species at the $PtSAFe_2O_3$ interface.

Zhu et al. [49] studied Pt_{SA} and small clusters with fewer than 10 atoms (91%) supported on Sn-containing mixed metal oxides. Using *in situ* FTIR for the sample with CO adsorbed showed there was no shift in the absorption band position with the decrease in CO coverage. Such bands are expected to change in frequency with CO coverage due to the interactions of CO adsorbed on neighboring Pt, and the lack

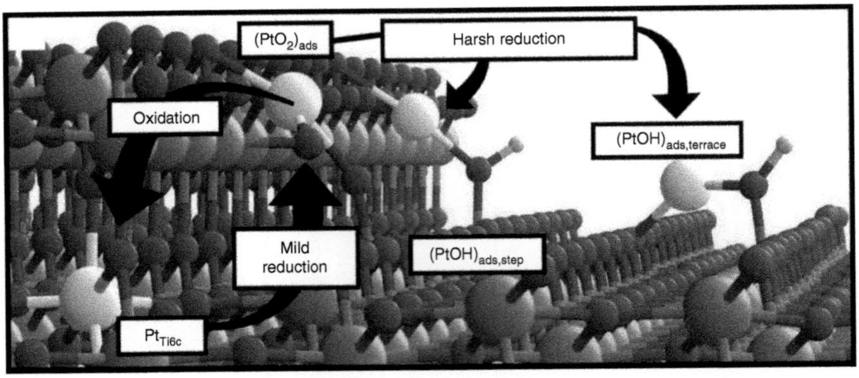

Figure 6.7 Schematic showing the proposed dynamic evolution of Pt_{SA}/TiO_2 catalysts following oxidation, mild reduction, and harsh reduction (shown left to right, respectively). Source: DeRita et al. [30]. Reproduced with permission of Springer Nature.

of shift suggests a lack of interaction between adsorbed CO molecules, consistent with the presence of single atoms or small clusters.

Jones et al. [46] investigated Pt_{SA}/CeO_2 catalysts, and DRIFTS measurements during CO oxidation reaction indicate that the 2095 cm^{-1} band due to CO on ionic Pt is the only observed feature, indicating that Pt_{NP} do not form under the conditions used.

Resasco et al. [32] used *in situ* IR spectroscopy during CO oxidation to study Pt_{SA} species produced on high surface area CeO_2 supports. Although the 2095 cm^{-1} band is observed, this is assigned to adsorption of CO on small oxidized Pt clusters. Instead, it is concluded that stable Pt_{SA} on CeO_2 does not absorb CO.

DeRita et al. [50] used IR spectroscopy to identify unique vibrational signatures of CO adsorbed on various Pt structures (Pt_{SA}, oxidized Pt clusters, and metallic Pt clusters) supported on TiO_2. No visible stretching vibrations associated with CO adsorbed on Pt_{SA} were observed under steady-state CO oxidation reaction conditions.

Liu et al. [35] followed the structural transformation of Pt_{SA} into Pt_{NP} on different oxides by *in situ* IR spectroscopy. The evolution of the bands (2075 and 2050 cm^{-1}) associated with CO adsorption is consistent with a surface reconstruction or disintegration of Pt_{NP} into smaller particles occurring during the CO oxidation reaction at high reaction temperature.

Nie et al. [47] used *in situ* IR to study supported bimetallic Pt–Pd catalysts dispersed on Al_2O_3 support. During chemisorption of CO followed by *in situ* oxidation on monometallic Pt catalyst, there was evidence of a small vibrational band due to strongly bound CO on a small fraction of Pt present as single atoms.

Liu et al. [51] used *in situ* DRIFTS to study Pt_{SA} prepared by reducing H_2PtCl_6 in liquid polydimethylsiloxane-polyethylene glycol (PDMS-PEG). The results show that the IR peak due to CO adsorption on Pt (band at 2084 cm^{-1}) does not change as a function of the CO pressure, as expected when there is only Pt_{NP} in the system. This study combined DRIFTS, NMR, and XPS techniques with DFT modeling of hypothetical atom sites to conclude that the Pt_{SA} has high activity in olefin hydrosilylation due to being present as an electron-deficient $Pt_{SA}{}^{\delta+}$ site in an octahedral $(R_1OR_2)_2Pt^{(0)}Cl_2H_2$ center with weak oxygen coordination.

6.3.3 *In situ/Operando* IR Spectroscopy of SACs Consisting of Metal Atoms Other than Platinum

Zhou et al. [60] studied atomically dispersed Co–P_3 species on CdS nanorods for photocatalytic activity in the dehydrogenation of formic acid to hydrogen. They used *in situ* attenuated total reflection (ATR) infrared spectroscopy to follow the effective dissociative adsorption of formic acid on the Co–P_3 species (evidenced by C–H bands at 2924 and 2851 cm^{-1}), which was completely decomposed after light irradiation (evidenced by disappearance of O–H bands at 3100–3600 cm^{-1} and growth of C=O band at 2342 cm^{-1}).

Abdel-Mageed et al. [15] used *operando* DRIFTS during CO oxidation at 250 °C on a SAC prepared by a covalent attachment of Cu atoms to the defect sites at the zirconium oxide clusters of the UiO-66 MOF. At the start of the reaction, a band at 2113 cm^{-1} becomes established, and later this is replaced by a band at 2132 cm^{-1}. Both are attributed to monocarbonyls adsorbed on Cu^{1+} sites, and the results show the Cu^{1+} sites are first activated and then transformed during the reaction. (There was no observation of bands attributed to Cu^{2+} sites.)

Qiu et al. [52] studied pure siliceous zeolite-supported Ru_{SA} active sites for ammonia synthesis. DRIFTS of adsorbed CO was carried out before and after being exposed to ammonia synthesis conditions, and the bands observed (2060 and 2000 cm^{-1}) were consistent with dicarbonyl species on Ru being present as SAs.

Cao et al. [18] used *operando* synchrotron radiation IR spectroscopy to study acidic OER for atomically dispersed Ru_{SA}-N_4 sites anchored on nitrogen-doped carbon support. A new band at 764 cm^{-1} appeared reversibly with change in potential from 1.2 to 1.6 V, and DFT simulations were used to identify the intermediate in the OER process as being the dynamic adsorption of single oxygen atoms on the Ru site under working potentials. The combination of *in situ* EXAFS (indicating the initial state of Ru_{SA} coordinated to four N atoms, and the *in situ* oxygen species adsorption on Ru_{SA}) and IR spectroscopy (indicating the O–Ru–N intermediate) were highly effective for developing the detailed schematic of the OER mechanism (see Figure 6.8).

Zhang et al. [61] studied Rh_{SA} anchored on an inert SiO_2 support, which are active in reducing nitric oxide with carbon monoxide, via two reaction pathways that are different from those on Rh_{NP}, as identified by *in situ/operando* IR studies in NO gas using a diffuse reflectance reaction cell. This was demonstrated via the replacement at room temperature of a band at 2058 cm^{-1} due to pristine CO adsorption on Rh_{SA} with a band at 2088 cm^{-1} attributed to coadsorption of NO molecules. Upon heating to 350 °C, a new band appeared at 2197 cm^{-1} attributed to the Rh-NCO intermediate.

Hülsey et al. [20] studied Rh_{SA} on phosphotungstic acid supports using *in situ* DRIFTS. The IR spectra were measured under CO with temperature increasing from 300 to 473 K. The results showed a growth in bands due to $Rh(CO)_2$, and a frequency shift consistent with a shift from Rh^{3+} to Rh^+, which was attributed to the creation of O_v on the support.

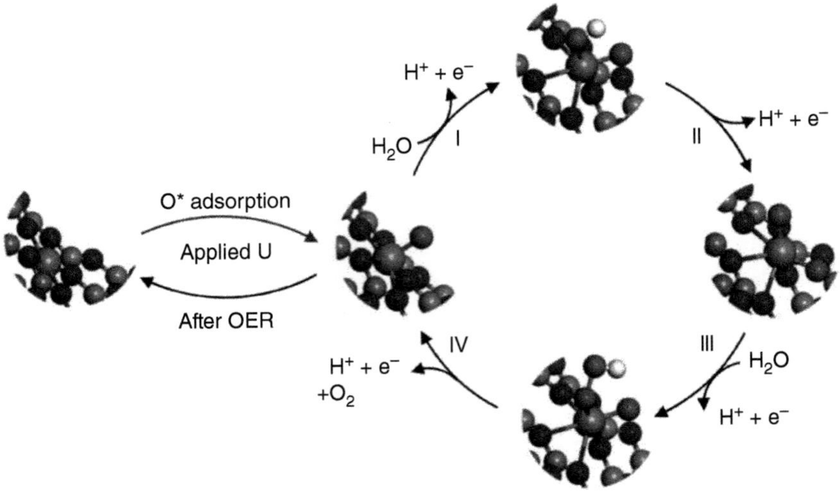

Figure 6.8 Schematic of the whole OER mechanism on Ru–N–C catalyst in the acidic electrolyte. All potentials are normalized to RHE. Source: Cao et al. [18]. Springer Nature. CC BY 4.0.

Piernavieja-Hermida et al. [22] studied Pd SACs using in FTIR spectroscopy. The catalysts had been prepared using ALD on alumina, and the *in situ* results were used to understand how to stabilize Pd_{SA} during synthesis. The results during CO adsorption showed bands at 2160, 2116, and 2094 cm^{-1} (respectively) that were used to distinguish different contributions between Pd^{2+}, Pd^{+}, and Pd^{0}, and between Pd_{SA} (band at 2160 cm^{-1}) and Pd_{NP} (bands at 2116 and 2094 cm^{-1}).

Hou et al. [62] performed *in situ* DRIFTS experiments of benzene oxidation over a $Pd_{SA}Co_{SA}/Al_2O_3$ catalyst. The observed bands indicating that the adsorbed benzene was oxidized to a series of organic intermediates in the presence of oxygen.

Pei et al. [53] used *in situ* FTIR to study CO chemisorption on Ag-alloyed Pd SACs for selective hydrogenation of acetylene under ethylene-rich conditions. Measurements showed that the surface possesses solely Pd_{SA} able to linearly bind CO, without any assemblies of neighboring Pd atoms.

Lu et al. [40] used *in situ* DRIFTS to investigate the adsorption of CO on $Ir/MgAl_2O_4$ catalysts, finding evidence of $Ir(CO)_2$.

Liang et al. [63] used *in situ* DRIFTS to study the mechanism for the WGSR on an Ir_{SA}/FeO_x SAC. After introducing CO and H_2O, there is a strong increase in absorption at high frequencies, which can be attributed to the production of O_v and the reduction of Fe^{3+} to Fe^{2+}.

6.3.4 *In situ/Operando* UV–vis and Mössbauer Spectroscopy, and XPS of SACs

In situ UV–vis was used by Shang et al. [55] on an Al_2O_3-supported Pd SAC, which is seen to have higher activity and selectivity compared to Pd_{NP} for the oxidation of cinnamyl alcohol. They observed a red shift in the cinnamyl alcohol absorption band when in contact with Pd_{SA}/Al_2O_3, owing to the effective expansion of the π-conjugation induced by the charge or energy transfer between the adsorbed cinnamyl alcohol and Pd_{SA}.

Operando ^{57}Fe Mössbauer spectroscopy was developed for the first time by Li et al. [64] to study single-atom Fe_{SA} catalysts with well-controlled site density and well-defined N-coordination environment. The catalysts were synthetized by a solvent-assisted linker-exchange method as model systems for electronic-level understanding of the ORR mechanism. The Mössbauer spectra were measured under ORR conditions in O_2-saturated acidic and alkaline electrolytes. All spectra present features assigned to low-spin Fe^{2+} in FeN_4C_{12}, intermediate-spin Fe^{2+} in FeN_4C_{10}, and high-spin Fe^{2+} in N-FeN_4C_{10}, and their dynamic evolution indicates the formation of O_2^--FeN_5 intermediate near the onset potential of ORR and the formation of O_2^-–FeN_4 intermediate at higher biases.

Pei et al. [53] used *in situ* XPS on Ag-alloyed Pd SACs for selective hydrogenation of acetylene under ethylene-rich conditions as a function of increasing reduction temperature. Samples were reduced *in situ* at different temperatures in the analyzer chamber under H_2 pressures of up to 2×10^{-7} Torr. The results showed that the binding energies of both Ag 3d and Pd 3d electrons decrease with increasing pre-reduction temperature. This is interpreted as due to an increased electron transfer from Ag to Pd, which would be consistent with Pd having an increased number of Ag neighbors.

Schilling et al. [56] carried out *in situ* studies using ambient-pressure (AP) XPS to study the effect of small amounts of atomically dispersed Pt on the reduction of Cu_2O. In particular, the XPS results for Cu 2p, O 1s, and Pt 4f electrons show that the oxidation results in both surface and subsurface Pt in a Cu_2O thin film (see Figure 6.6), and that the role of Pt_{SA} in catalyzing the reduction of Cu_2O is prevented if surface Pt is first saturated with CO.

Pereira-Hernández et al. [65] used near-ambient pressure XPS to study the CO oxidation performance of Pt SACs on a ceria support. Combining these measurements with CO temperature-programmed reduction, they found that high reactivity at low temperatures can be related to the improved reducibility of the CeO_2 support. The XPS results for Pt 4f electrons show that the as-synthesized catalysts are predominantly Pt^{2+}, and this changes to Pt^0 with exposure to CO. The XPS results for Ce 3d electrons show that activation involves the creation of Ce^{3+}, and this then returns to Ce^{4+} when CO_2 is formed.

Ambient-pressure XPS has been used by Nguyen et al. [26] to study *in situ* the surface chemistry of Pt and Pd atoms in singly dispersed bimetallic sites $Pt_{SA}Co_m$ and $Pd_{SA}Co_n$. The XPS results for Co 2p and Pt 4d electrons show that both elements are present in cationic state and remain in this state during reduction of nitric oxide with hydrogen in the temperature regime explored.

Hülsey et al. [20] studied via an integrated *operando/in situ* XPS, XAS, and DRIFTS study rhodium SAs on phosphotungstic acid for CO oxidation. *In situ* DRIFT spectroscopy consolidates the oxidation state and CO adsorption of Rh. To investigate the effect of the support, they used *in situ* XPS to follow the oxidation state of tungsten during the CO oxidation, finding that *W* is reduced close to room temperature.

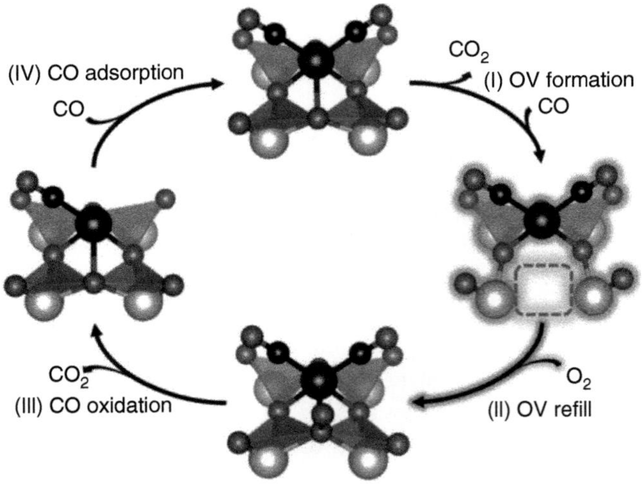

Figure 6.9 Key steps of the proposed reaction mechanism based on *in situ* spectroscopy; the steady-state catalyst structure and the corresponding rate-controlling step are highlighted. Source: Hüsley et al. [20]. Springer Nature CC BY 4.0.

This result allows the authors to conclude that the Rh_{SA}-catalyzed CO oxidation follows a rather unconventional Mars van Krevelen mechanism (see Figure 6.9) with the direct contribution of the heteropoly acid support. Rh undergoes a cycle between Rh^{3+} and Rh^{+}, and the structure with O_v is the dominant working state of the catalyst with the reoxidation of the support as the rate-controlling step.

6.4 *In situ/Operando* Electron Microscopy

6.4.1 State of the Art

TEM, also in scanning mode (STEM), can provide with direct visualization of SACs. The development of aberration correctors allowed S/TEM to routinely reach sub-angstrom resolution by the end of 1990s, [66–68] thus granting the necessary high spatial resolution required to understand whether they consist of single-metal atoms, and providing with insights about their location with respect to their support. S/TEM-related spectroscopies, energy-dispersive X-ray spectroscopy (EDS), and electron energy loss spectroscopy (EELS) can be further exploited for probing the chemical features of both metal atoms and their support.

According to literature, aberration-corrected STEM emerges as the technique of choice among the diverse kinds of TEM-based imaging techniques that can be used to investigate SACs, due to its very high local resolution, which permits to image and chemically investigate these kinds of materials with its sub-angstrom electron probe. Indeed, STEM imaging in HAADF geometry has been proven a very powerful tool, capable of revealing dispersed metal active sites and determining how they coordinate to the support and can be successfully used in different ways, such as studying *in situ* the formation of SACs or their behavior before and after catalytic reactions.

An aberration-corrected S/TEM can be schematized as a column constituted by an electron source, usually of field emission type, two or more condenser lenses followed by an aberration corrector, an objective lens (possibly with a dedicated aberration corrector), and a specimen holder located in the middle portion of the objective lens. Further into the column are positioned the intermediate and projector lenses, finally followed by an image-recording device, usually constituted by a charge-coupled device (CCD) or a complementary metal–oxide semiconductor (CMOS) camera. Further devices, such as EDS detectors and electron energy-loss spectrometers/filters, are frequently added to this basic configuration to provide additional analytical capabilities.

In STEM mode, the electron beam converges to form a very small probe, which rasters a region of the sample, thus forming a final magnified image that corresponds to the scanned surface area. The presence of a STEM (probe) aberration corrector allows the formation of an electron probe with sub-angstrom size, which is capable of acquiring images with the corresponding lateral resolution. The electrons pass through the sample, are scattered by its atoms, and collected by dedicated detectors to form an image that is generated synchronously with the beam scanning the sample. More in detail, the electrons elastically scattered upon a sufficient angle (75–150 mrad) are collected by an annular detector to form a HAADF image, with a scattering cross-section roughly proportional to $Z^{1.8}$. Due to this reason, the HAADF imaging is also defined as Z-contrast imaging [69]. In the case of SACs, usually there is a large difference in atomic number between the heavy metal catalysts and lighter elements of the support material (e.g. carbon, silica, and alumina), which in principle allows an easy observation of the metal atoms by Z-contrast imaging [70, 71], and which is the reason why this technique has become largely popular to image and chemically study SACs [72–74].

Along with its numerous advantages, this imaging technique has also some important limitations. The first one, which is inherent to any highly spatially resolved microscopy, concerns the statistical relevance of areas observed at a high level of resolution, which usually also implies the investigation of spatially limited areas. In other words, since the achievement of high resolution comes at the cost of the width of areas analyzed, the limited amount of single observations obtained in performing aberration-corrected STEM imaging poses a challenge about the representativity of the real degree of dispersion of heavy metal atoms achieved within the whole sample. The second limitation concerns the sensitivity of the single atoms to a very intense, energetic, and small-sized electron beam, as the one expected in an aberration-corrected STEM: in principle, such a beam can modify the location and spatial distribution of single atoms during its scanning of the sample surface [75].

A similar problem can also affect the stability of the support upon which the catalyst atoms are located, although it can manifest in two opposite ways, the latter depending on the nature of the support, both related to the two well-known kinds of beam damage [76]. The first one is the so-called knock-on effect, i.e. the atomic displacement suffered by an atom interacting with an electron beam of sufficient energy to remove it from the structure where it belongs. In this case, the only way to overcome the problem is to lower the electron energy, which automatically implies a loss in S/TEM resolution. The second possible effect of interaction between sample and electron beam is radiolysis, i.e. a beam-induced bonds breaking that, contrary to the knock-on effect, becomes less probable with increasing the energy of electron beam.

The third main limitation in using aberration-corrected STEM occurs in the cases where the interpretation of Z contrast imaging is not straightforward. A typical example of this circumstance is given by the case of lighter single atoms supported by heavy metal oxide [77, 78]. Here, further and diverse characterization methods are needed to assess the presence of light element on the heavier support. Indeed, the use of aberration-corrected HAADF–STEM imaging in conjunction with its related spectroscopies shows its strength and its capability in providing key information on catalytic sites [79].

In any case, it's of pivotal importance to highlight once again the very local nature of aberration-corrected STEM imaging, which can lead to misinterpret the overall nature of the investigated samples containing SACs. To overcome this intrinsic limitation, images with lower magnification can be acquired, aiming to investigate sufficiently large areas to definitively demonstrate the homogeneous dispersion of SAs over the support. However, since the low magnification Z-contrast cannot display clearly enough the differences in intensity ascribable to the higher or lower presence of SAs, concomitant EDS mapping can be exploited to show their overall presence even at that low magnification [73, 80–82]. Due to these general limitations inherent to STEM imaging, spectroscopic techniques such as FTIR, XAS, and XPS, which average information from much larger portion of the sample, should be coupled with the results coming from the very highly resolved aberration-corrected S/TEM imaging at different magnification. This allows to gain a more general and consistent view of the whole sample that clearly demonstrate how these different kinds of information (very local and hugely averaged) are in agreement.

Extending STEM imaging of SACs to *in situ/operando* modes adds a further level of complexity caused by the introduction of an additional volume of gases in the path of the electron beam. In principle, the use of an *in situ/operando* approach to aberration-corrected S/TEM should provide with real-time information over the evolution of the SACs + gas system, either during the formation of SACs or on the activation/degradation of SACs upon reaction conditions. Indeed, the dynamic behavior exhibited by SACs under those conditions can be hardly investigated in full by only using averaged spectroscopy techniques, since a very homogenous coordination environment would be mandatory to deconvolve different contributions [83].

Since the *in situ/operando* approach opens the door to the observation of dynamic processes, the time resolution of image acquisition becomes a parameter of paramount importance to effectively follow the evolution of the system under analysis. In this regard, the recent development of devices with ever-increasing time resolution, such as direct electron detectors for TEM and faster scanning detectors for STEM, permits to image fast processes with a time resolution unattainable few years ago and represents a major advantage for the observation of dynamic processes. However, it has to be clearly specified that the ultrafast TEM detectors developed in the last decade are much more capable to provide high frame-rate acquisition than those reachable using the STEM fastest scanning mode; thus, working in TEM mode currently is still the most effective solution in terms of temporal resolution. That being said, the second choice in performing *in situ* S/TEM experiments is represented by using either a dedicated environmental TEM (ETEM) or dedicated *in situ/operando* holders on "regular" TEMs. ETEM utilizes a series of differential pumping apertures to maintain the volume where the sample is located upon a certain degree of low pressure, while the rest of the TEM column is kept upon high vacuum conditions. However, this microscope configuration has some intrinsic limitations that limit the maximum gaseous pressure that can be maintained around the sample to values lower than atmospheric pressure, thus strongly limiting its capability to replicate benchtop reactions conditions. The only alternative to perform *in situ/operando* experiments with gaseous reactants close to atmospheric pressure is represented by using dedicated sample holders based on micro-electro-mechanical system (MEMS). These holders can provide different stimuli, such as heat or light, to trigger the reaction, but they also have some inherent limitations. Indeed, while *in situ* aberration-corrected S/TEM investigations performed in high vacuum conditions can usually be performed while keeping a very high spatial resolution, just a limited number of reactions involving SACs, such as their observation, synthesis, and formation, can be performed upon those vacuum conditions, which do not correspond to the real environmental conditions upon which many catalytic reactions occur. Due to all these circumstances, the ability to replicate by *in situ/operando* experiments the actual catalytic conditions (i.e. first of all, having the sample in gaseous environments with approximatively atmospheric pressure, and observing it by an electron beam, which has to also pass through a double membrane isolating that gaseous environment) unavoidably brings to a severe loss in spatial resolution and beam intensity, which reflects in lower signal-to-noise ratios in any spectral acquisition, either EDS or EELS. As a consequence, while recent technical advancements put aberration-corrected S/TEM as the preferred choice for direct observation of SACs, their characterization by *in situ/operando* approaches is not as widely diffused due to its more technically challenging nature.

6.4.2 *In situ* Imaging During the Synthesis of SACs

In recent years, different groups have already performed experiments devoted to image directly the formation or catalytic behavior of SACs by *in situ/operando* approaches or by combining *in situ* and *ex situ* observations. In this framework, Chen et al. published one of the earliest works that presents *in situ* HRTEM observation of noble-metal NPs being converted into atomically dispersed catalysts, [84] namely the dispersion of Ag_{NP} on the surface of hollandite manganese oxide. A thermal surface-mediated diffusion process enables the preparation of the material, but the formation of isolated metal atom catalytic sites as observed in the electron

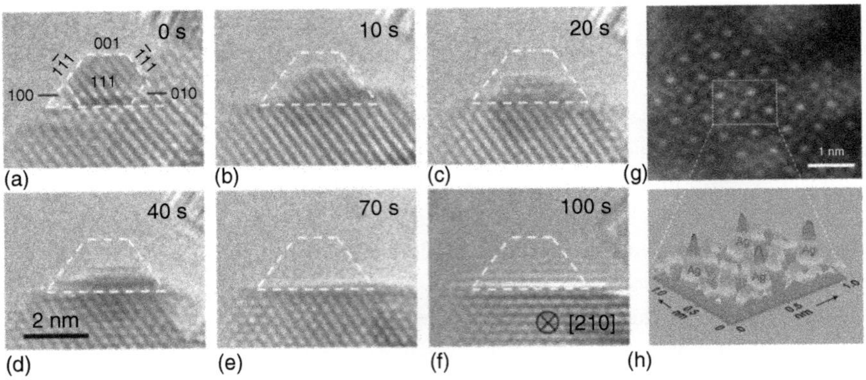

Figure 6.10 *In situ* disintegration process of an Ag_{NP} supported on the HMO surface. (a–f) *In situ* HRTEM images illustrating the shrinking process. (a) Zone-axis notations of the Ag_{NP} are given in italics. The dashed white trapezoid in (a–f) shows the size of the pristine Ag_{NP}. (g) *Ex situ* imaging of the isolated Ag atoms on the hollandite manganese oxide surface by aberration-corrected HAADF–STEM. (h) Three-dimensional projected image of the dashed rectangle in panel (g), showing the difference of contrast between Ag, Mn, and the background (each peak corresponds to an atom, and the height of the peaks is dependent on the atomic STEM contrast). Source: Chen et al. [84]. Adapted with permission of John Wiley and Sons.

microscope was simply due to the effect of the electron beam rather than a heating treatment. This is illustrated in Figure 6.10a, showing a truncated octahedral Ag_{NP} supported on the hollandite manganese oxide surface, which progressively shrank under the electron beam at first preserving its shape (Figure 6.10b–d), and eventually disappearing (Figure 6.10e,f), reflecting the diffusion of Ag atoms. The Ag_{SA} anchored on the oxide surfaces were further imaged *ex situ* by HAADF–STEM imaging and paired by EDS mappings (Figure 6.10g,h). Intensity variations were used to operate the distinction between the Ag atoms and the Mn/O atoms. Overall, a combination of electron microscopy images with X-ray absorption spectra established that the silver atoms were anchored on fivefold oxygen-terminated cavities on the surface of the support to form highly dense isolated metal active sites, leading to excellent reactivity in catalytic oxidation at low temperature.

More recently, Wei et al. performed a real *in situ* study of the dispersion of Pd/Pt/Au NPs initially trapped into ZIF-8 nanocrystals [85]. This was actually a follow-up of Yadong Li's work that developed a general host–guest strategy to fabricate various metal SACs on nitrogen-doped carbon (M_{SA}/N–C, M = Pt, Ir, Pd, Ru, Mo, Ga, Cu, Ni, Mn) [72], and is a convenient top-down route to obtain SACs from supported NPs (Figure 6.11a). For the current *in situ* TEM study, the initial material precursor was prepared by encapsulating Pd_{NP} during the synthesis of ZIF-8 nanocrystals (Pd_{NP}@ZIF-8). The evolution from Pd_{NP} to Pd_{SA} was followed by *in situ* ETEM in the temperature window from 100 to 1000 °C under an Ar atmosphere of 0.6 mbar. Images acquired at different temperatures show that the average diameter of Pd_{NP} steadily increases from room temperature to 900 °C, while the number of Pd_{NP} gradually decreases, indicating the co-existence of sintering and atomization processes (Figure 6.11b). As the temperature stabilizes at 1000 °C, emission of Pd

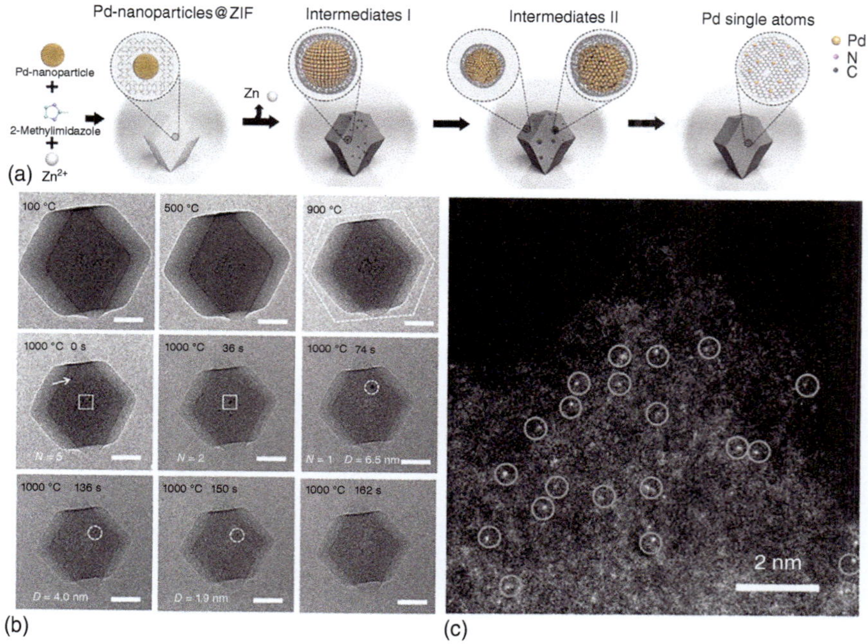

Figure 6.11 (a) Scheme of the transformation of nanoparticles to single atoms and structural characterizations of Pd_{SA}. (b) Representative movie images acquired at various temperatures and times of Pd_{NP}@ZIF-8 pyrolyzed *in situ* with ETEM under an Ar atmosphere. (c) *Ex situ* HAADF–STEM image of Pd_{SA} distributed on the nitrogen-doped carbon support derived from the pyrolysis of ZIF-8. Source: Wei et al. [85]. Adapted with permission of Springer Nature.

atoms becomes the major phenomenon and the latter are subsequently captured by nitrogen defects from the carbon support. Interestingly, the authors also proposed a comparison with a precursor material displaying only Pd_{NP} on the surface of ZIF-8 crystals (instead of being encapsulated inside). With the latter sample, the largest Pd_{NP} are not redispersed, thus suggesting that a thermal motion of large Pd particles inside the nitrogen-doped carbon is a necessary condition to enable the subsequent NP atomization. However, the Pd_{SA} could not be revealed by *in situ* ETEM and were instead characterized *ex situ* by HAADF–STEM imaging (Figure 6.11c). The study was further supported by DFT calculations and XAS to investigate the metal dispersion mechanism: high-temperature conversion from NP to SAs was found to be driven by the formation of the thermodynamically stable Pd-N_4 structure. Those thermally stable Pd-N_4 species exhibited higher activity and selectivity than their nanoparticles counterpart for the semi-hydrogenation of acetylene.

In 2021, Liu et al. elaborated an alternative strategy to the use of N-doped carbon as the anchoring support allowing the stabilization of isolated SAs via nitrogen coordination [86]. Figure 6.12a depicts the top-down fabrication processes followed by the authors to synthesize Fe_{SA} through pyrolysis. The initial precursor is a nanocomposite material made of 7 nm Fe_3O_4 nanoparticles distributed on

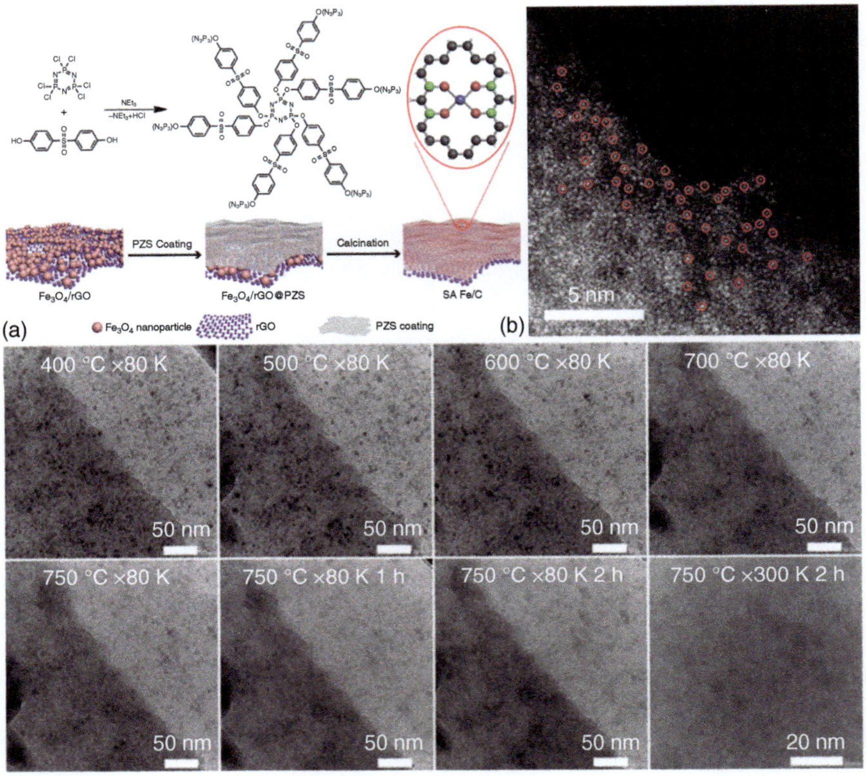

Figure 6.12 (a) Schematic illustrating the preparation processes of Fe_{SA} on carbon support. (b) *Ex situ* HAADF–STEM image of Fe_{SA} distributed on carbon (red dots indicate Fe_{SA}). c) Representative TEM images obtained at different temperatures during the *in situ* pyrolysis of Fe_3O_4/rGO@PZS under vacuum atmosphere. Source: Liu et al. [86]. Adapted with permission of John Wiley and Sons.

reduced graphene oxide (rGO) nanosheets and impregnated by a cross-linked poly(cyclotriphosphazene-*co*-4,4′-sulfonyldiphenol) (PZS). After pyrolysis at 750 °C under argon for two hours, the atomic dispersion of Fe_{SA} and their coordinated structures were characterized by aberration-corrected STEM (Figure 6.12b) and XAS. In addition, the conversion from Fe_3O_4 nanoparticles to Fe_{SA} was followed by *in situ* TEM from 400 to 750 °C under vacuum (Figure 6.12c). In the temperature window of 400–750 °C, Fe_3O_4 nanoparticles became gradually smaller and fewer. After two hours at 750 °C, all the NPs effectively vanished. Nevertheless, *in situ* HAADF–STEM imaging of the resulting single atoms was not provided. The authors state that the mechanism leading the atomic dispersion of metal is not entirely clear, but it is speculated that the polymer PZS plays a major role through the formation of PH_3 gases during pyrolysis allowing the extraction of Fe atoms from the Fe_3O_4 nanoparticles. The as-prepared Fe_{SA} exhibited high activity and superior selectivity for catalytic oxidation of benzene to phenol with hydrogen peroxide. Finally, the preparation method was further extended to Ni, Co, and Mn metals, and metal loadings could also be increased over 10 wt%.

Another example reported by Liu et al. makes use of a strong covalent metal–support interaction of Ru with a $MgAl_{1.2}Fe_{0.8}O_4$ spinel to stabilize Ru_{SA} [87]. What is quite interesting with their synthesis method is that the starting RuO_2 is a commercial powder simply mechanically mixed with the $MgAl_{1.2}Fe_{0.8}O_4$. This would allow the perspective of large-scale manufacture of thermally stable SACs for industrial applications, although the $MgAl_{1.2}Fe_{0.8}O_4$ spinel remains synthesized in the laboratory using a solvothermal synthesis approach. RuO_2 powders were found to disperse into single atoms upon high-temperature heat treatment under an oxidizing or inert atmosphere. Thanks to a control experiment using the $MgAl_2O_4$ spinel (i.e. without iron dopant) as a support, iron doping was revealed to be the major determinant to achieve the atomization of RuO_2 on the spinel surface through a strong covalent metal–support interaction effect. The actual dispersion process was directly visualized by *in situ* HAADF–STEM, a MEMS heating holder, and a secondary electron detection. As shown in Figure 6.13a–h, a large RuO_2 aggregate in the initial physical mixture had its size and morphology unchanged at temperature <900 °C. By maintaining the latter temperature, a melting phenomenon occurred, with the RuO_2 aggregate contracting by about 50% in all dimensions over the next 100 seconds. The authors also point out that *in situ* electron microscopy was not able to record the movement of individual Ru atoms under such conditions. *Ex situ* HAADF–STEM imaging was used instead to confirm the formation of uniformly dispersed Ru_{SA} (Figure 6.13i,j).

A similar strategy was also followed by Lang et al. isolating Pt atoms through a strong covalent metal–support interaction associated with Fe_2O_3 surface

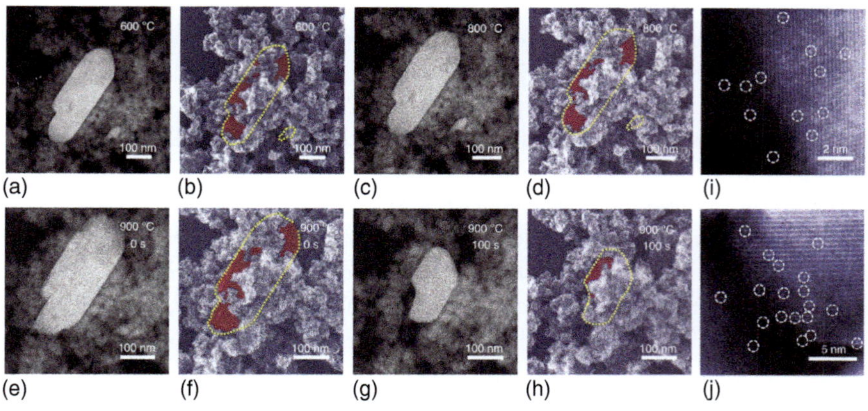

Figure 6.13 *In situ* imaging of RuO_2 dispersion: (a, c, e, g) HAADF–STEM images and (b, d, f, h) corresponding SE images of a $RuO_2 + MgAl_{1.2}Fe_{0.8}O_4$ physical mixture after calcination at 600, 800, and 900 °C (0, 100 seconds) under flowing O_2 (2 ml/min and 3.5 Pa). Yellow dashed lines in the SE images silhouette the RuO_2 aggregate, and red regions indicate exposed RuO_2 surfaces. (i, j) *Ex situ* HAADF–STEM images of $Ru_{SA}/MgAl_{1.2}Fe_{0.8}O_4$ sample demonstrating the atomic dispersion of Ru atoms. Source: Liu et al. [87]. Springer Nature / CC BY-SA 4.0.

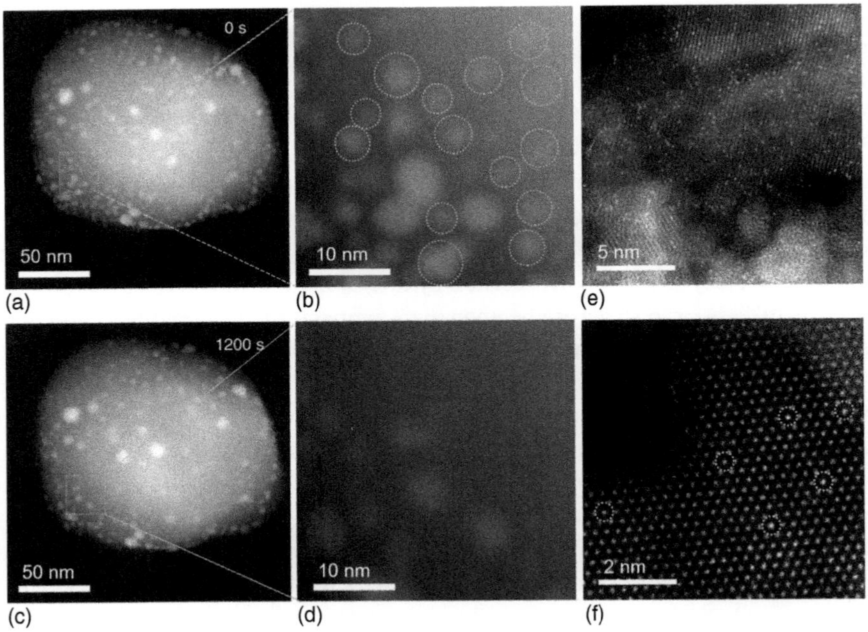

Figure 6.14 *In situ* characterization of Pt_{NP} oxidative dispersion. (a, b) HAADF–STEM images of 1 wt% Pt/Fe_2O_3 before, and (c, d) after *in situ* calcination at 800 °C under 1 bar flowing O_2 for 20 minutes: the white squares in panels a and c show the same sample area. White circles in panel b highlight the particles missing after calcination for 20 minutes; (e, f) *ex situ* HAADF–STEM images of Pt_{SA} distributed over an FeO_x support. Source: Lang et al. [88]. Springer Nature / CC BY-SA 4.0.

defects [88]. A 1 wt% Pt/Fe_2O_3 precursor was first prepared by co-precipitation resulting in the distribution of Pt_{NP} with an average particle size of about 3 nm. Disintegration of these Pt_{NP} during high-temperature calcination was directly visualized by *in situ* HAADF–STEM, using a MEMS heating holder while flowing pure O_2 at 1 bar pressure. After heating to 800 °C, the smaller Pt_{NP} shrank and/or vanished completely (Figure 6.14a,b vs. c,d). *Ex situ* HAADF–STEM imaging with sub-angstrom resolution combined with XAS were applied to highlight the Pt_{SA} dispersed on the FeO_x support (Figure 6.14e) and their average short-range order. Additional imaging on zone axis evidenced that some Pt atoms are exactly aligned with the Fe atomic columns (Figure 6.14f), while computational modeling revealed that iron oxide reducibility is crucial to anchor isolated Pt atoms.

6.4.3 *In situ* Observation of Catalysis Reactions at Single Atoms in Motion

The above-reported studies all focus on *in situ* imaging during the SAC synthesis itself, but it is worth stressing once more that their individual movements could not be captured, likely due to: (i) the inherent difficulty for the interpretation of phase contrast in TEM mode, (ii) the limited time resolution of STEM, (iii) the beam-induced damage of sample, and (iv) the increased atomic Brownian motion at

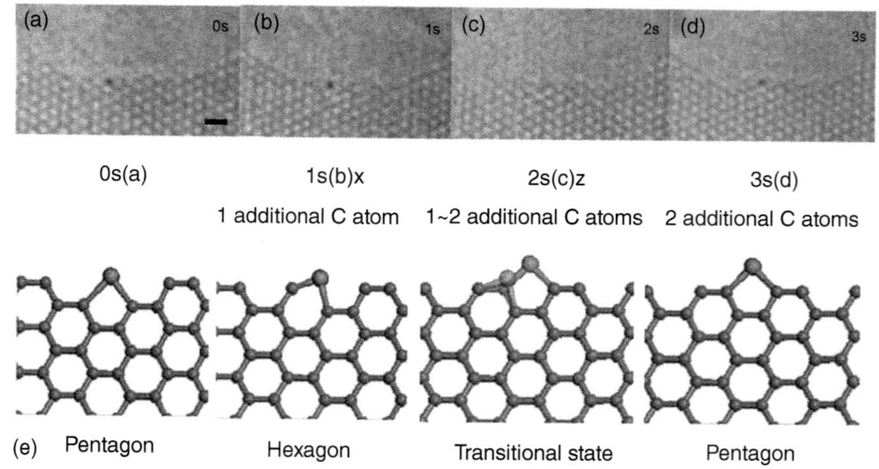

Figure 6.15 One cycle of catalytic growth of graphene edge. (a–d) A series of HRTEM images for four seconds; the Fe atom is highlighted as a larger dot, whereas the nearby carbon atoms are depicted as smaller dots. The dark shadow line, which is highlighted by the black circle in (c), is because of the motion of the Fe atom during exposure time. (Scale bar: 0.5 nm.) (e) The corresponding atomic structures for (a–d). Source: Fiedorow et al. [91]. Adapted with permission of United States National Academy of Sciences.

high temperature. However, these difficulties and limitations have been overcome under certain conditions, and the motions of individual binding sites could be probed by atomic-resolution imaging and/or *in situ* studies using experimental conditions relevant to catalysis science.

Wang et al. presented the interaction between Au_{SA} and graphene edges investigated via low-voltage (60 kV), aberration-corrected, and monochromated TEM. Those observations were performed in vacuum and at room temperature, so that the movement of atoms was only induced through knock-on displacements by electron irradiation. A collective motion of the Au atom and the nearby carbon atoms was observed in transition between various energy-favorable configurations. Most trapping and detrapping processes were found assisted by the nearby dangling carbon atoms [89]. In the same spirit, Zhao et al. performed atomically resolved TEM *in situ* investigations of Fe_{SA} at graphene edges [90]. A remarkable result of this work concerns the capture of the catalytic cycle involved in the growth of graphene through the removal and addition of sp^2 carbon by a single Fe atom (induced here by electron irradiation). As shown in Figure 6.15, the Fe atom was seen to move along the graphene edge, forming pentagonal and hexagonal rings as nearby carbon adatoms are incorporated into the graphene. The diffusion of Fe_{SA} at graphene edges is edge-dependent (zigzag and armchair), with sub-diffusion being shown for armchair edge termination and super-diffusion being shown for zigzag termination. The work was further supported by theoretical calculations explaining that this difference was due to different diffusion barriers between the edge terminations.

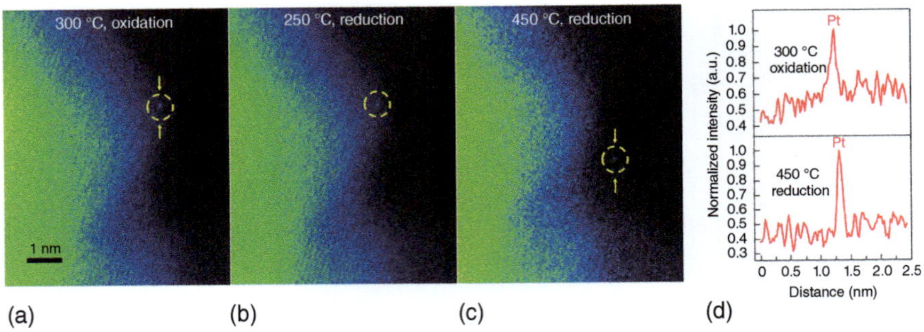

Figure 6.16 *In situ* STEM characterization of Pt_{SA}/TiO_2. (a–c) STEM images of Pt_{SA}/TiO_2 after 30 minutes at different annealing conditions: 300 °C, 1 bar of O_2 for 30 minutes (a); 250 °C, 1 bar of 5% H_2 (balanced with Ar) for 30 minutes (b); 450 °C, 1 bar of 5% H_2 (balanced with Ar) for 30 minutes (c). The yellow circles identify the same Pt_{SA}. A false-coloring scale was used in (a–c) to enhance contrast on the Pt atom. (d) Intensity profile of a line scan (along the yellow arrows) shown in (a) and (c) normalized to the identified Pt atom. Source: DeRita et al. [30]. Reproduced with permission of Springer Nature.

Further progress in observing single atoms in relevant experimental conditions was established by DeRita et al. [30]. Here, catalysts were synthesized via the adsorption of $[Pt(NH_3)_4]^{2+}$ complexes onto 5-nm-diameter anatase TiO_2 nanocrystals at a weight loading of ~0.025%. Using such low metal loading mitigates the formation of Pt clusters. *In situ* aberration-corrected STEM at atmospheric pressure was performed to examine the mobility of Pt atoms on the TiO_2 support (MEMS heating holder). To minimize beam irradiation, a very small beam current (<10 pA) was used for imaging. Considering the imaging condition (a pixel dwell time of 12 μs/pixel and a pixel size of 0.08 Å^2) for the STEM observation, these conditions correspond to an electron dose lower than 1×10^4 e/Å^2. Line-scan intensity analysis and *in situ* STEM experiments as a function of time were used to substantiate the identification of Pt_{SA} and demonstrate that the electron beam did not induce Pt motion or damage the beam-sensitive TiO_2 support. Experimental conditions investigated were a sequential exposure to 1 bar of O_2 at 300 °C, 1 bar of 5% H_2/Ar at 250 °C, and 1 bar of 5% H_2/Ar at 450 °C (Figure 6.16). These observations indicate that the Pt_{SA} are bounded covalently at similar sites on the support surface during the oxidation and the mild reduction pretreatments. When the stronger reduction treatment is applied (450 °C in H_2), the Pt_{SA} are reduced almost to a metallic state while being displaced to a different binding site on the TiO_2 support. The results were further completed by a combination of CO-FTIR spectroscopy, XAS, and *ex situ* STEM imaging along zone axis to derive possible structural models of the local coordination of Pt atoms. Those were further used as input of *ab initio* calculations to identify stable structures on the (101) terrace and (145) stepped surfaces of anatase TiO_2.

Moving onto a different support, Dessal et al. directed their efforts to the industrially relevant Pt/γ-Al_2O_3 catalyst [28]. Their catalyst was prepared as atomically dispersed SAs using a similar impregnation method as in DeRita's study, which involves the $[Pt(NH_3)_4]^{2+}$ complex as a starting Pt precursor. Resulting powders were calcined in air at 300 °C for two hours after the impregnation step. The atomic

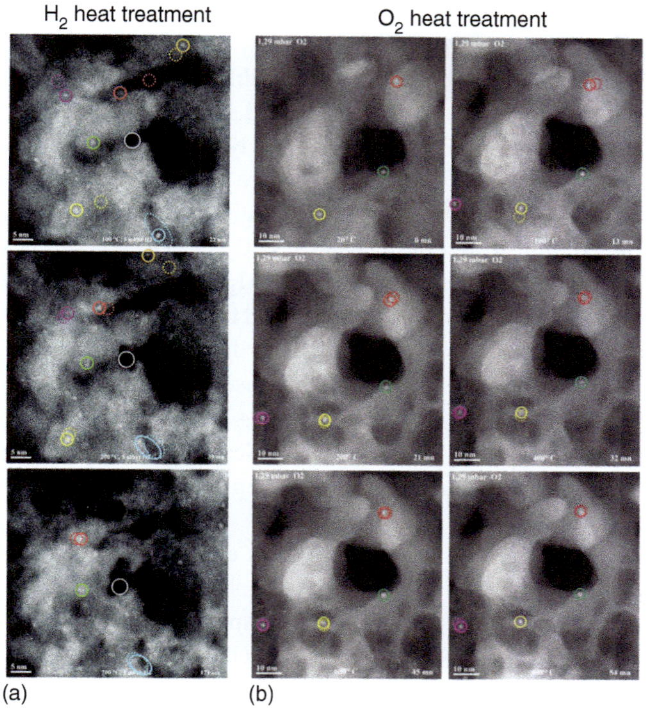

Figure 6.17 (a) ESTEM-ADF images recorded at increasing times (0, 22, 29, 39, 125, and 171 minutes) and temperatures (50, 100, 150, 200, 450, and 700 °C) for the 0.5 wt% Pt/γ-Al_2O_3 under 500 Pa H_2. Colored circles show individual clusters, while the blue oval shows a group of clusters. The central white circle shows the location used for beam focus tuning. (b) ESTEM-ADF image of Pt cluster stability in oxidizing conditions, recorded at increasing times (0, 13, 21, 32, 45, and 54 minutes) and temperatures (20, 100, 200, 400, 600, and 800 °C) for 0.5 wt% Pt/γ-Al_2O_3 under 130 Pa O_2. Colored circles show individual clusters. From one frame to the next, an identical location is indicated as a dotted circle with the same color both for (a) and (b). Source: Dessal et al. [28]. Reproduced with permission of Royal Society of Chemistry.

dispersion of platinum on alumina was confirmed after synthesis both by XAS and STEM imaging. DFT calculations revealed that oxygen from the gas phase directly contributes to the metal–support adhesion. Then, the dynamics of Pt atoms coordination under H_2 was investigated by environmental STEM (ESTEM) and XAS from RT up to 800 °C (up to 300 °C only for XAS). Figure 6.17 reveals a significant mobility even at temperatures lower than c. 200 °C due to the formation of clusters stable in size (0.88 ± 0.07 nm). Only a slight gradual increase in size above c. 400 °C is distinguished. Calculations of the sintering free energies through the series Pt_1H_2, Pt_2H_6, $Pt_{13}H_{32}$ show that sintering under H_2 is a favorable process from a thermodynamic point of view. However, the transition from supported $Pt_{13}H_{32}$ cluster up to bulk platinum is estimated as high as −145 kJ/(mol at$_{Pt}$), which means that sintering could go far beyond 1 nm particles in the presence of H_2. However, those particle sizes were hardly exceeding 1 nm as highlighted by ESTEM imaging. Because of the former experimental work from Fiedorow et al. and their own thermodynamic calculations, the authors expected to be able to redisperse the Pt species under O_2 atmosphere [91]. Thus, the Pt clusters and single atoms were also monitored by ESTEM under 130 Pa O_2 (Figure 6.17b). Both types of Pt species were quasi-immobile between RT and 800 °C. In the temperature range of 800–1000 °C, the metallic clusters were found to disappear. But this might not be directly only related to the effect of the O_2 heat treatment since the γ-Al_2O_3 support was severely restructured due to a probable phase transition toward α-Al_2O_3. Eventually, it was shown that oxygen atoms originating from gas-phase O_2 and from the Al_2O_3 surface serve as ligands that stabilize the Pt_{SA}. On the contrary, with the H_2 atmosphere, single atoms aggregate into subnanometric clusters because of the strong Pt—H bonds, which are forming at the expense of the Pt—O—Al bonds. This study clearly demonstrates that depending on the catalytic application, the as-prepared metal SAs may not be stable and can limit their practical applications for industrial processes.

The latter problematic was similarly investigated by Liu et al. using even more various experimental conditions [92]. Prior to their ESTEM study, the same authors reported a strategy to directly generate subnanometric Pt species (Pt single atoms and clusters) in the microporosity of MCM-22 zeolite using relatively low metal loading (<0.2 wt%) [93]. The dynamic structural transformation of those subnanometric Pt species was followed by *in situ* ESTEM with O_2 oxidation/H_2 reduction treatments and during $CO + O_2$, $CO + H_2O$, $NO + CO$, and $NO + H_2$ reactions. The authors were also careful to minimize the effects of beam irradiation on the sample using a very low beam current (2 pA). Their results with a reduction under H_2 flow were quite similar to the ones reported by Dessal et al. just above. With their catalyst, most Pt species are confined within the microporosity of MCM-22 crystallites, so that their agglomeration into Pt_{NP} is hindered during the reductive treatments leading instead to the formation of subnanometric clusters. On the other hand, O_2 oxidation at 550 °C also led to a completely different result, namely a quantitative redispersion of Pt clusters that had not been observed by Dessal et al. with the alumina support. When the loading of Pt in purely siliceous MCM-22 zeolite is maintained low (<0.1 wt%), subnanometric Pt species can disintegrate into Pt_{SA} in O_2 under relatively mild conditions. Furthermore, oxidative treatment with NO instead of O_2 was also found to be more effective in redispersing Pt clusters. This feature was used for the preparation of Pt@MCM-22 catalyst with higher metal loading up to 0.3 wt%, thanks to a subsequent NO treatment at 200–300 °C enabling the formation of subnanometric Pt species. Comparison with the O_2 post-treatment shows that the same state of dispersion could only be achieved using temperature as high as 650 °C. Then, the authors proceeded with experimental conditions approaching those used for various catalytic reactions at near-ambient pressure. A summary of the evolution of subnanometric Pt species under different reaction conditions is presented in Figure 6.18. Under reductive atmosphere ($CO + O_2$ and $CO + H_2O$), atomically dispersed Pt species coalesce into Pt clusters at a relatively low temperature

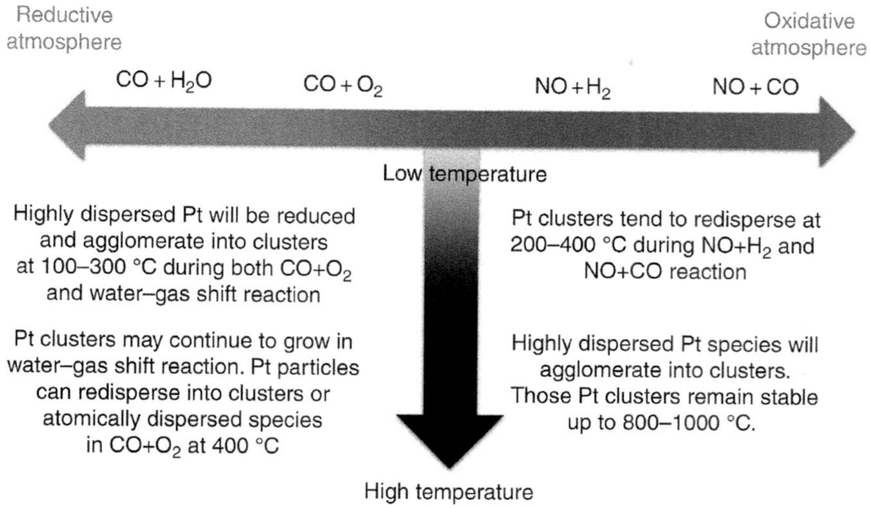

Figure 6.18 Comparison of the evolution of Pt species under different reaction conditions. The evolution of subnanometric Pt species is related with the reactants and reaction temperature. Source: Liu et al. [92]. Springer Nature CC BY 4.0.

(100–300 °C). However, in an oxidative atmosphere ($NO + H_2$ and NO + CO), subnanometric Pt clusters will disintegrate into atomically dispersed Pt species in the 200–400 °C temperature range. At higher temperature, the behavior of Pt species is more complex and dependent on the exact nature of the atmosphere. Comparing with Pt_{NP}, the behavior of Pt_{SA} is much more sensitive to the presence of reactants since dynamic and reversible transformation between single atoms, clusters, and nanoparticles has been observed for the same reaction. Therefore, the further development of *in situ* or even *operando* imaging of monoatomic metal catalysts is absolutely necessary when trying to identify the active species for a specific reaction.

The interplay between SA and NP was also highlighted by Xi et al. in the context of methane pyrolysis [94]. They used a catalyst that is a nanoporous gold material synthesized by dealloying Au–Ag alloy films. HRTEM image presented in Figure 6.19a shows that the as-synthesized nanoporous gold is well crystalline with clean surfaces. After CH_4 pyrolysis operated *ex situ*, the authors noticed that the pore size increased and an amorphous carbon layer was deposited expectedly on the gold surface (Figure 6.19b). The discovery that the carbon layer had a substantial amount of Au_{SA} smeared over its surface (Figure 6.19c) motivated the authors to study *in situ* the transformation of the catalyst. The CH_4 pyrolysis reaction was thus followed by HRTEM with high spatial–temporal resolutions and using a MEMS heating holder. Examination of Figure 6.19d–f evidenced the continuous growth of amorphous carbon, while some gold domains were either reducing or increasing their size overtime, while still retaining a well-defined long-range order (Figure 6.16d–f). The extensive emergence of Au_{SA} distributed on amorphous carbon was thus attributed to the dynamic disintegration/reconstitution of the gold crystals over the course of the reaction. Importantly, Figure 6.19g–i highlights an amorphous carbon region distant from the nanoporous gold surface, where a NP with the Au crystal structure is transiently formed and then quickly disappeared before 0.4125 seconds. This phenomenon imaged by *in situ* HRTEM suggests that both Au_{SA} and the Au_{NP} co-catalyzed the CH_4 pyrolysis reaction, which was further supported by subsequent DFT calculations. This work manifests that under certain conditions, there exists an equilibrium between nanocatalysts and SACs with a continuous migration/diffusion of atoms over the course of the catalytic reaction.

6.5 Summary and Conclusions

This chapter provides clear evidence that *in situ* and *operando* techniques are playing already an important role in the study of SACs, starting from their synthesis and going into their use in important catalytic processes. Combination of different *in situ/operando* techniques is sometime advantageously used, and in most cases one *in situ/operando* technique is used in combination with other *ex situ* techniques, to demonstrate that the study of SACs does need a multi-technique approach. The breadth of the studies covers main aspects, including the interaction with the support, which frequently dictates the formation of SACs.

This chapter has reviewed many recent examples of *in situ/operando* characterization of SACs. Among spectroscopy techniques, XAS is most widely used due to being element selective and capable of focusing on metal atoms in SACs. The XAS techniques routinely provide information on the oxidation state, and bonding of metal atoms to ligands. A limited number of studies have used advanced XAS techniques of PCA to quantify the conversion of initial to final states, and quantum mechanics simulation to scrutinize in hypothetical metal atom sites. Among IR and UV–vis spectroscopies, and XPS spectroscopy, the most widely used is IR spectroscopy. The latter has an overwhelming focus on bands associated with absorption of CO whose frequency is sensitive to the oxidation state and dispersion of the metal atom. The increasing use of XPS, which is also element selective, has been enabled by the recent development of ambient pressure XPS techniques.

There is great appeal to using electron microscopy techniques to probe supported SACs on an atomic scale, whether by imaging of structure using HRTEM and HAADF-STEM imaging, or by chemical analysis using EDS or EELS. Such studies require care to avoid electron-beam damage to single-atom sites. This chapter has reviewed recent *in situ/operando* electron microscopy studies, which are limited in number due to the additional challenge of applying these techniques when the sample is not in a vacuum. Meeting this challenge depends on the development of environmental electron microscopy techniques. Although limited in number, such studies are impressive for probing on an atomic scale the dynamics of catalytic processes involving supported SACs.

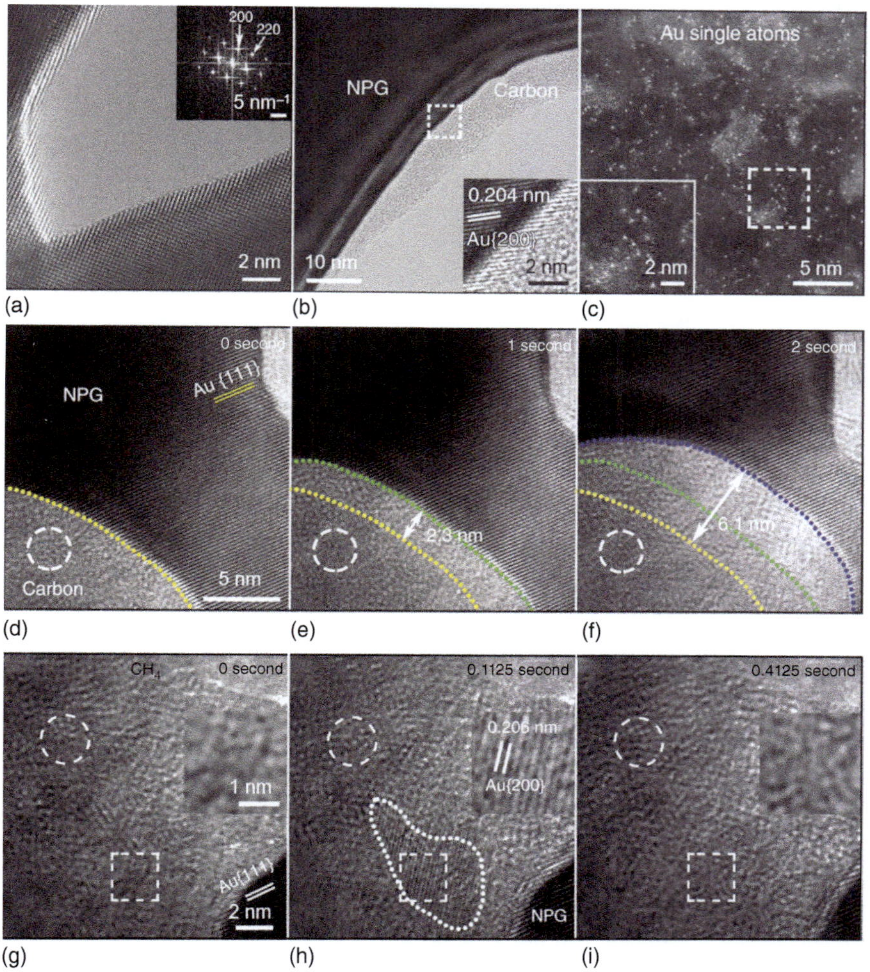

Figure 6.19 Dynamic processes of the catalytic CH_4 pyrolysis on nanoporous gold (NPG). *Ex situ* HRTEM image and its fast Fourier transform pattern (a) before and (b) after the reaction. (c) HAADF–STEM imaging of a carbon region on the nanoporous gold sample after CH_4 pyrolysis. (d–f) HRTEM images at three different moments during the reaction on a surface region. The images have the same scale bar. The yellow, green, and blue dashed lines indicate the positions of the surface at 0, 1, and 2 seconds, respectively. The white dashed circles denote the positions of a randomly selected carbon region, arrows indicate the thickness of the amorphous carbon at 0, 1, and 2 seconds. (g–i) HRTEM images of an amorphous carbon layer at three different moments during the reaction with all the conditions unchanged. The boxes in (g–f) indicate the positions of a nanoparticle that temporarily formed and disappeared. The insets in (g–i) are the close-up views of the boxed regions. Source: Xi et al. [94]. Springer Nature / CC BY-SA 4.0.

References

1 Kawai, J. (2000). Absorption techniques in X-ray spectrometry. In: *Encyclopedia of Analytical Chemistry* (ed. R.A. Meyers), 13288–13315. Chichester: Wiley.

2 Fernandez-Garcia, M. (2002). XANES analysis of catalytic systems under reaction conditions. *Catalysis Reviews* 44: 59–121.

3 Gu, J., Hsu, C.-S., Bai, L. et al. (2019). Atomically dispersed Fe^{3+} sites catalyze efficient CO_2 electroreduction to CO. *Science* 364 (6445): 1091–1094.

4 Zitolo, A., Ranjbar-Sahraie, N., Mineva, T. et al. (2017). Identification of catalytic sites in cobalt-nitrogen-carbon materials for the oxygen reduction reaction. *Nature Communications* 8: 957.

5 Koningsberger, D.C., Mojet, B.L., Van Dorssen, G.E. et al. (2000). XAFS spectroscopy; fundamental principles and data analysis. *Topics in Catalysis* 10: 143–155.

6 Feng, K., Zhang, H., Gao, J. et al. (2020). Single atoms or not? *The limitation of EXAFS. Applied Physics Letters* 116: 191903.

7 Soldatov, M.A., Martini, A., Bugaev, A.L. et al. (2018). The insights from X-ray absorption spectroscopy into the local atomic structure and chemical bonding of Metal–organic frameworks. *Polyhedron* 155: 232–253.

8 Jia, Q., Ramaswamy, N., Hafiz, H. et al. (2015). Experimental observation of redox-induced Fe–N switching behavior as a determinant role for oxygen reduction activity. *ACS Nano* 9 (12): 12496–12505.

9 Wang, L., Liu, X., Cao, L. et al. (2020). Active sites of single-atom iron catalyst for electrochemical hydrogen evolution. *The Journal of Physical Chemistry Letters* 11: 6691–6696.

10 Genovese, C., Schuster, M.E., Gibson, E.K. et al. (2018). Operando spectroscopy study of the carbon dioxide electro-reduction by iron species on nitrogen-doped carbon. *Nature Communications* 9: 935.

11 Nakatsuka, K., Yoshii, T., Kuwahara, Y. et al. (2017). Controlled synthesis of carbon-supported Co catalysts from single-sites to nanoparticles: characterization of the structural transformation and investigation of their oxidation catalysis. *Physical Chemistry Chemical Physics* 19: 4967–4974.

12 Zhang, Z., Xiao, J., Chen, X.-J. et al. (2018). Reaction mechanisms of well-defined metal-N_4 sites in electrocatalytic CO_2 reduction. *Angewandte Chemie International Edition* 57 (50): 16339–16342.

13 Jiang, K., Siahrostami, S., Zheng, T. et al. (2018). Isolated Ni single atoms in graphene nanosheets for high-performance CO_2 reduction. *Energy & Environmental Science* 11: 893–903.

14 Yang, H.B., Hung, S.-F., Liu, S. et al. (2018). Atomically dispersed Ni(I) as the active site for electrochemical CO_2 reduction. *Nature Energy* 3: 140–147.

15 Abdel-Mageed, A.M., Rungtaweevoranit, B., Parlinska-Wojtan, M. et al. (2019). Highly active and stable single-atom Cu catalysts supported by a metal–organic framework. *Journal of the American Chemical Society* 141 (13): 5201–5210.

16 Aitbekova, A., Wu, L., Wrasman, C.J. et al. (2018). Low-temperature restructuring of CeO_2-supported Ru nanoparticles determines selectivity in CO_2 catalytic reduction. *Journal of the American Chemical Society* 140 (42): 13736–41375.

17 Li, P., Wang, M., Duan, X. et al. (2019). Boosting oxygen evolution of single-atomic ruthenium through electronic coupling with cobalt-iron layered double hydroxides. *Nature Communications* 10: 1711.

18 Cao, L., Luo, Q., Chen, J. et al. (2019). Dynamic oxygen adsorption on single-atomic Ruthenium catalyst with high performance for acidic oxygen evolution reaction. *Nature Communications* 10: 4849.

19 Shan, J., Li, M., Allard, L.F. et al. (2017). Mild oxidation of methane to methanol or acetic acid on supported isolated rhodium catalysts. *Nature* 551: 605–608.

20 Hülsey, M.J., Zhang, B., Ma, Z. et al. (2019). In situ spectroscopy-guided engineering of rhodium single-atom catalysts for CO oxidation. *Nature Communications* 10: 1330.

21 Peterson, E.J., DeLaRiva, A.T., Lin, S. et al. (2014). Low-temperature carbon monoxide oxidation catalysed by regenerable atomically dispersed palladium on alumina. *Nature Communications* 5: 4885.

22 Piernavieja-Hermida, M., Lu, Z., White, A. et al. (2016). Towards ALD thin film stabilized single-atom Pd_1 catalysts. *Nanoscale* 8: 15348–15356.

23 Yan, H., Lv, H., Yi, H. et al. (2018). Understanding the underlying mechanism of improved selectivity in Pd_1 single-atom catalyzed hydrogenation reaction. *Journal of Catalysis* 366: 70–79.

24 He, Q., Lee, J.H., Liu, D. et al. (2020). Accelerating CO_2 electroreduction to CO over Pd single-atom catalyst. *Advanced Functional Materials* 30 (17): 2000407.

25 Bugaev, A.L., Skorynina, A.A., Braglia, L. et al. (2019). Evolution of Pt and Pd species in functionalized UiO-67 metal-organic frameworks. *Catalysis Today* 336: 33–39.

26 Nguyen, L., Zhang, S., Wang, L. et al. (2016). Reduction of nitric oxide with hydrogen on catalysts of singly dispersed bimetallic sites Pt_1Co_m and Pd_1Co_n. *ACS Catalysis* 6: 840–850.

27 Yang, M., Liu, J., Lee, S. et al. (2015). A common single-site $Pt(II)–O(OH)_x$ – species stabilized by sodium on "Active" and "Inert" supports catalyzes the water–gas shift reaction. *Journal of the American Chemical Society* 137 (10): 3470–3473.

28 Dessal, C., Sangnier, A., Chizallet, C. et al. (2019). Atmosphere-dependent stability and mobility of catalytic Pt single atoms and clusters on γ-Al_2O_3. *Nanoscale* 11: 6897–6904.

29 Dessal, C., Len, T., Morfin, F. et al. (2019). Dynamics of single Pt atoms on alumina during CO oxidation monitored by operando X-ray and infrared spectroscopies. *ACS Catalysis* 9 (6): 5752–5759.

30 DeRita, L., Resasco, J., Dai, S. et al. (2019). Structural evolution of atomically dispersed Pt catalysts dictates reactivity. *Nature Materials* 18: 746–751.

31 Yoo, M., Yu, Y.-S., Ha, H. et al. (2010). A tailored oxide interface creates dense Pt single-atom catalysts with high catalytic activity. *Energy & Environmental Science* 2020 (13): 1231–1239.

32 Resasco, J., DeRita, L., Dai, S. et al. (2020). Uniformity is key in defining structure–function relationships for atomically dispersed metal catalysts: the case of Pt/CeO_2. *Journal of the American Chemical Society* 142 (1): 169–184.

33 Wei, H., Ren, Y., and Wang, A. (2017). Remarkable effect of alkalis on the chemoselective hydrogenation of functionalized nitroarenes over high-loading Pt/FeO_x catalysts. *Chemical Science* 8 (7): 5126.

34 Zhang, S., Shan, J.-j., Zhu, Y. et al. (2013). WGS catalysis and in situ studies of CoO_{1-x}, $PtCo_n/Co_3O_4$, and $Pt_mCo_{m'}/CoO_{1-x}$ nanorod catalysts. *Journal of the American Chemical Society* 135 (22): 8283–8293.

35 Liu, L., Meira, D.M., Arenal, R. et al. (2019). Determination of the evolution of heterogeneous single metal atoms and nanoclusters under reaction conditions: which are the working catalytic sites? *ACS Catalysis* 9 (12): 10626–10639.

36 Li, Z., Qi, Z., Wang, S. et al. (2019). In situ formed Pt_3Ti nanoparticles on a two-dimensional transition metal carbide (MXene) used as efficient catalysts for hydrogen evolution reactions. *Nano Letters* 19 (8): 5102–5108.

37 Fang, S., Zhu, X., Liu, X. et al. (2020). Uncovering near-free platinum single-atom dynamics during electrochemical hydrogen evolution reaction. *Nature Communications* 11: 1029.

38 Kim, I.S., Li, Z., Zheng, J. et al. (2018). Sinter-resistant platinum catalyst supported by metal-organic framework. *Angewandte Chemie International Edition* 57 (4): 909–913.

39 Jiang, K., Luo, M., Peng, M. et al. (2020). Dynamic active-site generation of atomic iridium stabilized on nanoporous metal phosphides for water oxidation. *Nature Communications* 11: 2701.

40 Lu, Y., Wang, J., Yu, L. et al. (2019). Identification of the active complex for CO oxidation over single-atom Ir-on-$MgAl_2O_4$ catalysts. *Nature Catalysis* 2: 149–156.

41 Lebedev, D., Ezhov, R., Heras-Domingo, J. et al. (2020). Atomically dispersed iridium on indium tin oxide efficiently catalyzes water oxidation. *ACS Central Science* 6: 1189–1198.

42 Hoffman, A.S., Sokara, D., Zhang, S. et al. (2017). High-energy-resolution X-ray absorption spectroscopy for identification of reactive surface species on supported single-site iridium catalysts. *Chemistry – A European Journal* 23 (59): 14760–14768.

43 Guo, L.-W., P-P, D., Fu, X.-P. et al. (2016). Contributions of distinct gold species to catalytic reactivity for carbon monoxide oxidation. *Nature Communications* 7: 13481.

44 Li, X., Yang, X., Zhang, J. et al. (2019). In situ/operando techniques for characterization of single-atom catalysts. *ACS Catalysis* 9 (3): 2521–2531.

45 Ryczkowski, J. (2001). IR spectroscopy in catalysis. *Catalysis Today* 68: 263–381.

46 Jones, J., Xiong, H., DeLaRiva, A.T. et al. (2016). Thermally stable single-atomplatinum-on-ceria catalysts via atom trapping. *Science* 353 (6295):–150.

47 Nie, H., Howe, J.Y., Lachkov, P.T. et al. (2019). Chemical and structural dynamics of nanostructures in bimetallic Pt–Pd catalysts, their inhomogeneity, and their roles in methane oxidation. *ACS Catalysis* 9: 5445–5461.

48 Yang, K., Liu, Y., Deng, J. et al. (2019). Three-dimensionally ordered mesoporous iron oxide-supported single-atom platinum: highly active catalysts for benzene combustion. *Applied Catalysis B* 244: 650–659.

49 Zhu, Y., An, Z., and He, J. (2016). Single-atom and small-cluster Pt induced by Sn (IV) sites confined in an LDH lattice for catalytic reforming. *Journal of Catalysis* 341: 44–54.

50 DeRita, L., Dai, S., Lopez-Zepeda, K. et al. (2017). Catalyst architecture for stable single atom dispersion enables site-specific spectroscopic and reactivity measurements of CO adsorbed to Pt atoms, oxidized Pt clusters, and metallic Pt clusters on TiO_2. *Journal of the American Chemical Society* 139 (40): 14150–14165.

51 Liu, K., Hou, G., Mao, J. et al. (2019). Genesis of electron deficient $Pt_1(0)$ in PDMS-PEG aggregates. *Nature Communications* 10: 996.

52 Qiu, J.-Z., Hu, J., Lan, J. et al. (2019). Pure siliceous zeolite-supported Ru single-atom active sites for ammonia synthesis. *Chemistry of Materials* 31: 9413–9421.

53 Pei, G.X., Liu, X.Y., Wang, A. et al. (2015). Ag alloyed Pd single-atom catalysts for efficient selective hydrogenation of acetylene to ethylene in excess ethylene. *ACS Catalysis* 5: 3717–3725.

54 Schoonheydt, R.A. (2010). UV-VIS-NIR spectroscopy and microscopy of heterogeneous catalysts. *Chemical Society Reviews* 39 (12): 5051–5066.

55 Shang, Q., Tang, N., Qi, H. et al. (2020). A palladium single-atom catalyst toward efficient activation of molecular oxygen for cinnamyl alcohol oxidation. *Chinese Journal of Catalysis* 41: 1812–1817.

56 Schilling, A.C., Groden, K., Simonovis, J.P. et al. (2020). Accelerated Cu_2O reduction by single Pt atoms at the metal-oxide interface. *ACS Catalysis* 10: 4215–4226.

57 Corcoran, C.J., Tavassol, H., Rigsby, M.A. et al. (2010). Application of XPS to study electrocatalysts for fuel cells. *Journal of Power Sources* 195: 7856–7879.

58 Schnadt, J., Knudsen, J., and Johansson, N. (2020). Present and new frontiers in materials research by ambient pressure x-ray photoelectron spectroscopy. *Journal of Physics: Condensed Matter* 32: 413003.

59 Moses-DeBusk, M., Yoon, L.F., Allard, D.R. et al. (2013). CO oxidation on supported single Pt atoms: experimental and ab initio density functional studies of CO interaction with Pt atom on θ-$Al_2O_3(010)$ surface. *Journal of the American Chemical Society* 135 (34): 12634–12645.

60 Zhou, P., Zhang, Q., Xu, Z. et al. (2019). Atomically dispersed Co–P_3 on CdS nanorods with electron-rich feature boosts photocatalysis. *Advanced Materials* 32 (7): 1904249.

61 Zhang, S., Tang, Y., Nguyen, L. et al. (2018). Catalysis on singly dispersed Rh atoms anchored on an inert support. *ACS Catalysis* 8 (1): 110–121.

62 Hou, Z., Dai, L., Liu, Y. et al. (2021). Highly efficient and enhanced sulfur resistance supported bimetallic single-atom palladium–cobalt catalysts for benzene oxidation. *Applied Catalysis B: Environmental* 285 (119844).

63 Liang, J.-X., Lin, J., Liu, J. et al. (2020). A redox mechanism for the water–gas shift reaction. *Angewandte Chemie International Edition* 59: 12868–12875.

64 Li, X., Cao, C.-S., Hung, S.-F. et al. (2020). Identification of the electronic and structural dynamics of catalytic centers in single-Fe-atom material. *Chem* 6: 3440–3454.

65 Pereira-Hernández, X.I., DeLaRiva, A., Muravev, V. et al. (2019). Tuning Pt-CeO_2 interactions by high-temperature vapor-phase synthesis for improved reducibility of lattice oxygen. *Nature Communications* 10: 1358.

66 Haider, M., Uhlemann, S., Schwan, E. et al. (1998). Electron microscopy image enhanced. *Nature* 392: 768–769.

67 Krivanek, O.L., Dellby, N., and Lupini, A.R. (1999). Towards sub-Å electron beams. *Ultramicroscopy* 78: 1–11.

68 Krivanek, O.L., Lovejoy, T.C., Dellby, N. et al. (2014). Vibrational spectroscopy in the electron microscope. *Nature* 514: 209–214.

69 Zeitler, E. (2011). *Scanning Transmission Electron Microscopy*. New York: Springer.

70 Zhang, Q., He, X., Shi, J. et al. (2017). Atomic-resolution imaging of electrically induced oxygen vacancy migration and phase transformation in $SrCoO_{2.5-\sigma}$. *Nature Communications* 8: 104.

71 Kim, H., Zhang, J.Y., Raghavan, S. et al. (2016). Direct observation of Sr vacancies in $SrTiO_3$ by quantitative scanning transmission electron microscopy. *Physical Review X* 6: 041063.

72 Li, Z., Chen, Y., Ji, S. et al. (2020). Iridium single-atom catalyst on nitrogen-doped carbon for formic acid oxidation synthesized using a general host–guest strategy. *Nature Chemistry* 12: 764–772.

73 Fei, H., Dong, J., Arellano-Jiménez, M.J. et al. (2015). Atomic cobalt on nitrogen-doped graphene for hydrogen generation. *Nature Communications* 6: 8668.

74 Yan, D., Chen, J., and Jia, H. (2020). Temperature-induced structure reconstruction to prepare a thermally stable single-atom platinum catalyst. *Angewandte Chemie International Edition* 59 (32): 13562–13567.

75 Colliex, C., Gloter, A., March, K. et al. (2012). Capturing the signature of single stoms with the tiny probe of a STEM. *Ultramicroscopy* 123: 80–89.

76 Egerton, R.F. (2012). Mechanisms of radiation damage in beam-sensitive specimens, for TEM accelerating voltages between 10 and 300 kV. *Microscopy Research and Technique* 75 (11): 1550–1556.

77 Song, H., Li, Y., Lou, Z. et al. (2015). Synthesis of Fe-doped WO_3 nanostructures with high visible-light-driven photocatalytic activities. *Applied Catalysis B: Environmental* 166–167: 112–120.

78 Ngigi, E.M., Nomngongo, P.N., and Ngila, J.C. (2019). Synthesis and application of Fe-doped WO_3 nanoparticles for photocatalytic degradation of methylparaben using visible–light radiation and H_2O_2. *Catalysis Letters* 149: 49–60.

79 Guan, E., Ciston, J., Bare, S.R. et al. (2020). Supported metal pair-site catalysts. *ACS Catalysis* 10 (16): 9065–9085.

80 Wang, X., Chen, Z., Zhao, X. et al. (2018). Regulation of coordination number over single Co sites: triggering the efficient electroreduction of CO_2. *Angewandte Chemie International Edition* 130 (7): 1962–1966.

81 Liu, D., Li, J.C., Shi, Q. et al. (2019). Atomically isolated iron atom anchored on carbon nanotubes for oxygen reduction reaction. *ACS Applied Materials and Interfaces* 11 (43): 39820–39826.

82 Li, J.C., Maurya, S., Kim, Y.S. et al. (2020). Stabilizing single-atom iron electrocatalysts for oxygen reduction via ceria confining and trapping. *ACS Catalysis* 10 (4): 2452–2458.

83 Baares, M.A. (2007). Operando spectroscopy. *Catalysis Today* 126 (1–2): 1–264.

84 Chen, Y., Kasama, T., Huang, Z. et al. (2015). Highly dense isolated metal atom catalytic sites: dynamic formation and in situ observations. *Chemistry – A European Journal* 21 (48): 17397–17402.

85 Wei, S., Li, A., Liu, J.-C. et al. (2018). Direct observation of noble metal nanoparticles transforming to thermally stable single atoms. *Nature Nanotechology* 13: 856–861.

86 Liu, J., Cao, C., Liu, X. et al. (2021). Direct observation of metal oxide nanoparticles being transformed into metal single atoms with oxygen-coordinated structure and high-loadings. *Angewandte Chemie International Edition* 60 (28): 15248–15253.

87 Liu, K., Zhao, X., Ren, G. et al. (2020). Strong metal-support interaction promoted scalable production of thermally stable single-atom catalysts. *Nature Communications* 11: 1263.

88 Lang, R., Xi, W., Liu, J.-C. et al. (2019). Non defect-stabilized thermally stable single-atom catalyst. *Nature Communications* 10: 234.

89 Wang, H., Li, K., Cheng, Y. et al. (2012). Interaction between single gold atom and the graphene edge: A study via aberration-corrected transmission electron microscopy. *Nanoscale* 4: 2920–2925.

90 Zhao, J., Deng, Q., Avdoshenko, S.A. et al. (2014). Direct in situ observations of single Fe atom catalytic processes and anomalous diffusion at graphene edges. *Proceedings of the National Academy of Sciences of the United States of America* 111 (44): 15641–15646.

91 Fiedorow, R.M.J., Chahar, B.S., and Wanke, S.E. (1978). The sintering of supported metal catalysts II. Comparison of sintering rates of supported Pt, Ir, and Rh catalysts in hydrogen and oxygen. *Journal of Catalysis* 51 (2): 193–202.

92 Liu, L., Zakharov, D.N., Arenal, R. et al. (2018). Evolution and stabilization of subnanometric metal species in confined space by in situ TEM. *Nature Communications* 9: 574.

93 Liu, L., Díaz, U., Arenal, R. et al. (2017). Generation of subnanometric platinum with high stability during transformation of a 2D zeolite into 3D. *Nature Materials* 16: 132–138.

94 Xi, W., Wang, K., Shen, Y. et al. (2020). Dynamic co-catalysis of Au single atoms and nanoporous Au for methane pyrolysis. *Nature Communications* 11: 1919.

7

Contribution of Theoretical Calculations to Supported Metal Single-Atom Catalysis

Javier Navarro-Ruiz[1], Romuald Poteau[1], Iann C. Gerber[1], and Iker del Rosal[1]

[1] *LPCNO, Université de Toulouse, INSA-UPS-CNRS (UMR 5215), Institut National des Sciences Appliquées, 135 Avenue de Rangueil, F-31077 Toulouse, France*

7.1 Introduction

Single-atom catalysts (SACs) have sparked new interest in catalysis due to their high catalytic activity, stability, selectivity, and 100% atom utilization, as pointed out in previous chapters (Chapters 2, 3, 5 and 6) on their preparation and subsequent characterization. Also, see Refs. [1] and [2] for additional reviews with specific interest in innovative syntheses and characterization techniques for SACs with a focus on their electrochemical applications in the oxygen reduction reaction (ORR)/oxygen evolution reaction (OER), hydrogen evolution reaction (HER), and hydrocarbon conversion reactions for fuel cells (electro-oxidation of methanol, ethanol, and formic acid). SACs also represent interesting systems from a theoretical point of view, since the models used are very close to the real atomic configuration. In this sense, density functional theory (DFT) is a valuable tool to study electronic structures, by means of orbital analyses and density of states (DOS), as well as to rationalize and predict catalytic results thanks to the definition of appropriate descriptors [3, 4]. Advanced computational schemes allow to study the reaction pathways in detail, determining the transition states and energy barriers, with the aim of finding rate-limiting steps. In all cases, the investigations of the relationship between the structural parameters and the properties of the electronic structure of the SAC embedded or deposited on a support based on oxide or carbon are the first outcomes of such theoretical studies. The first attempt to describe the coordination of metal atomic species (Ag and Ti) on the surface of MgO using DFT was reported in 1992 [5]. Only 10 years later, the pioneering experimental study by Flytzani-Stephanopoulos and coworkers appeared, providing convincing evidence that ionic species of Au or Pt bind strongly to the surface of ceria, and not metal nanoparticles (NPs), were responsible for the activity observed in the water–gas shift reaction [6]. Theoretical works on the catalytic activity of a single-metal atom/site on the oxide surface were reported in 2006 in the context of olefin polymerization or alkane metathesis [7, 8]. The use of Au-SACs for CO oxidation was theoretically proposed in 2009 [9]. For recent comprehensive

Supported Metal Single Atom Catalysis, First Edition. Edited by Philippe Serp and Doan Pham Minh.
© 2022 WILEY-VCH GmbH. Published 2022 by WILEY-VCH GmbH.

reviews on the use of oxide-supported SACs, see the complementary studies reported in Refs. [10] and [11].

Rivera-Cárcamo and Serp have recently reviewed the SAC preparation focusing on carbon-based materials [12], where explanations of their catalytic performances are achieved, as well as a comprehensive understanding of metal–support interactions thanks to experimental and theoretical work. Furthermore, both defects and heteroatoms tend to strengthen the bonding between the transition metal (TM) atom and the carbon support. The preparation of a SACs requires specific sites on the support that will act as traps for the metal atoms. In that context, a pristine sp^2 carbon surface does not seem adequate to obtain stable SACs, even if for some metals on graphene, theoretical calculations have shown strong adsorption energy, high diffusion barrier, and significant charge transfer. The first complete DFT description of metal atoms embedded in a defective graphene layer was published in 2009 [13], followed by the work of Lu et al. which was devoted to investigating the first step of CO oxidation reaction mechanism catalyzed by the "Au-embedded graphene" [14]. After a review considering the description of carbon-based support models available in the literature, emblematic cases of SAC reactivity embedded in such supports are presented. In addition, theoretical works on the hydrogen spillover mechanism in those systems are also discussed due to its importance in several reactions. Finally, typical case studies on oxide surfaces are discussed.

7.2 Carbon-Based Support Models

Since the 1960s [15], it is widely accepted that the catalytic activity and selectivity of carbon catalysts are strongly influenced by their surface chemistry and electronic properties. Pure carbon supports for catalysis are usually classified depending on their dimensions: (i) carbon black, fullerene or onion-like structures in 0D; (ii) nanotubes in 1D; (iii) graphene and few-layer graphene (FLG) in 2D; and (iv) more complex structures such as mesoporous carbon or graphite in 3D. Describing theoretically the interactions between supported metallic SACs on all of these various supports is a huge task, the available literature being mostly limited to some specific cases due to the drastic increase in computational complexity when increasing dimensions. All these works are based on the fact that to anchor metal single atoms (SAs) on carbon supports, one needs specific sites. The next subsection (Section 7.2.1) presents important results concerning the description of those anchoring sites at the atomic scale, when in the second time the interaction with metallic SAs will be specified. We recall that obtaining experimental direct atomic-scale pictures of SAC systems is challenging. By combining advanced characterization techniques such as FTIR, Raman, XANES, EXAFS, HR-TEM, XPS, STEM-EELS, HAADF-STEM, and theoretical works mostly using DFT calculations on realistic models, useful insights of the metal–carbon-based support interactions can be obtained.

7.2.1 Anchoring Sites on Carbon Materials

Obviously, physicochemical properties of surface carbon materials dictate SA adsorption. Since most of the considered supports are made of sp^2-type carbon, basal and edge carbon atoms are usually considered [16, 17], as well as point defects essentially represented by structural carbon vacancies or non-aromatic rings [18]. For instance, the presence of such defects along graphene layer edges, or even incorporated in the graphene layer, is at the origin of their high reactivity, since many unpaired electrons are indeed localized at these specific sites [19]. Two types of carbon appear then, some carbene type on zigzag sites, or carbyne-like for armchair configurations [17]. Another possibility is to absorb or even substitute carbon atoms by various species (H, B, O, N, P, S, …), providing very stable surface compounds, which induce a more complex surface chemistry, as soon as structural point defects are also involved. In this respect, nitrogen is very interesting, since it is a versatile element, and it can dope the structure even by a single substitution [20–22]. Compared with pristine carbon-supported metal catalysts, these catalysts normally show superior catalytic performances in many reactions because of specific metal–support interactions resulting from N-doping [23]. Interestingly, the presence of surface groups [24], mainly oxygenated ones, can also reinforce the overall adsorption capacity of sp^2 and $sp^{2+\delta}$ carbon materials, for which the adsorption sites are mainly on basal planes, or on inner or outer surface in the case of carbon nanotubes (CNTs). Thus, it is of primary importance to be able to describe accurately, by means of computational approaches based on quantum chemistry, the effects of substituting/functionalizing the carbon support on the grafting of SAs, and of rationalizing the local electronic structure of the active site, to understand/predict their catalytic behavior [25]. Supports with a high level of graphitization, such as CNTs and graphene, have been extensively considered, since it is commonly accepted that the higher the graphitization's level, the better the stability due to a lower density of edges and defect sites. At the same time, those systems can always be viewed as simple models for more complex supports as nanofibers, hierarchical carbon, aerogels, and others ordered mesoporous carbon. We focus, in the following, on the interaction of single (non-)noble metal atom with various carbon nanostructures employed for catalytic purposes, namely 0D (fullerene-like), 1D (nanotubes), and 2D (graphene-like) systems.

7.2.1.1 SAs-Fullerene

Historically, the early significant atomic-scale investigations of the interaction between TM atoms and fullerenes dated from the beginning of this century, mainly in the context of hydrogen storage [26], with the aim of building polymerized structures. Metal fullerides have been the subject of intensive investigations over the past 20 years, and mostly compounds made of C_{60} with alkali and alkaline metals have been investigated in the superconductivity context [27]. In contrast, only few theoretical works investigating fullerides or metal–fullerene complexes

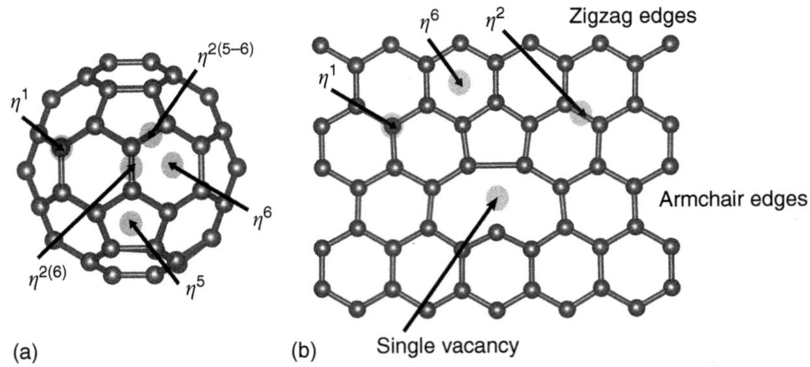

Figure 7.1 Typical adsorption modes on: (a) C_{60}, and (b) graphene layer presenting also unsaturated armchair and zigzag edges as well as a single reconstructed vacancy. Source: I. C. Gerber.

with TM have been reported [28–38], usually dealing with the interaction of a single TM atom adsorbed on a single C_{60} molecule, viewed as a versatile ligand due to various hapticities [24, 39].

Essentially, those works report the binding energy, the discrimination between the most stable adsorption sites either hollow (η^6), pentagonal (η^5), bridge on a single bond ($\eta^{2(5-6)}$) or on a double bond ($\eta^{2(6)}$), or even in top position, see Figure 7.1. For instance, Sc and Ti atoms prefer hollow sites with binding energies of 48 kcal/mol, while V and Cr stand for $\eta^{2(6)}$ sites with binding energies of 30 and 19 kcal/mol, respectively [26]. Pd_{SA} and Pt_{SA} bind strongly, with 34 and 64 kcal/mol, respectively, in a $\eta^{2(6)}$ coordination mode [40, 41], when Ni atoms also possess a strong affinity with C_{60}. We recall that this kind of study is only ideal in the sense that further molecular adsorption on the metallic center will occur, due to the presence of other species (H_2, CO, and water), as well as ligands associated with the metallic precursor's decomposition [42] or even other C_{60} [37]. In the specific case of Ru fullerides, DFT calculations have shown that Ru_{SA} can adopt a $\eta^{2(6)} - \eta^6$ coordination mode between two fullerenes, corroborated by EXAFS experiments. Charge transfer from the metal to the C_{60}, known as a good electron acceptor, has been estimated and confirmed by Raman spectroscopy. For a Pd_{SA} embedded in a C_{60}-polymeric 1/1 ratio, the preferential coordination is $\eta^{2(6)} - \eta^{2(6)}$ and remains the same, even in a 3D organization [38].

7.2.1.2 SAs-CNT

From a theoretical point of view, the interaction of single TM atoms with CNT started to be studied almost 15 years ago. Because the electronic structure of CNTs is crucially dependent on their diameter and helicity, metal-SA adsorption is expected to be highly dependent on these two properties [43, 44]. The anchorage of TM single atoms on pristine or even functionalized nanotube sidewalls has been widely investigated, mainly in hydrogen storage context, using the support of strong orbital analyses [45–48]. The curvature of the surface can affect the type of binding sites for the metal [49, 50], and the surface diffusion of metallic adatoms [51–53]. Defects, such as Stone–Wales (SW) or vacancies, which are inherently present or even induced by various physicochemical treatments, play a crucial role in the adsorption or surface diffusion of TM atoms on CNTs. Single vacancies, as depicted in Figure 7.1b, increase significantly the TM–carbon bonding, doubling the adsorption energy values for instance [54], due to the presence of σ-bonding of the metal to the surrounding carbon atoms around the defect edge. Because of bonding character differences with C atoms, adsorption energies can vary considerably from one 3d TM atom to the other and in return can be used to modulate the pristine electronic structure of CNTs. Most of the time, endohedral coordination is much less favorable thermodynamically [55], for instance as in the Fe@(6,6)CNT case [56].

7.2.1.3 SAs-Graphene

Graphene is thought to meet all the requirements of an ideal electrocatalyst support, since it possesses high electronic conductivity and excellent mechanical properties. DFT studies of 3d, 4d, and 5d TM atom adsorption on a pristine graphene surface are numerous and compilations of complete series of adsorption energies are even provided in some cases [57–60], the Figure 7.1 recalling the most favorable adsorption sites on graphene. Generally, TM atoms stand in hollow (η^6) sites, with a strong hybridization between the TM d orbitals and the π orbital of the graphene layer, inducing spin-state changes in the Co case [61, 62], with typical E_{ads} around 23 kcal/mol. Bridge (η^2) or top sites are supposedly less favorable, since it corresponds to a physisorption process that occurs mainly for half-filled and completely filled d orbital atoms. In these cases, the energy differences between different adsorption configurations are thus rather small.

The adsorption on pristine bilayer graphene (BLG) and on graphite has been theoretically studied [63, 64]. Generally, when more graphene layers are stacked, the binding energy is increased [65]. Edge sites are also found to be good anchoring sites, with large binding energies from 60 kcal/mol for armchair to 120 kcal/mol for zigzag edges [66], due to the presence of dangling bonds [67, 68]. However, due to experimental conditions, one could argue that most of those edge sites are passivated by hydrogen or possibly by other species. In this case, it has been shown that for a Fe atom adsorbed on a H-terminated site, the binding energy drops off but remains still larger than the adsorption on pristine graphene by 10 kcal/mol [69].

Intrinsic line of defects, made of vacancies essentially, or even grain boundaries as well as extrinsic, such as impurities defects, are also good candidates for a strong anchoring of TM atoms due to their high reactivity [70]. The presence of such defects thanks to the versatile nature of carbon atoms embedded in the graphene layer allows easy reconstruction by forming non-hexagonal rings. See Ref. [71] for a complete review of the numerous energetic aspects of graphene atomic structure changes. The simplest defect in any graphene-based materials, i.e. the single vacancy (SV), and thus the most studied one [72–75] has been confirmed experimentally (TEM and STM) despite its high formation energy around 170 kcal/mol [76, 77]. See Figure 7.2a for the corresponding atomic structure induced by the formation of a five-membered aromatic ring [78], leaving one highly reactive dangling bond on the opposite C atom. Interestingly, a small energy separation between the singlet and

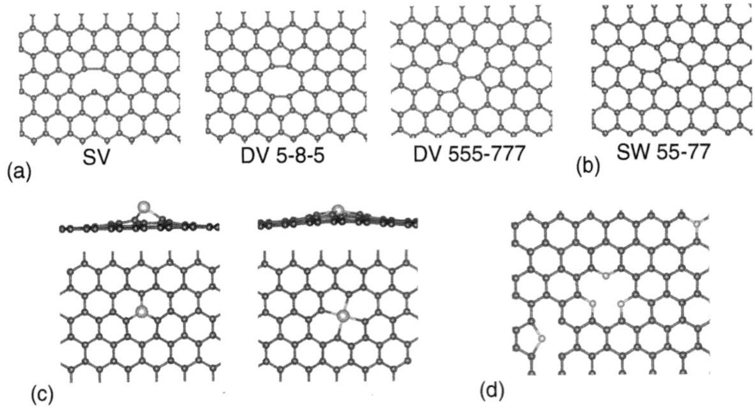

Figure 7.2 (a) Single vacancy, 5-8-5 vs. 555-777 divacancy atomic structures, (b) Stone–Wales (55-77) defect, (c) adsorption of Pt atom in SV and 5-8-5 DV, and (d) various nitrogen doping configurations: graphitic, pyridinic, and pyrrolic. Source: I. C. Gerber.

triplet state is obtained, denoting the carbene character of those mobile point defects [79], since the migration barrier was estimated to be approximately 30 kcal/mol for graphene as well as for small diameter single-walled CNTs [76, 77]. Moreover, when two SVs meet, a divacancy (DV) can be formed to stabilize the conformation, presenting a 5-8-5 structure, as it can be seen in Figure 7.2a. Upon a rotation of one of the C–C bonds in the octagon, the so-called 555-777 DV structure (Figure 7.2a) is obtained [80], which is more stable but with a large activation energy barrier estimated to be 115 kcal/mol [81]. DV defects appear to be immobile due to their high-energy migration barriers, justifying the use of 5-8-5 DV as model in several studies.

Localized defects induce strain extending for at least 2 nm away from the defect [82], thus attracting TM atoms. As a consequence, straining graphene increases its reactivity, anchoring metal atoms much more strongly than on pristine graphene. It has been reported that metal atoms usually bind more strongly to vacancies than to pristine carbon-support, by forming covalent bonds [13, 83]. Those extensive studies have shown that the binding energies for a TM atom in a SV are in the typical range of 45–190 kcal/mol, similar for adsorption in DV (see Figure 7.2c for atomic configurations). Since TM atomic radii are larger than the carbon one, the TM atom tends to be displaced outward from the graphene plane, as depicted in Figure 7.2c, more strongly for SV than for DV, which induces a long-range buckling of the graphene support.

Stone–Wales (SW 55-77) defects can be viewed as interesting point defects for further atomic adsorption [84], since they do not involve C atom removal. It corresponds to a single 90°-tilt of C–C bond, which transforms four hexagons into two pentagons and two heptagons, as shown in Figure 7.2b. Its formation energy is rather considerable, between 70 and 115 kcal/mol depending on the computational method [85–87]. Again, with the presence of more reactive bonds, TM atoms bind more strongly to SW defects than a pristine graphene layer, i.e. by almost 10 kcal/mol for a Co_{SA} [88].

One-dimensional defects have been observed in several experimental studies on graphene, see Ref. [18] for a review. These line defects are tilt boundaries with a tilt axis normal to the plane, which separate two graphene domains with different lattice orientations. These defect lines can be viewed as a line of reconstructed point defects presenting or not some dangling bonds. In the case of FLG, the layers can interconnect either by direct atomic bonding, forming grain boundaries, or by overlapping, holding by van der Waals interactions [89]. Because of its low dimensionality and the rigid bonding structure of graphene, the structural variety along these lines is typically governed by the presence of 7-5 rings and 8-5 ring defects [90]. Grain boundaries in graphene have been studied extensively and are predicted to have specific electronic, magnetic, and thus chemical properties [91–93]. Their activity allows for very localized metal deposition, and the presence of such line defects induces a very localized strain, which provides better reactivity than pristine graphene [94], acting as very localized anchoring sites, with stronger adsorption energies on the line defects.

Substitutional foreign atoms is another type of point defects usually used to anchor TM atoms [95]. Nitrogen, see Ref. [21] for a review of synthesis and characterization of N-doped graphene (N-G), boron (B-G) [96], and in a lesser extent others elements like S or P, have been studied [97]. Embedding a single heteroatom, as B, in the graphene layer typically induced important out-of-plane distortions, as well as introducing a hole, acting thus as a p-dopant [98]. As a consequence, the incorporation of boron into graphene substantially enhances the adsorption strength. Nitrogen doping can be achieved by several methods [99], leading to three common bonding configurations of N atoms in graphene: pyrrolic, pyridinic, and substitutional N (also denoted quaternary or graphitic) [100], the atomic configurations being shown in Figure 7.2d. Thus, the corresponding modulation of the local density state around the Fermi level of N-doped graphitic carbons certainly affect adsorption and electronic TM atom properties, providing a promising approach for optimizing SACs [101, 102]. The structural stability of N-G with graphitic-N, pyridinic-N, and pyrrolic-N, and TM atoms (from Sc to Ni) embedded into N-G has been investigated, revealing that the TM atoms embedded in an N_4-DV, by substituting four C atoms in a DV site with N atoms forming an N_4-centered structure, have binding energies larger than 160 kcal/mol [102]. When increasing the concentration of nitrogen, graphitic carbon nitride (g-C_3N_4), a porous material presenting a honeycomb-like structure with six-membered carbon–nitrogen rings and a band gap in its electronic structure, could form and stabilize single TM atoms due to the presence of defects [103]. DFT calculations of the adsorption, stabilization, and diffusion of metals established the stable location close to the sixfold interstices between heptazine units composing the g-C_3N_4 [104–107].

Graphene oxide (GO) is a complex non-stoichiometric material possessing a layered structure, with physicochemical properties highly dependent on synthesis procedures and post-synthesis treatments, since the O/C ratio depends much on the different chemical oxidation processes [108, 109]. Due to the use of strong oxidants under acidic conditions during GO synthesis, graphene surface and edges are fully functionalized by a wide range of groups such as hydroxyl, epoxy, and carboxyl, making: (i) the O/C ratio the critical parameter that defines the atomic structure, usually extracted from XPS experiments [110, 111], and (ii) GO modeling a challenging task [112]. In the literature, several structural models exist for GO. All of them present epoxy, mainly oxygen in η^2 [113], or hydroxyl (–OH) groups adsorbed on graphene basal plane [112, 114, 115], less with –COOH groups despite their fundamental roles in reactivity [116, 117]. Most of the possible applications of GO as support imply its reduction, to restore partially graphene's structure and physicochemical properties. Depending on the reduction process, various properties of the reduced graphene oxide (rGO) can be obtained, and thus different performances achieved [118]. It has been suggested that by removing –OH and –COOH groups, the sp^2 character of the carbon atoms is restored [119]. Only few theoretical studies have addressed the adsorption of metals (or other elements) on GO or rGO [120–123].

As expected, those theoretical works have pointed to stronger binding energies when GO is used as support compared to pristine graphene. In the case of a Fe_{SA}, the most favorable anchoring mode gives a binding energy of 160 kcal/mol, quite similar to defective graphene values. Combined with a high migration barrier of 75 kcal/mol, GO should ensure strong metal stability and avoid metal clustering. The presence of oxygen atoms allows for the anchoring of a single Pt atom [124]. The introduction of epoxy and hydroxyl groups on a graphene layer modifies its chemical properties, mainly due to the deformation of the basal plane, which depends on the type and the concentration of these functional groups. It can even result in the formation of surface dangling bonds, which can greatly stabilize Pt_{SA} adsorption [125].

7.2.2 Physicochemical Properties of the SAs upon Anchorage

Here, the theoretical structures of those SAC on carbon support are discussed, developing significant examples for various TM species. After the seminal theoretical work of Krasheninnikov et al. [13] on the description of TM atoms embedded in a graphene monolayer, numerous alternative DFT studies were published to describe the geometric and electronic structures, as well as to rationalize magnetic and electronic properties of incorporated TM atoms using simple models. Special mention to Ref. [126] must be made in this respect. Their discussion is based on the hybridization between the d states of the TM and the defect levels associated with the unreconstructed D_{3h} carbon vacancy (see Figures 7.3 and 7.4).

Magnetic states can thus be obtained for V, Cr, Mn, Co, Fe, Cu, Ag, and Au embedded atoms. In Ref. [127], the study of the electronic structures of embedded TMs in the case of N-doped SV as well as DV defects has been extended. It has been shown that Co–N_3, Fe–N_3, and Fe–N_4 sites are predicted to show ferromagnetic states with large magnetic moments and stabilization energies, resulting in a better stability as expressed by favorable binding energies. N_4-DV defects have been considered for various TMs from VII–IX groups [128]. The reported results have stated that the adsorption behavior of TMs is not intrinsic, since it can be severely altered by changes in the local geometry of the active site, the nearest neighbors' chemical nature, and the oxidation states. In this respect, the study of DOS for the localization of the d-band center of the TM atom is fundamental [129].

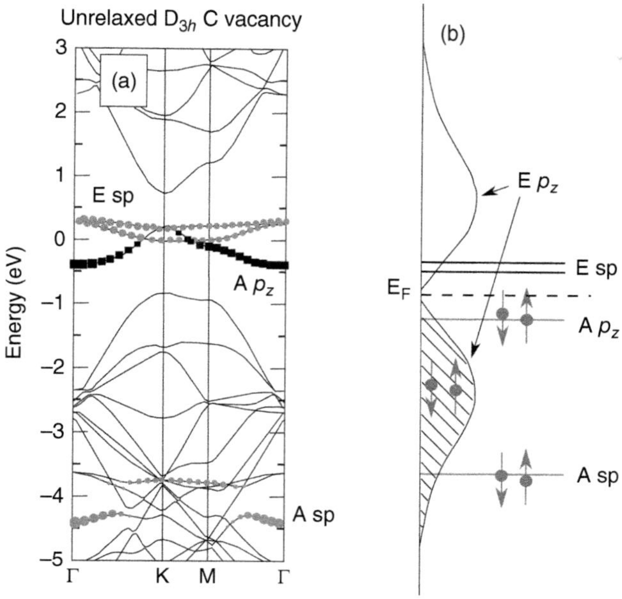

Figure 7.3 (a) Spin-compensated calculation of the band structure of a non-relaxed carbon vacancy (D_{3h} symmetry) in a 4 × 4 supercell of graphene. The electronic structure near E_F (Fermi level) is dominated by a fully symmetric p_z level (A p_z) and two defect levels with E-symmetry and sp character (E sp). (b) Approximate scheme of the electronic structure of the spin-compensated D_{3h} C vacancy, indicating the character and symmetry of the different levels and their occupations. Source: Santos et al. [126]. Reproduced with permission of the IOP Publishing.

7.2.2.1 Platinum

The combination of highly reactive Pt atoms and SV sites in graphene makes the Pt-embedded graphene a superior monodispersed atomic catalyst for CO oxidation, since the strong interfacial interaction is able to tune the energy level of Pt d-states for O_2 activation, by promoting the formation and dissociation of peroxide-like intermediates [130]. Besides, when a pyridinic N-doped site is close to the Pt atom, the CO oxidation efficiency and the resistance to poisoning is enhanced, due to the combined presence of a new spin-down channel below the Fermi level, as well as a spin-up channel above the Fermi level [131]. Since the formation of Pt–C bonds at graphene edges results of strong metal–carbon interaction, Pt atoms transfer electronic density to the support, thus forming $Pt^{\delta+}$ species, which appear to be efficient for the 3-nitrostyrene hydrogenation [132].

By attempting to tune the electronic properties of Pt_{SA} catalysts absorbed on pristine graphene sheets, thus adopting a η^2 coordination with the presence of various oxygenated groups, it was shown that the strengthening of the Pt–C interactions by the carbon support with different oxygen concentrations is more due to charge transfer than frontier-orbital hybridization. In general, the larger the concentration of oxygen-containing groups and the closer they are to the Pt atom, the stronger the binding energy of Pt to the carbon support. In the case of CO adsorption, by downshifting the d-band center of the Pt atom, due to the presence of epoxy

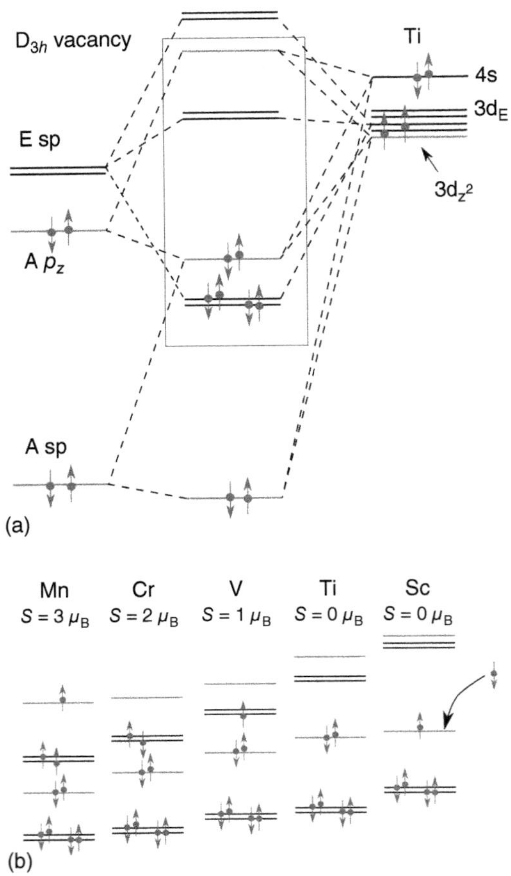

Figure 7.4 (a) Schematics of the hybridization between the 3d levels of Ti and the localized impurity levels of the D_{3h} C vacancy. (b) Schematic representation of the evolution of the electronic structure near E_F for several substitutional TM in graphene. The spin moment (S) is also indicated. Source: Santos et al. [126]. Reproduced with permission of the IOP Publishing.

groups, the coupling between the Pt d-band and the CO $2\pi^*$ state is reduced, thus decreasing the adsorption energy and at the same time the catalytic activity [133].

When comparing the adsorption states and diffusion behavior of a Pt_{SA} on pristine and B-, N-, O-, Si-, P-, and S-doped graphene by means of DFT calculations with a thorough study of the corresponding Pt electronic structures, Hasegawa et al. have shown that (i) localized orbitals, which can trap Pt atoms, are systematically created in the vicinity of dopants, and (ii) explained the nature of the bonding and its strength [134]. When analyzing the detailed projected density of states (pDOS) depending on the dopants and the position of the corresponding states available for further adsorption, it appears that O- and P-doping are particularly efficient to trap Pt atoms, preventing desorption as well as limiting diffusion, with a large (>50 kcal/mol) diffusion barrier compared to pristine graphene (3 kcal/mol). The O-doped graphene presents dangling bonds, which allows Pt atoms to chemically bond with C atoms.

7.2.2.2 Palladium

When embedded in a graphene sheet, through its adsorption in a SV, the strong interaction between the Pd atom and neighboring C atoms can be confirmed by the presence of overlapping peaks in the DOS. There are several Pd-4d and C-2p peaks emerging near the Fermi level, which implies the high activity of the Pd–graphene system. These states are available and ready to activate O_2 molecules for a further CO oxidation [135].

The N–Pd interaction in N-doped CNT, with N-graphitic or pyridinic sites, was investigated experimentally by XPS and NEXAFS and supported by a theoretical study [136]. The interaction of those nitrogen species with a Pd_{SA} has been characterized and corresponds to (a σ-type donation from the filled π-orbital of the N atom to the empty d-orbital of the Pd atom and a π back-donation from the filled Pd d-orbital to the π^* anti-bonding orbital of the N atom. On the one hand, the interaction of Pd atom with pyridinic site is mainly covalent with a partial ionic character, consistent with the chemical shift observed in the Pd 3d core level of divalent Pd. On the other hand, Pd on a graphitic site is covalently bound without any charge redistribution.

Decorated defective nanodiamond–graphene hybrids by Pd_{SA} have shown remarkable performance for the selective hydrogenation of acetylene to ethylene. The Pd–C anchoring mode does not allow for unselective subsurface hydrogen species presence and at the same time ensures the facile desorption of ethylene against the over-hydrogenation of undesired ethane, which is the key for the outstanding selectivity of the catalyst [137]. Upon the adsorption of Pd_{SA} on the sidewall of a single-walled CNT, the bonding character of PdH_2 has been described in detail [138]. It turns out that the role of s, p, and d orbitals on the bonding mechanism for all adsorbates and substrates have been addressed, and intermolecular donor–acceptor C–Pd and Pd–H delocalization after adsorption have been evidenced.

7.2.2.3 Other TMs

The graphitic and pyridinic N-sites, created after pyrolysis of metal organic frameworks followed by an annealing treatment under N_2, have appeared to strongly anchor Co_{SA} with very precise N coordination; mainly Co–N_2 and Co–N_4 sites were evidenced by EXAFS experiments [139]. Those Co–N_2 SACs have been identified to be active sites for ORR thanks to DFT calculations. Indeed, Co–N_2 has stronger binding energies toward various ORR intermediates than Co–N_4 due to the availability of more d-states in the Fermi level's vicinity. Stronger O_2 binding energy allows Co–N_2 to more efficiently capture the O_2 molecules from the solution than Co–N_4, satisfying the prerequisite for the ORR in proton exchange membrane fuel cell (PEMFC). An extensive theoretical study to investigate the potential of TM atoms embedded in buckled g-C_3N_4 monolayer (ML) has demonstrated that V, Cr, Mn, and Cu are much less susceptible to form NPs than the other TM atoms under investigation, except Ti and Sc, for which clustering is thermodynamically almost impossible. In the context of CO oxidation, Cr and Mn can be considered as promising SACs, since the metallic center adsorbs relatively weakly CO and O_2 and the activation of the latter is reasonable energetically, of about 14–17 kcal/mol [140].

Atomic Co incorporated in N-G has also been identified as active sites for the reduction of water to hydrogen, thanks to EXAFS [141]. Recently, Co_{SA}/N-G SACs have also been considered for dye-sensitized solar cells applications as counter electrode [142]. Electrochemical measurements have revealed that Co–N_4 sites embedded in N-G nanosheets are highly active, which also make the composite stable as counter electrode for the interconversion of the redox couple I^-/I_3^-. Co–N_4 sites have been shown to adsorb iodine rather weakly, leading to a good balance between adsorption and desorption processes. In the electrocatalytic reduction of CO_2 to CO (CO_2RR), both experiments and DFT calculations have shown that Ni on N-doped carbon are highly efficient catalysts due to the weak binding of CO on such catalysts that limits CO poisoning [143]. Sahoo et al. have investigated theoretically graphene-supported single TMs such as Cr, Mn, Fe, Co, and Cu atoms for the activation of methane and identified catalytically active centers through C–H bond cleavage [144]. Low activation barriers for both TM-adsorbed and embedded systems have been obtained, compared to that of free TM–methane systems. In this study, it is stated that the interaction of the three d_{xz}, d_{yz}, and d_{z^2} orbitals with methane, in terms of their relative positions and occupancy, play a key role in governing the catalytic activity of supported TM systems. Embedding TM has globally a negative impact, and lower activation barriers are yielded for Co and Fe adsorbed on pure graphene in η^6 coordination mode.

7.3 Hydrogen Spillover

In the last years, H-spillover has emerged as one of the most promising techniques for the achievement of high-density hydrogen storage at close to ambient conditions within solid-state materials. Spillover is defined as the transport of active species, adsorbed or formed on a specific surface (or site), migrated onto another surface that does not adsorb or form this species under the same conditions [145]. Thus, the adsorbed species gain access to a different surface phase (accepting surface) that is in contact with the original adsorbing and activating surface, although gas-phase spillover has also been reported (no contact). In heterogeneous catalysis, spillover is critical in adsorption and as a mechanistic step, being only important for that of catalytically active species. As spillover of hydrogen (H-spillover) is a fast phenomenon, which can compete with the rate of a catalytic reaction, it is involved in many surface reactions and transport phenomena. Additionally, it may include other atoms than hydrogen, such as oxygen or molecules like CO.

To explore the hydrogen storage by spillover, many experimental studies are performed on a wide variety of materials, notably carbon materials (fullerene, CNTs, graphene, graphite, and nano-horn, among others) and metal oxides (TM and non-TM oxides). These studies not only inspected the hydrogen storage capacity of different adsorbents but also inspected the effects of surface modification, support, doping techniques, and catalyst size on H-spillover [146]. It has been probed that H-spillover from a metal surface to the surface of a defect-free carbon support is energetically unfavorable. For the specific case of SACs, since not all embedded TMs has the same ability to dissociate H_2 once activated, metal NPs have been proposed as a plausible source of atomic hydrogen due to their greater ease, being able to provide it to SAs by spillover. Furthermore, it is also possible that the spillover of

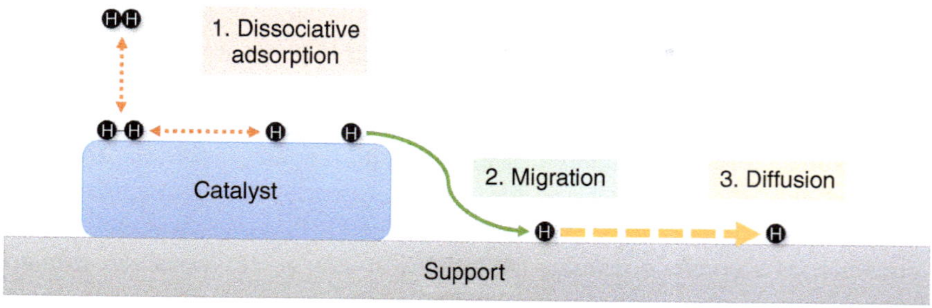

Figure 7.5 Mechanism of hydrogen spillover. Source: J. Navarro-Ruiz.

hydrogen atoms is from metal particles to a carbon support with defects or to a contaminated support. If the defects bind the atoms and if the distance between the defects is affordable, then the atoms might diffuse from the metal particles to the defect sites. As a result, these statements allowed a number of theoretical studies to be carried out to explore the mechanisms at the microscopic level [147].

H-spillover is a typical surface motion phenomenon, and hence a certain energetic barrier during the migration process needs to be surpassed. As it can be seen in Figure 7.5, H-spillover process mainly involves three primary steps: (i) activation and dissociation of gaseous hydrogen molecules on a TM catalyst in close contact with the support; (ii) migration of hydrogen atoms from the catalyst to the support; and (iii) diffusion of hydrogen atoms on the support surface [148]. In certain conditions, exploring a fourth step involving the hydrogen release process is mandatory. Additionally, different hydrogen species have to be considered for its spillover, such as hydrogen radicals ($H^{\bullet}$), protons (H^{+}), hydrides (H^{-}), and activated hydrogen (H^{*}), the latter being strongly influenced by the metallic presence halfway between a radical and an ionic species. The migration of H atoms from catalysts to supports, as well as H atom diffusion within the supports, may be mapped out by calculating minimum energy pathways.

In this section, the theoretical researches on H-spillover mechanism are summarized in the following three aspects: (i) the hydrogen adsorption and the subsequent dissociation on the metal catalyst; (ii) the hydrogen migration from the metal catalyst to the supports; and (iii) the hydrogen diffusion on the support surface. The support materials that will be discussed below for H-spillover are graphitic carbon materials, doped graphitic materials, and metal oxides.

7.3.1 Hydrogen Adsorption and Dissociation on the Metal Catalyst

The first elementary step in the spillover of hydrogen is to dissociate H_2 molecules on TM NPs. Of all the metals used in heterogeneous catalytic reactions, both Pd and Pt are among the most valuable catalysts in experiments and thus have been extensively investigated. From an experimental point of view, usually, the metal catalysts are well dispersed on support materials as nano-clusters with sizes of around 1 nm or even larger. In these conditions, the catalysts are able to facilitate H-spillover and significantly improve the hydrogen uptake. On the theory side, modeling 1 nm cluster deposited on a support start to be a challenging task, thus smaller aggregates are commonly used instead. Since in theory it is complex to model a 1 nm cluster, those that have four and six atoms are commonly used in computational studies.

The dissociative chemisorption of dihydrogen on Pt and Pd NPs is generally easy and is accompanied by the formation of strong metal–H bonds, in this case hydrogen acting as a hydride. Chen et al. studied the H_2 dissociative chemisorption on a Pt_6 cluster [149], their study determining that the H_2 dissociative chemisorption is governed by the charge transfer from the Pt cluster to H_2 molecules through the superposition of 5d orbitals of Pt_6 and σ^* orbital of H_2. Moreover, the cluster reaches its complete saturation with H atoms at the Pt/H ratio of 1 : 4 [149]. The H_2 dissociative chemisorption and H desorption on Pt_n clusters with different atomic numbers ($n = 2$–5, 7–9) have been studied by Cheng group and showed that the dissociated H atoms are strongly adsorbed on Pt clusters [150]. They reported a chemisorption energy of 18–42 kcal/mol and a desorption energy of an H atom of 58–74 kcal/mol, being the threshold of H sequential desorption energy in a range of 56–60 kcal/mol. Cheng group also studied the hydrogen dissociative chemisorption on Pd cluster (Pd_n, $n = 2$–9), and they obtained that the capacity of Pd clusters to adsorb H atoms is substantially less than that of Pt cluster, indicating that the catalytic efficiency of Pt NPs is higher than Pd ones [151]. Both studies show that the metal particle size of catalyst does not influence the H_2 dissociative chemisorption and the H desorption energy at full coverage value [149, 150], which is relevant to the realistic reaction conditions.

If the support is explicitly considered in the calculation cells, sp^2 carbon-based materials are commonly the most used in the theoretical study of H-spillover mechanism. On a Pt-doped graphite (0001) surface, where the Pt is on top forming two equivalent bridging bonds with carbon atoms, it was noted that a Pt_{SA} is capable of adsorbing up to 2 H_2 molecules and its dissociation does not require overcoming an energy barrier [152]. The binding energies were calculated to be 41 and 8 kcal/mol, respectively, one H_2 molecule being completely dissociated, while the other one exhibits Kubas binding (both H atoms are bound to the surface while their intramolecular bond is weakened). It was also found that it takes 88 kcal/mol to remove a single H atom from a carbon-supported PtH_4 moiety and 85 kcal/mol to remove a H atom from a PtH_2 moiety. On a Pt cluster, H_2 also dissociates effectively and spontaneously [152]. It was observed that up to 10 H atoms remained on the surface of the adsorbed Pt_4 cluster, and the average binding energy of hydrogen on the fully saturated Pt_4 cluster (Pt_4H_{10}) with respect to molecular H_2 was calculated to be 20 kcal/mol.

Others metals have been considered in the context of hydrogen storage. The feasibility of H-spillover on graphene in the presence of non-precious metals catalysts has been investigated. Molecular hydrogen dissociates into hydrogen atoms by interacting with the supported Ni_4 cluster [153], and it is found that it would be saturated by six or eight H_2 molecules depending on the configuration, yielding an average adsorption energy of 10 kcal/mol. The adsorbed H_2 on tetrahedral Ni_4 shows two types of adsorption states: an activated state with stretched H–H bonds (top Ni atoms) and a dissociated state with separated $H^{\delta-}$ atoms (bottom Ni atoms).

In order to fully understand the H-spillover mechanism on carbon materials, a series of four-atoms cluster TM catalyst have been used on penta-graphene [154], a theoretically proposed new 2D allotrope of carbon-based consisting exclusively of pentagons in a planar sheet geometry. The number of adsorbed H_2 molecules depends on the metal, and the calculated adsorption enthalpies of dihydrogen on Pt_4, Pd_4, Ni_4, and Ti_4 are 17 (6 H_2), 16 (7 H_2), 14 (6 H_2), and 18 (7 H_2) kcal/mol, respectively, which is larger than those on graphene except for Pt (20, 13, 10, and 12 kcal/mol for Pt_4, Pd_4, Ni_4, and Ti_4, respectively) [152, 153, 155]. On curved carbon surfaces such as carbon nano-horns, the binding energy is even higher: between 40 and 110 kcal/mol according to the surface coverage [156].

For the case of embedding a single heteroatom, such as boron, in the graphene layer, one generally observes significant out-of-plane distortions, as well as introducing a hole. As a consequence, the incorporation of boron atoms into graphene substantially enhances the adsorption strength for small metallic clusters, the gain for Pt_4 being larger than 23 kcal/mol [157]. This firmly bound catalytic metal, on B-doped graphene, can therefore effectively dissociate H_2 molecules into H atoms. The chemisorption energy per H atom for the saturated Pt_4 cluster on graphene and B-doped graphene sheet is 65 and 62 kcal/mol, respectively, indicating a strong H_2 dissociation ability for the metal cluster [157]. By combining the effect of B-doping and vacancy-defect on Pd-decorated graphene, the adsorption of hydrogen molecule is enhanced [158]. The adsorption energies values of H_2 on combined boron-and-vacancy graphene were larger than that of vacancy graphene, which ranged 1–30 kcal/mol compared to 1–10 kcal/mol. Interestingly, the B-doping decreased the interaction between the Pd_{SA} and the defected graphene but increased the stability of the adsorption of dissociated H_2.

Doping of the support with nitrogen or oxygen also affects significantly the Pd–H bonding energy. Rangel and Sansores studied the dihydrogen adsorption on small Pd clusters (Pd_n, $n = 1$–4) supported on pyridine-like nitrogen-doped graphene [159]. Up to four times higher is the binding energy between Pd clusters and N-doped graphene compared to pristine one, thus leading to different types of H_2 adsorption: (i) molecular hydrogen physisorbed, with almost negligible adsorption energies and each atom maintains its monoelectronic nature; (ii) activated state of H_2, with E_{ads} of about 11–22 kcal/mol and halfway between monoelectronic and hydride; and (iii) dissociated state, with E_{ads} around 12–36 kcal/mol, barriers below 13 kcal/mol, and where each species behaves as a hydride. In this sense, the influence that cluster nuclearity can exert on the molecular hydrogen interaction is limited, at least in ultra-small cluster ranges, since both activation and dissociation take place and can occur spontaneously at room temperature. On the other hand, Li et al. investigated the feasibility of H-spillover on single-layer GO in the presence of Ni_4 cluster and the role of oxygen-containing groups [153]. They found that supported Ni_4 attracts the hydroxyl nearby to form a Ni–OH bond, and molecular hydrogen can be readily activated, in which one H atom would gradually move toward the hydroxyl and would possibly combine with it to form a water molecule. Something similar happens with the epoxide group, both C–O bonds are broken, and the oxygen atom moves to the Ni–Ni bridge site, where it would react with the H atom of the activated hydrogen molecule to form Ni–OH bond and further bond with H to form H_2O molecules. Hence, doping effects could be an effective approach in the modification of graphitic surface for H-spillover studies.

Finally, reducible and non-reducible oxide supports have also been studied for H-spillover first step. The mechanism of H_2 adsorption on anatase titanium oxide (101) and γ-aluminum oxide (100) supports by using first-principles atomistic simulations showed that adsorption of a single hydrogen molecule is more favorable on the metal cluster than on the oxide (that is, about 35 kcal/mol on Pt_{13} vs. about 14 kcal/mol on TiO_2) [160]. A spontaneous dissociation of dihydrogen on the Pt_{13} cluster was also observed, independent of the support. On TiC(001), the reactivity of different supported clusters toward molecular hydrogen dissociation has been studied [161]. As a consequence of the substantial metal–carbide interactions, the adsorbed H_2 molecule becomes more difficult to dissociate on Pd_4/TiC and Pt_4/TiC systems, while for the Cu_4/TiC, Ag_4/TiC, and Au_4/TiC systems, the combination of some size effect and the polarization induced by the underlying carbide facilitates the dissociation of the hydrogen molecule. It is worth highlighting the SAC case of a Pd_{SA} on Fe_3O_4(001). It has been reported that, in this case, the H_2 dissociation is energetically favorable around 6–10 kcal/mol and occurs heterolytically between Pd (as a hydride) and surface oxygen (as a proton) [162].

7.3.2 Hydrogen Migration from the Metal Catalyst to the Support

The migration of H atom from the saturated metal cluster to the support is a key step in H-spillover, regardless of physisorption or chemisorption, which can be studied by performing the minimum energy pathway calculations. The previous studies have shown that the migration of H atoms from the fully saturated Pt_4 cluster to the defect-free graphitic support is rather endothermic and needs to overcome a large barrier (60–62 kcal/mol) [152, 157], which is close to the desorption energy of H atoms on Pt cluster, and it is unlikely to occur even at high hydrogen coverage (24 H atoms per Pt_4) [163]. Moreover, the migration of a unique H atom from Pt_{SA} to graphene layer evidenced an energy barrier of at least 60 kcal/mol [152], which is 50% lower than obtained for the non-catalyzed process. On Ni-loaded graphene, Li et al. evaluated the feasibility of H-spillover in the presence of Ni_4 cluster, and they indicate that the hydrogen migration is difficult to take place on the Ni/graphene interface due to the stronger Ni–H bond strength [153]. From an electrostatic point of view, the existing difficulties would also be justified, since the hydrogen atom must change the hydride character when it is bonded to metal toward one with a rather protonic tendency when it is bonded to carbon. These results indicate that the migration process is thermodynamically and kinetically very difficult and would not happen at normal conditions to any significant extent, leading to a very unfavorable starting point for H-spillover.

Singh et al. pointed out that the thermodynamic condition for H-spillover is that the H chemical potential (μ_H) on a metal catalyst must exceed the CH state energy level on a carbon substrate [164]. Their results show that the μ_H on a saturated Pd catalyst is lower than the energy of a single H atom on a pristine graphene, while it exceeds the CH state energy level in fully hydrogenated graphene. Thus, a H atom cannot migrate from the saturated Pd catalyst to the pristine graphene but can migrate to the hydrogenated graphene. Then, the energy barrier is further significantly reduced (16 kcal/mol) if a Pd_4 cluster is in the vicinity of a fully hydrogenated graphene phase [164]. Favorable energy barriers were also predicted for the spillover processes on fullerene C_{60} for Pt_{1-13} clusters [40]. Hydrogen transfer events between Pt and carbon regions on the $C_{60}Pt_n$ structures are highly dependent on the local atomic environment. When going from the Pt_{SA} to the small cluster regime, the spillover energy barriers vary between 16–37 kcal/mol, a result that is important to consider in order to more clearly understand the hydrogen storage in carbon nanostructures via chemical adsorption.

For CNTs, a lower barrier was also measured for the migration of two hydrogen atoms from a physisorbed Pt_6H_{24} cluster (11 kcal/mol per H atom) [165]. However, Juarez-Mosqueda et al. have found that the barrier for the simultaneous migration of two H atoms from the dodeca-hydrogenated Pt_4 cluster to the (10,10) CNT and large diameter CNT surface is 46 and 53 kcal/mol, respectively [166]. They have also observed that when bending the saturated Pt_4 cluster toward the support, the carbon structure is puckered down. This supports the fact that the hydrogen migration from the catalyst to the carbon is not a straightforward process, and even a rotational motion of the catalyst cluster does not lower the energy barrier. Therefore, curvature appears to have an important effect hampering the H migration, in addition to also proposing the Pt_4 migration as a key factor for the hydrogenation of the carbon supports [165]. Finally, on a curved carbon surface, DFT calculations have shown that the H atom migration from the hydrogenated Pt_4 cluster to the support surface is thermodynamically favorable at room temperature with low energy barrier and is highly exothermic [156].

Other TMs (e.g. Ti, Co, and Rh) can be experimentally used as catalysts for H-spillover. In previous studies, Guo et al. studied the H-spillover on penta-graphene by using the Pt_4, Pd_4, Ni_4, and Ti_4 as catalysts, the H migration barriers being 29, 25, 24, and 31 kcal/mol, respectively [154]. It can be seen that the catalytic effects of Ni and Ti in H-spillover is comparable with Pt and Pd, at lower cost. Subsequently, the same authors studied the catalytic properties of a tetrahedral bimetallic Ni/Pd, Ni/Pt, and Pt/Pd clusters for H-spillover also on penta-graphene [167]. The H migration barriers from the Ni_2/Pd_2, Ni_2/Pt_2, and Pt_2/Pd_2 stoichiometric bimetallic NP to penta-graphene are in the range of 15–22 kcal/mol, which are lower than that from the pure Pt, Pd, and Ni cluster. A plausible explanation for this significant decrease in the migration barrier may be related to the reduction of hydrogen adsorption energy on a mixed metal cluster.

As discussed in the previous section (Section 7.2), carbon supports may contain some defects or impurities. Their roles have been also investigated on H-spillover. For instance, Wu et al. reported that, by introducing boron atoms into graphene as carbon substitutes, the hydrogen migration barrier from the fully saturated Pt_4 cluster can be reduced from 63 to 44 kcal/mol [157]. The presence of nitrogen surface groups near the TM clusters also allows for a significant decrease in the migration energy of H atoms. Energy barriers of 18 (in the case of pyridinic group) and

12 kcal/mol (for pyrrolic group) were estimated in the case of the small Pd_4 tetrahedral cluster at full hydrogen saturation [168]. Psofogiannakis et al. reported that, on O-modified graphite surface, the migration of a H atom from Pt_4 or Pd_4 cluster to the nearest O atom is much easier, only crossing an energy barrier of 9 kcal/mol per H atom in the case of Pt [169, 170]. The reduced barrier may be due to the high binding strengths of H on C=O carbonyl groups on the periphery (85 kcal/mol) or on C–O–C epoxide groups (71 kcal/mol). Lastly, such a decrease was also reported for Ni_4 clusters in the vicinity of a hydroxyl. A H atom binds to the hydroxyl and desorbs with a relatively low energy barrier of 11 kcal/mol. On an epoxide group, a H atom combines with this oxygenated motif to form hydroxyl with a low-energy barrier of 7 kcal/mol [153], acting in all these cases as a species with a pronounced proton character. This low activation barrier suggests that the reaction is kinetically favored.

Concerning the transport of H atoms, it was previously proposed that hydrogen can migrate through the gas phase, revealed by DFT calculation, since a hydrogen atom can be in a physisorbed state (H^*) on a carbon-based support [152]. On supports based on metal oxides, Ahmed et al. studied the H-spillover process on a $Pt/\gamma\text{-}Al_2O_3(100)$ catalyst surface [171]. They found that the kinetic energy caused by the dissociation of the hydrogen bond on the hemispherical cuboctahedral Pt_{19} NP promotes the migration of the spillover H atom through the gas phase. The mechanism of hydrogen migration on anatase titanium oxide and the γ-aluminum oxide supports showed that, under experimental conditions, the barrier for spillover of dissociated hydrogen onto anatase is 10 kcal/mol, a process that is not influenced by the presence of water on the surface [160]. These results are in line with the ones of Sihag et al., where H-spillover has a tendency to take place on anatase (101) at medium hydrogen coverage (10 H atoms per Pt_4) from the perspectives of both thermodynamics and kinetics [163]. However, there is a strong effect of water on aluminum oxide, and hydrogen diffusion on its surface is much more restricted than on titanium oxide. A similar migration barrier was found for an H atom from a saturated Pt_6/H_{20} cluster to a terminal oxygen of $MoO_3(010)$ surface, in which an O–H bond (clearly hydrogen as a H^+) is formed with a low-energy barrier of 9 kcal/mol [172], indicating that hydrogen desorption energy from the Pt_6 cluster can be reduced significantly upon the cluster interaction with the support. At present, the influence of parameters such as temperature, H_2 pressure, and size of NP on the process of gas-phase H atom generation is not yet understood. Therefore, further researches need to be carried out.

7.3.3 Hydrogen Diffusion on the Support

The chemisorption of hydrogen on the surface of pristine graphene is an endothermic process from a thermodynamic view point and needs to reach an energy barrier from a kinetic point of view. We recall that the bond between a H atom and a graphene support (around 18 kcal/mol) is less strong that in a H_2 molecule (53 kcal/mol). There is also a local hindrance against the hybridization of the C–H bond from sp^2 to sp^3 [152]. Focusing on the interaction of atomic hydrogen with single-walled CNTs, the adsorption binding energy of 1 H atom ranges between 16 and 95 kcal/mol depending on H occupancy, tube diameter, and helicity (or chirality), as well as endohedral vs. exohedral binding [173]. An adsorption energy as low as 9 kcal/mol is obtained for 2 H atoms on a (5,0) semiconductor single-walled CNT. Thus, for the subsequent adsorption of hydrogen atoms, the first hydrogen chemisorption is very important. From experiments, the H diffusion on the support is essential for hydrogen storage by spillover. Chen et al. studied the H diffusion not only on graphene, but also on CNTs and polyaromatic carbon materials, and reported that the diffusion barriers are about 18–22, 25–37, 33–35 kcal/mol, respectively, which are relatively close to the adsorption energy [165]. However, they highlight the idea that the hydrogen atoms can be in a physical adsorption state ($H^{\bullet}$ or H^{*} depending on the interaction) on the carbon support, and this physisorbed H atoms can diffuse freely on the surfaces [165, 174]. So, theoretical results suggest that it could be difficult for chemisorbed H atoms (specifically H^{δ^-}) to move freely at temperatures close to room temperature, since diffusion requires dissociation of the C–H bond, with a comparatively high activation energy.

Taking into account the different existing carbon compounds, the H diffusion on the carbon sites having more hybridization of sp^3 nature is more difficult compared with that on sp^2 carbon atoms [175]. On the other hand, the curvature of the carbon materials is found to have a pronounced influence on the diffusion of H atoms with a higher barrier associated with carbon atoms of higher curvature, as in single-walled CNTs [165]. To overcome this drawback, Juarez-Mosqueda et al. found that a small Pt_4 cluster may move along the outer surface of the CNT with a very low energy barrier (4 kcal/mol), offering the possibility of complete hydrogenation of the nanotubes [166]. Thus, in the H-spillover process, the observed saturation of the CNT occurs through mobile Pt nano-clusters, which move on the support more easily than the chemisorbed hydrogens of the support and deposit or reattach hydrogens in the process. Other carbon nanostructures such as nano-horns have been proven, despite a high curvature, to be an interesting active area for accommodating migrated H atoms [156, 175]. The presence of hydrogen atoms on the TM catalyst can efficiently induce a hydrogen diffusion process through chemisorbed H atoms. The metal···H interaction significantly facilitates the hydrogen migration from C–H bonds nearby the active catalyst to its adjacent carbon atom, the reactions being spontaneous at room temperature and highly exothermic (energy barriers lower than 12 kcal/mol under ambient conditions).

Impurities or defects on the surface of the carbon support may also increase the binding strength of the hydrogen atoms without compromising the diffusion process. By introducing B impurities or hydroxyl (–OH) groups on the graphene surface, the H binding strength can be increased to 45 or 56 kcal/mol, respectively [157, 170]. It has been demonstrated, both theoretically and experimentally, the crucial role played by oxygenated surface groups to improve H-spillover by decreasing the H diffusion barrier [145, 169]. For instance, an energy barrier of only 8 kcal/mol has been reported for migration of a H atom from a hydroxyl group to an adjacent O (epoxide) atom on pure graphite [169]. Although the H diffusion barrier certainly depends on surface functionalization, the proximity of oxygen groups seems to facilitate the process. For the migration of a H atom between two C atoms that are both in the *ortho* position to –OH groups, the H diffusion barrier is 23 kcal/mol, very similar to the one on pristine graphene [170]. In addition, the stability of the oxygen surface groups is an important factor to take into account, since in several cases the desorption of water has been revealed due to the poor stability of some –OH groups against the approach of an H atom [153]. In this case, O-doped carbons can be readily hydrogenated in the presence of the metal catalyst, where both the hydroxyl and epoxide groups at the surface would inevitably desorb in the form of H_2O.

Lueking et al. also explored the influence of incorporating heteroatoms, such as boron, nitrogen, and oxygen atoms, for hydrogen diffusion on graphene, and reported that the –OH adducts and B substitutional dopants can facilitate significant hydrogen surface diffusion with an active barrier in the range of 12–24 kcal/mol [176]. Interestingly, the presence of polar hydride molecules that act as *shuttle gases* (H_2O, HF, or NH_3) has been proposed as an alternative to facilitate the migration of atomic H atoms adsorbed on graphene by co-adsorption [177]. So, a favorable catalytic effect (lowering the H migration barrier) and a favorable thermodynamic effect (activating the direct transition to the nearest-neighbor site) are provided.

Finally, the H-spillover mechanism on metal oxide surfaces should happen similarly to the process of H atom diffusion on oxidized carbon surfaces. The energy required to transfer a hydrogen atom among the different sites of titanium oxide lies between 14 and 16 kcal/mol, under experimental conditions [160]. This diffusion is not influenced by the presence of water on the surface or by the desorption of hydrogen, which requires about 56 kcal/mol. On aluminum oxide, the transfer between neighboring adsorption sites has an energy barrier ranging from 27 to 38 kcal/mol, depending on the hydration of the surface [160]. Furthermore, the hydrogen desorption energy of dehydrated aluminum oxide is 36 kcal/mol and can be as low as 9 kcal/mol at high water coverage. The hydrogen desorption rate is, therefore, faster or comparable to the surface diffusion, causing a gradient in hydrogen coverage of the surface. H-spillover on the $MoO_3(010)$ surface in the presence of a Pt catalyst revealed that the H atom is able to move nearly freely on the surface and diffuses into the bulk lattice at ambient temperatures, leading to the formation of hydrogen molybdenum bronze [172]. The high hydrogen diffusion is largely attributed to the massive H-bonding network in the MoO_3 lattice. On cerium oxide, the most important aspect is that there is no formation of OH^- ions, as the O–H distance is much larger than the bond length of the hydroxyl ion, and no water is formed on the catalyst because of interaction with hydrogen [178]. Thus, Pt/CeO_2 helps in production of active hydrogen and the H-spillover, increasing the hydrogen storage in the matrix with the H/Pt ratio much greater than 1. As a consequence, it appears that the O-doped carbon interface may not be able to facilitate the H-spillover as the oxide atom does in the metal–oxide interface.

7.4 Mechanistic Studies on C-SACs

Carbon-based SACs are widely investigated catalysts due to their appealing features that make them highly efficient for numerous important catalytic applications, such as oxidative cross-coupling, oxidation, hydrogenation, reduction, C–C coupling, photocatalysis, and electrocatalysis. Carbon supports are generally inexpensive, stable, exhibiting tunable morphologies, ordered porosity, and easy stabilization of various noble and non-noble metals. As will be seen in some selected examples, the successful adaptation of their morphology and composition provides an interesting basis for modulating electronic interactions and hence their catalytic activity. As such, DFT can shed light on these electronic properties and multistep reaction studies can contribute to the rational design of innovative SACs. Although several theoretical studies were recently cited in review articles [12, 179–182], mechanistic studies involving M_{SA}/graphene or even $M_{SA}N_4$/graphene catalysts (where M stands for metal) will be more particularly focused here. Reactions will be separated into two main domains, namely thermocatalysis and electrocatalysis. To the best of our knowledge, no remarkable photocatalytic theoretical studies were published, probably given the inherent difficulty in exploring reaction pathways that go through electronic excited states. The computational methods commonly used are generalized gradient approximation (GGA)-DFT methods, albeit they suffer from approximations that may lead to inaccuracies, especially in the case of strong spin-orbit coupling effects and strong on-site electron localization. A likely important approximation in the context of SACs is that GGA functionals incorporate no dispersion effects. Although it can be accounted for by appropriate corrections such as the popular D3 correction [183], such a contribution is neglected in most theoretical studies dealing with SACs stabilized by graphene or doped graphene. Furthermore, solvent effects are surprisingly not always accounted for in theoretical electrocatalytic studies, generally by means of the polarizable continuum model used to simulate the aqueous environment. Most of the studies reported hereafter incorporate free energy pathways calculated by DFT, sometimes with reaction intermediates only, and electronic structure analysis. Mechanistic studies are often relied on underlying microkinetic models, the Brønsted–Evans–Polanyi (BEP) rule, and volcano plots [184]. Electronic structure is generally analyzed using DOS diagrams, projected in relevant atomic orbitals (AOs). In addition to a fine understanding of the properties of SACs, the added value of some theoretical studies is the identification of relevant and simple descriptors, usually the adsorption free energy of a reaction intermediate involved in a rate-determining step. But it could also be an electronic factor, like the popular d-band center for TM heterogeneous catalysts [129].

7.4.1 Thermocatalysis

7.4.1.1 C–H Activation

These reactions are of paramount importance in several contexts, such as the direct conversion of methane to methanol [185] or the development of efficient methods for isotopic labeling with hydrogen isotopes [186]. The C–H bond activation is the critical and rate-determining step in methane conversion and as such some studies only focus on it. Sahoo et al. showed that the activation barrier is significantly reduced when the TM atom is adsorbed onto graphene compared to the TM atom occupying a vacancy (also known as M_{SA}/graphene-embedded) [144]. It was also shown that the interaction of the TM d_{xz}, d_{yz}, and d_{z^2} AO with methane, their relative position, and occupancy play a key role in the catalytic activity. Methane-to-methanol conversion is assumed to first go through the formation of metal-oxo species. C–H activation can then occur through a radical or non-radical pathway.

A comparative DFT study on M and MN_4 SA-embedded graphene (M = Ni and Si) performed within this assumption showed a relationship between the electron spin state and the reaction mechanism, with a higher activity and stability of a radical spin-triplet Eley–Rideal mechanism. It is interesting to note a recent joint experimental/theoretical study aiming at evaluating the activity toward the dehydrogenation of alkanes of a fluorenyl anion-Ni complex designed as a model for a pristine graphene-supported SACs [187]. This complex resembles a single-metal atom adsorbed on a Stone–Wales graphene defect. Its ability to dehydrogenate alkanes was investigated by DFT in the case of cyclohexane. This exothermic, kinetically accessible, two-step process first involves insertion of Ni at a C–H bond, followed by a β-hydride elimination. A computational study of the direct dehydrogenation of propane by $Pt_{SA}N_xB_y$/graphene catalysts (with $x+y=0$–3) revealed an intriguing tunable catalytic performance of Pt on nitrogen and boron doping [188]. Such a reaction is actually an important industrial reaction to produce propylene, and its main possible drawback is catalyst deactivation caused by coking and side reactions. Given the stability of SACs, exploration of the underlying dehydrogenation mechanism by supported SACs could play a role in the rational design of such catalysts. Interestingly, activation barriers and overall catalytic performance are discussed in terms of modulation of the electronic structure, assessed by the Pt_{SA} d-band center and in light of the BEP rule (see Figure 7.6). The main conclusions of

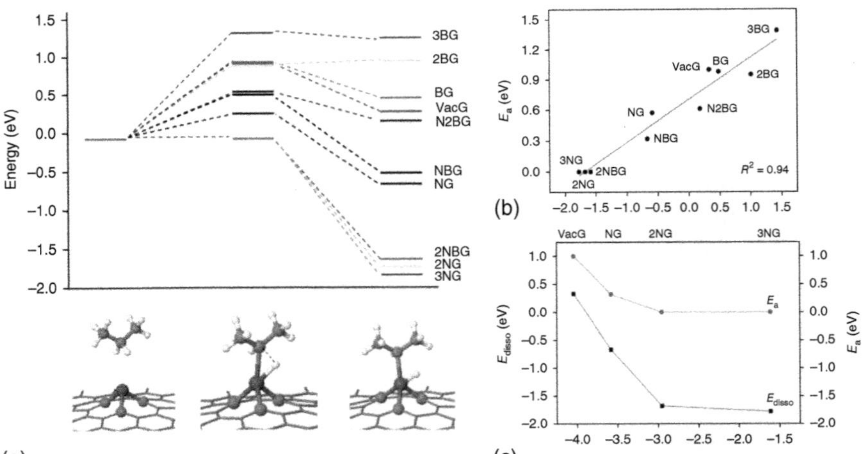

Figure 7.6 (a) Reaction pathway of the first C–H bond activation in propane molecule on the supported $Pt_{SA}N_xB_y$ catalyst adsorbed on graphene (dubbed xNyBG in the figure); (b) Illustration of the BEP relationship: linear relation between the calculated dissociation energy and barrier for the activation of the first C–H bond; (c) Relation between the d-band center of the Pt atom adsorbed on selected supports with the adsorption energy and the activation barrier. Source: Sun et al. [188]. Reproduced with permission of the American Chemical Society.

this study are that the supported Pt_{SA} are effective catalysts for the dehydrogenation of propane, that the pyridine nitrogen-doped support is predicted to be the most active, and that a good selectivity is expected, since the dehydrogenation in $Pt_{SA}N_3$ requires more energy than for desorption of propylene.

7.4.1.2 Hydrogenation

M_{SA}/graphene catalysts were also shown to be active and selective toward hydrogenation reactions. Yan et al. reported both experimentally and theoretically that Pd_{SA}–O/graphene selectively converted 1,2-butadiene to 1-butene, the first hydrogenation step being the rate-determining step [189]. It was confirmed by DFT calculations made on Pd_{SA}/graphene and Pd_{SA}–O/graphene models, where the oxygen atom bridges Pd_{SA} and the carbon support. The presence of the oxygen atom turns out to be not only crucial for anchoring and stabilizing Pd_{SA} during atomic layer deposition, but DFT calculations show that it also facilitates H_2 dissociation. Even worse, H_2 molecules can hardly be adsorbed onto oxygen-free Pd_{SA} atoms, suggesting that Pd_{SA}/graphene-embedded is catalytically inactive toward hydrogenation reactions. Atomically dispersed Pd [137] or Cu [190] atoms on nanodiamond/graphene hybrids were also reported to be efficient for hydrogenation of acetylene to ethylene, with high conversion, high selectivity, and good stability. Beyond these experimental observations, DFT mechanistic studies shed light on the differences between the activity of the two metal atoms and on the competition between hydrogenation reactions and desorption of partially hydrogenated species. While in Pd_{SA}/graphene the rate-limiting step is the H_2 dissociation, it turns out to be the $C_2H_2^* + H_{2(g)} \rightarrow C_2H_3^* + H^*$ reaction on Cu_{SA}/graphene catalysts. The key role of H_2 dissociation was later confirmed on M_{SA}/graphene catalysts (M = Ni, Pd, Pt), relative to the coordination number of M_{SA} [191]. Acetylene hydrochlorination by Au_{SA}/and $Au_{SA}N_n$/graphene catalysts was also studied by means of DFT calculations [192]. They are good co-adsorption substrates for HCl and C_2H_2 and have a catalytic effect on the hydrogenation of acetylene, with a significant decrease in the activation barrier with some N-doped catalysts. A more recent DFT study on $M_{SA}N_4$/graphene catalysts (with M = Ru, Pd, Au) revealed a shared feature, in the sense that C_2H_2 adsorption occurs prior to the HCl adsorption [193]. The energy barrier decreases in the order Pd > Ru > Au, and it is noteworthy that the highest barrier step for Au and Pd (the dissociation of the C_2H_3Cl molecule in the catalyst) is different from that of Ru (the formation of the desorption adduct). A recent mechanistic study also shed light on the potential of novel support graphdiyne in single Au catalysis [194].

7.4.1.3 CO_2 Hydrogenation

SAC have shown high performances regarding CO_2 hydrogenation and carbon monoxide oxidation reactions, which are another class of reactions of great environmental importance. CO_2 can be captured or converted into useful value-added products, such as formic acid, HCOOH. Since formic acid is formed by direct hydrogenation, it has a relatively high volumetric hydrogen density and low toxicity and is liquid under ambient conditions. It is sometimes considered that a viable

CO_2-to-HCOOH catalytic process may help to propel a virtuous hydrogen-based economy. Again, there is a need for a rational catalysts design and, as such, DFT is known to enhance current understanding of the CO_2 catalytic hydrogenation in a broader context than supported SAC [195]. Given all the possible reaction pathways, this is challenging for DFT. In 2016, Sirijaraensre and Limtrakul investigated two possible pathways on a Cu_{SA}/graphene catalyst [196]. It turned out that the first hydrogenation of CO_2 without activating H_2 goes through a high-energy barrier (~35 kcal/mol), whereas after an affordable H_2 activation (~20 kcal/mol), the CO_2 insertion into Cu–H followed by the second hydrogenation step involves barriers of less than ~15 kcal/mol. A more recent study revisited that seminal work and showed that a Ru_{SA}/graphene catalyst is also a promising candidate for designing a supported SAC capable of converting CO_2 to HCOOH [197]. With both catalysts, the direct CO_2 hydrogenation with H atoms from the H_2 molecule is not competitive, whereas the optimal pathways proceed via a preliminary H_2 activation, and later it could be assisted by the interaction of the second $H_{2(g)}$ with the HCOO intermediate.

7.4.1.4 CO Oxidation

The growing environmental problems caused by the emission of CO may be solved by the CO oxidation reaction, which is one of the most investigated reactions in heterogeneous catalysis [184]. It is catalyzed by noble metals such as Pt, Pd, Rh, Ir, and even Au. But these noble metal catalysts usually require a high reaction temperature for efficient processes. Although a nice review can be found in Ref. [182], a comprehensive description of the different reaction mechanisms of CO oxidation on graphene-based SAC, revealed by computational DFT studies, will now be provided. The standard Langmuir–Hinshelwood mechanism found in textbooks proceeds either by a direct dissociation of O_2 on the surface (Scheme 7.1a) or by a CO*-assisted O_2 dissociation pathway (Scheme 7.1b).

(a)

$$\begin{aligned} CO + * &\rightleftharpoons CO^* \\ O_2 + * &\longrightarrow O_2^* \\ O_2^* + * &\longrightarrow 2\,O^* \\ O^* + CO^* &\longrightarrow * + CO_2^* \\ CO_2^* &\longrightarrow * + CO_2 \end{aligned}$$

(b)

$$\begin{aligned} CO + * &\rightleftharpoons CO^* \\ O_2 + * &\rightleftharpoons O_2^* \\ O_2^* + CO^* &\longrightarrow CO_2^* + O^* \\ O^* + CO^* &\longrightarrow * + CO_2^* \\ 2\,CO_2^* &\longrightarrow 2* + 2\,CO_2 \end{aligned}$$

Scheme 7.1 Sequence of elementary steps for Langmuir–Hinshelwood CO oxidation on heterogeneous catalysts. (a) Direct O_2 dissociation; (b) CO*-assisted O_2 dissociation. Source: R. Poteau.

However, as summarized in Figure 7.7 by the work of Zhuo et al. [182], several DFT investigations suggested that three reaction pathways could be followed depending on M_{SA} and the graphene-coordination type (embedded in a single or double vacancy, locally doped or not) [14, 135, 198–203]. Interestingly, the highest barrier heights found in the literature do not exceed ~20 kcal/mol. The elementary step sequences **i**, **ii** and **ii′** are actually two-stage catalytic processes. The first stage in **i** follows a Langmuir–Hinshelwood mechanism with the final formation of a

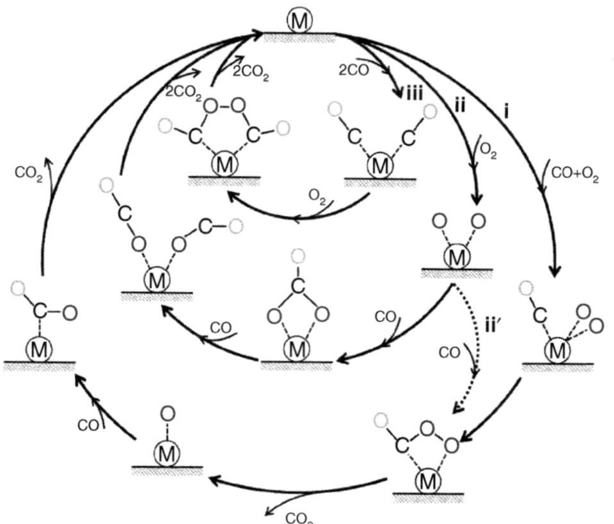

(a)

i

$$CO + O_2 + * \rightleftharpoons CO^* + O_2^*$$
$$O_2^* + CO^* \longrightarrow OCOO^*$$
$$OCOO^* \longrightarrow O^* + CO_2$$
$$CO + O^* \longrightarrow CO_2^*$$
$$CO_2^* \longrightarrow * + CO_2$$

ii

$$O_2 + * \rightleftharpoons O_2^*$$
$$CO + O_2^* \longrightarrow CO_3^*$$
$$CO_3^* + CO \longrightarrow 2\,CO_2^*$$
$$2\,CO_2^* \longrightarrow * + 2\,CO_2$$

ii′

$$O_2 + * \rightleftharpoons O_2^*$$
$$O_2^* + CO \longrightarrow OCOO^*$$
$$OCOO^* \longrightarrow O^* + CO_2$$
$$CO + O^* \longrightarrow CO_2^*$$
$$CO_2^* \longrightarrow * + CO_2$$

iii

$$2\,CO + * \rightleftharpoons 2\,CO^*$$
$$O_2 + 2\,CO^* \longrightarrow OCO{-}OCO^*$$
$$OCO{-}OCO^* \longrightarrow * + 2\,CO_2$$

(b)

Figure 7.7 (a) Proposed reaction mechanisms for CO oxidation on M_{SA}/graphene (**i**: Langmuir–Hinselwood and Eley–Rideal; **ii**: Eley–Rideal and Eley–Rideal; **iii**: trimolecule Eley–Rideal). Source: Zhuo et al. [182]. Reproduced with permission of the American Chemical Society. (b) Sequence of corresponding elementary steps.

peroxo-type OCOO* (which can be seen as a CO-assisted dissociation of O_2), while the second stage follows an Eley–Rideal mechanism where CO(g) reacts on a left O*. Two Eley–Rideal mechanisms are involved in the **ii** and **ii′** sequences. After preliminary adsorption of O_2, an insertion of CO(g) favors cleavage of O–O and leads to a carbonate-like CO_3 adsorbate. It then reacts with a second CO(g) to provide two released CO_2(g). The sequence **ii′** also begins with the adsorption of O_2 but then follows part of the **i** pathway, with the exception that CO(g) reacts in the outer sphere of O_2^*. The last possibility, **iii**, was found to occur on Pt_{SA}/N-graphene-supported catalyst via a "one-step trimolecule Eley–Rideal" mechanism [182]. O_2(g) reacts to form a bridge between two adsorbed CO molecules, resulting in the formation of an OCO–OCO* peroxo species. With a ~4 kcal/mol energy barrier for the OCO–OCO dissociation rate-limiting step, it can also be seen as a powerful CO assistance to O–O bond cleavage. MnN_3-doped graphene also proved to be a good candidate for CO oxidation, with better activity when coordinated in a DV [204]. The best reaction pathway was found to be the Langmuir–Hinshelwood mechanism (noted **i** in Figure 7.7). Interestingly, *ab initio* molecular dynamics (AIMD) simulations were also performed at several temperatures, suggesting that when adsorbed on a SV, CO_2 poisons MnN_3 SAC by remaining adsorbed, on the contrary to the DV catalyst. In summary, such a reaction is fairly well modeled and scrutinized using DFT calculations, which can play a role in the rational design of supported SACs. Such mechanistic knowledge, completed with the possible introduction of additional gas-phase reactants such as N_2O [199], could be used in microkinetic models or kinetic Monte-Carlo calculations to evaluate the possible outcome of these different pathways.

7.4.1.5 Other Reactions

In addition to these reactions, the mechanism for the synthesis of dimethyl carbonate by oxidation and carbonylation of methanol on a Cu_{SA}/graphene catalyst was investigated [205]. Calculations revealed the key role of a CH_3OCO intermediate, as well as a strong difference in activation energies of the rate-determining step between monovacancies and divacancies (~18 vs. ~46 kcal/mol). It is also interesting to mention that Cu(II) ions coordinated with nitrile-functionalized graphene were shown to perform well in O_2-mediated oxidative coupling of amines [206]. DFT was used both to characterize the coordination of Cu, its spin and charge density, and to address a probable reaction mechanism, which actually explains the superior performance of the catalyst. A captivating assumption, also indirectly supported by EPR spectroscopy, is the possible cooperation of two neighboring metal centers, favoring the formation of a cyclic intermediate (Figure 7.8). It is noteworthy, on the basis of this result, that the impact of the density of active sites on the catalytic activity of supported SAC could be taken into account more frequently, and particularly in mechanistic studies.

A thorough DFT kinetic and thermodynamic study of the direct catalytic reduction of NO on a SV graphene-based Fe_{SA} catalyst showed that it can be divided into two main stages [207]. The first, the reduction of two NO molecules to N_2O, is the rate-limiting step, while the second, the reduction of N_2O to N_2, is almost barrierless. With a low activation energy of ~19 kcal/mol, the Langmuir–Hinshelwood mechanism appears to be the dominant mechanism.

7.4.2 Electrocatalysis

Next, OER, HER, ORR, and CO_2RR will be the main reactions to focus on.

7.4.2.1 Water Splitting

Developing a long-term sustainable energy economy, based on cost-effective zero carbon footprint technologies, is one of the critical challenges we face today. In this regard, hydrogen production by water splitting is a promising technology. It is a very simple reaction in which water is decomposed into oxygen and hydrogen: $2H_2O = 2H_2 + O_2$. The electrochemical water splitting can be summarized as a two-electrode device in Figure 7.9 [208], in which on one side HER is performed when on the other electrode OER takes place.

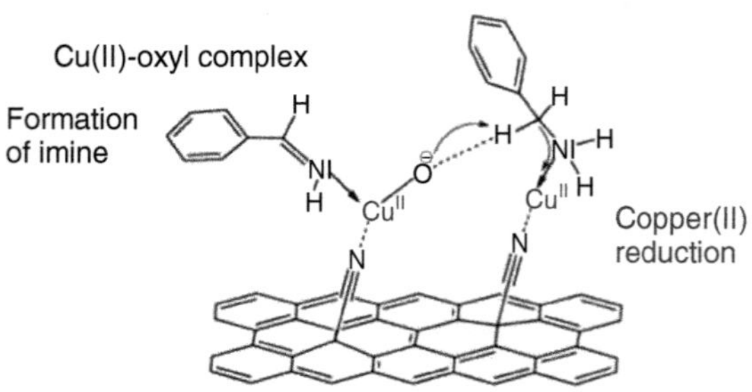

Figure 7.8 Possible cyclic intermediate between two Cu ions that could be involved in a O_2-mediated oxidative coupling of amines. Source: Bakandritsos et al. [206]. Reproduced with permission of Wiley-VCH.

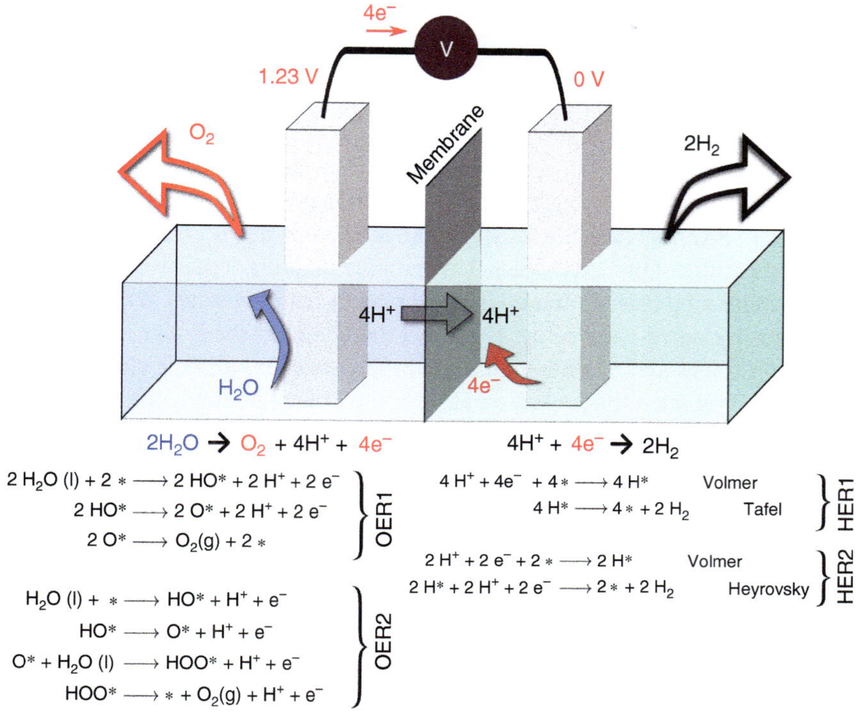

Figure 7.9 Schematic diagram of a photoelectrochemical cell for water splitting ($2H_2O = 2H_2 + O_2$), with reaction mechanisms in acidic medium. The equilibrium potentials are also provided. Source: R. Poteau.

HER is an electrocatalytic reaction that proceeds through the Volmer, Tafel, and/or Heyrovsky reaction mechanism (HER1 and HER2 in Figure 7.9). Although this part of the present chapter deals only with DFT-based multistep mechanistic studies, it was not possible not to mention DFT studies on HER even though most of them are based on a simple H adsorption energy descriptor [209]. In fact, zero hydrogen dissociative adsorption Gibbs energy, $\Delta_{\text{ads}}G^{\circ}_{\text{H}}$, corresponds to the top of the conventional volcano plot with the highest activity. There is a need to find alternative catalysts to the Pt electrodes, which lie on the top of the volcano, hindering their large-scale application due to their high cost. In this context, Lim et al. have calculated the hydrogen adsorption energy for 120 $M_{SA}X_n$/graphene candidates (X = C or N, $n = 3$ or 4) across all 3d, 4d, and 5d-block of TM [210]. Based on these DFT calculations, a superior HER activity of Fe_{SA}/graphene was observed, relative to the ionic bonding character between the hydrogen and the catalyst, rather than to the modulation of the position of the d AO of the TM. The lattice-confined Pt atoms in a nitrogen-containing porous carbon matrix (Pt@PCM) showed good electrocatalytic performance for HER [211], being rationalized by DFT calculations on $Pt_{SA}N_x$/graphene models. Interestingly, hydrogen adsorption has been assessed on both Pt and neighboring sites on the graphene surface, suggesting that confined Pt atoms and adjacent C and N atoms should be the active sites for HER. More recently, aniline-stacked graphene-anchored Pt atoms exhibited outstanding HER performance, with a mass current density 46 times higher than that of commercial 20 wt% Pt/C and a very good stability [212]. A low Gibbs energy of H adsorption has been found by DFT and has been related to the aniline modulation of the d-band center and DOS of the Pt atoms. The HER activity of a series of TM embedded in N-doped graphene has been studied using a combination of DFT calculations and electrochemical measurements. Co_{SA}–N_4/graphene was found to exhibit very good HER activity, relative to a low $\Delta_{ads}G^\circ_H$ (0.13 eV) and the electronic structure of the SAC, which was analyzed in terms of the average position of the d_{z^2}-containing orbitals and their resulting anti-bonding states, rather than in terms of the average d-band center [213]. All these studies show that maximizing HER activity consists of modulating the electronic structure and the charge of the TM. He et al. evaluated the impact on HER of the unique structure and electronic property of grain boundaries in graphene [214]. They anchored various TM at the so-called 558 grain boundaries and showed that with $\Delta_{ads}G^\circ_H$ lying at −0.13, −0.10, −0.08, and −0.01 eV respectively, Fe-, Co-, Ni-, and V-SACs offer a promising way to reduce the high cost of conventional HER catalysts (see also Figure 7.10).

Let us focus now on OER, HER's counterpart in electrochemical or photo (electro) chemical cells [215]. Two reaction mechanisms are possible, the direct pathway (OER1 in Figure 7.9) and the predominant OER2 pathway. The optimization of this four-step mechanism involving O*, HO*, and HOO* reaction intermediates often relies on a volcano relationship ($\Delta_{ads}G_O - \Delta_{ads}G_{HO}$) descriptor, a measure of the ability of surface oxygen atoms to form and break bonds with hydrogen and oxygen atoms [216, 217]. Most DFT-based studies consider only the thermodynamics of the reaction path along HO*, O*, and HOO* or simply consider this descriptor. An ideal OER catalyst actually requires that all four steps have free energies of the same magnitude at zero potential, i.e. 1.23 eV, equivalent to all reaction free energies being zero at equilibrium potential, 1.23 V. It turned out that no real oxide surface displays such

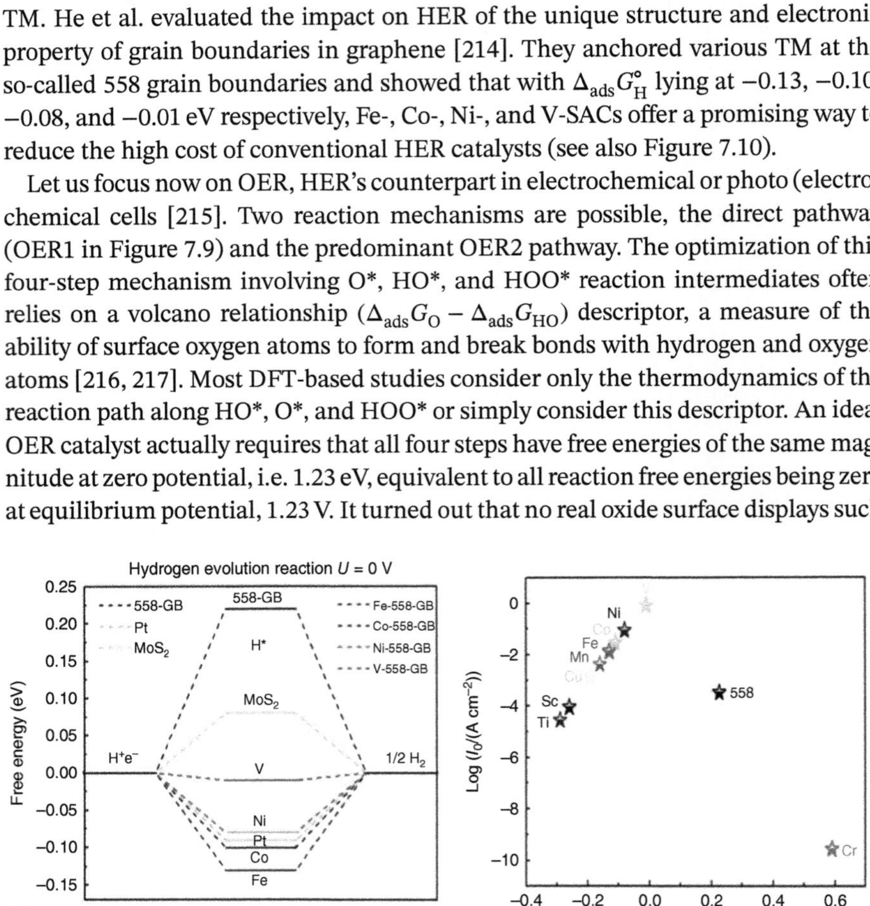

Figure 7.10 (a) DFT Gibbs energy diagram for HER at the equilibrium potential for Fe, Co, Ni, and V supported on a 558 grain boundary; (b) Volcano plot of the theoretical exchange current i_0 as a function of the Gibbs energy $\Delta_{ads}G_H$. The calculation of i_0 is based on Nørskov's model [209]: $i_0 = -ek_0/[1 + \exp(|\Delta_{ads}G_H|/k_BT)]$, where k_0 is related to the rate constant. Source: He et al. [214]. Reproduced with permission of Elsevier.

a property, the volcano optimum rather lying at *c.* 1.6 eV [216]. A recent DFT study performed in this context on several TM-decorated graphene sheets showed that they are good candidates for the OER, and in particular for the higher-coordinated TMs [218]. The support modification modulation of a Co-SAC in its electronic structure and in the $(\Delta_{ads}G_O - \Delta_{ads}G_{HO})$ descriptor has also been considered [219]. Results suggested improved performance of $Co_{SA}N_1$/graphene catalysts.

A mechanistic study involving four-coordination and three-coordination Ni–N–C configurations on a graphene support under alkaline media was also reported [220], considering the effect of a possible OH ligand binding on the TM. The results, summarized on the OER volcano diagram reported in Figure 7.11, show that some configurations could perform better than the RuO_2 reference catalyst.

7.4.2.2 Oxygen Reduction Reaction

Among the various electrochemical technologies, fuel cells require oxygen as the oxidant at the positive electrode, including direct methanol fuel cells. It may be interesting to refer to a recent review of ORR mechanisms for metal-free carbon-based electrocatalysts [221] to shed more light on possible reactions directly on graphene in SACs. Technical challenges and possible research directions are also given in Ref. [222], both for experimental and theoretical approaches, such as structure–activity relationships and catalytic reaction mechanisms at the atomic level. Yet, a recent general review has partially answered, underlining the role of theory in understanding the ORR mechanism and highlighting that a descriptor-based approach, namely $\Delta_{ads}G_{HO}$, can be used again to identify efficient catalysts [223], based on the $2e^-$ (ORR1), $4e^-$ (ORR2), and dissociative (ORR3) pathways reported in Scheme 7.2.

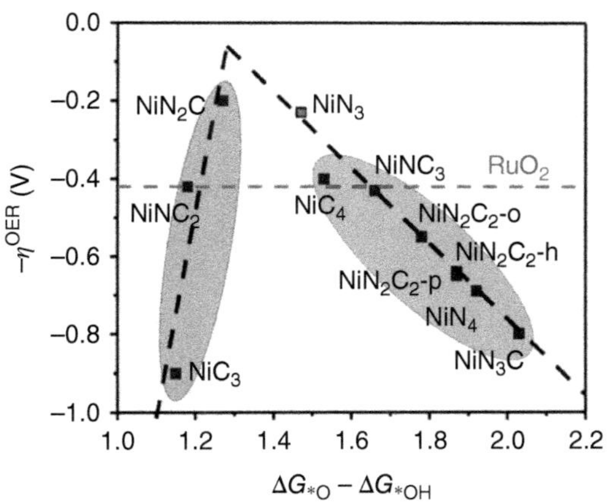

Figure 7.11 Volcano plots of OER catalytic activities on various types of Ni–N–C configurations on a graphene substrate. Some SACs exhibit a better theoretical overpotential, η^{OER}, than that of RuO_2 (the underlying reaction mechanism is a four-step process, each one being thermodynamically characterized by a free energy change, ΔG_i; η^{OER} (in V) is calculated as $\max(\Delta G_i) - 0.402$, where 0.402 V is the equilibrium reduction potential of an oxygen molecule in alkaline media). Source: Liang et al. [220]. Reproduced with permission of The Royal Society of Chemistry.

$$\left.\begin{array}{r}O_2(g) + * \longrightarrow O_2^* \\ O_2^* + H^+ + e^- + * \longrightarrow HOO^* \\ HOO^* + H^+ + e^- \longrightarrow H_2O_2 + *\end{array}\right\}\text{ORR1}$$

$$\left.\begin{array}{r}O_2(g) + * \longrightarrow O_2^* \\ O_2^* + H^+ + e^- + * \longrightarrow HOO^* \\ HOO^* + H^+ + e^- \longrightarrow H_2O\,(l) + O^* \\ O^* + H^+ + e^- \longrightarrow HO^* \\ HO^* + H^+ + e^- \longrightarrow H_2O\,(l)\end{array}\right\}\text{ORR2}$$

$$\left.\begin{array}{r}O_2(g) + 2* \longrightarrow 2\,O^* \\ 2\,O^* + 2\,H^+ + 2\,e^- \longrightarrow 2\,HO^* \\ 2\,HO^* + 2\,H^+ + 2\,e^- \longrightarrow 2\,H_2O\,(l)\end{array}\right\}\text{ORR3}$$

Scheme 7.2 ORR reaction pathways in acidic media, where ORR2 is the reverse process of OER2. The H_2O_2 intermediate in ORR1 may be further reduced to H_2O ($H_2O_2 + 2H^+ + 2e^- \rightarrow 2H_2O$) or undergo disproportionation to regenerate O_2 ($H_2O_2 \rightarrow 2H_2O + \frac{1}{2}O_2$). Source: R. Poteau.

Screening of the catalytic ORR performance of several M_{SA} has been considered at the DFT theory level by Chen et al. (M = Si, Ni, Fe, Co, Ag, and Au) [224]. This study consisted of evaluating the relative thermodynamic stability of the reaction intermediates HOO*, O*, and HO* along the ORR2 pathway. Given their progressive stability, Au, Co, and Ag SACs appeared to be good candidates as ORR catalysts. Thermodynamic stability and overpotentials were calculated for finite-size M_{SA}–C_3N_4 nanosheets (M = Mn, Fe, Co, Ni, Cu, Rh, Pd, Ag, Pt, Au), showing that most of them exhibit no ORR catalytic activity, except Ag and Pd [225]. But the OH-doping of the metal (M_{SA}–OH–C_3N_4) implied a decrease in the adsorption energies, with small theoretical overpotentials of 0.50 and 0.55 V for Ni and Cu, respectively. A systematic dispersion-corrected DFT study of the ORR in alkaline media of 34 SACs based on defective graphene (dubbed in the publication as M-SV or M-DV, M = B, N, O; Al, Si, P, S; Ga, Ge, As, Se; In, Sn, Sb, Te; Tl, Pb, Bi) was also reported [226]. It is noteworthy that, strictly speaking, no dissociation or bimolecular kinetic investigations are conducted in these articles, but the relative instability of some intermediates is sometimes referred to as a barrier height and the corresponding step is designated as a rate-limiting step, a somewhat hazardous shortcut. However, many studies are relied on it (a nice discussion in the one for HER, and probably valid in general for any electrocatalytic reaction, can be found in Ref. [227]). Nevertheless, the presence of defects seems to be in favor of ORR. Most of the DFT investigations are done in the context of the ORR2 mechanism and of the seminal work of Norskøv and coworkers [223]. Given that ORR2 and OER2 are inverse mechanistic processes, some authors even took advantage of the DFT adsorption energies to consider both descriptors and theoretical overpotential calculations [220]. The dissociative pathway (ORR3) was also studied in a series of TM_{SA}–N_4/graphene-embedded SACs (TM = Cr, Mn, Fe, Co, Ni, Cu, Mo, Tc, Ru, Rh, Pd, Ag, W, Re, Os, Ir, Pt, and Au), considering both the relative stability of the reaction intermediates and the O–O bond breaking transition state, i.e. species with an imaginary vibration frequency [228].

Barrier heights are strongly dependent on the metal, going from 0.19 eV for Mo to 1.97 eV for Co, which makes dissociation in the latter case a *true* rate-limiting step. Given the balance between kinetics and thermodynamics, MnN_4 SAs appeared to be the best graphene-embedded SACs. Although it did not involve TM-SACs, attention may be paid to the possible efficiency of P, N co-doped graphene. Geometric structure, stability, electronic properties, catalytic sites, and detailed ORR pathways were also studied by DFT on a 2D model [229]. These investigations showed that co-doping improves catalytic capacity due to a synergistic effect compared to single doping. Although the liquid phase can also play a role, the mechanistic studies cited above did not take into consideration the solvation effect. A noticeable exception is an AIMD study, aiming at comparing the specific mechanism of the $4e^-$ transfer of Mn_{SA}–N_4/graphene SACs in gas phase and in water, by using explicit water molecules [230]. Regarding the calculation of energetics in water, the choice to anneal to 0 K some chosen structures along the reaction path instead of performing a thermodynamic integration is questionable. It nevertheless offers interesting trends that deserve to be confronted in the future with more advanced simulations. It is found that the solvent environment can effectively promote charge transfer from the substrate to O_2, leading to the probable breaking of the O–O bond before being protonated.

7.4.2.3 Carbon Dioxide Reduction Reaction

Furthermore, it is considered as a green chemistry way of recycling carbon dioxide back to reusable forms of carbon. It is actually a complex process involving multiple e^-/H^+ transfers and numerous intermediates, and that competes with the HER side reaction (Figure 7.12). Given this complexity, to the best of our knowledge, there is no general volcano activity relationship based on simple adsorption energy descriptors. Therefore, it is necessary to conduct human and computational time-demanding multistep mechanistic studies. It is even possible to estimate the limiting potential of a catalyst (i.e. at $U = 0$ V), U_L, from the maximum free energy found along a reaction pathway, ΔG_{max}, using $U_L = -\Delta G_{max}/e$. But the best approach would probably be to couple the DFT calculations with kinetic Monte-Carlo approaches or microkinetic models. Some interesting trends can however emerge from basic DFT calculations. Under the same reaction conditions, an e^-/H^+ pair can be consumed by the HER side reaction, or feed the CO_2RR, thus forming formate OCHO* or carboxyl COOH* at the beginning of the reaction. Back et al. assumed that reactions with lower free energies are more selective [231]. They elegantly summed it up for a series of TM SACs as $\Delta_{ads}G_{COOH/OCHO}$ vs. $\Delta_{ads}G_H$ 2D diagram, which straightforwardly separates HER and CO_2RR selective SACs. Additionally, they also calculated the adsorption energies of several reaction intermediates involved in the CO_2RR. They found, in particular, that the predicted limiting potential for Pt-SACs is significantly less negative (−0.27 V) than for conventional TM catalysts (−0.7 to −0.8 V). A comprehensive analysis of electron density and DOS profiles suggests that TM-graphene orbital interactions and d-AO filling play an important role. A similar strategy was adopted in Ref. [232] for five TMs (Ag, Cu, Pd, Pt, and Co) embedded in a SV or DV of graphene. According

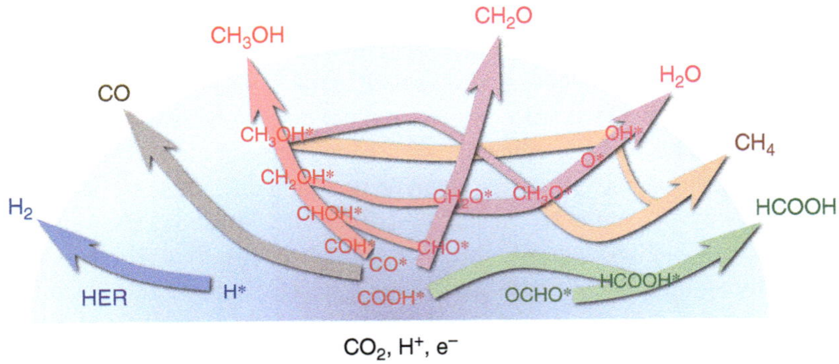

Figure 7.12 Competition between the HER and the reaction network of CO_2RR toward C_1 products. Source: R. Poteau.

to these DFT investigations, Ag has the lowest overpotential for CH_4 production (0.73 V), whereas Cu exhibits a preference for producing methanol (0.68 V). CO_2 to CO conversion is a valuable route in the context of the production of synthetic fuels using the Fischer–Tropsch process. A graphene-supported $Fe_{SA}N_5$-SAC has been experimentally shown to exhibit high efficiency in CO production, with a low overpotential of 0.35 V [233]. DFT calculations of free energy profiles, DOS, and charge density were performed to disclose the origin of this very good catalytic behavior. The catalytic activity of 24 TMs supported by a graphitic carbon nitride monolayer (g-CN) was found to be correlated with the adsorption free energies of two intermediates (OH and OCH) [234], themselves being related to two descriptors, namely the number of electrons in the valence d-shell and the enthalpy of vaporization of the TM@g-CN, in relation with the stability of the SAs. The descriptors were found to apply to a C_2N monolayer as well. Ni@g-CN, Cu@g-CN, and Co@C_2N were shown to be promising SACs for CO_2RR.

The axial pyrrolic N ligand of the FeN_5 model depletes the electron density of Fe 3d orbitals and reduces the π back-donation, resulting in a weaker Fe–CO bond strength for highly selective CO production. It is again an example where a coordination or environment effect modulates the adsorption energies by changing the electronic structure of the metal. It has also been achieved in Ref. [235] for the CO_2 to CO electroreduction. Experimental and DFT calculations revealed that the presence of a vacancy-defect in Ni–N_3 SACs can dramatically boost electrocatalytic activity. The study reported in Ref. [236], and illustrated in Figure 7.13 is a nice summary of the DFT-based strategy applied to this electrocatalytic reaction. Among other results, it comes from this study that all TM-C_3N_4 catalysts show better selectivity toward CO_2RR than toward HER and that Co SACs shows better activity and selectivity for the production of methanol. In this attempt to design more realistic simulations, recent work by Zhao and Liu should be taken into account [237], who explicitly considered a network of water molecules, used AIMD and a "slow-growth" sampling approach to calculate reaction kinetic barriers and that also accounted for the charge capacity of the TM, Ni in this study. Since this work highlights the crucial

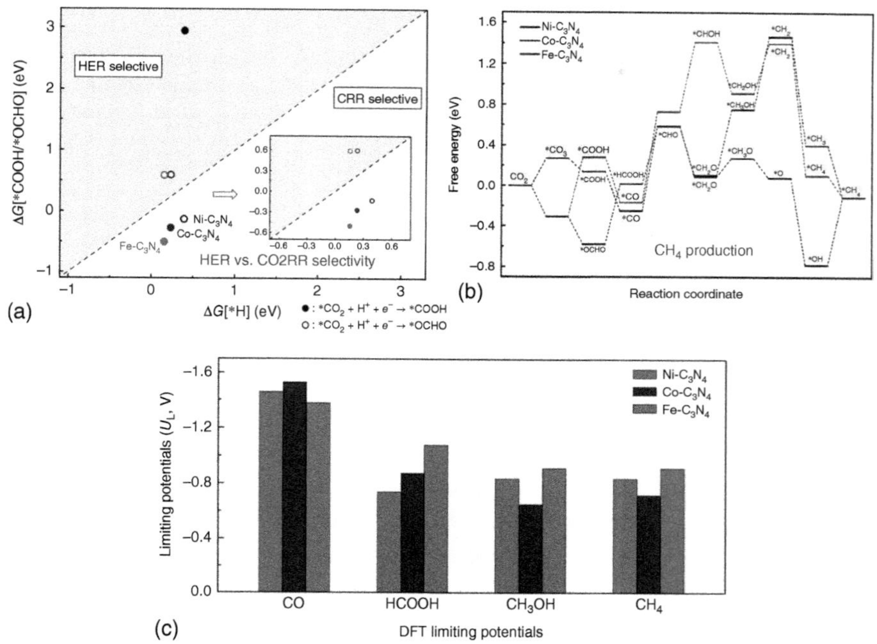

Figure 7.13 (a) Free-energy change for HER vs. CO_2RR selectivity, calculated from simple adsorption energy calculations (*vide supra*); (b) free-energy profile for CH_4 production of CO_2RR; (c) theoretical limiting potentials for the productions of CO, HCOOH, CH_3OH and CH_4. Source: Guo et al. [236]. Reproduced with permission of Wiley-VCH.

roles of charge capacity and hydrogen bonding, it could become a new state-of-the art methodology.

7.4.2.4 Other Reactions

They will not be described in much detail, but it should be mentioned that in recent years, graphene-supported SACs were also considered as possible interesting catalysts for the electrochemical reduction of CO [238], for the nitrogen reduction reaction [147, 239–244], for the NO reduction reaction [182], and for electrochemical conversion reactions in lithium batteries [245]. It is also noteworthy that the application of an electric field is an effective route to tune the performance of the catalysts, as shown theoretically for CO oxidation [246, 247].

7.5 Oxide Support Models

The doping of oxides has been widely studied since the 1970s. However, it is only recently, in the last two decades, that oxide-supported SACs have also been considered as active catalysts for different reactions. Describing theoretically both the interactions between SAs and oxide supports as well as their catalytic activity is a highly complex task. As a result, more and more theoretical as well as experimental studies appear regularly [11, 248–263]. The next subsections (Sections 7.5.1–7.5.7) present important results concerning the description of the coordination/interaction of a series of metal atoms spanning different columns and rows in the periodic table on several surface facets of different oxides (Al_2O_3, CoO and Co_3O_4, CeO_2, MgO and $MgAl_2O_4$, TiO_2, ZrO_2, and ZnO). A particular attention will be paid to the description of the theoretical studies carried out on the reactivity of these oxide-supported SACs.

7.5.1 Aluminum Oxide

Aluminum oxide (alumina, Al_2O_3) is one of the most versatile ceramics, utilized in an amazing range of structural and optical applications. Although most of the intrinsic properties of alumina are known, due to this versatility, many applied as well as fundamental studies appear each year. The main reason for this craze is undoubtedly the complexity (and utility) of alumina polymorphs. Alumina can crystallize in different crystal structures depending on the growing conditions. Based on literature, there are seven phases of alumina, also known as "transition aluminas," such as χ-, κ-, γ-, θ-, δ-, η-, and ρ-alumina. The most important, and common, polymorphs are denoted α, γ, θ, and κ. The κ phase is widely used as a wear-resistant coating due to its hardness and thermal stability, while the γ and θ phases find applications as catalysts or catalyst supports since they have large surface areas available for catalytic reactions. Although, the most common form of crystalline aluminum oxide is α-alumina. Also known as corundum, it does not belong to the transition alumina and constitutes the thermodynamically stable form. This crystalline structure is formed by increasing the temperature of the transition alumina to about 1050 °C. All the Al_2O_3 transitions are metastable but can be formed in growth experiments at low temperatures. However, they cannot be used in high-temperature applications, since diffusion mechanisms activate the irreversible transformation into the α phase. In contrast, α-alumina typically require high-growth temperatures (around 1000 °C), prohibiting the use of temperature sensitive substrates. For example, for transition aluminas derived from boehmite, γ-alumina is stable under a treatment temperatures range of 350–550 °C with a release of water molecules due to a dehydroxylation process. During the dehydroxylation step, the c parameter of the orthorhombic boehmite structure decrease, while the face-centered cubic (fcc) structure of the anionic sublattice remains unchanged. The octahedral Al^{3+} ions are ordered, while the tetrahedral Al^{3+} are disordered, leading to a distortion of the tetrahedron. At higher heat treatment temperatures around 800 °C, the γ-Al_2O_3 is transformed to δ-Al_2O_3 and then to θ-alumina at 1000 °C. The final α-alumina is formed when heating up boehmite at 1200 °C (see Figure 7.14). In the same way, if a gibbsite is treated at 300 °C a χ-alumina is formed, which at 970 °C evolves to κ-alumina before becoming α-alumina at around 1100 °C. When heating up bayerite, various transition aluminas are formed, with the transformation $\eta \rightarrow \theta$ taking place at 850 °C to finally reach the most stable α-alumina phase at 1200 °C. Interestingly, the oxygen sublattice in α-alumina is a hexagonal close-packed (hcp) structure with octahedral sites occupied with cations, while transition aluminas have a fcc

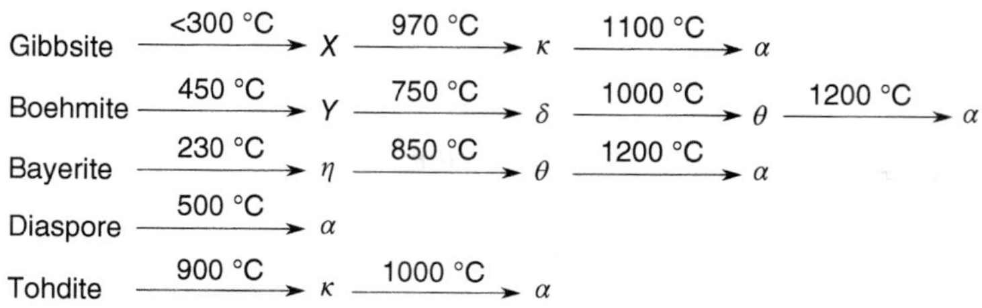

Figure 7.14 Phase transformation of alumina. Source: I. del Rosal.

arrangement of oxygen atoms and cations present in various proportions in both octahedral and tetrahedral sites. Due to the application of the metastable aluminas as catalysts or catalyst supports, there is an interest in stabilizing them with respect to the α phase. A natural course of action, taken by a number of researchers, would be to investigate the effects of dopants on the relative phase stability. Most dopants (d or f metals) tried in different works seem to retard the $\gamma \rightarrow \alpha$ transformation by hampering the diffusion mechanisms, but not to stop it [264–271].

The α-Al_2O_3 is a chemically inert, hard transparent, and uncolored material. Thus, α-Al_2O_3 thin films are often used as wear-resistant and high-temperature diffusion barrier coatings. This alumina phase is also used as insulator in electronics (due to the wide band gap of 8.8 eV) and in optics because it is completely transparent and stable at high temperature. This phase is also used not only in materials science, but also in gemology because, for example, rubies are nothing else than an α-Al_2O_3 doped with small amounts of Cr, and the sapphire gemstone an α-Al_2O_3 doped with Fe and Ti. The corundum structure can be described as a hcp oxygen sublattice, in which the Al^{3+} occupy two-thirds of the octahedral sites. In α-Al_2O_3, the basal (0001) surface plane is the lowest energy surface. On it, the metal-terminated surface is surprisingly stable, whereas the oxygen-terminated α-Al_2O_3/(0001) surface becomes stable only if hydrogen is present on the surface, i.e. if hydroxyl groups are formed.

The interaction energy of Cu atoms with α-Al_2O_3/(0001) has been theoretically studied using models of both clean (see Figure 7.15) and hydroxylated surfaces. At Cu coverage of 1 ML, Lodziana and Nørskov reported a strong dependency of the interaction strength related to the surface stoichiometry [273], the adsorption energy varying from 20 to 26 kcal/mol on a hydroxylated surface to 21 kcal/mol on a clean one. For non-stoichiometric surfaces, the interaction of Cu with the alumina surface varies from covalent-like on an aluminum-rich surface (75 kcal/mol) to ionic-like on an oxygen terminated surface (142 kcal/mol). Thus, according to the defect concentration created during the surface hydroxylation, the adsorption energy should be somewhere in between that for the stoichiometric and the non-stoichiometric surfaces. Likewise, their work shows that a strong hydroxyl coverage does not give rise to a stronger metal–aluminum oxide interaction. These results are partially in agreement with those of Kelber et al. published a year before [274, 275]. In this case, the calculations show that a similar adsorption energy is obtained on clean and highly hydroxylated surfaces. However, the presence of a small amount of hydroxyl

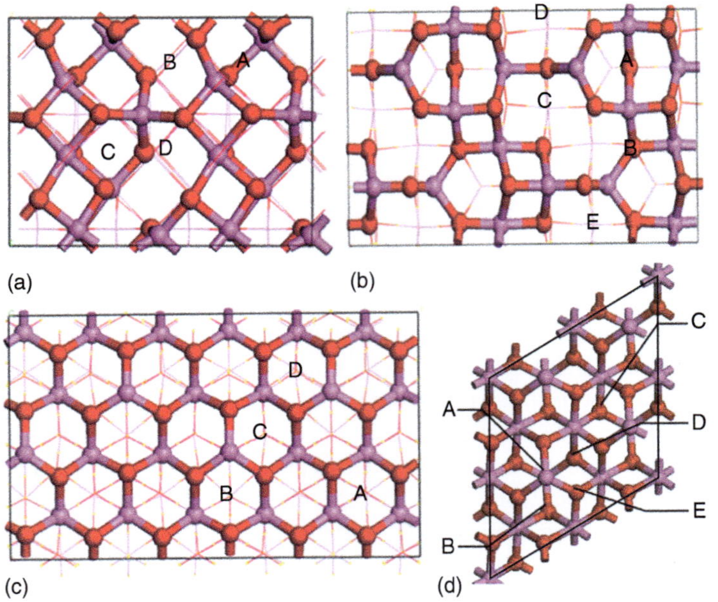

Figure 7.15 Chosen sites on the lowest energy termination of: (a) (100) γ-Al_2O_3, (b) (110) γ-Al_2O_3, (c) (111) γ-Al_2O_3, and (d) (0001) α-Al_2O_3. Sites are indicated with capital letters. Only top layers are shown by ball-and-stick, the atoms in the subsurface are shown by a wireframe. Pink atoms Al, and red atoms O. Source: Tan et al. [272]. Reproduced with permission of the American Chemical Society.

groups (only one OH group per surface unit area) increases the adsorption energy of Cu on the surface. At coverages less than 1/3 [276, 277], Cu atoms appear to reduce the surface Al atoms, with the formation of Cu^+ ions. The preferred surface coordination sites for these Cu^+ atoms are hollow sites close to oxygen atoms, with an Al atom underlying in the subsurface plane. Increasing the coverage to 2/3 leads to the formation of two sublayers with the coexistence of both Cu^+ and Cu^0 species forming zigzag chains. The preferred site for Ag is, as for Cu, the threefold hollow oxygen sites while Au lies on top of surface oxygen atoms, the adsorption energy going from 25 to 16 kcal/mol following the order Cu > Au > Ag [278–280]. This different behavior can be understood in terms of the ionization potential of these elements. For Cu and Ag, the largest contribution to the interaction energy arises from the charge transfer from the metal atom to the alumina surface while in the case of Au, the interaction is dominated by the polarization of the TM. The top coordination mode of Au is similar to that of Pd and Pt atoms, although the interaction energy of the latter is significantly larger (32 kcal/mol for Pd and 46 kcal/mol for Pt) [281–283]. As for Ag or Cu, the oxygen threefold vacant site is the preferential adsorption site for Ni atoms (E_{ads} = 41 kcal/mol) [283–285]. This strong interaction is attributed to the large overlapping between the d orbitals of Ni atoms and the p orbitals of the three oxygen atoms. Interestingly, Ni atom supported on an α-Al_2O_3/(0001) surface shows superior catalytic properties than Rh_{SA} or Pt_{SA} coordinated on the same surface for the methane activation reaction [286]. On the Ni SAC, the presence of an empty d orbital 1 eV lower than for Pd or Rh induces a better capability toward methane adsorption, i.e. a higher binding energy (5, 1, and <1 kcal/mol for Ni, Rh, and Pt, respectively). In the same way, the Ni atom also exhibits a low dissociation barrier of only 9 kcal/mol, which is 19 kcal/mol lower than for the two other metals.

The crystal structure of δ-Al_2O_3 has been studied for more than 50 years. Several crystallographic data showed that the structure is a complex intergrowth of two variants, identified as δ_1-Al_2O_3 and δ_2-Al_2O_3. These variants have identical structural motif, and only subtle differences in the rearrangement of a subset of octahedral and tetrahedral sites can distinguish them. Nonetheless, the intersection of these variants allows the creation of a structure, which has no long-range periodicity along the main direction. In contrast to the δ-Al_2O_3 structure, that has been debated and studied throughout the years, the structure of θ-Al_2O_3 has historically generated much less controversy. The θ-Al_2O_3 is a highly ionic and insulating phase (with a band gap of 7.4 eV) with a monoclinic symmetry (isostructural with β-Ga_2O_3 structure). In this structure, half of the Al^{3+} ions occupy octahedral sites and the other half are at tetrahedral sites.

The structure of γ-Al_2O_3 is traditionally a cubic-defect spinel type in which the oxygen atoms are arranged in a cubic close packing with the Al^{3+} atoms more or less randomly distributed between the octahedral and tetrahedral sites. Among the various phases, the γ-Al_2O_3 can be formed at low temperature, greater than 350 °C. This alumina phase is an important material in microelectronic processing due to its porous structure, high surface area, and high catalytic surface activity. γ-Al_2O_3 is undeniably the most relevant alumina phase. This phase is also an extremely powerful material in many industrial processes, acting as an adsorbent, a catalyst, and/or catalyst support. γ-Al_2O_3 exhibits at their surface a large variety of hydroxyl groups with several coordination modes. Their number and type depend on the pretreatment temperature: the higher the pretreatment temperature, the lower the amount of OH groups. In addition to other properties, the presence of these hydroxyl groups is one of the reasons why this surface is often used in the petroleum and petrochemical industries as catalyst supports for TM. The coordination surface chemistry involves the creation of single-metal atoms covalently or ionically bonded to a solid support. The reaction of the surface-OH groups with the appropriate molecular precursors forms well-defined surface species, which can then be post-treated to modify their environment according to the desired applications. Their synthesis usually involve the reaction of a surface-OH group with a generic L_n–Metal–X_x complex (X = R, NR_2, OR, Cl, …). The corresponding grafted surface–O–ML_nMX_{x-1} species are formed via a proton transfer to the M–X group and a release of HX. Thus, contrarily to the SACs, the supported coordination compounds retain at least part of their ligands after grafting [10]. On alumina, surface coordination chemistry studies have grown exponentially over the past 20 years. However, on this same support, the number of studies involving SAC catalysts remains, in comparison, quite low [287]. The (100) and (110) surfaces exposed mainly by the γ-Al_2O_3 exhibit two distinct behaviors regarding the dehydroxylation process. The (110) surface maintains a high degree of hydroxyl coverage even at high temperature, whereas the (100) surface is easily dehydrated at low temperature (see Figure 7.15 for clean surfaces). On the γ-Al_2O_3/(100) surface, the adsorption energies of a Pd_{SA} are in a range from −9 to −33 kcal/mol [7]. The most stable structure corresponding to an Al–Pd–O bridging mode in the vicinity of the penta-coordinated aluminum atoms. On the extremely corrugated (110) surface, the interaction energies for the Pd atom vary from −23 to −39 kcal/mol. These adsorption energies are reduced if a highly hydroxylated γ-Al_2O_3/(110) surface is considered (E_{ads} vary from −5 to −32 kcal/mol on this surface). This difference is due to the decreases in the Lewis acidity of the aluminum surface atoms and the Lewis basicity of the oxygen surface atoms when the number of hydroxyl groups increase. The hydroxyl coverage also has an impact on the hopping rate of Pd, because this one is greatly reduced when the number of OH groups increases. On this same surfaces, Rh atoms adsorb strongly than Pd atoms, with adsorption energies in the range −78 to −64 kcal/mol [288]. These results are in agreement with experimental decreasing of the diffusion of the Rh atom with respect to Pd one [289].

In the last years, the studies involving non-hydroxylated alumina surfaces were extended to a series of metal atoms spanning different columns and rows in the periodic table (Au, Cu, Ag, Pt, Pd, Ni, Rh, Sn, In, and Ir) coordinated on the two same γ-Al_2O_3/(100) and γ-Al_2O_3/(110) low-index surfaces as well as on the γ-Al_2O_3/(111) facet [272, 290]. On γ-Al_2O_3/(100), the preferred adsorption configuration for all metal atoms is a hollow site interacting with both two oxygen and two aluminum atoms. The DFT-calculated adsorption energies for Rh and Pd atoms are in agreement with the aforementioned theoretical results, with a stronger adsorption for Rh (−82 kcal/mol) than for Pd (−60 kcal/mol). On this same site, Pt (−107 kcal/mol) and Ir (−105 kcal/mol) are strongly coordinated than Rh, while Cu (−38 kcal/mol), Au (−22 kcal/mol) and Ag (−18 kcal/mol) will be more weakly coordinated than Pd. Ni (−82 kcal/mol) is found between Rh and Pd. At calcination conditions (300 °C, 0.2×10^5 Pa O_2), the Pt atom is strongly anchored onto the support due to the formation of three Pt–O bonds and one Pt–O–O–Al interaction. Thus, in these reaction conditions, gas-phase O_2 molecules can serve as stabilizing ligands for the Pt_{SA}, in addition to the alumina support [291]. On γ-Al_2O_3/(110), the strongest binding site for the metal atoms correspond to a hollow site between an Al^{3+} and an oxygen atom, as observed by Valero et al. [7]. On this surface, the adsorption energies are reduced from 10 to 20 kcal/mol for Pd, Pt, Ir, Rh, and Ni while those of Ag and Au increase by the same order of magnitude. Furthermore, the adsorption energies of Sn and In on this facet are more than two times larger than those of Pt, Pd, or Ag. The strong interaction of the Sn and In atoms with the oxide surface can be explained by the strong covalent contribution to this interaction. Indeed, the 5p orbitals of the Sn and In atoms extending over wider energy ranges, which allows a better interaction with the hybrid sp orbitals of Al atoms. On γ-Al_2O_3/(111), metals bind on a hollow site formed by three oxygen atoms and three aluminum atoms. On this surface, the adsorption energies are two to three times higher than those on the two other surfaces (between −190 and −87 kcal/mol). Interestingly, on the basis of their work, Tan et al. have shown, by combining DFT calculations and a statistical learning approach, that the adsorption energy can be estimated from the gas-phase metal–oxygen binding energy and the band gap of the oxide support [272].

The influence of TM dopants on the bond length, charge state, band edge, and redox properties of γ-Al_2O_3 is also a topic of constant interest. For example, Fu et al. have shown that both the structural and electronic properties as well as the reactivity of alumina can be modified upon metal doping [292]. Indeed, both the properties and the reactivity are related to the concentration of Al vacancies, which decreases with increasing atomic number of dopant TM due to a charge compensation effect. The substitution of a surface Al atoms by Cu is also thermodynamically accessible and leads to the increase in the dehydration temperature of the γ-Al_2O_3/(110) surface [293]. In the same way, the increase of the Ce^{3+} and Cu^{2+} concentration also causes an increase of the number of emitting ions [294].

7.5.2 Cobalt Oxides

For cobalt oxide surfaces, the simplest form is the rock salt monoxide CoO for which at 300 K the Co(II) atoms are octahedrally coordinated by lattice oxygen atoms [295] but slightly tetragonally distorted below 291 K [296]. In the same way, the Co_3O_4 spinel oxide is a thermodynamically stable form of the cobalt oxide at room temperature and standard oxygen partial pressure. In this form, the tetrahedral and octahedral interstices of the O^{2-} lattice are occupied, respectively, by Co^{2+} and Co^{3+} ions [297]. In 2015, Qiao et al. have experimentally demonstrated the high activity of a Au_{SA}/Co_3O_4 SAC for CO oxidation [298]. Following this experimental work, the adsorption and catalytic activity of Au atoms on Co_3O_4 has recently been studied theoretically by Lou et al. [299]. In this work, only the (100) facet is selected as surface model, since this facet is the most exposed active plane for Co_3O_4 systems. On the (100) facet, contrarily to the (110) surface, which only exhibits Co^{3+} sites or the (111) surface, which contains only Co^{2+} sites, both Co^{3+} and Co^{2+} sites are exposed. On this surface, the Au atom is strongly coordinated to two oxygen atoms (with a binding energy of 60 kcal/mol) forming a linear configuration over the Co_3O_4 surface. This favorable interaction is possible due to 5d and 6s orbital hybridization of Au atoms. The presence of the Au atom increases the amount of active oxygen atoms, which can explain the high activity of Au_{SA}/Co_3O_4 SAC for CO oxidation (see Figure 7.16).

Indeed, at the Au_{SA}/Co_3O_4 site, the coordination of a CO molecule to Co atoms is a thermodynamically favorable reaction (E_{ads} = 6 kcal/mol). In the same way, the C–O coupling between the CO molecule and one of the activated oxygen atoms of the surface is also a kinetically accessible process (activation barrier of 7 kcal/mol). The final CO_2 release induces the formation of an oxygen vacancy (O_v), who is occupied by the Au atom, leading to the formation of two Au–Co interactions (see Figure 7.17). On this new active site, CO and O_2 molecules can be adsorbed at the Au and Co sites, respectively. The formation of a CO_2 molecule from this adduct is an almost barrierless process, as evidenced by the low activation barrier of 3 kcal/mol. Finally, the release of the CO_2 molecule leads to the regeneration of the initial linear O–Au–O configuration. Thus, the high activity of Au_{SA}/Co_3O_4 SAC for CO

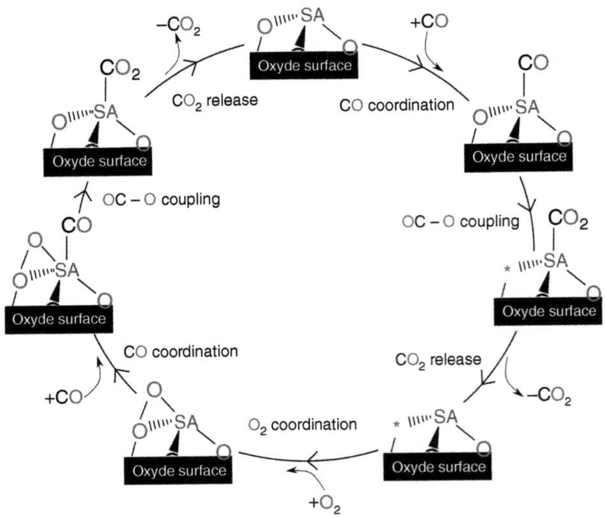

Figure 7.16 Scheme of the Mars–van Krevelen-type CO oxidation mechanism mediated by an oxide-supported single-atom catalyst. Source: I. del Rosal.

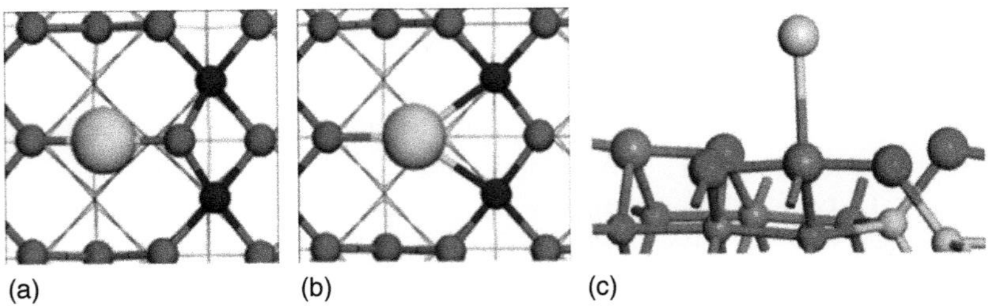

Figure 7.17 Coordination modes of Au atoms over Co_3O_4 during the CO oxidation reaction. (a) Initial coordination mode of Au atom on a (100) facet; (b) coordination mode of Au atom on a (100) facet after the first CO_2 release; (c) coordination mode of Au atom on a (110) facet. Source: I. del Rosal.

oxidation is attributed to the presence of this two highly active reaction pathways. A similar study was carried out by Yang et al. considering the coordination of a gold atom on the (110) surface rather than on the (100) surface [300]. Interestingly, the Au atom is more strongly coordinated on the (110) facet than on the (100) one ($E_{ads} = -87$ kcal/mol on the (110) facet vs. −60 kcal/mol on the (100) facet).

Unlike to the (100) facet, in the (110) one, the Au atom is coordinated in top position with respect to a Co^{3+} atom (see Figure 7.17). This coordination mode is attributed to the stronger overlap between the d orbitals of the Au and Co atoms compared to the overlap between the d orbitals of the Au atom and the p orbitals of the oxygen atoms. From a reactivity point of view, the CO oxidation takes place through an Eley–Rideal mechanism, with a CO molecule adsorbed on the Au atom and a O_2 molecule in the gas phase (see Figure 7.18). The rate-determining step of this reaction corresponds to the second step of the mechanism, i.e. the breaking

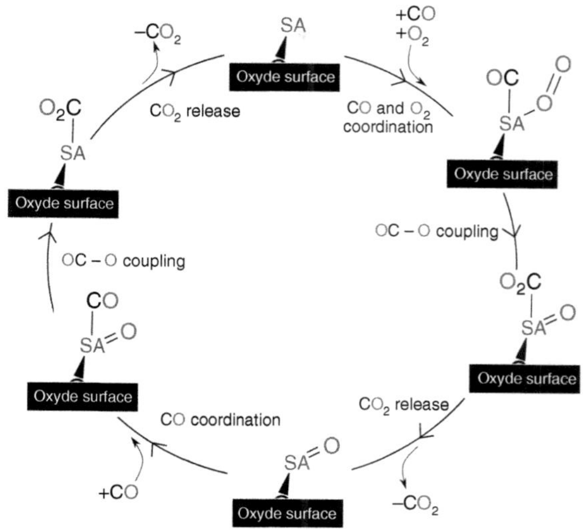

Figure 7.18 Scheme of the Eley–Rideal CO oxidation mechanism mediated by an oxide-supported single-atom catalyst. Source: I. del Rosal.

of the O–O and C–Au bonds leading to the release of the CO_2 molecule (activation barrier of 6 kcal/mol). The first step, corresponding to a C–O coupling, takes places overcoming an energy barrier of 3 kcal/mol. In the final product, an oxygen atom remains strongly coordinated to the Au atom. This oxygen can spontaneously react with a new CO molecule to form the second CO_2 molecule. After this second reaction, the catalytic cycle is completed and the initial Au_{SA}/Co_3O_4 SAC recovered. Thus, as experimentally observed [298], the CO oxidation mediated by Au_{SA}/Co_3O_4 SAC is a thermodynamically favorable and kinetically facile process independently of the facet on which the Au atom is coordinated.

7.5.3 Cerium Oxide

Among the different metal oxide support materials, cerium oxide (ceria, CeO_2) is one of the most common supports because it is a stable, abundant, and inexpensive rare earth material. CeO_2 has attracted considerable interest because of its applications in many areas, including chemical catalysis, fuel cell, and oxygen storage devices. In addition to their unique redox properties, CeO_2 can stabilize different precious metal on its surfaces, avoiding the growth and aggregation of metal particles, and thus maintaining and improving the activity of the SACs. Many experimental and theoretical investigations have reported the segregation, bonding, and migration behaviors of different metal and clusters on the ceria surface. In the same way, CeO_2 is widely considered to be not merely a support for different metals but also an energetic participant in the catalytic process: (i) ceria is able to contain numerous O_v and cerium ions, which induces an excellent oxygen diffusion, and (ii) its morphology with different preferentially exposed crystal faces turned out to be a critical factor in several reactions. CeO_2 is composed of three low-index surfaces, (111), (110), and (100) for which the surface energies increase in the order (111) > (110) > (100) (see Figure 7.19). The most stable (111) facet exposes threefold oxygen and sevenfold cerium atoms, whereas the (110) surface

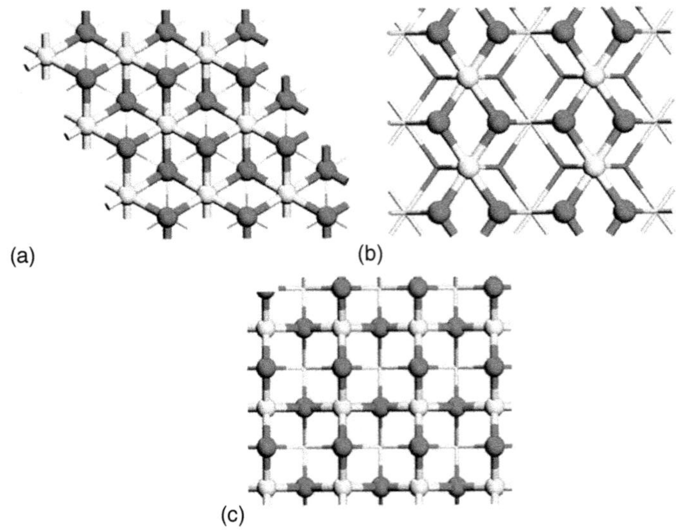

Figure 7.19 Top view of optimized surface structures of: (a) (111) CeO_2, (b) (110) CeO_2, and (c) (100) CeO_2. Source: Liu et al. [301]. Reproduced with permission of The Royal Society of Chemistry.

exposes threefold oxygen and sixfold cerium atoms and the (100) facet exposes twofold oxygen and sixfold cerium atoms. The formation of SACs on CeO_2(111) single crystals has been the subject of a high number of experimental and theoretical work [302]. However, much less information is available on the other surfaces [303].

One of the pioneering (theoretical) works on this topic concerns the coordination of the Pt atom on the (111) ceria surface [304]. The adsorption of Pt on both an unreduced and reduced CeO_2 surface shows that the adsorption on the reduced surface is stronger than on the unreduced one. Therefore, the O_v made the interaction between the Pt and ceria much stronger (over than 36 kcal/mol) than on the unreduced surface. In the same way, the presence of the Pt atom reduces the energy needed to create an O_v by as much as 8 kcal/mol. Interestingly, the DOS of the Pt atom on the reduced ceria has a very high main peak close to the Fermi level, which is much sharper and narrower than compared with that of a Pt on the unreduced surface. This metal induced gap states allow accommodating the additional electrons from the formation of the O_v, which facilitates the reduction. Several years later, Hatanka et al. have shown, by studying the coordination of a Pt atom on the (001), (011), and (111) CeO_2 facets, that if the ceria surface is equally composed of (001), (011), and (111) planes, five cerium atoms interact on average with Pt to form Pt anchoring sites [305, 306]. In 2016, Dvořák et al. showed that, on the (111) surface, the preferential Pt atoms adsorption sites are located on the step edges present on this facet at temperatures lower than 1000 K [307]. On these sites, the Pt atom interacts with four lattice oxygen atoms in a highly stable PtO_4 planar configuration. This study also excludes the stabilization of Pt atoms on the O_v, contrary to what had been postulated by Yang et al., the Pt coordination on the step edges being 50 kcal/mol more stable than on O_V [304]. More recently, Tang et al. performed the study of the coordination and catalytic activity of a Pt_{SA} catalysts dispersed on (111), (110), and (100) CeO_2 surfaces [308]. More precisely, the substitution of one cerium atom of each surface by a Pt has been considered. The coordination of the Pt atom is favorable on the three ceria surfaces from a thermodynamic point of view, with adsorption energies between 126 and 132 kcal/mol. On the (111) surface, the Pt atom takes the place of a Ce^{4+} atom, and its oxidation state can be reduced from IV to II in the presence of an O_V. On the (110) surface, a fourfold Pt atom is formed with nearly planar local structure, which implies a strong distortion of the coordination site due to the reduction of the Pt from a Pt^{4+} to Pt^{2+} by the oxidation of two surface oxygen atoms. Consequently, Pt substitution on ceria (110) surface is more stable than on ceria (111) and (100). Interestingly, this planar structure is similar to the structure proposed by Dvořák et al. on the (111) step edges [307]. From a reactivity view point, the CO oxidation mediated by the Pt atoms coordinated on the three surfaces has been considered. In all the cases, the reaction takes place thought a Mars–van Krevelen-type mechanism (see Figure 7.16). Thus, the reaction starts by the coordination of one CO molecule to the SAC. The CO can attack one of the oxygen atoms interacting with the SAC to produce the first CO_2 molecule. After CO_2 desorption, one O_V is formed and a O_2 molecule coordinates onto this vacancy. Finally, the second CO molecule comes to react with the adsorbed O_2, leading to the formation of the second CO_2 and regenerating the SAC. Based on calculated thermodynamic and kinetic criteria, the order of reactivity follows $Pt/CeO_2(110) < Pt/CeO_2(111) < Pt/CeO_2(100)$. This result is in good agreement with experiments, in which the polyhedral ceria exhibiting (111) and (100) facets and the nanorods exposing dominant (111) facets show similar (high) performance for CO oxidation, while the cube ceria exposing only (110) facets has the lowest activity [309]. On the other hand, considering the coordination of a Pt atom on the (111), (110), and the polar (100) ceria facets, and not the substitution of one cerium atom by a Pt, only the (100) surface can stabilize isolated Pt atoms [310]. These results agree with previous findings showing that on the (111) and (110) surfaces, NPs are formed instead of SAs [309, 311]. Interestingly, on the (100) surface, a high diversity of coordination modes are possible for the Pt atoms surrounded by a variable number of oxygen atoms between two and four. This is possible due to the low coordination of the surface oxygen atoms, which increases their diffusion [312]. In their work, Daelman et al. also showed that some oxidation and charge states might coexist due to a dynamic charge transfer between the metal and the oxide. The reactivity is also closely related to this dynamic behavior. The CO oxidation can take place on a Pt-4O site following a classical Mars–van Krevelen-type mechanism (see Figure 7.16). On this site, the reaction is kinetically accessible with the highest activation barrier around 23 kcal/mol (corresponding to the C–O coupling) and thermodynamically favorable (the final state is 28 kcal/mol below reactants). In this case, the formation of an O_V during the CO oxidation process can also lead to the formation of a Pt-2O site, thanks to the facile oxygen diffusion. On this site, the CO adsorption is relatively weak on Pt^0 but stronger on Pt^+ and Pt^{2+} (1 vs. 19 kcal/mol). Then, O_2 can adsorb, forming a percarbonate species with an increase of the O–O bond distance. Here again, the O_2 adsorption energy depends on the oxidation state of the Pt atom, with an energy difference that can be as large as 28 kcal/mol. Finally, the CO_2 formation take place, overcoming an activation barrier of 13 kcal/mol with the low-lying Pt(0) state and 26 kcal/mol with a Pt(I). Thus, on a Pt-2O site, the electronic structure of the material needs to be dynamic to allow both the adsorption of CO and O_2 (occurring on Pt^+ or Pt^{2+}) and the C–O coupling (occurring at the neutral state) steps.

As for Pt, the addition of Zr to ceria by the substitution of a cerium lattice position by a Zr atom alters the structural and electronic properties of CeO_2 [313]. From a structural point of view, the substitution of Zr into the ceria lattice results in a slight distortion of the local metal coordination environment independently of the considered surface ((100), (110), and (111)) due to the shorter metal–oxygen bonds around the Zr atom. The vacancy formation energy for Zr substituted ceria surfaces is lowered by, respectively, 10, 20, and 26 kcal/mol on the (100), (110), and (111) surface with respect to pure ceria but remains in all the cases endothermic. On these systems, the preferred adsorption configuration for dissociated methane is consistent with pure ceria surfaces, with both the methyl group and the H atom adsorbed on surface oxygen atoms coordinated to the Zr atom (see Figure 7.20). The adsorption energies are exothermic independently of the considered surface. Likewise, this reaction is between 4 and 9 kcal/mol more exothermic, depending on the exposed facet, for surfaces containing Zr than on pure ceria. DOS data suggest that for both the dissociation of methane and the vacancy formation, the reduction process results in the reduction of cerium atoms instead of Zr atoms. Thus, the incorporation of Zr atoms in the ceria lattice lowers the energy of O_v formation, as well as the reaction energy for the methane dissociative adsorption. Interestingly, the CO oxidation mediated by a (111) ceria surface in which one cerium has been substituted by a Zr is a kinetically accessible (activation barrier of 15 kcal/mol) and thermodynamically favorable reaction by 12 kcal/mol [314]. However, the same reaction on a (110) surface is endothermic by 72 kcal/mol. On this crystal surface, the binding energies of CO are rather high, leading to the poisoning of the surface.

Depending on the catalyst preparation, pretreatment, exposed facets, and reaction environment, Pd atoms can exist on the ceria surface with different local environments and oxidation states, each with particular catalytic properties. For example, Pd^{4+} atoms adopts an octahedral oxygen coordination environment characteristic of the d^6 metal centers, while Pd^{2+}, as the d^8 metal centers, prefer a square planar geometry. As in the case of Zr, Mayernick and Janik have considered the substitution of a cerium lattice position by a Pd atom over (111), (110), and (100) surfaces and analyze the impact of this substitution on the O_v formation and methane adsorption [313]. The Pd substitution into the (111) surface of ceria leads to a notable distortion of the surface structure around the Pd, especially when an O_v is formed. This distortion is due to the formation of a square planar ligand coordination around the Pd

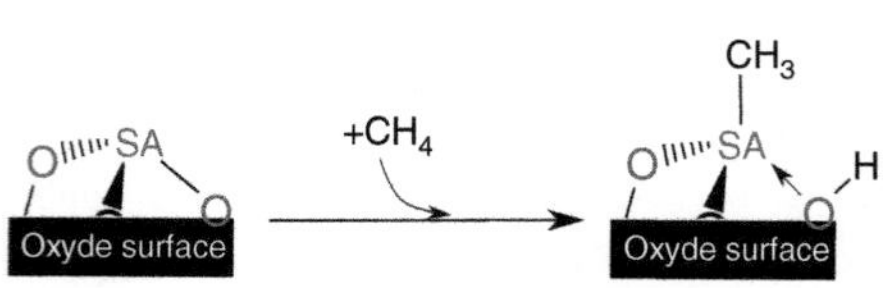

Figure 7.20 Scheme of the methane dissociative adsorption on an oxide-supported single-atom catalyst. Source: I. del Rosal.

atom, consistent with the coordination environment of a Pd^{2+} atom. For the (110) and (100) surfaces, no surface specific distortions around the Pd center are observed for intact surfaces. The formation of an O_v does not result in any significant distortion, since in both cases the Pd atom is in a four-coordinate environment. The vacancy formation energy for Pd-substituted ceria surfaces is lowered by, respectively, 53, 51, and 47 kcal/mol on the (100), (110), and (111) surface with respect to pure ceria. Thus, for Pd/CeO_2 surfaces, as for pure ceria, O_v formation is more exothermic for the (110) surface, followed by the (100) and endothermic for the (111) surface. The preferred adsorption configuration for dissociated methane involves the methyl and hydrogen adsorption to surface oxygen atoms coordinated to Pd (see Figure 7.20). The methane dissociation is an exothermic process on the three facets by more than 12 kcal/mol with respect to pure CeO_2. DOS analysis confirms the reduction of Pd as a result of the methane dissociation. Thus, the incorporation of Pd atoms in the ceria lattice lowers the energy of O_v formation, as well as the reaction energy for the methane dissociative adsorption. This study was completed a year later by the same authors in considering, on the (100), (110), and (111) crystal surfaces, Pd_{SA} as adsorbed onto the surface, substituted into the surface at Ce sites, and inserted into void spaces underneath the first ceria layer [315]. On the (111) surface, the Pd incorporation lead to the formation of a mixed surface oxide, which is expected to be the most stable form at room temperature and standard O_2 pressure. The coordination of the Pd_{SA} at the ceria surface is only possible at high temperature (973 K) and below 10^7 atm of O_2. On the (110) surface, the incorporation of the Pd_{SA} into the surface is possible at low oxygen pressure (below 10^{-5} atm) and high temperature (973 K). However, at room temperature and standard O_2 pressure, the Pd_{SA} remains at the ceria surface. Finally, on the (100) surface, the Pd_{SA} remains at the ceria surface independently of the considered temperature. The incorporation of the single atom on the ceria surface will only take place at oxygen pressures above 10^{13} atm and at 973 K. These results suggest that the surface of ceria (100) is the one that will most favor the formation of Pd SACs. The CO oxidation mediated by Pd SACs located on or in the (110) ceria surface was studied latter by Song et al. [316]. According to this work, the most stable adsorption site for a Pd_{SA} on the (110) ceria surface is a fourfold hollow site in which the Pd^{2+} cation is connected to four surface oxygen atoms. The adsorption energy on this site is −69 kcal/mol, i.e. 28 kcal/mol more stable than the coordination site considered by Mayernick and Janik [315], in which the Pd was coordinated in a bridging fashion between two surface oxygen atoms. On this fourfold Pd atom, the CO oxidation takes places through a classical Mars–van Krevelen-type mechanism (see Figure 7.16). The reaction is thermodynamically favorable by 47 kcal/mol for the formation of the first CO_2 molecule and by 47 kcal/mol for the second one. From a kinetic point of view, the rate-determining step corresponds to C–O coupling (activation barrier of 22 kcal/mol) for the first CO_2 formation and to the O–O activation (activation barrier of 12 kcal/mol) for the formation of the second CO_2 molecule. Interestingly, during the CO oxidation process, the adsorption of the first CO molecule leads to the migration of the Pd_{SA} to a bridge site between two oxygen atoms, and the migration of the Pd_{SA} back to the fourfold site only occurs after the release of the second CO_2 molecule. The CO oxidation mediated by (110) ceria surface in which one cerium atom is substituted by a Pd_{SA} has also been studied. In this case, the formation of the first CO_2 molecule is a facile reaction both from a kinetic and thermodynamic point of view. However, after the release of the CO_2 molecule, the presence of the O_v induces a reorganization of the ceria surface leading to the migration of the Pd_{SA} into the surface, and consequently to the deactivation of the catalyst. The methane activation mediated by a Pd_{SA} incorporated on (111) and (110) ceria surfaces has also been studied [317]. In complement to the study of Mayernick and Janik [313], Senftle et al. have shown that on (111) and (110) crystal surfaces, the formation of Pd^{2+} species is thermodynamically more stable under standard reaction conditions. Nevertheless, a lower barrier is obtained for the C–H bond activation with a Pd^{4+}, despite requiring an endergonic oxidation step prior to activation. This difference is due to the fact that methane activation induces the transfer of one hydrogen to a surface oxygen atom coordinated to the Pd_{SA}. With Pd^{4+}, the ion plays the role of reduction center while over the thermodynamically preferred Pd^{2+} cation, this role is played by a Ce^{4+}. Thus, the formation of Pd^{4+} plays a crucial role in CH_4 activation over Pd-doped ceria.

Both the stability and catalytic activity of Rh_{SA} on (111) CeO_2 surface were recently investigated by Amsler et al. [318]. In this case, the SAC is formed by the coordination of a $HRh(CO)_4$ complex on top of the ceria surfaces or at edges, where the Rh_{SA} is coordinated to one or two oxygen atoms, respectively. On flat (111) CeO_2 only one CO molecule is released, but for surfaces containing steps, such as (211) and (221) CeO_2, the final Rh system contains only two CO molecules ($HRh(CO)_2/CeO_2$). On ceria, the highest adsorption free energy was found for the stepped (211) and (221) surfaces, −18 and −16 kcal/mol, respectively. This is closer than two times the adsorption energy on flat (111) and (100) CeO_2 surfaces, 10 and 10 kcal/mol, respectively. Thus, Rh_{SA} are expected to adsorb only on the corrugated surfaces. The gas-phase hydroformylation of ethylene mediated by Rh/(111) CeO_2 was also studied in this work, the reaction taking place through the well-accepted mechanism proposed by Heck and Breslow (see Figure 7.21) [319, 320]. The mechanism starts by the coordination of an olefin to the metal center, followed by the insertion of the H–metal bond. Then, an additional CO molecule is coordinated to the metal center and subsequently inserted onto the olefin–metal bond. Finally, an oxidative addition of H_2 occurs, followed by the reductive elimination of the resulting aldehyde. On Rh/(111) CeO_2, the reaction is thermodynamically favorable by around 12 kcal/mol. The rate-determining step corresponds to the CO insertion into the Rh–C bond with an activation barrier of 19 kcal/mol. Thus, the Rh SAC is highly active for olefin hydroformylation, showing a catalytic activity comparable to that of molecular catalyst, while being stable toward Rh leaching in solution. More recently, the use of a one-pot combination of Ru_{SA}/CeO_2 and Rh_{SA}/CeO_2 SAC has been considered for the olefin isomerization-hydrosilylation through a tandem process [321]. The goal in this work is, as reported with homogeneous complexes [322], to integrate the first catalyst (Ru_{SA}/CeO_2) in order to do the olefin double-bond migration and the second one (Rh_{SA}/CeO_2) for the hydrosilylation of the olefin, all in a single-reaction medium. In that purpose, (211) CeO_2 surface has been considered by Dvořák et al. [307], for the Rh and Ru coordination, i.e. a step edge that connects

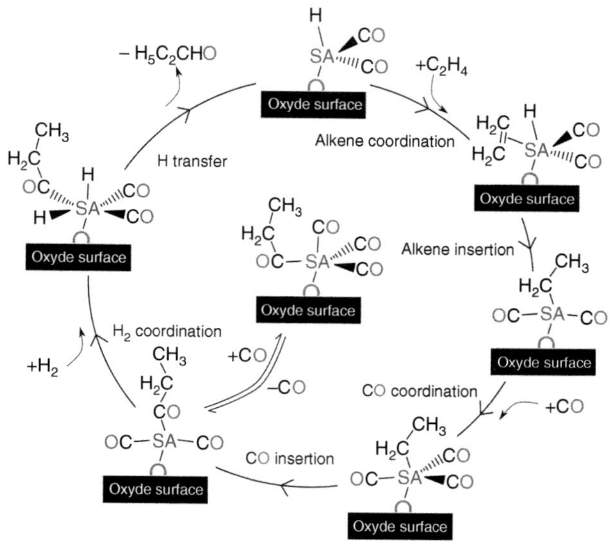

Figure 7.21 Scheme of the ethylene hydroformylation mediated by an oxide-supported single-atom catalyst. Source: I. del Rosal.

two (111) facets of ceria. In both cases, the SAC adopts a similar coordination mode, with the metal center in a bridging position between two surface oxygen atoms.

Along with experimental results, Ru SAC is more active than the Rh counterpart for the olefin isomerization. The activation barrier with Ru_{SA}/CeO_2 (9 kcal/mol) is 8 kcal/mol lower than that computed with the Rh SAC. This difference is attributed to the stronger olefin coordination on the Ru SAC than on the Rh SAC. Interestingly, the oxidative addition of Et_3SiH to the SAC, prior to the olefin isomerization, is fundamental due to a decreasing of more than 7 kcal/mol on the activation barriers. The second half of the tandem reaction, i.e. the hydrosilylation of the resulting olefin was also investigated with only Rh_{SA}/CeO_2. The reaction takes place through a Chalk–Harrod mechanism, in which the resting state corresponds to an energetically feasible Rh–H species (see Figure 7.22) [323]. The reaction starts by both the oxidative addition of Et_3SiH and the coordination of the olefin to Rh_{SA}–H/CeO_2, and continuing by the olefin insertion to the Rh–H bond. The last step corresponds to the alkane insertion to the Rh–Si bond, leading to the formation and releasing of the final product. In this case, the rate-determining step corresponds to the olefin insertion, with an activation barrier of 20 kcal/mol. The formation of the final product is thermodynamically favorable by 15 kcal/mol with respect to the entrance channel. These results are in agreement with the experimental observed high activity (reaction yield of 70%) and selectivity (greater than 92% for the terminal organosilane) of the 2-octene isomerization/hydrosilylation using the one-pot combination of 1.0 Ru_{SA}/CeO_2 and 1.0 Rh_{SA}/CeO_2 catalysts.

From an experimental point of view, several works have shown that highly dispersed Au atoms supported on metal oxides, such as ceria, are active catalysts for several reactions including CO oxidation [324–326]. In this way, the presence

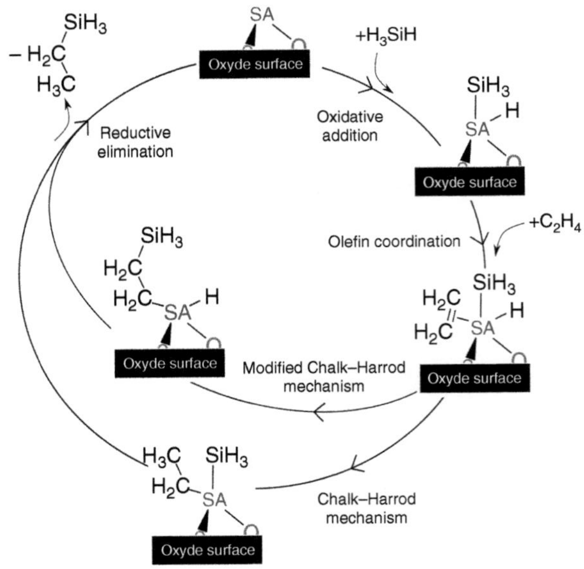

Figure 7.22 Scheme of the olefin isomerization-hydrosilylation mechanism mediated by an oxide-supported single-atom catalyst. Source: I. del Rosal.

of Au^{3+} atoms seems to be crucial to explain the high activity of this system, with a direct correlation between the concentration of Au^{3+} species and the catalytic activity of Au/CeO_2. In order to shed light on this experimental results, the CO oxidation mediated by Au/(111) CeO_2 has been studied by Shapovalov and Metiu [327]. More precisely, special attention has been paid to both the CO oxidation reaction pathway and the formation and stability of both carbonate intermediates and of the O_v. The results obtained with Au/(111) CeO_2 were then compared with those obtained with the pure oxide surface or Ag and Cu doped, i.e. Ag/(111) CeO_2 and Cu/(111) CeO_2. The formation of the Au SAC takes place by the substitution of one-surface cerium atoms by an Au one. The CO oxidation induces the exothermic formation of several carbonates by reacting with different pairs of oxygen atoms around the Au atom. From this reaction two carbonate populations are formed, the first one seems to be very stable and will be present on the surface during the catalytic reaction. The second one easily decomposes to produce CO_2 and form an O_v on the surface. Following a Mars–van Krevelen-type mechanism (see Figure 7.16), an O_2 molecule adsorbs onto the O_v and reacts with a new CO to form CO_2, recovering the active site. Interestingly, the energy needed to form one O_v follows the trend: Cu/(111) CeO_2 > Ag/(111) CeO_2 > Au/(111) CeO_2. This difference is explained by the fact that if a surface cerium atom is replaced by a metal atom with a lower valence, this creates an electron deficit in its neighborhood. Consequently, the bond between the oxide and the oxygen atoms in the neighborhood of the substituted metal atom is highly activated. This effect will be increased if the structure of the dopant oxide has a structure different from that of the host. The coordination of the same metal atoms on a pristine (111) CeO_2 surface was studied by Branda et al., predicting the coordination of both Ag and Cu atoms on a threefold site formed by three O atoms of the first ceria layer [328]. On this site, the Cu atom is more strongly coordinated than Ag (E_{ads} = −43 kcal/mol for Cu vs. 23 kcal/mol for Ag) but in both cases Cu^+ and Ag^+ species would be formed by the coordination on the ceria surface. For Au atom, according to the calculation method, Au prefers to adsorb on a top site, i.e. coordinated to only one oxygen surface atom or in a twofold site. This preference for a top coordination is in good agreement with similar calculations on the same crystal surface or on the (110) facet [329, 330]. Concerning the oxidation degree of the metal center, according to the calculation method, a neutral Au atom or an oxidized Au^+ can be obtained. In the same way, Castellani et al. predicted that upon adsorption on the regular (111) ceria surface, Au remains neutral [330]. However, a study conducted by Hernández et al. suggests that Au atoms at the (111) ceria surface can adopt different oxidation degrees according to the adsorption site: (i) on a top or bridge coordination sites an Au^+ is formed; (ii) on the hollow sites the Au atom remains neutral; and (iii) adsorption at an O_v site induces the formation Au^- atoms [331]. Thus, probably a dynamic distribution of neutral Au and Au^+ atoms takes place on the ceria surface [332]. The thermodynamic, structural, and electronic properties of Cu_{SA} on a (111) CeO_2 surface were also investigated by Szabová et al. [333].

The coordination of a Cu atom on the pristine ceria surface has also been considered. In this case, the most stable coordination mode (by 70 kcal/mol) is on a hollow site formed by three oxygen surface atoms. The calculated thermodynamics shows that adsorption on surface O_v is disfavored with respect to the coordination on the pristine surface. These results are in opposition to those obtained with Au atoms, for which Au adsorption at the cerium or O_v sites is strongly preferred to that on the pristine surface [329, 334–336]. A complementary study to that of Shapovalov and Metiu [327], concerning the CO oxidation reaction mediated by an Au atom adsorbed on a cerium vacancy site or on a stoichiometric (111) CeO_2 surface, was carried out by Camellone [9]. In addition to the study about the stability of the different intermediates involved in the CO oxidation reaction, the transition states connecting these intermediates were also located and analyzed in this work. On the Au_{SA}, the CO coordination induces the spontaneous diffusion of the metal center from bridge to top site. Then, the CO oxidation takes place through the tilted CO molecule followed by the C–O coupling with a lattice oxygen atom, and finally the CO_2 desorption with the concomitant diffusion of the Au atom into the formed O_v. The rate-limiting step of the overall reaction, with an activation energy of 20 kcal/mol, corresponds to the first step, i.e. the tilting of the CO molecule to bring it closer to the surface. Interestingly, the formation of the final product, in which the Au atom migrates from the O_v, induces the formation of an $Au^{\delta-}$ atom which does not allow the adsorption of other molecules neither CO nor O_2, essential for the catalyst regeneration. Consequently, after the first CO_2 formation, the catalyst deactivates. The substitution of one cerium atom by an Au one leads to a more active catalyst. These species can sustain a catalytic cycle in which the oxidation state of the Au atoms remains positive and can promote multiple CO oxidation without deactivating. In the case of Au_{SA} coordinated to a corrugated ceria surface such as the stepped (331) [337] or (110) facets [338], the rate-determining step corresponds to the C–O coupling, which requires a barrier of 19 kcal/mol. Thus, this reaction is competitive with respect to the same one with the Au atom coordinated on the (111) surface. However, from a kinetic point of view, the CO_2 desorption is a thermo-neutral reaction on the (331) surface (2 kcal/mol) but endothermic on the (111) surface (28 kcal/mol).

7.5.4 Magnesium Oxides

MgO, often considered as a prototype, is among the most studied metal oxides due to its simple geometric (Mg^{2+} and O^{2-} ions are octahedral) and electronic structure (7–8 eV band gap). As a compound with a rock salt structure, MgO typically exposes the (100), (110), and (111) facets (see Figure 7.23). The (100) and (110) surfaces are both neutral and type 1 in Tasker's classification [339] (neutral with equal numbers of anions and cations on each plane), whereas the (111) face is a type 3 surface (charged and has a dipole moment in the repeat unit perpendicular to the surface). In MgO, the surface energy of the (100) surface is much lower than other surfaces. Interestingly, unlike most of the currently used oxide supports, MgO is a barely non-reducible support [340]. Consequently, due to this limited redox chemistry, MgO supports allow addressing the intrinsic redox catalytic properties of the supported metal atoms. On the (100) surface, the metal interacts in a rather weak way with the acidic sites (above a Mg^{2+} cation) but a moderately large interaction is predicted above the basic sites (above an O^{2-} anion) [341]. In the specific case of Pd, this difference is due to the better overlap between the oxygen p orbitals and the Pd metal orbitals, which enhances the metal–oxygen bonding [342]. In general, the mixing of the d orbitals or the hybridized s–d orbitals with the oxygen 2p band leads to the formation of a covalent polar bond of moderate strength. However, if the

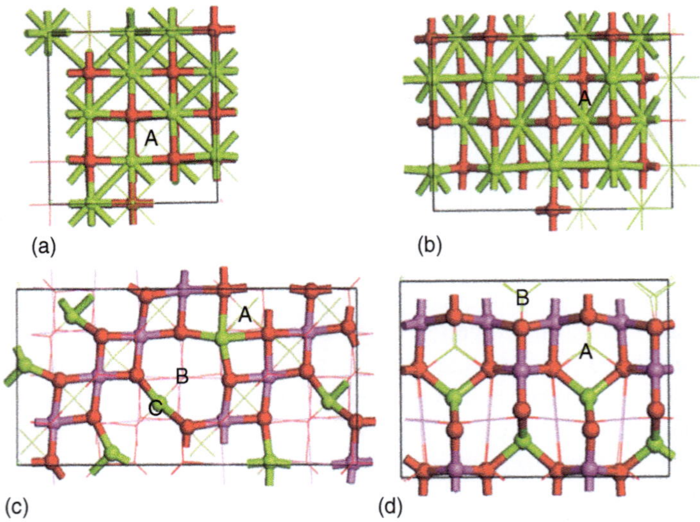

Figure 7.23 Chosen sites on the lowest energy termination of: (a) (100) MgO, (b) (110) MgO, (c) (100) $MgAl_2O_4$, and (d) (110) $MgAl_2O_4$. Sites are indicated with capital letters. Only top layers are shown by ball and stick, and the atoms in the subsurface are shown by a wireframe. Green atoms represent Mg, pink atoms Al, and red atoms O. Source: Tan et al. [272]. Reproduced with permission of the American Chemical Society.

metal binds via the s electrons, the interaction between the metal center and the support will be weak mainly due to polarization interactions and dispersion effects with little mixing with the support orbitals. Thus, according to several theoretical investigations, the coordination of the metal center above the basic sites of the (100) MgO surface, with an adsorption energy between 5 and 9 kcal/mol, is weak or very weak for Au [343], Ag [5, 343–346], Cu [343, 347, 348], Cr [343], and Mo [343] and, with an adsorption energy between 10 and 35 kcal/mol, is relatively strong for Ni [341, 343, 348], Pd [341, 343, 349], Pt [341, 343], W [343], and Fe [350]. Interestingly, all of these calculated adsorption energies, even for the strongest ones, are considerably smaller than the metal–metal bond strength in the corresponding bulk materials. Consequently, on this pristine surface, the cluster formation should probably be preferred over the SAC one.

However, due to the variety of conditions used for the MgO synthesis, several surface topological defects can be formed at the oxide surface, such as mono and diatomic steps, providing a rough surface topology which is expected to be favorable for the metal SA coordination/reactivity [351]. The adsorption energy of a metal center on the (100) MgO surface is considerably enhanced by the presence of defects [352]. In general, the electrons can be trapped on the vacancies formed by removing an oxygen atom from the MgO lattice surface. As a consequence, if the vacancy is filled by 2, 1, or 0 electrons the defect is called respectively F^0, F^+, or F^{2+}. The F^+ center is paramagnetic, whereas F^0 and F^{2+} centers are diamagnetic. For Cu, Ni, Ag, and Pd atoms, adsorption is found to be stronger on F sites by 23 to 55 kcal/mol compared with the coordination on basic sites [353]. On the F^+ site, the coordination of Cu and Ag atoms is stronger than on the F sites due to the coupling of the unpaired electron of the vacancy with the single valence electron of metal centers. However, for Ni and Pd, the coordination on the F^+ site is less favorable (by 12 and 30 kcal/mol, respectively) than on the F sites. For Pd, the presence of an F^{2+} vacancy stabilizes the adsorption energy by 53 kcal/mol with respect to the pristine (100) MgO surface [352]. In a step, the binding energy of Au atoms is 29 kcal/mol, around 40% higher than on the (100) MgO surface [354]. The adsorption energies on F^0 and F^+ centers of MgO become 73 and 92 kcal/mol, respectively, i.e. more than four times larger than on the regular sites. On F^0, this strong covalent polar bond is formed due to the easy polarizability of the electron density inside the cavity that can be transferred to the Au atom. On F^+, as for Cu and Ag atoms, the important adsorption energy can be explained by the coupling of the unpaired electron in the O_v with that of the Au atom, leading to the formation of a two-center two-electron bond. In the case of Ca atoms [355], they adsorb directly above the oxygen atom, with an adsorption energy of 19 kcal/mol. However, the calculated binding at both neutral and charged O_v is less favorable by 11 and 2 kcal/mol, respectively. The adsorption on O_v at steps are also weaker (30 kcal/mol on an F^0 site and 44 kcal/mol on an F^+ site) than on a regular step site (48 kcal/mol). The strong preference of Sn coordination on the basic site (44 kcal/mol) over the acidic one (12 kcal/mol) is in line with the aforementioned results made for TM atoms on (100) MgO surface [356]. The adsorption values for all sites vary in the order $F^0 < F^+ < Sn/O^{2-} < F^{2+}$, indicating from a thermodynamic view point the preference for the formation of the Sn/F^0 structure

(E_{ads} = 70 kcal/mol). Interestingly, the adsorption of Sn at the F^0 defect is more exothermic than the adsorption of all first-row TMs, except Ni, at this same site [357].

In the structure of magnesium aluminate spinel, $MgAl_2O_4$, the O^{2-} ions are considered to form a face-centered array, Mg^{2+} and Al^{3+} occupy the 1/8 tetrahedral and 1/2 octahedral positions, respectively. A common defect in spinel is the cation exchange, known as the anti-site defect, which has been shown to be enhanced by SA coordination [358]. This defect leads to an increase in the system disorder because certain tetrahedral sites will then be occupied by Al^{3+} ions and, conversely, octahedral sites will be occupied by Mg^{2+} ions. For the (100) surface, two termination modes can be obtained: a fully Mg-terminated surface or a O–Al–O termination exposing principally oxygen and aluminum atoms arranged diagonally between each other. Only the O–Al–O termination has been considered by Tan et al. in their work. On this surface, due to its low symmetry, the adsorption behavior of the different metals differs: Au, Mg, and Cu prefer a bridging coordination between two Mg atoms, whereas Pd, Ni, Rh, Ir, and Pt binds in hollow sites formed by Mg, Al, and O atoms. In these sites, the DFT-calculated metal adsorption energies are between those calculated on the (100) and (110) facets of MgO: −90 kcal/mol for Pt < Ir < Ni < Rh < Au < Cu < −28 kcal/mol for Ag. Along the (110) orientation, the atomic stacking sequence is $MgAlO_2$–AlO_2, also leading to two terminations: $MgAlO_2$ and AlO_2, albeit only the latter is considered in this work. On this facet, a clear preference for a bridging coordination between two nearby oxygen atoms on the surface is observed independently of the considered metal. DFT-calculated metal adsorption energies are between −169 kcal/mol for Ir and −72 kcal/mol for Au. Thus, for every metal, the adsorption is stronger on $MgAl_2O_4$ than on MgO (see Figure 7.23). This energy difference can be explained as a consequence of the surface restructuring induced by the metal adsorption on the (110) $MgAl_2O_4$ surface. More recently, both the stability and catalytic activity of Ru, Rh, Pd, Ir, and Pt dispersed at flat (001) and stepped MgO surface were investigated by Sarma et al. [359]. In this work, the (301) MgO surface has been used because it provides an extended (001) facet as well as a step edge. More recently, the coordination of Au, Cu, Ag, Pt, Pd, Ni, Rh, and Ir on several sites of the (100) and (110) facets of MgO and $MgAl_2O_4$ has been studied by Tan et al. [272]. The study of the (111) facet of these oxides was not performed in this work due to the aforementioned instability of the facet. In agreement with previous theoretical results, except for Rh and Ir atoms, which prefers to bind on a hollow site formed by both two oxygen and two magnesium atoms, all the SACs adsorb directly above an oxygen atom. However, in this case, the adsorption energies calculated by DFT are greater than the previously calculated ones with values ranging from −16 kcal/mol for Ag to −71 kcal/mol for Pt (Pt < Ir < Rh < Ni < Pd < Au < Cu < Ag). On the (110) facet, except for Pt that prefers to coordinate above an oxygen atom, the rest of the metal centers prefer to bind on a bridging position between two oxygen atoms. The DFT-calculated metal adsorption energy is exacerbated on this facet with values between −109 and −48 kcal/mol following the order: Ir < Pt < Ni < Rh < Cu < Pd < Au < Ag.

According to DFT results, in the absence of carbon oxide molecules, the preferential coordination mode for the isolated metal atoms on the (301) MgO surface is, regardless of the metal identity, on a high-confinement octahedral coordination lattice position below the MgO step edge. However, when CO is introduced, the metal atom remains at the surface with a CO molecule coordinated on it. For Pd and Pt, the most stable oxidation state is +II and, as expected for a d^8 metal, the system adopts a square planar coordination geometry. For Rh, Ir, and Ru atoms, the most stable oxidation state is +IV. The three atoms adopt an octahedral geometry, with four oxygen atoms of the surface in the equatorial positions and one oxygen atom of the second MgO layer and the CO molecule on the axial positions. The CO oxidation mechanism mediated by Pt_{SA}/MgO and Pd_{SA}/MgO SAC was also been studied. For Pt, the first CO_2 formation takes place by the coordination of an O_2 molecule to the metal center followed by the coupling of one of the oxygen atoms of O_2 with the CO molecule (see Figure 7.24). This step corresponds to the rate-determining step with an activation barrier of 20 kcal/mol close to the experimental value of 24 kcal/mol. A Mars–van-Krevelen mechanism (see Figure 7.16) has been also tested, but in this case the CO coupling with one of the oxygen atom of the MgO surface, with an activation barrier of 45 kcal/mol, is not competitive from a kinetic point of view. Comparatively, similar reaction energy barriers were obtained with the Pd SAC. In addition to this work, the stability and catalytic activity of a $HRh(CO)_4$ complex dispersed on the same (301) MgO surface were investigated by Amsler et al. [318]. While the ceria-supported Rh catalyst is found to be highly active, this is not the case for the Rh/MgO systems. On this occasion the SAC is formed, as on ceria, by the coordination of a $HRh(CO)_4$ complex to the MgO surface on top of the surfaces or at edges, where the Rh atom is coordinated to one or two oxygen, respectively. On flat (100) MgO surfaces, only one CO molecule is released, but for surfaces containing steps, such as (301) MgO, the final Rh system contains only two CO molecules ($HRh(CO)_2$/MgO). The stronger coordination is observed on stepped (301) MgO surface with an adsorption free energy over −30 kcal/mol. This is more than three times

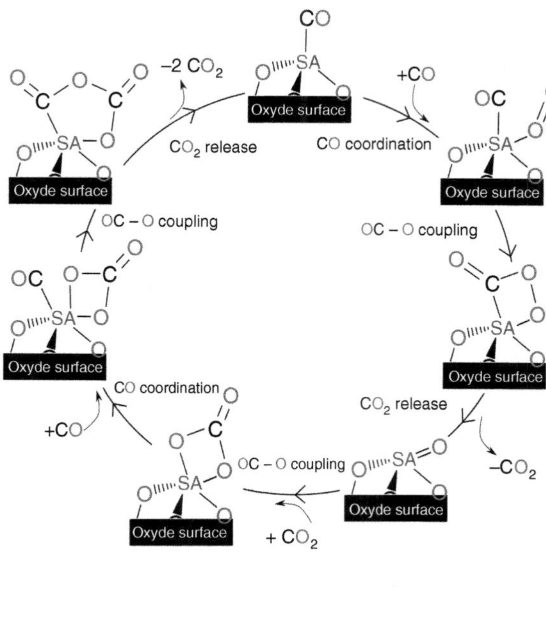

Figure 7.24 Scheme of the carbonate mediated CO-oxidation mechanism on an oxide-supported single-atom catalyst. Source: I. del Rosal.

the adsorption energy on flat (100) MgO (−10 kcal/mol). This adsorption energy is also higher than on the (111), (211), (221), and (100) ceria surfaces by more than 12 kcal/mol. Thus, Rh atoms are expected to adsorb only on the corrugated surfaces. The gas-phase hydroformylation of ethylene mediated by Rh/(100) MgO and Rh/(301) MgO was also studied by Amsler et al. The reaction takes places, as on Rh_{SA}/CeO_2, through the mechanism proposed by Heck and Breslow. A comparison between Rh/(100) MgO and molecular Rh-based catalysts shows that the stability of the different intermediates and the reaction barriers are similar, with the main difference being the high stability of the final aldehyde product. This difference is due to an attractive interaction between the carbonyl oxygen of the aldehyde and the support. Concerning the Rh/(301) MgO system, this complex is clearly most stable but also less active. On this system, the rate-determining step corresponds to the CO insertion onto the Rh–C bond, but in this case the activation barrier is 10 kcal/mol higher than for Rh/(100) MgO (20 vs. 29 kcal/mol). Thus, the bicoordination mode of the Rh complex on the MgO surface increases the stability of the system but also reduces the activity by limiting the flexibility of the catalyst.

7.5.5 Titanium Dioxide

Among the group of compounds of technological important belonging to the family of metal oxides, titanium dioxide occupies a privileged position as one of the most studied. TiO_2 is used in a wide range of applications, for example on self-cleaning and antimicrobial surfaces, food colorants, cosmetics, water or air purification, gas detection devices or as photocatalytic materials, as they have great potential to use solar energy. TiO_2 is naturally present in three main phases (see Figure 7.25): anatase (tetragonal, a = b = 3.782 Å, c = 9.502 Å), brookite (orthorhombic, a = 5.135 Å, b = 5.436 Å, c = 9.166 Å), and rutile (tetragonal, a = b = 4.584 Å, c = 2.953 Å). The most abundant phase in the natural state is rutile, which is the stable state at ordinary pressure and temperatures. The thermodynamically metastable phases,

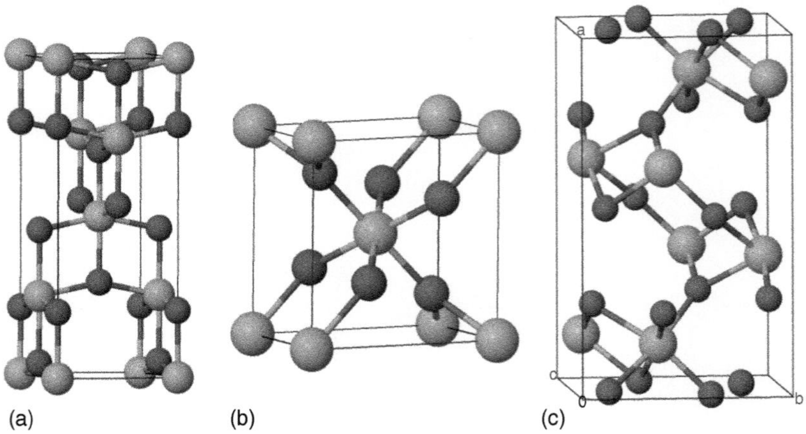

Figure 7.25 Crystal structure of (a) anatase, (b) rutile, and (c) brookite. Source: I. del Rosal.

i.e. anatase and brookite, are irreversibly transformed into rutile at elevated temperatures (between 600 and 700 °C for pure TiO_2).

Photocatalytically, despite the larger experimental anatase band gap of ~3.2 eV, compared to ~3.0 eV for rutile, the photocatalytic performance of anatase is generally considered superior to that of the more stable rutile. This improved performance of the anatase phase is attributed to a higher density of localized states, a higher mobility of electron–hole pairs, and an improved surface hydroxyl density. However, in both cases, TiO_2 poorly absorbs visible light due to its wide band gap, which restricts its application to the UV region. By reducing its band gap, TiO_2 can be used in various applications in the visible spectrum of solar light. Both theoretical and experimental investigations have shown that the band gap can be reduced by various methods, with doping (by metals or non-metals) being considered the simplest and most effective way to shift the band gap from TiO_2 to the visible light region. The electronic structure and photocatalytic properties of TiO_2 have been extensively studied from both a theoretical and experimental point of view. Particular attention has been paid to the doping effect of numerous elements of the fourth [360–377], fifth [369, 372, 375, 378–388], and sixth [369, 371, 372, 375, 378, 379, 381, 385–393] periods of the periodic table on these properties. These studies have shown that: (i) the standard DFT fails in describing most of the experimentally observed features of doped TiO_2 due to the self-interaction error inherent in standard DFT functions, so it is necessary to complement GGA with a "+ U" on-site Coulomb correction that allows a direct comparison with the experimental data [394, 395]; (ii) the presence of a donor dopant induces the introduction of unoccupied d states in the lower part of the conduction band that downshifts its edge and consequently accelerates the electron injection process; and (iii) in the presence of an acceptor dopant, the d states of the dopants hybridize with the O 2p and Ti 3d states to provide impurity energy levels, which either shifts the edge of the conduction band upward due to the insertion of d-occupied orbitals into this band and/or appear separately in the TiO_2 band gap. It was even shown that the enhanced visible light photocatalytic activity of TiO_2 can be achieved by Ti^{3+} self-doping [396, 397]. Interestingly, the presence of Ti^{3+} induces the introduction of occupied states in the band gap that extends the absorbance to the visible range. These occupied states are shallower in anatase than in rutile due to the greater distortion of the octahedral environment of Ti^{3+} ions in anatase compared to rutile. Rare earth metal doping of TiO_2 has also been shown to be an efficient way to reduce the TiO_2 band gap. Liang et al. analyzed the effect of 17 rare earths doping agents on the TiO_2 DOS [398], evidencing that the photocatalytic activity of these systems is enhanced with respect to pure TiO_2. In the same way, Lin et al. reported the increase in the absorption of visible light of the (N, La) co-doped TiO_2 due to the formation of several impurity states appearing in the band gap [399]. Similarly, Khan et al. studied the band structures of mono- or co-doped Yb or N TiO_2 [400]. They observed an increase in the optical response as a result of co-doping with N and Yb, compared to the corresponding mono-doped systems, due to the coupling of the N 2p, Yb 4f, and O 2p states that reduced the band gap without creating isolated states in it.

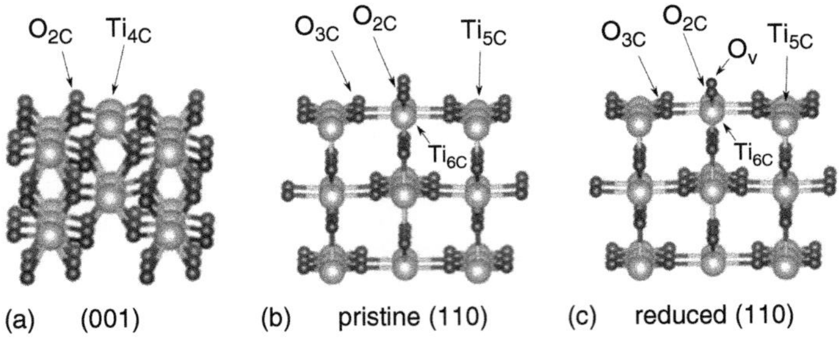

Figure 7.26 Side views of rutile (a) TiO_2 (001), (b) pristine, and (c) reduced (110) facets. Source: Wei et al. [401]. MDPI. CC BY 4.0.

Of the low-index rutile surfaces, the (110) surface (see Figure 7.26) is the lowest in energy and, consequently, the most studied. This facet contains two different types of Ti atoms: sixfold- (Ti_{6C}) and fivefold-coordinated (Ti_{5C}) Ti atoms. Two types of oxygen atoms are also present on this surface: threefold-coordinated (O_{3C}) within the main surface plane and twofold-coordinated bridging O atoms (O_{2C}). Among all these atoms, the O_{2C} and Ti_{5C} atoms are unsaturated. There are other rutile TiO_2 terminations that are experimentally accessible, such as the (100), (001), and (101) facets. It should be noted, as can be seen in Figure 7.26, facets (110) and (001) of rutile TiO_2 are roughly flat, which is not the case for (100) and (101) facets. From a theoretical point of view, only the (110) and (001) facets have been considered for the study of the coordination and catalytic activity of SAs. Thus, this section will only focus on these two surfaces. From a structural point of view, on the (001) surface (see Figure 7.26), all Ti atoms are fourfold-coordinated (Ti_{4C}), and all the O atoms twofold-coordinated (O_{2C}).

Between 2007 and 2008, Asaduzzaman and Krüger have reported the adsorption and diffusion of several TMs on a rutile TiO_2 (110) surface. For Mo_{SA} atoms [402], the threefold hollow site between two O_{2C} and one O_{3C} atoms is the most stable adsorption site. This coordination mode induces the partial insertion, on the TiO_2 surface, of the Ti atom closest to the adsorbed Mo in the direction of an empty interstitial octahedral oxygen site. If this atomic displacement is continued further, an even more favorable coordination mode is obtained (by −20 kcal/mol with respect to the previous coordination mode) in which Mo_{SA} has replaced the Ti atom and the latter has moved to the interstitial octahedral site. Interestingly, the same behavior is observed with V_{SA} [403], and Ti_{SA}, Mn_{SA}, and Cr_{SA} which form stable dioxides [404]. However, for late 3d elements such as Fe_{SA}, Co_{SA}, Ni_{SA}, and Cu_{SA} [404], the SA remains on the surface occupying a threefold hollow site. More recently, in 2019, Tang et al. have evidenced for Rh_{SA} on a rutile TiO_2 (110) surface that these atoms have no static local coordination but can switch from inactive to active structure according to the reaction conditions [405]. Under oxygen-rich conditions, the most favorable structure corresponds to the substitution of Ti_{6C} by Rh_{SA} whereas under oxygen-poor conditions, in which an O_v is formed, the most stable configuration for Rh_{SA} is above an O_v corresponding to a threefold subsurface oxygen. Upon submitting the system to a pressure of CO, thermodynamics tends to drive all Rh_{SA} to a bridging coordination mode between two O_{2C} atoms to allow the coordination of two CO molecules on the Rh_{SA}, regardless of the presence of O_v on the experimental conditions. This adaptive coordination is also observed under H_2 pressure. Thus, even if a preferred site for SA can be obtained, the availability of these sites on the reaction condition can induce the trapping of metal atoms at other sites by inducing a variety of coordination environments surrounding the SA at the oxide surface [406].

For the adsorption of Au_{SA} on a pristine (110) TiO_2 surface, Pillay and Hwang [407] and Vijay et al. [408] have reported that the most stable coordination sites are those between two O_{3C} and one O_{2C} atoms, and between two O_{2C} atoms, in this order. However, it is interesting to note that the presence of an O_v made the interaction between the Au_{SA} and TiO_2 much stronger (more than 23 kcal/mol) than on the pristine surface. Thus, the strong coordination of Au_{SA} on an O_v can serves as an initial nucleation site during sintering, as evidenced by the theoretical work of Li et al. [409]. From a catalytic activity point of view, atomically dispersed Au_{SA} on exposed (110) and (001) facets of rutile TiO_2 exhibits high efficiency for methanol dehydrogenation [410] and oxidation [411] processes. DFT mechanistic studies have been conducted to shed light on the higher activity of Au_{SA}/TiO_2 (110) compared to Au_{SA}/TiO_2 (001). At Au_{SA}/TiO_2 (110), the oxidation reaction of methanol begins with adsorption and deprotonation of the methanol molecule, followed by C–H activation and concomitant desorption of H_2O. This step leads to the formation of a formaldehyde intermediate in which the C atom interacts with the Au_{SA}, and the O atom with a Ti surface atom. Then, an O_2 molecule is adsorbed onto the reduced rutile surface, allowing the formation and desorption of formate. It is interesting to note that, prior to desorption, the formate molecule remains in interaction with the rutile support through the donation of a lone pair of one of the two O atoms to a Ti surface atom. On Au_{SA}/TiO_2 (001), the methanol O–H and C–H activation leads to the release of the formaldehyde molecule. This molecule is then coordinated to the Au atom (donation of a lone pair of one of the two O atoms to the metal) and leads to the formation of a very stable formate intermediate that makes Au_{SA}/TiO_2 (001) less active than Au_{SA}/TiO_2 (110).

For Pt_{SA}, as for Au_{SA}, the preferred coordination site on a rutile TiO_2 (110) surface is a hollow one formed by two O_{3C} and one O_{2C} atoms (with an adsorption energy of −49 kcal/mol). On this Pt SAC, the adsorption and initial decomposition of methylamine (CH_3NH_2) were studied by Lv et al. [412]. Methylamine is strongly adsorbed on Pt_{SA} via the lone pair of the N atom, by −58 kcal/mol. From this first adduct, the C–H, N–H, and C–N bond activation was also studied with a clear kinetic preference for the activation of the C–H bond followed by the activation of N–H and C–N, with an activation barrier of 30, 39, and 52 kcal/mol, respectively. From a thermodynamic point of view, for this first step, the formation of a CH_2NH_2 molecule and a H atom both coordinated to the Pt_{SA} is endothermic by 12 kcal/mol. Previously, Ammal and Heyden had also considered the coordination of a Pt_{SA} on the same rutile TiO_2 (110) surface [413]. In this case, with an adsorption energy of −49 kcal/mol, a threefold site formed by one in-plane O_{3C} and two Ti_{5C} atoms has been considered as a realistic coordination under water-gas shift (WGS) reaction conditions, see Figure 7.27. Thanks to DFT and microkinetic modeling, Ammal et al. studied the high activity of

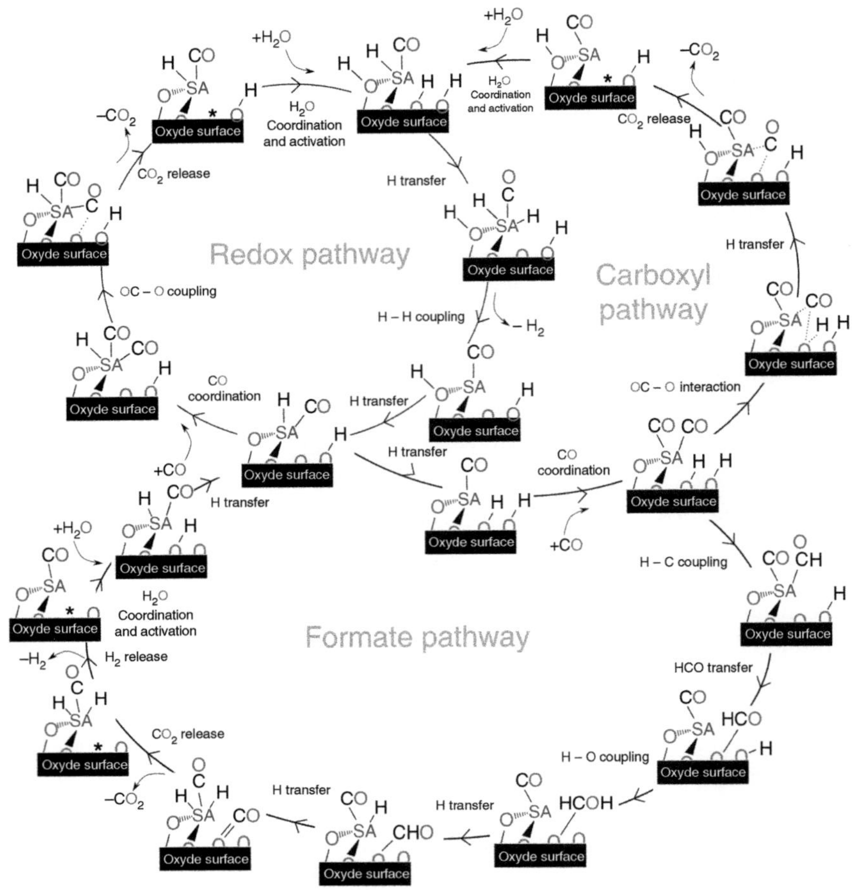

Figure 7.27 Reaction network of possible WGS reaction steps mediated by Pt_{SA} or Pt_{NP}/TiO_2. Source: I. del Rosal.

small Pt clusters (Pt_8) supported on TiO_2 (110) for the WGS reaction. More precisely, they have reported computational models for various WGS reaction mechanisms on corner and edge interface sites for rutile-supported Pt_8 clusters TiO_2 (110) at different temperatures [414–417]. Their work showed that the CO-promoted redox pathway dominates at temperatures below 623 K and the classical redox pathway becomes dominant at temperatures above 673 K. In the same way, associative pathways with formate or carboxyl intermediate are disqualified regardless of the temperature considered. These studies also suggest that H_2O dissociation and H diffusion contribute significantly to the overall rate, as well as the critical role of O_V on the TiO_2 surface for the WGS activity of Pt_8/TiO_2 catalysts. Using the same methodology, it has been confirmed that atomically dispersed Pt^{2+}_{SA}/TiO_2 exhibit very high activity for the WGS reaction at temperatures below 673 K [418]. In this case, Pt^{2+}_{SA} is connected to four surface O atoms, adopting a distorted square planar geometry. Regarding the Pt_8/TiO_2 catalysts, the rates calculated at different temperatures for the WGS reaction propose that the classical redox pathway is the dominant one in Pt_{SA}^{2+}/TiO_2 systems, as well as a similar activity between Pt_{SA}^{2+}/TiO_2 and Pt_8/TiO_2 at low temperatures. The catalytic activity of a Pt_{SA} adsorbed on an undoped $TiO_2(110)$ surface has also been considered by the same authors [419]. In that case, it was found that even if Pt_{SA}/TiO_2 can exhibit reasonably good activity for WGS, and the calculated rate is an order of magnitude lower than with Pt_{SA}^{2+}/TiO_2.

High-temperature AIMD simulations has been carried out by Humphrey et al. in 2020 to identify all possible coordination modes of a Pt_{SA} on the (110) surface of rutile TiO_2 [420]. This work highlights the presence of a wide range of coordination environments and binding energies for Pt_{SA}/TiO_2 with, in particular, a new almost linear O–Pt–O configuration on both pristine and reduced surfaces never observed in previous studies.

A Pt_{SA}^{4+} in a distorted square planar geometry substituting a Ti surface atom was also considered for the chemisorption and activation of methane by Fung et al. [421] and Xu et al. [422], respectively. Interestingly, methane is strongly chemisorbed on Pt_{SA}^{4+}/TiO_2 systems because the empty Pt d_{z^2} orbital is located in the band gap and can effectively accept σ-electron donation from the C–H methane bond. From the chemisorbed CH_4, the heterolytic C–H bond cleavage takes place, overcoming an accessible activation barrier of approximately 7 kcal/mol. The methane dissociation reaction leads to the formation of a Pt–C bond and the transfer of the H atom to an O_{2C} atom (and not to an O_{3C} atom as observed on other supports, see Figure 7.20).

On anatase, (101) facet is predicted to be the most thermodynamically stable surface, followed by (100) and (001) facets. As can be seen in Figure 7.28, (101) surface exhibits a sawtooth-like surface corrugation with surface ridges formed by Ti_{5C} and O_{2C} atoms. O_{3C} atoms can be found on the surface by interconnecting rows of Ti_{5C} and Ti_{6C} atoms located at the bottom of the ridges. On the (001) surface, only the Ti_{5C} atoms are present. Each of these is bonded to two raised O_{2C} and two lowered O_{3C} atoms. Finally, the anatase (001) surface exhibits exclusively Ti_{5C} atoms as well as O_{2C} and O_{3C} atoms.

On (101) anatase surface, Han et al. showed that Pt_{SA} prefers coordinately unsaturated step edge sites involving two O_{2C} atoms [423, 424], with an adsorption energy of between 66 and 68 kcal/mol [425, 426]. Coordination on a bridge site between an O_{2C} and a Ti_{5C} atom or on a threefold bridging site involving an edge O_{2C} atom and lower terrace Ti_{6C} and O_{3C} atoms were also considered but are less favorable

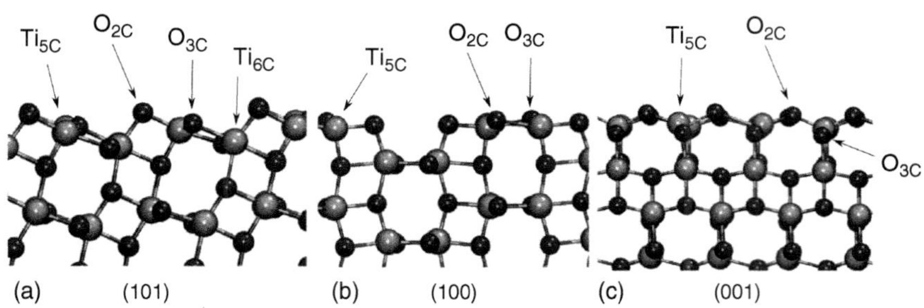

Figure 7.28 Side views of anatase TiO_2 (101) (a), (100) (b), and (001) (c) facets. Source: I. del Rosal.

by 19 and 14 kcal/mol, respectively. In the case of Pd_{SA}, the highest adsorption energy (50 kcal/mol) is obtained on the same site on the pristine (101) anatase surface [427], whereas for Au_{SA} a slight preference is observed for a coordination on top of a Ti_{5C} atom (adsorption energy of 8 kcal/mol) rather than on the bridging site (adsorption energy of 9 kcal/mol) [428]. The formation of an O_v when removing an O_{2C} atom induces an increase in the adsorption energy in all three cases by 11, 47, and 64 kcal/mol for Pd_{SA}, Pt_{SA}, and Au_{SA}, respectively. In comparison [429], the coordination (formation energy) of Ag_{SA} (−59 kcal/mol) on a bridge O_v of a (101) anatase TiO_2 surface is less favorable than for Pd_{SA} (−77 kcal/mol), Pt_{SA} (−139 kcal/mol) or Au_{SA} (−84 kcal/mol). However, a higher formation energy is obtained with Os_{SA} (−146 kcal/mol) and Ir_{SA} (−156 kcal/mol). In the same way, O_v formation becomes easier when Ti_{5C} or Ti_{6C} atoms are replaced by a Pt_{SA}, while the presence of a Pt_{SA} on a step edge site or substituting an O_{2C} atom shows no effect [430, 431]. Similar results were obtained with the Pd_{SA} coordination, which enhances the O_v formation energy by about 46 kcal/mol [432]. On (001) anatase TiO_2 surface, Pd_{SA}, Pt_{SA}, and Au_{SA} adopt a bridging coordination mode between an O_{2C} and a Ti_{5C} atoms [433]. At this site, Pt_{SA} and Pd_{SA} metal atoms are less coordinated (adsorption energy of 62 and 32 kcal/mol for Pt_{SA} and Pd_{SA}, respectively) than on the (101) surface, while Au_{SA} are more strongly coordinated (adsorption energy of 35 kcal/mol). In the case of Au_{SA}, it can also be strongly adsorbed (123 kcal/mol) on the (001) anatase TiO_2 surface by inserting the metal atom into a Ti_{5C}–O_{2C} bond, leading to the formation of a Ti_{5C}–Au_{SA}–O_{2C} species [434]. The CO coordination in these three SAs has also been studied. In this case, the adsorption energy of CO on Au_{SA}/TiO_2 (001) (13 kcal/mol) and Pt_{SA}/TiO_2 (001) (16 kcal/mol) are slightly stronger than on a clean TiO_2 (001) surface (6 kcal/mol [435]), whereas for Pd_{SA}/TiO_2 (001) (4 kcal/mol) the CO adsorption is weaker than on a Ti_{5C} surface atom. These studies were completed in 2018 by Iyemperumal et al. [436], modeling the adsorption of all 29 TM atoms on the TiO_2 anatase (101), and in 2020 by Wang et al. [437], studying the coordination of 12 TM atoms (Ag, Au, Co, Cr, Cu, Fe, Nb, Ni, Mn, Pd, Pt, and V) on anatase TiO_2(101) and (001) surfaces. Interestingly, relative to TiO_2(101), the presence of numerous unsaturated O_{2C} atoms on (001) provides an extremely strong ability to stabilize SAs, resulting in exceptional stability and interfacial bonding. Thus, (001) has been identified as an excellent substrate for loading SAs.

The alkene hydrosilylation reaction mediated by a Pt_{SA} anchored on a bridging site between two O_{2C} atoms on a TiO_2 anatase (101) surface was explored both experimentally and theoretically by Chen et al. [438]. As proposed by Chalk and Harrod (see Figure 7.22), the reaction mechanism begins with the oxidative addition of Si–H to Pt_{SA} and the alkene insertion into the Pt_{SA}–H bond, followed by the Si–C reductive elimination. In this case, the Chalk–Harrod mechanism is more favorable than the modified Chalk–Harrod mechanism, where the alkene is inserted into the Pt_{SA}–Si bond. For this reaction, the rate-determining step corresponds to the Si–C reductive elimination with an activation barrier of 39 kcal/mol. Interestingly, according to the experimental data, in the three elementary steps, the energy barriers for Pt_{SA}/TiO_2 are lower than those for Pt_{NP}/TiO_2 by 2 to 7 kcal/mol. This difference is attributed to the partially positive valence electronic structure of

Pt_{SA}/TiO_2 and its atomic dispersion. At this same coordination site, the formaldehyde oxidation mediated by Pt_{SA}/TiO_2 has been considered by Li et al. [439, 440] and Ding et al. [441]. In this system, the formaldehyde oxidation process is carried out through an Eley–Rideal mechanism that includes six elementary steps: (i) O_2 adsorption and dissociation leading to the formation of $Pt_{SA}{=}O$ and $Ti_{5C}{=}O$ bonds; (ii) the C–O coupling between the formaldehyde molecule and the activated O atom bonded to Pt_{SA}; (iii) the methylenedioxy dehydrogenation ($H_2CO_2{-}Pt_{SA} \rightarrow HCO_2{-}Pt_{SA} + HO{-}Ti_{5C}$); (iv) the formate dehydrogenation ($HCO_2{-}Pt_{SA} \rightarrow H{-}Pt_{SA} + CO_2$); (v) CO_2 desorption; and (vi) H_2O formation and release. In this pathway, the reaction between formaldehyde and the activated O atom is energetically more favorable than between formaldehyde and adsorbed O_2 (or lattice oxygen). From a thermodynamic point of view, this reaction is exothermic by 105 kcal/mol and the dehydrogenation of HCO_2 displays the highest activation energy barrier (55 kcal/mol). The decomposition of formic acid into H_2 and CO_2 has been investigated by Rezaei and Chermahini for a TiO_2 anatase (101) surface in which a Ti_{5C} atom is substituted by a Pt_{SA} [442]. The activity of this system was then compared to that of a pristine TiO_2 anatase (101) surface. For Pt_{SA}/TiO_2, the decomposition reaction starts with the coordination of *trans*-formic acid on Pt_{SA} by donation from a lone pair of the carbonyl O atom to the Pt_{SA}. From this adduct, the first step involves the formation of *cis*-formic acid followed by the coupling of the two H atoms, leading to the release of H_2 and CO_2. In this case, the final product is endothermic (by c. 28 kcal/mol) and the rate-determining step corresponds to the H–H coupling with an activation barrier of 31 kcal/mol. On the pristine TiO_2 anatase (101) surface, *cis*-formic acid interacts with a Ti_{5C} atom by lone pair donation of the hydroxyl O atom to the Ti_{5C} atom. In this case, the isomerization step corresponds to the rate-determining one with an activation barrier of 21 kcal/mol. Therefore, according to DFT calculations, the decomposition reaction of formic acid on the pristine TiO_2 surface is thermodynamically and kinetically more favorable than on Pt_{SA}/TiO_2.

Substitution of a Ti_{5C} atom of the TiO_2 anatase (001) surface by a Mn_{SA} was also considered in several DFT studies [443–445]. Recently, the CO oxidation reaction mediated by this Mn_{SA} system was explored by Mi et al. [445]. Different possible reaction mechanism were considered in this study, but the most favorable one follows the Mars–van-Krevelen mechanism (see Figure 7.16). Thus, the adsorbed CO reacts with a lattice oxygen to generate a CO_2 molecule. The reduced surface with an O_v can then be re-oxidized by an O_2 molecule, leading to the formation of two O atoms that can easily react with a new CO molecule to form a CO_2 molecule and regenerate the catalyst. Interestingly, the low-energy barriers (6 kcal/mol for the first CO oxidation, 4 kcal/mol for the re-oxidation process, and 12 kcal/mol for the second CO_2 formation) and the high exothermicity of the overall reaction (152 kcal/mol) make $Mn_{SA}/TiO_2(001)$ anatase a prospective low-temperature catalyst for CO oxidation.

In the case of $Pd_{SA}/TiO_2(101)$ anatase, Zhang et al. reported both experimentally and theoretically that singly dispersed Pd_{SA} are selective and highly active for more than 10 Sonogashira C–C coupling reactions ($R{-}C{\equiv}H + R'{-}X \rightarrow R{-}C{\equiv}C{-}R'$; $X = Br, I$; $R' =$ Aryl or Vinyl) [446]. According to experimental data, the Pd_{SA} adopts a distorted square planar geometry in which four added O atoms bind to Pd_{SA}

and Ti_{5C} atoms. From a mechanistic point of view, the DFT calculations suggest that Pd_{SA} acts as a site to chemisorb iodobenzene dissociatively to generate an intermediate, phenyl, which then couples with phenylacetylenyl attached to one of the added O atoms leading to the formation of the final product, diphenylacetylene. The rate-limiting step of this pathway corresponds to the dissociative chemisorption of iodobenzene with an activation barrier of 12 kcal/mol in reasonable agreement with the experimental value of 7 kcal/mol.

Cu_{SA} coordinated on a pristine TiO_2 anatase (101) surface, as aforementioned, prefers a bridging coordination mode between two O_{2C} atoms with an adsorption energy of c. −52 kcal/mol [447–450]. A similar adsorption energy is obtained in the presence of an O_v, −51 kcal/mol. In this case, the most stable adsorption is obtained when Cu_{SA} filled the O_v. A slightly lower adsorption energy (−49 kcal/mol) is observed when the Cu_{SA} atom is adsorbed near the vacancy but did not fully occupy it. This system was then used as catalysts for CO_2 reduction [450]. Activation of CO_2 occurs when curved CO_2 is formed by reducing it. To this end, Cu_{SA} atoms in/near an O_v were found to be good reducers for CO_2, in contrast to Cu_{SA} adsorbed on pristine TiO_2. The Cu_{SA}/O_v system can reduce the reaction barrier for CO_2 dissociation to very low values between 2 and 4 kcal/mol, which gives a high catalytic production of CO. Photoexcited electrons could significantly assist SA in this process, as evidenced by decreases in reaction energy (c. 23 kcal/mol) and barrier (6 kcal/mol) for the CO_2 reduction on pristine Cu_{SA}/TiO_2.

7.5.6 Zirconium Oxide

Zirconium dioxide, owing to its mechanical, thermal, and chemical properties, is an attractive ceramic material for various applications (fuel cells, gas sensing, and oxygen sensors, among others) and is frequently used as a catalyst or catalyst support. ZrO_2 have three polymorphic forms (see Figure 7.29): (i) at room temperature, zirconia has a monoclinic crystal structure (m-ZrO_2); (ii) on heating between around 1200 and 2370 °C, it transforms into a tetragonal phase (t-ZrO_2); and (iii) at a temperature above 2370 °C, the stable phase is fluorite, which has complete cubic symmetry (c-ZrO_2) [452]. The surface sites on zirconia include hydroxyl groups, O_v, coordinatively unsaturated Zr–O pairs, Lewis basic sites (O^{2-}), and Lewis acid sites (Zr^{3+}, Zr^{4+}). The concentration of these sites is different according to the crystal phase, which significantly influences their physical and thermomechanical properties [453]. For example, Pokrovski et al. reported that m-ZrO_2 exhibited a higher concentration and basicity of hydroxyl group [454], as well as the stronger Zr^{4+}/O^{2-} pairs compared to t-ZrO_2 [455]. For ZrO_2, as for silica or alumina, the concentration of the hydroxyl groups on the surface can be modulated by thermal treatment in vacuum or in different atmospheres. In recent years, computer modeling techniques have increasingly been applied to the investigation of the geometric and electronic structures of metal/oxide interfaces in order to provide an atomic-level understanding of their properties.

In the case of the zirconia support, at the beginning of the 2000s, the main focus has been on the simulation of the support interaction with Pt or Pd atoms. On

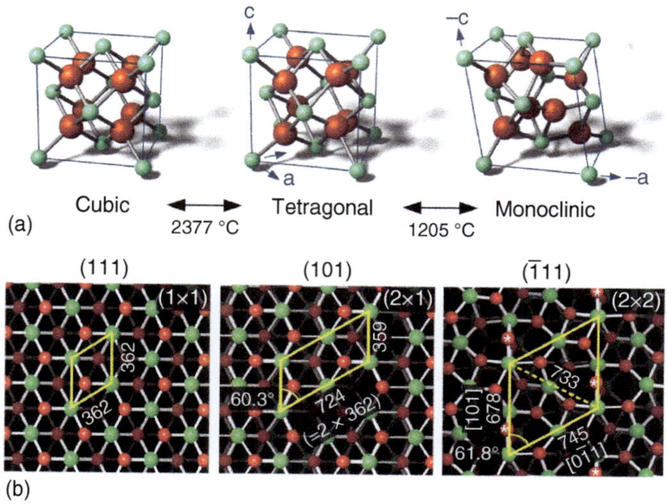

Figure 7.29 The three stable ambient-pressure bulk phases of zirconia: (a) bulk unit cells and (b) bulk-terminated surfaces equivalent to the (111) surface of cubic ZrO_2, with the surface unit cells marked in yellow. Oxygen (2^-) ions are depicted in red, zirconium (4^+) ions in green. Room-temperature cell sizes in (b) are given in pm. Source: Lackner et al. [451]. Reproduced with permission of Elsevier.

c-ZrO_2, both metal atoms show a higher adsorption energies on the (011) facet than on the (111) one, with an energy difference greater than 24 kcal/mol [456, 457]. In both surfaces, the Pt coordination is found to be thermodynamically more stable than the Pd coordination, suggesting a higher mobility for the Pd atom than the Pt one at the *c*-ZrO_2 surface. Likewise, independently of the considered facet, for Pt an onefold oxygen coordination is clearly favored while for Pd the adsorption energy is similar on onefold and threefold oxygen sites. On the (111) surface, Rh atoms, as Pt, show much stronger interaction with the zirconia surface than the Pd atoms (−42 kcal/mol for Rh and Pt vs. −32 kcal/mol for Pd) [458]. Interestingly, the adsorption energy on a defective (111) surface, i.e. with a neutral O_V, for both Pd and Pt metal atoms is three to four times more favorable than on the pristine *c*-ZrO_2/(111) surface (163 and 127 kcal/mol vs. 61 and 32 kcal/mol, respectively, for Pt and Pd) [459]. On a (310) surface, containing well-defined terrace sites, the adsorption energy is comparable with that for the oxygen-defective surfaces (149 and 123 kcal/mol for Pt and Pd, respectively). The Rh coordination on an *m*-ZrO_2 surface has also been investigated in order to clarify the single-atom effect on the zirconia reducibility [460]. This study shows that adsorbed Rh species reduces the formation energy of a nearby O_V by 16 kcal/mol (83 vs. 66 kcal/mol for pristine *m*-$ZrO_2(\bar{1}11)$ and Rh/*m*-$ZrO_2(\bar{1}11)$, respectively). This stabilization should be due to the localization of the electrons that remain after the oxygen removal by a partial hybridization of the electronic states of Rh with the valence band. The same effect is seen when a Pt atom is coordinated on the same surface. In this case, the presence of a Pt atom lowers the O_V formation energy so much that it becomes exothermic by −17 kcal/mol [262]. In the same way, on an *m*-ZrO_2

surface, the adsorption energy for Pt is always higher than that for Rh at the same site, which indicates a higher affinity of Pt toward the zirconia surface. For both Rh and Pt atoms, the formation energy is lower at the edge site, indicating that the metal atom prefers under-coordinated sites. The nature of an Rh_{SA} stabilized on the (101) most stable facet of *t*-ZrO_2 [461–464] has also been investigated [465]. The clean *t*-ZrO_2(101) surface is a regular arrangement of three-coordinated oxygen atoms and six-coordinated Zr atoms. Different coordination modes have been considered for the Rh_{SA}: an Rh atom bound to two or three oxygen atoms, an Rh atom bound to an oxygen atom and to an OH group or, as considered previously by Kwon et al., an Rh atom substituting a lattice Zr cation [466]. The strongest coordination modes correspond to the atom bounded to three oxygen atoms with an adsorption energy of 116 kcal/mol followed by the bridging mode containing an OH group (110 kcal/mol).

This kind of supported metal atoms can bind one or two CO molecules, forming single CO or geminal CO complexes. Based on the CO adsorption energies and CO frequencies, this study shows that an Rh(I) atom bridging two oxygen of the lattice or one oxygen atom and a hydroxyl group coexists at the zirconia surface. More recently, *ab initio* thermodynamic approaches were applied to show that the most stable structure on this surface corresponds to a five-coordinated Rh structure, in which the Rh atom would be oxidized by O_2 during the calcination process [467]. This structure, even if it is very stable, is not active for the oxidation of methane to methanol due to an over-oxidation of the CH_3 at the OCH_3 intermediate that prevents the methanol formation. In contrast, a less stable four-coordinated Rh species should be the active site for CH_3OH formation due to the high stabilization of the CH_3 group and the facile activation of H_2O_2 on this site. For the oxidation of methane to methanol, the dissociation of H_2O_2 on the catalyst surface is crucial to understand the catalytic activity of SACs (see Figure 7.30). This dissociation step leads to the formation of a free O_2 molecule and the hydrogenation of two surface oxygen atoms. After the adsorption of methane on this site, the C–H bond activation is followed by the coupling of the methyl radical with one of the formed hydroxyl groups to produce principally CH_3OH and CH_3OOH (as minor product) [468]. According to the reaction profiles for the conversion of methane mediated by Ru_{SA}/ZrO_2, Fe_{SA}/ZrO_2, and Rh_{SA}/ZrO_2, Fe_{SA}/ZrO_2 SAC appears to be the best candidate for converting methane to methanol selectively, although in all three cases the conversion can occur under mild conditions via a radical pathway. For Ru_{SA}/ZrO_2 SAC, the formation of CH_3OOH is, from a kinetic point of view, competitive with the CH_3OH formation. However, for Fe_{SA}/ZrO_2 SAC, the formation of CH_3OOH is discarded due to a high activation barrier of 64 kcal/mol. In the same way, methanol production at the Fe_{SA}/ZrO_2 site is more favorable than at the Rh site, the barriers of both the methyl radical formation and the methanol formation at the Fe_{SA}/ZrO_2 site being reduced by 11 and 3 kcal/mol, respectively, compared to Rh_{SA}/ZrO_2. Therefore, Fe_{SA}/ZrO_2 SAC is probably more efficient and selective than Ru_{SA}/ZrO_2 and Rh_{SA}/ZrO_2.

The Ru atom diffusion on the *t*-ZrO_2(101) surface has also been theoretically studied by Tosoni and Pacchioni [469]. As observed by Harrath et al. [468], on a defect-free *t*-ZrO_2(101) surface, Ru atoms form stable oxygen bridging adducts on several adsorption sites, with adsorption energies going up to −66 kcal/mol.

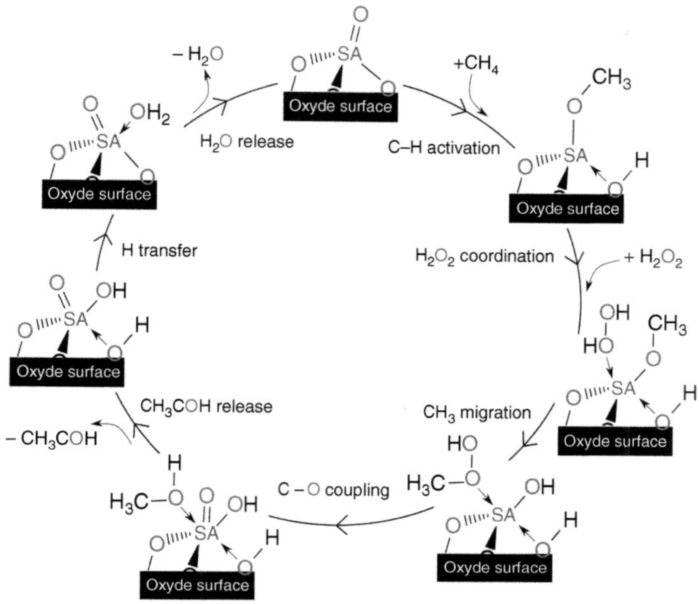

Figure 7.30 Scheme of the methane oxidation reaction mediated by an oxide-supported single-atom catalyst. Source: I. del Rosal.

The SAC diffusion between these different adsorption sites is a kinetically facile process, with activation barriers lower than 12 kcal/mol. The presence of surface O_v radically changes the diffusion landscape of Ru atoms. In presence of surface O_v, the Ru atom is strongly stabilized if adsorbed on the vacancy (−89 kcal/mol). The diffusion process induces the migration of both the vacancy and the metal through a complex mechanism that implies major surface rearrangements with a rather high activation energy. Thus, O_v on the zirconia surface act as trapping sites. In the presence of hydroxyl groups, the adsorption energy of Ru atoms is similar with those reported for the defect-free t-$ZrO_2(101)$ surface. However, the presence of the hydroxyl groups hinders the diffusion of the Ru atom, with kinetic barriers two times higher than on the dehydroxylated surface. If the Ru atoms are adsorbed on t-$ZrO_2(101)$ and m-ZrO_2 $(\bar{1}11)$ surfaces, based on the CO adsorption energies and CO frequencies, the most probable coordination modes are forming a RuO or RuO_2 surface species. These two forms are obtained by the interaction of a Ru atom with one or two surface hydroxyl groups with, respectively, the release of 1/2 or 1 molecule of H_2 [470]. Finally, catalysts consisting of Au, Ag, or Cu atoms supported at the (111) c-ZrO_2 surface have also been investigated [471], mainly due to their good activity and selectivity for the synthesis of methanol from carbon dioxide and hydrogen [472, 473]. In the three cases, the metal atom is coordinated in a bridging position between an oxygen atom and a surface Zr. The adsorption energies were found to vary in the order Cu (−24 kcal/mol) > Au (−19 kcal/mol) > Ag (−8 kcal/mol). The interaction between the metal atoms and the support results mainly from the overlap of the orbitals of the metal centers and those of the surface oxygen atoms, which implies an s–d hybridization of the metal orbitals.

7.5.7 Zinc Oxide

Noble metal catalysts supported on ZnO are highly active systems for CO oxidation. However, these metals suffer from their scarcity and high cost, so that SACs can maximize the metal atom efficiency. In this way, Rh, Au, and Pt SACs supported on ZnO nanowires, which mainly exposes $(10\bar{1}0)$ facets, were successfully developed and investigated experimentally and theoretically in recent years [474, 475]. The stability of these atoms onto the ZnO surface can be explained by their strong interaction with the coordinated lattice oxygen atoms around a Zn vacancy position. The possible substitution of a subsurface Zn lattice by these different atoms has also been studied from a theoretical point of view, highlighting an endothermic process by more than 10 kcal/mol. Thus, the substitution will take place only on the $(10\bar{1}0)$ ZnO surface. Interestingly, the presence of these surface embedded atoms has a great impact on the catalytic properties of these surfaces by changing dramatically: the adsorption sites, binding strength, reaction energetics, and barriers. For the methanol steam-reforming reaction [474], only the coordination of Pt and Au atoms has been considered. In these systems, the calculated binding energies for the reactants and intermediates increase between 4 and 23 kcal/mol on Pt_{SA}/ZnO, and between 4 and 42 kcal/mol on Au_{SA}/ZnO with respect to a pristine ZnO surface. This difference can be explained in part by the presence of a considerable number of states available around the Fermi level with embedded Pt and Au atoms compared to the pristine ZnO. In the same way, the formation of these stronger interactions can be due to the larger spatial extension of d-orbitals of Pt and Au atoms with respect to Zn, allowing more extensive charge transfer and redistribution. From a kinetic point of view, independently of the considered system, the dehydrogenation of methoxy to formaldehyde is the rate-determining step of the methanol steam-reforming reaction. For Pt_{SA}/ZnO, this transition state remains energetically lower than that on pristine ZnO, while for Au_{SA}/ZnO the activation energy becomes slightly higher. This implies that Pt_{SA}/ZnO would have a higher activity than ZnO while that of Au_{SA}/ZnO will be similar to that of ZnO. These theoretical results are in agreement with those obtained experimentally, for which the Pt_{SA}/ZnO systems have a TOF of over 1000 times higher than that of the pristine ZnO. The CO oxidation mediated by Rh, Au, and Pt atoms supported on ZnO nanowire has also been recently studied both from a theoretical and experimental point of view by Han et al. [475]. In this work, Rh_{SA}/ZnO systems showed much higher activity for the CO oxidation than other noble metals, which are generally considered as good candidates for the CO oxidation. The CO oxidation reaction takes place through a Mars–van Krevelen mechanism (see Figure 7.16). In this case, independently of the considered systems, the overall process is thermodynamically favorable by more than 70 kcal/mol. From a kinetic point of view, the O_2 dissociation on the O_v is the rate-limiting step for the three SAC systems. For this step, the calculated barriers are 19, 21, and 24 kcal/mol for Rh_{SA}/ZnO, Pt_{SA}/ZnO, and Au_{SA}/ZnO, respectively, suggesting that, in agreement with the experimental results, the activity of CO oxidation is in the order: Rh_{SA}/ZnO > Pt_{SA}/ZnO > Au_{SA}/ZnO.

7.6 Conclusions

DFT calculations have proven invaluable in elucidating SAC anchoring modes. Modifications of the electronic structure on the embedding of metal atoms by varying the type of surfaces, their orientations, the presence of various defects and/or heteroatoms, for instance based on the evaluation of DOS, the AO analyses and the d-band center, as well as charge transfer estimates, are precious tools for the rationalization of catalytic activity. In some cases, by defining appropriate descriptors, predictive results can be obtained. Many metals have already been tested, and for some of them (Pt, Pd, etc.) the reactivity has been thoroughly investigated. At times, DFT outcomes can be used as valuable inputs for large-scale approaches such as Monte-Carlo to determine the long-term evolution of the catalytic process, along with microkinetic modeling. Additionally, machine learning is recently receiving increased attention and its progress is significantly aiding in the rational design of SACs. However, in some cases, improvements could be achieved. For instance, in the case of hydrogenation reactions, spillover mechanism has not been entirely described so far, since rather large energy barriers are usually provided by DFT calculations and thermodynamics are not always favorable. Considering electrocatalytic reactions, the effects of solvents, as well as the application of an electrostatic potential, are not explicitly taken into account most of the time. In terms of future perspective, the influence of adding the second metal atom in the vicinity of a SAC is also a case that has not been addressed as much. Besides, little is known about the synergistic effect of the common presence of NPs, clusters and SAs, as it requires the use of more complex models, in addition to the effect of SA concentrations. Therefore, the theoretical works should definitively push toward those directions in the near future.

Acknowledgements

This work was supported by the Agence Nationale de la Recherche (COMET project ANR-19-CE07-0030), which is gratefully acknowledged.

References

1 Zhu, C., Fu, S., Shi, Q. et al. (2017). Single-atom electrocatalysts. *Angewandte Chemie International Edition* 56: 13944–13960.
2 Kim, J., Kim, H.E., and Lee, H. (2017). Single-atom catalysts of precious metals for electrochemical reactions. *ChemSusChem* 11: 104–113.
3 Nørskov, J.K., Abild-Pedersen, F., Studt, F. et al. (2011). Density functional theory in surface chemistry and catalysis. *Proceedings of the National Academy of Sciences of the United States of America* 108: 937–943.
4 Chizallet, C. and Raybaud, P. (2014). Density functional theory simulations of complex catalytic materials in reactive environments: beyond the ideal surface at low coverage. *Catalysis Science & Technology* 4: 2797–2813.

5 Schönberger, U., Andersen, O.K., and Methfessel, M. (1992). Bonding at metal-ceramic interfaces; Ab initio density-functional calculations for Ti and Ag on MgO. *Acta Metallurgica et Materialia* 40: S1–S10.

6 Fu, Q., Saltsburg, H., and Flytzani-Stephanopoulos, M. (2003). Active non-metallic Au and Pt species on ceria-based water-gas shift catalysts. *Science* 301: 935–938.

7 Valero, M.C., Digne, M., Sautet, P. et al. (2006). DFT study of the interaction of a single palladium atom with γ-alumina surfaces: the role of hydroxylation. *Oil & Gas Science and Technology* 61: 535–545.

8 Joubert, J., Delbecq, F., Sautet, P. et al. (2006). Molecular understanding of alumina supported single-site catalysts by a combination of experiment and theory. *Journal of the American Chemical Society* 128: 9157–9169.

9 Camellone, M.F. and Fabris, S. (2009). Reaction mechanisms for the CO oxidation on Au/CeO_2 catalysts: activity of substitutional Au^{3+}/Au^+ cations and deactivation of supported Au^+ adatoms. *Journal of the American Chemical Society* 131: 10473–10483.

10 Samantaray, M.K., D'Eia, V., Pump, E. et al. (2020). The comparison between single atom catalysis and surface organometallic catalysis. *Chemical Reviews* 120: 734–813.

11 Li, L., Chang, X., Lin, X. et al. (2020). Theoretical insights into single-atom catalysts. *Chemical Society Reviews* 49: 8156–8178.

12 Rivera-Cárcamo, C. and Serp, P. (2018). Single atom catalysts on carbon-based materials. *ChemCatChem* 10: 5058–5091.

13 Krasheninnikov, A.V., Lehtinen, P.O., Foster, A.S. et al. (2009). Embedding transition-metal atoms in graphene: structure, bonding, and magnetism. *Physical Review Letters* 102: 126807.

14 Lu, Y.H., Zhou, M., Zhang, C. et al. (2009). Metal-embedded graphene: a possible catalyst with high activity. *Journal of Physical Chemistry C* 113: 20156–20160.

15 Coughlin, R.W. (1969). Carbon as adsorbent and catalyst. *Product R&D* 8: 12–23.

16 Radovic, L.R. and Bockrath, B. (2005). On the chemical nature of graphene edges: origin of stability and potential for magnetism in carbon materials. *Journal of the American Chemical Society* 127: 5917–5927.

17 Acik, M. and Chabal, Y.J. (2011). Nature of graphene edges: a review. *Japanese Journal of Applied Physics* 50: 070101–070117.

18 Banhart, F., Kotakoski, J., and Krasheninnikov, A.V. (2011). Structural defects in graphene. *ACS Nano* 5: 26–41.

19 Radovic, L.R. (2021). Probing the "elephant": on the essential difference between graphenes and polycyclic aromatic hydrocarbons. *Carbon* 171: 798–805.

20 Ayala, P., Arenal, R., Rümmeli, M. et al. (2010). The doping of carbon nanotubes with nitrogen and their potential applications. *Carbon* 48: 575–586.

21 Wang, H., Maiyalagan, T., and Wang, X. (2012). Review on recent progress in nitrogen-doped graphene: synthesis, characterization, and its potential applications. *ACS Catalysis* 2: 781–794.

22 Lee, W.J., Maiti, U.N., Lee, J.M. et al. (2014). Nitrogen-doped carbon nanotubes and graphene composite structures for energy and catalytic applications. *Chemical Communications* 50: 6818–6830.

23 Cao, Y., Mao, S., Li, M. et al. (2017). Metal/porous carbon composites for heterogeneous catalysis: old catalysts with improved performance promoted by N-doping. *ACS Catalysis* 7: 8090–8112.

24 Axet, M.R., Dechy-Cabaret, O., Durand, J. et al. (2016). Coordination chemistry on carbon surfaces. *Coordination Chemistry Reviews* 308: 236–345.

25 Hughes, Z.E. and Walsh, T.R. (2015). Computational chemistry for graphene-based energy applications: progress and challenges. *Nanoscale* 7: 6883–6908.

26 Yildirim, T., Í niguez, J., and Ciraci, S. (2005). Molecular and dissociative adsorption of multiple hydrogen molecules on transition metal decorated C_{60}. *Physical Review B* 72: 153403. '

27 Rosseinsky, M.J. (1998). Recent developments in the chemistry and physics of metal fullerides. *Chemistry of Materials* 10: 2665–2685.

28 Andriotis, A.N., and Menon, M. (1999). Geometry and bonding in small $(C_{60})_nNi_m$ clusters. *Physical Review B* 60: 4521.

29 Ivanova, V.N. (2000). Fullerene compounds with transition metals MnC_{60}: preparation, structure, and properties. *Journal of Structural Chemistry* 41: 135–148.

30 Andriotis, A.N., Menon, M., and Froudakis, G.E. (2000). Contrasting bonding behaviors of 3D transition metal atoms with graphite and C_{60}. *Physical Review B* 62: 9867.

31 Ramaniah, L.M., Boero, M., and Laghate, M. (2004). Tantalum-fullerene clusters: a first-principles study of static properties and dynamical behavior. *Physical Review B* 70: 035411.

32 Shukla, M.K., Dubey, M., and Leszczynski, J. (2008). Theoretical investigation of electronic structures and properties of C_{60}-gold nanocontacts. *ACS Nano* 2: 227–234.

33 Voityuk, A.A. (2009). Thermochemistry of Pt-fullerene complexes: semiempirical study. *Journal of Physical Chemistry A* 113: 11801–11808.

34 Ramaniah, L.M. and Boero, M. (2010). Mobility of a single alkali metal atom on fullerene C_{60}: first principles molecular dynamical study. *Journal of Chemical Physics* 133: 134701.

35 Matsubara, M., Massobrio, C., Ramaniah, L.M. et al. (2010). Electronic structure and localization properties of C_{60}-Ta_n clusters (n=1,3): a first-principles study. *Physical Review B* 81: 195433.

36 Zeng, Q., Chu, X., Yang, M. et al. (2012). Spin-orbit coupling effect on Au-C_{60} interaction: a density functional theory study. *Chemical Physics* 395: 82–86.

37 Leng, F., Gerber, I.C., Lecante, P. et al. (2016). Synthesis and structure of ruthenium-fullerides. *RSC Advances* 6: 69135–69148.

38 Goclon, J., Winkler, K., and Margraf, J.T. (2017). Theoretical investigation of interactions between palladium and fullerene in polymer. *RSC Advances* 7: 2202–2210.

39 Soto, D. and Salcedo, R. (2012). Coordination modes and different hapticities for fullerene organometallic complexes. *Molecules* 17: 7151.

40 Méndez-Camacho, R. and Guirado-López, R.A. (2013). Adsorption and diffusion of hydrogen on C_{60}-supported Pt_n clusters. *Journal of Physical Chemistry C* 117: 10059–10069.

41 Pham, N.N.T. and Le, H.M. (2017). A density functional theory investigation of Ni_n, Pd_n, and Pt_n clusters (n=1–4) adsorbed on buckminsterfullerene. *ChemPhysChem* 18: 1376–1384.

42 Konarev, D.V., Khasanov, S.S., Nakano, Y. et al. (2014). Linear coordination fullerene C_{60} polymer $[Ni(Me_3P)_2(\mu\text{-}\eta_2,\eta_2\text{-}C_{60})]_\infty$ bridged by zerovalent nickel atoms. *Inorganic Chemistry* 53: 11960–11965.

43 Durgun, E., Dag, S., Ciraci, S. et al. (2004). Energetics and electronic structures of individual atoms adsorbed on carbon nanotubes. *Journal of Physical Chemistry B* 108: 575–582.

44 Umadevi, D. and Sastry, G.N. (2012). Metal ion binding with carbon nanotubes and graphene: effect of chirality and curvature. *Chemical Physics Letters* 549: 39–43.

45 Yildirim, T. and Ciraci, S. (2005). Titanium-decorated carbon nanotubes as a potential high-capacity hydrogen storage medium. *Physical Review Letters* 94: 175501.

46 Durgun, E., Ciraci, S., and Yildirim, T. (2008). Functionalization of carbon-based nanostructures with light transition-metal atoms for hydrogen storage. *Physical Review B* 77: 085405.

47 Valencia, H., Gil, A., and Frapper, G. (2010). Trends in the adsorption of 3*D* transition metal atoms onto graphene and nanotube surfaces: a DFT study and molecular orbital analysis. *Journal of Physical Chemistry C* 114: 14141–14153.

48 Valencia, H., Gil, A., and Frapper, G. (2015). Trends in the hydrogen activation and storage by adsorbed 3*D* transition metal atoms onto graphene and nanotube surfaces: a DFT study and molecular orbital analysis. *Journal of Physical Chemistry C* 119: 5506–5522.

49 Menon, M., Andriotis, A.N., and Froudakis, G.E. (2000). Curvature dependence of the metal catalyst atom interaction with carbon nanotubes walls. *Chemical Physics Letters* 320: 425–434.

50 Chen, G. and Kawazoe, Y. (2006). Interaction between a single Pt atom and a carbon nanotube studied by density functional theory. *Physical Review B* 73: 125410.

51 Shu, D.J. and Gong, X.G. (2001). Curvature effect on surface diffusion: the nanotube. *Journal of Chemical Physics* 114: 10922–10926.

52 Wu, M.C., Li, C.L., Hu, C.K. et al. (2006). Curvature effect on the surface diffusion of silver adatoms on carbon nanotubes: deposition experiments and numerical simulations. *Physical Review B* 74: 125424.

53 Teddy, J., Falqui, A., Corrias, A. et al. (2011). Influence of particles alloying on the performances of Pt-Ru/CNT catalysts for selective hydrogenation. *Journal of Catalysis* 278: 59–70.

54 Zhuang, H.L., Zheng, G.P., and Soh, A.K. (2008). Interactions between transition metals and defective carbon nanotubes. *Computational Materials Science* 43: 823–828.

55 Chen, Y.K., Liu, L.V., Tian, W.Q. et al. (2011). Theoretical studies of transition-metal-doped single-walled carbon nanotubes. *Journal of Physical Chemistry C* 115: 9306–9311.

56 Yu, L., Li, W.X., Pan, X. et al. (2012). In- and out-dependent interactions of iron with carbon nanotubes. *Journal of Physical Chemistry C* 116: 16461–16466.

57 Chan, K.T., Neaton, J.B., and Cohen, M.L. (2008). First-principles study of metal adatom adsorption on graphene. *Physical Review B* 77: 235430.

58 Liu, X., Wang, C.Z., Hupalo, M. et al. (2013). Metals on graphene: interactions, growth morphology, and thermal stability. *Crystals* 3: 79.

59 Manadé, M., Vi nes, F., and Illas, F. (2015). Transition metal adatoms on graphene: a systematic density functional study. *Carbon* 95: 525–534.

60 Manadé, M., Vi nes, F., Gil, A. et al. (2018). On the H_2 interaction on transition metal adatoms supported on graphene: a systematic density functional study. *Physical Chemistry Chemical Physics* 20: 3819–3830.

61 Rudenko, A.N., Keil, F.J., Katsnelson, M.I. et al. (2012). Adsorption of cobalt on graphene: electron correlation effects from a quantum chemical perspective. *Physical Review B* 86: 075422.

62 Donati, F., Dubout, Q., Autès, G. et al. (2013). Magnetic moment and anisotropy of individual Co atoms on graphene. *Physical Review Letters* 111: 236801.

63 Appy, D., Lei, H., Wang, C.Z. et al. (2014). Transition metals on the (0001) surface of graphite: fundamental aspects of adsorption, diffusion, and morphology. *Progress in Surface Science* 89: 219–238.

64 Tang, Y., Zhang, H., Shen, Z. et al. (2017). The electronic and diffusion properties of metal adatoms on graphene sheets: a first-principles study. *RSC Advances* 7: 33208–33218.

65 Hardcastle, T.P., Seabourne, C.R., Zan, R. et al. (2013). Mobile metal adatoms on single layer, bilayer, and trilayer graphene: an Ab initio DFT study with van der Waals corrections correlated with electron microscopy data. *Physical Review B* 87: 195430.

66 Zhao, J., Deng, Q., Avdoshenko, S.M. et al. (2014). Direct in situ observations of single Fe atom catalytic processes and anomalous diffusion at graphene edges. *Proceedings of the National Academy of Sciences of the United States of America* 111: 15641–15646.

67 Wang, H., Li, K., Cheng, Y. et al. (2012). Interaction between single gold atom and the graphene edge: a study via aberration-corrected transmission electron microscopy. *Nanoscale* 4: 2920–2925.

68 Yamazaki, K., Maehara, Y., Lee, C.C. et al. (2018). Atomic structure and local electronic states of single Pt atoms dispersed on graphene. *Journal of Physical Chemistry C* 122: 27292–27300.

69 Yang, Y.-E., Xiao, Y., Yan, X.H., and Dai, C.J. (2015). Transport properties of zigzag graphene nanoribbons adsorbed with single iron atom. *Chinese Physics B* 24: 117204.

70 Terrones, H., Lv, R., Terrones, M. et al. (2012). The role of defects and doping in 2D graphene sheets and 1D nanoribbons. *Reports on Progress in Physics* 75: 062501.

71 Skowron, S.T., Lebedeva, I.V., Popov, A.M. et al. (2015). Energetics of atomic scale structure changes in graphene. *Chemical Society Reviews* 44: 3143–3176.

72 Ma, Y., Lehtinen, P.O., Foster, A.S. et al. (2004). Magnetic properties of vacancies in graphene and single-walled carbon nanotubes. *New Journal of Physics* 6: 68.

73 Lehtinen, P.O., Foster, A.S., Ma, Y. et al. (2004). Irradiation-induced magnetism in graphite: a density functional study. *Physical Review Letters* 93: 187202.

74 Yazyev, O.V. and Helm, L. (2007). Defect-induced magnetism in graphene. *Physical Review B* 75: 125408.

75 Valencia, A.M. and Caldas, M.J. (2017). Single vacancy defect in graphene: insights into its magnetic properties from theoretical modeling. *Physical Review B* 96: 125431.

76 El-Barbary, A.A., Telling, R.H., Ewels, C.P. et al. (2003). Structure and energetics of the vacancy in graphite. *Physical Review B* 68: 144107.

77 Krasheninnikov, A.V., Lehtinen, P.O., Foster, A.S. et al. (2006). Bending the rules: contrasting vacancy energetics and migration in graphite and carbon nanotubes. *Chemical Physics Letters* 418: 132–136.

78 Berber, S. and Oshiyama, A. (2006). Reconstruction of mono-vacancies in carbon nanotubes: atomic relaxation vs. spin polarization. *Physica B: Condensed Matter* 376–377: 272–275.

79 Gerber, I., Oubenali, M., Bacsa, R. et al. (2011). Theoretical and experimental studies on the carbon-nanotube surface oxidation by nitric acid: interplay between functionalization and vacancy enlargement. *Chemistry: A European Journal* 17: 11467–11477.

80 Lee, G.D., Wang, C.Z., Yoon, E. et al. (2005). Diffusion, coalescence, and reconstruction of vacancy defects in graphene layers. *Physical Review Letters* 95: 205501.

81 Wu, L., Hou, T., Li, Y. et al. (2013). First-principles study on migration and coalescence of point defects in monolayer graphene. *Journal of Physical Chemistry C* 117: 17066–17072.

82 Krasheninnikov, A.V. and Nieminen, R.M. (2011). Attractive interaction between transition-metal atom impurities and vacancies in graphene: a first-principles study. *Theoretical Chemistry Accounts* 129: 625–630.

83 Bhattacharya, A., Bhattacharya, S., Majumder, C. et al. (2010). Transition-metal decoration enhanced room-temperature hydrogen storage in a defect-modulated graphene sheet. *Journal of Physical Chemistry C* 114: 10297–10301.

84 Stone, A.J. and Wales, D.J. (1986). Theoretical studies of icosahedral C_{60} and some related species. *Chemical Physics Letters* 128: 501–503.

85 Li, L., Reich, S., and Robertson, J. (2005). Defect energies of graphite: density-functional calculations. *Physical Review B* 72: 184109.

86 Ertekin, E., Chrzan, D.C., and Daw, M.S. (2009). Topological description of the Stone-Wales defect formation energy in carbon nanotubes and graphene. *Physical Review B* 79: 155421.

87 Ma, J., Alfè, D., Michaelides, A. et al. (2009). Stone-Wales defects in graphene and other planar sp^2-bonded materials. *Physical Review B* 80: 033407.

88 Harman, A.J. and Raelynn, L. (2014). Density functional theory study of metal adatoms at or near a Stone-Wales defect in graphene. *Procedia Engineering* 93: 2–7.

89 Robertson, A.W., Bachmatiuk, A., Wu, Y.A. et al. (2011). Atomic structure of interconnected few-layer graphene domains. *ACS Nano* 5: 6610–6618.

90 Liu, Y. and Yakobson, B.I. (2010). Cones, pringles, and grain boundary landscapes in graphene topology. *Nano Letters* 10: 2178–2183.

91 Červenka, J. and Flipse, C.F.J. (2009). Structural and electronic properties of grain boundaries in graphite: planes of periodically distributed point defects. *Physical Review B* 79: 195429.

92 Malola, S., Häkkinen, H., and Koskinen, P. (2010). Structural, chemical, and dynamical trends in graphene grain boundaries. *Physical Review B* 81: 165447.

93 Kou, L., Tang, C., Guo, W. et al. (2011). Tunable magnetism in strained graphene with topological line defect. *ACS Nano* 5: 1012–1017.

94 Bissett, M.A., Konabe, S., Okada, S. et al. (2013). Enhanced chemical reactivity of graphene induced by mechanical strain. *ACS Nano* 7: 10335–10343.

95 Liu, H., Liu, Y., and Zhu, D. (2011). Chemical doping of graphene. *Journal of Materials Chemistry* 21: 3335–3345.

96 Panchakarla, L.S., Subrahmanyam, K.S., Saha, S.K. et al. (2009). Synthesis, structure, and properties of boron- and nitrogen-doped graphene. *Advanced Materials* 21: 4726–4730.

97 Paraknowitsch, J.P. and Thomas, A. (2013). Doping carbons beyond nitrogen: an overview of advanced heteroatom doped carbons with boron, Sulphur and phosphorus for energy applications. *Energy & Environmental Science* 6: 2839–2855.

98 Faccio, R., Fernández-Werner, L., Pardo, H. et al. (2010). Electronic and structural distortions in graphene induced by carbon vacancies and boron doping. *Journal of Physical Chemistry C* 114: 18961–18971.

99 Batzill, M. (2012). The surface science of graphene: metal interfaces, CVD synthesis, nanoribbons, chemical modifications, and defects. *Surface Science Reports* 67: 83–115.

100 Wei, D., Liu, Y., Wang, Y. et al. (2009). Synthesis of N-doped graphene by chemical vapor deposition and its electrical properties. *Nano Letters* 9: 1752–1758.

101 Feng, H., Ma, J., and Hu, Z. (2010). Nitrogen-doped carbon nanotubes functionalized by transition metal atoms: a density functional study. *Journal of Materials Chemistry* 20: 1702–1708.

102 Yang, M., Wang, L., Li, M. et al. (2015). Structural stability and O_2 dissociation on nitrogen-doped graphene with transition metal atoms embedded: a first-principles study. *AIP Advances* 5: 067136.

103 Chen, Z., Mitchell, S., Vorobyeva, E. et al. (2017). Stabilization of single metal atoms on graphitic carbon nitride. *Advanced Functional Materials* 27: 1605785.

104 Zhang, Y., Wang, Z., and Cao, J. (2014). Prediction of magnetic anisotropy of 5*D* transition metal-doped g-C_3N_4. *Journal of Materials Chemistry C* 2: 8817–8821.

105 Ghosh, D., Periyasamy, G., Pandey, B. et al. (2014). Computational studies on magnetism and the optical properties of transition metal embedded graphitic carbon nitride sheets. *Journal of Materials Chemistry C* 2: 7943–7951.

106 Du, J., Xia, C., Xiong, W. et al. (2016). Tuning the electronic structures and magnetism of two-dimensional porous C_2N via transition metal embedding. *Physical Chemistry Chemical Physics* 18: 22678–22686.

107 Zheng, Y., Jiao, Y., Zhu, Y. et al. (2017). Molecule-level g-C_3N_4 coordinated transition metals as a new class of electrocatalysts for oxygen electrode reactions. *Journal of the American Chemical Society* 139: 3336–3339.

108 Chen, D., Feng, H., and Li, J. (2012). Graphene oxide: preparation, functionalization, and electrochemical applications. *Chemical Reviews* 112: 6027–6053.

109 Mao, S., Pu, H., and Chen, J. (2012). Graphene oxide and its reduction: modeling and experimental progress. *RSC Advances* 2: 2643–2662.

110 Yang, D., Velamakanni, A., Bozoklu, G. et al. (2009). Chemical analysis of graphene oxide films after heat and chemical treatments by X-ray photoelectron and micro-Raman spectroscopy. *Carbon* 47: 145–152.

111 Krishnamoorthy, K., Veerapandian, M., Yun, K. et al. (2013). The chemical and structural analysis of graphene oxide with different degrees of oxidation. *Carbon* 53: 38–49.

112 Zhou, S. and Bongiorno, A. (2014). Density functional theory modeling of multilayer "Epitaxial" graphene oxide. *Accounts of Chemical Research* 47: 3331–3339.

113 Saxena, S., Tyson, T.A., and Negusse, E. (2010). Investigation of the local structure of graphene oxide. *Journal of Physical Chemistry Letters* 1: 3433–3437.

114 Lahaye, R.J.W.E., Jeong, H.K., Park, C.Y. et al. (2009). Density functional theory study of graphite oxide for different oxidation levels. *Physical Review B* 79: 125435.

115 Boukhvalov, D.W., Dreyer, D.R., Bielawski, C.W. et al. (2012). A computational investigation of the catalytic properties of graphene oxide: exploring mechanisms by using DFT methods. *ChemCatChem* 4: 1844–1849.

116 Dimiev, A.M., Alemany, L.B., and Tour, J.M. (2013). Graphene oxide. Origin of acidity, its instability in water, and a new dynamic structural model. *ACS Nano* 7: 576–588.

117 Sun, Y., Yang, S., Chen, Y. et al. (2015). Adsorption and desorption of U(VI) on functionalized graphene oxides: a combined experimental and theoretical study. *Environmental Science and Technology* 49: 4255–4262

118 Pei, S. and Cheng, H.M. (2012). The reduction of graphene oxide. *Carbon* 50: 3210–3228.

119 Tung, V.C., Allen, M.J., Yang, Y. et al. (2008). High-throughput solution processing of large-scale graphene. *Nature Nanotechnology* 4: 25–29.

120 Wang, L., Lee, K., Sun, Y.Y. et al. (2009). Graphene oxide as an ideal substrate for hydrogen storage. *ACS Nano* 3: 2995–3000.

121 Li, F., Zhao, J., and Chen, Z. (2011). Fe-Anchored graphene oxide: a low-cost and easily accessible catalyst for low-temperature CO oxidation. *Journal of Physical Chemistry C* 116: 2507–2514.

122 Dobrota, A.S., Pas˘ ti, I.A., and Skorodumova, N.V. (2015). Oxidized graphene as an electrode material for rechargeable metal-ion batteries – a DFT point of view. *Electrochimica Acta* 176: 1092–1099.

123 Esrafili, M.D., Sharifi, F., and Nematollahi, P. (2016). Al- or Si-Decorated graphene oxide: a favorable metal-free catalyst for the N_2O reduction. *Applied Surface Science* 387: 454–460.

124 Groves, M.N., Malardier-Jugroot, C., and Jugroot, M. (2012). Improving platinum catalyst durability with a doped graphene support. *Journal of Physical Chemistry C* 116: 10548–10556.

125 Dobrota, A.S., Pas˘ ti, I.A., Mentus, S.V. et al. (2016). A general view on the reactivity of the oxygen-functionalized graphene basal plane. *Physical Chemistry Chemical Physics* 18: 6580–6586.

126 Santos, E.J.G., Ayuela, A., and Sánchez-Portal, D. (2010). First-principles study of substitutional metal impurities in graphene: structural, electronic and magnetic properties. *New Journal of Physics* 12: 053012.

127 Kattel, S., Atanassov, P., and Kiefer, B. (2012). Stability, electronic and magnetic properties of in-plane defects in graphene: a first-principles study. *Journal of Physical Chemistry C* 116: 8161–8166.

128 Calle-Vallejo, F., Martínez, J.I., and Rossmeisl, J. (2011). Density functional studies of functionalized graphitic materials with late transition metals for oxygen reduction reactions. *Physical Chemistry Chemical Physics* 13: 15639–15643.

129 Hammer, B. and Nørskov, J.K. (1995). Electronic factors determining the reactivity of metal surfaces. *Surface Science* 343: 211–220.

130 Liu, X., Sui, Y., Duan, T. et al. (2014). CO oxidation catalyzed by Pt-embedded graphene: a first-principles investigation. *Physical Chemistry Chemical Physics* 16: 23584–23593.

131 Wang, C.S., Wang, H., Wu, R. et al. (2018). Evaluating the stability of single-atom catalysts with high chemical activity. *Journal of Physical Chemistry C* 122: 21919–21926.

132 Lou, Y., Wu, H., and Liu, J. (2019). Nanocarbon-edge-anchored high-density Pt atoms for 3-nitrostyrene hydrogenation: strong metal-carbon interaction. *iScience* 13: 190–198.

133 Mahmoodinia, M., Åstrand, P.O., and Chen, D. (2017). Tuning the electronic properties of single-atom Pt catalysts by functionalization of the carbon support material. *Journal of Physical Chemistry C* 121: 20802–20812.

134 Hasegawa, S., Kunisada, Y., and Sakaguchi, N. (2017). Diffusion of a single platinum atom on light-element-doped graphene. *Journal of Physical Chemistry C* 121: 17787–17795.

135 Xu, G., Wang, R., Yang, F. et al. (2017). CO oxidation on single Pd atom embedded defect-graphene via a new termolecular Eley–Rideal mechanism. *Carbon* 118: 35–42.

136 Arrigo, R., Schuster, M.E., Xie, Z. et al. (2015). Nature of the N–Pd interaction in nitrogen-doped carbon nanotube catalysts. *ACS Catalysis* 5: 2740–2753.

137 Huang, F., Deng, Y., Chen, Y. et al. (2018). Atomically dispersed Pd on nanodiamond/graphene hybrid for selective hydrogenation of acetylene. *Journal of the American Chemical Society* 140: 13142–13146.

138 López-Corral, I., Irigoyen, B., and Juan, A. (2014). Bonding in PdH_2 and Pd_2H_2 systems adsorbed on carbon nanotubes: implications for hydrogen storage. *International Journal of Hydrogen Energy* 39: 8780–8790.

139 Yin, P., Yao, T., Wu, Y. et al. (2016). Single cobalt atoms with precise N-coordination as superior oxygen reduction reaction catalysts. *Angewandte Chemie International Edition* 128: 10958–10963.

140 Li, S.L., Yin, H., Kan, X. et al. (2017). Potential of transition metal atoms embedded in buckled monolayer g-C_3N_4 as single-atom catalysts. *Physical Chemistry Chemical Physics* 19: 30069–30077.

141 Fei, H., Dong, J., Arellano-Jiménez, J. et al. (2015). Atomic cobalt on nitrogen-doped graphene for hydrogen generation. *Nature Communications* 6: 8668.

142 Cui, X., Xiao, J., Wu, Y. et al. (2016). A graphene composite material with single cobalt active sites: a highly efficient counter electrode for dye-sensitized solar cells. *Angewandte Chemie International Edition* 128 (23): 6820–6824.

143 Möller, T., Ju, W., Bagger, A. et al. (2019). Efficient CO_2 to CO electrolysis on solid Ni–N–C catalysts at industrial current densities. *Energy & Environmental Science* 12: 640–647.

144 Sahoo, S., Suib, S.L., and Alpay, S.P. (2018). Graphene supported single atom transition metal catalysts for methane activation. *ChemCatChem* 10: 3229–3235.

145 Psofogiannakis, G.M. and Froudakis, G.E. (2011). Fundamental studies and perceptions on the spillover mechanism for hydrogen storage. *Chemical Communications* 47: 7933–7943.

146 Gerber, I.C. and Serp, P. (2020). A theory/experience description of support effects in carbon-supported catalysts. *Chemical Reviews* 120: 1250–1349.

147 Guo, J.H., Li, S.J., Su, Y. et al. (2020). Theoretical study of hydrogen storage by spillover on porous carbon materials. *International Journal of Hydrogen Energy* 45: 25900–25911.

148 Cheng, H., Chen, L., Cooper, A.C. et al. (2008). Hydrogen spillover in the context of hydrogen storage using solid-state materials. *Energy & Environmental Science* 1: 338–354.

149 Chen, L., Cooper, A.C., Pez, G.P. et al. (2007). Density functional study of sequential H_2 dissociative chemisorption on a Pt_6 cluster. *Journal of Physical Chemistry C* 111: 5514–5519.

150 Zhou, C., Wu, J., Nie, A. et al. (2007). On the sequential hydrogen dissociative chemisorption on small platinum clusters: a density functional theory study. *Journal of Physical Chemistry C* 111: 12773–12778.

151 Zhou, C., Yao, S., Wu, J. et al. (2008). Hydrogen dissociative chemisorption and desorption on saturated subnano palladium clusters (Pd_n, n = 2–9). *Physical Chemistry Chemical Physics* 10: 5445–5451.

152 Psofogiannakis, G.M. and Froudakis, G.E. (2009). DFT study of the hydrogen spillover mechanism on Pt-doped graphite. *Journal of Physical Chemistry C* 113: 14908–14915.

153 Li, Q., Wang, H., Xia, H. et al. (2014). Density functional study of hydrogen adsorption and diffusion on Ni-loaded graphene and graphene oxide. *International Journal of Quantum Chemistry* 114: 879–884.

154 Guo, J.H., Liu, D.D., Li, X.D. et al. (2018). Pt_4, Pd_4, Ni_4, and Ti_4 catalyzed hydrogen spillover on penta-graphene for hydrogen storage: the first-principles and kinetic Monte Carlo study. *International Journal of Hydrogen Energy* 43: 2247–2255.

155 Ramos-Castillo, C.M., Reveles, J.U., Cifuentes-Quintal, M.E. et al. (2016). Ti_4- and Ni_4-doped defective graphene nanoplatelets as efficient materials for hydrogen storage. *Journal of Physical Chemistry C* 120: 5001–5009.

156 Rungnim, C., Faungnawakij, K., Sano, N. et al. (2018). Hydrogen storage performance of platinum supported carbon nanohorns: a DFT study of reaction mechanisms, thermodynamics, and kinetics. *International Journal of Hydrogen Energy* 43: 23336–23345.

157 Wu, H.Y., Fan, X., Kuo, J.L. et al. (2011). DFT study of hydrogen storage by spillover on graphene with boron substitution. *Journal of Physical Chemistry C* 115: 9241–9249.

158 Zhou, Q., Wang, C., Fu, Z. et al. (2015). Hydrogen adsorption on palladium anchored defected graphene with B-doping: a theoretical study. *International Journal of Hydrogen Energy* 40: 2473–2483.

159 Rangel, E. and Sansores, E. (2014). Theoretical study of hydrogen adsorption on nitrogen doped graphene decorated with palladium clusters. *International Journal of Hydrogen Energy* 39: 6558–6566.

160 Karim, W., Spreafico, C., Kleibert, A. et al. (2017). Catalyst support effects on hydrogen spillover. *Nature* 541: 68–71.

161 Gomez, T., Florez, E., Rodriguez, J.A. et al. (2011). Reactivity of transition metals (Pd, Pt, Cu, Ag, Au) toward molecular hydrogen dissociation: extended surfaces versus particles supported on TiC(001) or small is not always better and large is not always bad. *Journal of Physical Chemistry C* 115: 11666–11672.

162 Doudin, N., Yuk, S.F., Marcinkowski, M.D. et al. (2019). Understanding heterolytic H_2 cleavage and water-assisted hydrogen spillover on Fe_3O_4(001)-supported single palladium atoms. *ACS Catalysis* 9: 7876–7887.

163 Sihag, A., Xie, Z.L., Thang, H.V. et al. (2019). DFT insights into comparative hydrogen adsorption and hydrogen spillover mechanisms of Pt_4/graphene and Pt_4/anatase (101) surfaces. *Journal of Physical Chemistry C* 123: 25618–25627.

164 Singh, A.K., Ribas, M.A., and Yakobson, B.I. (2009). H-Spillover through the catalyst saturation: an Ab initio thermodynamics study. *ACS Nano* 3: 1657–1662.

165 Chen, L., Cooper, A.C., Pez, G.P. et al. (2007). Mechanistic study on hydrogen spillover onto graphitic carbon materials. *Journal of Physical Chemistry C* 111: 18995–19000.

166 Juarez-Mosqueda, R., Mavrandonakis, A., Kuc, A.B. et al. (2015). Theoretical analysis of hydrogen spillover mechanism on carbon nanotubes. *Frontiers in Chemistry* 3: 2.

167 Guo, J.H., Li, X.D., Cheng, X.L. et al. (2018). The theoretical study of the bimetallic Ni/Pd, Ni/Pt and Pt/Pd catalysts for hydrogen spillover on penta-graphene. *International Journal of Hydrogen Energy* 43: 19121–19129.

168 Rangel, E., Sansores, E., Vallejo, E. et al. (2016). Study of the interplay between N-graphene defects and small Pd clusters for enhanced hydrogen storage via a spill-over mechanism. *Physical Chemistry Chemical Physics* 18: 33158–33170.

169 Psofogiannakis, G.M. and Froudakis, G.E. (2009). DFT study of hydrogen storage by spillover on graphite with oxygen surface groups. *Journal of the American Chemical Society* 131: 15133–15135.

170 Psofogiannakis, G.M., Steriotis, T.A., Bourlinos, A.B. et al. (2011). Enhanced hydrogen storage by spillover on metal-doped carbon foam: an experimental and computational study. *Nanoscale* 3: 933–936.

171 Ahmed, F., Alam, M.K., Suzuki, A. et al. (2009). Dynamics of hydrogen spillover on Pt/γ-Al_2O_3 catalyst surface: a quantum chemical molecular dynamics study. *Journal of Physical Chemistry C* 113: 15676–15683.

172 Chen, L., Cooper, A.C., Pez, G.P. et al. (2008). On the mechanisms of hydrogen spillover in MoO_3. *Journal of Physical Chemistry C* 112: 1755–1758.

173 Yang, F.H., Lachawiec, A.J., and Yang, R.T. (2006). Adsorption of spillover hydrogen atoms on single-wall carbon nanotubes. *Journal of Physical Chemistry B* 110: 6236–6244.

174 Sha, X., Knippenberg, M.T., Cooper, A.C. et al. (2008). Dynamics of hydrogen spillover on carbon-based materials. *Journal of Physical Chemistry C* 112: 17465–17470.

175 Yodsin, N., Rungnim, C., Promarak, V. et al. (2018). Influence of hydrogen spillover on Pt-decorated carbon nanocones for enhancing hydrogen storage capacity: a DFT mechanistic study. *Physical Chemistry Chemical Physics* 20: 21194–21203.

176 Lueking, A.D., Psofogiannakis, G., and Froudakis, G.E. (2013). Atomic hydrogen diffusion on doped and chemically modified graphene. *Journal of Physical Chemistry C* 117: 6312–6319.

177 Han, S.S., Kim, H., and Park, N. (2011). Effect of shuttling catalyst on the migration of hydrogen adatoms: a strategy for the facile hydrogenation of graphene. *Journal of Physical Chemistry C* 115: 24696–24701.

178 Dutta, G., Waghmare, U.V., Baidya, T. et al. (2007). Hydrogen spillover on CeO_2/Pt: enhanced storage of active hydrogen. *Chemistry of Materials* 19: 6430–6436.

179 Zhang, H., Liu, G., Shi, L. et al. (2017). Single-atom catalysts: emerging multifunctional materials in heterogeneous catalysis. *Advanced Energy Materials* 8: 1701343.

180 Yang, Y., Yang, Y., Pei, Z. et al. (2020). Recent progress of carbon-supported single-atom catalysts for energy conversion and storage. *Matter* 3: 1442–1476.

181 Gawande, M.B., Fornasiero, P., and Zbořil, R. (2020). Carbon-based single-atom catalysts for advanced applications. *ACS Catalysis* 10: 2231–2259.

182 Zhuo, H.Y., Zhang, X., Liang, J.X. et al. (2020). Theoretical understandings of graphene-based metal single-atom catalysts: stability and catalytic performance. *Chemical Reviews* 120: 12315–12341.

183 Goerigk, L., Hansen, A., Bauer, C. et al. (2017). A look at the density functional theory zoo with the advanced GMTKN55 database for general main group thermochemistry, kinetics and noncovalent interactions. *Physical Chemistry Chemical Physics* 19: 32184–32215.

184 Chorkendorff, I. and Niemantsverdriet, J.W. (2003). *Concepts of Modern Catalysis and Kinetics*. Wiley-VCH.

185 Zakaria, Z. and Kamarudin, S.K. (2016). Direct conversion technologies of methane to methanol: an overview. *Renewable and Sustainable Energy Reviews* 65: 250–261.

186 Lepron, M., Daniel-Bertrand, M., Mencia, G. et al. (2021). Nanocatalyzed hydrogen isotope exchange. *Accounts of Chemical Research* 54: 1465–1480.

187 Borrome, M. and Gronert, S. (2019). Gas-phase dehydrogenation of alkanes: C-H activation by a graphene-supported nickel single-atom catalyst model. *Angewandte Chemie International Edition* 58: 14906–14910.

188 Sun, X.Y., Han, P., Li, B. et al. (2018). Tunable catalytic performance of single Pt atom on doped graphene in direct dehydrogenation of propane by rational doping: a density functional theory study. *Journal of Physical Chemistry C* 122: 1570–1576.

189 Yan, H., Lv, H., Yi, H. et al. (2018). Understanding the underlying mechanism of improved selectivity in Pd_1 single-atom catalyzed hydrogenation reaction. *Journal of Catalysis* 366: 70–79.

190 Huang, F., Deng, Y., Chen, Y. et al. (2019). Anchoring Cu_1 species over nanodiamond-graphene for semi-hydrogenation of acetylene. *Nature Communications* 10: 4431.

191 Zhuo, H.Y., Yu, X., Yu, Q. et al. (2020). Selective hydrogenation of acetylene on graphene-supported non-noble metal single-atom catalysts. *Science China Materials* 63: 1741–1749.

192 Gong, W., Zhao, F., and Kang, L. (2018). Novel nitrogen-doped Au-embedded graphene single-atom catalysts for acetylene hydrochlorination: a density functional theory study. *Computational and Theoretical Chemistry* 1130: 83–89.

193 Zhou, X., Zhu, M., and Kang, L. (2019). Single-atom X/g-C_3N_4(X = Au_1, Pd_1, and Ru_1) catalysts for acetylene hydrochlorination: a density functional theory study. *Catalysts* 9: 808.

194 Ali, S., Lian, Z., and Li, B. (2021). Density functional theory study of a graphdiyne-supported single Au atom catalyst for highly efficient acetylene hydrochlorination. *ACS Applied Nano Materials* 4: 6152–6159.

195 Podrojková, N., Sans, V., Oriňak, A. et al. (2020). Recent developments in the modelling of heterogeneous catalysts for CO_2 conversion to chemicals. *ChemCatChem* 12: 1802–1825.

196 Sirijaraensre, J. and Limtrakul, J. (2016). Hydrogenation of CO_2 to formic acid over a Cu-embedded graphene: a DFT study. *Applied Surface Science* 364: 241–248.

197 Sredojević, D.N., S˘ ljivančanin, v.Z., Brothers, E.N. et al. (2018). Formic acid synthesis by CO_2 hydrogenation over single-atom catalysts based on Ru and Cu embedded in graphene. *ChemistrySelect* 3: 2631–2637.

198 Li, Y., Zhou, Z., Yu, G. et al. (2010). CO catalytic oxidation on iron-embedded graphene: computational quest for low-cost nanocatalysts. *Journal of Physical Chemistry C* 114: 6250–6254.

199 Wannakao, S., Nongnual, T., Khongpracha, P. et al. (2012). Reaction mechanisms for CO catalytic oxidation by N_2O on Fe-embedded graphene. *Journal of Physical Chemistry C* 116: 16992–16998.

200 Zhang, X., Lu, Z., Xu, G. et al. (2015). Single Pt atom stabilized on nitrogen doped graphene: CO oxidation readily occurs via the tri-molecular Eley–Rideal mechanism. *Physical Chemistry Chemical Physics* 17: 20006–20013.

201 Liu, Z., He, T., Liu, K. et al. (2017). Structural, electronic and catalytic performances of single-atom Fe stabilized by divacancy-nitrogen-doped graphene. *RSC Advances* 7: 7920–7928.

202 Tang, Y., Chen, W., Shen, Z. et al. (2017). Nitrogen coordinated silicon-doped graphene as a potential alternative metal-free catalyst for CO oxidation. *Carbon* 111: 448–458.

203 Li, D., Li, W., and Zhang, J. (2019). Catalytic CO oxidation by Fe doped penta-graphene: a density functional study. *Molecular Catalysis* 470: 48–55.

204 Luo, M., Liang, Z., Liu, C. et al. (2020). Theoretical calculation of different reaction mechanisms for CO oxidation on MnN_3-doped graphene. *ACS Omega* 5: 21203–21210.

205 Sun, W., Shi, R., Wang, X. et al. (2017). Density-functional theory study of dimethyl carbonate synthesis by methanol oxidative carbonylation on single-atom Cu_1/graphene catalyst. *Applied Surface Science* 425: 291–300.

206 Bakandritsos, A., Kadam, R.G., Kumar, P. et al. (2019). Mixed-valence single-atom catalyst derived from functionalized graphene. *Advanced Materials* 31: 1900323.

207 Yang, W., Gao, Z., Liu, X. et al. (2019). Directly catalytic reduction of NO without NH_3 by single atom iron catalyst: a DFT calculation. *Fuel* 243: 262–270.

208 Li, X., Zhao, L., Yu, J. et al. (2020). Water splitting: from electrode to green energy system. *Nano-Micro Letters* 12: 131.

209 Nørskov, J.K., Bligaard, T., Logadottir, A. et al. (2005). Trends in the exchange current for hydrogen evolution. *Journal of the Electrochemical Society* 152: J23–J26.

210 Lim, J., Back, S., Choi, C. et al. (2018). Ultralow overpotential of hydrogen evolution reaction using Fe-doped defective graphene: a density functional study. *ChemCatChem* 10: 4450–4455.

211 Zhang, H., An, P., Zhou, W. et al. (2018). Dynamic traction of lattice-confined platinum atoms into mesoporous carbon matrix for hydrogen evolution reaction. *Science Advances* 4: eaao6657.

212 Ye, S., Luo, F., Zhang, Q. et al. (2019). Highly stable single Pt atomic sites anchored on aniline-stacked graphene for hydrogen evolution reaction. *Energy & Environmental Science* 12: 1000–1007.

213 Hossain, M.D., Liu, Z., Zhuang, M. et al. (2019). Rational design of graphene-supported single atom catalysts for hydrogen evolution reaction. *Advanced Energy Materials* 9: 1803689.

214 He, T., Zhang, C., and Du, A. (2019). Single-atom supported on graphene grain boundary as an efficient electrocatalyst for hydrogen evolution reaction. *Chemical Engineering Science* 194: 58–63.

215 Narayanan, H., Viswanathan, B., Krishnamurthy, K.R. et al. (2019). Hydrogen from photo-electrocatalytic water splitting. In: *Solar Hydrogen Production* (ed. F. Calise, M.D. D'Accadia, M. Santarelli et al.), 419–486. Elsevier.

216 Man, I.C., Su, H.Y., Calle-Vallejo, F. et al. (2011). Universality in oxygen evolution electrocatalysis on oxide surfaces. *ChemCatChem* 3: 1159–1165.

217 Dickens, C.F., Kirk, C., and Nørskov, J.K. (2019). Insights into the electrochemical oxygen evolution reaction with Ab initio calculations and microkinetic modeling: beyond the limiting potential volcano. *Journal of Physical Chemistry C* 123: 18960–18977.

218 Gao, G., Bottle, S., and Du, A. (2018). Understanding the activity and selectivity of single atom catalysts for hydrogen and oxygen evolution via Ab initial study. *Catalysis Science & Technology* 8: 996–1001.

219 Chen, C., Chen, G., and Kong, X. (2018). Enhanced oxygen evolution reaction for single atomic Co catalyst via support modification: a density functional theory design predication. *Inorganic Chemistry* 57: 13020–13026.

220 Liang, Z., Luo, M., Chen, M. et al. (2020). Exploring the oxygen electrode Bi-functional activity of Ni–N–C-doped graphene systems with N, C Co-ordination and OH ligand effects. *Journal of Materials Chemistry A* 8: 20453–20462.

221 Ma, R., Lin, G., Zhou, Y. et al. (2019). A review of oxygen reduction mechanisms for metal-free carbon-based electrocatalysts. *npj Computational Materials* 5: 78.

222 Liu, M., Wang, L., Zhao, K. et al. (2019). Atomically dispersed metal catalysts for the oxygen reduction reaction: synthesis, characterization, reaction mechanisms and electrochemical energy applications. *Energy & Environmental Science* 12: 2890–2923.

223 Kulkarni, A., Siahrostami, S., Patel, A. et al. (2018). Understanding catalytic activity trends in the oxygen reduction reaction. *Chemical Reviews* 118: 2302–2312.

224 Chen, X., Chen, S., and Wang, J. (2016). Screening of catalytic oxygen reduction reaction activity of metal-doped graphene by density functional theory. *Applied Surface Science* 379: 291–295.

225 Chen, X. and Hu, R. (2019). DFT-based study of single transition metal atom doped g-C_3N_4 as alternative oxygen reduction reaction catalysts. *International Journal of Hydrogen Energy* 44: 15409–15416.

226 Zhang, X., Xia, Z., Li, H. et al. (2019). The mechanism and activity of oxygen reduction reaction on single atom doped graphene: a DFT method. *RSC Advances* 9: 7086–7093.

227 Lindgren, P., Kastlunger, G., and Peterson, A.A. (2019). A challenge to the $G \sim 0$ interpretation of hydrogen evolution. *ACS Catalysis* 10: 121–128.

228 Li, L., Huang, R., Cao, X. et al. (2020). Computational screening of efficient graphene-supported transition metal single atom catalysts toward the oxygen reduction reaction. *Journal of Materials Chemistry A* 8: 19319–19327.

229 Liang, Z., Liu, C., Chen, M. et al. (2019). Oxygen reduction reaction mechanism on P, N Co-doped graphene: a density functional theory study. *New Journal of Chemistry* 43: 19308–19317.

230 Cao, H., Xia, G.J., Chen, J.W. et al. (2020). Mechanistic insight into the oxygen reduction reaction on the Mn–N_4/C single-atom catalyst: the role of the solvent environment. *Journal of Physical Chemistry C* 124: 7287–7294.

231 Back, S., Lim, J., Kim, N.Y. et al. (2017). Single-atom catalysts for CO_2 electroreduction with significant activity and selectivity improvements. *Chemical Science* 8: 1090–1096.

232 He, H. and Jagvaral, Y. (2017). Electrochemical reduction of CO_2 on graphene supported transition metals – towards single atom catalysts. *Physical Chemistry Chemical Physics* 19: 11436–11446.

233 Zhang, H., Li, J., Xi, S. et al. (2019). A graphene-supported single-atom FeN_5 catalytic site for efficient electrochemical CO_2 reduction. *Angewandte Chemie International Edition* 58: 14871–14876.

234 Yuan, H., Li, Z., Zeng, X.C. et al. (2020). Descriptor-based design principle for two-dimensional single-atom catalysts: carbon dioxide electroreduction. *The Journal of Physical Chemistry Letters* 11: 3481–3487.

235 Rong, X., Wang, H.J., Lu, X.L. et al. (2020). Controlled synthesis of a vacancy-defect single-atom catalyst for boosting CO_2 electroreduction. *Angewandte Chemie International Edition* 59: 1961–1965.

236 Guo, C., Zhang, T., Deng, X. et al. (2019). Electrochemical CO_2 reduction to C_1 products on single nickel/cobalt/iron-doped graphitic carbon nitride: a DFT study. *ChemSusChem* 12: 5126–5132.

237 Zhao, X. and Liu, Y. (2020). Unveiling the active structure of single nickel atom catalysis: critical roles of charge capacity and hydrogen bonding. *Journal of the American Chemical Society* 142: 5773–5777.

238 Kirk, C., Chen, L.D., Siahrostami, S. et al. (2017). Theoretical investigations of the electrochemical reduction of CO on single metal atoms embedded in graphene. *ACS Central Science* 3: 1286–1293.

239 Ling, C., Ouyang, Y., Li, Q. et al. (2018). A general two-step strategy–based high-throughput screening of single atom catalysts for nitrogen fixation. *Small Methods* 3: 1800376.

240 Riyaz, M. and Goel, N. (2019). Single-atom catalysis using chromium embedded in divacant graphene for conversion of dinitrogen to ammonia. *ChemPhysChem* 20: 1954–1959.

241 Liu, P., Fu, C., Li, Y. et al. (2020). Theoretical screening of single atoms anchored on defective graphene for electrocatalytic N_2 reduction reactions: a DFT study. *Physical Chemistry Chemical Physics* 22: 9322–9329.

242 Qian, Y., Liu, Y., Zhao, Y. et al. (2020). Single vs double atom catalyst for N_2 activation in nitrogen reduction reaction: a DFT perspective. *EcoMat* 2: e12014.

243 Sahoo, S.K., Heske, J., Antonietti, M. et al. (2020). Electrochemical N_2 reduction to ammonia using single Au/Fe atoms supported on nitrogen-doped porous carbon. *ACS Applied Energy Materials* 3: 10061–10069.

244 Ji, Y., Li, Y., Dong, H. et al. (2020). Ruthenium single-atom catalysis for electrocatalytic nitrogen reduction unveiled by grand canonical density functional theory. *Journal of Materials Chemistry A* 8: 20402–20407.

245 Wang, J., Jia, L., Zhong, J. et al. (2019). Single-atom catalyst boosts electrochemical conversion reactions in batteries. *Energy Storage Materials* 18: 246–252.

246 Xu, X.Y., Guo, H., and Zhao, C. (2017). Probing the electric field effect on the catalytic performance of Mn-doped graphene to CO oxidation. *Journal of Physical Chemistry C* 121: 27983–27991.

247 Xu, X.Y., Xu, H., Guo, H. et al. (2020). Mechanism investigations on CO oxidation catalyzed by Fe-doped graphene: a theoretical study. *Applied Surface Science* 523: 146496.

248 Pacchioni, G. (2008). Modeling doped and defective oxides in catalysis with density functional theory methods: room for improvements. *Journal of Chemical Physics* 128: 182505.

249 Yang, X.F., Wang, A., Qiao, B. et al. (2013). Single-atom catalysts: a new frontier in heterogeneous catalysis. *Accounts of Chemical Research* 46: 1740–1748.

250 Pacchioni, G. (2013). Electronic interactions and charge transfers of metal atoms and clusters on oxide surfaces. *Physical Chemistry Chemical Physics* 15: 1737–1757.

251 Yongyong, J., Panpan, H., Jun, R. et al. (2015). Single atom catalysis: concept, method and application. *Progress in Chemistry* 27: 1689–1704.

252 Peters, B. and Scott, S.L. (2015). Single atom catalysts on amorphous supports: a quenched disorder perspective. *Journal of Chemical Physics* 142: 104708.

253 Tang, Y., Zhao, S., Long, B. et al. (2016). On the nature of support effects of metal dioxides MO_2 (M = Ti, Zr, Hf, Ce, Th) in single-atom gold catalysts: importance of quantum primogenic effect. *Journal of Physical Chemistry C* 120: 17514–17526.

254 Wang, L., Huang, L., Liang, F. et al. (2017). Preparation, characterization and catalytic performance of single-atom catalysts. *Chinese Journal of Catalysis* 38: 1528–1539.

255 Parkinson, G.S. (2017). Unraveling single atom catalysis: the surface science approach. *Chinese Journal of Catalysis* 38: 1454–1459.

256 Tan, K. (2017). Predicting metal–support interactions in oxide-supported single-atom catalysts. PhD thesis, B. S. Material Chemistry. University of Science and Technology of China.

257 Li, H., Zhang, H.X., Yan, X.L. et al. (2018). Carbon-supported metal single atom catalysts. *New Carbon Materials* 33: 1–11.

258 Liu, J.C., Tang, Y., Wang, Y.G. et al. (2018). Theoretical understanding of the stability of single-atom catalysts. *National Science Review* 5: 638–641.

259 O'Connor, N.J., Jonayat, A.S.M., Janik, M.J. et al. (2018). Interaction trends between single metal atoms and oxide supports identified with density functional theory and statistical learning. *Nature Catalysis* 1: 531–539.

260 Lai, W. (2019). Highly efficient single atom based materials in energy conversion. PhD thesis. University of Wollongong.

261 Dietze, E.M. and Plessow, P.N. (2019). Predicting the strength of metal–support interaction with computational descriptors for adhesion energies. *Journal of Physical Chemistry C* 123: 20443–20450.

262 Kauppinen, M.M., Melander, M.M., and Honkala, K. (2020). First-principles insight into CO hindered agglomeration of Rh and Pt single atoms on m-ZrO_2. *Catalysis Science & Technology* 10: 5847–5855.

263 Gutić, S.J., Dobrota, A.S., Fako, E. et al. (2020). Hydrogen evolution reaction-from single crystal to single atom catalysts. *Catalysts* 10: 290.

264 Bye, G.C. and Simpkin, G.T. (1974). Influence of Cr and Fe on formation of α-Al_2O_3 from γ-Al_2O_3. *Journal of the American Chemical Society* 57: 367–371.

265 Ozawa, M., Kimura, M., and Isogai, A. (1990). Thermal stability and characterization of γ-Al_2O_3 modified with lanthanum or cerium. *Journal of Materials Science Letters* 9: 709–711.

266 Verdozzi, C., Jennison, D.R., Schultz, P.A. et al. (1999). Sapphire (0001) surface, clean and with *d*-metal overlayers. *Physical Review Letters* 82: 799.

267 Bogicevic, A. and Jennison, D.R. (1999). Variations in the nature of metal adsorption on ultrathin Al_2O_3 films. *Physical Review Letters* 82: 4050.

268 Ragan, D.D., Mates, T., and Clarke, D.R. (2003). Effect of yttrium and erbium ions on epitaxial phase transformations in alumina. *Journal of the American Ceramic Society* 86: 541–545.

269 Fu, L., Li, X., Liu, M. et al. (2013). Insights into the nature of Cu doping in amorphous mesoporous alumina. *Journal of Materials Chemistry A* 1: 14592–14605.

270 Limmer, K.R., Neupane, M.R., Brennan, R.E. et al. (2016). Rare-earth dopant effects on the structural, energetic, and magnetic properties of alumina from first principles. *Journal of the American Chemical Society* 99: 4007–4012.

271 Somjit, V. and Yildiz, B. (2019). Doping α-Al_2O_3 to reduce its hydrogen permeability: thermodynamic assessment of hydrogen defects and solubility from first principles. *Acta Materialia* 169: 172–183.

272 Tan, K., Dixit, M., Dean, J. et al. (2019). Predicting metal–support interactions in oxide-supported single atom catalysts. *Industrial and Engineering Chemistry Research* 58: 20236–20246.

273 Łodziana, Z. and Nørskov, J.K. (2001). Adsorption of Cu and Pd on α-Al_2O_3(0001) surfaces with different stoichiometries. *Journal of Chemical Physics* 115: 11261–11267.
274 Kelber, J.A., Niu, C., Shepherd, K. et al. (2000). Copper wetting of α-Al_2O_3(0001): theory and experiment. *Surface Science* 446: 76–88.
275 Niu, C., Shepherd, K., Martini, D. et al. (2000). Cu interactions with α-Al_2O_3(0001): effects of surface hydroxyl groups versus dehydroxylation by ar-ion sputtering. *Surface Science* 465: 163–176.
276 Hernández, N.C. and Sanz, J.F. (2002). First principles study of Cu atoms deposited on the α-Al_2O_3(0001) surface. *Journal of Physical Chemistry B* 106: 11495–11500.
277 Sanz, J.F. and Hernández, N.C. (2005). Mechanism of Cu deposition on the α-Al_2O_3(0001) surface. *Physical Review Letters* 94: 016104.
278 Hernández, N.C. and Sanz, J.F. (2004). First principles simulations of Cu and Au deposition on α-Al_2O_3(0001) surface. *Applied Surface Science* 238: 228–232.
279 Hernández, N.C., Graciani, J., Márquez, A. et al. (2005). Cu, Ag and Au atoms deposited on the α-Al_2O_3(0001) surface: a comparative density functional study. *Surface Science* 575: 189–196.
280 Nigam, S. and Majumder, C. (2010). Growth pattern of Ag_n($n = 1$–8) clusters on the α-Al_2O_3(0001) surface: a first principles study. *Langmuir* 26: 18776–18787.
281 Gomes, J.R.B., Illas, F., Hernández, N.C. et al. (2002). Surface model and exchange-correlation functional effects on the description of Pd/α-Al_2O_3(0001). *Journal of Chemical Physics* 116: 1684–1691.
282 Rivanenkov, V.V., Nasluzov, V.A., Shor, A.M. et al. (2003). Adsorption of Pd and Pt atoms on α-Al_2O_3(0001): density functional study of cluster models embedded in an elastic polarizable environment. *Surface Science* 525: 173–183.
283 Briquet, L.G.V., Catlow, C.R.A., and French, S.A. (2008). Comparison of the adsorption of Ni, Pd, and Pt on the (0001) surface of α-alumina. *Journal of Physical Chemistry C* 112: 18948–18954.
284 Ma, Q., Klier, K., Cheng, H. et al. (2001). Interaction between catalyst and support. 2. Low coverage of Co and Ni at the alumina surface. *Journal of Physical Chemistry B* 105: 2212–2221.
285 Zhang, M., Chen, J., Yu, Y. et al. (2013). DFT study on the structure of Ni/α-Al_2O_3 catalysts. *Applied Surface Science* 287: 97–107.
286 Gao, F., Gao, S., and Meng, S. (2017). Screening single-atom catalysts for methane activation: α–Al_2O_3 (0001)–supported Ni. *Physical Review Materials* 1: 035801.
287 Larmier, K., Chizallet, C., and Raybaud, P. (2015). Tuning the metal–support interaction by structural recognition of cobalt-based catalyst precursors. *Angewandte Chemie International Edition* 54: 6824–6827.
288 Shi, X.R. and Sholl, D.S. (2012). Nucleation of Rh_n ($n= 1$–5) clusters on γ-Al_2O_3 surfaces: a density functional theory study. *Journal of Physical Chemistry C* 116: 10623–10631.
289 Bäumer, M., Frank, M., Heemeier, M. et al. (2000). Nucleation and growth of transition metals on a thin alumina film. *Surface Science* 454–456: 957–962.

290 Gao, H. (2016). DFT study of the adsorption properties of single Pt, Pd, Ag, In and Sn on the γ-Al_2O_3(110) surface. *Chemical Physics Letters* 657: 11–17.

291 Dessal, C., Sangnier, A., Chizallet, C. et al. (2019). Atmosphere-dependent stability and mobility of catalytic Pt single atoms and clusters on γ-Al_2O_3. *Nanoscale* 11: 6897–6904.

292 Fu, L., Yang, H., Hu, Y. et al. (2017). Tailoring mesoporous γ-Al_2O_3 properties by transition metal doping: a combined experimental and computational study. *Chemistry of Materials* 29: 1338–1349.

293 Shi, L., Huang, Y., Lu, Z.H. et al. (2021). Surface property of the Cu doped γ-Al_2O_3: a density functional theory study. *Applied Surface Science* 535: 147651.

294 Mulwa, W.M. and Dejene, F.B. (2018). γ-Al_2O_3:$Ce^{3+}Cu^{2+}$ as a Phosphor material; DFT+U and experimental approach. *Physical Sciences Reviews* 3: 20170165.

295 Greenwald, S. (1953). The antiferromagnetic structure deformations in CoO and MnTe. *Acta Crystallographica* 6: 396–398.

296 Jauch, W., Reehuis, M., Bleif, H.J. et al. (2001). Crystallographic symmetry and magnetic structure of CoO. *Physical Review B* 64: 052102.

297 Smith, W.L. and Hobson, A.D. (1973). The structure of cobalt oxide, Co_3O_4. *Acta Crystallographica* B29: 362–363.

298 Qiao, B., Lin, J., Wang, A. et al. (2015). Highly active Au_1/Co_3O_4 single-atom catalyst for CO oxidation at room temperature. *Chinese Journal of Catalysis* 36: 1505–1511.

299 Lou, Y., Cai, Y., Hu, W. et al. (2020). Identification of active area as active center for CO oxidation over single Au atom catalyst. *ACS Catalysis* 10: 6094–6101.

300 Yang, C., Zhao, Z.Y., and Liu, Q.J. (2021). Theoretical study of CO oxidation on Au_1/Co_3O_4(110) single atom catalyst using density functional theory calculations. *Materials Science in Semiconductor Processing* 123: 105578.

301 Liu, B., Li, W., Song, W. et al. (2018). Carbonate-mediated Mars–van Krevelen mechanism for CO oxidation on cobalt-doped ceria catalysts: facet-dependence and coordination-dependence. *Physical Chemistry Chemical Physics* 20: 16045–16059.

302 Bera, P. and Hegde, M.S. (2015). Noble metal ions in CeO_2 and TiO_2: synthesis, structure and catalytic properties. *RSC Advances* 5: 94949–94979.

303 Nolan, M. (2012). Charge transfer and formation of reduced Ce^{3+} upon adsorption of metal atoms at the ceria (110) surface. *Journal of Chemical Physics* 136: 134703.

304 Yang, Z., Lu, Z., and Luo, G. (2007). First-principles study of the Pt/CeO_2(111) interface. *Physical Review B* 76: 075421.

305 Hatanaka, M., Takahashi, N., Takahashi, N. et al. (2009). Reversible changes in the Pt oxidation state and nanostructure on a ceria-based supported Pt. *Journal of Catalysis* 266: 182–190.

306 Hatanaka, M., Takahashi, N., Tanabe, T. et al. (2010). Ideal Pt loading for a Pt/CeO_2-based catalyst stabilized by a Pt–O–Ce bond. *Applied Catalysis B: Environmental* 99: 336–342.

307 Dvořák, F., Camellone, M.F., Tovt, A. et al. (2016). Creating single-atom Pt-ceria catalysts by surface step decoration. *Nature Communications* 7: 10801.

308 Tang, Y., Wang, Y.G., and Li, J. (2017). Theoretical investigations of $Pt_1@CeO_2$ single-atom catalyst for CO oxidation. *Journal of Physical Chemistry C* 121: 11281–11289.

309 Jones, J., Xiong, H., DeLaRiva, A.T. et al. (2016). Thermally stable single-atom platinum-on-ceria catalysts via atom trapping. *Science* 353: 150–154.

310 Daelman, N., Capdevila-Cortada, M., and López, N. (2019). Dynamic charge and oxidation state of Pt/CeO_2 single-atom catalysts. *Nature Materials* 18: 1215–1221.

311 Bruix, A., Lykhach, Y., Matolínová, I. et al. (2014). Maximum noble-metal efficiency in catalytic materials: atomically dispersed surface platinum. *Angewandte Chemie International Edition* 53: 10525–10530.

312 Capdevila-Cortada, M. and López, N. (2016). Entropic contributions enhance polarity compensation for CeO_2(100) surfaces. *Nature Materials* 16: 328–334.

313 Mayernick, A.D. and Janik, M.J. (2008). Methane activation and oxygen vacancy formation over CeO_2 and Zr, Pd substituted CeO_2 surfaces. *Journal of Physical Chemistry C* 112: 14955–14964.

314 Yang, Z., Fu, Z., Zhang, Y. et al. (2010). Direct CO oxidation by lattice oxygen on Zr-doped ceria surfaces. *Catalysis Letters* 141: 78–82.

315 Mayernick, A.D. and Janik, M.J. (2009). Ab initio thermodynamic evaluation of Pd atom interaction with CeO_2 surfaces. *Journal of Chemical Physics* 131: 084701.

316 Song, W., Su, Y., and Hensen, E.J.M. (2015). A DFT study of CO oxidation at the Pd–CeO_2(110) interface. *Journal of Physical Chemistry C* 119: 27505–27511.

317 Senftle, T.P., van Duin, A.C.T., and Janik, M.J. (2015). Role of site stability in methane activation on $Pd_xCe_{1-x}O_\delta$ surfaces. *ACS Catalysis* 5: 6187–6199.

318 Amsler, J., Sarma, B.B., Agostini, G. et al. (2020). Prospects of heterogeneous hydroformylation with supported single atom catalysts. *Journal of the American Chemical Society* 142: 5087–5096.

319 Breslow, D.S. and Heck, R.F. (1960). Mechanism of the hydroformylation of olefins. *Chemistry and Industry* 17: 467.

320 Heck, R.F. and Breslow, D.S. (1961). The reaction of cobalt hydrotetracarbonyl with olefins. *Journal of the American Chemical Society* 83: 4023–4027.

321 Sarma, B.B., Kim, J., Amsler, J. et al. (2020). One-pot cooperation of single-atom Rh and Ru solid catalysts for a selective tandem olefin isomerization-hydrosilylation process. *Angewandte Chemie International Edition* 59: 5806–5815.

322 Jia, X. and Huang, Z. (2015). Conversion of alkanes to linear alkylsilanes using an iridium–iron-catalysed tandem dehydrogenation–isomerization–hydrosilylation. *Nature Chemistry* 8: 157–161.

323 Chalk, A.J. and Harrod, J.F. (1965). Homogeneous catalysis. II. The mechanism of the hydrosilation of olefins catalyzed by group VIII metal complexes. *Journal of the American Chemical Society* 87: 16–21.

324 Carrettin, S., Concepción, P., Corma, A. et al. (2004). Nanocrystalline CeO_2 increases the activity of Au for CO oxidation by two orders of magnitude. *Angewandte Chemie International Edition* 43: 2538–2540.

325 Carrettin, S., Corma, A., Iglesias, M. et al. (2005). Stabilization of Au(III) on heterogeneous catalysts and their catalytic similarities with homogeneous Au(III) metal organic complexes. *Applied Catalysis A: General* 291: 247–252.

326 Guzman, J., Carrettin, S., and Corma, A. (2005). Spectroscopic evidence for the supply of reactive oxygen during CO oxidation catalyzed by gold supported on nanocrystalline CeO_2. *Journal of the American Chemical Society* 127: 3286–3287.

327 Shapovalov, V. and Metiu, H. (2007). Catalysis by doped oxides: CO oxidation by $Au_xCe_{1-x}O_2$. *Journal of Catalysis* 245: 205–214.

328 Branda, M.M., Hernández, N.C., Sanz, J.F. et al. (2010). Density functional theory study of the interaction of Cu, Ag, and Au atoms with the regular CeO_2(111) surface. *Journal of Physical Chemistry C* 114: 1934–1941.

329 Chen, Y., Hu, P., Lee, M.H. et al. (2008). Au on (111) and (110) surfaces of CeO_2: a density-functional theory study. *Surface Science* 602: 1736–1741.

330 Castellani, N.J., Branda, M.M., Neyman, K.M. et al. (2009). Density functional theory study of the adsorption of Au atom on cerium oxide: effect of low-coordinated surface sites. *Journal of Physical Chemistry C* 113: 4948–4954.

331 Hernández, N.C., Grau-Crespo, R., de Leeuw, N.H. et al. (2009). Electronic charge transfer between ceria surfaces and gold adatoms: a GGA+U investigation. *Physical Chemistry Chemical Physics* 11: 5246–5252.

332 Branda, M.M., Castellani, N.J., Grau-Crespo, R. et al. (2009). On the difficulties of present theoretical models to predict the oxidation state of atomic Au adsorbed on regular sites of CeO_2(111). *Journal of Chemical Physics* 131: 094702.

333 Szabová, L., Camellone, M.F., Huang, M. et al. (2010). Thermodynamic, electronic and structural properties of Cu/CeO_2 surfaces and interfaces from first-principles DFT+U calculations. *Journal of Chemical Physics* 133: 234705.

334 Nolan, M., Soto Verdugo, V., and Metiu, H. (2008). Vacancy formation and CO adsorption on gold-doped ceria surfaces. *Surface Science* 602: 2734–2742.

335 Zhang, C., Michaelides, A., King, D.A. et al. (2008). Structure of gold atoms on stoichiometric and defective ceria surfaces. *Journal of Chemical Physics* 129: 194708.

336 Zhang, C., Michaelides, A., King, D.A. et al. (2009). Anchoring sites for initial Au nucleation on $CeO_2\{111\}$: O vacancy versus Ce vacancy. *Journal of Physical Chemistry C* 113: 6411–6417.

337 Liu, J.C., Wang, Y.G., and Li, J. (2017). Toward rational design of oxide-supported single-atom catalysts: atomic dispersion of gold on ceria. *Journal of the American Chemical Society* 139: 6190–6199.

338 Ziemba, M. and Hess, C. (2020). Influence of gold on the reactivity behaviour of ceria nanorods in CO oxidation: combining operando spectroscopies and DFT calculations. *Catalysis Science & Technology* 10: 3720–3730.

339 Tasker, P.W. (1979). The stability of ionic crystal surfaces. *Journal of Physics C: Solid State Physics* 12: 4977–4984.

340 Puigdollers, A.R., Schlexer, P., Tosoni, S. et al. (2017). Increasing oxide reducibility: the role of metal/oxide interfaces in the formation of oxygen vacancies. *ACS Catalysis* 7: 6493–6513.

341 López, N. and Illas, F. (1998). Ab initio modeling of the metal–support interface: the interaction of Ni, Pd, and Pt on MgO(100). *Journal of Physical Chemistry B* 102: 1430–1436.

342 Goniakowski, J. (1998). Adsorption of palladium on the MgO(100) surface: dependence on the metal coverage. *Physical Review B* 58: 1189.

343 Yudanov, I., Pacchioni, G., Neyman, K. et al. (1997). Systematic density functional study of the adsorption of transition metal atoms on the MgO(001) surface. *Journal of Physical Chemistry B* 101: 2786–2792.

344 Li, C., Wu, R., Freeman, A.J. et al. (1993). Energetics, bonding mechanism, and electronic structure of metal-ceramic interfaces: Ag/MgO(001). *Physical Review B* 48: 8317.

345 Smith, J.R., Hong, T., and Srolovitz, D.J. (1994). Metal-ceramic adhesion and the Harris functional. *Physical Review Letters* 72: 4021.

346 Heifets, E., Kotomin, E.A., and Orlando, R. (1996). Hartree-Fock simulation of the Ag/MgO interface structure. *Journal of Physics: Condensed Matter* 8: 6577–6584.

347 Li, Y., Langreth, D.C., and Pederson, M.R. (1995). Copper adsorption potentials of MgO(001). *Physical Review B* 52: 6067.

348 Pacchioni, G. and Rösch, N. (1996). Supported nickel and copper clusters on MgO(100): a first-principles calculation on the metal/oxide interface. *Journal of Chemical Physics* 104: 7329–7337.

349 Yudanov, I.V., Vent, S., Neyman, K. et al. (1997). Adsorption of Pd atoms and Pd_4 clusters on the MgO(001) surface: a density functional study. *Chemical Physics Letters* 275: 245–252.

350 Li, C. and Freeman, A.J. (1991). Giant monolayer magnetization of Fe on MgO: a nearly ideal two-dimensional magnetic system. *Physical Review B* 43: 780.

351 Chizallet, C., Costentin, G., Che, M. et al. (2006). Revisiting acido-basicity of the MgO surface by periodic density functional theory calculations: role of surface topology and ion coordination on water dissociation. *Journal of Physical Chemistry B* 110: 15878–15886.

352 Giordano, L., Goniakowski, J., and Pacchioni, G. (2001). Characteristics of Pd adsorption on the MgO(100) surface: role of oxygen vacancies. *Physical Review B* 64 (7): 075417.

353 Matveev, A.V., Neyman, K.M., Yudanov, I.V. et al. (1999). Adsorption of transition metal atoms on oxygen vacancies and regular sites of the MgO(001) surface. *Surface Science* 426: 123–139.

354 Del Vitto, A., Pacchioni, G., Delbecq, F. et al. (2005). Au atoms and dimers on the MgO(100) surface: a DFT study of nucleation at defects. *Journal of Physical Chemistry B* 109: 8040–8048.

355 Zhu, J., Farmer, J.A., Ruzycki, N. et al. (2008). Calcium adsorption on MgO(100): energetics, structure, and role of defects. *Journal of the American Chemical Society* 130: 2314–2322.

356 Matczak, P. (2018). Theoretical study of Sn adsorbed on the MgO(100) surface with defects. *Comptes Rendus Chimie* 21: 669–675.

357 Fernandez, S., Markovits, A., and Minot, C. (2008). Adsorption of the first row of transition metals on the perfect and defective MgO(100) surface. *Chemical Physics Letters* 463: 106–111.

358 Izumi, K., Mizokawa, T., and Hanamura, E. (2007). Optical response and electronic structure of Zn-doped $MgAl_2O_4$. *Journal of Applied Physics* 102: 053109.

359 Sarma, B.B., Plessow, P.N., Agostini, G. et al. (2020). Metal-specific reactivity in single-atom catalysts: CO oxidation on 4*D* and 5*D* transition metals atomically dispersed on MgO. *Journal of the American Chemical Society* 142: 14890–14902.

360 Park, M.S., Kwon, S.K., and Min, B.I. (2002). Electronic structures of doped anatase TiO_2:ti_1-xM_xO_2 (M=Co,Mn, Fe, Ni). *Physical Review B* 65: 161201(R).

361 Umebayashi, T., Yamaki, T., Itoh, H. et al. (2002). Analysis of electronic structures of 3D transition metal-doped TiO_2 based on band calculations. *Journal of Physics and Chemistry of Solids* 63: 1909–1920.

362 Geng, W.T. and Kim, K.S. (2003). Structural, electronic, and magnetic properties of a ferromagnetic semiconductor: Co-doped TiO_2 rutile. *Physical Review B* 68 (12): 125203.

363 Weng, H., Yang, X., Dong, J. et al. (2004). Electronic structure and optical properties of the co-doped anatase TiO_2 studied from first principles. *Physical Review B* 69: 125219.

364 Gao, G.Y., Yao, K.L., Liu, Z.L. et al. (2007). Magnetism and electronic structure of Cr-doped rutile from first-principles calculations. *Journal of Magnetism and Magnetic Materials* 313: 210–213.

365 Shao, G. (2008). Electronic structures of manganese-doped rutile TiO_2 from first principles. *The Journal of Physical Chemistry C* 112: 18677–18685.

366 Zuo, Z., Huang, W., Han, P. et al. (2009). A DFT study on the interaction of Co with an anatase TiO_2 (001)-(1×4) surface. *Journal of Natural Gas Chemistry* 18: 78–82.

367 Long, R. and English, N.J. (2010). First-principles calculation of electronic structure of V-doped anatase TiO_2. *ChemPhysChem* 11: 2606–2611.

368 Körner, W. and Elsässer, C. (2011). Density functional theory study of dopants in polycrystalline TiO_2. *Physical Review B* 83: 205315.

369 García-Mota, M., Vojvodic, A., Abild-Pedersen, F. et al. (2012). Electronic origin of the surface reactivity of transition-metal-doped TiO_2 (110). *The Journal of Physical Chemistry C* 117: 460–465.

370 Çelik, V. and Mete, E. (2013). Electronic and optical properties of Cr and Cr–N doped anatase TiO_2 from screened coulomb hybrid calculations. *Journal of Physics: Condensed Matter* 25: 365502.

371 Hao, L., Bai, F.Q., Kong, C.P. et al. (2015). Theoretical studies of heteroatom-doping in TiO_2 to enhance the electron injection in dye-sensitized solar cells. *RSC Advances* 5: 79868–79873.

372 Pan, J.W., Li, C., Zhao, Y.F. et al. (2015). Electronic properties of TiO_2 doped with Sc, Y, La, Zr, Hf, V, Nb and Ta. *Chemical Physics Letters* 628: 43–48.

373 Saini, M., Kumar, M., and Som, T. (2017). Ab initio study of 3D transition metal-doping effects in rutile- TiO_2: role of bandgap tunability in conductivity behaviour. *Applied Surface Science* 418: 302–307.

374 Li, X., Shi, J., Chen, H. et al. (2017). A DFT study on the modification mechanism of (Cr, C) co-doping for the electronic and optical properties of anatase TiO_2. *Computational Materials Science* 129: 295–303.

375 Tada, K., Koga, H., Hayashi, A. et al. (2017). Effects of halogens on interactions between a reduced TiO_2 (110) surface and noble metal atoms: a DFT study. *Applied Surface Science* 411: 149–162.

376 Li, X., Xie, Q., and Tian, Z. (2018). A DFT study of electronic structures and optical properties of nickel, nitrogen doped rutile TiO_2. *Chemical Physics Letters* 710: 143–146.

377 Wang, R.Y., Wang, J.X., Jia, J. et al. (2021). The growth pattern and electronic structures of $cu_n(n = 1–14)$ clusters on rutile TiO_2 (110) surface. *Applied Surface Science* 536: 147793.

378 Murata, Y., Fukuta, S., Ishikawa, S. et al. (2000). Photoelectrochemical properties of TiO_2 rutile microalloyed with 4D and 5D transition elements. *Solar Energy Materials and Solar Cells* 62: 157–165.

379 Morgan, B.J., Scanlon, D.O., and Watson, G.W. (2009). Small polarons in Nb- and Ta-doped rutile and anatase TiO_2. *Journal of Materials Chemistry* 19: 5175–5178.

380 Yu, X., Li, C., Tang, H. et al. (2010). First principles study on electronic structures and properties of Sn-doped rutile TiO_2. *Computational Materials Science* 49: 430–434.

381 Long, R. and English, N.J. (2011). Electronic structure of cation-codoped TiO_2 for visible-light photocatalyst applications from hybrid density functional theory calculations. *Applied Physics Letters* 98 (14): 142103.

382 Iwaszuk, A. and Nolan, M. (2011). Electronic structure and reactivity of Ce- and Zr-doped TiO_2: assessing the reliability of density functional theory approaches. *The Journal of Physical Chemistry C* 115: 12995–13007.

383 Zhao, W., Lin, H., Li, Y. et al. (2012). Growth mechanism of palladium clusters on rutile TiO_2(110) surface. *Journal of Natural Gas Chemistry* 21: 544–555.

384 Song, K., Han, X., and Shao, G. (2013). Electronic properties of rutile TiO_2 doped with 4D transition metals: first-principles study. *Journal of Alloys and Compounds* 551: 118–124.

385 Lamrani, A.F., Ouchri, M., Belaiche, M. et al. (2014). Half metallic antiferromagnetic behavior in doped TiO_2 rutile with double impurities (Os, Mo) from Ab initio calculations. *Thin Solid Films* 570: 45–48.

386 Yang, K., Dai, Y., Huang, B. et al. (2014). First-principles GGA+U study of the different conducting properties in pentavalent-ion-doped anatase and rutile TiO_2. *Journal of Physics D: Applied Physics* 47: 275101.

387 Jin, C., Dai, Y., Wei, W. et al. (2017). Effects of single metal atom (Pt, Pd, Rh and Ru) adsorption on the photocatalytic properties of anatase TiO_2. *Applied Surface Science* 426: 639–646.

388 Li, D.H., Jiang, Z.Y., Lin, Y.M. et al. (2021). Theoretical studies of photocatalytic behaviors of isoelectronic Sn/Pb-doped TiO_2: DFT+U. *Applied Surface Science* 555: 149714.

389 Zhang, X., Liu, F., Huang, Q.L. et al. (2011). Dye-sensitized W-doped TiO_2 solar cells with a tunable conduction band and suppressed charge recombination. *The Journal of Physical Chemistry C* 115: 12665–12671.

390 Khan, M., Cao, W., Chen, N. et al. (2013). Influence of tungsten doping concentration on the electronic and optical properties of anatase TiO_2. *Current Applied Physics* 13: 1376–1382.

391 Jha, P.K., Gupta, S.K., and čević, I.L. (2013). Electronic structure, photocatalytic properties and phonon dispersions of X-doped (X = N, B and Pt) rutile TiO_2 from density functional theory. *Solid State Sciences* 22: 8–15.

392 Belosˇ ević-Čavor, J., Koteski, V., Umićević, A. et al. (2018). Effect of 5D transition metals doping on the photocatalytic properties of rutile TiO2. *Computational Materials Science* 151: 328–337.

393 Tada, K., Koga, H., Okumura, M. et al. (2018). Clarification of the interaction between au atoms and the anatase TiO_2 (112) surface using density functional theory. *Surface Science* 670: 23–32.

394 Dudarev, S.L., Botton, G.A., Savrasov, S.Y. et al. (1998). Electron-energy-loss spectra and the structural stability of nickel oxide: an LSDA+U study. *Physical Review B* 57: 1505–1509.

395 Morgan, B.J. and Watson, G.W. (2007). A DFT+U description of oxygen vacancies at the TiO_2 rutile (110) surface. *Surface Science* 601: 5034–5041.

396 Wang, Z., Wen, B., Hao, Q. et al. (2015). Localized excitation of Ti^{3+} ions in the photoabsorption and photocatalytic activity of reduced rutile TiO_2. *Journal of the American Chemical Society* 137: 9146–9152.

397 Wen, B., Hao, Q., Yin, W.J. et al. (2018). Electronic structure and photoabsorption of Ti^{3+} ions in reduced anatase and rutile TiO_2. *Physical Chemistry Chemical Physics* 20: 17658–17665.

398 Liang, B., Mianxin, S., Tianliang, Z. et al. (2009). Band gap calculation and photo catalytic activity of rare earths doped rutile TiO_2. *Journal of Rare Earths* 27: 461–468.

399 Lin, Y., Su, Y., Zhu, Y. et al. (2017). The electronic structure and optical absorption of rutile TiO_2 with La and N dopants from first-principles calculation. *Computational Materials Science* 131: 178–186.

400 Khan, M., Zeng, Y., and Zhenghua, L. (2019). Defect analysis of TiO_2 doped with ytterbium and nitrogen by Ab initio calculations. *Theoretical Chemistry Accounts* 138: 22.

401 Wei, B., Tielens, F., and Calatayud, M. (2019). Understanding the role of rutile TiO_2 surface orientation on molecular hydrogen activation. *Nanomaterials* 9: 1199.

402 Asaduzzaman, A.M. and Krüger, P. (2007). Adsorption and diffusion of a molybdenum atom on the TiO_2 (110)surface: a first-principles study. *Physical Review B* 76: 115412.

403 Asaduzzaman, A.M. and Krüger, P. (2008). Adsorption and cluster growth of vanadium on TiO_2 (110) studied by density functional theory. *The Journal of Physical Chemistry C* 112: 4622–4625.

404 Asaduzzaman, A.M. and Krüger, P. (2008). Adsorption of 3D transition elements on a TiO_2 (110) surface. *The Journal of Physical Chemistry C* 112: 19616–19619.

405 Tang, Y., Asokan, C., Xu, M. et al. (2019). Rh single atoms on TiO_2 dynamically respond to reaction conditions by adapting their site. *Nature Communications* 10: 4488

406 Christopher, P. (2019). Single-atom catalysts: are all sites created equal? *ACS Energy Letters* 4: 2249–2250.

407 Pillay, D. and Hwang, G.S. (2005). Growth and structure of small gold particles on rutile TiO_2 (110). *Physical Review B* 72: 205422.

408 Vijay, A., Mills, G., and Metiu, H. (2003). Adsorption of gold on stoichiometric and reduced rutile TiO_2 (110) surfaces. *The Journal of Chemical Physics* 118: 6536–6551.

409 Li, L., Li, W., Zhu, C. et al. (2021). A DFT+U study about agglomeration of Au atoms on reduced surface of rutile TiO_2 (110). *Materials Chemistry and Physics* 271: 124944.

410 Wang, J., Zhang, W., Zhu, W. et al. (2020). Rutile TiO_2 supported single atom Au catalyst: a facile approach to enhance methanol dehydrogenation. *Molecular Catalysis* 482: 110670.

411 Wang, J., Lei, M., Wang, Z. et al. (2021). Methanol oxidation over rutile Au1@TiO_2 catalyst: importance of facets and oxygen vacancy. *Applied Surface Science* 542: 148541.

412 Lv, C.Q., Liu, J.H., Guo, Y. et al. (2016). DFT+U investigation on the adsorption and initial decomposition of methylamine by a Pt single-atom catalyst supported on rutile (110)TiO_2. *Applied Surface Science* 389: 411–418.

413 Ammal, S.C. and Heyden, A. (2011). Nature of Pt_n/TiO_2(110) interface under water–gas shift reaction conditions: a constrained Ab initio thermodynamics study. *The Journal of Physical Chemistry C* 115: 19246–19259.

414 Ammal, S.C. and Heyden, A. (2013). Origin of the unique activity of Pt/TiO_2 catalysts for the water–gas shift reaction. *Journal of Catalysis* 306: 78–90.

415 Walker, E., Ammal, S.C., Terejanu, G.A. et al. (2016). Uncertainty quantification framework applied to the water–gas shift reaction over Pt-based catalysts. *The Journal of Physical Chemistry C* 120: 10328–10339.

416 Ammal, S.C. and Heyden, A. (2014). Water–gas shift catalysis at corner atoms of Pt clusters in contact with a TiO_2 (110) support surface. *ACS Catalysis* 4: 3654–3662.

417 Walker, E.A., Mitchell, D., Terejanu, G.A. et al. (2018). Identifying active sites of the water–gas shift reaction over titania supported platinum catalysts under uncertainty. *ACS Catalysis* 8: 3990–3998.

418 Ammal, S.C. and Heyden, A. (2016). Water–gas shift activity of atomically dispersed cationic platinum versus metallic platinum clusters on titania supports. *ACS Catalysis* 7: 301–309.

419 Ammal, S.C. and Heyden, A. (2017). Titania-supported single-atom platinum catalyst for water–gas shift reaction. *Chemie Ingenieur Technik* 89: 1343–1349.

420 Humphrey, N., Bac, S., and Sharada, S.M. (2020). Ab initio molecular dynamics reveals new metal-binding sites in atomically dispersed Pt1/TiO_2 catalysts. *The Journal of Physical Chemistry C* 124: 24187–24195.

421 Fung, V., Hu, G., Tao, F.F. et al. (2019). Methane chemisorption on oxide-supported Pt single atom. *ChemPhysChem* 20: 2217–2220.

422 Xu, X., Wang, X., and Jiang, D.E. (2021). Band gap as a novel descriptor for the reactivity of 2D titanium dioxide and its supported Pt single atom for methane activation. *The Journal of Physical Chemistry Letters* 12: 2484–2488.

423 Han, Y., Liu, C.J., and Ge, Q. (2006). Interaction of Pt clusters with the anatase TiO_2(101) surface: a first principles study. *The Journal of Physical Chemistry B* 110: 7463–7472.

424 Han, Y., Liu, C.J., and Ge, Q. (2007). Effect of surface oxygen vacancy on Pt cluster adsorption and growth on the defective anatase TiO_2 (101) surface. *The Journal of Physical Chemistry C* 111: 16397–16404.

425 Yoo, M., Yu, Y.S., Ha, H. et al. (2020). A tailored oxide interface creates dense Pt single-atom catalysts with high catalytic activity. *Energy & Environmental Science* 13: 1231–1239.

426 Thang, H.V., Pacchioni, G., deRita, L. et al. (2018). Nature of stable single atom Pt catalysts dispersed on anatase TiO_2. *Journal of Catalysis* 367: 104–114.

427 Zhang, J., Zhang, M., Han, Y. et al. (2008). Nucleation and growth of palladium clusters on anatase TiO_2 (101) surface: a first principle study. *The Journal of Physical Chemistry C* 112: 19506–19515.

428 Vittadini, A. and Selloni, A. (2002). Small gold clusters on stoichiometric and defected TiO_2 anatase (101) and their interaction with CO: a density functional study. *The Journal of Chemical Physics* 117: 353–361.

429 Lin, L., Shi, Z., Huang, J. et al. (2020). Molecular adsorption properties of CH_4 with noble metals doped onto oxygen vacancy defect of anatase TiO_2 (101) surface: first-principles calculations. *Applied Surface Science* 514: 145900.

430 Wang, X., Zhang, L., Bu, Y. et al. (2021). Interplay between invasive single atom Pt and native oxygen vacancy in anatase TiO_2 (101) surface: a theoretical study. *Applied Surface Science* 540: 148357.

431 Chen, Y., Ji, S., Sun, W. et al. (2019). Engineering the atomic interface with single platinum atoms for enhanced photocatalytic hydrogen production. *Angewandte Chemie International Edition* 59: 1295–1301.

432 Wang, X., Rui, Z., Zeng, Y. et al. (2017). Synergetic effect of oxygen vacancy and Pd site on the interaction between Pd/anatase TiO_2 (101) and formaldehyde: a density functional theory study. *Catalysis Today* 297: 151–158.

433 Wanbayor, R. and Ruangpornvisuti, V. (2012). A periodic DFT study on binding of Pd, Pt and Au on the anatase TiO_2 (001) surface and adsorption of CO on the TiO_2 surface-supported Pd, Pt and Au. *Applied Surface Science* 258: 3298–3301.

434 Sun, C. and Smith, S.C. (2012). Strong interaction between gold and anatase TiO_2 (001) predicted by first principle studies. *The Journal of Physical Chemistry C* 116: 3524–3531.

435 Scaranto, J. and Giorgianni, S. (2009). A DFT study of CO adsorbed on clean and hydroxylated anatase TiO_2 (001) surfaces. *Molecular Physics* 107: 1997–2003.

436 Iyemperumal, S.K., Pham, T.C., Bauer, J. et al. (2018). Quantifying support interactions and reactivity trends of single metal atom catalysts over TiO_2. *The Journal of Physical Chemistry C* 122: 25274–25289.

437 Wang, T., Qiu, S., Dai, Z. et al. (2020). Exploration of TiO_2 as substrates for single metal catalysts: a DFT study. *Applied Surface Science* 533: 147362.

438 Chen, Y., Ji, S., Sun, W. et al. (2018). Discovering partially charged single-atom Pt for enhanced anti-Markovnikov alkene hydrosilylation. *Journal of the American Chemical Society* 140: 7407–7410.

439 Li, S., Lu, X., Guo, W. et al. (2012). Formaldehyde oxidation on the Pt/TiO_2(101) surface: a DFT investigation. *Journal of Organometallic Chemistry* 704: 38–48.

440 Li, Z., Wang, X., Jia, L. et al. (2017). Reduction of HCHO with OH- on Pt loading anatase TiO_2 (001) surface: a DFT calculation. *Catalysis Communications* 92: 23–26.

441 Ding, J., Yang, Y., Liu, J. et al. (2020). Catalytic reaction mechanism of formaldehyde oxidation by oxygen species over Pt/TiO_2 catalyst. *Chemosphere* 248: 125980.

442 Rezaei, M. and Chermahini, A.N. (2020). A DFT study on production of hydrogen from biomass-derived formic acid catalyzed by Pt-TiO_2. *International Journal of Hydrogen Energy* 45: 20993–21003.

443 Boronat, M. and Corma, A. (2011). Generation of defects on oxide supports by doping with metals and their role in oxygen activation. *Catalysis Today* 169: 52–59.

444 Wei, L., Cui, S., Guo, H. et al. (2018). Study on the role of MN species in low temperature SCR on $MnOx/TiO_2$ through experiment and DFT calculation. *Molecular Catalysis* 445: 102–110.

445 Mi, T.G., Wu, Y.W., Xu, M.X. et al. (2021). Theoretical insights into the roles of active oxygen species in heterogeneous oxidation of CO over Mn/TiO_2 catalyst. *Applied Catalysis A: General* 616: 118104.

446 Zhang, X., Sun, Z., Wang, B. et al. (2018). C-C coupling on single-atom-based heterogeneous catalyst. *Journal of the American Chemical Society* 140: 954–962.

447 Seriani, N., Pinilla, C., and Crespo, Y. (2015). Presence of gap states at Cu/TiO_2 anatase surfaces: consequences for the photocatalytic activity. *The Journal of Physical Chemistry C* 119: 6696–6702.

448 Alghannam, A., Muhich, C.L., and Musgrave, C.B. (2017). Adatom surface diffusion of catalytic metals on the anatase TiO_2 (101) surface. *Physical Chemistry Chemical Physics* 19: 4541–4552.

449 Iyemperumal, S.K. and Deskins, N.A. (2017). Activation of CO_2 by supported cu clusters. *Physical Chemistry Chemical Physics* 19: 28788–28807.

450 Chen, J., Iyemperumal, S.K., Fenton, T. et al. (2018). Synergy between defects, photoexcited electrons, and supported single atom catalysts for CO_2 reduction. *ACS Catalysis* 8: 10464–10478.

451 Lackner, P., Zou, Z., Mayr, S. et al. (2019). Surface structures of ZrO_2 films on Rh(111): from two layers to bulk termination. *Surface Science* 679: 180–187.

452 Christensen, A. and Carter, E.A. (1998). First-principles study of the surfaces of zirconia. *Physical Review B* 58: 8050.

453 Stefanovich, E.V., Shluger, A.L., and Catlow, C.R.A. (1994). Theoretical study of the stabilization of cubic-phase ZrO_2 by impurities. *Physical Review B* 49: 11560.

454 Pokrovski, K., Jung, K.T., and Bell, A.T. (2001). Investigation of CO and CO_2 adsorption on tetragonal and monoclinic zirconia. *Langmuir* 17: 4297–4303.

455 Tanabe, K. and Yamaguchi, T. (1994). Acid-base bifunctional catalysis by ZrO_2 and its mixed oxides. *Catalysis Today* 20: 185–197.

456 Alfredsson, M. and Catlow, C.R.A. (2001). Modelling of Pd and Pt supported on the {111} and {011} surfaces of cubic-ZrO_2. *Physical Chemistry Chemical Physics* 3: 4129–4140.

457 Alfredsson, M. and Catlow, C.R.A. (2002). A comparison between metal supported c-ZrO_2 and CeO_2. *Physical Chemistry Chemical Physics* 4: 6100–6108.

458 Jung, C., Ishimoto, R., Tsuboi, H. et al. (2006). Interfacial properties of ZrO_2 supported precious metal catalysts: a density functional study. *Applied Catalysis A: General* 305: 102–109.

459 Alfredsson, M. and Catlow, C.R.A. (2004). Predicting the metal growth mode and wetting of noble metals supported on c-ZrO_2. *Surface Science* 561: 43–56.

460 Bazhenov, A.S., Kauppinen, M.M., and Honkala, K. (2018). DFT prediction of enhanced reducibility of monoclinic zirconia upon rhodium deposition. *Journal of Physical Chemistry C* 122: 6774–6778.

461 Orlando, R., Pisani, C., Ruiz, E. et al. (1992). Ab initio study of the bare and hydrated (001) surface of tetragonal zirconia. *Surface Science* 275: 482–492.

462 Morterra, C., Cerrato, G., Ferroni, L. et al. (1994). Surface characterization of yttria-stabilized tetragonal ZrO_2 Part 1. Structural, morphological, and surface hydration features. *Materials Chemistry and Physics* 37: 243–257.

463 Gennard, S., Corà, F., and Catlow, C.R.A. (1999). Comparison of the bulk and surface properties of ceria and zirconia by Ab initio investigations. *Journal of Physical Chemistry B* 103: 10158–10170.

464 Hofmann, A., Clark, S.J., Oppel, M. et al. (2002). Hydrogen adsorption on the tetragonal ZrO_2(101) surface: a theoretical study of an important catalytic reactant. *Physical Chemistry Chemical Physics* 4: 3500–3508.

465 Thang, H.V. and Pacchioni, G. (2020). On the real nature of Rh single-atom catalysts dispersed on the ZrO_2 surface. *ChemCatChem* 12: 2595–2604.

466 Kwon, Y., Kim, T.Y., Kwon, G. et al. (2017). Selective activation of methane on single-atom catalyst of rhodium dispersed on zirconia for direct conversion. *Journal of the American Chemical Society* 139: 17694–17699.

467 Zhao, Q., Liu, B., Xu, Y. et al. (2020). Insight into the active site and reaction mechanism for selective oxidation of methane to methanol using H_2O_2 on a Rh_1/ZrO_2 catalyst. *New Journal of Chemistry* 44: 1632–1639.

468 Harrath, K., Yu, X., Xiao, H. et al. (2019). The key role of support surface hydrogenation in the CH_4 to CH_3OH selective oxidation by a ZrO_2-supported single-atom catalyst. *ACS Catalysis* 9: 8903–8909.

469 Tosoni, S. and Pacchioni, G. (2017). Influence of surface hydroxylation on the Ru atom diffusion on the ZrO_2(101) surface: a DFT study. *Surface Science* 664: 87–94.

470 Viet Thang, H., Tosoni, S., Fang, L. et al. (2018). Nature of sintering-resistant, single-atom Ru species dispersed on zirconia-based catalysts: a DFT and FTIR study of CO adsorption. *ChemCatChem* 10: 2634–2645.

471 Grau-Crespo, R., Hernández, N.C., Sanz, J.F. et al. (2007). Theoretical investigation of the deposition of Cu, Ag, and Au atoms on the ZrO_2(111) surface. *Journal of Physical Chemistry C* 111: 10448–10454.

472 Koeppel, R.A., Baiker, A., Schild, C. et al. (1991). Carbon dioxide hydrogenation over Au/ZrO_2 catalysts from amorphous precursors: catalytic reaction mechanism. *Journal of the Chemical Society, Faraday Transactions* 87: 2821–2828.

473 Baiker, A., Kilo, M., Maciejewski, M. et al. (1993). Hydrogenation of CO_2 over copper, silver and gold/zirconia catalysts: comparative study of catalyst properties and reaction pathways. *Studies in Surface Science and Catalysis* 75 1257–1272.

474 Gu, X.K., Qiao, B., Huang, C.Q. et al. (2014). Supported single Pt_1/Au_1 atoms for methanol steam reforming. *ACS Catalysis* 4: 3886–3890.

475 Han, B., Lang, R., Tang, H. et al. (2019). Superior activity of Rh_1/ZnO single-atom catalyst for CO oxidation. *Chinese Journal of Catalysis* 40: 1847–1853.

8

Supported Metal Single Atom Thermocatalysts for Selective Hydrogenation

Eva Castillejos[1]*, Ana B. Dongil*[2]*, Inmaculada Rodríguez-Ramos*[2]*, and Antonio Guerrero-Ruiz*[1]

[1] *UNED, Departamento de Química Inorgánica y Química Técnica, Facultad de Ciencias, Edificio Las Rozas, Av. De Esparta s/n, 28232 Las Rozas, Madrid, Spain*
[2] *Instituto de Catálisis y Petroleoquímica, CSIC, C/ Marie Curie 2, 28049 Madrid, Spain*

8.1 Introduction

Single-atom catalysts (SACs), which consist of isolated metal atoms dispersed on solid supports [1], have emerged as a new concept in the development of catalytic materials, and among other purposes, they can aid to the rational design of heterogeneous catalysts. In the previous chapters of this book, different aspects of SACs, mainly related to their preparation and characterization, have been presented. Of particular relevance is the definition of SAC, which should be related to their application for heterogeneous catalytic processes. Very recently, an extended description of SAC has been introduced, including many situations of isolated atoms, and comprising practically all the elements of the Periodic Table with relevance as catalytic ingredients [2, 3]. Beyond the enlarged definition of SACs given by these authors, it should be included in the SAC definition categories species such as metal complexes or organometallic compounds, typically used as homogeneous catalysts, surface organometallic catalysts, metal-organic frameworks with an isolated element, or even metalloenzymes, etc. However, in this chapter, we will assume that the SAC definition only includes isolated free metal atoms (typically transition metals), which means free of chemically bonded ligand (see *Chapter 1*). These metallic SACs are usually supported/stabilized over solid materials. A reason for this restriction is that catalytic hydrogenation reactions involve as principal catalytic ingredient-specific transition metals (in zero or with low oxidation state), on which the substrate to be hydrogenated should be chemisorbed. The ambiguity of the definition of SAC and the relationships with concepts of surface organometallic chemistry (SOMC) has been well documented [4]. Briefly, the overlap between both concepts, SAC and SOMC, can be easily understood when the single-atom metal in SAC interacts with reactants to form surface organometallic species, being the first step for all catalytic processes.

Supported Metal Single Atom Catalysis, First Edition. Edited by Philippe Serp and Doan Pham Minh.
© 2022 WILEY-VCH GmbH. Published 2022 by WILEY-VCH GmbH.

For application in homogeneous catalysis, many organometallic compounds can also be considered as isolated metal atoms. In this case, each individual active site or the principal species involved in catalytic cycles can be perfectly identified, for instance, when used as homogeneous hydrogenation catalysts [5]. However, in the case of heterogeneous catalytic reactions on solid catalysts, including supported metal clusters, the surface catalytic centers are probably constituted by an ensemble of atoms of the same element (i.e. in a metallic cluster, for which a defined number of surface atoms will be required to efficiently perform the chemisorption of reactants), or a combination of the metal atoms with promoters (typically non-metal elements) or metal atoms placed in an alloy or interacting with support surface species, etc. Thus, it is necessary, in each case, to identify the nature and working mechanism of this ensemble of surface active sites [6]. Obviously, this multicomponent definition of the catalytic surface sites implies a certain contradiction with the SAC, where a single metal atom could be considered as representative of the active sites, but also opens the possibility of including, in the interpretation of catalytic properties, effects associated with the different support materials. In summary, the adsorption of the substrate to be hydrogenated should be associated with the surface properties of SACs to understand the catalytic properties, selectivity, and specific activity. In addition, these SAC studies have to be analyzed comparatively with the properties of metallic cluster catalysts.

A decisive difference of SACs vs. nanoparticles (NPs) when applied as hydrogenation catalysts is related to their ability or not to activate/dissociate di-hydrogen, this being an essential step in the reaction mechanism of many hydrogenation processes. Notice that we are not excluding from the present chapter many possible reactions of hydrogenation using transfer hydrogen co-reactants or alternative hydrogen sources (hydrazine, formic acid, alcohols, etc.). However, as the initial step for performing the hydrogenation reaction, when hydrogen is the reactant, it should be chemisorbed and dissociated before the incorporation of the hydrogen atoms into the hydrogenated products. Only a few studies of hydrogen activation have been reported over SAC. Of particular relevance is the study [7], where Pt single atoms (Pt_{SA}) deposited on the surface of Ni nanocrystals are studied comparing them with Pt_{SA} supported on activated carbon, TiO_2, SiO_2, and ZSM-5. All these Pt SACs were tested for the hydrogenation of 3-nitrostyrene, observing that the obtained catalytic activity (expressed as turnover frequency, TOF) is much higher for Pt_{SA}/Ni nanocrystals. The enhanced hydrogen supply in this latter catalyst due to the spontaneous dissociation of H_2 on Pt_{SA} and on Ni surface as well as the facile diffusion of H atoms over the Ni nanocrystals is considered responsible of its improved behavior. The same arguments were used in a study on acetylene hydrogenation with Pd_{SA} deposited on the surface of a Cu(111) single crystal [8]. In this case, Pd_{SA} can activate hydrogen, which then populates the Cu surface where it is weakly bound and effective for selective hydrogenation to proceed. Density functional theory (DFT) calculations have provided evidence allowing to attribute the high activity of Pd_{SA} anchored into the cavities of polymeric graphitic carbon nitride (g-C_3N_4), for both alkynes and nitroarenes hydrogenation, to the facile hydrogen activation (heterolytic dissociation) and hydrocarbon adsorption on atomically dispersed Pd sites [9].

A critical property to be achieved with the design of new catalysts is the control of selectivity. In this sense, a recent study using SAC consisting of Pt_{SA} diluted within Cu_{NP} (single atom alloy, SAA), has evidenced that hydrogenation with atomic hydrogen of C=O bonds in unsaturated aldehydes requires Pt sites to promote the initial activation of molecular hydrogen [10]. However, the definition of the active surface sites is more complex. In the surface of this catalyst are identified Pt_{SA}, which indirectly modify electronic properties of neighboring Cu atoms, a thin Cu oxide layer covering the metallic Cu_{NP}, as well as Cu^0 surface sites. In conclusion, it was determined that the hydrogenation selectivity could be explained by the preferential adsorption of the unsaturated aldehyde molecules via the terminal oxygen atom over the copper support. In addition, aspects as competitive adsorption among the reactants and products could contribute to obtaining the desired selectivity. Another situation also interesting in relation to the activity and selectivity of SACs as hydrogenation catalysts is the case of partially reducible metal oxide supports. For example, Ni_{SA} on ceria cannot act directly as active species but rather as promoters. This is the case reported in [11]. In this work, it is demonstrated that the ceria surface contains the active surface sites for the selective hydrogenation of alkynes. Based on DFT studies, a mechanism of heterolytic dissociation of H_2 over oxygen vacancies of CeO_2 (111) is proposed, in which the catalytic centers consist of Lewis pairs containing separated O and Ce surface sites. The resulting O–H and Ce–H species seem to catalyze the hydrogenation of acetylene. Thus, although X-ray photoelectron spectroscopy (XPS) analysis reveals the presence of a minority of Ni^0 species in the spent Ni/CeO_2 catalyst, the role of Ni_{SA} seems to be just as the promoter to create oxygen vacancies in CeO_2. Therefore, the Ni_{SA} are not directly involved in the catalytic hydrogenation reaction. Contrarily, for noble metal containing SAC supported over metallic surfaces, or for SAA, hydrogen activation and spillover have been demonstrated for numerous systems [12]. A key parameter for these multiple metallic catalytic surfaces, in relation to hydrogenation reactions, should be the competitive adsorption on the surface of both reactants: substrate molecules to be hydrogenated and hydride species. This competition can modify not only the proportion of both reactants adsorbed at the SAC surface but also the desorption of the intermediate species. Finally, as a result of all these dynamic processes taking place at the metallic catalyst surfaces, the reaction selectivity is directed to a selected product [13].

Another important aspect to be considered in relation to the hydrogen activation and diffusion during the hydrogenation reactions over SACs is the possibility of hydrogen species spillover and migration over the surface of non-metallic supports. This phenomenon described as hydrogen spillover, which is particularly known for supports that contain surface hydroxyl species, is also recognized to operate in the case of SACs due to the extended interaction between the anchored single metal atoms and the support. Thus, the phenomena associated with the support properties play a major role in determining the catalytic performances. This is the case of the Pt_{SA}/FeO_x system, where the strong metal-support interaction (SMSI) seems to lead to an extensive hydrogen spillover, which promotes the hydrogenation of 3-nitrostyrene to 3-aminostyrene, achieving high specific activities and selectivity values close to 99% [14]. The effect associated with hydrogen spillover species, migration of hydrogen atoms from metal particles to support materials, depends on the reducibility of the support, so H atoms are transformed to protons and electrons on reducible supports, such as MoO_3, TiO_2, or WO_3, reducing surface cations and protonating neighboring oxygen anions. On unreducible supports, the presence of surface defects can also provide anchoring centers for the migration of H atoms from the metal to the support. The role of hydrogen spillover in the hydrogenation of alkenes or aromatic substrates has been critically reviewed, stressing that adsorbed H atoms may simply act as a hydrogen source in these reactions [15].

In active and selective SACs, the specific interactions with the support should also be maximized because of the high surface free energy of the isolated metal atoms, which can lead to sintering. This opens the possibility of strong electronic metal-support interactions. As an example, in a comparative study of Pt SACs supported over Co_3O_4, CeO_2, ZrO_2, and graphene, detailed spectroscopic characterization and theoretical calculations revealed that these interactions modify the unoccupied 5d state of Pt atoms, which modulates their adsorption properties [16]. This effect is enhanced on the Co_3O_4 support. This electronic promotion was also demonstrated on Pd/Co_3O_4, which showed high activity and stability for hydrogenation reactions.

To date, there have been various reviews on SACs [2, 17–20], and some of them include the hydrogen transfer reactions. Hydrogen transfer reactions are important reactions in the chemical industry as they are used for the synthesis of active pharmaceutical materials, dyes, or fine chemicals [18]. Thus, it is not surprising that the hydrogenation reaction is one of the most intensively investigated topics in catalysis since 25% of chemical transformations include at least one hydrogenation step [20]. Indeed, in the synthesis of many products, there are various functional groups such as —C≡C, —C=O, $—NO_2$, —C≡N, —COOH(R), and $—CONH_2$ that have to be selectively reduced by H_2 to their corresponding alkenes, alcohols, and amine. These products are key intermediates for the fine chemical, polymer, agrochemical, and pharmaceutical industries. Lately, the new field of SACs has emerged as the finest alternative for heterogeneous catalysts used in these catalytic applications.

In the following sections of this chapter, we are going to review the recent applications of SACs as hydrogenation catalysts. For this, we are going to group them depending on the nature of the support on which the single metallic atoms are located.

8.2 Hydrogenation Reactions Catalyzed by Single-Atom Supported on Carbon Materials

Carbon-supported metal SACs are considered as great candidates for hydrogenation reactions. Notably, carbon-based supports are a vast family of materials, which have been explored over the past decade as one of the most common supports (or catalysts) because of their unusual features such as chemical and mechanical stability, tunable porosity and surface properties, good thermal and electrical conductivity, high specific surface area, variable structural and morphological combinations, and easy handling. Carbon-based SACs are widely used because of their unusual

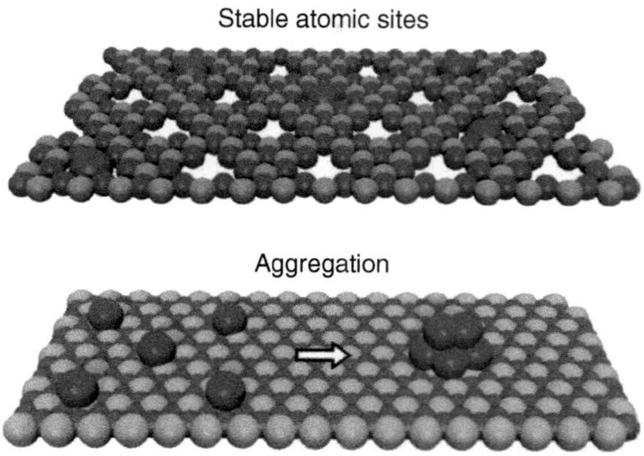

Figure 8.1 Illustration presenting Pd_{SA} on graphitic carbon nitride (top) and alumina (bottom). While the atoms on the Al_2O_3 tend to aggregate, forming a Pd cluster, this does not occur with g-C_3N_4. Source: Vilé et al. [9]. Reproduced with permission of Wiley VCH.

features, including porosity, tunable morphologies, or/and simple immobilization, making them highly attractive SACs for hydrogenation reactions and many other important catalytic applications. However, due to the low binding energy between metal atoms and carbon structures, SACs on carbon supports are frequently synthesized at high annealing temperatures and feature low metal loading. In general, appropriate supports for SACs are limited to functionalized carbon structures, on which the concentration and type of oxygen groups can be controlled. In addition, the presence of defects such as vacancies will significantly increase the adsorption energy of metal SACs (e.g. Pd, Ru, Ni, or Cu atoms) on carbon materials (*Chapter 3*). Finally, to overcome the tendency of metal agglomeration (Figure 8.1), extensive efforts have been devoted to developing effective synthetic strategies via either generating SMSI or/and introducing appropriate anchoring sites, as later will be well documented for an extensive range of materials [2].

To further arbitrate the coordination environment of SACs supported on carbon materials, C-doping with specific heteroatoms presenting higher binding energy with the metal is another efficient strategy [21]. Nitrogen doping can easily tailor the properties of these carbon supports, thanks to its chemical versatility and to the availability of a wide range of nitrogen-doped carbons produced by a single N/C substitution [22]. Compared with carbon-supported SACs, N-doped catalysts generally show superior catalytic performances in many reactions since strong metal nitrogen bonds enhance the specific metal-support interactions (MSIs). Three types of N atoms doped in the graphene layer have been identified: pyrrolic, pyridinic, and graphitic [23], as shown schematically in Figure 8.2.

These catalysts offer unrivaled opportunities for efficient catalytic transformations since the interactions are strong enough with the reactants and weak with the reaction products.

Enormous efforts have been dedicated toward the design of materials with tailored metal particle sizes to improve the catalytic performance in hydrogenation reactions. Until now, metal single atoms that are anchored onto heteroatom doped carbon materials such as nitrogen, oxygen, sulfur, or phosphorous showed excellent catalytic performance for catalytic reactions such as selective hydrogenations. The sites that isolated metal atoms anchored on carbon supports are capable of

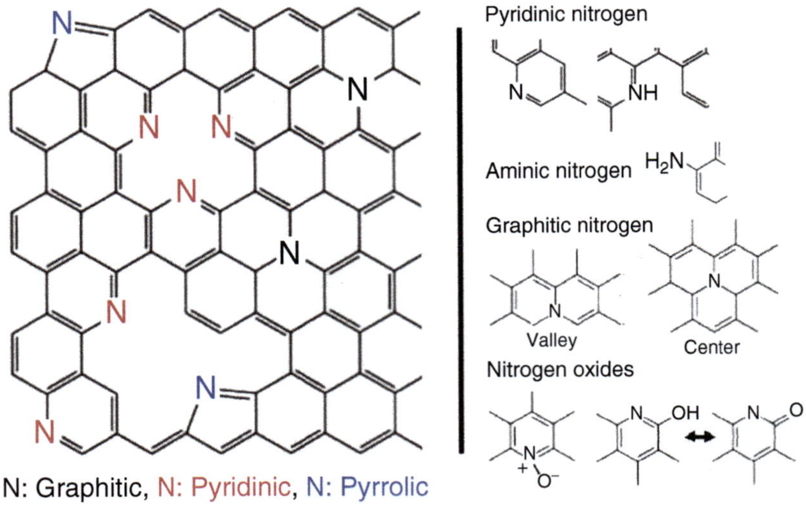

Figure 8.2 Various configurations of N atoms introduced in a graphene layer. Source: Inagaki et al. [23]. Reproduced with permission of Elsevier.

activating the H_2 molecule, catalyzing hydrogenation reactions. Furthermore, SACs are suitable for adsorbing the hydrogen molecule generating needed species leading to the hydrogenation reaction. In a systematic approach, the dihydrogen activation is divided into the steps of splitting of H_2 and H-atom transfers with homolytic and heterolytic splitting at the active centers. Here, we aim to report on the development achieved in researching carbon-based SACs, including principally noble metals such as Pd, Pt, Ru, and Au. Subsequently, we continue toward Co, Cu, Ni, and Fe, and their application to hydrogenation reactions.

8.2.1 Noble-Metal Single-Atom Catalysts

Pd-based catalysts are commonly employed in hydrogenation reactions due to their exceptional activity. However, Pd_{SA} synthesis on carbon materials is challenging due to the lack of stable coordination sites, resulting in the formation of Pd_{NP} across aggregation. Despite this difficulty, Pd SACs with low Pd loadings, high hydrogenation activity and selectivity, as well as good durability could be developed. In this sense, atomically dispersed Pd_{SA} on graphene exhibited high conversion and high selectivity in the selective hydrogenation of 1,3-butadiene to butenes [24]. The authors proposed that the 1,3-butadiene adsorption on graphene confining Pd_{SA} could take place via mono-π-adsorption mode since the di-π adsorption usually requires a larger particle size than a single atom on the surface (Figure 8.3). Then, the selectivity was caused by a favorable steric effect and the change of 1,3-butadiene adsorption mode on the Pd_{SA}. This adsorption mode avoids the complete hydrogenation of 1,3-butadiene to butane, favoring the 1-butene production.

Furthermore, excellent durability against deactivation via aggregation of metal atoms or by carbonaceous deposits has been achieved with these catalytic materials [24]. The graphene (2D material) confining metal single-atom described above

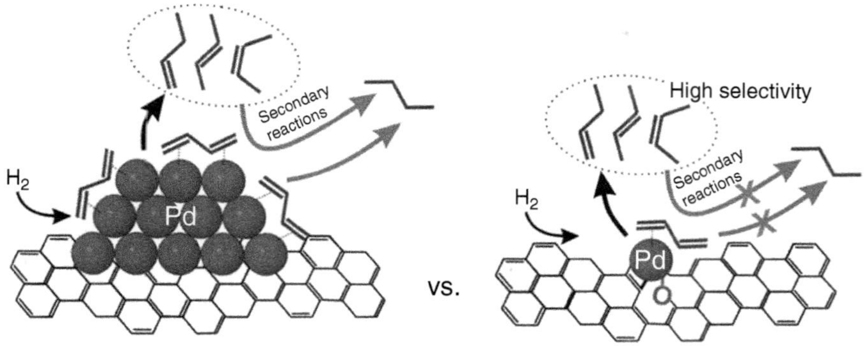

Figure 8.3 Schematic example of butene selectivity on Pd_{SA}/graphene catalyst. Source: Yan et al. [24]. Reproduced with permission of American Chemical Society.

should be differentiated from metal NP inside 2D carbon materials and single metal atom catalysts in graphite or mesoporous carbon. The confining means that the SA can be stabilized by the coordination environment of 2D structures through strong covalent bonds.

The outstanding selectivity of atomically dispersed Pd atoms on graphene was described by Huang et al. [25] for acetylene to ethylene reaction in the presence of an excess of ethylene. This catalyst blocks the unselective subsurface species and ensures the ethylene desorption against the over hydrogenation to ethane. The over hydrogenation of ethylene to ethane is associated with the lack of unselective subsurface hydrogen and with the existence of an energetically favored path for ethylene desorption. Atomically dispersion of Pd atoms on graphene through Pd—C bond anchoring is the key to understand the catalyst selectivity.

Also, the anchoring of Pd atoms into the cavities of mesoporous polymeric g-C_3N_4 led to high activity and selectivity in comparison with commercial catalysts based on NP [9]. Graphitic carbon nitride is a class of N-doped carbon material that can exist in graphene and graphite carbon structures. As far as dihydrogen activation is concerned, the single metal atoms are usually positively charged and therefore act as a Lewis acid, while the nearby nitrogen atoms from the N-doped carbon can act as a Lewis base; the combination of the two facilitates the dissociation of hydrogen via a heterolytic pathway [22]. Again, the facile H_2 activation and hydrocarbon adsorption on atomically dispersed Pd sites illustrate the benefits of the single-atom catalysis.

An improvement toward coke resistance was obtained with a similar catalyst (g-C_3N_4 support) tested on the acetylene hydrogenation [26]. Following the behavior described above, single-atom sites adsorb reactants and hydrogen but do not adsorb products because of steric hindrance. The catalyst presented a low activity but a high selectivity. It is important to note that the distinction between Pd graphite and Pd carbon nitride performance is mainly related to catalytic selectivity. The selectivity of these two materials might be attributed to the porosity, acidity of support, and the variation of charge induced by the Pd-support interaction. In brief, there is a remarkable support effect on the catalytic properties of Pd SACs, which leads to new knowledge into hydrogenation reactions.

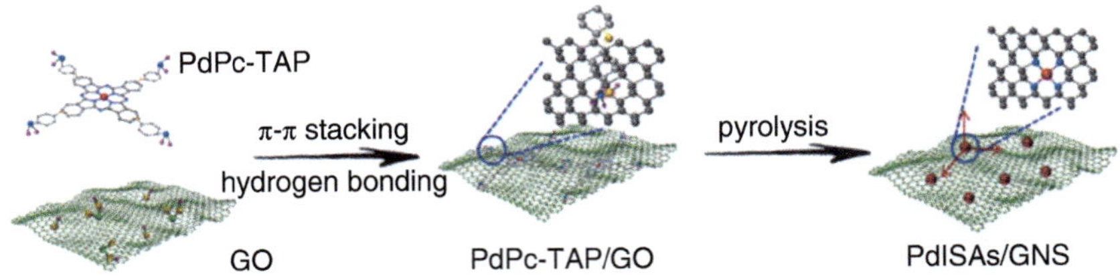

Figure 8.4 Schematic "*in situ*" immobilization of Pd–N on GO. Source: Liu et al. [27]. Reproduced with permission of Royal Society of Chemistry.

Another innovative strategy has been developed by Tang's group [27] to overcome challenges in the synthesis of SACs, such as reduced stability of the active sites and/or the difficulty in stabilizing high contents of single atoms. Thus, Pd_{SA} were successfully anchored onto graphene oxide via an "*in-situ*" immobilization, growth, and pyrolysis post-treatment (Figure 8.4). This catalyst exhibited outstanding catalytic activity and stability toward the rhodamine hydrogenation, being 21 times higher than that of a commercial Pd/C catalyst. The atomically dispersed Pd atoms assure the homogeneity of active sites and maximize the atomic utilization. The coordinated $Pd–N_x$ species not only serve as a draft to anchor the Pd_{SA} but also facilitate the mass transfer and diffusion of H* species on graphene.

Other form of Pd SACs, where Pd_{SA} are anchored on the inner walls of mesoporous nitrogen-doped carbon nanospheres, was synthesized via a hard-template Lewis acid doping strategy [28]. The dispersion of Pd_{SA} on the support, with a higher specific surface area, enables more Pd atoms to participate in acetylene semihydrogenation. The catalyst showed both high Pd utilization efficiency and excellent ethylene selectivity. The application of N-doped carbons as host materials for SACs is increasing because of the strong binding affinity of the heteroatom with transition metals. Understanding the relation between the properties of N-doped carbons, their interaction with metals, and the performance of the resulting catalysts is crucial to guide the design of more effective SACs.

Pt_{SA} catalysts are much less active for hydrogenation reactions than the ones containing Pd_{SA}, NPs, or clusters, which is probably related, in the case of NPs or clusters, with the need of ensembles of atoms to perform the reactant adsorptions. Despite that, it has been shown that the formation at the edges of the graphene layers of a platinum carbon bond results in electronic strong metal–carbon interaction. Subsequently, the Pt atoms transfer electrons to the support forming $Pt^{\delta+}$ species. These species are very efficient for the 3-nitrostyrene hydrogenation to 3-vinylaniline with a turnover 20 times higher than that of the best catalyst for such hydrogenation reactions reported in the literature [29]. As discussed above, the H_2 dissociation over Pt_{SA} may follow a heterolytic dissociation pathway to yield H species at the Pt-carbon interface, significantly enhancing the catalytic activity. Moreover, the spillover phenomenon of the activated H species over the carbon support cannot be rejected. However, the influence of such spillover H species on hydrogenation reactions is not clear and needs to be further investigated.

The authors cited above proposed that the strong Pt_{SA}-carbon interaction significantly modifies the electronic structure of the carbon-edge-anchored Pt_{SA}, and correspondingly their catalytic properties for H_2 activation. Notably, the surface physicochemical properties of the carbon support play a decisive role via this SMSI, modifying the catalytic behaviors of supported metal species.

A facile and versatile precursor-dilution strategy was developed to synthesize a series of SACs on N-doped carbon supports. The Pt/N–C catalyst obtained with this strategy exhibited excellent chemo and regioselectivity in the hydrogenation of terminal alkynes in 1-nitro-4-ethynylbenzene and 1-ethynyl-4-vinylbenzene to alkenyl groups while keeping —NO_2 and —C=CH_2 undamaged. In the catalysis of complex molecules, SACs are able to reach great selectivity comparable to homogeneous and enzyme catalysts [30].

Interestingly, an effective strategy to optimize the selectivity of Pt_{SA} can be efficiently tuned by isolating Pt contiguous atoms and forming Pt–Zn intermetallic NP supported on hollow N-doped carbon nanotubes. This Pt–Zn intermetallic catalyst promotes the hydrogenation of 4-nitrophenylacetylene to 4-aminophenylacetylene with a much higher selectivity (99%) and conversion (99%) than the comparison samples with Pt_{SA} and Pt_{NP} (Pt/CN, CN = carbon nanospheres). DFT calculations reveal that Pt–Zn intermetallic NPs promote the hydrogenation of the nitro group, whereas the positive Zn atoms assist the adsorption of the nitro group [31].

Ru_{NP} are also known to be catalytically active for hydrogenation reactions. However, Ru_{SA} are poor catalysts for this kind of reaction because of their inability to efficiently dissociate dihydrogen compared to other metals such as Pd. DFT calculations have shown that electro-deficient Ru_{SA} and/or few atom Ru clusters are not performant for the H_2 activation compared to larger clusters ($n \geq 13$) during nitrobenzene hydrogenation [32]. For the more crowded substrate 2,3-dimethyl-2-butene, DFT calculations have shown that this is due to steric hindrance. Importantly, these reactions can be used as tests to distinguish samples containing metallic sub-nanometer NPs. DFT was also used to examine the catalytic performance of Ru and Cu single atoms embedded on defects in graphene for the selective hydrogenation of CO_2 into formic acid (FA)[33]. Three different pathways for the reaction were considered. The first relies on CO_2 direct hydrogenation with protons from the H_2 molecule. However, this reaction path does not depict a positive route for FA synthesis. The other two reaction mechanisms start with the H_2 dissociative adsorption. The CO_2 hydrogenation happens with the H atoms from the dissociated H_2 at Ru catalyst, while the Cu catalyst favors the proton transfer from an additional H_2 co-adsorbed with CO_2 on hydrogenated SAC.

Remarkably, it has been reported that synergistic effects occur in Ru/C catalysts for nitro-compound hydrogenation by using mixtures of SAs and NPs. First, Ru_{NP} activated dihydrogen, and hydrogen spillover provided hydride to Ru_{SA} that can hydrogenate the substrate [34].

In general, for a single metal atom and given carbon support, high electronic MSI is expected. Oppositely, the efficiency of MSI tuning rapidly decayed with increasing nanoparticle size. In heterogeneous catalysis, strong MSI can meaningfully influence the chemical reactions, being one of the few tools able to improve the catalytic performances in terms of selectivity, activity, and stability. As an example, charge transfer between supports (carbon nanotubes or defective TiO_2) and ruthenium (metallic surface electron-rich species) produced methane in the CO_2 selective hydrogenation, while the electron-deficient ones were selective for the reverse water–gas shift reaction [35]. Therefore, a direct correlation exists between the electronic density on the metal and the selectivity. This work opens perspectives for the design of SACs on carbon materials for CO_2 selective hydrogenation.

To finally tackle noble metals, single gold atoms were used to tune the structural and the electronic properties of the resulting catalysts [36]. It is accepted that well-defined architectures are showing the possibility for deriving structure–performance relationships. The approach is based on the surface diffusion of gold atoms on different cavities of N-doped carbon. The Au_{SA} can be tailored by controlling the coordination neighbors, the activation temperature, and the oxidation state. DFT relates the strong interaction between the substrate to be hydrogenated (acetylene, 1-hexyne, 2-methyl-3-butyn-2-ol, and 4-pentyl-1-ol) and specific gold atoms on cavities, revealing that semihydrogenation on the gold SAC is substrate-dependent. While unfunctionalized alkynes and alkynols with γ-OH are structure insensitive, alkynols with α-OH are sensitive, and the catalytic responses can be related to the strong interaction between these latter substrates and specific gold-cavity ensembles (mimicking a molecular recognition site).

8.2.2 Non-Noble Metal Single-Atom Catalysts

Despite notable achievements with noble metals, the replacement of precious noble metal catalysts with non-noble metal catalysts for hydrogenation reactions is receiving worldwide attention. The noble metal catalysts, due to their specific electronic structure, are highly active but may have difficulty in differentiating competitive functional groups in many cases. There is a great need to develop the design of inexpensive and active/selective non-noble metal catalysts. Non-noble SACs have caught more and more attention and are rapidly developing among heterogeneous catalysts. The non-noble metal catalysts supported on N-doped carbon have also shown super catalytic activity toward a wide range of reactions, including hydrogenations. As mentioned, N-doped carbon is ideally suited to stabilize noble or non-noble metal single atoms.

Gascon and coworkers [37] have reported mesoporous N-doped carbon cobalt catalyst with dispersed Co (3.5 wt% Co) sites in the carbon matrix. This heterogeneous catalyst displays high activity and selectivity (>99%) for the selective hydrogenation of nitrobenzene to aniline under mild conditions (~360 K, ~2 MPa). When more challenging substrates (functionalized nitroarenes) are hydrogenated, the catalyst displays excellent chemoselectivity to the corresponding substituted anilines. The presence of mesoporosity and the high surface area of this catalyst improve the mass transport of reactants and/or products and the accessibility to the active $Co–N_x$ sites and greatly reduce deactivation due to fouling.

Regarding the selective hydrogenation of nitrobenzene, Co_{SA} supported on N-doped carbon showed a much better performance compared with Co_{NP}. The specific activity was 5.4 and 32 times higher than those of Co_{NP} supported on

N-doped carbon and active carbon, respectively [38]. Another example of the use of Co_{SA} as catalysts for the chemoselective hydrogenation of nitroarenes to azoxy compounds under mild conditions has been recently described [39]. In this regard, the Co_{SA} are coordinated with four pyridinic nitrogen atoms in the graphitic layer, while two oxygen molecules are weakly adsorbed on Co atoms perpendicular to the $Co–N_4$ plane. This catalyst presented an excellent performance in the above-mentioned reaction.

Common strategies to prevent the formation of metal NP include reducing the metal loading to a low level. Interestingly, an atomic layer deposition (ALD) approach allowed tuning the density of the supported Co_{SA} on graphene from 0.4% up to 2.5% without the formation of any Co_{NP}. In respect to this strategy, ALD was used to ensure that each Co precursor molecule was anchored on an active site of the graphene support. These active sites can be regenerated by ozone treatment in the second pulse of each ALD cycle, permitting the loading of another batch of Co_{SA}. Therefore, the Co_{SA} loading can be tuned by monitoring the ALD cycle number [40]. The catalyst exhibited remarkably high selectivity toward azoxy compounds in the hydrogenation of nitrobenzenes. This can be ascribed to the electronic coupling between adjacent oxygen and Co atoms, which results in a positively charged catalytic site. Subsequently, the adsorption of electron-deficient azoxy compounds is weaker, preserving the full nitroarenes hydrogenation. This research has opened up an avenue to control the loading of single metal atoms.

In contrast to some conventional metal catalysts, Zhou et al. [41] described an active, selective, stable, and low-cost Co/N-doped carbon catalyst for the hydrogenation of nitrobenzene into aniline under mild conditions. The non-precious metal catalyst showed full nitrobenzene conversion and high aniline selectivity (97%).

In general, SACs possessing particular electronic structures and advantages of maximum atom efficiency are expected to be efficient catalysts. Thus, Co_{SA} anchored on N-doped ultrathin carbon nanosheets have demonstrated an outstanding catalytic activity for selective transfer hydrogenation of nitroarenes to amines with formic acid as hydrogen donor [42].

In practical industry applications, larger molecule sizes and more complex structures are often used. In this sense, anchored Co_{SA} on N-doped carbon have been designed and prepared with excellent properties for selective hydrogenation of several pharmaceutical molecules with large sizes, such as nimodipine and/or 2-(3,4methylenedioxyphenylethyl) quinoline [43]. These single atoms have also been encapsulated in a hierarchical porous N-doped carbon. Anchored Co_{SA} on N-doped carbon displays significantly higher catalytic activity than encapsulated Co in this N-doped carbon for these reactions. These materials have turned out to be ideal catalysts for hydrogenation reactions since they combine improved mass transfer ability and the advantages of the high activity of SACs.

Otherwise, earth-abundant, inexpensive, and non-toxic metals, such as copper-based catalysts, have also been developed and evaluated for hydrogenation reactions. Atomically dispersed Cu catalysts supported on graphitic carbon shells have exhibited interesting performances for acetylene hydrogenation [44]. The structural feature of these Cu atoms anchored over graphene through Cu—C bonds ensures the acetylene activation and easy ethylene desorption, which could be the key to the outstanding activity, selectivity, and stability of the catalyst.

The rational design of heterogeneous non-noble metal SACs includes Ni as a highly efficient and selective metal for hydrogenation of nitroarenes with hydrogen as a reducing agent [45]. The strategy for achieving atomic dispersion of Ni atoms involves also using N-doped porous carbon. The Ni_{SA} anchored on N-doped porous carbon by bonding with nitrogen–carbon seem to be the active sites. Notably, the $Ni–N_3$ species seems to contribute more to the catalyst activity compared with $Ni–N_2$ and/or $Ni–N_4$ ones. DFT calculation also showed that the $Ni–N_3$ species exhibits the highest activity according to the lowest adsorption energy and the longest elongation of N—O bonds of nitrobenzene, originated from an induced charge transfer between the active sites (N or Ni in $Ni–N_3$ and N or O in nitrobenzene).

$Ni_{SA}–N_x$ species have also been used to hydrogenate α,β-unsaturated aldehydes. Carbon-based materials, such as nanotubes and nanosheets, in which confined $Ni(0)_{NP}$ and $Ni_{SA}–N_x$ species exist, showed high activity and selectivity for the hydrogenation of α,β-unsaturated aldehydes [46]. This catalytic performance could be due to the synergistic effect between $Ni(0)_{NP}$ and single atomic $Ni–N_x$ species via a two-step mechanism: adsorption and activation of H_2 molecules on $Ni(0)_{NP}$ active sites and adsorption and activation of cinnamaldehyde on $Ni–N_x$ active sites. It is worth mentioning again that nanocarbon confined Ni means that the Ni atoms are stabilized by the coordination through strong covalent bonds via nitrogen doping.

A durable Ni SAC has also been investigated on N-doped carbon for the hydrogenation of various biomass-bearing unsaturated groups, including nitroarenes (—NO_2), aldehyde (—C=O), styrene (—C=C—), phenylacetylene (—C≡C—), and quinoline (—C=N—) [47]. During H_2 activation, the dihydrogen molecule approaches Ni_{SA} and subsequently, it is heterolytically dissociated via interaction with Ni and the adjacent pyridinic N atoms. Taking advantage of computational modeling and integrated spectroscopy characterization, the active site structure was identified as $(Ni–N_4)\cdots N$, where isolated Ni cation and the adjacent uncoordinated pyridinic N atom form a frustrated Lewis pair for the heterolytic dihydrogen activation.

Ni atoms embedded in N-doped carbon catalysts with the assistance of metal organic frameworks also showed great potential as an alternative to noble metal catalysts. The catalytic performance of the Ni-based catalysts was analyzed for acetylene hydrogenation. The selectivity for ethylene seems to be due to the uniform coordination of N atoms around the Ni centers, adjusting the d-orbitals of the Ni atoms [48].

In search of advanced, effective, and sustainable SACs, other systems have been studied in the area of heterogeneous catalysis, such as the double atoms catalysts (DACs, *Chapter 15*). In view of this, Ni DACs supported on N-doped porous carbon have shown very good selectivity toward the reverse water–gas shift reaction and unusually better catalytic activity compared to single atom Ni_{SA}/N-carbon without excessive hydrogenated products such as C_2H_6, CH_3OH, and HCOOH, even under H_2 rich conditions [49]. Further studies for exploring other DAC systems and their uses are underway at present.

In the same way, a combination of Fe_{SA} and Fe_2O_3 clusters dispersed on N-doped carbon showed high activity and chemoselectivity for the hydrogenation of nitro-compounds under mild conditions with hydrazine as reductant [50]. The

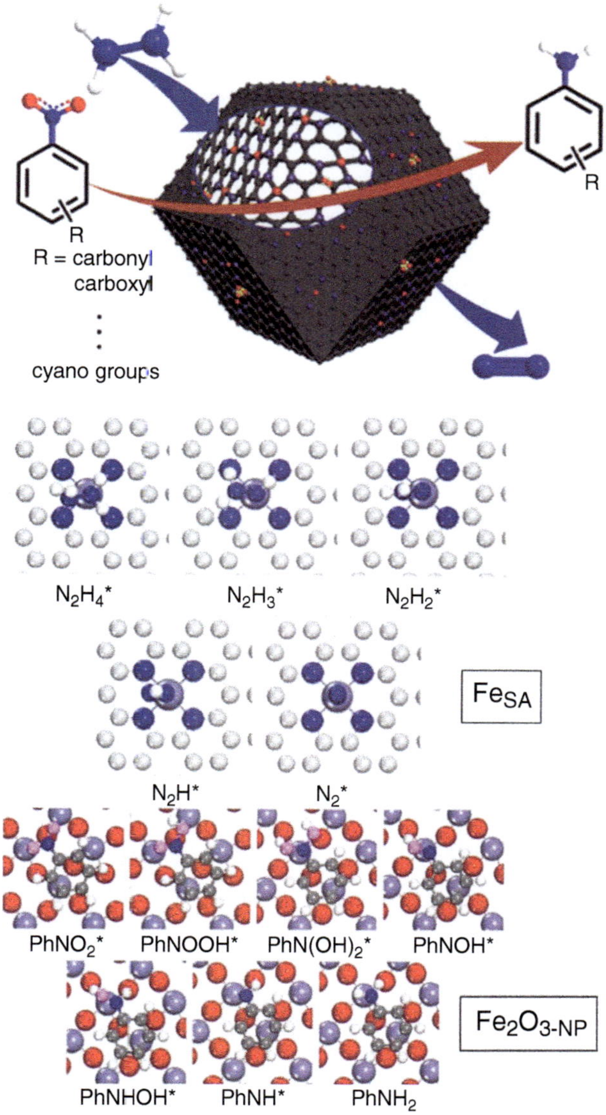

Figure 8.5 Optimized structures of intermediates on the FeN_4 site embedded in graphene and the Fe_2O_3 surface in hydrazine dehydrogenation and nitrobenzene reduction processes, respectively. Source: Yun et al. [50]. Reproduced with permission of American Chemical Society.

authors suggested that the reaction took place quickly since the support facilitates catching and enrichment of the reactants. Moreover, the surface area and the pore size facilitate the substrate to easily arrive at the catalytically active sites. Finally, the H atoms produced from hydrazine hydrate are produced by Fe_{SA}, while the Fe_2O_3 clusters mainly favor the hydrogenation of the nitro compounds (Figure 8.5).

In Table 8.1, we summarize the recent studies in the field of supported SACs on carbon developed for hydrogenation reactions.

8.3 Hydrogenation Reactions Catalyzed by SACs Supported on Unreducible Metal Oxides

SACs are usually stabilized by diverse surface defects. The so-called SMSI is often seen as critical to sustaining high catalytic activity under demanding catalyst operation conditions (i.e. high temperature, high water vapor pressure, etc.). Indeed, SMSI has been directly linked to the presence of electronic defects that can be prepared with ease on the surfaces of reducible oxides (e.g. CeO_2 and TiO_2). Electronic defects, however, are not present on γ-Al_2O_3, an unreducible oxide that is one of the most commonly used catalyst supports for practical applications. However, a study by Kwak et al. [51] provided a new understanding of the nature of strong interactions between catalytic phases and oxide support materials. They used a combination of the ultrahigh magnetic field, solid-state magic-angle spinning nuclear magnetic resonance spectroscopy, and high-angle annular dark-field scanning transmission electron microscopy (HAADF-STEM) coupled with DFT calculations to reveal the nature of anchoring sites of a catalytically active phase of platinum on the surface of γ-Al_2O_3. The results obtained show that the coordinatively unsaturated pentacoordinated Al^{3+} (Al^{3+} penta) centers present on the (100) facets of the γ-Al_2O_3 surface anchor the Pt_{SA}. Thereafter, active research focused on the synthesis of alumina with a large amount of pentacoordinated Al^{3+} sites has been carried out to effectively prepare atomically dispersed metal catalysts.

A highly stable, atomically dispersed Pt catalyst supported on well-aligned mesoporous Al_2O_3 (m-Al_2O_3) with a 0.2 wt% Pt loading was prepared by a modified sol–gel solvent vaporization self-assembly method, followed by calcination in air and reduction with H_2 [52]. Characterization of the catalyst by aberration-corrected HAADF-STEM and by Extended X-ray absorption fine structure (EXAFS) clearly demonstrated the existence of isolated Pt atoms (coordinates with four oxygen atoms) with high number density on the support (Figure 8.6).

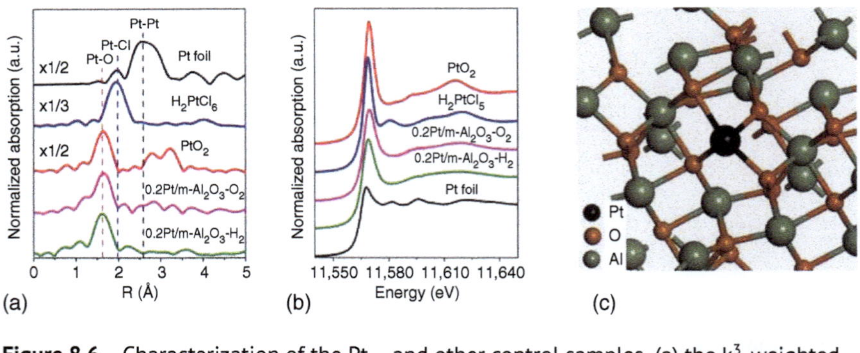

Figure 8.6 Characterization of the Pt_{SA} and other control samples. (a) the k^3-weighted Fourier transform of EXAFS spectra derived from EXAFS, $\Delta = 3.0–12.0\,Å^{-1}$, (b) normalized XANES spectra at the Pt L_3-edge of Pt foil, H_2PtCl_6, and PtO_2, 0.2 Pt/m-Al_2O_3-O_2, and 0.2 Pt/m-Al_2O_3-H_2, (c) schematic illustration of individual Pt atom located on the surface of m-Al_2O_3 for sample 0.2 Pt/m-Al_2O_3-H_2. Source: Zhang et al. [52]. Springer Nature. CC BY 4.0.

Table 8.1 Summary of carbon-supported metal single atoms for hydrogenation reactions.

MetalSA wt%	Support	Synthetic method	Reaction	Catalytic performance	References
Pd 0.25	Graphene	Atomic layer deposition	Butadiene to butenes	100% butene selectivity at 95% conversion under mild reaction condition, excellent durability against deactivation	[24]
Pd 0.11	Nanodiamond-graphene hybrid	Wet chemistry methods	Acetylene to ethylene	High conversion (100%), high selectivity (90%), and good stability	[25]
Pd 0.50	Polymeric g-C_3N_4	Wet chemistry methods	Alkynes and nitroarenes to alkenes and aminoarenes	High catalyst activity and high selectivity (>90%)	[9]
Pd 0.5	g-C_3N_4	Atomic layer deposition	Acetylene to ethylene	Much higher selectivity than the Pd_{NP} catalysts at near-complete conversions, although the activity was lower. High resistance to coke formation	[26]
Pd 1.9	Graphene	Immobilization and thermal post-treatments	Rhodamine	Outstanding catalytic activity and stability (21.3 times higher than that of the commercial Pd/C)	[27]
Pd 0.04	N-doped carbon nanosphere	Synthetic-template doping	Acetylene to ethylene	Higher activity than that of its non-mesoporous counterpart by a factor of 4, excellent C_2H_4 selectivity (90%), and good long-term stability	[28]
Pt 0.08	Hollow nanocarbons	Electrostatic adsorption	Nitrostyrene to vinylaniline	TOF > 31 000 h^{-1}, 20 times higher than that of the best catalyst reported in the literature. Selectivity of 78% under mild conditions	[29]

(Continued)

Table 8.1 (Continued)

MetalSA wt%	Support	Synthetic method	Reaction	Catalytic performance	References
Pt 0.43	N-doped carbon	Thermal method	1-nitro-4-ethynylbenzene and 1-ethynyl-4-vinylbenzene to alkenyl groups	High conversion, high chemo- and regioselectivity toward terminal alkynes and good stability	[30]
Pt, Pt/Zn 0.8	N-doped carbon nanotubes	Precipitation method and thermal post-treatments	4-nitrophenylacetylene to 4-aminophenylacetylene	Pt/Zn much higher conversion (>99%) and selectivity (99%) than the comparison samples with Pt_{SA} and Pt_{NP}	[31]
Ru 9–20	C_{60}	Schlenk techniques	Nitrobenzene or butenes	Low hydrogenation activity	[32]
Ru/Cu	Graphene	Atomic model	CO_2 to formic acid	—	[33]
Ru 1	Carbon nanotubes	Wet chemistry methods	Nitrobenzene to amines	High catalytic activity, high stability, and robustness	[34]
Ru 0.5, Ni 1	Carbon nanotubes and TiO_2	Schlenk techniques/wet chemistry and thermal post-treatments	CO_2 to CH_4 or CO	Catalysts supported on TiO_2 are more active and sometimes more stable than those deposited on CNT. Ni catalysts are very selective for CO, and Ru catalysts are more selective for CH_4	[35]
Au 0.5	N-doped carbon	Wet chemistry and thermal post-treatments	Alkynes to alkenes	Higher activity of Au_{SA} than Au_{NP}. Full selectivity to alkenes	[36]
Co 3.5	N-doped carbon	Wet chemistry methods	Nitrobenzene to amines	High activity and selectivity (>99%) under mild conditions (~360 K, ~2 MPa)	[37]
Co 1.3	N-doped carbon	Wet chemistry and thermal post-treatments	Nitrobenzene to amines	100% conversion and >99% selectivity after four hours of reaction. Activity 5.4 and 32.0 times higher than that of Co_{NP}/NC and Co_{NP}/AC, respectively	[38]

Co 3.6	N-doped carbon	Atomic layer deposition	Nitroarenes to azo compounds	Excellent activity (TOF 35.9 h^{-1}), chemoselectivity and stability under optimized reaction conditions (80 °C, 3 MPa H_2, 1.5 hours)	[39]
Co 0.4–2.5	Graphene	Atomic layer deposition	Nitroarenes to azoxy compounds	Exceptional activity (TOF > 3 s^{-1}) and remarkable high selectivity	[40]
Co 4.6	N-doped carbon nanosheets	Wet chemistry methods	Nitroarenes to azo compounds	Good activity and excellent selectivity. TOF of 110.6 h^{-1}, 20 times higher than the best results of Co_{NP} reported in the literature under similar reaction conditions.	[42]
Co 1.2	N-doped carbon	Wet chemistry methods	Large-sized pharmaceutical molecules	Un-encapsulated SA exhibits significantly higher catalytic activity than encapsulated SA	[43]
Cu 0.25	Graphitic carbon shells	Wet chemistry methods	Acetylene to ethylene	High conversion (95%), high selectivity (98%), and good stability at 200 °C (for more than 60 hours)	[44]
Ni 4.4	N-doped carbon	Wet chemistry and thermal post-treatments	Nitroarenes to azo compounds	High activity, chemoselectivity, and reusability. TON 84 and TOF 8.4 h^{-1}.	[45]
Ni 2.1	Carbon-based nanocomposite	Wet chemistry and thermal post-treatment	α,β-Unsaturated aldehyde to aldehydes	Superior selective hydrogenation activity and stability	[46]

(Continued)

Table 8.1 (Continued)

MetalSA wt%	Support	Synthetic method	Reaction	Catalytic performance	References
Ni 7.5	N-doped carbon	Wet chemistry and thermal post-treatments	Nitroarenes, aldehyde, styrene, phenylacetylene, and quinoline to polyols	High activity, nearly 100% yields under mild conditions, and unprecedented durability under very harsh conditions (245 °C, 60 bar H_2, presence of tungstic acid in hot water)	[47]
Ni	N-doped carbon	Wet chemistry and thermal post-treatments	Acetylene to ethylene	Excellent conversion and high selectivity (both over 90%) even at a high temperature of 200 °C	[48]
Ni 0.09	N-doped carbon	Wet chemistry and thermal post-treatment	CO_2 to CO or CH_4	97.0% selectivity with 22.7% conversion of CO_2 at H_2-rich condition (4 H_2: 1 CO_2), 773 K and 1 atm	[49]
Fe 1.22	N-doped carbon	Wet chemistry and thermal post-treatment	Nitrobenzene to amine	Excellent catalytic performance (TOF 1923 h^{-1}, much higher than those of the previously reported catalysts), chemoselectivity, and tolerance under room temperature	[50]

It is found that the complex effect between the metal and support precursors enables the Pt cation to be strongly anchored on the inner surface of the support, likely stabilized by coordinatively unsaturated pentahedral Al^{3+} centers. Additionally, the template used in the synthesis facilitates the formation of highly stable mesoporous structures enriched with unsaturated pentahedral Al^{3+} centers, further enhancing MSI.

Next, selective hydrogenation of 1,3-butadiene was studied over this Pt SAC, which transformed 1,3-butadiene into butenes with 99% selectivity at 30 °C, without affecting the co-feed propylene molecule. Remarkably, the 0.2 $Pt_{SA}/m\text{-}Al_2O_3$ catalyst exhibited slightly increased activity and near-quantitative selectivity toward butenes after a long-term treatment under hydrogen at 200 °C for 24 hours. The preservation of the isolated Pt active sites was confirmed by HAADF-STEM after the reaction.

In a similar manner, Tang et al. [53] synthesized an Al_2O_3 rich in pentahedral Al^{3+} centers (named as rp-Al_2O_3), which provided enough "defect sites" to anchor atomically dispersed Ru species and to modulate the geometric and electronic features of Ru species. Individual Ru atoms with fine dispersion were distinguished in the HAADF-STEM images. Additional information about the interaction of surface species was obtained by ^{27}Al MAS NMR, Raman, XPS, and CO-DRIFT analyses. It was revealed that the geometric and electronic features of Ru are significantly changed through SMSI, and this must affect the catalytic performance seriously. With the strong interactions between Ru species and pentacoordinated Al^{3+} sites, the $Ru_{SA}/rp\text{-}Al_2O_3$ catalyst exhibits extremely high catalytic activity. Catalytic performance of $Ru_{SA}/rp\text{-}Al_2O_3$ for hydrogenation of benzene was tested and compared with Ru/commercial-Al_2O_3 (and with data reported previously in the scientific literature). We can see in Table 8.2 that the $Ru_{SA}/rp\text{-}Al_2O_3$ shows a fairly high catalytic activity in benzene hydrogenation, with TOF of 5180 h^{-1} within one hour at 80 °C under 3 MPa H_2. In contrast, the Ru/commercial-Al_2O_3 catalyst shows low catalytic activity, with TOF of 417 h^{-1} under the same reaction conditions.

Notably, Ru/rp-Al_2O_3 can be recovered simply by filtration and reused five times without activity decay at conversions of 100% and 60%. Other arene hydrogenations were also investigated. Ru/rp-Al_2O_3 was active for various monosubstituted and disubstituted arenes, such as toluene, *p*-xylene, *m*-xylene, and *o*-xylene, and it showed high selectivity to decalin in the hydrogenation of naphthalene.

Table 8.2 Catalytic performance of various catalysts for benzene hydrogenation to cyclohexane at 80 °C and 3 MPa of H_2.

Catalyst	*t* (h)	Yield (%)[a)]	TOF (h^{-1})[b)]
Ru/rp-Al_2O_3	1	100	5180
Ru/commercial-Al_2O_3	1	18.5	417

a) Yield to cyclohexane, determined by GC.
b) Turnover frequency defined as a mole of benzene converted per mole of total metal loading per hour at a conversion lower than 30%. Catalyst (20 mg), benzene (400 mg).

Source: Tang et al. [53]. Reproduced with permission of American Chemical Society.

In a comparative work [54], the formation of dispersed Pt_{SA} on different supports (TiO_2, Al_2O_3, and CeO_2), their evolution to NP, and their catalytic performance in hydrogenation reactions were studied. The authors prepared a series of Pt/Al_2O_3 samples containing Pt ranging from SA to clusters and NPs. First, a sample containing Pt_{SA} (0.2 Pt_{SA}/Al_2O_3) was synthesized. Then, by reduction of this catalyst under hydrogen at increasing temperatures, from 200 to 450 °C, the Pt_{SA} agglomerated into Pt_{NP}. The series of Pt/Al_2O_3 catalysts were tested for the hydrogenation of 3-nitrostyrene and, similar to the situation of Pt/TiO_2 catalysts, the 0.2 Pt_{SA}/Al_2O_3 sample shows negligible activity for the hydrogenation of 3-nitrostyrene to 3-aminostyrene. The ranking of activity for the hydrogenation of 3-nitrostyrene indicates that Pt clusters and Pt_{NP} show higher reactivity than Pt_{SA}. The authors also tested the reactivity of Pt/Al_2O_3 catalysts for the hydrogenation of styrene to check whether Pt_{SA} can be used for the hydrogenation of C=C bonds. They found that Pt_{SA} also shows negligible activity toward the hydrogenation of C=C bonds. The Pt/Al_2O_3 catalysts with agglomerated Pt species showed much higher activity, and the activity increased with the percentage of Pt_{NP} in the catalyst, indicating that Pt_{NP} are more active than Pt_{SA} and clusters for the hydrogenation of C=C bonds. These results indicate that singly dispersed Pt atoms supported on Al_2O_3 should be much less active than Pt_{NP} in the hydrogenation of —NO_2 and C=C groups.

In line with the above work is the recent observation by Shi et al. on Cu-catalyzed hydrogenation reactions [55]. They [55] prepared a series of size-controllable Cu/Al_2O_3 catalysts from Cu_{SA} (0.5 Cu_{SA}/Al_2O_3) to 9.3 nm Cu_{NP} (2.0 Cu_{NP}/Al_2O_3 sample) using ALD. They aimed at developing a low-cost non-noble metal catalyst with high selectivity and high coking resistance, which can replace the industrially used Pd-based catalysts for the selective hydrogenation of acetylene in excess of ethylene. These catalysts were characterized by HAADF-STEM, X-ray absorption spectroscopy (XAS),*in situ* XPS, and *in situ* thermogravimetry analysis (TGA) to confirm the dominant presence of isolated Cu atoms in the SAC sample. In addition, the Cu_{SA} remained very stable and had +1 valence state after reduction at 300 °C in hydrogen. The catalysts were studied in the semihydrogenation of acetylene in excess of ethylene, showing that a decrease in the Cu particle size reduces the activity considerably but gradually improves both the ethylene selectivity and stability. In particular, the Cu SAC exhibited the highest ethylene selectivity of 91% at complete conversion along with excellent long-term stability for at least 40 hours, in sharp contrast with the rapid deactivation of Cu_{NP} catalysts. *In situ* thermogravimetry measurements further reveal that coke formation on Cu SACs is significantly suppressed by up to ~89% compared to that on the 9.3 nm Cu_{NP} catalysts. These results demonstrated that Cu SAC on Al_2O_3 showed lower activity for the selective C=C hydrogenation reactions but higher selectivity and coking resistance.

Kwak et al. [56] prepared a series of Ru on alumina catalysts with Ru loadings that assured a dispersion range from atomic to 3D clusters and tested their CO_2 reduction performances. Atomically dispersed (as evidenced by STEM) Ru on alumina initially produced CO exclusively by CO_2 reduction. This is in complete contrast with the catalytic behavior of 3D Ru clusters supported on alumina (formed at higher Ru loadings) that are efficient methanation catalysts. However, catalysts with low metal loading (SACs) are unstable under reaction conditions of CO_2 reduction and form large metal clusters. This clustering is accompanied by a large increase in CH_4 selectivity and a drop in CO formation selectivity. On the basis of these results, the authors proposed that CO can be produced by different reaction mechanisms on active Ru centers with different particle sizes. Furthermore, CO may not be a simple reaction intermediate in the path of CO_2 hydrogenation to CH_4.

Following the approach of confinement of active metal NPs in micro- or mesopores of the support to construct single metal atom catalysts, Moliner et al. [57] performed direct synthesis of a nanocrystalline (20–50 nm) high-silica CHA zeolite containing encapsulated Pt_{NP} of ~1 nm. They used the combination of *N,N,N*-trimethyl-1-adamantammonium (TMAda), and Pt-mercapto complex as organic structure-directing agent (OSDA) and Pt precursor, respectively. In comparison with Pt/SiO_2 or Pt-containing Al-rich zeolites, this material showed enhanced stability toward metal sintering in a variety of conditions, including H_2, O_2, and H_2O thermal treatment at high temperatures. EXAFS and HAADF-STEM analyses highlighted the reversible interconversion of the supported metal structures from ~1 nm NP to site-isolated Pt_{SA} in reducing and oxidizing atmospheres. Finally, to demonstrate the successful encapsulation of Pt within the CHA structure, the catalyst was studied in the hydrogenation of olefins of different molecular sizes. Ethylene (0.39 Å) and propylene (0.45 Å) were selected because at mild temperatures, the former diffuses relatively fast through the zeolite, whereas the latter diffuses very slowly. The Pt-CHA-2 catalyst converted more than 80% of ethylene (to ethane) and only 2% of propylene (to propane) under identical experimental conditions. In drastic contrast, a Pt/SiO_2 sample consisting of 2 nm (average size) Pt_{NP} produced similar conversion for both olefins under the same reaction conditions, as one would expect in the absence of internal diffusion limitations. The very low catalytic activity observed for the propylene using Pt-CHA-2 unambiguously confirms the effective encapsulation of the Pt_{NP} within the cavities of the high-silica CHA material.

Another effective way to achieve highly dispersed noble metal catalysts is through the interaction between the lattice-confined component in layered double hydroxides (LDHs) and the supported metal precursors. Recently, it has been reported a simple approach to synthesize highly dispersed Pt catalysts supported on ZnSnAl mixed metal oxides (MMOs) using the induction effect of the electron-rich Sn sites confined in ZnSnAl-LDH [58]. This ZnSnAl-MMO supported Pt catalyst showed enhanced activity and selectivity to 2-pentenol in the selective hydrogenation reaction of 2-pentenal under mild conditions. The authors showed by HAADF-STEM that the induction of the lattice-confined Sn sites in LDHs enhanced the dispersion of Pt in the catalyst. In addition, the strong electronic interaction between Pt_{NP} and MMO support was revealed by XPS analyses. The catalytic evaluation results showed that this electronic effect promoted the performance of Pt by favoring the activation of the C=O bond; consequently, the selectivity to 2-pentenol increased with an increasing amount of Sn and decreased with increasing temperature.

In a similar manner, the anchoring of a Ru precursor onto the surface of LDH in a basic medium successfully generated a Ru SAC with high efficiency for selective hydrogenation of CO_2 to formic acid under mild reaction conditions [59]. A structural study by EXAFS and HAADF-STEM indicated that the catalyst was constituted by isolated single-atomic Ru species with octahedral coordination geometry bearing

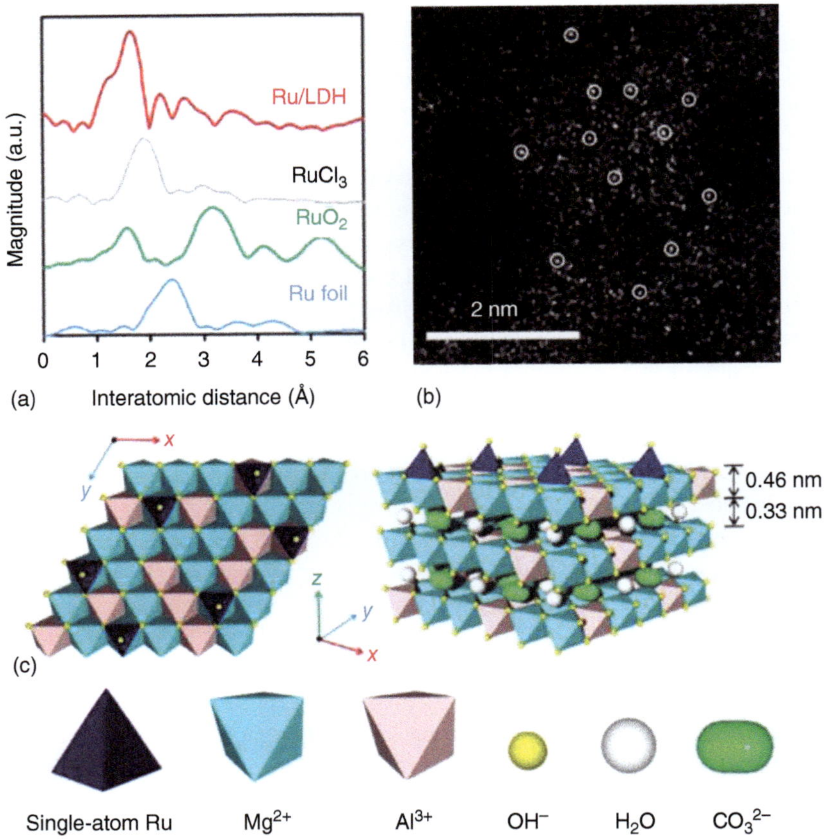

Figure 8.7 (a) Ru K-edge FT-EXAFS spectra, (b) HAADF-STEM image of Ru/LDH (the circles are drawn around some of the isolated Ru atoms), and (c) schematic illustration of Ru/LDH (top and side view). Source Mori et al. [59]. Reproduced with permission of American Chemical Society.

one hydroxyl and two water ligands grafted onto a triad of O atoms originating from the basic hydroxyl groups on the LDH surface (Figure 8.7). The strong basicity of the OH ligands surrounding the active Ru centers has a concentration effect of adsorbed CO_2 near the neighboring Ru center, boosting its catalytic activity.

In Table 8.3, we summarize the recent studies in the field of supported SAC on unreducible metal oxides developed for hydrogenation reactions.

8.4 Hydrogenation Reactions Catalyzed by SACs Supported on Reducible Metal Oxide CeO_2 and TiO_2

The hydrogenation of several nitroarenes with different functional groups was reported by Lu et al. [60] over Pd/CeO_2 prepared by impregnation with different Pd weight loadings (1.0%, 2.0%, 2.5%, 3.0%, 5.0%) in the absence of solvent or additives.

Table 8.3 Summary of unreducible metal oxide supported metal single atoms for hydrogenation reactions.

Support	SAC	Synthetic method	Reaction	Catalytic performance	References
Al_2O_3	Pt	sol–gel solvent vaporization self-assembly method	1,3-butadiene to butenes	Increased activity and near-quantitative selectivity toward butenes	[52]
Al_2O_3	Ru	Impregnation	Benzene to cyclohexane	TOF of 5180 h^{-1} within one hour at 80 °C under 3 MPa H_2	[53]
Al_2O_3	Pt	Impregnation	3-nitrostyrene to 3-aminostyrene	Negligible activity	[54]
Al_2O_3	Cu	Atomic layer deposition	Acetylene to ethylene	Low activity, selectivity of 91%, and long-term stability for at least 40 hours	[55]
Al_2O_3	Ru	Impregnation	CO_2 to CO	Low activity, selectivity, and stability (clustering during the reaction)	[56]
CHA zeolite	Pt	One-pot with organic structure-directing agent (OSDA)	Ethylene to ethane	Reversible interconversion from ~1 nm NP to Pt_{SA} in reducing and oxidizing atmospheres	[57]
LDH	Pt	Impregnation	2-pentenal to 2-pentenol	Enhanced activity and selectivity to 2-pentenol	[58]
LDH	Ru	Impregnation	CO_2 to formic acid	High efficiency for selective hydrogenation of CO_2 to formic acid under mild reaction conditions	[59]

The presence of Pd_{SA} was suggested from the absence of diffractions peaks or particles on the microscopy images. However, the observation of Pd–O vibrations by Raman spectroscopy could actually point to the presence of PdO_x NP at least coexisting with the Pd_{SA}. The authors tried to show that a higher concentration of Pd_{SA} estimated as the ratio of Pd^{+2} to Pd^0 obtained by XPS enhanced the yield in the nitroarenes hydrogenation.

Karim et al. [61] evaluated the hydrogenation of acetylene over Pt supported on oxygen-deficient TiO_2 anatase. Experimentally, the Pt/TiO_2 catalysts were prepared by strong electrostatic adsorption (SEA) and incipient wetness impregnation with weight loadings of 0.025%, 0.1%, and 1% to vary the Pt nuclearity from SAs to sub-nanometer clusters and NPs. From STEM images, the presence of Pt_{SA} was verified on the samples with 0.025 and 0.1 wt%, for which 100% and 35% of Pt_{SA} were estimated, respectively. Unfortunately, the conversion achieved with the catalyst containing exclusively Pt_{SA} was negligible to be detected by the analytical instrument, and the authors were only able to establish a correlation between the Pt particle size when clusters were formed, and the activity and selectivity of the acetylene hydrogenation.

To further investigate these systems, the authors also studied Pt_n/TiO_2 (n = number of atoms per catalytic site = 1, 2, 4, and 8), by DFT and microkinetics approaches [62]. DFT study is indeed one of the best alternatives to study these challenging systems. The authors obtained data about the adsorption modes and strength of several species and suggested that as the number of Pt atoms of the cluster increases from 1 to 8, the energetic barrier of the hydrogenation is decreased, which can be directly correlated with the activity. Based on these findings, the authors speculated that the energy barrier could be related to the different adsorption modes that can occur on Pt clusters compared to the lower degree of freedom of adsorbents on Pt_{SA}. Overall, the study concluded that while the activity on the acetylene hydrogenation was enhanced by increasing the number of Pt atoms, the selectivity to ethylene decreases.

The activation of C—O bond has been evaluated using several SAC. The idea behind the use of single metal atoms for C—O activation of biomass compounds is that due to side effects, the ring moiety is not adsorbed on the metal, and the C—O activation proceeds selectively. Bearing in mind this strategy, Fu et al. [63] employed Pt_{SA} on TiO_2 prepared by impregnation and tested the catalysts in the furfuryl alcohol conversion to 2-methylfuran. A combination of HAADF-STEM, EXAFS, and CO-FTIR shown in Figure 8.8, seemed to prove that the catalysts with 0.04 wt% Pt consisted of Pt_{SA} along with areas where two to three atoms could be observed as well as larger clusters, which were more obvious as the Pt content increased to 1 wt% (Figure 8.8b).

The Pt SACs were more active for the hydrodeoxygenation (HDO) than the bare support by 31-fold, which was ascribed to the larger number of oxygen vacancies close to the Pt atoms. In addition, no hydrogenation of the ring was observed, apparently due to the absence of Pt clusters where the reactant could be adsorbed parallel to the metal, i.e. ensemble effect. Upon increasing the Pt loading up to 1 wt%, the HDO rate was below the theoretical value if the entire Pt was found as single atoms, while the ring saturation increased as well.

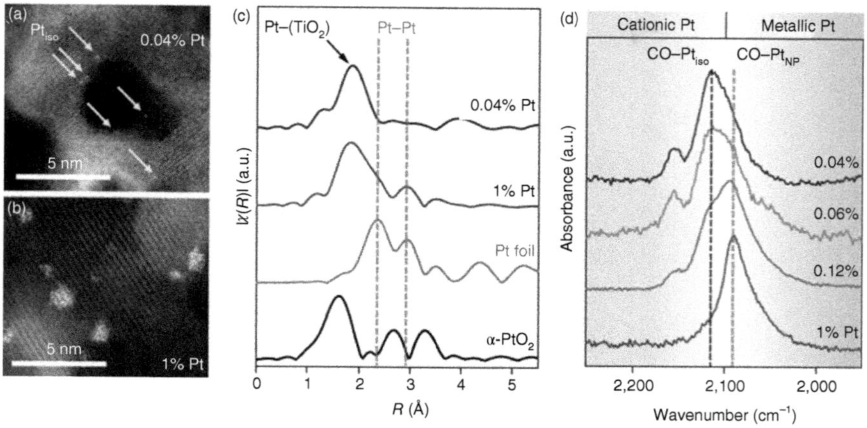

Figure 8.8 (a) HAADF-STEM image of 0.04 wt% Pt on TiO_2, (b) HAADF-STEM image of 1 wt% Pt on TiO_2, (c) EXAFS spectra of Pt/TiO_2 catalysts and Pt standards at the Pt L_3 edge, and (d) CO-FTIR adsorption spectra of *x* wt% Pt/TiO_2. Source: Fu et al. [63]. Reproduced with permission of Springer Nature.

In contrast to this work, Resasco et al. [64] recently found that Pt_{SA} supported on TiO_2 displayed significantly less activity on the *m*-cresol HDO compared to small Pt_{NP}. The SACs were prepared with 0.025 wt% loading and compared with catalysts synthesized by conventional incipient wetness impregnation using 1 wt%. Due to the difficulty in characterizing the active centers, the authors supported their findings with kinetic measurements combined with some common characterization techniques and theoretical modeling. The Pt SAC supported on TiO_2, seemed to be stable, as suggested by CO-FTIR spectra, which showed the same adsorption band in the fresh and spent catalyst. The stability was tentatively ascribed to the low metal loading, which would result in Pt located on the most stable TiO_2 sites. As mentioned, the catalysts with Pt_{SA} displayed significantly lower activity compared to those obtained with Pt clusters. The lower ability of isolated Pt atoms to activate hydrogen was indeed confirmed using cyclohexene hydrogenation as a test reaction since the TOF was again much higher for the supported Pt_{NP} than for supported Pt_{SA}, which barely promoted the formation of cyclohexane.

The experimental observations were further compared with DFT calculations. The systems were modeled considering that Pt_{SA} consisted of PtO_2 over anatase TiO_2 (101) terrace and Pt_{SA} coordinated to two lattice-oxygen at a bridging position on the TiO_2 (101). These two structures showed that despite the dissociative adsorption of H_2 on the PtO_2 structure is exothermic (−282 kJ/mol), the energetically preferred adsorption sites are oxygen atoms from the support adjacent to the Pt atom. The adsorption of H atoms on these oxygen atoms would create hydroxyl groups that would not be active for hydrogenation in contrast to H atoms adsorbed on Pt. The DFT calculations also proved that cresol adsorption on Pt_{SA} is favored through the oxygen atom, which would increase the selectivity to the target HDO reaction. Therefore, a strategy combining both sites, as depicted in Figure 8.9 for

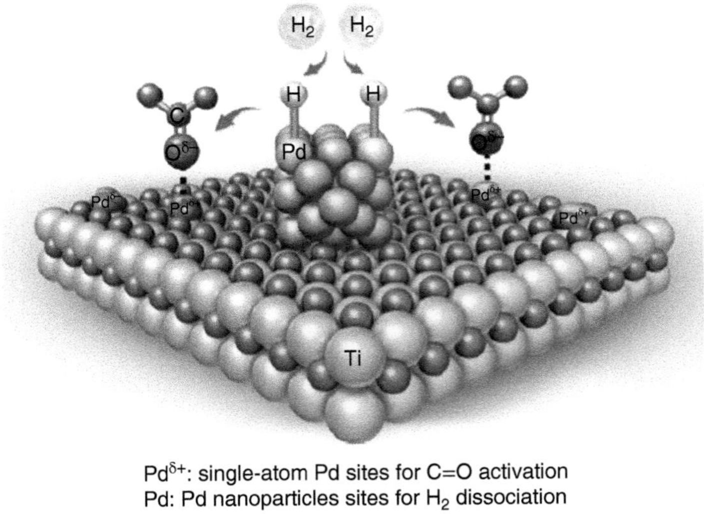

Figure 8.9 Schematic representation of bifunctional Pd^0/Pd_{SA} on TiO_2. Source: Kuai et al. [65]. Springer Nature. CC BY 4.0.

Pd-based catalysts, could lead to catalyst optimization for HDO in terms of activity and selectivity [65].

Kuai et al. [65] also raised the problem of single metal atoms for the hydrogenation reaction of C—O bonds. In this case, they designed catalysts with two different active sites for the hydrogenation of C=O bonds and proposed to use the spillover mechanism to overcome the lower activation ability of H_2 on single metal atoms. The dissociation of H_2 is limited on the Pd_{SA} since it can only take place heterolytically, this being energetically not favored. Hence, the catalysts contained both Pd_{NP} and Pd_{SA} supported on mesoporous TiO_2 (Figure 8.9). The C=O bond activation could take place on the Pd_{SA}, while H_2 might be activated on the Pd_{NP} and the H atoms migrate to the Pd_{SA} where C=O bonds are adsorbed. Synthetically, these catalysts were prepared by a modified large-scaled spray-assisted method, and conversion of several substrates reached 100% after one hour. The coexistence of Pd_{NP} and Pd_{SA} was verified by aberration-corrected HAADF-STEM and the Fourier transform X-ray absorption fine structure spectra (FT-EXAFS) as shown on Figure 8.10.

Among C=O reduction reactions, the hydrogenation of CO_2 has received increasing attention due to its potential for valorization of a greenhouse gas to valuable products such as methanol. Recently, using DFT calculation, Ma et al. [66] studied the hydrogenation of CO_2 over Rh isolated atoms supported on the model TiO_2 (101) surface. The simulations revealed that H_2 and CO_2 can be adsorbed on Rh_{SA} and that CO_2 reacts with dissociated H_2 to get COOH intermediate species, which is subsequent transformed into CO and H_2O. Further hydrogenation of CO seems to be limited by the strong adsorption of this molecule onto the Rh_{SA}, which was taken as the explanation for the high selectivity toward CO. The same methodology was used to study Ru, Pd, and Pt single atoms supported on TiO_2 (101) surface.

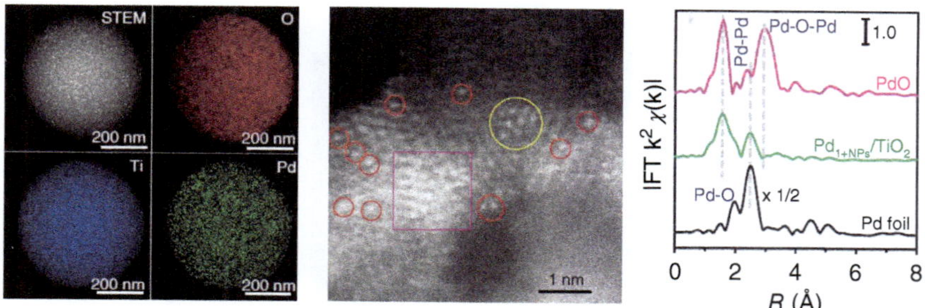

(a) (b) (c)

Figure 8.10 (a) HAADF-STEM mapping of Pd/TiO_2, (b) HAADF of Pd/TiO_2, and (c) R-spaced Pd K-edge FT-EXAFS spectra of $Pd^{1+}NPs/TiO_2$ referred to bulk Pd foil and PdO. Source: Kuai et al. [65]. Springer Nature/CC BY 4.0.

The results showed that Pd displayed similar behavior to Rh and that it could co-adsorb both H_2 and CO_2, leading to high selectivity to CO. On the other hand, Ru might be able to co-adsorb H_2, CO_2, and CO, which leads to the formation of other products, while Pt seemed to be inactive toward CO_2 hydrogenation since co-adsorption of H_2 with CO_2 was not favored.

Wang et al. also [67] studied the CO_2 reduction using single-atom 0.05 wt% Pt/CeO_2 [verified by HAADF-STEM (Figure 8.11)], and nano-clustered 2 wt% Pt/CeO_2. Pt_{SA} supported on ceria proved to be more active toward CO_2 hydrogenation than Pt_{NP}, i.e. by a factor of 7, selectively forming CO, unlike the theoretical behavior described in Ref. [66] with Pt_{SA} on TiO_2. The difference with TiO_2 is the activation of CO_2 by CeO_2, which was observed by *in situ* DRIFTs, while H_2 is activated on the metal in both systems. Similar to the conclusions of Ref. [66] for Rh and Pt_{SA} on TiO_2, it seems that CO adsorption on Pt is weak, which inhibits further hydrogenation to CH_4 and prevents poisoning.

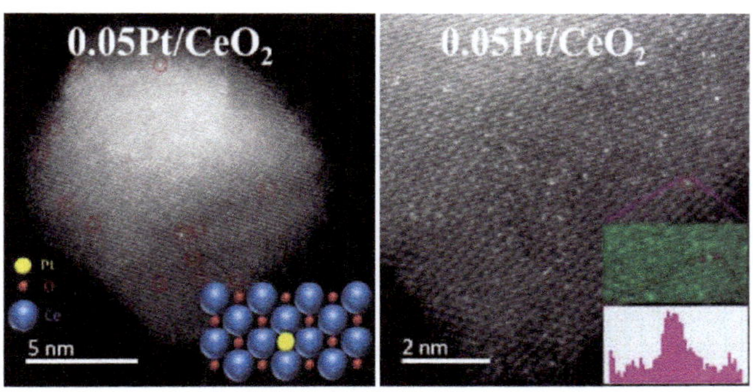

(a) (b)

Figure 8.11 HAADF-STEM images of 0.05 Pt_{SA}/CeO_2 (inset: line- scanning intensity profile obtained for the zoomed in area in (b)). Source: Wang et al. [67]. Reproduced with permission of American Chemical Society.

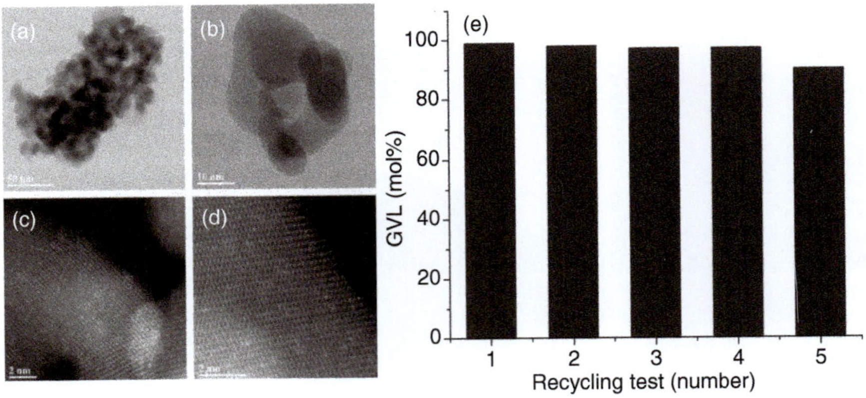

Figure 8.12 STEM of the fresh Ru/ZrO_2 (a) low- and (b) high-magnification; (c, d) HAADF-STEM images, (e) yield to GVL in successive runs at 30 bar of H_2 and 150 °C in dioxane as solvent 10 wt% LA; LA/Ru = 1000. Source: Ftouni et al. [68]. Reproduced with permission of American Chemical Society.

Ftouni et al. [68] studied the hydrogenation of levulinic acid (LA) to γ-valerolactone (GVL) using 1 wt% Ru_{SA} supported on the surface of TiO_2, ZrO_2, and active carbon. The HAADF-STEM images of Ru/ZrO_2, shown in Figure 8.12 , proved that this catalyst displayed mostly Ru single atoms on the surface. The yield to GVL obtained with Ru/ZrO_2 was the highest among the tested catalysts. However, the specific activity was not estimated. Also, recycling tests indicated that the catalyst was quite stable as regards yield levels, although the HAAF-STEM images of the spent catalyst showed some Ru_{NP} of average size 2.8 nm after the fifth recycling test.

Cao et al. [69] synthetized a catalyst with 0.6 wt% of Ir using a metal-organic framework (UiO-66) as a template followed by pyrolysis as shown in Figure 8.13. This material displayed single iridium atoms, as observed in Figure 8.14. While the activity of this catalyst was lower than that measured for Ir_{NP} supported on carbon and ZrO_2 with 2.7 and 0.6 wt% iridium loading, it showed remarkable stability even after 7 runs and over 99% selectivity to GVL during the hydrogenation of LA. The stability was ascribed to the enhanced interactions between the atomic iridium and the

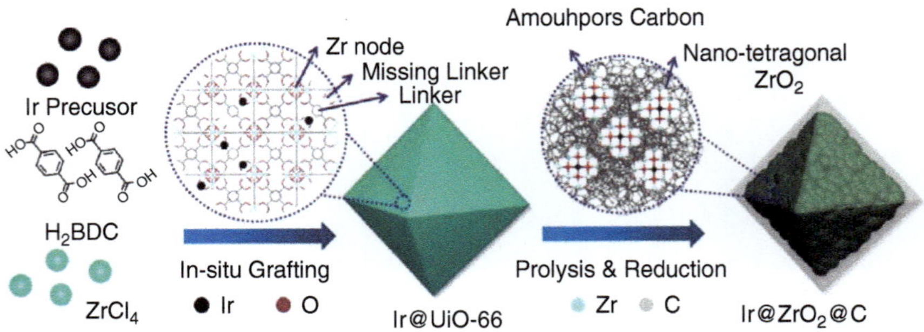

Figure 8.13 Synthesis of Ir@ZrO_2@C using UiO-66 as a template. Source: Cao et al. [69]. Reproduced with permission of Elsevier.

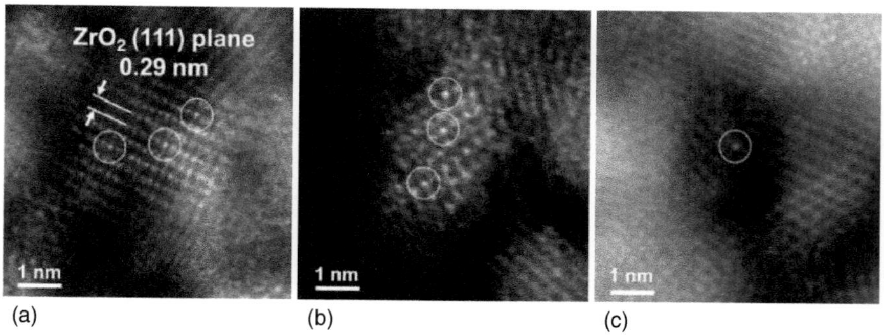

Figure 8.14 (a–c) HAADF-STEM images of Ir@/ZrO_2@C. Source: Cao et al. [69]. Reproduced with permission of Elsevier.

ZrO_2 support, as well as the confinement of the amorphous carbon, this avoiding the metal leaching observed for the other tested catalysts.

Kim et al. [70], also synthesized Pt_{SA} on a metal-organic framework based on zirconia by ALD. The formation of Pt_{SA} was verified by CO adsorption followed by *in situ* IR during the ALD synthesis. Vibration at 2090 and 2040 cm^{-1} corresponding to Pt_{SA} and a few-atom Pt clusters were observed, respectively. This technique also showed that the synthesis temperature during ALD was critical since the proportion of Pt_{SA} was significantly higher when the synthesis was performed at 115 °C (LT-Pt-ALD) compared to 160 °C (HT-Pt-ALD). The synthesis temperature also influences the actual Pt loading, which was 0.15 and 2.5 Pt atoms per Zr_6 node, respectively. The catalysts were tested in the ethylene hydrogenation reaction, showing high conversion of ethylene even at 25 °C. The authors used the dispersion estimated from CO chemisorption to calculate the TOF to account for the accessible Pt atoms. However, the TOF was indeed slightly lower for the sample with a larger fraction of Pt_{SA}, 0.64 vs. 0.81 s^{-1}. The result was not conclusive since CO poisoning during the ALD synthesis was observed on the Pt_{SA}, and this would result in higher activity for LT-Pt-ALD if only accessible atoms were considered.

In Table 8.4, we summarize the recent studies in the field of supported SAC on reducible metal oxides developed for hydrogenation reactions.

8.5 Hydrogenation Reactions Catalyzed by SACs Supported on Metallic Surfaces

Another common strategy for obtaining and evaluating SAC is using a non-noble metal as support for noble metal single metal atoms, this enhancing H_2 dissociation as it can take place on both metals. Applying this methodology, Peng et al. [7] supported Pt_{SA} on Ni surfaces and tested it in the hydrogenation of 3-nitrostyrene. The actual existence of Pt_{SA} atoms was verified by HAADF-STEM and XANES, as shown in Figure 8.15. This catalyst displayed much higher activity, c. 1800 h^{-1} compared to Pt_{SA} supported on active carbon, TiO_2, SiO_2, and ZSM-5 whose values were in the range 60–160 h^{-1}. This superior activity was ascribed to the activation of H_2 on

Table 8.4 Summary of reducible metal oxide supported metal single atoms for hydrogenation reactions.

Support	SAC	Synthetic method	Reaction	Catalytic performance	References
CeO_2	Pd	Impregnation	Nitroarene hydrogenation	100% selectivity, TOF >10 000 h^{-1}	[60]
TiO_2	Pt	Electrostatic adsorption	Acetylene hydrogenation	Selectivity with Pt_{SA} 100% vs. 50% with Pt_{NP}. TOF with Pt_{SA} 1–2 orders of magnitude lower	[61]
TiO_2	Pt	—	Acetylene hydrogenation	Selectivity with Pt_{SA} 100% vs. 50% with Pt_{NP}. TOF with Pt_{SA} 3 orders of magnitude lower	[62]
TiO_2	Pt	Impregnation	Furfuryl alcohol to 2-methylfuran	Conversion rate and ring hydrogenation increased with Pt loading by 1–2 orders of magnitude depending on Pt loading	[63]
TiO_2	Pt	Impregnation	*m*-Cresol HDO	Pt_{SA} no activity	[64]
TiO_2	Pd	Spray-assisted method	Aromatic aldehyde hydrogenation	Selectivity to alcohol 100% TOF >4000 h^{-1}.	[65]
TiO_2	Rh, Pd, Pt, Ru	—	CO_2 reduction	—	[66]
CeO_2	Pt	Impregnation	CO_2 reduction	100% selectivity to CO using SAC vs. 30–70% with Pt_{NP}. CO_2 conversion rate increased by a factor of 5–13 using SAC depending on the temperature	[67]
ZrO_2	Ru	Impregnation	Levulinic acid to γ-valerolactone	Recyclable in 4 successive runs at GVL yields of c. 50%	[68]
ZrO_2	Ir	Impregnation	Levulinic acid to γ-valerolactone	Selectivity to GVL >99%. Recyclable in 7 successive runs at conversion levels of c. 25%	[69]
ZrO_2	Pt	Atomic layer deposition	Ethylene hydrogenation	Activity of 0.8 s^{-1}	[70]

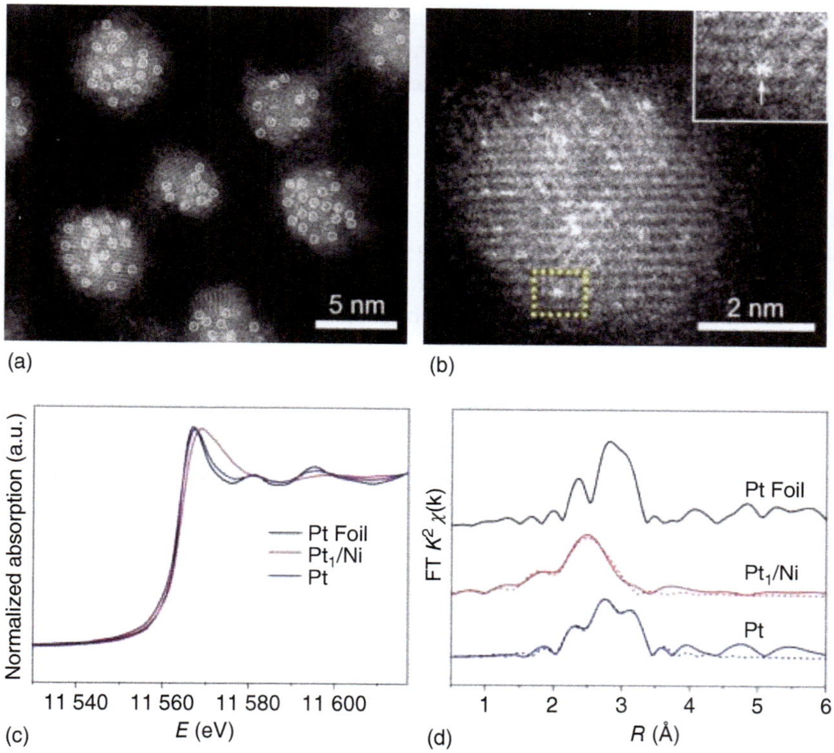

Figure 8.15 (a, b) HAADF-STEM images of Pt single atoms on Ni surface. (c) Pt L_3-edge XANES profiles of Pt foil, Pt_1/Ni, and Pt/C and (d) Pt L_3-edge EXAFS spectra in R space. Pt foil was used as the reference. Source: Peng et al. [7]. Reproduced with permission of American Chemical Society.

both metals along with the spillover mechanism. In addition, the synergetic action between Pt_{SA} and Ni surface leads to the optimal 3-nitrostyrene adsorption mode to obtain selectively 3-vinylaniline.

This strategy was also used by Abdel-Rahman et al. [71] who studied the gas-phase hydrogenation of propyne to propene using Pd_{SA} (2 wt%) supported on a Cu surface, and the results were compared with those obtained with Cu (111). The formation of propene was favored on the Pd/Cu system and took place starting at 383 K preceded by propyne adsorption on the surface at 300 K to form a di–σ/di–π structure.

To date in the open literature, there were only these two studies mentioned above, which have been reported on the development of supported SACs on metallic surfaces for hydrogenation reactions.

8.6 Summary and Conclusions

This chapter presents a revision of recent researches studies concerning the application of supported single metal atoms as heterogeneous catalysts for a variety of hydrogenation reactions. Among the substrates that have been hydrogenated using SACs, there are many organic compounds such as alkynes, di-olefins, aromatics, nitroaromatics, nitroarenes, unsaturated aldehydes, furfuryl alcohol, or cresol. Emerging reactions such as the reductive transformation of CO_2 into CO or formic acid using SACs have also been explored (see also Chapter 10). In this chapter, the SACs have been categorized as a function of the support materials: supported on carbon materials (emphasizing on N-doped carbons), supported on unreducible and reducible metal oxide (mainly CeO_2 and TiO_2), or supported on metallic surfaces (typically noble metals over Cu or Ni). The metals used to build SACs are both noble metals such as Pd, Pt, Ru, Au, or Rh, and non-noble metals as Cu, Ni, or Co. Very relevant catalytic concepts have been managed for the interpretation of the intermediate species and to postulate reaction mechanisms. Among these, it can be underlined: the oxidation state of single metal atoms anchored on each support, the dihydrogen activation and mobility by spillover, the adsorption modes of the reactants on SACs, or at the interphases SAC-support, and the chemical or electronic MSIs.

Some contradictory evidence are inferred when data of catalytic performances in hydrogenation reactions with supported single metal atoms are comparatively evaluated. For some reactions, and with a certain type of supports or metal atoms, these new materials seem to be superior in terms of catalytic performances, improved specific activities, or higher selectivity to desired products, in comparison with classical supported metal NP or metal clusters. However, for other metal atoms and other hydrogenation reactions, supported single metals are practically inactive. A possible reason for these findings is the very low amount of metal incorporated on the supports for obtaining single atoms, which makes the characterization of materials with isolated atoms difficult and the catalytic reaction studies hard to be performed. However, the design of selective catalytic sites for hydrogenation reactions based on supported single metal atoms is at present an exciting field of research. In addition, today, the very large majority of catalytic studies addressing the subject are limited to a comparison of the reactivity of isolated atoms with that of metallic particles, even though these two types of species must coexist in many catalysts. Interestingly, some studies have already shown that cooperative catalysis between single atoms and nanoparticles can operate for hydrogenation reactions [72]. In that context, the control of the single atom/particle ratio of the catalysts is also challenging.

Thus, even if developments of catalytic materials based on supported single metal atoms do not have large technological implementations up to now, the management of concepts and ideas should conduce shortly to improved catalytic materials because the rational design of improved new catalysts will be based on rigorous scientific studies, both experimental one and theoretical analysis, developed using supported SACs. For these reasons, the steps for the application of supported SACs for selective hydrogenation reactions are being given.

Acknowledgments

This work was supported by the Spanish Agencia Estatal de Investigación (AEI) and EU (FEDER) (projects CTQ2017-89443-C3-1-R and CTQ2017-89443-C3-3-R), which are gratefully acknowledged.

References

1 Liu, J. (2017). Catalysis by supported single metal atoms. *ACS Catalysis* 7 (1): 34–59. https://doi.org/10.1021/acscatal.6b01534.

2 Kaiser, S.K., Chen, Z., Faust Akl, D. et al. (2020). Single-atom catalysts across the periodic table. *Chemical Reviews* 120 (21): 11703–11809. https://doi.org/10.1021/acs.chemrev.0c00576.

3 Mitchell, S. and Pérez-Ramírez, J. (2020). Single atom catalysis: a decade of stunning progress and the promise for a bright future. *Nature Communications* 11 (1): 10–12. https://doi.org/10.1038/s41467-020-18182-5.

4 Samantaray, M.K., D'Elia, V., Pump, E. et al. (2020). The comparison between single atom catalysis and surface organometallic catalysis. *Chemical Reviews*: 734–813. https://doi.org/10.1021/acs.chemrev.9b00238.

5 Osborn, J.A., Jardine, F.H., Young, J.F. et al. (1966). The preparation and properties of tris(triphenylphosphine)halogenorhodium(I) and some reactions thereof including catalytic homogeneous hydrogenation of olefins and acetylenes and their derivatives. *Journal of the Chemical Society A: Inorganic, Physical, Theoretical*: 1711–1732. https://doi.org/10.1039/j19660001711.

6 Parkinson, G.S. (2019). Single-atom catalysis: how structure influences catalytic performance. *Catalysis Letters* 149 (5): 1137–1146. https://doi.org/10.1007/s10562-019-02709-7.

7 Peng, Y., Geng, Z., Zhao, S. et al. (2018). Pt single atoms embedded in the surface of Ni nanocrystals as highly active catalysts for selective hydrogenation of nitro compounds. *Nano Letters* 18 (6): 3785–3791. https://doi.org/10.1021/acs.nanolett.8b01059.

8 Kyriakou, G., Boucher, M.B., Jewell, A.D. et al. (2012). Isolated metal atom geometries as a strategy for selective heterogeneous hydrogenations. *Science* 335 (6073): 1209–1212. https://doi.org/10.1126/science.1215864.

9 Vilé, G., Albani, D., Nachtegaal, M. et al. (2015). A stable single-site palladium catalyst for hydrogenations. *Angewandte Chemie – International Edition* 54 (38): 11265–11269. https://doi.org/10.1002/anie.201505073.

10 Cao, Y., Chen, B., Guerrero-Sánchez, J. et al. (2019). Controlling selectivity in unsaturated aldehyde hydrogenation using single-site alloy catalysts. *ACS Catalysis* 9 (10): 9150–9157. https://doi.org/10.1021/acscatal.9b02547.

11 Riley, C., Zhou, S., Kunwar, D. et al. (2018). Design of effective catalysts for selective alkyne hydrogenation by doping of ceria with a single-atom promotor. *Journal of the American Chemical Society* 140 (40): 12964–12973. https://doi.org/10.1021/jacs.8b07789.

12 Hannagan, R.T., Giannakakis, G., Flytzani-Stephanopoulos, M. et al. (2020). Single-atom alloy catalysis. *Chemical Reviews* 120 (21): 12044–12088. https://doi.org/10.1021/acs.chemrev.0c00078.

13 Jiang, L., Liu, K., Hung, S.-F. et al. (2020). Facet engineering accelerates spillover hydrogenation on highly diluted metal nanocatalysts. *Nature Nanotechnology* 15 (10): 848–853. https://doi.org/10.1038/s41565-020-0746-x.

14 Wei, H., Liu, X., Wang, A. et al. (2014). FeO_x-supported platinum single-atom and pseudo-single-atom catalysts for chemoselective hydrogenation of functionalized nitroarenes. *Nature Communications* 5 (1): 5634. https://doi.org/10.1038/ncomms6634.

15 Prins, R. (2012). Hydrogen spillover. Facts and fiction. *Chemical Reviews* 112 (5): 2714–2738. https://doi.org/10.1021/cr200346z.

16 Li, J., Guan, Q., Wu, H. et al. (2019). Highly active and stable metal single-atom catalysts achieved by strong electronic metal-support interactions. *Journal of the American Chemical Society* 141 (37): 14515–14519. https://doi.org/10.1021/jacs.9b06482.

17 Rivera-Cárcamo, C. and Serp, P. (2018). Single atom catalysts on carbon-based materials. *ChemCatChem* 10 (22): 5058–5091. https://doi.org/10.1002/cctc.201801174.

18 Gawande, M.B., Fornasiero, P., and Zbořil, R. (2020). Carbon-based single-atom catalysts for advanced applications. *ACS Catalysis* 10 (3): 2231–2259. https://doi.org/10.1021/acscatal.9b04217.

19 Wang, Y., Mao, J., Meng, X. et al. (2019). Catalysis with two-dimensional materials confining single atoms: concept, design, and applications. *Chemical Reviews* 119 (3): 1806–1854. https://doi.org/10.1021/acs.chemrev.8b00501.

20 Zhang, L., Zhou, M., Wang, A. et al. (2020). Selective hydrogenation over supported metal catalysts: from nanoparticles to single atoms. *Chemical Reviews* 120 (2): 683–733. https://doi.org/10.1021/acs.chemrev.9b00230.

21 Li, X., Rong, H., Zhang, J. et al. (2020). Modulating the local coordination environment of single-atom catalysts for enhanced catalytic performance. *Nano Research* 13 (7): 1842–1855. https://doi.org/10.1007/s12274-020-2755-3.

22 Gerber, I.C. and Serp, P. (2020). A theory/experience description of support effects in carbon-supported catalysts. *Chemical Reviews* 120 (2): 1250–1349. https://doi.org/10.1021/acs.chemrev.9b00209.

23 Inagaki, M., Toyoda, M., Soneda, Y. et al. (2018). Nitrogen-doped carbon materials. *Carbon* 132: 104–140. https://doi.org/10.1016/j.carbon.2018.02.024.

24 Yan, H., Cheng, H., Yi, H. et al. (2015). Single-atom Pd_1/graphene catalyst achieved by atomic layer deposition: remarkable performance in selective hydrogenation of 1,3-butadiene. *Journal of the American Chemical Society* 137 (33): 10484–10487. https://doi.org/10.1021/jacs.5b06485.

25 Huang, F., Deng, Y., Chen, Y. et al. (2018). Atomically dispersed Pd on nanodiamond/graphene hybrid for selective hydrogenation of acetylene. *Journal of the American Chemical Society* 140 (41): 13142–13146. https://doi.org/10.1021/jacs.8b07476.

26 Huang, X., Xia, Y., Cao, Y. et al. (2017). Enhancing both selectivity and coking-resistance of a single-atom Pd1/C3N4 catalyst for acetylene hydrogenation. *Nano Research* 10 (4): 1302–1312. https://doi.org/10.1007/s12274-016-1416-z.

27 Liu, Q., Wang, J., Zhang, J. et al. (2020). In situ immobilization of isolated Pd single-atoms on graphene by employing amino-functionalized rigid molecules and their prominent catalytic performance. *Catalysis Science and Technology* 10 (2): 450–457. https://doi.org/10.1039/c9cy02110h.

28 Feng, Q., Zhao, S., Xu, Q. et al. (2019). Mesoporous nitrogen-doped carbon-nanosphere-supported isolated single-atom Pd catalyst for highly efficient semihydrogenation of acetylene. *Advanced Materials* 31 (36): 1–7. https://doi.org/10.1002/adma.201901024.

29 Lou, Y., Wu, H., and Liu, J. (2019). Nanocarbon-edge-anchored high-density Pt atoms for 3-nitrostyrene hydrogenation: strong metal-carbon interaction. *IScience* 13: 190–198. https://doi.org/10.1016/j.isci.2019.02.016.

30 He, X., He, Q., Deng, Y. et al. (2019). A versatile route to fabricate single atom catalysts with high chemoselectivity and regioselectivity in hydrogenation. *Nature Communications* 10 (1): 3663. https://doi.org/10.1038/s41467-019-11619-6.

31 Han, A., Zhang, J., Sun, W. et al. (2019). Isolating contiguous Pt atoms and forming Pt–Zn intermetallic nanoparticles to regulate selectivity in 4-nitrophenylacetylene hydrogenation. *Nature Communications* 10 (1): 3787. https://doi.org/10.1038/s41467-019-11794-6.

32 Rivera-Cárcamo, C., Leng, F., Gerber, I.C. et al. (2020). Catalysis to discriminate single atoms from subnanometric ruthenium particles in ultra-high loading catalysts. *Catalysis Science and Technology* 10 (14): 4673–4683. https://doi.org/10.1039/D0CY00540A.

33 Sredojević, D.N., Šljivančanin, Ž., Brothers, E.N. et al. (2018). Formic acid synthesis by CO_2 hydrogenation over single-atom catalysts based on Ru and Cu embedded in graphene. *ChemistrySelect* 3 (9): 2631–2637. https://doi.org/10.1002/slct.201702836.

34 Fernandes, D.M., Rocha, M., Rivera-Cárcamo, C. et al. (2020). Ru single atoms and nanoparticles on carbon nanotubes as multifunctional catalysts. *Dalton Transactions* 49 (29): 10250–10260. https://doi.org/10.1039/D0DT02096F.

35 Rivera-Cárcamo, C., Scarfiello, C., García, A.B. et al. (2020). Stabilization of metal single atoms on carbon and TiO_2 supports for CO_2 hydrogenation: the importance of regulating charge transfer. *Advanced Materials Interfaces* 2001777 https://doi.org/10.1002/admi.202001777.

36 Lin, R., Albani, D., Fako, E. et al. (2019). Design of single gold atoms on nitrogen-doped carbon for molecular recognition in alkyne semi-hydrogenation. *Angewandte Chemie – International Edition* 58 (2): 504–509. https://doi.org/10.1002/anie.201805820.

37 Sun, X., Olivos-Suarez, A.I., Osadchii, D. et al. (2018). Single cobalt sites in mesoporous N-doped carbon matrix for selective catalytic hydrogenation of nitroarenes. *Journal of Catalysis* 357: 20–28. https://doi.org/10.1016/j.jcat.2017.10.030.

38 Wang, H., Wang, Y., Li, Y. et al. (2020). Highly efficient hydrogenation of nitroarenes by N-doped carbon-supported cobalt single-atom catalyst in ethanol/water mixed solvent. *ACS Applied Materials and Interfaces* 12 (30): 34021–34031. https://doi.org/10.1021/acsami.0c06632.

39 Liu, W., Zhang, L., Yan, W. et al. (2016). Single-atom dispersed Co-N-C catalyst: structure identification and performance for hydrogenative coupling of nitroarenes. *Chemical Science* 7 (9): 5758–5764. https://doi.org/10.1039/c6sc02105k.

40 Yan, H., Zhao, X., Guo, N. et al. (2018). Atomic engineering of high-density isolated Co atoms on graphene with proximal-atom controlled reaction selectivity. *Nature Communications* 9 (1): 3197. https://doi.org/10.1038/s41467-018-05754-9.

41 Zhou, P., Jiang, L., Wang, F. et al. (2017). High performance of a cobalt–nitrogen complex for the reduction and reductive coupling of nitro compounds into amines and their derivatives. *Science Advances* 3 (2): 1–11. https://doi.org/10.1126/sciadv.1601945.

42 Li, H., Cao, C., Liu, J. et al. (2019). Cobalt single atoms anchored on N-doped ultrathin carbon nanosheets for selective transfer hydrogenation of nitroarenes. *Science China Materials* 62 (9): 1306–1314. https://doi.org/10.1007/s40843-019-9426-x.

43 Huang, R., Cao, C., Liu, J. et al. (2020). Integration of metal single atoms on hierarchical porous nitrogen-doped carbon for highly efficient hydrogenation of large-sized molecules in the pharmaceutical industry. *ACS Applied Materials and Interfaces* 12 (15): 17651–17658. https://doi.org/10.1021/acsami.0c03452.

44 Huang, F., Deng, Y., Chen, Y. et al. (2019). Anchoring Cu1 species over nanodiamond-graphene for semi-hydrogenation of acetylene. *Nature Communications* 10 (1): 4431. https://doi.org/10.1038/s41467-019-12460-7.

45 Yang, F., Wang, M., Liu, W. et al. (2019). Atomically dispersed Ni as the active site towards selective hydrogenation of nitroarenes. *Green Chemistry* 21 (3): 704–711. https://doi.org/10.1039/c8gc03664k.

46 Ning, L., Liao, S., Li, H. et al. (2019). Carbon-based materials with tunable morphology confined Ni(0) and Ni-Nx active sites: highly efficient selective hydrogenation catalysts. *Carbon* 154: 48–57. https://doi.org/10.1016/j.carbon.2019.07.099.

47 Liu, W., Chen, Y., Qi, H. et al. (2018). A durable nickel single-atom catalyst for hydrogenation reactions and cellulose valorization under harsh conditions. *Angewandte Chemie – International Edition* 57 (24): 7071–7075. https://doi.org/10.1002/anie.201802231.

48 Dai, X., Chen, Z., Yao, T. et al. (2017). Single Ni sites distributed on N-doped carbon for selective hydrogenation of acetylene. *Chemical Communications* 53 (84): 11568–11571. https://doi.org/10.1039/c7cc04820c.

49 Chen, H., Zhang, Y., He, Q. et al. (2020). A facile route to fabricate double atom catalysts with controllable atomic spacing for the r-WGS reaction. *Journal of Materials Chemistry A* 8 (5): 2364–2368. https://doi.org/10.1039/c9ta13192b.

50 Yun, R., Zhan, F., Li, N. et al. (2020). Fe single atoms and Fe_2O_3 clusters liberated from N-doped polyhedral carbon for chemoselective hydrogenation under mild conditions. *ACS Applied Materials and Interfaces* 12 (30): 34122–34129. https://doi.org/10.1021/acsami.0c09124.

51 Kwak, J.H., Hu, J., Mei, D. et al. (2009). Coordinatively unsaturated Al^{3+} centers as binding sites for active catalyst phases of platinum on γ-Al_2O_3. *Science* 325 (5948): 1670–1673. https://doi.org/10.1126/science.1176745.

52 Zhang, Z., Zhu, Y., Asakura, H. et al. (2017). Thermally stable single atom Pt/m-Al_2O_3 for selective hydrogenation and CO oxidation. *Nature Communications* 8 (1): 16100. https://doi.org/10.1038/ncomms16100.

53 Tang, N., Cong, Y., Shang, Q. et al. (2017). Coordinatively unsaturated Al^{3+} sites anchored subnanometric ruthenium catalyst for hydrogenation of aromatics. *ACS Catalysis* 7 (9): 5987–5991. https://doi.org/10.1021/acscatal.7b01816.

54 Liu, L., Meira, D.M., Arenal, R. et al. (2019). Determination of the evolution of heterogeneous single metal atoms and nanoclusters under reaction conditions: which are the working catalytic sites? *ACS Catalysis* 9 (12): 10626–10639. https://doi.org/10.1021/acscatal.9b04214.

55 Shi, X., Lin, Y., Huang, L. et al. (2020). Copper catalysts in semihydrogenation of acetylene: from single atoms to nanoparticles. *ACS Catalysis* 10 (5): 3495–3504. https://doi.org/10.1021/acscatal.9b05321.

56 Kwak, J.H., Kovarik, L., and Szanyi, J. (2013). CO_2 reduction on supported Ru/Al_2O_3 catalysts: cluster size dependence of product selectivity. *ACS Catalysis* 3 (11): 2449–2455. https://doi.org/10.1021/cs400381f.

57 Moliner, M., Gabay, J.E., Kliewer, C.E. et al. (2016). Reversible transformation of Pt nanoparticles into single atoms inside high-silica chabazite zeolite. *Journal of the American Chemical Society* 138 (48): 15743–15750. https://doi.org/10.1021/jacs.6b10169.

58 Wang, H., Lan, X., Wang, S. et al. (2020). Selective hydrogenation of 2-pentenal using highly dispersed Pt catalysts supported on ZnSnAl mixed metal oxides derived from layered double hydroxides. *Catalysis Science & Technology* 10 (4): 1106–1116. https://doi.org/10.1039/C9CY02200G.

59 Mori, K., Taga, T., and Yamashita, H. (2017). Isolated single-atomic Ru catalyst bound on a layered double hydroxide for hydrogenation of CO_2 to formic acid. *ACS Catalysis* 7 (5): 3147–3151. https://doi.org/10.1021/acscatal.7b00312.

60 Shi, X., Wang, X., Shang, X. et al. (2017). High performance and active sites of a ceria-supported palladium catalyst for solvent-free chemoselective hydrogenation of nitroarenes. *ChemCatChem* 9 (19): 3743–3751. https://doi.org/10.1002/cctc.201700631.

61 Kuo, C.-T., Lu, Y., Kovarik, L. et al. (2019). Structure sensitivity of acetylene semi-hydrogenation on Pt single atoms and subnanometer clusters. *ACS Catalysis* 9 (12): 11030–11041. https://doi.org/10.1021/acscatal.9b02840.

62 Ma, H.Y. and Wang, G.C. (2020). Selective hydrogenation of acetylene on Ptn/TiO_2 ($n = 1, 2, 4, 8$) surfaces: structure sensitivity analysis. *ACS Catalysis* 10 (9): 4922–4928. https://doi.org/10.1021/acscatal.0c00190.

63 Fu, J., Lym, J., Zheng, W. et al. (2020). C–O bond activation using ultralow loading of noble metal catalysts on moderately reducible oxides. *Nature Catalysis* 3 (5): 446–453. https://doi.org/10.1038/s41929-020-0445-x.

64 Resasco, J., Yang, F., Mou, T. et al. (2020). Relationship between atomic scale structure and reactivity of Pt catalysts: hydrodeoxygenation of m-Cresol over

isolated Pt cations and clusters. *ACS Catalysis* 10 (1): 595–603. https://doi.org/10.1021/acscatal.9b04330.

65 Kuai, L., Chen, Z., Liu, S. et al. (2020). Titania supported synergistic palladium single atoms and nanoparticles for room temperature ketone and aldehydes hydrogenation. *Nature Communications* 11 (1): 48. https://doi.org/10.1038/s41467-019-13941-5.

66 Ma, S., Song, W., Liu, B. et al. (2016). Elucidation of the high CO_2 reduction selectivity of isolated Rh supported on TiO_2: a DFT study. *Catalysis Science and Technology* 6 (15): 6128–6136. https://doi.org/10.1039/c5cy02158h.

67 Wang, Y., Arandiyan, H., Scott, J. et al. (2018). Single atom and nanoclustered Pt catalysts for selective CO_2 reduction. *ACS Applied Energy Materials* 1 (12): 6781–6789. https://doi.org/10.1021/acsaem.8b00817.

68 Ftouni, J., Muñoz-Murillo, A., Goryachev, A. et al. (2016). ZrO_2 is preferred over TiO_2 as support for the Ru-catalyzed hydrogenation of levulinic acid to γ-valerolactone. *ACS Catalysis* 6 (8): 5462–5472. https://doi.org/10.1021/acscatal.6b00730.

69 Cao, W., Lin, L., Qi, H. et al. (2019). In-situ synthesis of single-atom Ir by utilizing metal-organic frameworks: an acid-resistant catalyst for hydrogenation of levulinic acid to Γ-valerolactone. *Journal of Catalysis* 373: 161–172. https://doi.org/10.1016/j.jcat.2019.03.035.

70 Kim, I.S., Li, Z., Zheng, J. et al. (2018). Sinter-resistant platinum catalyst supported by metal-organic framework. *Angewandte Chemie* 130 (4): 921–925. https://doi.org/10.1002/ange.201708092.

71 Abdel-Rahman, M.K. and Trenary, M. (2020). Propyne hydrogenation over a Pd/Cu(111) single-atom alloy studied using ambient pressure infrared spectroscopy. *ACS Catalysis* 10 (17): 9716–9724. https://doi.org/10.1021/acscatal.0c02475.

72 Serp, P. (2021). Cooperativity in supported metal single atom catalysis. *Nanoscale* 13 (12): 5985–6004. https://doi.org/10.1039/D1NR00465D.

9

Supported Metal Single-Atom Thermocatalysts for Oxidation Reactions

Laurent Piccolo[1], Stéphane Loridant[1], and Phillip Christopher[2]

[1] *University of Lyon, IRCELYON, 2 Avenue Albert Einstein, F-69626, Villeurbanne, France*
[2] *University of California, Department of Chemical Engineering, Engineering II Building, Santa Barbara, CA 93106-5080, USA*

9.1 Introduction

Single-atom catalysts (SACs) have been the subject of intense and increasing interest for a decade [1–11]. They generally differ from the so-called "single-site heterogeneous catalysts" or heterogenized homogeneous catalysts – typically mononuclear metal complexes grafted on silica – since they avoid or at least minimize the use of labile organic ligands [12–14]. By design, metal atoms in SACs coordinate to atoms of solid supports or hosts [2]. Beyond nanocatalysis, the current fascination for SAC in materials research parallels a significant advance in the exploration of the subnanometric world. With respect to conventional supported catalysts, SAC provide both metal cost savings (especially interesting in the case of noble metals) and potential for novel catalytic properties owing to unique active-site electronic structure. In particular, SACs potentially cumulate the site-specific selectivity of homogeneous catalysts and the advantages of heterogeneous catalysts, i.e. their easy handling and recovery [2].

Since the 1950s, there have been a number of early reports on heterogeneous catalysis by supported single metal atoms, identified mostly by spectroscopic characterization, e.g. for noble metals on Al_2O_3 [15–17], MgO [18–20], CeO_2 [21–23], and ZrO_2 [24], as well as for early transition metals anchored on SiO_2 and other oxides [12]. From the 2011 paper by Zhang and coworkers [25], the term "single atoms" (SAs) has been widely used, especially for late transition metals, whereas equivalent terms such as "isolated" and "individually dispersed" metal atoms or species, as well as "single sites" and "mononuclear species," were previously preferred – and are sometimes still in use (the term "atomically dispersed" is also frequently employed, but this may also cover, e.g. 2D raft-like or subnanometric 3D clusters [26]).

Though conventional heterogeneous metal-based catalysts often contain SAs in addition to clusters and nanoparticles (NPs) [1], it is only recently that atomic dispersion of the metal can be accurately controlled and assessed, in particular through visualization by aberration-corrected scanning transmission electron microscopy

Supported Metal Single Atom Catalysis, First Edition. Edited by Philippe Serp and Doan Pham Minh.
© 2022 WILEY-VCH GmbH. Published 2022 by WILEY-VCH GmbH.

(aberration-corrected STEM) [27]. The possibility to "see" individual atoms has stimulated research into synthesis, characterization, and reactivity of SACs, especially for late transition metal atoms on oxide supports. The pioneering works on so-identified SACs in the early 2010s were devoted to noble metal atoms supported on high-surface-area metal oxides, with CO oxidation (in absence or presence of hydrogen) [25, 28] or water–gas shift (WGS) [29, 30] as the main test-reactions. Since then, the scope of SACs has considerably expanded to all domains of heterogeneous catalysis, including electrocatalysis and photocatalysis, in parallel to the increasing use of non-noble metals and carbon/nitrogen-containing hosts. Regarding thermal oxidation reactions, noble metals supported on oxides are still by far the most popular systems among recent studies, but oxophillic 5B-6B group metals isolated on transition metal oxides, i.e. supported monomeric/isolated oxometallates discovered decades ago [31–33], also have to be mentioned.

This chapter covers total oxidation reactions (including CO oxidation), WGS, and selective oxidations on most types of SAC, including so-called single-atom alloys (SAAs) [34]. Although some reforming reactions such as methane steam reforming include an oxidation step, they are not addressed in this chapter, and the reader is referred to the dedicated Chapter 11. While gas-phase oxidations have been mostly carried out on oxide-supported SACs, liquid-phase oxidations have often relied on carbon-hosted SACs. In the following, we have chosen to mostly focus on experimental investigations providing clear evidence of metal site isolation, though in some cases "pseudo SAC" [35] in the form of loosely packed clusters, rafts, multimers, or other atomically dispersed metal species (present by choice, lack of control, or through restructuring [26]) may be considered as well.

9.2 Oxide-Supported Single-Atom Catalysts

In their pioneering article, Zhang and coworkers reported on an investigation of Pt/FeO_x SACs for CO oxidation [25]. Though the current SAC literature is dominated by carbon-based materials (especially for electrocatalytic applications), most of the works on SAC in the early 2010s and before, used transition metal oxides as the support, and this field is still very active [36, 37]. In addition to recent SAC studies on late transition metals, it has been known for decades that various selective (mild) oxidation reactions are well catalyzed by oxide-supported molecular structures consisting of early transition metals coordinated to oxygen atoms, i.e. oxometallates. At low coverage of active early transition metals, the corresponding solids can be considered as SAC.

Due to their chemical and mechanical properties, metal oxides are the most widely employed supports in heterogeneous catalysis. The first role of the oxide support is to efficiently stabilize metal NPs through iono-covalent interactions. The strength of metal-support interaction strongly depends on the nature of both entities, and typically increases with the metal oxophilicity and the oxide support reducibility, including in the case of SACs [38–41]. Transition metal oxides are particularly important for oxidation reactions, where they can play an active role owing to their acid–base properties, the direct involvement of lattice oxygen and oxygen vacancies (O_v – Mars-van Krevelen, MvK, mechanism [42]), and the surface diffusion of reactive intermediates between active metal species (spillover).

The first and main part of this section is devoted to CO oxidation, as this reaction is by far the most investigated for SAC, mostly based on noble metal active centers. The subsequent sections concern SAC applications to preferential CO oxidation (PROX) in the presence of dihydrogen, WGS, total oxidations, and selective oxidations. A few examples of atomically dispersed metal active sites in zeolites are also included in this section.

9.2.1 CO Oxidation

Carbon monoxide oxidation ($2CO + O_2 \rightarrow 2CO_2$) catalyzed by platinum-group metals (PGM) has been deeply investigated and used as a model catalytic reaction for a century [43]. CO oxidation has become especially important with the development of automotive catalytic converters in the 1970s to meet emission control regulations [44]. Nowadays, three-way catalysts on gasoline-powered vehicles typically consist of Pt, Pd, or Rh NPs, and sometimes SAs, dispersed on a porous oxide for converting CO as well as unburnt hydrocarbons and nitrogen oxides [44–46]. Because three-way catalysts represent the largest commercial use of these precious metals, significant research efforts have focused on SACs for exhaust pollution mitigation, including CO oxidation [47–50]. In Sections 9.2.1.1–9.2.1.5, we highlight results from the recent literature associated with CO oxidation over late transition metal atoms anchored on non-reducible (including Al_2O_3) and reducible (including FeO_x, TiO_2, and CeO_2) oxides.

9.2.1.1 PGM on Alumina

As early as in the 1950s, Yang and Garland ascribed CO-FTIR features at 2027 and 2095 cm^{-1} to Rh_{SA} dicarbonyls adsorbed on alumina and investigated their reaction with O_2, H_2, and H_2O, making their 2 wt% Rh/Al_2O_3 sample one of the first reported examples of SAC [15]. More recently, Yates et al. confirmed the previous IR assignment for $Rh^I(CO)_2$ on alumina [51], and could photochemically convert these species, in the presence of O_2, to $Rh^I(CO)(O)_x$ species and CO_2 at sub-ambient temperature [16]. From *ab initio* molecular dynamics simulations – based on density functional theory, DFT – applied to CO oxidation on $Rh/\gamma\text{-}Al_2O_3(110)$, Ghosh and Nair proposed a two-step (thermal) reaction scheme where O_2 dissociates on the Rh atom, CO reacts with chemisorbed O, producing CO_2 and leaving an O atom coordinated to Rh and Al atoms [52]. Then, a second CO molecule enters the cycle, followed by a second O_2 molecule, leading to $Rh(O)_2$ and CO_2, which terminates the cycle (Figure 9.1a, path A). The breaking of the Rh—C bond constitutes the rate-determining step. An alternative pathway (Figure 9.1a, path B) was also proposed. Notably, the alumina support is continuously involved in the reaction path, in order to accommodate an O atom also bonded to the Rh atom.

In contrast, from their experimental and theoretical investigation of CO oxidation over a 0.18 wt% $Pt/\theta\text{-}Al_2O_3$ SAC, Moses-DeBusk et al. introduced a reaction scheme

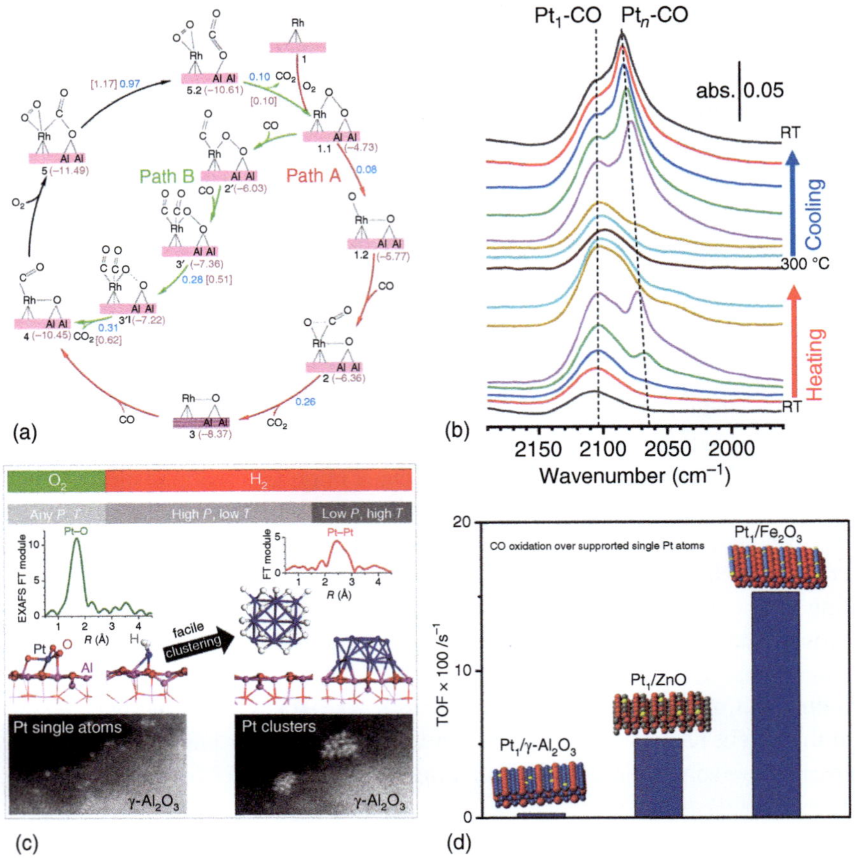

Figure 9.1 (a) Proposed reaction schemes for CO oxidation on $Rh_{SA}/\gamma\text{-}Al_2O_3$ based on DFT calculations. Source: Ghosh et al. [52]. Reproduced with permission from Wiley-VCH. (b) DRIFTS spectra in the C–O stretching region for an atomically-dispersed 0.3 wt% $Pt/\gamma\text{-}Al_2O_3$ catalyst under $CO{-}O_2$ mixture (2 : 2 vol%). Source: Dessal et al. [53]. Reproduced with permission from American Chemical Society. (c) Influence of the reactive atmosphere (O_2 vs. H_2) on Pt nuclearity and atom coordination. From top to bottom: EXAFS, DFT models, and environmental STEM images. Source: Dessal et al. [54]. Reproduced with permission of Royal Society of Chemistry. (d) Comparison of the CO oxidation activities of single Pt atoms supported on $\gamma\text{-}Al_2O_3$, ZnO, and Fe_2O_3. Source: Lou and Liu [55]. Reproduced with permission from American Chemical Society.

without direct involvement of the support [28]. O_2, then CO, would adsorb on the Pt_{SA}, forming a bidentate Pt carbonate, followed by the release of a CO_2 molecule. In a second step, CO adsorbs on Pt(O), which leads to the production of CO_2 and the regeneration of the Pt_{SA}. However, the first CO_2 formation step – from $Pt(CO_3)$ – was found highly endergonic (2.22 eV), suggesting that the SAC would be covered with carbonates at room temperature. The existence of the $Pt(CO_3)$ species was supported by a diffuse reflectance infrared spectroscopy (DRIFTS) study of CO adsorption, showing an absorption feature at 1659 cm^{-1}.

Newton et al. investigated an industrial 5 wt% Pt/γ-Al_2O_3 catalyst by DRIFTS and X-ray absorption spectroscopy (XAS) combined with mass spectrometry under alternate CO/O_2 pulses [56, 57]. The authors suggested that periodic CO_2 formation at room temperature is mediated by Pt carbonates (IR feature at ~1690 cm^{-1}) associated to single Pt cations coexisting with Pt_{NP}. Under CO, CO_2 would be released through the $Pt(CO_3) \rightarrow Pt(O) + CO_2$ reaction, as in the work by Moses-DeBusk et al. However, under O_2, Pt(O) would convert to $Pt(O)_2$ thanks to the neighboring presence of Pt_{NP} that enable O_2 dissociation, as well as CO storage. $Pt(O)_2$ is then converted back to $Pt(CO_3)$ under the next CO pulse.

Also using DRIFTS of adsorbed CO, Ding et al. investigated catalysts with various Pt SAs/NPs ratios and various supports, including HZSM-5 mesoporous zeolite, SiO_2, γ-Al_2O_3, TiO_2, and ZrO_2 [58]. For alumina, IR bands centered at ~2100 and ~2050 cm^{-1} were assigned to CO adsorbed on SA and NP, respectively. Upon replacing CO with O_2 – or H_2O for the WGS reaction – at 100 °C, only the CO/NP band vanished while the CO/SA band remained unchanged. This indicates that, for both reactions and all oxides, only NPs are active at this temperature, while Pt_{SA} would act as spectator species owing to the strong binding of CO molecules (carbonates were not mentioned).

The above report, mostly based on infrared (IR) measurements at low temperature and under model conditions, seriously questioned the CO oxidation efficiency of oxide-supported noble metal-based SACs. Using *operando* DRIFTS and XAS coupled with online mass spectrometry, Dessal et al. investigated CO oxidation over a 0.3 wt% Pt/γ-Al_2O_3 (157 m^2/g) SAC prepared by a conventional impregnation-calcination method [53]. Both techniques revealed a gradual Pt aggregation-activation process during heating/cooling cycles. Figure 9.1b shows a series of DRIFTS spectra recorded under CO–O_2 mixture during a temperature cycle, showing the appearance of a C–O stretching band (~2070 cm^{-1}) corresponding to *in situ* formed nanometric Pt clusters at 100 °C, which vanishes at higher temperatures due to full CO conversion (as monitored by mass spectrometry). Comparatively, CO species vibrating at ~2100 cm^{-1}, assigned to CO on Pt_{SA}, were strongly adsorbed and appeared much less reactive, in agreement with the work by Ding et al. [58]. Figure 9.1c shows the respective influences of O_2 and H_2 adsorption on Pt nuclearity, as seen from *in situ* extended X-ray absorption fine structure (EXAFS), environmental STEM, and DFT calculations [54]. While Pt—O_{ads}—Al bonds form under pure O_2 and stabilize the Pt_{SA} in place, the adsorption of H_2 – just like that of CO, even under excess O_2 – induces Pt clustering and reduction. The resulting H-covered subnanometric Pt clusters are mobile on the surface. Overall, evidence supports the conclusions that alumina-supported cationic Pt_{SA} are prone to clustering in the presence of reducing gases, and they are less active for CO oxidation than their Pt cluster or NP counterparts.

While the question of the intrinsic activity of Pt/Al_2O_3 SACs and their operation mechanism is still open, the influence of the support composition has been analyzed by Lou and Liu [55]. The authors compared γ-Al_2O_3, ZnO, and Fe_2O_3 solids loaded with Pt at ultralow content (0.03–0.04 wt%) deposited by a strong electrostatic adsorption method. The CO oxidation turnover frequency (TOF) at

140 °C ranges from 0.003 to 0.053 and 0.15 s^{-1} for Pt/γ-Al_2O_3, Pt/ZnO, and Pt/Fe_2O_3 SACs, respectively (Figure 9.1d). This suggests that the SAC activity increases with the oxide reducibility, and that the support directly influences the nature of the active sites. However, as discussed further below, the increasing reactivity of Pt SACs with increasing support reducibility does not always hold true (e.g. for CeO_2) because the support reducibility also controls how strongly the Pt atom coordinates to the support.

As mentioned above, the stabilization of SACs in general, and of single noble-metal atoms on irreducible oxides in particular, is a major challenge in SAC research, especially when reducing molecules such as CO and H_2 are present in the reactant feed [26, 59, 60]. However, strategies have been developed to address this issue for Al_2O_3 supports. For example, Zhang et al. succeeded in the synthesis of thermally stable 0.2 wt% Pt/Al_2O_3 SACs by anchoring the Pt atoms in the internal surface of mesoporous alumina (c. 200 m^2/g) through a one-pot sol–gel method (self-assembly complexing of H_2PtCl_6, Al isopropoxide and P123 polymer) followed by air calcination and H_2 reduction [61]. The SAC was found active not only for CO oxidation but also for H_2-involving reactions. The Pt cations are thought to coordinate with unsaturated pentacoordinate Al^{3+} centers [62] via four O bridges. Another anchoring strategy relies on dopants such as La^{3+} and Ba^{2+} [63, 64].

9.2.1.2 PGM on Iron Oxide

In 2011, Qiao et al. popularized the concept of SA catalysis through an analysis of platinum supported on iron oxide (290 m^2/g) [25]. Using a coprecipitation method with chloroplatinic acid as the metal precursor, they prepared a SAC with 0.17 wt% Pt loading. The catalyst was characterized by STEM, XAS and *in situ* Fourier transform infrared spectroscopy of adsorbed CO (CO-FTIR). This work initiated the widespread use of aberration-corrected STEM imaging to complement traditionally used spectroscopic techniques to reveal or confirm the isolated nature of supported metal atoms, as well as their oxidation state [27]. The SAC exhibited a c. twice superior TOF, both for CO oxidation at 27 °C and PROX at 27 and 80 °C, with respect to its Pt cluster-containing counterpart, which was prepared from the same method though with a much higher Pt loading (2.5 wt%). Based on DFT calculations, this result was ascribed to the cationic nature of Pt atoms, which would reduce both the CO–Pt binding strength and the reaction barriers. The authors also suggested that Pt atoms occupy the positions of surface Fe atoms and coordinate to c. three O atoms. The same group later reported similar investigations of Ir/FeO_x SACs, which were found less active than the Pt/FeO_x SAC [65] and Ir subnanometer clusters supported on FeO_x [66].

In a recent investigation of Pt supported on amorphous Fe_2O_3 nanosheets, Chen et al. found an even larger superiority of the SACs compared to cluster-based catalysts for CO oxidation around 70 °C [67]. This result was ascribed to the positive role of the amorphous support structure on the presence of Pt-stabilizing defects with decreased CO adsorption energy, and the abundance of O_v prone to activate O_2. The scanning tunneling microscopy (STM) study of Parkinson and coworkers

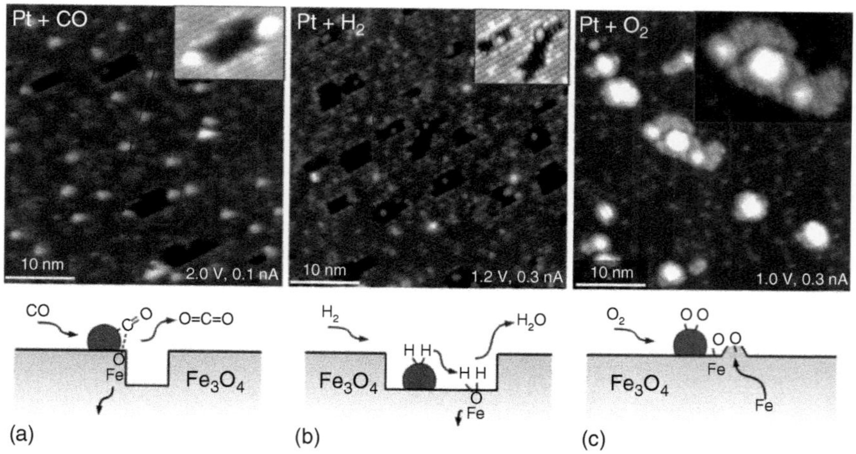

Figure 9.2 STM images acquired following exposure of the Pt/Fe_3O_4(001) model catalyst to 1×10^{-7} mbar of CO (a), H_2 (b), or O_2 (c) at 277 °C, and illustration of the corresponding surface processes. Source: Bliem et al. [68]. Adapted with permission of Wiley-VCH.

on a $Pt_{1\text{-}6}/Fe_3O_4$(001) model catalyst (resulting from the CO-induced partial sintering of initially single Pt atoms) provides a striking atomic-scale visualization of CO-induced lattice O extraction at the perimeter of Pt clusters, which often sit at the edge of or inside monolayer holes in the oxide surface (Figure 9.2a) [68]. The corresponding formation of gaseous CO_2 and O_v constitutes the first step of the MvK mechanism. The second step involves the replenishment of O_v from gas-phase O_2 (absent in the case of Figure 9.2a). Islands of O_v were observed to increase in size under exposure of the model catalyst to H_2, which leads to water evolution via Pt-induced hydrogen spillover and OH group formation (Figure 9.2b). In contrast, exposure to O_2 leads to the filling of O_v and the creation of Fe_3O_4(001) islands through oxygen spillover and reaction with bulk Fe atoms (Figure 9.2c).

In a recent study of the model Rh/Fe_3O_4(001) SAC using STM, X-ray photoelectron spectroscopy (XPS) and temperature-programmed desorption (TPD), the same group identified an O_2-induced Rh oxidation/clustering process, whereas CO strongly adsorbs and stabilizes Rh adatom isolation [69]. Under these high-vacuum conditions, while RhO_x clusters catalyze CO oxidation at sub-ambient temperature through the classical Langmuir–Hinshelwood (LH) mechanism, Rh carbonyls can only catalyze the reaction through the MvK mechanism at higher temperature.

The existing works on Pt and Rh SACs on FeO_x supports are consistent in proposing MvK as the primary reaction mechanism, although the reactivity of the SAC seems to be controlled by the nature of the FeO_x support (phase and surface structure).

9.2.1.3 Noble Metals on Titania

TiO_2 has long served as a model oxide that imparts distinct catalytic chemistry onto supported metal active sites. For example, TiO_2 is the classic support that enables strong metal-support interactions with Pt-group metals, and one that provides

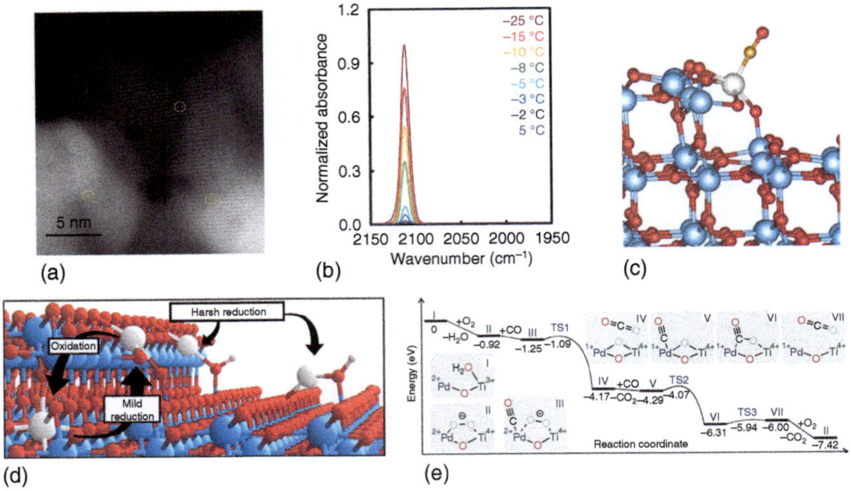

Figure 9.3 (a) Aberration-corrected STEM image of Pt atoms (identified by yellow circles) on c. 5 nm diameter anatase TiO_2 particles. Source: DeRita et al. [72]. Adapted with permission of American Chemical Society. (b) CO probe-molecule FTIR spectra of a Pt/TiO_2 SAC as a function of temperature during a TPD measurement. (c) DFT-calculated structure of CO bound to PtO_2 adsorbed at a TiO_2(145) step edge. Source: (b) and (c) Thang et al. [73]. Reproduced with permission from Elsevier. (d) Schematic showing the structural transformation of isolated Pt species as a function of environmental conditions. Source: DeRita et al. [74]. Reproduced with permission from Springer Nature. (e) DFT-calculated reaction pathway for CO oxidation on a Pd/TiO_2 SAC that was supported by spectroscopic and kinetic analyses. Source: Liu et al. [75]. Reproduced with permission from Elsevier.

size-dependent reactivity to supported Au particles [70, 71]. TiO_2 has similarly served as a model system for understanding the structure, dynamic behavior, and reactivity of SAC active sites. Particularly, small anatase crystallites have been demonstrated as supports that allow for the synthesis of atomically dispersed Pt species with uniform local coordination environments rivaling single-crystal oxides and zeolite supports. The uniformity of Pt active sites enables unambiguous spectroscopic fingerprinting, structural analysis, and the development of structure–function relationships.

SACs have been synthesized using small (∼5 nm diameter) anatase TiO_2 crystallites, low Pt loadings (<0.1 wt%), and principles of strong electrostatic adsorption. Aberration-corrected STEM imaging was used to demonstrate the existence of solely Pt_{SA} in these samples (Figure 9.3a) [72–74]. Probe-molecule CO-FTIR spectroscopy and TPD were used to chemically characterize the Pt active sites (Figure 9.3b). It was observed that following 250 °C reduction of the catalyst in H_2, bound CO exhibited a vibrational wavenumber of ∼2112 cm^{-1} with a full-width at half maximum lower than 10 cm^{-1}.

This suggested that Pt existed in a +2 oxidation state with a uniform coordination environment across the support [72, 76, 77]. TPD measurements showed an approximate CO binding energy of 0.9 eV (∼87 kJ/mol). Detailed comparisons to many different Pt coordination environments on TiO_2 (101) and (145) surfaces allowed the identification of adsorbed PtO_2 species at terrace or step edges as the most representative coordination environment of the atomically dispersed Pt active sites (Figure 9.3c) [73]. This assignment was further substantiated through site-resolved STEM imaging and EXAFS analyses, which showed quantitative and visual correlation between the DFT-predicted structure and the experimental analysis [74]. Consistency between the chemical (oxidation state, CO adsorption energy, and CO vibrational frequency) and structural (STEM imaging and EXAFS analysis) analyses from experiments and theory was only feasible because of the highly uniform Pt local coordination environment on the anatase TiO_2 support.

While the Pt active-site structure on anatase TiO_2 was well resolved after a particular pre-treatment, it has been further demonstrated that the Pt local coordination environment is adaptive to variations in temperature, oxidizing or reducing chemical species, and photochemical stimulation [74, 78–82]. For example, it was inferred from a combination of microscopy, spectroscopy, and theory that Pt prefers to reside in a cation-replaced position in the TiO_2 lattice with sixfold coordination to oxygen following oxidizing pre-treatment; an adsorbed PtO_2 state following mild reducing pre-treatment; and as a mobile PtOH species under harsher reducing conditions (Figure 9.3d) [74]. Further analysis under much harsher reducing environments has shown that the coordination environment of Pt species could include hydrogen atoms and unique sites on the support [80, 81]. Finally, it has been shown that extended exposure to CO oxidation reaction conditions and photochemical environments can sometimes lead to the formation of Pt clusters [78, 79, 82]. The adaptive nature of atomically dispersed Pt on TiO_2 suggests that the reactivity of these active sites can likely be tuned by controlling the coordination environment, and also that the active-site structure may be different following pre-treatment and exposure to reactive environments. Interestingly, it has also been observed that other atomically dispersed metals, such as Cu and Rh, exhibit similar dynamic physical and electronic structures as a function of environmental conditions [83, 84].

The reactivity of Pt/TiO_2 SACs have been assessed both from chemical (how strongly species bind to Pt sites) and catalytic (rates of reactions) perspectives. CO-FTIR and TPD measurements correlated to DFT calculations showed that CO binds more strongly to c. 1 nm diameter metallic or oxidized Pt clusters on TiO_2, as compared to atomically dispersed Pt sites. This is particularly interesting because CO bound to Pt_{SA} and small PtO_x clusters exhibited almost identical stretching frequencies (~2110 cm^{-1}) but binding energy differences superior to 100 kJ/mol, due to the under-coordination of Pt in partially reduced PtO_x clusters. This highlights an important challenge in using CO-FTIR to characterize Pt SACs, and makes a strong point that the co-existence of Pt_{SA} and small oxidized Pt clusters on a support can be easily confused.

It has been shown under multiple reaction conditions that Pt_{SA}/TiO_2 catalysts exhibit higher CO oxidation activity on per Pt mass and per Pt site bases as compared to small Pt NPs; this is consistent with what has been observed for Pt/FeO_x [72, 85]. Furthermore, it has been shown that coordination environments, which resulted in close to neutral atomically dispersed Pt, exhibited stronger binding to CO and higher CO oxidation activity as compared to coordination environments where Pt existed in a +2 oxidation state [74]. However, there have also been reports claiming that Pt_{NP} on TiO_2 are actually more reactive than Pt_{SA} on TiO_2 [79]. The differences in conclusions regarding the relative activities of atomically dispersed Pt and Pt_{NP} on TiO_2 likely point to the critical role of the support structure, Pt precursor (for example H_2PtCl_6 can lead to Cl poisoning of active sites), and Pt coordination environment in controlling the catalytic performance.

The use of TiO_2 as a support for atomically dispersed metals is not limited to Pt. Reports of CO oxidation on Au/TiO_2 and Pd/TiO_2 have provided further evidence of high activity from atomically dispersed metals on TiO_2 [75, 86]. For example, it was reported that Pd_{SA}/TiO_2 catalysts exhibit the highest activity for CO oxidation compared to all previous reports on Pd-based catalysts. It is clear from our review of the literature above that various groups have measured enhanced CO oxidation activity from atomically dispersed precious metals on TiO_2 as compared to NP of the same metal. Reports countering this conclusion also exist, which highlights the challenge in studying this class of materials. Finally, it has been observed that the reactivity of atomically dispersed Pt species could be further promoted by localizing Pt at CeO_2–TiO_2 interfaces [87]. This observation again points to the critical role that local coordination plays in dictating the reactivity of atomically dispersed Pt, and suggests that supports likely participate directly in the CO oxidation reaction.

Mechanistic analysis of CO oxidation over Pt/TiO_2 and Pd/TiO_2 through DFT calculations, kinetic measurements coupled to microkinetic modeling, and spectroscopic analyses have provided a consistent picture [74, 75, 85]. Generally, it has been proposed that the reaction proceeds via an MvK mechanism, with formation of an O_v at the Pt(Pd)-TiO_2 interface, followed by O_2 adsorption and dissociation, and then CO adsorption and oxidation (Figure 9.3e). Consistency in each of the proposed mechanisms is found in the observation of less CO poisoning on single metal atoms as compared to metal clusters of the same composition. This derives from the cationic oxidation state and strong support coordination of the metal SA. However, details regarding rate-determining steps, reaction orders, and apparent barriers are specific to each case, where small differences in support (for example anatase vs. rutile, or defect concentration) and metal (coordination environment, site uniformity) are critical for dictating reactivity.

9.2.1.4 Late Transition Metals on Ceria

Cerium oxide (CeO_2) has been extensively investigated for many reactions, both as a catalyst and a support [88–91]. The distinct catalytic performances of CeO_2 derive from unique physical properties, including high reducibility while remaining in its fluorite structure, easy electron transfer between Ce^{4+} and Ce^{3+} cations, high mobility of surface oxygen species, and high oxygen storage capacity [88–90, 92, 93]. Furthermore, defects such as O_v and Ce^{3+} can be stabilized in nanocrystallites contained in ceria supports, leading to sub-stoichiometric oxides [94, 95]. CeO_2 also contains at its surface both acid and basic sites in close proximity [89]. In particular, a high proportion of strong basic sites can be present, leading to easy carbonation, which can influence the catalytic reaction [96]. The redox properties of ceria have made it an ideal PGM support for automotive catalytic converters, which is currently the main application of this material; in particular, Pt, Pd, and Rh NP supported on CeO_2 exhibit excellent CO oxidation performances [97, 98].

M/CeO_2 SACs with M = Pt, Rh, Cu, and Au have been prepared by various methods including classical incipient wetness impregnation [99], atomic layer deposition [100], and gas phase atom trapping where volatile MO_x species adsorb and are stabilized on some support surfaces [101]. For example, atomically dispersed metals have been shown to be stabilized on certain facets of CeO_2 by an oxidizing treatment at 800 °C [99–102]. In the case of Pd_{SA}/CeO_2, high-temperature atom trapping by calcination at 800 °C increased the amount of O_v, resulting in decreased Pd–O (lattice oxygen from CeO_2) coordination in square planar Pd_1O_4 [103]. Defect sites can serve to stabilize noble metal catalysts with atomic dispersion by adsorption of ascorbic acid to generate Ce^{3+} cations [104], or by doping with Ga^{3+} or Zr^{4+} cations [105, 106]. Similarly, a CeO_2/Al_2O_3 support containing smaller CeO_2 particles and more surface defects led to more atomically dispersed Pt and abundant Pt–O–Ce structures [107]. In fact, the defect-induced trapping of metal cations such as Au^+ or Au^{3+} [108], or mobile metal oxide species such as PtO_2 [109], have been proposed to lead to the exclusive formation of atomically dispersed metals.

However, the exact location and coordination of Pt_{SA}, as well as the potential existence of small oxidized PtO_x clusters in samples previously reported as SACs, is still debated. The challenge with providing conclusive evidence of Pt site isolation on CeO_2 stems from the heavy mass of Ce, which minimizes contrast with Pt in STEM imaging, and the fact that XAS is a sample-averaged technique. Early on it was proposed that Pt_{SA} prefer to sit in fourfold coordinated symmetric $Pt–O_4$ CeO_2 micro-environments, and be rendered inactive for molecular adsorption by the formation of a highly stable Pt d^8 structure [110]. However, various experimental reports have claimed to identify Pt SAs on CeO_2 through CO-FTIR based on a stretching band at 2095 cm^{-1}, which is stable at high temperature [101, 109, 111, 112].

The location of these Pt atoms on CeO_2 has been claimed to range from {111} ceria steps or {111}, {110}, or {100} ceria facets, to surface or bulk Ce substitutes forming $Ce_{1-x}Pt_xO_{2-y}$ solid solutions (Ref. [113] and references therein). In an attempt to clarify the structure of Pt single sites determined by DFT calculations, experimental EXAFS data were fitted with theoretical ones by Maurer et al. [113]. As shown in Figure 9.4a–l, the fit substantially improves for several exposed facets, going from adsorbed to surface-substituted Pt species. The agreement is quite good within the radial distance range 1.1–3.4 Å for models with Pt^{2+} substituted at {110} facets, which corresponds to fourfold hollow sites. Within the best-fit model, adsorbed CO is predicted to have a stretching frequency of ~2095 cm^{-1}, consistent with experimental measurements. However, it has been noted that using CO vibrational frequency to correlate DFT-calculated structures and experimental measurements can lead to misassignment, and that additionally comparing the CO adsorption energy is critical for substantiating a DFT-based structural assignment [73, 114]. Based on this issue and detailed microscopy and spectroscopy analysis, Resasco et al. proposed that the 2095 cm^{-1} signature found in Pt/CeO_2 catalysts stems from the existence of CO adsorbing to a sub-population of small Pt_xO_y clusters that are challenging to detect by STEM or EXAFS [115]. Thus, it seems that there is still a

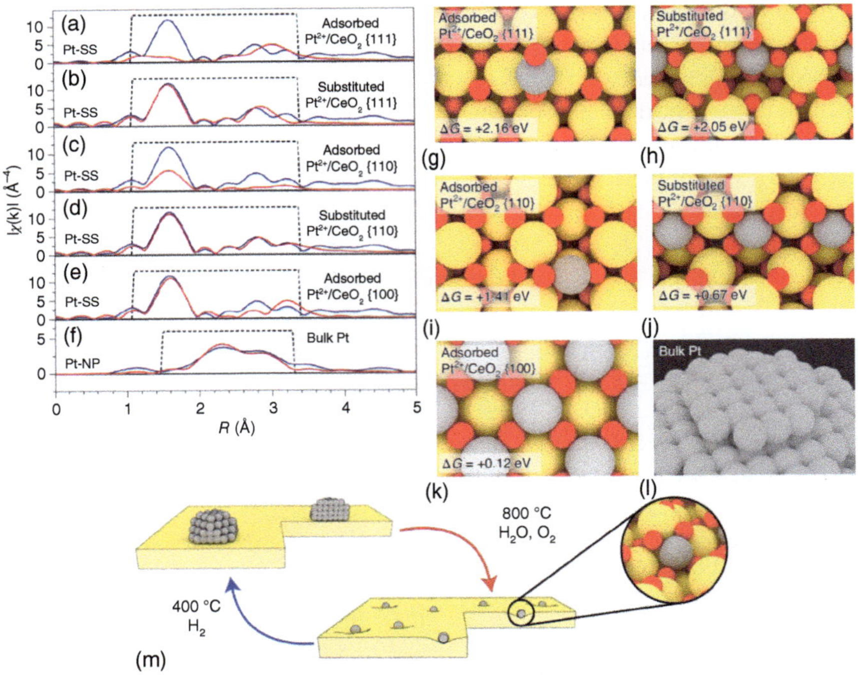

Figure 9.4 EXAFS analysis of a Pt/CeO_2 SAC. (a–f) Fourier transform k^3-weighted EXAFS data for Pt single sites and Pt NPs recorded at the Pt L_3 edge. The experimental data (blue) were fitted (red) in the range 1.1–3.4 Å (dashed box) using five possible locations of Pt^{2+}. Bulk Pt was used to fit Pt NP. (g–l) The corresponding structures (grey, Pt; yellow, Ce; red, O), as determined by DFT calculations, showing the free-energy values. (m) Illustration of Pt nanoparticle redispersion and Pt-induced restructuring of the CeO_2 support at high temperatures under oxidizing conditions as well as particle formation from the atomically dispersed state under reducing conditions. Source: Maurer et al. [113]. Reproduced with permission from Springer Nature.

disagreement in literature on spectroscopic assignments associated with atomically dispersed Pt species on CeO_2, which highlights the interesting and challenging nature of studying SACs.

The catalytic properties of M_{SA}/CeO_2 catalysts depend on the oxidation state of the metal. For instance, oxidized Pt_{SA} presented poorer activity than Pt_{NP}, whereas metallic Pt_{SA} showed higher activity for the oxidation of CO, NO, and CH_4, which is consistent with observations for Pt/TiO_2 [116]. However, the common assignment of a fixed oxidation state of metal atoms in SACs is oversimplified. Daelman et al. identified several well-defined charge states that are dynamically interconnected by combining DFT calculations and first-principle molecular dynamics on Pt_{SA} adsorbed on the CeO_2(100) surface [117]. A similar analysis was published simultaneously by Alexopoulos et al. for Pd/Al_2O_3, which also demonstrated the dynamic charge state of SACs during CO oxidation [118].

The dynamic behavior suggests that *in situ* measurements of SA oxidation states probe the state of the most abundant species in the catalytic cycle. The study of electron transfer between the metal and CeO_2 determined by the relative positions of the Ce 4f levels with respect to those of the noble metal is crucial to understand the nature of the active site and the associated reaction mechanism. Pt—O—Ce bond formation is favored by the presence of defects such as O_v associated with Ce^{3+} cations [109, 119, 120], and their amount was related to the distortion degree of the interface. In this regard, Pt_{SA}/CeO_2 sites were constructed at continuously-twisted surfaces in a monolith by directly growing porous CeO_2 single crystals and confining Pt in the lattice [121]. It was shown that these sites not only contribute to the chemisorption of CO but also effectively activate the lattice oxygen linked to Pt ions for CO oxidation. In fact, the distortion of Pt—O—Ce bonds, which is energetically favorable through strong metal-support interaction, balances the Fermi energy and the charge density between the Pt species and the CeO_2 support, leading to variations in the valence state and the electronic structure of Pt_{SA}/CeO_2 [120]. Note that a correlation between the strength of the Pt-oxide/oxide-support interaction and the electron density of oxygen in oxide supports was reported [122]. The latter, as determined from the binding energy of the O 1s electron, was found maximal for CeZrY mixed oxide and CeO_2, and minimal for Al_2O_3 and SiO_2.

Furthermore, the local coordination environment of SAs determines their catalytic activity. For example, a hydrothermal treatment was suggested by Nie et al. to create a new type of thermally stable $O_{lattice}H$ active site in the vicinity of Pt^{2+}, leading to enhanced catalytic activity for CO oxidation [111]. Besides, the addition of phosphorus (an electron acceptor) into CeO_2 was reported by Ma et al. to considerably decrease the electron density of Pt atoms [123]. Their higher valence state gave rise to an activity enhancement by up to 10 times in the hydrogenation of styrene, cyclohexene, phenylacetylene, and nitrobenzene. It was also shown that phosphorus favors reactant adsorption strength and hydrogen spillover, both playing an important role in this enhancement.

The catalytic performances of SACs depend on the considered reaction, and this is particularly true when they are supported on CeO_2. For instance, Au_{SA}/CeO_2 was found much more active for CO oxidation than its NP counterpart, but showed poor activity for H_2 oxidation [124]. This unique feature would be beneficial to the PROX of CO in H_2-rich gas stream (Section 9.2.2). CO oxidation on Pt_{SA}/CeO_2 [125] and Rh_{SA}/CeO_2 [126] was found to follow the MvK mechanism: adsorbed CO molecules react with O species to form CO_2 and O_v, which are then replenished by adsorption of O_2. On the Rh/CeO_2 SAC, the measured CO and O_2 reaction orders were 0.2 and −0.03, respectively, i.e. totally different from those obtained for the nanocatalyst counterpart (−1 and 1, respectively) that favors an LH mechanism [126]. CO oxidation on Pt/CeO_2 [112] and Au/CeO_2 [127] SAC is promoted by the presence of H_2O. For Pt_{SA}/CeO_2, CO was proposed to easily react with the hydroxyl from dissociated water to yield a carboxyl intermediate, which subsequently dehydrogenates with the help of a lattice hydroxyl to generate CO_2 and water (water-mediated mechanism) [112]. For Au_{SA}/CeO_2, the promoting effect was attributed to the local atomic and electronic structure of SACs that facilitates an efficient reaction channel of CO with OH [127]. CO oxidation on NP periphery sites would be much less promoted by the presence of H_2O.

Controversies on the intrinsic activity of M_{SA}/CeO_2 catalysts can be found in the literature [110, 111, 113, 128–131]. Beyond discrepancies due to the preparation method and reaction conditions, the use of *in situ* techniques is crucial to compare the performances of SACs with those of cluster or NP catalysts [26, 113, 132]. It is especially true considering the dynamic structural behavior of metal/ceria catalysts, showing e.g. the formation of clusters or NPs from SAs under reducing conditions and redispersion under oxidizing conditions (Figure 9.4m), even at low temperature (<300 °C) [133–135]. Recently, the onset of CO oxidation was found connected to the migration of Pt_{SA} from fourfold hollow sites of ceria, forming few-atom clusters [113]. This suggests that the latter are the active sites or that the dynamics of Pt_{SA}/CeO_2 contributes to CO oxidation.

9.2.1.5 Other Catalysts

A number of other oxides have been employed as SA supports. In 2001, Abbet et al. reported an investigation of Pd atoms anchored to MgO(100) surface O_v (F-centers) using temperature-programmed reaction of O_2 with preadsorbed CO at −183 °C, IR spectroscopy, and DFT calculations [20]. Low-temperature (−13 °C) and high-temperature (227 °C) reaction routes were identified, with $Pd(CO)_2O_2$ and $PdCO_3CO$ found as precursors, respectively. However, the process leads to the filling of the O_v, and consequently to the migration and clustering of CO-carrying Pd atoms, as observed for Pt/Al_2O_3 [53]. Recently, Sarma et al. systematically investigated CO oxidation over M/MgO SAC (M = Ru, Rh, Pd, Ir, Pt) using XAS, FTIR, and DFT [136]. The metal cations were shown to occupy octahedral positions of the MgO lattice under step edges. O_2-lean CO + O_2 conditions lead to the partial deconfinement of the metal atoms and the formation of bidentate metal carbonate species, which would constitute the reactive intermediates. Also similar to the results obtained for alumina, the involvement of the (irreducible) MgO support was found limited, and the SAC activities were found lower than those of the supported cluster counterparts. Though Pt leads to the lowest apparent activation energy, the latter was found in the range of 96 ± 19 kJ/mol for the whole set of metals (see also Section 9.2.1.6).

Copper oxides have been used as SA supports in several studies. Though Zhou et al. detected no CO oxidation activity for SA in a model Pt/CuO/Cu(110) catalyst due to the weakness of CO adsorption [137], Therrien et al. reported that the reaction proceeds at low temperatures on a $Pt/Cu_2O/Cu(111)$ SAC [138]. In the latter case, from STM, TPD, IR, XPS, and DFT, the reaction was shown to involve lattice O atoms and neighboring charge-neutral Pt_{SA} through a MvK mechanism. CO oxidation was also shown to proceed through this mechanism over a model Au/CuO/Cu(110) SAC, in which Au_{SA} are negatively charged [139]. In a recent study by Wang et al. Fe_1O_3 motifs were grown on a $Cu_2O(100)$ single crystal by atomic layer deposition and characterized by near-ambient-pressure XPS, STM, and DFT [140]. The densely distributed but isolated Fe centers, with an oxidation state close to +3, were found active for CO oxidation at 200 °C. Whether Pt or Fe, the single metal atoms were shown to diffuse into the Cu_2O subsurface under reaction conditions, leading to deactivation of the model catalyst [138, 140]. As pointed out by Kropp et al. from DFT calculations in the cases of Pt and Pd SACs supported on MO(001) (M = Fe, Mg, Mn, Ni), catalyst deactivation may also be caused by metal sintering due to the O_2-induced healing of the O_v that stabilize the single metal atoms (in a negatively charged state) [141].

Various other oxides, accompanied with specific synthesis strategies, have been employed as supports for CO oxidation-active SACs, e.g. Ir on $MgAl_2O_4$ spinel [142], Au [143] and Pd [144] on cobalt oxides, Pd on $(CeZrHfTiLa)O_x$ high-entropy fluorite oxides [145], and Pt on $Cr_{1.3}Fe_{0.7}O_3$ [146]. These studies point out the importance of surface O_v and/or structural defects such as steps for the SAC stabilization. Remarkably, a 0.067 wt% Au_{SA}/Co_3O_4 catalyst exhibited a high activity (TOF = 3.1 s^{-1}) at −75 °C, which was ascribed from DFT calculations and isotopic experiments to a predominant LH pathway over an active site composed of a Au_{SA} and neighboring Co and O/O_{vac} species [143].

Using FTIR, *operando* XAS, kinetic measurements, and DFT calculations to investigate a model $Ir/MgAl_2O_4$ SAC with an ultralow metal loading (0.0025 wt%), Lu et al. have proposed an original CO oxidation mechanism involving Ir(CO) species as the active supported complex and Ir(CO)(O) as the most stable intermediate [142]. The latter would be the SAC resting state involved in the rate-determining step following an Eley–Rideal mechanism: $Ir(CO)(O) + CO \rightarrow Ir(CO) + CO_2$. Thus, CO would act as a ligand rather than a poison to (cationic) Ir atoms. Notably, the SAC exhibited reaction orders toward CO and O_2 different from those of supported Ir_{NP}, corresponding to different reaction mechanisms. This makes possible the quantification of surface site fractions in catalysts containing both SAs and NPs [147].

9.2.1.6 Discussion

A few ideas emerge from the above review of the literature: controversies regarding the characterization, local coordination environment, reactivity, and catalytic mechanisms of oxide-supported late-transition-metal atoms in CO oxidation catalysts are prevalent in literature.

The first issue is the characterization of SACs with the aims of substantiating the complete dispersion of metals as site-isolated atoms (ions) and identifying the primary local coordination of the metal species. Aberration-corrected STEM imaging executed through a combination of large-area and local measurements can effectively demonstrate the primary existence of atomically dispersed metals. However, this becomes challenging when the support oxide contains cations that have similar or higher atomic number as compared to the metal atoms (e.g. Pt and Ce, or Rh and Ce). In these cases, additional characterization is required to assess the primary metal structure. The most common tools are XAS and CO-FTIR. Again, results from these characterization techniques can be challenging to interpret. For example, single cationic Pt atoms on an oxide support can show similar XAS and CO-FTIR spectra as compared to small non-uniform PtO_x clusters on the same support [72, 115]. Thus, it has become apparent that chemical probes – for example reducibility, measurements of the adsorption energy of a probe molecule, or reactivity of a probe reaction such as simple hydrogenation chemistry – may be best suited to analyze the relative dispersion of metals across a sample [73, 115, 148].

Once the latter is assessed, the uniformity of the local coordination environment should then be considered prior to assigning a dominant local coordination environment [76].

While apparent discrepancies exist in the characterization of SACs, a few common themes have emerged. First, for reducible supports, single late-transition-metal atoms tend to adopt cation-replaced position in the oxide lattice following oxidative treatments [25, 74, 83, 84, 110, 117]. Exposure to reducing environments results in reduction of the oxidation state of the metal atom (ion) and likely a reduction in the M–O coordination number. Finally, harsher reduction results in the production of mobile atomically dispersed species in close to neutral states, which serve as the precursors for cluster formation. Non-reducible oxide supports such as Al_2O_3 [53, 54] and MgO [136], which have softer interaction with metal atoms as compared to reducible oxides, maximize reduction and clustering phenomena in the presence of CO or H_2.

The second common theme, in the case of reducible oxide supports, is that Pt SACs tend to exhibit lower CO adsorption energies as compared to metallic Pt clusters. This stems from the cationic oxidation state of Pt in SACs, and the strong coordination to the support. A result of the decreased CO binding energy on Pt SACs is the reduced CO coverage on the Pt active site during CO oxidation [72].

Finally, as previously mentioned, two main types of CO oxidation pathways have been proposed on oxide-supported SACs: the MvK and the LH mechanisms [44, 48]. The former is favored on reducible oxides, while the latter necessarily involves carbonate-like species, as illustrated in Figure 9.5a. According to the recent literature analysis by Beniya and Higashi of CO oxidation over PGM and Au catalysts supported on various oxides, Pt SAC are overall more efficient than Pt cluster catalysts above c. 100 °C, while supported Au_{NP} are superiors to SAC at low temperature [44]. However, as previously discussed in the chapter, this issue is still strongly debated, especially for the Pt/CeO_2 and Pt/Al_2O_3 systems in relation to their actual active site, owing to the dynamic evolution of the Pt atom nuclearity and oxidation state under reaction conditions.

Few experimental attempts have been made to compare the effect of the metal or the support identity on the SAC performance. Comparing PGM atomically dispersed on MgO for CO oxidation, Sarma et al. recently reported that Ru, Pd, and Pt are more active than Rh and Ir, and that metallic clusters are overall more active than the single cations (Figure 9.5b) [136]. Computational studies can help in the design of SAC and rationalize their *modus operandi*. Still in the case of the MgO support, using DFT calculations and microkinetic modeling, Xu et al. predicted from Brønsted–Evans–Polanyi relationships that Ag and Cu SACs should perform better than Ni, Pd, and Pt SACs for CO oxidation (Figure 9.5c) [149]. The LH mechanism was found to be more favorable than the MvK and Eley–Rideal pathways. Such an example can motivate future research toward the design of SACs based on non-conventional metals.

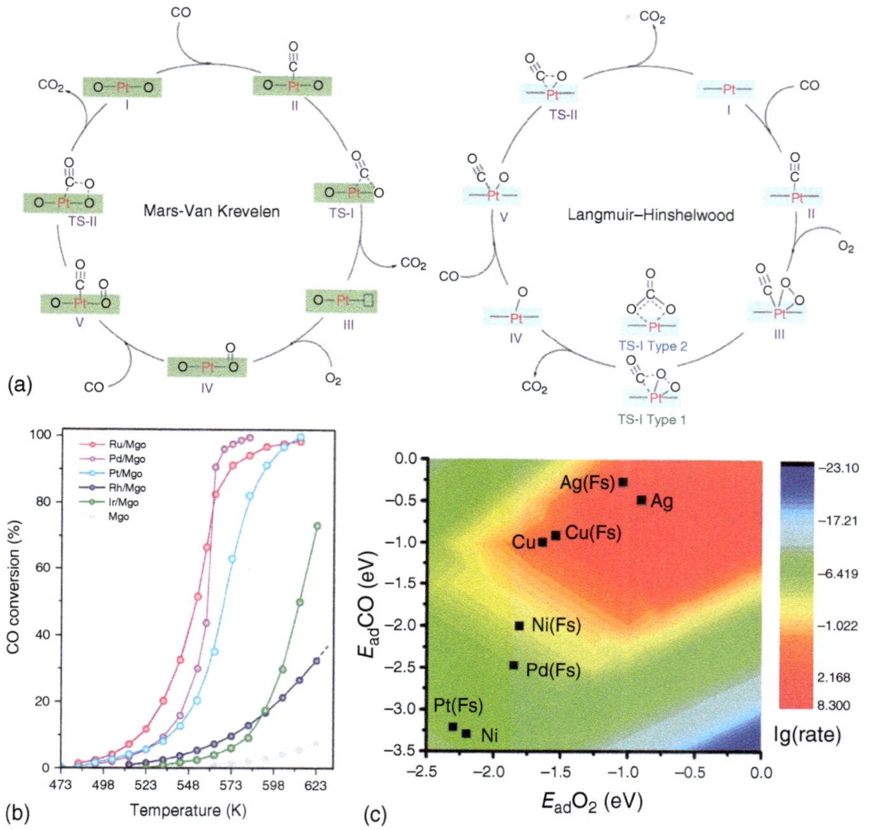

Figure 9.5 (a) Typical reaction pathways of Pt SACs via the MvK or the LH mechanism. Source: Lu et al. [48]. Reproduced with permission from Wiley-VCH. (b) Evolution of CO conversion with the reaction temperature for M/MgO SAC (0.1 M_{at} nm^{-2} metal content, 1 bar, CO : O_2 = 1 : 20 vol%). Source: Sarma et al. [136]. Reproduced with permission from American Chemical Society. (c) Contour plot of the Sabatier activity with respect to the CO and O_2 adsorption energies on different SAC under typical experimental conditions for low-temperature CO oxidation ($T = 0\,°C$, $P_{O2} = 0.21$ bar, $P_{CO} = 0.01$ bar). (Fs) represents M/Fs-defect MgO. Source: Xu et al. [149]. Reproduced with permission from Royal Society of Chemistry.

9.2.2 Preferential CO Oxidation in Hydrogen (PROX)

Preferential CO oxidation is useful for the removal of CO impurities present in hydrogen produced from hydrocarbon reforming. Such a pure hydrogen is then able to feed proton-exchange-membrane fuel cells [97]. Like for conventional CO oxidation, noble metals (especially Pt and Au) supported on reducible oxides are the most efficient catalysts for this reaction, i.e. they exhibit high activity and selectivity at low temperature (<100 °C). The extension of this knowledge to SACs is challenging owing to the highly reducing hydrogen-rich atmosphere, potentially favoring metal atom clustering [54]. This may explain the scarcity of studies related to PROX on SACs.

Zhang and coworkers reported the PROX performances (reactant mixture $CO : O_2 : H_2 = 1 : 1 : 40$ vol%) of Pt and Ir SAs supported on iron oxide. The Ir_{SA}/FeO_x catalyst was found much less efficient than both the Pt/FeO_x SAC [65] (consistently with nanocatalysts [97]) and an Ir/FeO_x catalyst based on subnanometric Ir clusters [66]. In contrast, the Pt_{SA}/FeO_x catalyst was reported to be more active than its cluster-based counterpart [25]. The same group also investigated the PROX reaction over 0.05 and 0.3 wt% Au_{SA}/CeO_2 prepared by a simple adsorption method [124]. The catalyst was found to be stable, highly active and fully selective at 70–120 °C. The high performance at relatively high temperature was ascribed to inhibited H_2 dissociation on SACs, which would disfavor the unselective route. Gan et al. recently reported a dry ball milling method able to produce, from acetate precursors, kilogram-scale 0.1 wt% Au_{SA}/CeO_2 for PROX [150].

Notably, several types of highly dispersed catalysts containing hydroxide species were reported to exhibit excellent PROX performances. Zhang and coworkers prepared subnanometric-cluster $Ir/Fe(OH)_x$ [151] and $Rh/Fe(OH)_x$ [152] catalysts by a co-precipitation method, with high CO conversion efficiency near room temperature and tolerance to CO_2 and H_2O poisoning. L. Cao et al. adopted a reverse strategy by selectively and atomically dispersing $Fe(OH)_x$ species at the surface of silica-supported Pt_{NP} through atomic layer deposition [153]. The catalyst achieved complete and 100% selective CO removal from −73 to 107 °C. From XAS, XPS, and DFT investigations, the PROX mechanism was suggested to involve $Pt\text{-}Fe_1(OH)_3$ interfacial sites, with COOH (resulting from $CO + Fe_1(OH)_3$ reaction) dehydrogenation to CO_2 as the rate-determining step. Note that this catalyst may also be seen as an SAA (Section 9.3.2) with dilute Fe atoms deposited onto Pt_{NP}. So do oxide-supported Pt–Fe nanoalloy catalysts prepared by Zhang et al., who also measured high PROX performance for this bimetallic system [154].

Even more recently, S. Cao et al. reported the preparation of a Pt SAC by incipient wetness impregnation of silica or alumina with an aqueous solution containing $H_2Pt(OH)_6$ as well as KOH or CsOH [155]. The alkali ions were assumed to stabilize $Pt\text{-}O(OH)_x$ species at the support surface, with good PROX performance around 100 °C. Here again, similar to what was previously proposed for supported gold nanocatalysts [156], surface hydroxyl groups (regenerated by $O_2 + H_2$ reaction) are believed to play a prominent role as they open a faster CO oxidation pathway in the presence of H_2. This can be put in relation to the strong promoting effect of water reported for CO oxidation over a $Pt/Cr_{1.3}Fe_{0.7}O_3$ SAC [146] and ceria-supported SACs (Section 9.2.1.4). However, in these cases, the relevant reaction to be considered may be the WGS.

9.2.3 Water–Gas Shift Reaction (WGSR)

The water–gas shift reaction (WGSR) ($CO + H_2O \leftrightarrow CO_2 + H_2$) is an important industrial process to produce CO-free hydrogen or to adjust the H_2/CO ratio required for Fischer–Tropsch and methanol synthesis [157–160]. The WGSR is thermodynamically more favorable at lower temperatures ($\Delta H_{298\,K} = -41.2$ kJ/mol) and kinetically favored at high temperatures. Therefore, the reaction is industrially performed in several stages with different catalysts: the high-temperature shift reaction is performed at 350–450 °C on iron-oxide-based catalysts, while the low temperature shift reaction is performed at 190–250 °C on copper-zinc-oxide-based catalysts. In the past two decades, noble-metal-based catalysts have been widely investigated for the WGSR due to their high stability and activity, and among them SACs have aroused great attention (see also Chapter 10).

In 2003, the Flytzani–Stephanopoulos group reported the catalytic role of Au and Pt cations in WGSR [21]. Interestingly, the authors used a sodium cyanide leaching method to remove metallic NPs contained in conventional Au/CeO_2 and Pt/CeO_2 catalysts. They found that their activity was almost unchanged, and concluded that only isolated metal cations or clusters strongly bonded to the support are the active sites. This has been debated since the activity of such catalysts could be modified by the presence of sodium cyanide (formation of Na[Au$(CN)_2$]), and Au/CeO_2 catalysts prepared by this method were less active than commercial Cu/ZnO/Al_2O_3, contrarily to other catalysts prepared by the deposition–precipitation method [161–164]. Furthermore, a precise comparison of the activity of the SA and NP active sites is not obvious.

The same group later showed that the introduction of alkali ions (sodium or potassium) can stabilize, at appreciable loading (c. 1 wt%), atomically dispersed Pt/Au–O_x(OH)-S sites (where S is the oxide support), which are able to catalyze the WGSR at low temperature (<250 °C) [29, 165–168]. Alkali cations were reported by Ammal and Heyden to behave similarly as reducible oxide supports by supplying OH species for the WGSR and activating H_2O molecules [169]. In a study by the Stair group, high-angle annular dark field (HAADF) STEM images confirmed the higher dispersion of Pt in Pt–Na/SiO_2 as compared to Pt/SiO_2. However, CO-FTIR data revealed that the promotional effect of Na^+ in the WGSR mainly originates from an alteration of the properties of Pt_{NP}, not from Pt_{SA} [58].

Au_{SA} and Pt_{SA} supported on CeO_2 as well as TiO_2, SiO_2, FeO_x, MnO_x, and inert KLTL zeolites, were later proposed by several groups to be the active sites in the WGSR [170, 171]. As shown in Figure 9.6, the TOF are similar on different supports, and only specific to the nature of the metal cation of the active site, i.e. Au(I) (45 ± 5 kJ/mol) or Pt(II) (70 ± 10 kJ/mol) [172]. However, by using CO-FTIR spectroscopy to distinguish Pt_{SA} from Pt_{NP}, the Stair group concluded that Pt_{SA} in Pt/H-ZSM5 and Pt/SiO_2 catalysts behave as spectators for the WGSR, unlike Pt_{NP} [58]. Their low catalytic activity was attributed to the strong binding of CO molecules.

The role of support nanostructuring was also investigated. Ceria nanorods exhibiting {110} and {100} crystal surfaces stabilize gold as single atoms and clusters (<1 nm), while Au_{NP} (~3 nm) were present on the {100} facets of ceria nanocubes. The activities of Au/CeO_2 nanorods were found at least 1 order of magnitude higher for the WGSR and steam reforming of methanol at low temperatures [173]. Furthermore, hollow mesoporous CeO_2 microspheres with a high surface area and a mesoporous structure were shown to stabilize single Au atoms catalyzing

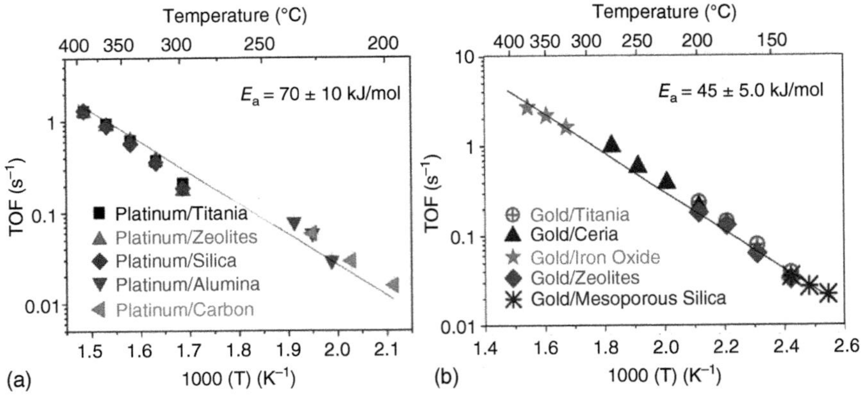

Figure 9.6 (a) TOF of the WGSR on Na-containing Pt catalysts on various supports in a simulated reformate-gas mixture ($CO : H_2O : CO_2 : H_2 = 11 : 26 : 7 : 26$ vol%); (b) TOF plot for the WGSR over atomically dispersed gold catalysts in the same mixture. Source: Yang and Flytzani-Stephanopoulos [172]. Reproduced with permission from Elsevier.

the WGSR with higher performances than those obtained using ceria synthesized through co-precipitation and sol–gel methods [174].

High WGSR performances were also reported for SACs containing other elements than Au or Pt. For instance, the catalytic activity of a 0.37 wt% Rh_{SA}/TiO_2 catalyst was around four times higher than that of a Rh/TiO_2 nanocatalyst, without any methanation at 300 °C, even under CO_2- and H_2-rich WGSR stream. The suppression of methanation was attributed to the absence of H_2 dissociative adsorption on Rh_{SA}/TiO_2 [175]. The activities of Ir/FeO_x [30] and Pd/FeO_x [176] SACs were 1 order of magnitude higher than those of their cluster or NPs counterparts. This better performance was ascribed to the enhanced reducibility of the FeO_x support, and the generation of O_v that promote the dissociation of H_2O to form H_2 and atomic oxygen.

Environmental STEM in controlled WGS atmospheres revealed the formation of clusters of only a few gold atoms resulting from SA dynamics, and the catalytic effect of low-coordination surface sites [177]. Conversely, Pd can be predominantly dispersed as isolated atoms onto TiO_2 during the reverse-WGSR at 400 °C. The thermodynamic stability of Pd was associated with Pd–Ti coordination (as evidenced by EXAFS), which manifests upon O_v formation [178].

A major difference in reaction mechanisms between SAs and NPs was highlighted for Pt/FeO_x catalysts [179]. On Pt_{NP}, CO is strongly adsorbed and reacts with OH groups, leading to formates, which then decompose to simultaneously produce CO_2 and H_2. On Pt_{SA}, adsorbed O species formed by dissociation of H_2O on O_v of FeO_x combine with weakly adsorbed CO to produce CO_2. The activation energy of this process is only 33 kJ/mol, vs. 61 kJ/mol for Pt_{NP}/FeO_x catalysts. A similar redox mechanism was reported for Pd_{SA}/FeO_x, with also a low activation energy of c. 30 kJ/mol [176]. A specific redox mechanism, involving electron transfer from Fe^{3+}-O $\cdots$ Ir^{2+}-O_{vac} to Fe^{2+}-O_{vac} $\cdots$ Ir^{3+}-O (dual metal active site), was proposed for Ir_{SA}/FeO_x from a theoretical and experimental study: H_2O dissociates to OH on Ir_{SA}, and to H on the first-neighbor O atom bonded with an Fe site [180]. CO adsorbed on Ir_{SA} reacts with another adjacent O species to produce CO_2, yielding an O_v. H_2 is then formed by migration of H from adsorbed OH toward Ir_{SA}, and its subsequent reaction with another H.

9.2.4 Total Oxidation of Hydrocarbons

Catalytic reactions consisting of hydrocarbon total oxidation are useful in a range of energy generation devices such as gas turbines and heating systems, and environmental applications such as the purification of indoor and outdoor air, and vehicle exhaust. In 1999, Iwasawa and coworkers reported an EXAFS investigation of atomically dispersed Pt and $PtMo_6$ species on MgO prepared by impregnation-calcination, forming a spinel-like distorted structure [18]. These catalysts were found to be as active as MgO-supported metallic Pt particles for propane combustion. Single Pt ions could be converted into Pt_6 clusters upon a propane reducing treatment, and regenerated back upon O_2 calcination at 450 °C. Recently, Wang et al. showed that a Pt/La-Al_2O_3 SAC was active for the propene combustion [63]. The authors found that the Pt atoms remained isolated if, and only if, oxidic Ba species were additionally introduced in the preparation. A number of SAC studies have also been devoted to methane oxidation. Tang et al. prepared Pt-doped CeO_2 catalysts by several methods and found that the rate of methane partial oxidation with CO_2 (to CO) and total oxidation with O_2 (to CO_2) increases with the fraction of ionic Pt sites [23]. From DFT calculations, it was suggested that the dissociative adsorption of methane is the rate-determining step, which is favored at the surface in the vicinity of Pt dopants, where O atoms are more labile. In contrast to these results, Jeong et al. recently found that highly oxidized Pt SACs are less active for methane oxidation as well as CO and NO oxidation than Pt_{NP} catalysts, which are themselves less active than metallic Pt SACs [116]. The Pt oxidation state, which would be barely changed upon the reactions, could be controlled by using as support defective ceria islands at an alumina surface, and by varying the catalyst prereduction temperature. Yan et al. reported an original strategy to synthesize a thermally stable Pt_{SA}/Mn_2O_3 catalyst for methane combustion [181]. Platinum was preloaded as SA on Mn_3O_4 through a redox precipitation method, and the resulting SAC precursor was calcined in humid air (3 vol% H_2O) at 800 °C for five days. This led to the strong anchoring of single Pt^{4+} atoms onto the reconstructed Mn_2O_3 phase, which was further optimized by H_2O_2 etching of the oxide surface. Besides Pt-based SACs, Pd atomically dispersed on oxides was also found efficient for methane and toluene combustion [75, 182]. Overall, it is often pointed out that the supports play a role as important as that of the metal atoms [75], which in some cases only have a promoting effect on the inherent oxidation activity of the oxides [181].

Manganese oxide was also used as a SAC support by Tang and coworkers in the form of microporous hollandite Mn oxide (HMO) nanorods. Silver atoms could be anchored at the pore openings of HMO(001) facets, and the resulting catalyst was active for the combustion of volatile organic compounds such as formaldehyde [183] and benzene [184]. Surprisingly, the replacement of Ag with Na located at similar sites of HMO led to enhanced formaldehyde abatement performance, which was ascribed to an increased electron density of neighboring surface lattice oxygen atoms with respect to Ag/HMO [185]. Hence, the role of the alkali metal could be more than that of a promoter, as it was previously believed from the high performance of an atomically dispersed Na-promoted Pt/TiO_2 catalyst for the total oxidation of formaldehyde [186]. Potassium-loaded hollandite-type MnO_2 and TiO_2 structures were also found efficient for diesel soot oxidation [187]. Still with Mn oxide, Zhang et al. prepared a SAC consisting of Pt_{SA} on defective MnO_2 nanosheets through a one-pot hydrothermal method for low-temperature toluene oxidation [188]. Original strategies were also employed for the preparation of SACs for the oxidative removal of benzene, whether on $Pd/CuO/\gamma\text{-}Al_2O_3$ prepared by galvanic replacement [189] or on Pt-loaded ordered mesoporous Fe_2O_3 [190]. For the latter, the authors found enhanced activity and stability with respect to supported NP. The superiority of Pt_{SA} was also reported for the total oxidation of methanol over Pt/Co_3O_4 [191] and of butanone on Pt/WO_3 [192].

9.2.5 Selective Oxidation Reactions

9.2.5.1 Early Transition Metals on Oxides

Site isolation, which corresponds to the spatial separation of active sites at the surface of heterogeneous catalysts, is a key parameter in selective oxidation (SELOX) reactions to avoid deep oxidation (formation of CO_x) and obtain high selectivity to partial oxidation products [193–195]. This concept was formulated in the early 1950s by Grasselli, and was defined as one of the seven pillars in SELOX [194]. Since then, it has been widely applied both to bulk oxides and supported early transition metals such as V, Mo, Ti, and Cr.

Various methods, such as impregnation at low metal loading, co-condensation [196], thermolytic molecular precursor method [197–200], cation immobilization by amino groups [201], and flame pyrolysis [202], have been developed to isolate supported active sites, also called monomeric sites. This has led to high selectivities in various reactions such as oxidative dehydrogenation of methanol [197], epoxidation of cyclohexene [198, 203], hydroxylation of benzene [202], and photo-assisted epoxidation of styrene [204]. In particular, VO_x/SiO_2 catalysts were applied to challenging reactions such as (photo)catalytic oxidation of methane to methanol and formaldehyde [196, 200, 205, 206] and oxidative dehydrogenation of propane [202, 207–209], during which CO_x species are very easily produced.

The molecular structure of these isolated sites was characterized by vibrational spectroscopies (especially Raman for iono-covalent oxides) [196, 198, 200, 201, 210, 211], UV-Vis spectroscopy [198–201, 204, 206, 207, 212, 213], electron spin resonance spectroscopy [196, 201, 207, 212], and EXAFS [197, 202]. The need for *in situ* techniques quickly became evident since the structure of supported oxometallates strongly depends on their environment. For instance, polyoxometalates supported on silica, which are stable in hydrated conditions (e.g. in ambient air), decompose upon dehydration at elevated temperature, leading to isolated species [210, 214]. The role of bridging bonding with the support and of the support itself was also investigated. Khaliullin and Bell reported that the TOF of methanol oxidation to formaldehyde is more than 100 times greater for VO_x/TiO_2 and VO_x/ZrO_2, than for VO_x/SiO_2 [215]. The authors proposed that formaldehyde formation from methoxy groups may require pairs of adjacent VO_4 groups or V_2O_5 dimers. For propane oxidative dehydrogenation to propylene, the TOF of VO_x/ZrO_2 catalysts was found almost independent on the surface vanadia coverage (i.e. similar for monomeric and polymeric vanadia species) [216]. This suggests that the reaction requires only one surface VO_4 group. However, the propylene selectivity increases with increasing surface vanadia loading, which was explained by the removal of unselective Zr–OH groups upon vanadia deposition. Finally, the role of support defects was underlined. For example, Launay et al. proposed that hydrolyzed D2 defects in SiO_2 would act as preferential anchoring sites for monomeric hydroxylated species, which are the most efficient ones for the SELOX of methane [210, 211].

Progress in electron microscopy has allowed the direct visualization of monomeric and polymeric species [217–220], as illustrated in Figure 9.7 by HAADF-STEM images of WO_x/ZrO_2, which is a solid well known for its acidic properties, but is also efficient for some SELOX reactions [221, 222]. It has also led to a change of terminology in the community working on selective oxidation. For instance, one can find the term "single Cr atoms" in the recent literature [223].

9.2.5.2 Late Transition Metals on Oxides

SACs such as Pd_{SA}/silicate [224] and Rh_{SA} on ZrO_2 [225], TiO_2 [226], ZSM-5 zeolite [226], or CeO_2 [227], were shown to be efficient for the SELOX of methane in aqueous solution using either H_2O_2 or O_2 as oxidant. For Rh_{SA}/ZrO_2, DFT calculations indicate that four-coordinated Rh atoms stabilize CH_3 species by suppressing further dehydrogenation, unlike five-coordinated Rh atoms [225, 228]. Rh_{SA} also facilitate the activation of H_2O_2 and the formation of the CH_3OOH intermediate. The spontaneous dissociative adsorption of H_2O_2 on Rh_{SA}/ZrO_2 was reported to form an O_2Rh active site and to hydrogenate the surrounding ZrO_2 surface [229]. After the adsorption of CH_4 on O_2Rh/ZrO_2-2H, the C—H bond is activated by H abstraction with a barrier of 1.23 eV, leading to a methyl radical and the hydrogenated HOO-Rh species. Adsorbed CH_3 can further react with HOO-Rh to produce CH_3OH with a barrier of 0.30 eV. After desorption of CH_3OH, a second methane molecule can be adsorbed on the ORh/ZrO_2-2H site, which is easily activated (barrier of 0.60 eV) to form CH_3. The second CH_3OH molecule is then produced with a barrier of 0.39 eV. In the case of Rh_{SA}/ZSM-5, the oxidative addition pathway for the activation of methane occurs at coordinately unsaturated Rh atoms, with a very low barrier of 0.07 eV for Rh(CO) [230]. The rate-determining step corresponds to the formation of a C—OH bond, which is promoted by CO coordinated to Rh. Finally, water was shown to prevent poisoning by CO, and to protonate an intermediate RhOOH species.

As compared to their NPs counterparts, several SACs exhibited a higher activity for the SELOX of alcohols with molecular oxygen, which was ascribed to their maximal number of interfacial sites [231]. Their higher selectivity was explained by the activation of lattice oxygen at the interface. Similarly, Au_{SA}/SiO_2 catalysts prepared by cyanide leaching were found to be very active for the solvent-free aerobic oxidation

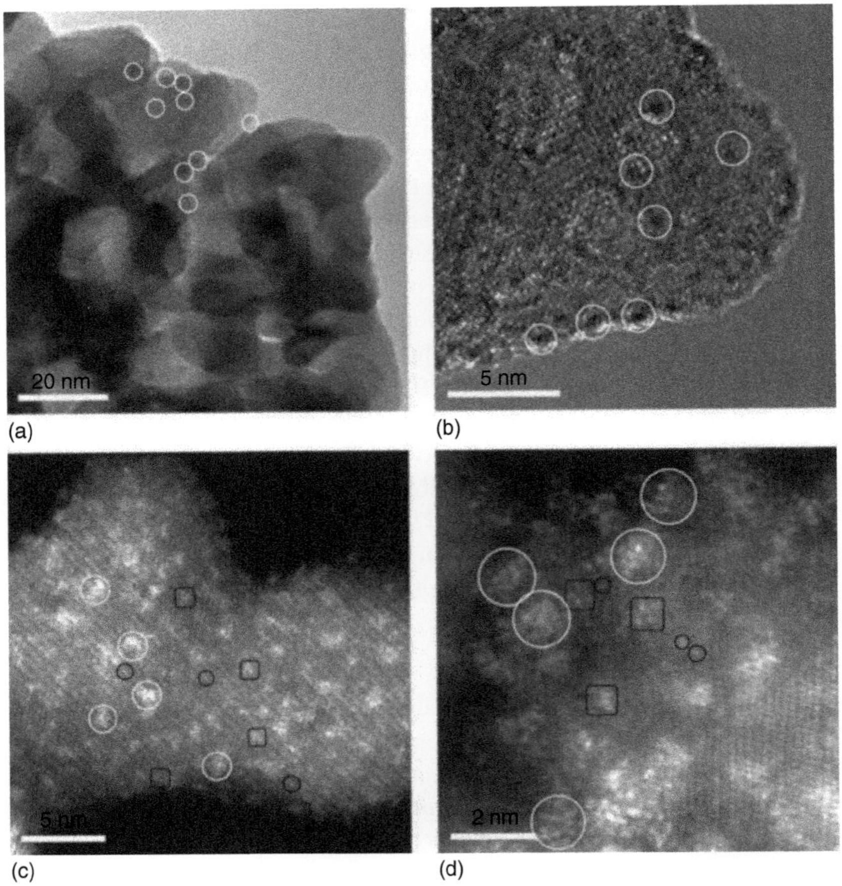

Figure 9.7 TEM (a), HRTEM (b), and HAADF-STEM (c, d) images of a WO_x/ZrO_2 sample. Black circles indicate the presence of single W atoms corresponding to surface mono-tungstate species. Black squares indicate surface poly-tungstate species with several W atoms linked by oxygen bridging bonds. White circles indicate WO_x clusters with diameters of ~0.8–1.0 nm. Source: Zhou et al. [217]. Reproduced with permission of Springer Nature.

of benzyl alcohol to benzaldehyde [232]. Their activities were measured to be identical to those of the unleached catalysts, and turnover-number values higher than $4 \times 10^5\ h^{-1}$ were obtained based on the Au content. 100% selectivity below 180 °C and high stability for 10 hours was reported for methanol dehydrogenation to produce methyl formate and H_2 on Na^+-stabilized Au_{SA}/TiO_2 prepared by incipient wetness impregnation of Au_{SA}–O_x–Na_9–$(OH)_y$ solutions [233]. Pd/SiO_2 and Cu/SiO_2 catalysts containing metal NPs are unselective and inactive, respectively, for this reaction. Interestingly, 92% selectivity to methyl formate at 65% conversion and high stability for 10 hours were obtained with a $Pd_{0.01}Cu/SiO_2$ SAA catalyst (Section 9.3.2) prepared by adding a small amount of Pd onto Cu NPs [234]. Pd_{SA}/Al_2O_3 exhibited quite high performance in the aerobic SELOX of cinnamyl and crotyl alcohols under mild conditions [17], though with a lower selectivity to the α,β-unsaturated carbonyl compound (~90%) than state-of-the-art supported Au catalysts [235]. In the case of SELOX of cinnamyl alcohol, the TOF of Pd^{II} isolated sites is 4400 h^{-1} at 60 °C, against 538 h^{-1} at 120 °C for Au/CeO_2 catalysts. For the SELOX of crotyl alcohol, the TOF is 1 order of magnitude higher than that of surface PdO_x species, while a high selectivity is retained [17]. In another study, a Pd_{SA}/Al_2O_3 catalyst showed higher activity and selectivity as compared to Pd_{NP} for the oxidation of cinnamyl alcohol [236]. This work also highlighted the key role of coordinately unsaturated Al^{3+} sites in trapping Pd atoms and tuning their electronic properties. It was also concluded that oxygen species formed by the interaction of O_2 with Pd_{SA}/Al_2O_3 oxidize the partially dehydrogenated intermediates to cinnamaldehyde. Pd_{SA}/TiO_2 was reported to catalyze the epoxidation of light olefins (except ethylene) in the presence of O_2 and H_2 at room temperature [237]. However, one can note that the yields were quite low (e.g. 1–2% for propene to propylene oxide). Ninety-one percent conversion of 5-hydroxymethylfurfural and 81% selectivity to 2,5-diformylfuran for two hours at 110 °C were obtained with a Ru_{SA}/NiO catalyst prepared by a ball-milling method [238]. Finally, Rh_{SA}/ZSM-5 was found efficient to convert CH_4 to acetic acid and methanol through coupling of CH_4, CO and O_2 in solution below 150 °C [239]. Computational studies suggested that CH_4 is activated by Rh_1O_5 anchored on the wall of micropores of the ZSM-5 zeolite. The resulting CH_3 then couples with CO and OH to produce acetic acid with a low activation barrier.

9.3 Single-Atom Catalysts Supported on Carbon and Other Materials

SAC supports are not limited to oxides and zeolites, and support materials such as metal-organic frameworks [240, 241] or sulfides [242] have been investigated for selective oxidation. Here we briefly report on two important catalyst types, namely carbon-based SACs and SAAs.

9.3.1 Carbon and Nitrogen-Hosted SAC

Besides oxide-supported SACs for gas-phase thermal oxidation reactions, carbon-based materials are by far the most widely employed SAC supports or hosts for applications in liquid-phase electrocatalysis, photocatalysis, and hydrogenation or coupling thermocatalysis [243]. Among other advantages, one can cite the beneficial properties of carbon (graphite, graphene, carbon nanotubes (CNTs), activated carbon, etc.) in terms of specific surface area, electrical and thermal conductivities, and tunable surface functionalization (see Chapter 3). Doping of carbon with nitrogen or using a carbon nitride such as graphitic C_3N_4, allows for the stabilization of noble and non-noble metals in an isolated form through several metal–nitrogen bonds [2]. A number of studies have investigated the catalytic thermal oxidation properties of carbon-supported SAC, most often under mild liquid-phase conditions.

9.3.1.1 Selective Oxidation of Alcohols

Li et al. reported the preparation of a Co/N-G (N-doped graphene) SAC by heat treatment of cobalt salts and graphene oxide in ammonia atmosphere [244]. The catalyst showed high stability and atom efficiency for the selective aerobic oxidation of benzyl alcohol and derivatives to the corresponding aldehydes. For benzyl alcohol, the achieved conversion was 94.8% with 97.5% benzaldehyde selectivity. Xie et al. further reported that all M/N–C (M = Fe, Co, Ni, Cu, Cr) SACs – synthesized by pyrolysis of metal nitrate and nicarbazin – are able to catalyze benzyl alcohol oxidation to benzaldehyde, Cu/N–C being the most efficient system [245]. In addition, Fe/N–C was found active for the aqueous-phase aerobic oxidation of 5-hydroxymethylfurfural to 2,5-diformylfuran. In comparison, the oxidation of aliphatic alcohols (ethanol, 1,6-hexanediol, glycerol) was found much less efficient.

Ding et al. designed a SAC consisting of an Fe-based polymerized ionic liquid monolayer on CNTs, which exhibited more than 99% regioselectivity for phenol hydroxylation to catechol (1,2-dihydroxybenzene) with hydrogen peroxide, and an activity superior to that of its homogeneous free ion counterpart [246]. As suggested by experimental and theoretical investigations, the Fe atom would coordinate to four oxygen atoms at the CNT surface, and the reaction mechanism would involve the non-radical addition of an OH group at the *ortho*-carbon atom of phenol through the formation of a ferric-hydroperoxo complex.

9.3.1.2 Selective Oxidation of Hydrocarbons

Liu et al. prepared Fe/N–C SAC by an MgO template-sacrificial pyrolytic approach, which exhibited high room-temperature activity, selectivity, and reusability for the C—H bond oxidation in a number of substrates – including ethylbenzene toward acetophenone – dissolved in an aqueous solution of *tert*-butyl hydroperoxide [247]. Using STEM, XPS, XAS, electron spin resonance spectroscopy, Mössbauer spectroscopy and KSCN titration experiments, the authors showed that the pyrolysis temperature (600–800 °C under N_2 atmosphere) controls the relative concentrations of FeN_x species ($x = 4$–6) coexisting in the SAC, which in turn critically affect the reaction TOF in this order: $Fe^{III}N_5 > Fe^{III}N_6 > Fe^{II}N_4$.

Bakandritsos et al. reported the beneficial influence of mixed valence for Cu atoms anchored onto graphene functionalized with nitrile groups (cyanographene) to reach near 100% conversions and selectivities for the aerobic oxidation of benzylic C—H bonds and the oxidative coupling of amines toward pharmaceutical synthons [248]. These high performances were explained by a short-distance cooperative synergy between Cu(II) and Cu(I) species (cyanographene allows the partial reduction of Cu[II] to Cu[I]), as observed in metalloenzymes).

Acetophenone could also be synthesized with high yield and regioselectivity from aryl alkenes – including styrene – through the Wacker-type aerobic oxidation using isopropanol as the hydrogen source and a Co–N–C SAC prepared by impregnation-pyrolysis [249]. The corresponding industrial process, which uses catalytic $PdCl_2$ and stoichiometric $CuCl_2$, has a number of limitations, which makes the SAC an elegant alternative.

Zhang et al. anchored Cu atoms in Cu–N_3 coordination onto porous hollow graphitic carbonitride spheres synthesized using a template-free preassembly strategy, which exhibited high conversion and higher selectivity and stability than Cu_{NP} for the oxidation of benzene to phenol [250]. The Fe/N–C system was also reported to be efficient for this reaction, while Fe_{NP} counterparts showed lower conversion and selectivity [251, 252]. Noticeably, Deng et al. were able to directly reveal the atomic structure of FeN_4 centers embedded in graphene by combining STEM and STM methods [251].

9.3.1.3 Other Reactions

Another reaction for which a SAC was found much more efficient than its homogeneous or nanoparticulate counterparts is the gold-catalyzed selective oxidation of silanes to silanols using water, as reported by Chen et al. for Au_{SA} anchored through three bonds onto mesoporous polymeric graphitic carbon nitride (mpg-C_3N_4) [253].

The use of SACs was also found beneficial to reactions involving hydrogen peroxide, such as the direct synthesis of H_2O_2 from H_2 and O_2 over atomically dispersed $PdCl_x$/C [254], and the peroxone reaction between O_3 and H_2O_2 – for hydroxyl radical generation in acid solution – over Mn–N_4 sites dispersed on g-C_3N_4 [255]. Similarly, a Cu/C_3N_4 SAC was recently found efficient for the activation of H_2O_2 to hydroxyl radicals (Fenton reaction) and subsequent oxidative degradation of rhodamine B, in the context of organic wastewater treatment [256]. SACs such as Fe/ and Cu/N–C were also reported to exhibit enzyme-like activity ("SA nanozymes") in e.g. (per)oxidase-catalyzed reactions [257, 258].

The excellent activity as well as steam and sulfur-resistance of the emblematic Fe/N–C system – consisting here of Fe–N_4 centers confined in polymeric carbon nitride through a copolymerization approach – for gas-phase H_2S oxidation into elemental sulfur was recently reported by Lei et al. [259]. Based on DFT calculations, the authors propose H_2S and O_2 dissociative adsorptions at (probably separate) Fe–N_4 centers, leading to HS–Fe–N_4 and O* radical species, which in turn exothermally generate S and H_2O (Figure 9.8a).

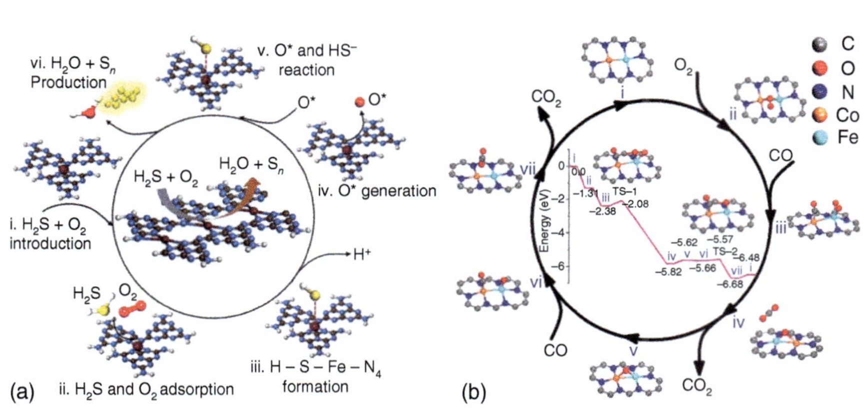

Figure 9.8 DFT calculation-based reactions schemes. (a) H_2S oxidation on Fe–N_4-carbon nitride. Source: Lei et al. [259]. Reproduced with permission from Wiley-VCH. (b) CO oxidation on Fe–Co–N–C. Source: Adapted with permission from Wang et al. [260]. Reproduced with permission from American Chemical Society.

Finally, CO oxidation, the usual gas-phase test reaction, was also carried out on carbon-based SACs. In particular, these support materials are ideally suited to elaborate bimetallic "double-atom catalysts" (see Chapter 15) and explore the catalytic performance of neighboring metal elements. Zhou et al. reported the stabilization of Pt–Ru dimers by forming Pt—C and Ru—N bonds in the triangular cavities of N-deficient g-C_3N_4 through an icing-assisted photocatalytic reduction method, leading to improved CO oxidation activity with respect to monometallic SA and double-atom counterparts [261]. Wang et al. synthesized a single-site catalyst consisting of N-coordinated Fe–Co dimers in a carbon matrix from the pyrolysis of an Fe-impregnated Zn/Co metal-organic framework, leading to CO oxidation activity at a temperature as low as −73 °C [260]. From *in situ* XAS, pulse-adsorption microcalorimetry, and DFT, the authors concluded that the reaction proceeds through the LH mechanism, with CO and O_2 preferentially adsorbing on Co and Fe atoms, respectively (Figure 9.8b).

9.3.2 Single-Atom Alloy Catalysts

SAAs [34, 262], which were previously referred to as site/atom-isolated or ultralow-loading alloys, can also be considered as SACs. They consist of isolated atoms of a metal at the surface of another metal, which is either in the form of an extended single crystal or of (supported) NP. The metal phase can be an intermetallic compound such as $Al_{13}Fe_4$ [263], or an alloy such as Ag–Cu dilute in Cu [264]. Although the SAA strategy is interesting to isolate active centers and tune their electronic structure, it has rarely been applied – at the notable exception of Au–Pd alloys – to oxidation reactions. One of the main reasons is the potential poisoning of the more reactive dopant atom in oxidative environments, which may render the dopant atom inactive. Several surface-science or powder-catalyst studies focused on CO oxidation over Au–Pd alloys dilute in Pd [265–268]. A matter of debate is the need for contiguous Pd surface atoms in the dissociative adsorption of O_2 and its reaction with CO through the LH mechanism. Ouyang et al. examined hydrogen peroxide synthesis from hydrogen oxidation over Au–Pd/TiO_2 catalysts [269]. Isolated Pd atoms were assumed to be active and selective sites for H_2O_2 formation, while contiguous Pd ensembles would be unselective, i.e. catalyze H_2O formation. Zhang et al. reported enhanced glucose oxidation performances for colloidal Au–Pd NPs presenting dilute, low-coordinated and negatively charged surface Au atoms [270]. Sun et al. found that photodepositing 0.05 wt% site-isolated Pd at the surface of 0.94 wt% Au_{NP} on TiO_2 was enough to reach a maximum activity for the aerobic solvent-free oxidation of benzyl alcohol to benzaldehyde [271]. The reader is referred to Chapter 4 of this book for more information on SAA catalysts.

9.4 Summary and Conclusions

A number of thermal oxidation reactions have been investigated over SACs, CO oxidation being by far the most studied one. A large part of this chapter has attempted to review the knowledge on the structure, performance, and CO oxidation mechanism of late transition metal atoms supported on various oxides. The latter part of the chapter highlighted interesting properties of SACs in PROX, WGS, and hydrocarbon combustion as well. In addition, oxometallate monomers supported on oxides and single metal atoms stabilized in N-doped carbon or carbon nitride are efficient for gas-phase and liquid-phase selective oxidations, respectively.

For the important aspect of SAC stability (both in terms of atom isolation and catalytic performance), it is possible to rank the metal/oxide systems as a function of the metal element across the periodic table and the reducibility of the oxide. While early transition metals supported on reducible oxides will typically remain isolated owing to stabilizing oxygen bridges, noble metals supported on non-reducible oxides are prone to diffusion and aggregation. Moreover, highly oxidizing reaction conditions are much more favorable to metal atom anchoring and isolation than reducing ones. In order to avoid metal atom clustering, a frequent approach has initially consisted in the use of very low metal loadings, which does not avoid surface diffusion but considerably reduces the probability of multimer formation by reducing the metal atom collision frequency. Complementary approaches use tiny support particles or strongly anchoring oxide patches onto a high-surface area support (e.g. CeO_2/Al_2O_3) for trapping SAs. In addition, recent SAC design strategies rely on the stabilization of SAs on their support or in their host through defect engineering or doping, as in the case of N-doped carbons.

Provided the stability criterion is fulfilled, another important question arises concerning the oxidation activity of SACs compared to nanocatalysts. The previous review of the literature for metal/oxide SACs shows that the answer not only depends on the considered catalytic system, but sometimes also on the initial oxidation state of the SAs. In some cases, SACs have been reported to be inactive, whereas their supported clusters or NPs counterparts are efficient. Thus, SACs should not be considered as the universal panacea. Beyond the possible decrease in metal content, prominent advantages of supported SACs lie in their potential to replace homogeneous catalysts (heterogenization) or noble-metal-based heterogeneous catalysts (metal substitution) in selected reactions. Promising results have been reported in that regard, and plausible reaction mechanisms – often involving O_v and the MvK mechanism for oxide-supported SACs – have been suggested. However, to go further, we believe that the SAC community urgently needs more systematic studies, e.g. by comparing several metals, both in the SA and NP forms, for a given support and reaction, together with in-depth characterization at least before and after reaction, if not *operando* (which is particularly relevant as the state of metal species dynamically responds to the chemical environment). This will guide theoretical modeling, which in turn will be in position to predict efficient SACs. This holds especially true for non-noble transition metals, which have been comparatively little investigated owing to their challenging preparation and characterization.

A final challenge faced in developing a predictive theory of SAC performance with variation in the metal and support composition is metal and support-dependent differences in the active site coordination environment. For example, if one assumes that Pt^{2+} prefers to exist in a fourfold square planar coordination to an oxide support, then the question is whether the support surface can allow this. For CeO_2 the answer seems to be yes, while for TiO_2 the geometry forms via additional OH species creates a strained fourfold coordination. Thus, the preferred coordination environment of each metal and each support is likely to be different, based on well-known coordination preferences from inorganic chemistry and the accessibility of these structures on a specific surface, making the existence of clear trends more challenging than for extended metal surfaces. This increased complexity, however, also presents increased opportunity for controlling catalytic functionality!

References

1 Liu, J. (2017). Catalysis by supported single metal atoms. *ACS Catalysis* 7: 34–59.

2 Mitchell, S., Vorobyeva, E., and Pérez-Ramírez, J. (2018). The multifaceted reactivity of single-atom heterogeneous catalysts. *Angewandte Chemie International Edition* 57: 15316–15329.

3 Zhang, H., Liu, G., Shi, L., and Ye, J. (2018). Single-atom catalysts: emerging multifunctional materials in heterogeneous catalysis. *Advanced Energy Materials* 8: 1701343.

4 Yan, H., Su, C., He, J., and Chen, W. (2018). Single-atom catalysts and their applications in organic chemistry. *Journal of Materials Chemistry A* 6: 8793–8814.

5 Cui, X., Li, W., Ryabchuk, P. et al. (2018). Bridging homogeneous and heterogeneous catalysis by heterogeneous single-metal-site catalysts. *Nature Catalysis* 1: 385–397.

6 Wang, A., Li, J., and Zhang, T. (2018). Heterogeneous single-atom catalysis. *Nature Reviews Chemistry* 2: 65–81.

7 Zhang, L., Ren, Y., Liu, W. et al. (2018). Single-atom catalyst: a rising star for green synthesis of fine chemicals. *National Science Review* 5: 653–672.

8 Liu, J., Bunes, B.R., Zang, L., and Wang, C. (2018). Supported single-atom catalysts: synthesis, characterization, properties, and applications. *Environmental Chemistry Letters* 16: 477–505.

9 Li, X., Yang, X., Huang, Y. et al. (2019). Supported noble-metal single atoms for heterogeneous catalysis. *Advanced Materials* 31: 1902031.

10 Sykes, E.C.H. and Christopher, P. (2020). Recent advances in single-atom catalysts and single-atom alloys: opportunities for exploring the uncharted phase space in-between. *Current Opinion in Chemical Engineering* 29: 67–73.

11 Kaiser, S.K., Chen, Z., Faust Akl, D. et al. (2020). Single-atom catalysts across the periodic table. *Chemical Reviews* 120: 11703–11809.

12 Thomas, J.M., Raja, R., and Lewis, D.W. (2005). Single-site heterogeneous catalysts. *Angewandte Chemie International Edition* 44: 6456–6482.

13 Copéret, C., Comas-Vives, A., Conley, M.P. et al. (2016). Surface organometallic and coordination chemistry toward single-site heterogeneous catalysts: strategies, methods, structures, and activities. *Chemical Reviews* 116: 323–421.

14 Samantaray, M.K., D'Elia, V., Pump, E. et al. (2020). The comparison between single atom catalysis and surface organometallic catalysis. *Chemical Reviews* 120: 734–813.

15 Yang, C. and Garland, C.W. (1957). Infrared studies of carbon monoxide chemisorbed on rhodium. *The Journal of Physical Chemistry* 61: 1504–1512.

16 Wovchko, E.A. and Yates, J.T. (1998). Activation of O_2 on a photochemically generated RhI site on an Al_2O_3 surface: low-temperature O_2 dissociation and CO oxidation. *Journal of the American Chemical Society* 120: 10523–10527.

17 Hackett, S.F.J., Brydson, R.M., Gass, M.H. et al. (2007). High-activity, single-site mesoporous Pd/Al_2O_3 catalysts for selective aerobic oxidation of allylic alcohols. *Angewandte Chemie International Edition* 46: 8593–8596.

18 Asakura, K., Nagahiro, H., Ichikuni, N., and Iwasawa, Y. (1999). Structure and catalytic combustion activity of atomically dispersed Pt species at MgO surface. *Applied Catalysis A: General* 188: 313–324.

19 Abbet, S., Sanchez, A., Heiz, U. et al. (2000). Acetylene cyclotrimerization on supported size-selected Pdn clusters ($1 \leq n \leq 30$): one atom is enough! *Journal of the American Chemical Society* 122: 3453–3457.

20 Abbet, S., Heiz, U., Häkkinen, H., and Landman, U. (2001). CO oxidation on a single Pd atom supported on magnesia. *Physical Review Letters* 86: 5950–5953.

21 Fu, Q., Saltsburg, H., and Flytzani-Stephanopoulos, M. (2003). Active non-metallic Au and Pt species on ceria-based water-gas shift catalysts. *Science* 301: 935–938.

22 Aguilar-Guerrero, V. and Gates, B.C. (2008). Kinetics of CO oxidation catalyzed by highly dispersed CeO_2-supported gold. *Journal of Catalysis* 260: 351–357.

23 Tang, W., Hu, Z., Wang, M. et al. (2010). Methane complete and partial oxidation catalyzed by Pt-doped CeO_2. *Journal of Catalysis* 273: 125–137.

24 Zhang, X., Shi, H., and Xu, B.-Q. (2005). Catalysis by gold: isolated surface Au^{3+} ions are active sites for selective hydrogenation of 1,3-butadiene over Au/ZrO_2 catalysts. *Angewandte Chemie International Edition* 44: 7132–7135.

25 Qiao, B., Wang, A., Yang, X. et al. (2011). Single-atom catalysis of CO oxidation using Pt_1/FeO_x. *Nature Chemistry* 3: 634–641.

26 Piccolo, L. (2021). Restructuring effects of the chemical environment in metal nanocatalysis and single-atom catalysis. *Catalysis Today* 373: 80–97.

27 Liu, Q. and Zhang, Z. (2019). Platinum single-atom catalysts: a comparative review towards effective characterization. *Catalysis Science & Technology* 9: 4821–4834.

28 Moses-DeBusk, M., Yoon, M., Allard, L.F. et al. (2013). CO oxidation on supported single Pt atoms: experimental and ab initio density functional studies of CO interaction with Pt atom on θ-Al_2O_3(010) surface. *Journal of the American Chemical Society* 135: 12634–12645.

29 Yang, M., Allard, L.F., and Flytzani-Stephanopoulos, M. (2013). Atomically dispersed $Au-(OH)_x$ species bound on titania catalyze the low-temperature water-gas shift reaction. *Journal of the American Chemical Society* 135: 3768–3771.

30 Lin, J., Wang, A., Qiao, B. et al. (2013). Remarkable performance of Ir_1/FeOx single-atom catalyst in water gas shift reaction. *Journal of the American Chemical Society* 135: 15314–15317.

31 Mastikhin, V.M., Lapina, O.B., Balzhinimaev, B.S. et al. (1987). Catalytically active complexes and influence of SiO_2 on the catalytic properties of the active component of vanadium catalysts for SO_2 oxidation. *Journal of Catalysis* 103: 160–169.

32 Hardcastle, F.D., Wachs, I.E., Horsley, J.A., and Via, G.H. (1988). The structure of surface rhenium oxide on alumina from laser Raman spectroscopy and X-ray absorption near-edge spectroscopy. *Journal of Molecular Catalysis* 46: 15–36.

33 Hardcastle, F.D. and Wachs, I.E. (1990). Raman spectroscopy of chromium oxide supported on alumina, titania, and silica: a comparative study. In: *Olefin Metathesis and Polymerization Catalysts: Synthesis, Mechanism and Utilization*, NATO ASI Series (eds. Y. İmamoğlu, B. Zümreoğlu-Karan and A.J. Amass), 545–545. Dordrecht: Springer Netherlands.

34 Hannagan, R.T., Giannakakis, G., Flytzani-Stephanopoulos, M., and Sykes, E.C.H. (2020). Single-atom alloy catalysis. *Chemical Reviews* 120: 12044–12088.

35 Wei, H., Liu, X., Wang, A. et al. (2014). FeO_x-supported platinum single-atom and pseudo-single-atom catalysts for chemoselective hydrogenation of functionalized nitroarenes. *Nature Communications* 5: 5634.

36 Zhou, J., Xu, Z., Xu, M. et al. (2020). A perspective on oxide-supported single-atom catalysts. *Nanoscale Advances* 2: 3624–3631.

37 Lang, R., Du, X., Huang, Y. et al. (2020). Single-atom catalysts based on the metal–oxide interaction. *Chemical Reviews* 120: 11986–12043.

38 Hemmingson, S.L. and Campbell, C.T. (2017). Trends in adhesion energies of metal nanoparticles on oxide surfaces: understanding support effects in catalysis and nanotechnology. *ACS Nano* 11: 1196–1203.

39 O'Connor, N.J., Jonayat, A.S.M., Janik, M.J., and Senftle, T.P. (2018). Interaction trends between single metal atoms and oxide supports identified with density functional theory and statistical learning. *Nature Catalysis* 1: 531–539.

40 Dietze, E.M. and Plessow, P.N. (2019). Predicting the strength of metal–support interaction with computational descriptors for adhesion energies. *Journal of Physical Chemistry C* 123: 20443–20450.

41 Tan, K., Dixit, M., Dean, J., and Mpourmpakis, G. (2019). Predicting metal–support interactions in oxide-supported single-atom catalysts. *Industrial and Engineering Chemistry Research* 58: 20236–20246.

42 Mars, P. and van Krevelen, D.W. (1954). Oxidations carried out by means of vanadium oxide catalysts. *Chemical Engineering Science* 3: 41–59.

43 Langmuir, I. (1922). The mechanism of the catalytic action of platinum in the reactions $2CO + O_2 = 2CO_2$ and $2H_2 + O_2 = 2H_2O$. *Transactions of the Faraday Society* 17: 621–654.

44 Beniya, A. and Higashi, S. (2019). Towards dense single-atom catalysts for future automotive applications. *Nature Catalysis* 2: 590–602.

45 Lambert, C.K. (2019). Current state of the art and future needs for automotive exhaust catalysis. *Nature Catalysis* 2: 554–557.

46 Getsoian, A.(Bean), Theis, J.R., Paxton, W.A. et al. (2019). Remarkable improvement in low temperature performance of model three-way catalysts through solution atomic layer deposition. *Nature Catalysis* 2: 614–622.

47 Zhang, H., Fang, S., and Hu, Y.H. (2021). Recent advances in single-atom catalysts for CO oxidation. *Catalysis Reviews: Science and Engineering*, in press. https://doi.org/10.1080/01614940.2020.1821443.

48 Lu, Y., Zhang, Z., Lin, F. et al. (2020). Single-atom automobile exhaust catalysts. *ChemNanoMat* 6: 1659–1682.

49 Datye, A.K. and Votsmeier, M. (2021). Opportunities and challenges in the development of advanced materials for emission control catalysts. *Nature Materials*, 20: 1049–1059. https://doi.org/10.1038/s41563-020-00805-3

50 Doherty, F., Wang, H., Yang, M., and Goldsmith, B.R. (2020). Nanocluster and single-atom catalysts for thermocatalytic conversion of CO and CO_2. *Catalysis Science & Technology* 10: 5772–5791.

51 Yates, J.T., Duncan, T.M., Worley, S.D., and Vaughan, R.W. (1979). Infrared spectra of chemisorbed CO on Rh. *The Journal of Chemical Physics* 70: 1219–1224.

52 Ghosh, T.K. and Nair, N.N. (2013). Rh1/γ-Al_2O_3 single-atom catalysis of O_2 activation and CO oxidation: mechanism, effects of hydration, oxidation state, and cluster size. *ChemCatChem* 5: 1811–1821.

53 Dessal, C., Len, T., Morfin, F. et al. (2019). Dynamics of single Pt atoms on alumina during CO oxidation monitored by operando X-ray and infrared spectroscopies. *ACS Catalysis* 9: 5752–5759.

54 Dessal, C., Sangnier, A., Chizallet, C. et al. (2019). Atmosphere-dependent stability and mobility of catalytic Pt single atoms and clusters on γ-Al_2O_3. *Nanoscale* 11: 6897–6904.

55 Lou, Y. and Liu, J. (2017). CO oxidation on metal oxide supported single Pt atoms: the role of the support. *Industrial and Engineering Chemistry Research* 56: 6916–6925.

56 Newton, M.A., Ferri, D., Smolentsev, G. et al. (2015). Room-temperature carbon monoxide oxidation by oxygen over Pt/Al_2O_3 mediated by reactive platinum carbonates. *Nature Communications* 6: 8675.

57 Newton, M.A., Ferri, D., Smolentsev, G. et al. (2016). Kinetic studies of the Pt carbonate-mediated, room-temperature oxidation of carbon monoxide by oxygen over Pt/Al_2O_3 using combined, time-resolved XAFS, DRIFTS, and mass spectrometry. *Journal of the American Chemical Society* 138: 13930–13940.

58 Ding, K., Gulec, A., Johnson, A.M. et al. (2015). Identification of active sites in CO oxidation and water-gas shift over supported Pt catalysts. *Science* 350: 189–192.

59 Li, H., Wang, M., Luo, L., and Zeng, J. (2019). Static regulation and dynamic evolution of single-atom catalysts in thermal catalytic reactions. *Advanced Science* 6: 1801471.

60 Liu, L. and Corma, A. (2020). Evolution of isolated atoms and clusters in catalysis. *Trends in Chemistry* 2: 383–400.

61 Zhang, Z., Zhu, Y., Asakura, H. et al. (2017). Thermally stable single atom Pt/m-Al_2O_3 for selective hydrogenation and CO oxidation. *Nature Communications* 8: 16100.

62 Kwak, J.H., Hu, J., Mei, D. et al. (2009). Coordinatively unsaturated Al^{3+} centers as binding sites for active catalyst phases of platinum on γ-Al_2O_3. *Science* 325: 1670–1673.

63 Wang, H., Dong, J., Allard, L.F. et al. (2019). Single-site Pt/La-Al_2O_3 stabilized by barium as an active and stable catalyst in purifying CO and C_3H_6 emissions. *Applied Catalysis B: Environmental* 244: 327–339.

64 Peterson, E.J., DeLaRiva, A.T., Lin, S. et al. (2014). Low-temperature carbon monoxide oxidation catalysed by regenerable atomically dispersed palladium on alumina. *Nature Communications* 5: 4885.

65 Liang, J.-X., Lin, J., Yang, X.-F. et al. (2014). Theoretical and experimental investigations on single-atom catalysis: Ir_1/FeO_x for CO oxidation. *Journal of Physical Chemistry C* 118: 21945–21951.

66 Lin, J., Chen, Y., Zhou, Y. et al. (2017). More active Ir subnanometer clusters than single-atoms for catalytic oxidation of CO at low temperature. *AIChE Journal* 63: 4003–4012.

67 Chen, W., Ma, Y., Li, F. et al. (2019). Strong electronic interaction of amorphous Fe_2O_3 nanosheets with single-atom Pt toward enhanced carbon monoxide oxidation. *Advanced Functional Materials* 7: 1904278.

68 Bliem, R., van der Hoeven, J., Zavodny, A. et al. (2015). An atomic-scale view of CO and H_2 oxidation on a Pt/Fe_3O_4 model catalyst. *Angewandte Chemie International Edition* 54: 13999–14002.

69 Jakub, Z., Hulva, J., Ryan, P.T.P. et al. (2020). Adsorbate-induced structural evolution changes the mechanism of CO oxidation on a Rh/Fe_3O_4(001) model catalyst. *Nanoscale* 12: 5866–5875.

70 Tauster, S.J., Fung, S.C., and Garten, R.L. (1978). Strong metal-support interactions. Group 8 noble metals supported on titanium dioxide. *Journal of the American Chemical Society* 100: 170–175.

71 Haruta, M. (1997). Size- and support-dependency in the catalysis of gold. *Catalysis Today* 36: 153–166.

72 DeRita, L., Dai, S., Lopez-Zepeda, K. et al. (2017). Catalyst architecture for stable single atom dispersion enables site-specific spectroscopic and reactivity measurements of CO adsorbed to Pt atoms, oxidized Pt clusters, and metallic Pt clusters on TiO_2. *Journal of the American Chemical Society* 139: 14150–14165.

73 Thang, H.V., Pacchioni, G., DeRita, L., and Christopher, P. (2018). Nature of stable single atom Pt catalysts dispersed on anatase TiO_2. *Journal of Catalysis* 367: 104–114.

74 DeRita, L., Resasco, J., Dai, S. et al. (2019). Structural evolution of atomically dispersed Pt catalysts dictates reactivity. *Nature Materials* 18: 746.

75 Liu, P., Zhao, Y., Qin, R. et al. (2018). A vicinal effect for promoting catalysis of Pd_1/TiO_2: supports of atomically dispersed catalysts play more roles than simply serving as ligands. *Science Bulletin* 63: 675–682.

76 Hoffman, A.S., Fang, C.-Y., and Gates, B.C. (2016). Homogeneity of surface sites in supported single-site metal catalysts: assessment with band widths of metal carbonyl infrared spectra. *Journal of Physical Chemistry Letters* 7: 3854–3860.

77 Hadjiivanov, K.I. and Vayssilov, G.N. (2002). Characterization of oxide surfaces and zeolites by carbon monoxide as an IR probe molecule. In: *Advances in Catalysis*, vol. 47, 307–511. Academic Press.

78 Dessal, C., Martínez, L., Maheu, C. et al. (2019). Influence of Pt particle size and reaction phase on the photocatalytic performances of ultradispersed Pt/TiO_2 catalysts for hydrogen evolution. *Journal of Catalysis* 375: 155–163.

79 Liu, L., Meira, D.M., Arenal, R. et al. (2019). Determination of the evolution of heterogeneous single metal atoms and nanoclusters under reaction conditions: which are the working catalytic sites? *ACS Catalysis* 9: 10626–10639.

80 Humphrey, N., Bac, S., and Mallikarjun Sharada, S. (2020). Ab initio molecular dynamics reveals new metal-binding sites in atomically dispersed Pt_1/TiO_2 catalysts. *Journal of Physical Chemistry C* 124: 24187–24195.

81 Han, B., Guo, Y., Huang, Y. et al. (2020). Strong metal–support interactions between Pt single atoms and TiO_2. *Angewandte Chemie International Edition* 132: 11922–11927.

82 Piccolo, L., Afanasiev, P., Morfin, F. et al. (2020). Operando X-ray absorption spectroscopy investigation of photocatalytic hydrogen evolution over ultradispersed Pt/TiO_2 catalysts. *ACS Catalysis* 10: 12696–12705.

83 Lee, B.-H., Park, S., Kim, M. et al. (2019). Reversible and cooperative photoactivation of single-atom Cu/TiO_2 photocatalysts. *Nature Materials* 18: 620–626.

84 Tang, Y., Asokan, C., Xu, M. et al. (2019). Rh single atoms on TiO_2 dynamically respond to reaction conditions by adapting their site. *Nature Communications* 10: 4488.

85 Hoang, S., Guo, Y., Binder, A.J. et al. (2020). Activating low-temperature diesel oxidation by single-atom Pt on TiO_2 nanowire array. *Nature Communications* 11: 1062.

86 Shi, J.L., Zhao, X.J., Zhang, L.Y. et al. (2017). An oxidized magnetic Au single atom on doped TiO_2(110) becomes a high performance CO oxidation catalyst due to the charge effect. *Journal of Materials Chemistry A* 5: 19316–19322.

87 Yoo, M., Yu, Y.-S., Ha, H. et al. (2020). A tailored oxide interface creates dense Pt single-atom catalysts with high catalytic activity. *Energy & Environmental Science* 13: 1231–1239.

88 Agarwal, S., Mojet, B.L., Lefferts, L., and Datye, A.K. (2015). Ceria nanoshapes—structural and catalytic properties. In: *Catalysis by Materials with Well-Defined Structures*, Chapter 2 (eds. Z. Wu and S.H. Overbury), 31–70. Amsterdam: Elsevier.

89 Mann, A.K.P., Wu, Z., and Overbury, S.H. (2015). The characterization and structure-dependent catalysis of ceria with well-defined facets. In: *Catalysis by Materials with Well-Defined Structures*, Chapter 3 (eds. Z. Wu and S.H. Overbury), 71–97. Amsterdam: Elsevier.

90 Montini, T., Melchionna, M., Monai, M., and Fornasiero, P. (2016). Fundamentals and catalytic applications of CeO_2-based materials. *Chemical Reviews* 116: 5987–6041.

91 Nguyen, T.-S., Postole, G., Loridant, S. et al. (2014). Ultrastable iridium-ceria nanopowders synthesized in one step by solution combustion for catalytic hydrogen production. *Journal of Materials Chemistry A* 2: 19822–19832.

92 Li, P., Chen, X., Li, Y., and Schwank, J.W. (2019). A review on oxygen storage capacity of CeO_2-based materials: influence factors, measurement techniques, and applications in reactions related to catalytic automotive emissions control. *Catalysis Today* 327: 90–115.

93 Martin, D. and Duprez, D. (1996). Mobility of surface species on oxides. 1. Isotopic exchange of $^{18}O_2$ with ^{16}O of SiO_2, Al_2O_3, ZrO_2, MgO, CeO_2, and CeO_2-Al_2O_3. Activation by noble metals. Correlation with oxide basicity. *The Journal of Physical Chemistry* 100: 9429–9438.

94 Paier, J., Penschke, C., and Sauer, J. (2013). Oxygen defects and surface chemistry of ceria: quantum chemical studies compared to experiment. *Chemical Reviews* 113: 3949–3985.

95 Ma, Y., Gao, W., Zhang, Z. et al. (2018). Regulating the surface of nanoceria and its applications in heterogeneous catalysis. *Surface Science Reports* 73: 1–36.

96 Lafaye, G., Barbier, J., and Duprez, D. (2015). Impact of cerium-based support oxides in catalytic wet air oxidation: conflicting role of redox and acid–base properties. *Catalysis Today* 253: 89–98.

97 Nguyen, T.-S., Morfin, F., Aouine, M. et al. (2015). Trends in the CO oxidation and PROX performances of the platinum-group metals supported on ceria. *Catalysis Today* 253: 106–114.

98 Morfin, F., Nguyen, T.-S., Rousset, J.-L., and Piccolo, L. (2016). Synergy between hydrogen and ceria in Pt-catalyzed CO oxidation: an investigation on Pt–CeO_2 catalysts synthesized by solution combustion. *Applied Catalysis B: Environmental* 197: 2–13.

99 Pereira-Hernández, X.I., DeLaRiva, A., Muravev, V. et al. (2019). Tuning Pt-CeO_2 interactions by high-temperature vapor-phase synthesis for improved reducibility of lattice oxygen. *Nature Communications* 10: 1358.

100 Ye, X., Wang, H., Lin, Y. et al. (2019). Insight of the stability and activity of platinum single atoms on ceria. *Nano Research* 12: 1401–1409.

101 Jones, J., Xiong, H., DeLaRiva, A.T. et al. (2016). Thermally stable single-atom platinum-on-ceria catalysts via atom trapping. *Science* 353: 150–154.

102 Yu, W.-Z., Wang, W.-W., Li, S.-Q. et al. (2019). Construction of active site in a sintered copper–ceria nanorod catalyst. *Journal of the American Chemical Society* 141: 17548–17557.

103 Jiang, D., Wan, G., García-Vargas, C.E. et al. (2020). Elucidation of the active sites in single-atom Pd_1/CeO_2 catalysts for low-temperature CO oxidation. *ACS Catalysis* 10: 11356–11364.

104 Chen, J., Wanyan, Y., Zeng, J. et al. (2018). Surface engineering protocol to obtain an atomically dispersed Pt/CeO_2 catalyst with high activity and stability for CO oxidation. *ACS Sustainable Chemistry & Engineering* 6: 14054–14062.

105 Feng, Y., Wan, Q., Xiong, H. et al. (2018). Correlating DFT calculations with CO oxidation reactivity on Ga-doped Pt/CeO_2 single-atom catalysts. *Journal of Physical Chemistry C* 122: 22460–22468.

106 Vieira, L.H., Assaf, J.M., and Assaf, E.M. (2020). Stabilization of atomically dispersed rhodium sites on ceria-based supports under reaction conditions probed by in situ infrared spectroscopy. *Materials Letters* 277: 128354.

107 Tan, W., Alsenani, H., Xie, S. et al. (2020). Tuning single-atom Pt_1-CeO_2 catalyst for efficient CO and C_3H_6 oxidation: size effect of ceria on Pt structural evolution. *ChemNanoMat* 6: 1797–1805.

108 Camellone, M.F. and Fabris, S. (2009). Reaction mechanisms for the CO oxidation on Au/CeO_2 catalysts: activity of substitutional Au^{3+}/Au^+ cations and deactivation of supported Au^+ adatoms. *Journal of the American Chemical Society* 131: 10473–10483.

109 Kunwar, D., Zhou, S., DeLaRiva, A. et al. (2019). Stabilizing high metal loadings of thermally stable platinum single atoms on an industrial catalyst support. *ACS Catalysis* 9: 3978–3990.

110 Bruix, A., Lykhach, Y., Matolínová, I. et al. (2014). Maximum noble-metal efficiency in catalytic materials: atomically dispersed surface platinum. *Angewandte Chemie International Edition* 53: 10525–10530.

111 Nie, L., Mei, D., Xiong, H. et al. (2017). Activation of surface lattice oxygen in single-atom Pt/CeO_2 for low-temperature CO oxidation. *Science* 358: 1419–1423.

112 Wang, C., Gu, X.-K., Yan, H. et al. (2017). Water-mediated mars–Van Krevelen mechanism for CO oxidation on ceria-supported single-atom Pt_1 catalyst. *ACS Catalysis* 7: 887–891.

113 Maurer, F., Jelic, J., Wang, J. et al. (2020). Tracking the formation, fate and consequence for catalytic activity of Pt single sites on CeO_2. *Nature Catalysis* 3: 824–833.

114 Aleksandrov, H.A., Neyman, K.M., Hadjiivanov, K.I., and Vayssilov, G.N. (2016). Can the state of platinum species be unambiguously determined by the stretching frequency of an adsorbed CO probe molecule? *Physical Chemistry Chemical Physics* 18: 22108–22121.

115 Resasco, J., DeRita, L., Dai, S. et al. (2020). Uniformity is key in defining structure–function relationships for atomically dispersed metal catalysts: the case of Pt/CeO_2. *Journal of the American Chemical Society* 142: 169–184.

116 Jeong, H., Shin, D., Kim, B.-S. et al. (2020). Controlling the oxidation state of Pt single atoms for maximizing catalytic activity. *Angewandte Chemie International Edition* 59: 20691–20696.

117 Daelman, N., Capdevila-Cortada, M., and López, N. (2019). Dynamic charge and oxidation state of Pt/CeO_2 single-atom catalysts. *Nature Materials* 18: 1215–1221.

118 Alexopoulos, K., Wang, Y., and Vlachos, D.G. (2019). First-principles kinetic and spectroscopic insights into single-atom catalysis. *ACS Catalysis* 9: 5002–5010.

119 Lee, J., Ryou, Y., Kim, J. et al. (2018). Influence of the defect concentration of ceria on the Pt dispersion and the CO oxidation activity of Pt/CeO_2. *Journal of Physical Chemistry C* 122: 4972–4983.

120 Jiang, Z., Jing, M., Feng, X. et al. (2020). Stabilizing platinum atoms on CeO_2 oxygen vacancies by metal-support interaction induced interface distortion: mechanism and application. *Applied Catalysis B: Environmental* 278: 119304.

121 Xie, K., Xiao, Y., and Li, H. (2021). Activating lattice oxygen at twisted surface in mesoporous CeO_2 single crystal for efficient and durable catalytic CO oxidation. *Angewandte Chemie International Edition* 60: 5240–5244.

122 Nagai, Y., Hirabayashi, T., Dohmae, K. et al. (2006). Sintering inhibition mechanism of platinum supported on ceria-based oxide and Pt-oxide–support interaction. *Journal of Catalysis* 242: 103–109.

123 Ma, Y., Chi, B., Liu, W. et al. (2019). Tailoring of the proximity of platinum single atoms on CeO_2 using phosphorus boosts the hydrogenation activity. *ACS Catalysis* 9: 8404–8412.

124 Qiao, B., Liu, J., Wang, Y.-G. et al. (2015). Highly efficient catalysis of preferential oxidation of CO in H_2-rich stream by gold single-atom catalysts. *ACS Catalysis* 5: 6249–6254.

125 Tang, Y., Wang, Y.-G., and Li, J. (2017). Theoretical investigations of Pt_1@CeO_2 single-atom catalyst for CO oxidation. *Journal of Physical Chemistry C* 121: 11281–11289.

126 Han, B., Li, T., Zhang, J. et al. (2020). A highly active Rh_1/CeO_2 single-atom catalyst for low-temperature CO oxidation. *Chemical Communications* 56: 4870–4873.

127 Zhao, S., Chen, F., Duan, S. et al. (2019). Remarkable active-site dependent H_2O promoting effect in CO oxidation. *Nature Communications* 10: 3824.

128 Hegde, M.S. and Bera, P. (2015). Noble metal ion substituted CeO_2 catalysts: electronic interaction between noble metal ions and CeO_2 lattice. *Catalysis Today* 253: 40–50.

129 Kibis, L.S., Kardash, T.Y., Derevyannikova, E.A. et al. (2017). Redox and catalytic properties of $Rh_xCe_{1-x}O_{2-\delta}$ solid solution. *Journal of Physical Chemistry C* 121: 26925–26938.

130 Wang, H., Liu, J.-X., Allard, L.F. et al. (2019). Surpassing the single-atom catalytic activity limit through paired Pt-O-Pt ensemble built from isolated Pt_1 atoms. *Nature Communications* 10: 3808.

131 Meunier, F.C., Cardenas, L., Kaper, H. et al. (2021). Synergy between metallic and oxidized Pt sites unravelled during room temperature CO oxidation on Pt/ceria. *Angewandte Chemie International Edition* 60: 3799–3805.

132 Guo, L.-W., Du, P.-P., Fu, X.-P. et al. (2016). Contributions of distinct gold species to catalytic reactivity for carbon monoxide oxidation. *Nature Communications* 7: 13481.

133 Gänzler, A.M., Casapu, M., Vernoux, P. et al. (2017). Tuning the structure of platinum particles on ceria in situ for enhancing the catalytic performance of exhaust gas catalysts. *Angewandte Chemie International Edition* 56: 13078–13082.

134 Gänzler, A.M., Casapu, M., Maurer, F. et al. (2018). Tuning the Pt/CeO_2 interface by in situ variation of the Pt particle size. *ACS Catalysis* 8: 4800–4811.

135 Ferré, G., Aouine, M., Bosselet, F. et al. (2020). Exploiting the dynamic properties of Pt on ceria for low-temperature CO oxidation. *Catalysis Science & Technology* 10: 3904–3917.

136 Sarma, B.B., Plessow, P.N., Agostini, G. et al. (2020). Metal-specific reactivity in single-atom catalysts: CO oxidation on 4d and 5d transition metals atomically dispersed on MgO. *Journal of the American Chemical Society* 142: 14890–14902.

137 Zhou, X., Yang, W., Chen, Q. et al. (2016). Stable Pt single atoms and nanoclusters on ultrathin CuO film and their performances in CO oxidation. *Journal of Physical Chemistry C* 120: 1709–1715.

138 Therrien, A.J., Hensley, A.J.R., Marcinkowski, M.D. et al. (2018). An atomic-scale view of single-site Pt catalysis for low-temperature CO oxidation. *Nature Catalysis* 1: 192–198.

139 Zhou, X., Shen, Q., Yuan, K. et al. (2018). Unraveling charge state of supported Au single-atoms during CO oxidation. *Journal of the American Chemical Society* 140: 554–557.

140 Wang, C., Tissot, H., Stenlid, J.H. et al. (2019). High-density isolated Fe_1O_3 sites on a single-crystal $Cu_2O(100)$ surface. *Journal of Physical Chemistry Letters* 10: 7318–7323.

141 Kropp, T., Lu, Z., Li, Z. et al. (2019). Anionic single-atom catalysts for CO oxidation: support-independent activity at low temperatures. *ACS Catalysis* 9: 1595–1604.

142 Lu, Y., Wang, J., Yu, L. et al. (2019). Identification of the active complex for CO oxidation over single-atom Ir-on-$MgAl_2O_4$ catalysts. *Nature Catalysis* 2: 149.

143 Lou, Y., Cai, Y., Hu, W. et al. (2020). Identification of active area as active center for CO oxidation over single Au atom catalyst. *ACS Catalysis* 10: 6094–6101.

144 Xu, Z., Zhang, Y., Qin, L. et al. (2020). Crystal facet induced single-atom Pd/Co_xO_y on a tunable metal–support interface for low temperature catalytic oxidation. *Small* 16: 2002071.

145 Xu, H., Zhang, Z., Liu, J. et al. (2020). Entropy-stabilized single-atom Pd catalysts via high-entropy fluorite oxide supports. *Nature Communications* 11: 3908.

146 Wang, T., Xing, J.-Y., Jia, A.-P. et al. (2020). CO oxidation over Pt/$Cr_{1.3}Fe_{0.7}O_3$ catalysts: enhanced activity on single Pt atom by H_2O promotion. *Journal of Catalysis* 382: 192–203.

147 Lu, Y., Kuo, C.-T., Kovarik, L. et al. (2019). A versatile approach for quantification of surface site fractions using reaction kinetics: the case of CO oxidation on supported Ir single atoms and nanoparticles. *Journal of Catalysis* 378: 121–130.

148 Resasco, J., Yang, F., Mou, T. et al. (2020). Relationship between atomic scale structure and reactivity of Pt catalysts: hydrodeoxygenation of m-cresol over isolated Pt cations and clusters. *ACS Catalysis* 10: 595–603.

149 Xu, H., Xu, C.-Q., Cheng, D., and Li, J. (2017). Identification of activity trends for CO oxidation on supported transition-metal single-atom catalysts. *Catalysis Science & Technology* 7: 5860–5871.

150 Gan, T., He, Q., Zhang, H. et al. (2020). Unveiling the kilogram-scale gold single-atom catalysts via ball milling for preferential oxidation of CO in excess hydrogen. *Chemical Engineering Journal* 389: 124490.

151 Lin, J., Qiao, B., Liu, J. et al. (2012). Design of a highly active $Ir/Fe(OH)_X$ catalyst: versatile application of Pt-group metals for the preferential oxidation of carbon monoxide. *Angewandte Chemie International Edition* 51: 2920–2924.

152 Guan, H., Lin, J., Li, L. et al. (2016). Highly active subnano $Rh/Fe(OH)_x$ catalyst for preferential oxidation of CO in H_2-rich stream. *Applied Catalysis B: Environmental* 184: 299–308.

153 Cao, L., Liu, W., Luo, Q. et al. (2019). Atomically dispersed iron hydroxide anchored on Pt for preferential oxidation of CO in H_2. *Nature* 565: 631.

154 Zhang, H., Liu, X., Zhang, N. et al. (2016). Construction of ultrafine and stable PtFe nano-alloy with ultra-low Pt loading for complete removal of CO in PROX at room temperature. *Applied Catalysis B: Environmental* 180: 237–245.

155 Cao, S., Zhao, Y., Lee, S. et al. (2020). High-loading single Pt atom sites [$Pt\text{-}O(OH)_X$] catalyze the CO PROX reaction with high activity and selectivity at mild conditions. *Science Advances* 6: eaba3809.

156 Quinet, E., Piccolo, L., Morfin, F. et al. (2009). On the mechanism of hydrogen-promoted gold-catalyzed CO oxidation. *Journal of Catalysis* 268: 384–389.

157 Pal, D.B., Chand, R., Upadhyay, S.N., and Mishra, P.K. (2018). Performance of water gas shift reaction catalysts: a review. *Renewable and Sustainable Energy Reviews* 93: 549–565.

158 Saeidi, S., Fazlollahi, F., Najari, S. et al. (2017). Hydrogen production: perspectives, separation with special emphasis on kinetics of WGS reaction: a state-of-the-art review. *Journal of Industrial and Engineering Chemistry* 49: 1–25.

159 Zhu, M. and Wachs, I.E. (2016). Iron-based catalysts for the high-temperature water-gas shift (HT-WGS) reaction: a review. *ACS Catalysis* 6: 722–732.

160 Gradisher, L., Dutcher, B., and Fan, M. (2015). Catalytic hydrogen production from fossil fuels via the water gas shift reaction. *Applied Energy* 139: 335–349.

161 Haruta, M. (2011). Spiers memorial lecture. *Faraday Discussions* 152: 11–32.

162 Sakurai, H., Akita, T., Tsubota, S. et al. (2005). Low-temperature activity of Au/CeO_2 for water gas shift reaction, and characterization by ADF-STEM, temperature-programmed reaction, and pulse reaction. *Applied Catalysis A: General* 291: 179–187.

163 Oyama, S.T., Gaudet, J., Zhang, W. et al. (2010). Platinum-like catalytic behavior of Au^+. *ChemCatChem* 2: 1582–1586.

164 Gaudet, J., Bando, K.K., Song, Z. et al. (2011). Effect of gold oxidation state on the epoxidation and hydrogenation of propylene on Au/TS-1. *Journal of Catalysis* 280: 40–49.

165 Zhai, Y., Pierre, D., Si, R. et al. (2010). Alkali-stabilized Pt-OH_x species catalyze low-temperature water-gas shift reactions. *Science* 329: 1633–1636.

166 Flytzani-Stephanopoulos, M. (2014). Gold atoms stabilized on various supports catalyze the water-gas shift reaction. *Accounts of Chemical Research* 47: 783–792.

167 Yang, M., Li, S., Wang, Y. et al. (2014). Catalytically active Au-$O(OH)_X$- species stabilized by alkali ions on zeolites and mesoporous oxides. *Science* 346: 1498–1501.

168 Yang, M., Liu, J., Lee, S. et al. (2015). A common single-site Pt(II)–$O(OH)_X$– species stabilized by sodium on "active" and "inert" supports catalyzes the water-gas shift reaction. *Journal of the American Chemical Society* 137: 3470–3473.

169 Ammal, S.C. and Heyden, A. (2019). Understanding the nature and activity of supported platinum catalysts for the water-gas shift reaction: from metallic nanoclusters to alkali-stabilized single-atom cations. *ACS Catalysis* 9: 7721–7740.

170 Wang, C., Yang, M., and Flytzani-Stephanopoulos, M. (2016). Single gold atoms stabilized on nanoscale metal oxide supports are catalytic active centers for various reactions. *AIChE Journal* 62: 429–439.

171 Kuai, L., Liu, S., Cao, S. et al. (2018). Atomically dispersed Pt/metal oxide mesoporous catalysts from synchronous pyrolysis–deposition route for water-gas shift reaction. *Chemistry of Materials* 30: 5534–5538.

172 Yang, M. and Flytzani-Stephanopoulos, M. (2017). Design of single-atom metal catalysts on various supports for the low-temperature water-gas shift reaction. *Catalysis Today* 298: 216–225.

173 Yi, N., Si, R., Saltsburg, H., and Flytzani-Stephanopoulos, M. (2010). Active gold species on cerium oxide nanoshapes for methanol steam reforming and the water gas shift reactions. *Energy & Environmental Science* 3: 831–837.

174 Xiang, Y., He, J., Sun, N. et al. (2020). Hollow mesoporous CeO_2 microspheres for efficient loading of Au single-atoms to catalyze the water-gas shift reaction. *Microporous and Mesoporous Materials* 308: 110507.

175 Guan, H., Lin, J., Qiao, B. et al. (2017). Enhanced performance of Rh_1/TiO_2 catalyst without methanation in water-gas shift reaction. *AIChE Journal* 63: 2081–2088.

176 Sun, X., Lin, J., Zhou, Y. et al. (2017). FeO_x supported single-atom Pd bifunctional catalyst for water gas shift reaction. *AIChE Journal* 63: 4022–4031.

177 Gai, P.L., Yoshida, K., Ward, M.R. et al. (2016). Visualisation of single atom dynamics in water gas shift reaction for hydrogen generation. *Catalysis Science & Technology* 6: 2214–2227.

178 Nelson, N.C., Chen, L., Meira, D. et al. (2020). In situ dispersion of palladium on TiO_2 during reverse water-gas shift reaction: formation of atomically dispersed palladium. *Angewandte Chemie International Edition* 59: 17657–17663.

179 Chen, Y., Lin, J., Li, L. et al. (2018). Identifying size effects of Pt as single atoms and nanoparticles supported on FeO_X for the water-gas shift reaction. *ACS Catalysis* 8: 859–868.

180 Liang, J.-X., Lin, J., Liu, J. et al. (2020). Dual metal active sites in an Ir_1/FeO_X single-atom catalyst: a redox mechanism for the water-gas shift reaction. *Angewandte Chemie International Edition* 59: 12868–12875.

181 Yan, D., Chen, J., and Jia, H. (2020). Temperature-induced structure reconstruction to prepare a thermally stable single-atom platinum catalyst. *Angewandte Chemie International Edition* 59: 13562–13567.

182 Duan, Q., Zhang, C., Sun, S. et al. (2020). Atomically dispersed palladium-based catalysts obtained via constructing a spatial structure with high performance for lean methane combustion. *Journal of Materials Chemistry A* 8: 7395–7404.

183 Huang, Z., Gu, X., Cao, Q. et al. (2012). Catalytically active single-atom sites fabricated from silver particles. *Angewandte Chemie International Edition* 51: 4198–4203.

184 Chen, Y., Huang, Z., Zhou, M. et al. (2017). Single silver adatoms on nanostructured manganese oxide surfaces: boosting oxygen activation for benzene abatement. *Environmental Science & Technology* 51: 2304–2311.

185 Chen, Y., Gao, J., Huang, Z. et al. (2017). Sodium rivals silver as single-atom active centers for catalyzing abatement of formaldehyde. *Environmental Science & Technology* 51: 7084–7090.

186 Zhang, C., Liu, F., Zhai, Y. et al. (2012). Alkali-metal-promoted Pt/TiO_2 opens a more efficient pathway to formaldehyde oxidation at ambient temperatures. *Angewandte Chemie International Edition* 51: 9628–9632.

187 Liu, T., Li, Q., Xin, Y. et al. (2018). Quasi free K cations confined in hollandite-type tunnels for catalytic solid (catalyst)-solid (reactant) oxidation reactions. *Applied Catalysis B: Environmental* 232: 108–116.

188 Zhang, H., Sui, S., Zheng, X. et al. (2019). One-pot synthesis of atomically dispersed Pt on MnO_2 for efficient catalytic decomposition of toluene at low temperatures. *Applied Catalysis B: Environmental* 257: 117878.

189 Xu, L., Chen, D., Qu, J. et al. (2018). Replacement reaction-based synthesis of supported palladium catalysts with atomic dispersion for catalytic removal of benzene. *Journal of Materials Chemistry A* 6: 17032–17039.

190 Yang, K., Liu, Y., Deng, J. et al. (2019). Three-dimensionally ordered mesoporous iron oxide-supported single-atom platinum: highly active catalysts for benzene combustion. *Applied Catalysis B: Environmental* 244: 650–659.

191 Jiang, Z., Feng, X., Deng, J. et al. (2019). Atomic-scale insights into the low-temperature oxidation of methanol over a single-atom Pt_1-Co_3O_4 catalyst. *Advanced Functional Materials* 29: 1902041.

192 Zhang, Q., Qin, X.-X., Duan-Mu, F.-P. et al. (2018). Isolated platinum atoms stabilized by amorphous tungstenic acid: metal–support interaction for synergistic oxygen activation. *Angewandte Chemie International Edition* 57: 9351–9356.

193 Grasselli, R.K. (2001). Genesis of site isolation and phase cooperation in selective oxidation catalysis. *Topics in Catalysis* 15: 93–101.

194 Grasselli, R.K. (2002). Fundamental principles of selective heterogeneous oxidation catalysis. *Topics in Catalysis* 21: 79–88.

195 Grasselli, R.K. (2014). Site isolation and phase cooperation: two important concepts in selective oxidation catalysis: a retrospective. *Catalysis Today* 238: 10–27.

196 Nguyen, L.D., Loridant, S., Launay, H. et al. (2006). Study of new catalysts based on vanadium oxide supported on mesoporous silica for the partial oxidation of methane to formaldehyde: catalytic properties and reaction mechanism. *Journal of Catalysis* 237: 38–48.

197 Shannon, I.J., Maschmeyer, T., Oldroyd, R.D. et al. (1998). Metallocene-derived, isolated Mo^{VI} active centres on mesoporous silica for the catalytic dehydrogenation of methanol. *Journal of the Chemical Society, Faraday Transactions* 94: 1495–1499.

198 Bradley, C.A., McMurdo, M.J., and Tilley, T.D. (2007). Selective catalytic cyclohexene oxidation using titanium-functionalized silicone nanospheres. *Journal of Physical Chemistry C* 111: 17570–17579.

199 Jarupatrakorn, J. and Tilley, T.D. (2002). Silica-supported, single-site titanium catalysts for olefin epoxidation. A molecular precursor strategy for control of catalyst structure. *Journal of the American Chemical Society* 124: 8380–8388.

200 Ruddy, D.A., Ohler, N.L., Bell, A.T., and Tilley, T.D. (2006). Thermolytic molecular precursor route to site-isolated vanadia–silica materials and their catalytic performance in methane selective oxidation. *Journal of Catalysis* 238: 277–285.

201 Lee, C.-H., Lin, T.-S., and Mou, C.-Y. (2003). $(VO)^{2+}$ ions immobilized on functionalized surface of mesoporous silica and their activity toward the hydroxylation of benzene. *The Journal of Physical Chemistry B* 107: 2543–2551.

202 Rossetti, I., Mancini, G.F., Ghigna, P. et al. (2012). Spectroscopic enlightening of the local structure of VOX active sites in catalysts for the Odh of propane. *Journal of Physical Chemistry C* 116: 22386–22398.

203 Thomas, J.M. (2008). How far is the concept of isolated active sites valid in solid catalysts? *Topics in Catalysis* 50: 98–105.

204 Wang, G., Zhang, S., Huang, Y. et al. (2012). Styrene epoxidation over V-SBA-15 with alkaline-earth metal ion promotion under photo-assisted conditions. *Appl. Catal. A* 413–414: 52–61.

205 Du, G., Lim, S., Yang, Y. et al. (2006). Catalytic performance of vanadium incorporated MCM-41 catalysts for the partial oxidation of methane to formaldehyde. *Applied Catalysis A: General* 302: 48–61.

206 Hu, Y., Nagai, Y., Rahmawaty, D. et al. (2008). Characteristics of the photocatalytic oxidation of methane into methanol on V-containing MCM-41 catalysts. *Catalysis Letters* 124: 80.

207 Chalupka, K., Thomas, C., Millot, Y. et al. (2013). Mononuclear pseudo-tetrahedral V species of VSiBEA zeolite as the active sites of the selective oxidative dehydrogenation of propane. *Journal of Catalysis* 305: 46–55.

208 Liu, J., Mohamed, F., and Sauer, J. (2014). Selective oxidation of propene by vanadium oxide monomers supported on silica. *Journal of Catalysis* 317: 75–82.

209 Rozanska, X., Fortrie, R., and Sauer, J. (2007). Oxidative dehydrogenation of propane by monomeric vanadium oxide sites on silica support. *Journal of Physical Chemistry C* 111: 6041–6050.

210 Launay, H., Loridant, S., Pigamo, A. et al. (2007). Vanadium species in new catalysts for the selective oxidation of methane to formaldehyde: specificity and molecular structure dynamics with water. *Journal of Catalysis* 246: 390–398.

211 Launay, H., Loridant, S., Nguyen, D.L. et al. (2007). Vanadium species in new catalysts for the selective oxidation of methane to formaldehyde: activation of the catalytic sites. *Catalysis Today* 128: 176–182.

212 Nozaki, C., Lugmair, C.G., Bell, A.T., and Tilley, T.D. (2002). Synthesis, characterization, and catalytic performance of single-site iron(III) centers on the surface of SBA-15 silica. *Journal of the American Chemical Society* 124: 13194–13203.

213 Su, W., Wang, S., Ying, P. et al. (2009). A molecular insight into propylene epoxidation on Cu/SiO_2 catalysts using O_2 as oxidant. *Journal of Catalysis* 268: 165–174.

214 Wachs, I.E. (1999). In situ Raman spectroscopy studies of catalysts. *Topics in Catalysis* 8: 57–63.

215 Khaliullin, R.Z. and Bell, A.T. (2002). A density functional theory study of the oxidation of methanol to formaldehyde over vanadia supported on silica, titania, and zirconia. *The Journal of Physical Chemistry B* 106: 7832–7838.

216 Gao, X., Jehng, J.-M., and Wachs, I.E. (2002). In Situ UV–Vis–NIR diffuse reflectance and Raman spectroscopic studies of propane oxidation over ZrO_2-supported vanadium oxide catalysts. *Journal of Catalysis* 209: 43–50.

217 Zhou, W., Ross-Medgaarden, E.I., Knowles, W.V. et al. (2009). Identification of active Zr–WO_X clusters on a ZrO_2 support for solid acid catalysts. *Nature Chemistry* 1: 722–728.

218 Kim, T., Burrows, A., Kiely, C.J., and Wachs, I.E. (2007). Molecular/electronic structure–surface acidity relationships of model-supported tungsten oxide catalysts. *Journal of Catalysis* 246: 370–381.

219 Ross-Medgaarden, E.I., Knowles, W.V., Kim, T. et al. (2008). New insights into the nature of the acidic catalytic active sites present in ZrO_2-supported tungsten oxide catalysts. *Journal of Catalysis* 256: 108–125.

220 Ross-Medgaarden, E.I., Wachs, I.E., Knowles, W.V. et al. (2009). Tuning the electronic and molecular structures of catalytic active sites with titania nanoligands. *Journal of the American Chemical Society* 131: 680–687.

221 Figueras, F., Palomeque, J., Loridant, S. et al. (2004). Influence of the coordination on the catalytic properties of supported W catalysts. *Journal of Catalysis* 226: 25–31.

222 Kourieh, R., Bennici, S., and Auroux, A. (2012). Acid and redox properties of tungstated zirconia catalysts. *Reaction Kinetics, Mechanisms and Catalysis* 105: 101–111.

223 Shen, Q., Cao, C., Huang, R. et al. (2020). Single chromium atoms supported on titanium dioxide nanoparticles for synergic catalytic methane conversion under mild conditions. *Angewandte Chemie International Edition* 59: 1216–1219.

224 Huang, W., Zhang, S., Tang, Y. et al. (2016). Low-temperature transformation of methane to methanol on Pd_1O_4 single sites anchored on the internal surface of microporous silicate. *Angewandte Chemie International Edition* 55: 13441–13445.

225 Kwon, Y., Kim, T.Y., Kwon, G. et al. (2017). Selective activation of methane on single-atom catalyst of rhodium dispersed on zirconia for direct conversion. *Journal of the American Chemical Society* 139: 17694–17699.

226 Shan, J., Li, M., Allard, L.F. et al. (2017). Mild oxidation of methane to methanol or acetic acid on supported isolated rhodium catalysts. *Nature* 551: 605–608.

227 Bai, S., Liu, F., Huang, B. et al. (2020). High-efficiency direct methane conversion to oxygenates on a cerium dioxide nanowires supported rhodium single-atom catalyst. *Nature Communications* 11: 954.

228 Zhao, Q., Liu, B., Xu, Y. et al. (2020). Insight into the active site and reaction mechanism for selective oxidation of methane to methanol using H_2O_2 on a Rh_1/ZrO_2 catalyst. *New Journal of Chemistry* 44: 1632–1639.

229 Harrath, K., Yu, X., Xiao, H., and Li, J. (2019). The key role of support surface hydrogenation in the CH_4 to CH_3OH selective oxidation by a ZrO_2-supported single-atom catalyst. *ACS Catalysis* 9: 8903–8909.

230 Bunting, R.J., Thompson, J., and Hu, P. (2020). The mechanism and ligand effects of single atom rhodium supported on ZSM-5 for the selective oxidation of methane to methanol. *Physical Chemistry Chemical Physics* 22: 11686–11694.

231 Li, T., Liu, F., Tang, Y. et al. (2018). Maximizing the number of interfacial sites in single-atom catalysts for the highly selective, solvent-free oxidation of primary alcohols. *Angewandte Chemie International Edition* 57: 7795–7799.

232 Thomas, A., He, Q., and Edwards, J.K. (2011). Preparation of ultra low loaded Au catalysts for oxidation reactions. *Faraday Discussions* 152: 381–392.

233 Cao, S., Yang, M., Elnabawy, A.O. et al. (2019). Single-atom gold oxo-clusters prepared in alkaline solutions catalyse the heterogeneous methanol self-coupling reactions. *Nature Chemistry* 11: 1098–1105.

234 Shan, J., Giannakakis, G., Liu, J. et al. (2020). PdCu single atom alloys for the selective oxidation of methanol to methyl formate at low temperatures. *Topics in Catalysis* 63: 618–627.

235 Liu, L. and Corma, A. (2018). Metal catalysts for heterogeneous catalysis: from single atoms to nanoclusters and nanoparticles. *Chemical Reviews* 118: 4981–5079.

236 Shang, Q., Tang, N., Qi, H. et al. (2020). A palladium single-atom catalyst toward efficient activation of molecular oxygen for cinnamyl alcohol oxidation. *Chinese Journal of Catalysis* 41: 1812–1817.

237 Hikazudani, S., Mochida, T., Yano, K. et al. (2011). Mono-atomically dispersed Pd on TiO_2 as a catalyst for epoxidation of light olefins at low temperatures in the presence of H_2 and O_2. *Catalysis Communications* 12: 1396–1400.

238 Liu, Y., Gan, T., He, Q. et al. (2020). Catalytic oxidation of 5-hydroxymethylfurfural to 2,5-diformylfuran over atomically dispersed ruthenium catalysts. *Industrial and Engineering Chemistry Research* 59: 4333–4337.

239 Tang, Y., Li, Y., Fung, V. et al. (2018). Single rhodium atoms anchored in micropores for efficient transformation of methane under mild conditions. *Nature Communications* 9: 1231.

240 He, W.-L., Yang, X.-L., Zhao, M., and Wu, C.-D. (2018). Suspending ionic single-atom catalysts in porphyrinic frameworks for highly efficient aerobic oxidation at room temperature. *Journal of Catalysis* 358: 43–49.

241 Babucci, M., Guntida, A., and Gates, B.C. (2020). Atomically dispersed metals on well-defined supports including zeolites and metal–organic frameworks: structure, bonding, reactivity, and catalysis. *Chemical Reviews* 120: 11956–11985.

242 Chen, Z., Liu, C., Liu, J. et al. (2019). Cobalt single-atom-intercalated molybdenum disulfide for sulfide oxidation with exceptional chemoselectivity. *Advanced Materials* 32: 1906437.

243 Gawande, M.B., Fornasiero, P., and Zbořil, R. (2020). Carbon-based single-atom catalysts for advanced applications. *ACS Catalysis* 10: 2231–2259.

244 Li, M., Wu, S., Yang, X. et al. (2017). Highly efficient single atom cobalt catalyst for selective oxidation of alcohols. *Applied Catalysis A: General* 543: 61–66.

245 Xie, J., Yin, K., Serov, A. et al. (2017). Selective aerobic oxidation of alcohols over atomically-dispersed non-precious metal catalysts. *ChemSusChem* 10: 359–362.

246 Ding, Y., Zhang, P., Xiong, H. et al. (2020). Tuning regioselective oxidation toward phenol via atomically dispersed iron sites on carbon. *Green Chemistry* 22: 6025–6032.

247 Liu, W., Zhang, L., Liu, X. et al. (2017). Discriminating catalytically active FeN_x species of atomically dispersed Fe–N–C catalyst for selective oxidation of the C—H bond. *Journal of the American Chemical Society* 139: 10790–10798.

248 Bakandritsos, A., Kadam, R.G., Kumar, P. et al. (2019). Mixed-valence single-atom catalyst derived from functionalized graphene. *Advanced Materials* 31: 1900323.

249 Huang, G., Wang, L., Luo, H. et al. (2020). Isopropanol as a hydrogen source for single atom cobalt-catalyzed wacker-type oxidation. *Catalysis Science & Technology* 10: 2769–2773.

250 Zhang, T., Zhang, D., Han, X. et al. (2018). Preassembly strategy to fabricate porous hollow carbonitride spheres inlaid with single $Cu–N_3$ sites for selective oxidation of benzene to phenol. *Journal of the American Chemical Society* 140: 16936–16940.

251 Deng, D., Chen, X., Yu, L. et al. (2015). A single iron site confined in a graphene matrix for the catalytic oxidation of benzene at room temperature. *Science Advances* 1: e1500462.

252 Wu, K., Zhan, F., Tu, R. et al. (2020). Dopamine polymer derived isolated single-atom site metals/N-doped porous carbon for benzene oxidation. *Chemical Communications* 56: 8916–8919.

253 Chen, Z., Zhang, Q., Chen, W. et al. (2018). Single-site Au^I catalyst for silane oxidation with water. *Advanced Materials* 30: 1704720.

254 Ledendecker, M., Pizzutilo, E., Malta, G. et al. (2020). Isolated Pd sites as selective catalysts for electrochemical and direct hydrogen peroxide synthesis. *ACS Catalysis* 10: 5928–5938.

255 Guo, Z., Xie, Y., Xiao, J. et al. (2019). Single-atom $Mn–N_4$ site-catalyzed peroxone reaction for the efficient production of hydroxyl radicals in an acidic solution. *Journal of the American Chemical Society* 141: 12005–12010.

256 Xu, J., Zheng, X., Feng, Z. et al. (2021). Organic wastewater treatment by a single-atom catalyst and electrolytically produced H_2O_2. *Nature Sustainability* 4: 233–241.

257 Huang, L., Chen, J., Gan, L. et al. (2019). Single-atom nanozymes. *Science Advances* 5: eaav5490.

258 Lin, Z., Zheng, L., Yao, W. et al. (2020). A facile route for constructing Cu–N–C peroxidase mimics. *Journal of Materials Chemistry B* 8: 8599–8606.

259 Lei, G., Tong, Y., Shen, L. et al. (2020). Highly active and sulfur-resistant Fe–N_4 sites in porous carbon nitride for the oxidation of H_2S into elemental sulfur. *Small* 16: 2003904.

260 Wang, J., You, R., Zhao, C. et al. (2020). N-coordinated dual-metal single-site catalyst for low-temperature CO oxidation. *ACS Catalysis* 10: 2754–2761.

261 Zhou, P., Hou, X., Chao, Y. et al. (2019). Synergetic interaction between neighboring platinum and ruthenium monomers boosts CO oxidation. *Chemical Science* 10: 5898–5905.

262 Kyriakou, G., Boucher, M.B., Jewell, A.D. et al. (2012). Isolated metal atom geometries as a strategy for selective heterogeneous hydrogenations. *Science* 335: 1209–1212.

263 Gaudry, É., Chatelier, C., Loffreda, D. et al. (2020). Catalytic activation of a non-noble intermetallic surface through nanostructuration under hydrogenation conditions revealed by atomistic thermodynamics. *Journal of Materials Chemistry A* 8: 7422–7431.

264 Greiner, M.T., Jones, T.E., Beeg, S. et al. (2018). Free-atom-like d states in single-atom alloy catalysts. *Nature Chemistry* 10: 1008–1015.

265 Maroun, F., Ozanam, F., Magnussen, O.M., and Behm, R.J. (2001). The role of atomic ensembles in the reactivity of bimetallic electrocatalysts. *Science* 293: 1811–1814.

266 Gao, F., Wang, Y., and Goodman, D.W. (2009). CO oxidation over AuPd(100) from ultrahigh vacuum to near-atmospheric pressures: the critical role of contiguous Pd atoms. *Journal of the American Chemical Society* 131: 5734–5735.

267 Ward, T., Delannoy, L., Hahn, R. et al. (2013). Effects of Pd on catalysis by Au: CO adsorption, CO oxidation, and cyclohexene hydrogenation by supported Au and Pd–Au catalysts. *ACS Catalysis* 3: 2644–2653.

268 Luneau, M., Shirman, T., Filie, A. et al. (2019). Dilute Pd/Au alloy nanoparticles embedded in colloid-templated porous SiO_2: stable Au-based oxidation catalysts. *Chemistry of Materials* 31: 5759–5768.

269 Ouyang, L., Da, G., Tian, P. et al. (2014). Insight into active sites of Pd–Au/TiO_2 catalysts in hydrogen peroxide synthesis directly from H_2 and O_2. *Journal of Catalysis* 311: 129–136.

270 Zhang, H., Watanabe, T., Okumura, M. et al. (2012). Catalytically highly active top gold atom on palladium nanocluster. *Nature Materials* 11: 49–52.

271 Sun, J., Han, Y., Fu, H. et al. (2017). Au@Pd/TiO_2 with atomically dispersed Pd as highly active catalyst for solvent-free aerobic oxidation of benzyl alcohol. *Chemical Engineering Journal* 313: 1–9.

10

Supported Metal Single-Atom Thermocatalysts for the Activation of Small Molecules

Marcos G. Farpón, Wilson Henao*, and Gonzalo Prieto*

ITQ Instituto de Tecnología Química, Universitat Politècnica de València-Consejo Superior de Investigaciones Científicas (UPV-CSIC), Av. Los Naranjos s/n, 46022, Valencia, Spain

10.1 Introduction

Small molecules, such as H_2O, CH_4, CO_x, and N_2, are ubiquitous in major natural bio-geochemical cycles. Owing to their abundance and delocalized availability, either as part of natural resources or as (side-)products of large-scale chemical processes, these simple molecules can play an important role as (renewable) feedstocks and/or versatile intermediates at the root of various production chains toward commodity chemicals and energy carrier compounds. However, they are generally rather recalcitrant and unreactive molecules. Therefore, their activation requires typically overcoming significant kinetic barriers and, when successful, activation typically leads to high-energy transition-state and intermediate species, which make it challenging to selectively steer the overall conversion pathways. While nature has developed highly optimized enzymatic catalysts for the selective activation of such small molecules under very mild conditions, e.g. photosystem II for water cleavage, oxidoreductases for CO_2 reduction, or monooxygenases for selective methane oxidation, the development of synthetic catalysts capable of activating and driving a selective transformation of these small molecules into value-added compounds remains a very active field of research. In particular, the design of solid catalysts to this end represents both a challenging and a potentially highly rewarding task in view of the distinctive benefits they present compared to their molecular counterparts, i.e. high thermal and mechanical stabilities, technically uncomplicated recovery from the reaction media, and therefore amenability to continuous conversion processes.

This chapter aims to provide an overview of the rationale behind, the advances achieved, as well as the open challenges remaining in the development of single-atom solid catalysts for the selective transformation of small molecules. Emphasis has been placed on three C_1 building-block molecules, i.e. CH_4, CO_2, and CO, as showcases. The development of energy- and material-efficient routes for the catalytic conversion of these compounds into added-value chemicals and fuels

* Marcos G. Farpón and Wilson Henao have contributed equally.

Supported Metal Single Atom Catalysis, First Edition. Edited by Philippe Serp and Doan Pham Minh.
© 2022 WILEY-VCH GmbH. Published 2022 by WILEY-VCH GmbH.

is expected to be central to the transition toward a chemical industry disconnected from crude oil as the universal feedstock, and therefore with a lower carbon footprint. For each of these molecules, the fundamentals underlying their activation are first succinctly discussed. This helps to understand the interest in the design and tuning of monoatomic metal centers as active catalyst sites. Finally, selected examples from the recent scientific literature are surveyed, which illustrate the progress achieved as well as the hurdles identified in the application of this booming class of catalyst materials in the area of small-molecule valorization.

10.2 Methane Conversion on Single-Atom Catalysts

Methane, as the main component of natural gas reservoirs, as well as shale and clathrate deposits, represents an abundant, largely scattered, and unconventional carbon-based resource. A selective and cost-effective conversion of methane is regarded as an opportunity to provoke a switch in the intensive chemical industry, from the current localized and globally dwindling crude oil as the central feedstock to an alternative raw material with intrinsically lower carbon footprint. Despite its vast availability, the use of methane as a chemical feedstock remains largely unexploited. For instance, associated natural gas is often flared to locally produce energy, contributing significantly to the global greenhouse gases emissions, owing to the lack of an effective catalytic technology to valorize methane on site [1]. The catalytic valorization of methane into fuels and oxygenates has been long studied [2–4]. However, efforts have traditionally revitalized in a cyclic fashion, often triggered by spiking oil prices. More recently, environmental concerns about the high carbon footprint associated with conventional oil refining have again turned the spotlight onto alternative chemical routes based on C_1 building blocks, of which CH_4 is a clear exponent. As shown in Figure 10.1a, methane can be converted into a variety of industrially relevant added-value chemicals and fuel compounds either by direct routes or indirectly, i.e. via syngas ($H_2 + CO$) as a versatile intermediate.

The conversion of syngas into different compounds is thermodynamically downhill, as water ($\Delta G_f{}^{\circ}{}_{(l)} = -237$ kJ/mol) is typically a major reaction side product. This makes indirect methane conversion pathways more attractive to achieve high yields and contributes to broaden the array of thermodynamically feasible end products. However, the energetically demanding reforming of methane into syngas makes these processes to rely on the so-called *economy of scale*, i.e. they are profitable only at very large scales. Therefore, syngas-mediated methane conversion routes are hardly compatible with intensified and miniaturized processes, as required for the valorization of commonly small-scale and delocalized unconventional natural gas reservoirs. For these reasons, direct transformation routes of methane into chemicals and fuels at low temperatures, which eliminate the expenses of intermediate syngas generation, are highly sought after.

Direct methane conversion routes might be classified into oxidative and non-oxidative depending on whether or not an oxidant is applied as co-reactant and hydrogen scavenger. Non-oxidative paths suffer from thermodynamic limitations

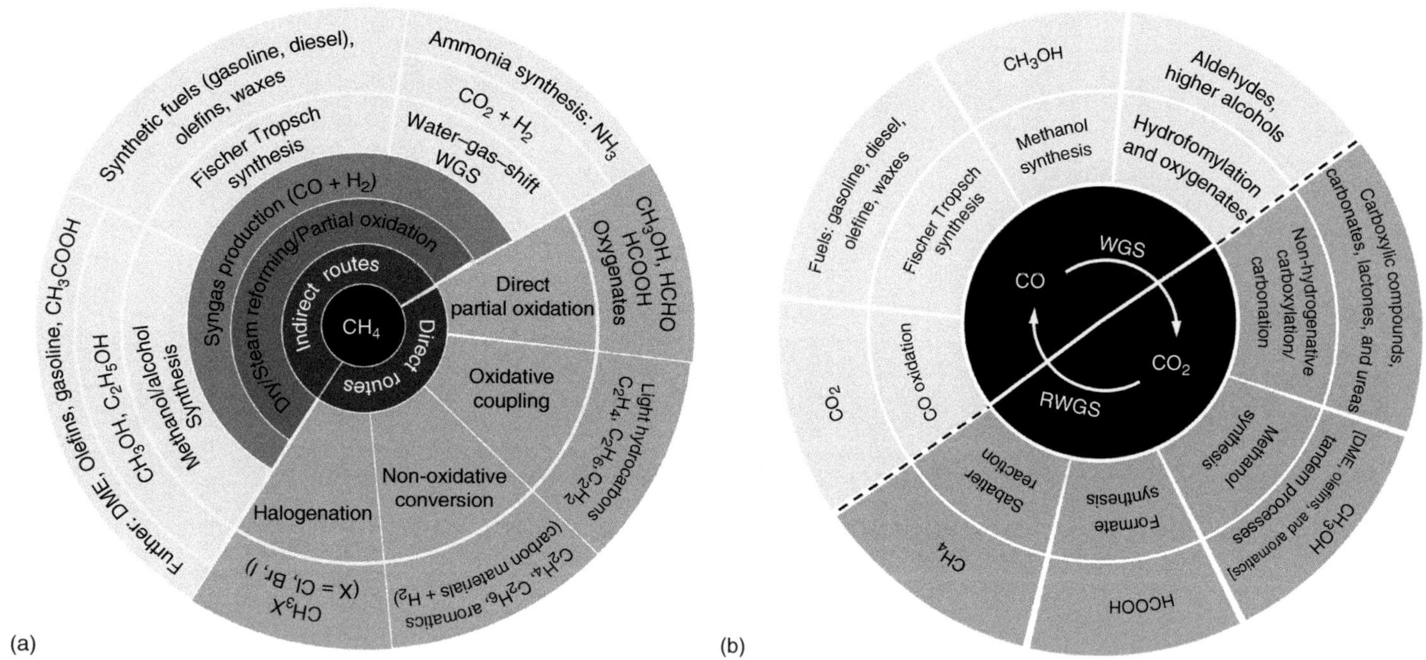

Figure 10.1 Major catalytic valorization routes for (a) methane and (b) carbon dioxide into industrially relevant commodity chemicals and energy carriers. Source: Farpón, M.G., Henao,W., Prieto, G.

as the reaction equilibria lay toward the reactants at the relevant operation temperatures, due to the rejection of hydrogen as molecular H_2. On the contrary, the selective partial methane oxidation into oxygenated compounds (CH_3OH, HCHO, HCOOH) is thermodynamically favorable, and the use of oxidants such as H_2O_2, O_2, and NO_x promotes the reaction to proceed at low temperatures [5]. As a matter of fact, in nature, methane mono-oxygenase enzymes based on Fe–O–Fe metal clusters can oxidize CH_4 selectively into methanol using O_2 as the oxygen source even at room temperature [6]. While such biocatalysts are difficult to scale up, the nature of their active sites serves as inspiration for the development of synthetic solid catalysts containing low-atomicity metal clusters, down to the limit of isolated metal atoms stabilized on oxide supports. Nonetheless, despite significant research work in the past decades, per-reactor-pass yields and selectivities achieved thus far do not suffice to make direct methane conversion routes based on heterogeneous catalysis attractive from a technological standpoint. Hence, further innovations are necessary.

10.2.1 Methane Activation: Mechanistic Considerations

The activation and selective functionalization of the methane C—H bond, the simplest and the least reactive among all hydrocarbons, is one of the foremost challenges in modern chemistry. Difficulties in methane activation lay in the substantial energy required to complete both homo- and heterolytic cleavages of the first C—H bond. The full T_d symmetry, with four equivalent C—H bonds, set the dipole moment of methane to be essentially null (2.84×10^{-40} C^2 m^2/J) [7]. Hence, the molecule requires a relatively high local electric field to be polarized, as a first requirement to further undergo either a nucleophilic or an electrophilic attack. Furthermore, methane is an extremely weak acid with a very low proton affinity, which hampers its activation by conventional acid–base chemistry. In the molecule ground state, all electrons occupy binding molecular orbitals in a $(1a_1)^2(2a_1)^2(1t_2)^6$ electron configuration. Due to the very large gap between the lowest unoccupied molecular orbital (LUMO) and the highest occupied molecular orbital (HOMO) (10.6 eV at the B3LYP level of theory) [8], its ionization energy is high and its electron affinity very poor. Therefore, electron donation to the LUMO from nucleophiles and electron removal from the HOMO by electrophiles are energetically difficult and sterically hindered [9]. All these features give rise to a high C—H bond dissociation energy (BDE) of 439.3 kJ/mol, which makes the first C—H bond cleavage to often be the rate-determining step in catalytic conversion routes [10].

An additional limitation of direct methane conversion routes into higher added-value chemicals is the fact that the C—H BDE for $-CH_x$ intermediates ($x \leq 2$) is typically lower (422 – 267 kJ/mol) than that in the starting methane molecule. This renders these monomeric species more reactive and thereby makes it difficult to steer their conversion path. When it comes to oxidative conversion routes, most of the intermediates are more prone to undergo oxidation than methane itself, with over-oxidation into CO or CO_2 being thermodynamically most favorable [4]. In the case of oxygenated species such as CH_xO_y, over-oxidation is further promoted by their polar nature and their high sticking coefficients on the typically oxophilic surfaces of solid catalysts.

Most of the current fundamental knowledge on C–H activation mechanisms for methane has emerged from studies with enzymes and well-defined synthetic molecular organometallic complexes, which operate in solution. However, this conceptual framework is certainly insightful to understand the site requirements for methane activation processes on supported single-atom catalysts (SACs), which, owing to their monoatomicity and availability to undergo multifold coordination, most closely resemble their molecular counterparts. The interaction of transition metals with d-orbitals (dz^2, dxz, and dyz) can enhance the C—H bond reactivity in methane by changing the relative energies of the hydrocarbon molecular orbitals or the molecule's polarity [7]. For instance, it has been realized that a strong interaction with coordinatively unsaturated metal centers of Fe and Cu can alter the T_d symmetry structure of methane into less symmetric C_{3V} or D_{2d} structures, as a result of the alteration of the local electric field via the Stark effect [8]. This distortion of the methane symmetry changes the relative position of the LUMO and HOMO of the C—H bond, facilitating its activation via electron transfer from/to the d-orbitals of the transition metal center. The charge transfer (CT) might take place from the occupied dπ orbital of the metal to the σ^* orbital of the coordinated C—H bond (reverse CT), as well as from the back-donation of the filled σ(C—H) bond to the empty dσ orbital of the metal (forward CT), as shown in Figure 10.2a. Both processes cooperatively contribute to destabilize the C—H bond, and their relative strength defines the electronic nature of the bond cleavage. Typically, in electron-deficient late transition metals, the forward CT dominates over the reverse CT, due to their low-energy dσ and dπ electrons, and therefore the C—H bond is activated via an electrophilic attack. Conversely, electron-rich transition metals possess high-energy dσ and dπ orbitals, and the dominant frontier orbital interaction is the reverse CT, favoring a nucleophilic C–H activation [11].

Works by Shilov [13], Labinger [14], and others [12, 15, 16] have significantly added to the current mechanistic understanding on C—H bond activation by transition metal centers. Activation might proceed through pathways involving the formation of transient M—C bonds, where the metal complex directly achieves the cleavage of the first C—H bond in methane, resulting in a methyl ligand. Alternatively, activation might take place via a free-radical mechanism in which the abstraction of hydrogen from methane leads to free $^{\cdot}CH_3$ radicals. The mechanism involved depends on the steric and electronic properties of the metal center. In M—C bond-mediated mechanisms, C—H bond activation begins with the σ-coordination of methane to the inner sphere of the transition metal center, followed by the scission of the C—H bond, forming a M—CH_3 intermediate with a lower BDE and, consequently, more reactive than the pristine methane molecule. As summarized in Figure 10.2b, the C—H bond cleavage might proceed by oxidative addition, electrophilic activation, or σ-bond metathesis paths. In the case of oxidative addition, the C–H linkage is cleaved via a nucleophilic attack driven by the electron back-donation of the dπ-orbitals of low-valence transition metals to the σ^* C–H orbital. Following cleavage, the metal atom is in this case

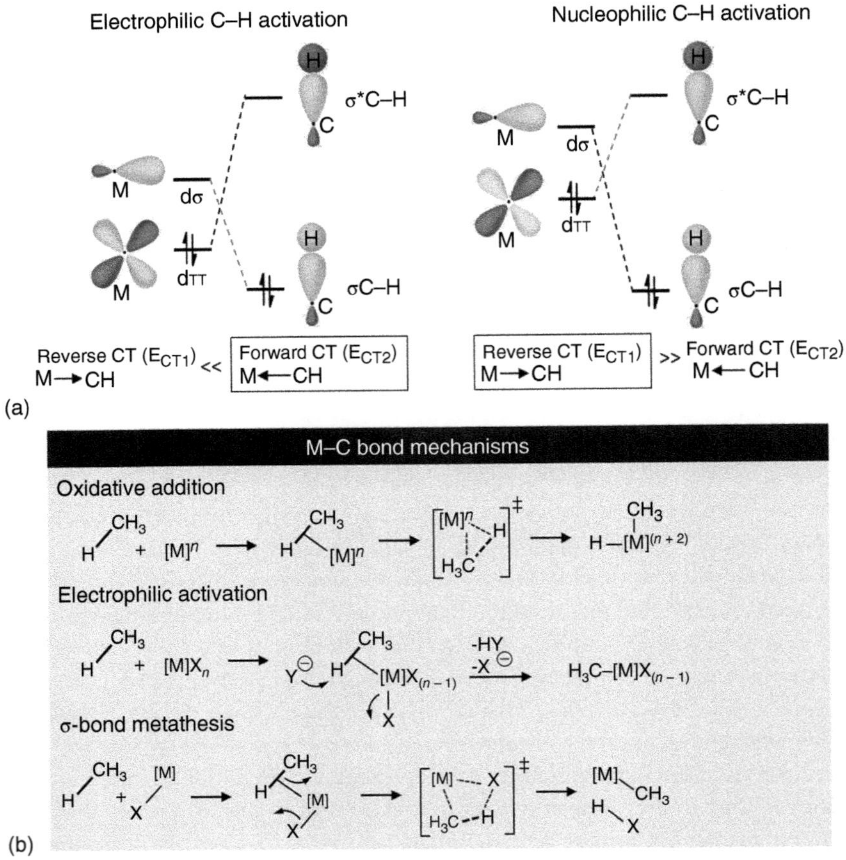

Figure 10.2 (a) Frontier orbital interactions for electrophilic and nucleophilic C—H bond activation involving transition metal centers. Source: Roudesy et al. [11]. Reproduced with permission of Elsevier. (b) C—H bond cleavage mechanisms in the catalyzed activation of methane. Source: Ravi et al. [12]. Reproduced with permission of Wiley-VCH GmbH.

covalently bonded to the methyl and the hydride fragments, resulting in an increase in its formal oxidation state by 2 units. Conversely, electrophilic activation involves electron-deficient late transition metals in relatively high oxidation states. In this mechanism, the metal cation attacks methane in an electrophilic manner, in which the σ-bond donation is stronger than the π-back-donation. The organometallic M–C species generated exist only as a transient intermediate, and the heterolytic C–H cleavage is usually conducted by an external anion Y^-. Finally, particularly in cases where the metal center exists in a d^0 configuration (Sc, lanthanides, and actinides), and therefore oxidative addition is forbidden, the C—H bond might be cleaved via a σ-bond metathesis. In this case, activation proceeds as the C—H bond in methane is σ-coordinated to the metal center, leading to a four-centered transition state from which a new M—C bond and a H–X linkage are formed, without the involvement of any metal hydride species. Commonly, X is another ligand previously linked to the metal center.

Compared to the two former mechanisms, σ-bond metathesis has been claimed to better describe the methane activation pathway in heterogeneous systems, where active sites display neighboring M–X acid–base pairs. Typically, $M^{\delta+}$ is a coordinatively unsaturated Lewis acidic metal site that polarizes the C—H bond and promotes its heterolytic dissociation, whereas $X^{\delta-}$ is a neighboring surface oxygen from the oxide support that helps to stabilize reaction intermediates and products. A representative example of σ-bond metathesis activation has been put forth by Tang et al. [17] to rationalize the activity exhibited by Rh_{SA} anchored within the channels of a mordenite framework inverted (MFI) zeolite (ZSM-5) for the oxidative carboxylation of methane with O_2/CO (Figure 10.3a). According to the proposed mechanism, Rh_{SA}/ZSM-5 isolated metal centers first react with O_2-forming $Rh_{SA}O_5$ active sites. Then, the C—H bond of methane is activated via a four-centered transition state $[CH_3\text{-}H\text{-}O\text{-}Rh_{SA}O_4]$, which leads to the formation of Rh–CH_3 and Rh–O–H species on the Rh_{SA}. In a next elementary step, carboxylation takes place via the insertion of a CO molecule into the Rh—O bond of Rh—O—H, forming a Rh-bounded COOH group that finally couples to Rh–CH_3 to form a CH_3COOH molecule, eventually desorbing from the active site. The remaining Rh=O activates the C—H bond of a second CH_4 molecule to form Rh–CH_3 and Rh–OH, which after a second carboxylation event leads to the formation and desorption of a second CH_3COOH molecule, turning the monoatomic center over.

Finally, metal centers might be not directly but rather indirectly involved in the C—H bond scission, in reactions which proceed via a Fenton-type pathway typically with peroxides (H_2O_2) but also with molecular O_2 as the oxidant sources. In this case, metal centers are involved in the genesis of reactive ·OH and/or ·OOH radical species, which are responsible for hydrogen abstraction from CH_4 in a subsequent step, away from the metal center. This mechanism was proposed to describe the selective activation of CH_4 on a Rh_{SA}/CeO_2 SAC in solution to produce CH_3OH and CH_3OOH (Figure 10.3b) [18]. According to this mechanistic proposal, the participation of coordinatively unsaturated metal sites from the CeO_2 support is important to facilitate the radical decomposition of H_2O_2 on the isolated Rh centers.

10.2.2 Methane Conversion on Single-Atom Catalysts: State of the Art and Challenges Ahead

10.2.2.1 Oxidative Routes

Given their capacity to strongly interact with both methane and oxidant molecules, SACs have been primarily explored as candidates for the oxidative conversion of methane under mild conditions [19, 20]. With the exception of group 11 (coinage) metals (Ag, Au), which show very low binding energies for carbon species and thus suffer from high methane dehydrogenation energy barriers [21], extended metallic surfaces of 4d and 5d block metals such as Pd, Rh, Pt, Ir, and Mo are active for C—H bond cleavage reactions. However, successive dehydrogenation of –CH_x intermediate species, thus extensive oxidation into CO_x, typically prevents high selectivities to oxygenates. In contrast, coordinatively isolated metal atoms have been

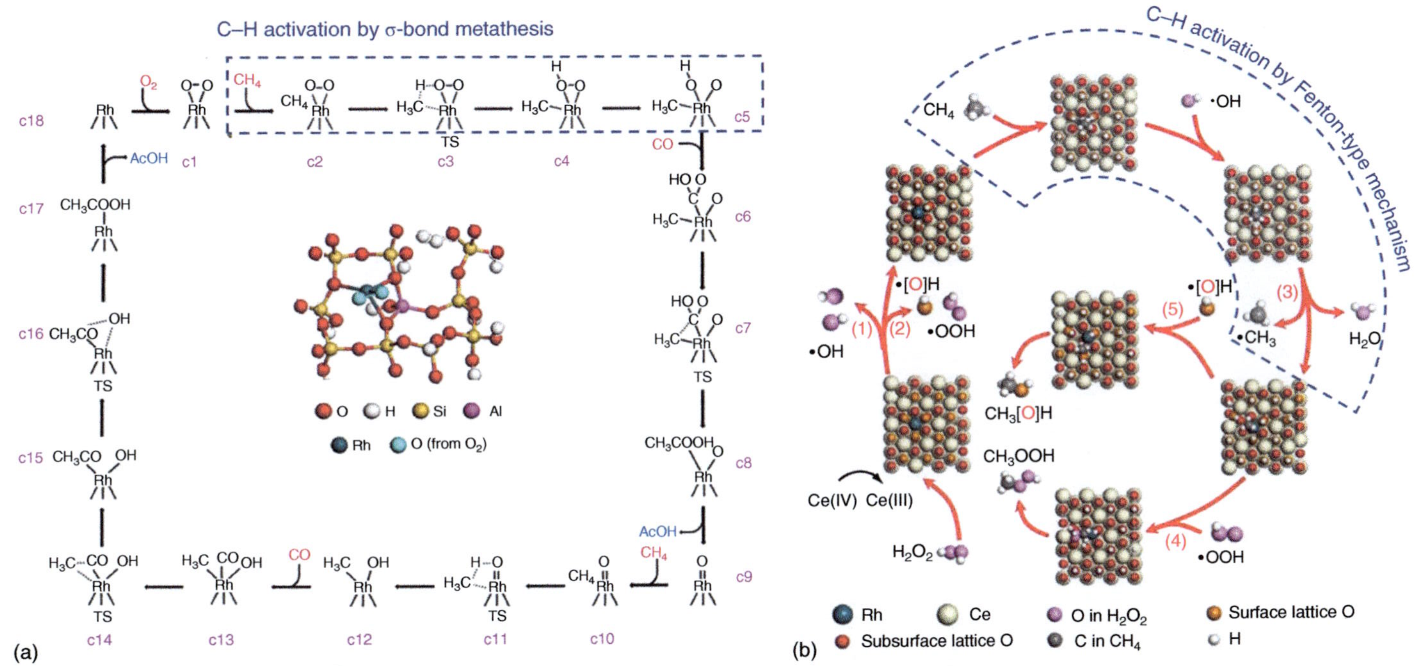

Figure 10.3 Examples of C—H bond activation of methane with oxide-supported single-atom catalysts conducted by (a) σ-bond metathesis on Rh_{SA}/ZSM-5. Source: Tang et al. [17], https://www.nature.com/articles/s41467-018-03235-7. Springer Nature. CC BY 4.0., and (b) a Fenton-type radical pathway on Rh/CeO_2. Source: Bai et al. [18], https://www.nature.com/articles/s41467-020-14742-x (*last accessed 27 May 2021*). CC BY-SA 4.0.

predicted, on the basis of density functional theory (DFT) calculations, to be more suitable candidate sites to inhibit successive dehydrogenations and thus stabilize $-CH_3$ intermediate species, which can therefore engage in selective oxidation pathways [22].

A number of examples have been reported with Rh_{SA} stabilized on various oxide supports such as H-ZSM-5 (MFI) zeolite [17, 23], TiO_2 [23], CeO_2 [18], or ZrO_2 [19]. These catalysts have shown activity for the direct conversion of methane to C_{1-2} oxygenates under remarkably mild reaction conditions (i.e. T = 50–150 °C, $P(CH_4)$ = 5–30 bar CH_4). Moreover, literature discloses a significant impact of the oxide support on the catalytic performance. For instance, Flytzani-Stephanopoulos and coworkers [23] showed that the acidity of the support may serve as a design parameter to steer product selectivity toward either C_1 methanol or C_2 acetic acid when methane is converted in aqueous phase under O_2/CO gaseous atmospheres at 150 °C. Isolated Rh atoms stabilized on the mildly Lewis acidic TiO_2 led exclusively to methanol as the oxidation product, whereas the use of the Brønsted acidic H-ZSM-5 zeolite as support led to a significant share of acetic acid within the oxidation products. Low reaction temperatures (<170 °C) and low partial pressures of O_2 (≤2 bar) were desirable to suppress the over-oxidation of the products into formic acid and CO_2, although the yield to oxygenates was found to decrease significantly, hinting at an oxygen-limited conversion. Methanol and acetic acid yields of up to 2 and 22 mmol/g_{cat} after three hours of reaction, respectively, were reported under the screened reaction conditions. In an independent work, Tang et al. [17] proposed a mechanism based on a σ-bond metathesis activation of methane at the isolated Rh centers within the ZSM-5 zeolite microporous architecture (Figure 10.3a). Huang et al. [24] applied a similar Pd_{SA}/H-ZSM-5 SAC for the selective oxidation of methane to methanol with H_2O_2 as oxidant in aqueous media, using CuO as co-catalyst, at temperatures below 100 °C. Under optimized conditions (T = 70 °C, $P(CH_4)$ = 30 bar, 0.5 M H_2O_2), the selectivity to methanol exceeded 85%, with a Pd-based *turnover frequency* (TOF, number of reactant molecules converted per unit (metal) active center and time) of 2.78 s^{-1} when 2.0 wt% CuO is used as a co-catalyst.

By atomically dispersing Rh on CeO_2 nanowires (NWs), Bai et al. [18] proposed the CeO_2 support to play an important role in the formation of $^{\cdot}OOH$ and $^{\cdot}OH$ radical species in an aqueous H_2O_2 medium, as an important step in a Fenton-like mechanism for the selective oxidation of CH_4 to CH_3OH and CH_3OOH. Working with Rh_{SA}/ZrO_2, Kwon et al. [19] showed that the reaction pathway may be directed toward either methanol or ethane in H_2O_2 aqueous or gas-phase O_2 reaction media, respectively. Stabilization of $-CH_3$ intermediates on the low-coordination Rh_{SA} was deemed to be critical to avoid successive dehydrogenations, which dominated on the extended Rh surfaces of nanoparticle (NP)-based analogue catalysts, leading to full oxidation to CO_2. Rhodium showed the highest alcohol production rate at a temperature of 70 °C, among various investigated metals (Rh, Pd, Pt, and Ir). In the aqueous H_2O_2 medium, DFT calculations pointed to a mechanism starting by C—H bond scission to $-CH_3$ followed by the formation of a methyl hydroperoxide ($-CH_3OOH$) intermediate, which is further deoxygenated into CH_3OH. In contrast, a mechanistic route involving the migration of a first $-CH_3$ fragment from the Rh atom onto a vicinal O^{2-} ion from the ZrO_2 support, followed by dissociation of a second CH_4 molecule on the metal center, was proposed to stabilize two adjacent $-CH_3$ groups, leading to the C–C coupling activity toward ethane in gas-phase catalysis with O_2 as the oxidant.

First-row transition metals such as Cr, Mn, Fe, Co, and Cu have also been explored as active species in SACs for the oxidative activation of methane, although less extensively compared to precious metals [25, 26]. This might be in part because it is more challenging to stabilize these oxophilic metals as isolated atoms on oxide supports while preventing the formation of sub-surface mixed compounds with the support. Alternative to oxide supports, carbon-based materials are also considered as platforms to stabilize atomically dispersed metals [27, 28]. These supports typically preclude the formation of thermodynamically stable mixed compounds, while their low e-contrast facilitates electron microscopy characterization of the supported metal species, often used as a proof for the atomic dispersion. FeN_4 sites anchored on graphene nanosheets (FeN_4/GNs) converted methane into methanol and other oxygenates ($HOCH_2OOH$ and HCOOH) at room temperature following a radical pathway [26]. The conversion was proposed to start via activation of FeN_4 with H_2O_2 to form oxygenated metal species O–FeN_4–O (with release of H_2O). The resulting center was surmised to be active for the homolytic cleavage of the methane C—H bond into $^{\cdot}CH_3$ radicals, which combine easily with $^{\cdot}OH$ and $^{\cdot}OOH$ groups to form CH_3OH and CH_3OOH. Compared to other metals (M = Cr, Mn, and Co), the Gibbs free energy of formation (ΔG_f) of the O–MN_4–O active site was found to be moderate in the case of iron, leading to a comparatively lower reaction energy barrier for C–H activation (c. 75 kJ/mol).

Despite the significant progress achieved in short time, the low-temperature selective methane oxidation using SACs is still at an incipient stage of development. Figure 10.4 summarizes the STY and selectivity to added-value C_1 and C_2 oxygenates reported in representative studies. To provide a qualitative benchmark, Figure 10.4 includes also the same performance parameters for two mature industrial processes for the production of methanol, i.e. methanol synthesis from syngas mixtures using Cu/ZnO/Al_2O_3 solid catalysts [33], and acetic acid, i.e. the Cativa process for methanol carbonylation catalyzed by iridium/ruthenium carbonyl molecular catalysts [34]. As observed, product selectivity levels reported for SACs using H_2O_2 or even O_2/CO as oxidants are comparable to those of the established industrial processes. However, productivity levels remain notably lower at present. Therefore, improvements in space–time yield (STY) might be required to trigger industrial interest for the SAC-catalyzed direct and selective methane conversion to oxygenate chemicals. Perhaps a more direct comparison might be established to the so-called Periana-Catalytica process [35] for the selective oxidation of methane to methanol, via methyl bisulfate as intermediate, in highly concentrated H_2SO_4 media catalyzed by Pt(II) salts and complexes. Rooted in the landmark article by Roy Periana and coworkers [36], this process has thus far not reached industrial implementation, a major technical hurdle being product recovery from concentrated H_2SO_4 or oleum reaction media.

As shown in Figure 10.4, comparison of productivity levels for SA catalysis with, to our best knowledge, the finest results achieved with Pt(II) catalysts in oleum media [29, 30] shows that the former remain generally much lower, with the only exception of methane oxo-carbonylation catalyzed by Rh/ZSM-5 SAC in aqueous media, for which the reported STY productivity is only about a factor of two lower. Several other important considerations need to be taken into account to assess process profitability, including: (i) the need for highly corrosive media, e.g. concentrated H_2SO_4 in the Periana-Catalytica concept or iodide promoters in the Cativa™ process, (ii) the cost of the oxidant, which is particularly high for H_2O_2, (iii) energy and cost-intensive syngas production in the case of state-of-the-art methanol synthesis, and (iv) the technical feasibility of product recovery and concentration. Nonetheless, it stems from the comparison presented above those improvements in product throughput are still required with SAC-based systems to approach the performance of the most effective molecular catalyst systems for the direct and selective methane oxidation.

10.2.2.2 Non-oxidative Routes

Next to oxidative routes, methane can also be converted under non-oxidative conditions to light alkenes and aromatics. In these cases, the reactions involved are reversible and endothermic. This poses thermodynamic constraints to the reaction yield per reactor pass. Therefore, integrated designs of catalysts and reactor concepts, e.g. to combine methane conversion with a selective removal of the product H_2 from the reaction medium, as well as high operation temperatures (typically >700 °C), are often applied to pull the reaction equilibrium toward the products [37]. However, the latter result also in high rates of coke deposition. Moreover, the high operation temperatures and the overall reducing reaction atmospheres are aspects that are known to promote reductive agglomeration of metal catalysts, and thus *a priori* limit the SAC applicability.

In a landmark paper, Bao and coworkers [38] reported that Fe_{SA} partially embedded on a silica matrix (Fe_{SA}/SiO_2) are effective catalysts for CH_4 activation under oxygen-free conditions and its conversion with high selectivity toward ethylene and aromatics. The absence of metal ensembles with Fe—Fe bonds suppressed iron carbidization and C–C coupling events, thus inhibiting the major catalyst deactivation pathway under the non-oxidative reaction conditions, i.e. coke deposition, which was found to proceed intensively on polynuclear $Fe(C_x)$ nanocrystals. Already in this first study, a 48.1% methane conversion was achieved, with a combined carbon selectivity to ethylene and aromatics >99%, and no significant deactivation after 60 hours on stream at 1090 °C. Mechanistically, authors proposed SiO_2-embedded FeC_2 sites to dissociate methane into $-CH_3$ and –H ad-species, which become adsorbed onto Fe and C sites, respectively, followed by the release of free $^{\cdot}CH_3$ radicals to the gas phase where these monomers undergo further coupling, dehydrogenation, and cyclization steps to ethylene, benzene, and naphthalene via non-catalytic thermochemical processes (Figure 10.5a). Nonetheless, it remained unclear what the driving force for the desorption of $-CH_3$ species strongly adsorbed on the Fe sites is, and how homogeneous-phase free-radical mechanisms could

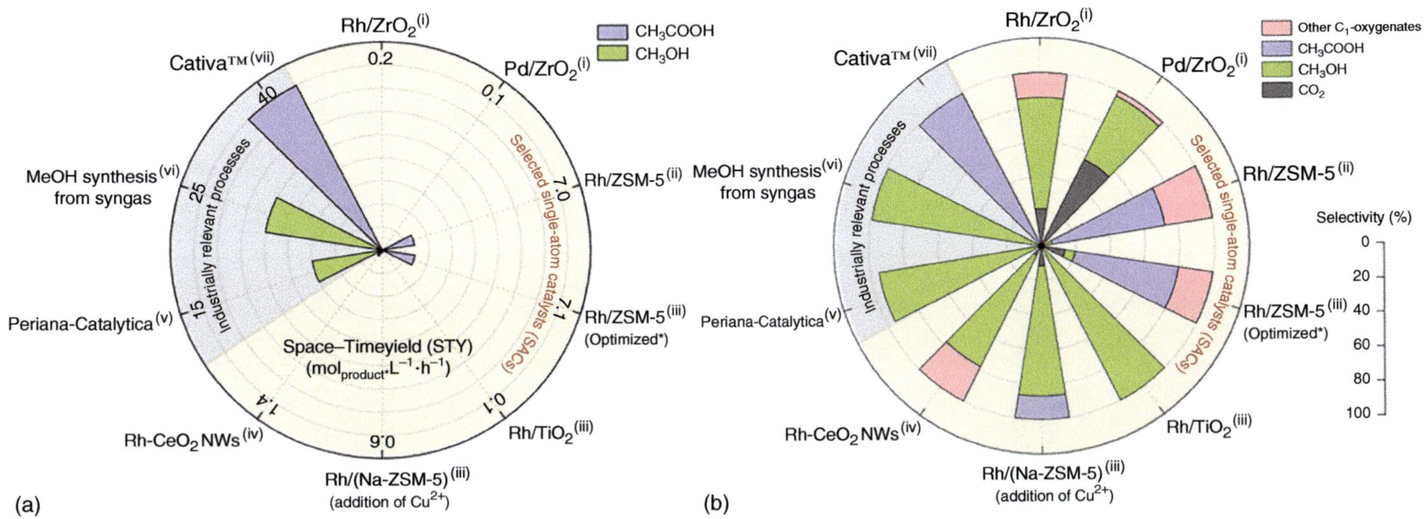

Figure 10.4 Windrose diagrams showing a comparison of (a) the productivity (space–time yield (STY), $mol_{product}/(l\,h)$) and (b) selectivity (%) performance parameters among selected SACs reported for the selective oxidation of methane into C_1 and C_2 added-value oxygenates. For the sake of benchmarking, indicative state-of-the-art performance values are also included for two established full-industrial-scale processes for the production of methanol (methanol synthesis) and acetic acid (Cativa™ methanol carbonylation process), as representative C_1 and C_2 oxygenates, respectively, as well as the Periana-Catalytica process concept of direct catalytic methane oxidation in concentrated sulfuric acid media. Legend for reaction conditions and literature references: (i) Rh/ZrO_2, Pd/ZrO_2: 30 bar CH_4, 70 °C, 30 mg catalyst, 10 ml 0.5 M H_2O_2, 30 minutes [19]; (ii) Rh/ZSM-5: 50 bar CH_4, 10 bar CO and 8 bar O_2, 150 °C, 28 mg catalyst, 10 ml H_2O, 12 hours [17]; (iii) Rh/ZSM-5 (optimized*), Rh/TiO_2, Rh/(Na-ZSM-5) (addition of Cu^{2+}): 20 bar CH_4, 5 bar CO and 2 bar O_2, 150 °C, 20 mg catalyst, 20 ml H_2O, 3 hours, *additional washing and filtration steps performed, 1 hour reaction [23]; (iv) Rh/CeO_2 NW: 5 bar CH_4, 50 °C, 10 mg catalyst, 20 ml 1 M H_2O_2, 1 hour [18]; (v) Periana-Catalytica: data reported for optimized reaction conditions in oleum using K_2PtCl_4 as catalyst, 15 ml 20% oleum, 65 bar CH_4 (72 bar total pressure), 215 °C, 15 ml 20% oleum, 600 μM K_2PtCl_4 as catalyst, 2 hour, CH_4 conversion 20–30% [29, 30]; (vi) methanol synthesis from syngas: 200–300 °C and 35–100 bar syngas pressure [31]; (vii) Cativa™ methanol carbonylation process, 150–200 °C, 30–60 bar [32]. Source: Farpón, M.G., Henao,W., Prieto, G.

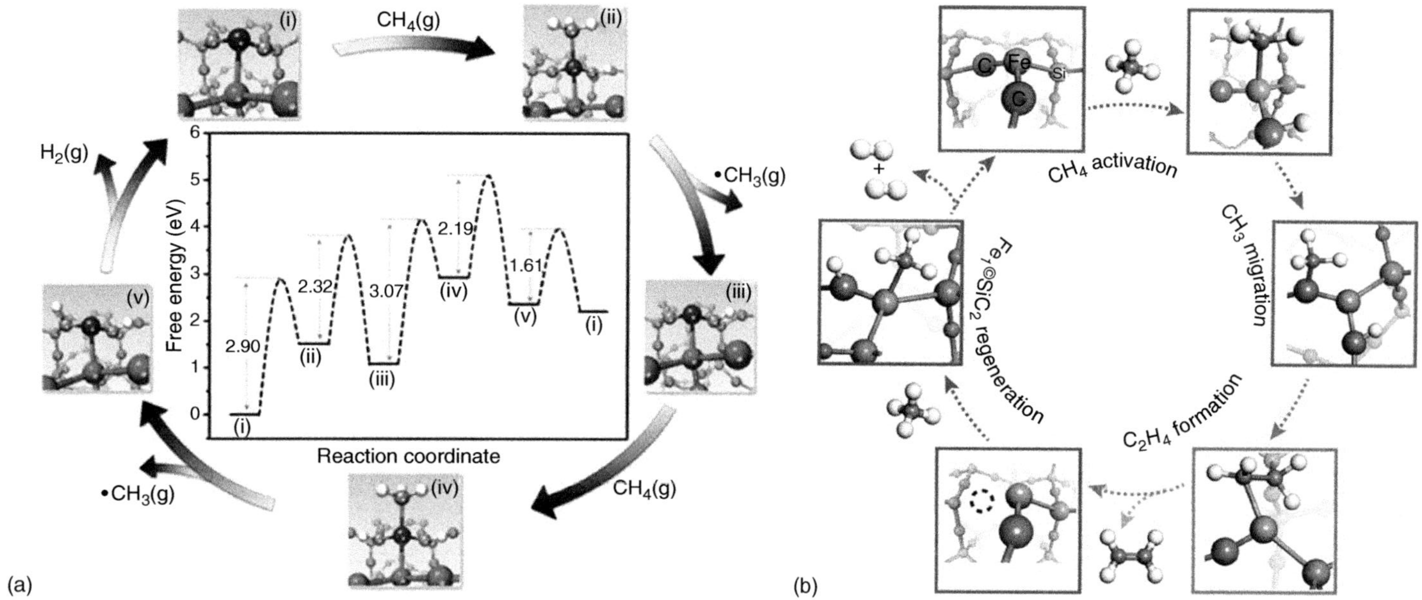

Figure 10.5 (a) Scheme of proposed free-radical mechanism for activation of methane with Fe_{SA}/SiO_2 under oxygen-free conditions. Source: Schwach et al. [2]. Reproduced with permission of American Chemical Society. (b) Illustration of the quasi-MvK mechanism proposed for the non-oxidative methane conversion at an Fe_{SA}/SiC_2 single-atom active center. Source: Liu et al. [39]. Reproduced with permission of Wiley-VCH GmbH.

account for the outstanding selectivity observed. Hence, alternative mechanistic proposals have also been presented. In a recent work, Liu et al. [39] suggested that the transfer of $-CH_3$ fragments to an adjacent carbon site of the FeC_2 active center might be energetically more favorable than desorbing into the gas phase, triggering a *quasi* Mars van Krevelen (MvK) surface reaction mechanism that involves extracting and refilling surface C atoms coordinated to the isolated Fe_{SA} (Figure 10.5b).

Since the pioneering report by Dalian researchers, a good deal of research interest has been devoted to this atomically dispersed iron catalyst system for methane conversion. In particular, efforts have focused on devising reaction concepts enabling high yield to ethylene, by shifting thermodynamics and preventing extensive oligomerization and cyclization as well as carbon deposition [40]. Oh et al. [41] developed a millisecond contact-time reactor by fusing Fe/SiO_2 powder into the inner wall of an open quartz tube, reaching a 11.3% CH_4 conversion per reactor pass with 91% C_{2+} selectivity at 1000 °C after 50 hours of reaction. Sakbodin et al. [42] achieved up to 30% CH_4 conversion and 99% selectivity to C_2 (ethylene and acetylene) and aromatic (benzene and naphthalene) products, by integrating the Fe_{SA}/SiO_2 catalyst in a high-temperature tubular $SrCe_{0.7}Zr_{0.2}Eu_{0.1}O_{3-\delta}$ hydrogen-permeation membrane. The selectivity toward C_2 hydrocarbons or aromatics could be adjusted by either adding or removing H_2 from the membrane reactor feed and permeate gas streams, respectively. Exploiting a three-heating-zone oven, Postma and Lefferts [43] found that the axial temperature profile and residence time upstream and downstream of an Fe_{SA}/SiO_2 catalyst bed influenced not only the CH_4 conversion and product distribution, but also the formation of carbonaceous deposits. A shallow catalyst bed followed by a significant gas residence time at high temperature increased methane conversion and inhibited coking.

In a similar vein, Xie et al. [44] tested Pt_{SA}/CeO_2 for the non-oxidative conversion of methane. Under optimized reaction conditions, a c. 15% methane conversion could be achieved at 975 °C, with 74.6% selectivity to C_2 products (C_2H_6, C_2H_4, and C_2H_2). Similar to the case of the original Fe-based catalyst system, no significant coke deposition was observed with the atomically dispersed catalyst, in contrast to the significant carbon buildup noticed with supported Pt_{NP}. As predicted by DFT calculations, low-coordination Pt sites can prevent the successive dehydrogenation of methane by stabilizing $Pt-CH_3$ intermediates [45]. Although the primary role of the support in the Pt_{SA}/CeO_2 catalysts is the stabilization of the single-atom metal species at elevated temperatures [46], synergetic effects at the interface of CeO_2 and Pt_{SA} may also be responsible for the CH_4 conversion with relatively low C–H activation barriers [47].

The suppression of carbon/coke deposition on the catalyst surface by reducing the atomicity of the catalytic metal species to the limit of SAs has been exploited also for other methane conversion processes [22, 38, 44, 45, 48]. Marcinkowski et al. [22] reported that Pt atoms isolated within Cu extended surfaces, so-called Pt/Cu single-atom alloy catalysts, exhibit superior coke resistance compared to a neat Pt surface, due to the weak binding of carbonaceous adsorbates offered by the low carbophilicity of copper. Tang et al. [49] reported that the integration of Ni_{SA} and Ru_{SA} on a common CeO_2 support led to higher activities for the dry reforming of CH_4 with CO_2 than the monometallic analogue materials, and ascribed the superior performance to complementary CO_2 and CH_4 activation as well as hydrogen adatom recombination on the two separate metal sites on the basis of DFT calculations. Also, under methane dry reforming conditions, Akri et al. [50] proposed that the atomic dispersion of the metal underlaid the remarkable stability and negligible coke buildup observed with nickel catalysts dispersed on hydroxyapatite supports. Comparative catalysts based on supported Ni_{NP} in their as-reduced (activated) state suffered from a notorious deactivation associated with high coke contents in the "spent" catalyst. The results provide a clear link between the initial atomic metal dispersion and coking inhibition. However, it is not possible to unequivocally rule out from the reported data that small, low-atomicity Ni clusters, for which low carbon deposition rates have been predicted [48], develop under the rather severe reaction conditions as the true active sites. Along this reasoning, part of the same authors reported in a follow-up study that significant sintering of the initially atomically dispersed Ni metal had taken place under the dry reforming reaction conditions, leading to the clear observation of Ni_{NP} on the surface of the hydroxyapatite support in the "spent" catalyst [51]. These findings illustrate that small metal clusters or nanoparticles derived from atomically dispersed precursors might also display unique catalytic properties [52]. Moreover, they underscore the importance of ascertaining, preferably via various independent physicochemical methods, the persistence of atomic metal dispersion prior, during, and after catalytic action, before performance might be unequivocally associated with single-atom sites.

10.3 CO_2 Conversion on Single-Atom Catalysts

Owing to their proven causality for climate change, the mitigation and even reversion of net anthropogenic carbon dioxide emissions to the atmosphere is currently a topic of major scientific and technological interest [53]. Catalysis is expected to play a central role in this endeavor, as it not only may contribute to the reduction of the overall carbon footprint of the intensive chemical industry and transportation, but it also provides means for a chemical recycling of CO_2 into chemicals and energy vectors. Figure 10.1b summarizes some of the major catalytic routes to convert CO_2 into added-value compounds. Direct CO_2 fixation is possible via its engagement into the synthesis of organic derivatives, such as carbonates, carboxylates, lactones, or ureas. While these routes benefit from catalysis, the hurdles they typically face are most often related to thermodynamic rather than kinetic constraints. Perhaps broader is the array of reductive CO_2 conversion routes, i.e. those which entail the use of a reducing agent to lower the formal oxidation state of carbon, which is C(IV) in CO_2.

Among the reductive conversion pathways, hydrogenation routes show the benefits of being connected to the well-established array of industrial processes for the valorization of CO, given that the (reverse) water–gas-shift reaction (rWGSR) interrelates both carbon oxides. Under certain scenarios, which include the availability of cost-effective renewable hydrogen as a low-carbon footprint reducing agent as well as point sources of waste CO_2 of sufficient concentration, CO_2 hydrogenation is considered an attractive route for its chemical recycling [54]. The contribution of CO_2 hydrogenation to a net reduction of the overall carbon emissions is higher when the CO_2 is disconnected from power generation, and particularly if it has a renewable origin, as in the purification of biogas [55]. On the contrary, it is less appealing if the CO_2 is connected to power generation [56]. Next to CO, a number of light oxygenates and hydrocarbons might be accessed via direct CO_2 hydrogenation, including formic acid and formates [57], methanol via methanol synthesis [58, 59], and methane through the Sabatier reaction [60]. An even broader array of commodity chemicals and liquid fuels is accessed via first partial reduction of CO_2 to the more reactive, thus versatile, carbon monoxide.

10.3.1 CO_2 Activation: Mechanistic Considerations

Carbon dioxide is thermodynamically a very stable molecule ($\Delta G^0{}_f = -394$ kJ/mol), which makes most of the CO_2 conversion routes energetically uphill in a broad range of temperatures. It is a symmetric molecule with equivalent and oppositely oriented C=O bonds ($\Delta H_{bond} \sim 800$ kJ/mol), and hence it holds an essentially null net dipole moment, factors that contribute to its inertness [61]. The induction of a net dipole moment, via bending of the molecule, is one of the mechanisms of which nature takes advantage for the activation of CO_2 by means of enzymes such as carbon monoxide dehydrogenase based on Ni/Fe active sites, or formate dehydrogenase (FDHs) bearing Cu/Mo active centers, which achieve the reduction of CO_2 to CO and formates, respectively [62]. Super-alkali chemicals have also been previously reported to facilitate the activation of CO_2 via bending of the molecule [63].

Electron transfer to the CO_2 molecule causes the bending of the O—C—O bond away from its ground-state linear geometry. Bending has in turn implications for key electronic properties, notably a shift of the LUMO ($2\pi_u$ molecular orbital) toward lower energy values, and a reduction of the HOMO-LUMO energy gap [62, 64, 65], which facilitate electron acceptance [66, 67]. This variation in the energy levels of the molecular orbitals of CO_2 as a function of the O—C—O bond bending degree is qualitatively described in the Walsh diagram derived from DFT calculations and shown in Figure 10.6a [64]. Upon acceptance of an electron, the molecule transforms into the $^{\cdot}CO_2^-$ radical, a species which is metastable in the gas phase but might be stabilized on polar surfaces and in solvent media [69, 70]. The formation of this radical anion contributes making the central carbon atom more electrophilic as well as to the elongation and thus weakening of the C—O bond (from 1.16 to 1.24 Å), increasing the reactivity with respect to the starting linear-shaped CO_2 [62, 64]. Acceptance of negatively charged hydride (H^-), hydroxyl (OH^-), and oxygen ion (O^{2-}) species leads to the formation of formate ($HCOO^-$), bicarbonate (HCO_3^-), and carbonate ions (CO_3^{2-}), respectively, which are commonly identified with *in situ* and *operando* Fourier-transform infrared (FTIR) spectroscopy as surface (intermediate) species

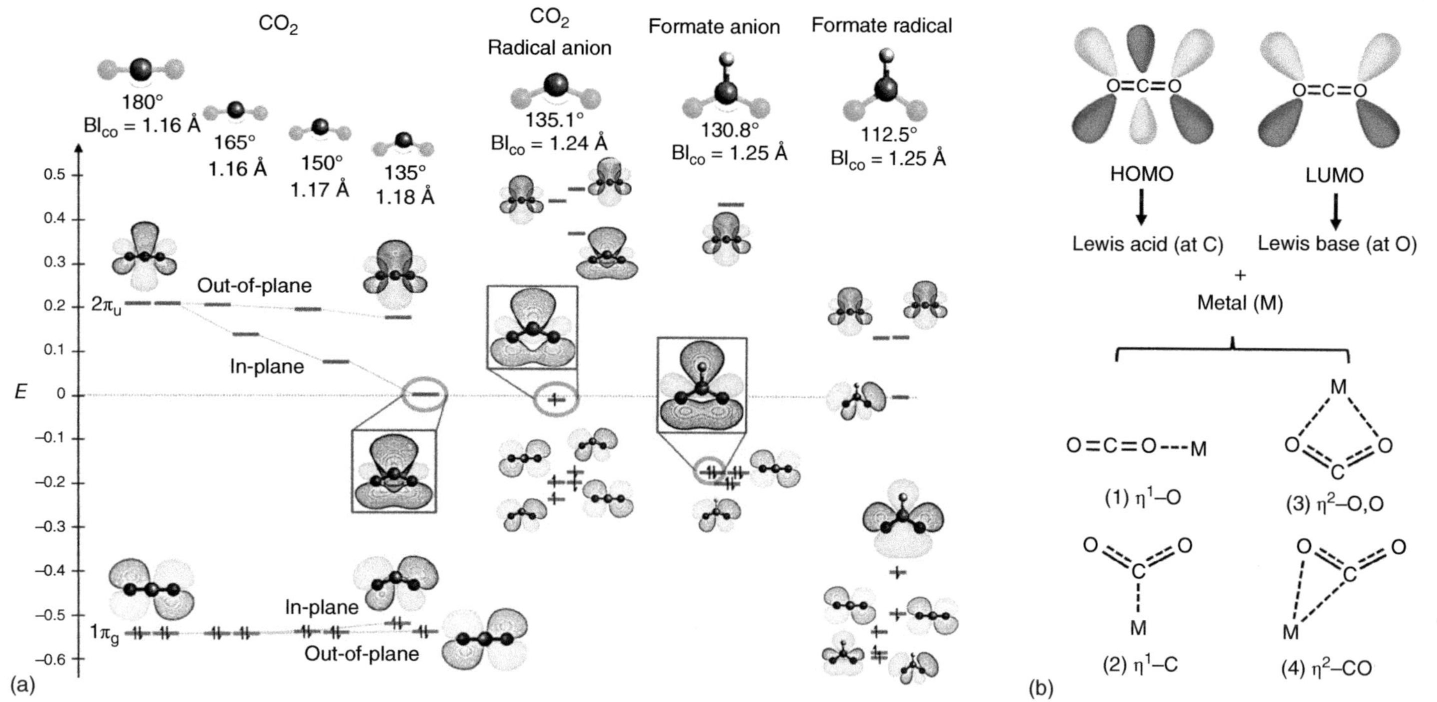

Figure 10.6 (a) Walsh diagrams (*y*-axis, energy in Hartree, E_h) showing the evolution of the energy of molecular orbitals with the characteristic O−C−O angle and C−O bond length of CO_2 in the linear (left) and its different bent forms: $\cdot CO_2^-$ radical, formate ion, and formate radical (right). Source: Álvarez et al. [64]. Reproduced with permission of Wiley-VCH GmbH. (b) Schematic illustration of (1) η^1-O, (2) η^1-C, (3) η^2 O, O, and (4) η^2 C,O interactions of CO_2 with a mononuclear metal center. Source: Mascetti et al. [68]. Reproduced with permission of Wiley-VCH GmbH.

upon the activation of CO_2 on metal (oxide) catalysts, and which all display a bent molecular configuration [71, 72].

Extended metal surfaces are known to adsorb and dissociate CO_2 into ·CO and ·O surface fragments [73]. However, particularly in the presence of hydrogen as reducing agent, a variety of adsorbate species might be generated (formate, formyl, carboxylate, methoxy, etc.), the further reactivity of which becomes dictated not only by the nature of the metal surface but also by the relative surface coverages of various ad-species. The interaction with monoatomic metal complexes, on the other hand, is typically better defined, hence preferred for a highly selective CO_2 conversion. As summarized schematically in Figure 10.6b, CO_2 can interact with isolated metal centers in three major fashions, i.e. via the electrophilic carbon atom (η^1-C) with electron-rich metals in a low oxidation state, via a nucleophilic (Lewis basic) oxygen atom (η^1-O or η^2-O,O) with electron-poor metal centers in high oxidation state, or through η^2 C,O bond [74, 75].

A number of metal complexes bounded to CO_2 have been isolated and characterized, shedding light into the configuration of CO_2 upon interaction with the corresponding mononuclear metal sites [76–79]. In all cases, evidence could be retrieved that the CO_2 molecule is bent. The energy required to bend CO_2 is over-balanced by the interaction energy with the metal center, primarily electron back-donation from the metal to the π^* orbital of the adsorbate, with a minimum binding energy reached at an O–C–O bending angle of c. 138° [80]. The π-backbonding component decreases among the different CO_2 coordination modes in the order η^2-C,O > η^1-C > η^1-O. In cases where the $M^{\delta+}/C^{\delta+}$ repulsive interaction is great, for example at positively charged metal centers, the end-on mode η^1-O is stabilized over the other two. On the other hand, binding through the C atom (η^1-C) might be anticipated in cases where both the metal center is electron-rich and the side-on coordination is not allowed, for some reasons such as coordination number. CO_2 can also insert into M—H and M—C bonds at monoatomic metal centers. Particularly the former is of significance for reductive CO_2 conversions to formic acid/formates. These have been successfully demonstrated with a number of organometallic molecular catalysts based on Ru and Ir, reaching TOF as high as 150 000 h^{-1} [76–78]. Remarkable performances have also been reported with mononuclear organometallic complexes for the rWGSR under mild conditions [79]. The performances observed with these mononuclear molecular catalysts have stimulated the exploration of solid SAC for CO_2 hydrogenation catalysis, as a means to pair the aforementioned performances with technically advantageous solid catalysts.

10.3.2 CO_2 Hydrogenation on Single-Atom Catalysts: State of the Art, Advantages, and Limitations

Given the known high activity of molecular complex catalysts, the CO_2 hydrogenation to formic acid/formates, i.e. to a carbon formal oxidation state of +II, seems an obvious first candidate reaction for SACs. Formic acid is an important base chemical for the formulation of preservatives, as well as in the textile and fine chemical industries. Mori et al. [81] reported a catalyst consisting of Ru_{SA} dispersed on a layered double-hydroxide support for the conversion of CO_2 into formates in an alkaline aqueous medium. More recently, Shao et al. [57] developed Ir-based SACs inspired in a ligand-stabilized molecular catalyst. Among the set of catalysts developed, the one consisting of Ir_{SA} dispersed on a porous organic polymer architecture exposing aminopyridine functionalities delivered a *turnover number* (TON, cumulative number of reactant molecules converted per unit metal atom) of c. 25 200. This is, to our best knowledge, the highest reaction rate reported to date for the conversion of CO_2 to formate using solid catalysts [76].

Further reduction to the level of methanol, i.e. a formal carbon oxidation state of −II, has received less attention in the field of SACs. Methanol and other light alcohols are very important compounds in the chemical industry as solvents, precursors for aldehyde and carboxylic acid derivatives, as well as prospective condense energy carriers [82–84]. In spite of the interest in developing alternative catalytic systems to conventional solid catalysts based on Cu_{NP} interfaced with oxides such as ZnO, ZrO_2, or Ga_2O_3 [85–87], and capable of driving the reaction at milder temperatures – thus benefiting from the higher methanol yields enabled by thermodynamics at operation temperatures <200 °C – only few mononuclear molecular catalysts are known for the direct reduction of CO_2 to methanol. In those few cases reported to date, the engineering of typically multidentate organic ligands has proven to be essential. However, even under optimized reaction conditions, limited turnover numbers (TON < 500) have been achieved [88, 89].

The generally poor performance evidenced for mononuclear molecular catalysts in solution might have contributed to curtailing the interest in developing supported SACs for the direct hydrogenation of CO_2 to methanol. Nonetheless, Pérez-Ramirez and coworkers demonstrated the exploitation of an atomically dispersed metal to enhance the catalytic performance of oxides with intrinsic activity for CO_2 hydrogenation to methanol. Following their earlier reports on the use of In_2O_3 as catalyst for methanol synthesis from CO_2 [90], they found that further surface decoration of this oxide with Pd_{SA} enhanced the methanol production rate by a factor of 3.5. Extensive characterization both prior to and after catalytic action revealed that Pd was metallic under reaction conditions, likely forming very-low-atomicity clusters, with less than four Pd atoms, aggregated from the single-atom precursor. These tiny clusters were proposed to be responsible for an enhanced hydrogen activation, and their promotional effect could not be paired by comparative materials displaying larger Pd clusters on the In_2O_3 support. Undesirably, the latter combined the enhanced hydrogen dissociation with the development of new metallic sites for the rWGSR, lowering the overall selectivity to methanol [91]. In a yet more recent article, to the best of our knowledge the first of its kind, Ye and coworkers [92] have extended the hydrogenation of CO_2 into the C_2 alcohol, ethanol, over Ir_{SA}/In_2O_3. In this case, catalysis proceeds in liquid water phase. The introduction of atomically dispersed surface Ir centers shifted entirely the alcohol selectivity from methanol, the only alcohol produced with the neat In_2O_3, to ethanol. Remarkably, this shift in alcohol selectivity was progressively lost as the nuclearity of the Ir species increased with increasing the metal surface loading, suggesting that ethanol production is exclusive of isolated metal sites. A combination of isotopic labeling, *in situ* diffuse reflectance infrared

Fourier-transform (DRIFT) spectroscopy experiments, and DFT calculations led to the proposal of vicinal $^{\cdot}CO–Ir^{\delta+}$ metal carbonyl and $^{\cdot}CH_3O–O_v$ methoxy group at an oxygen vacancy as key intermediates to achieve C–C coupling [92]. It was inferred that higher nuclearity Ir species (nanoclusters) favored a higher surface availability of dissociated hydrogen and thus premature hydrogenation to methanol. This first example of the successful combination of CO_2 partial reduction with C–C coupling activity to produce C_{2+} oxygenates, which resembles activities only observed hitherto with homologation molecular carbonyl catalysts in solution [93], opens up an interesting niche of application for SACs in the context of CO_2 valorization.

Perhaps the most studied catalytic route for CO_2 hydrogenation is the Sabatier reaction, i.e. its full hydrogenation to CH_4 (carbon formal oxidation state of –IV). This reaction is known to be catalyzed by extended surfaces of metals showing an intermediate CO_2 binding strength, such as Ru and Ni [94]. Hence, NP-based catalysts of these and other metals, supported on suitable oxide carriers, are the standard catalysts for CO_2 methanation [95]. Methanation typically competes with the rWGSR, which leads to CO as a side product. It has compellingly been established in the literature that the selectivity of the process to CH_4 depends on the metal NP size [96–99]. Wu et al. [100] addressed the impact of metal dispersion on the product distribution for a series of Ni/SiO_2 catalysts. They observed that the selectivity to methane increased with increasing the Ni_{NP} size and proposed two competing mechanisms to account for these particle size effects: (i) a consecutive mechanism, in which CO_2 is first partially reduced to CO via a formate intermediate, followed by further dissociation of CO and hydrogenation to CH_4 via CH_x intermediates; and (ii) a parallel mechanism, in which both CO and CH_4 products are derived from the dissociation or hydrogenation, respectively, of a common surface formate ($HCOO^-$) intermediate. More recently, Vogt et al. [60] proposed similar mechanisms also for a series of SiO_2-supported Ni catalysts synthesized with different metal NP sizes. In their case, an optimal Ni_{NP} size, in the range of 2–3 nm, was found to maximize the catalyst activity. Common to these different studies is the observation that the selectivity to CO progressively increases as the Ni_{NP} size is decreased down to the sub-nm range. The results thus suggest that metal clusters of low atomicity (in the limit, single-atom sites) are only capable to catalyze the $2e^-$ redox cycle necessary to obtain CO via the rWGSR, whereas metal ensembles of a minimum size are required to catalyze the complete ($8e^-$) redox cycle required for full hydrogenation to CH_4 [58]. These findings might be rationalized on the basis of a consecutive mechanism, in which CO may act as an intermediate, given that CO dissociation, which is mandatory for methanation, is known to require multi-atom ensembles on the surface of metal NP of a certain size to proceed (*vide infra*, Section 10.3.1).

In light of the above findings for nanoparticulate catalysts with different particle sizes, it is not surprising that atomically dispersed metal catalysts, i.e. at the limit of monoatomicity, have been found to be selective to the rWGSR under CO_2 hydrogenation conditions. Amal and coworkers [101] compared the performance of atomically dispersed Pt_{SA}/CeO_2 with a NP-based counterpart. Under identical reaction conditions, the former delivered an approximately 2.5–7.5 times greater reaction rate per unit surface Pt atom than the catalyst containing Pt agglomerates. Moreover,

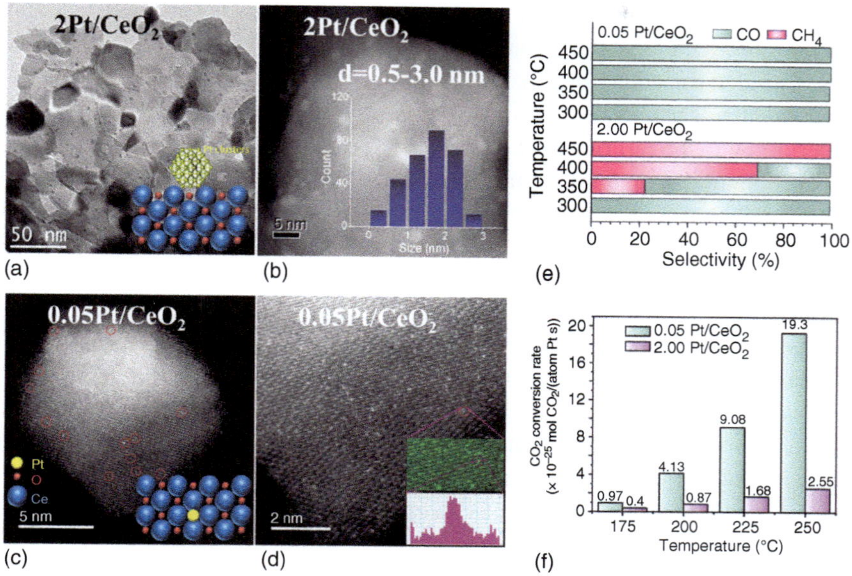

Figure 10.7 (a, b) Transmission electron microscopy (TEM) and HAADF-STEM micrographs of a clustered 2% Pt/CeO_2 catalyst. Structural model of supported Pt nanoclusters on the CeO_2 surface (inset to (a)) and Pt size distribution (inset to (b)). (c, d) HAADF-STEM micrographs of an atomically dispersed 0.05% Pt/CeO_2 catalyst. Red circles indicate atom-sized high Z-contrast motifs interpreted as isolated Pt atoms. Structural model of Pt_{SA} stabilized on CeO_2 (inset to c) and Pt K-line EDS line-scanning intensity profile obtained from the zoomed area (inset to d). (e) Selectivity distributions for 0.05% Pt/CeO_2 and 2% Pt/CeO_2 at different reaction temperatures and (f) CO_2 conversion rates expressed as mol CO_2 converted per exposed atom of Pt and unit time. Catalytic tests have been performed at atmospheric pressure with a 5% CO_2, 62.5% H_2, 32.5% N_2 gas feed mixture. Source: Wang et al. [101]. Adapted with permission of American Chemical Society.

isolated Pt sites led to a 100% selectivity to CO, whereas methane was also observed with the catalyst additionally exposing extended Pt metal surfaces to the reactants, Figure 10.7. Based on *in situ* DRIFT spectroscopy results, a mechanism was proposed, which starts with the initial activation of CO_2 on oxygen vacancies on the CeO_2 surface to then diverge depending on the metal agglomeration degree, regardless of the nuclearity of the metal species. Platinum atomic dispersion was proposed to lead to a comparatively weaker binding of CO (and related formyl intermediates) restricting its further hydrogenation and minimizing rate inhibition effects by CO poisoning, which is commonly observed on metallic clusters and nanoparticles. In another study, Kwak et al. [97] analyzed the role of metal nanoparticle size with a battery of Ru/Al_2O_3 catalysts. They identified atomically dispersed Ru as those metal centers responsible for a 100% CO selectivity (rWGSR), whereas Ru clusters and Ru_{NP} were found to be accountable for the formation of methane. The markedly different selectivity pattern offered by SACs and their metal agglomerate counterparts has indeed been exploited to track the stability of atomically dispersed metals under reaction conditions. Similar observations were made by Aitbekova et al. [98] who reported that oxidative metal redispersion of a Ru_{NP}/CeO_2 catalyst in an

O_2 atmosphere results in an abrupt change in selectivity under CO_2 hydrogenation conditions, from methanation to water–gas-shift reaction (WGSR). Frei and coworkers [58] studied the evolution of solid–solution Ni/MgO SACs under reaction conditions. At temperatures below 300 °C, CO was observed as the main product, i.e. the performance fingerprint for SACs or very-low-atomicity metal species. However, once the reaction temperature was raised above 300 °C, the original selectivity could not be recovered, and methane was observed among the products, even after restoring the initial lower temperatures. These non-reversible changes in selectivity were taken as an indication for the occurrence of permanent metal clustering. Taken together, these observations underscore the significance of assessing the occurrence of performance *hysteresis* effects upon cycling the operation conditions, as a means to validate the stability of SAC during their exposure to relevant operation conditions.

So far, the routes discussed for the valorization of CO_2 involve reduction events in the formal oxidation state of the carbon atom. Nonetheless, there is a wide variety of industrially relevant, non-reductive CO_2 conversion routes into valuable chemicals such as acrylates, epoxides, or carbonates among others. These processes have attracted interest since they do not require a hydrogen source, and they typically operate under milder conditions than reductive routes [102]. Recently, research in the context of SACs has been extended to these non-reductive routes. For instance, Zhao et al. studied the N-formylation of amines with CO_2 as formylation agent using a SAC based on atomically dispersed Pt on titanium carbide with Ti defects ($Ti_{3-x}C_2T_y$, where T stands for O, OH, or F) [103]. Due to the high reducibility of the support even at very mild conditions, the formation of Ti defects could be exploited to accommodate the Pt species. According to their observations, it is the positive charge of the Pt atoms that seems to play a pivotal role for the superior catalytic performance of the Pt_{SA} in contrast with Pt_{NP}. In another work, Wang and coworkers [104] reported a catalytic system comprising Zn_{SA} dispersed on a N-doped graphene matrix evaluated in the cycloaddition of several epoxides with CO_2. The Zn-based SAC achieved high values of TOF and TON (2889 and 8666h^{-1}, respectively), although they showed certain limitations for the activation of other substrates (i.e. internal epoxides such as cyclohexene oxide) as well as regarding catalyst reusability.

10.4 CO Conversion on Single-Atom Catalysts

Carbon monoxide is a very versatile intermediate compound in the chemical industry. It might be derived, as component of syngas mixtures ($H_2 + CO + (CO_2)$), from essentially any carbon-based raw material, e.g. via reforming or partial oxidation of (unconventional) natural gas or via gasification of coal or biomass resources. In pure form, carbon monoxide is accessed via the gasification of coal with CO_2 into CO-rich *producer gas*, via the Boudouard reaction ($CO_2 + C_{(s)} \rightarrow 2CO$), the direct partial oxidation of coal in defect of oxygen ($O_2 + 2C_{(s)} \rightarrow 2CO$), or as side product from the reduction of metal oxide ores with carbon ($xC_{(s)} + M_yO_{x(s)} \rightarrow xCO_{(g)} + yM_{(s)}$).

As part of syngas mixtures, CO acts as carbon precursor/building block for important chemicals such as methanol [105], higher alcohol and aldehyde oxygenates via the (reductive) hydroformylation of olefins [106], as well as synthetic hydrocarbon fuels through the Fischer–Tropsch synthesis (FTS) [107]. As a pure compound, carbon monoxide is important as reactant in carbonylation processes such as the Reppe and Koch reactions [108], in the synthesis of phosgene as precursor for isocyanates, polycarbonates, and polyurethanes, as well as in the purification of metals mediated by volatile metal carbonyl compounds [109]. Additionally, catalytic conversion of CO is also relevant in the context of its removal from syngas mixtures, in processes aiming at the purification of hydrogen streams.

10.4.1 CO Activation: Fundamental Considerations

Carbon monoxide is a linear-shaped diatomic molecule held together by a strong multiple C—O bond ($\Delta H_{C—O} = 1077$ kJ/mol), which has been described as either double or triple bond depending on various considerations about the molecule's physical properties and the Lewis resonant form adopted [110]. The electronegativity difference between the C and the O atoms, alongside the dative bond from O to C, results in a small permanent dipole moment (c. 0.125 D). As shown in Figure 10.8a, the high-energy HOMO 5σ molecular orbital in CO holds a directional character, with the electronic density localized on the C atom, enabling a rather strong interaction of the molecule with metal atoms via σ-donation into the unoccupied d-orbitals of the latter. Moreover, the 2π* LUMO, which is also localized on the C atom, facilitates electron back-donation from the metal.

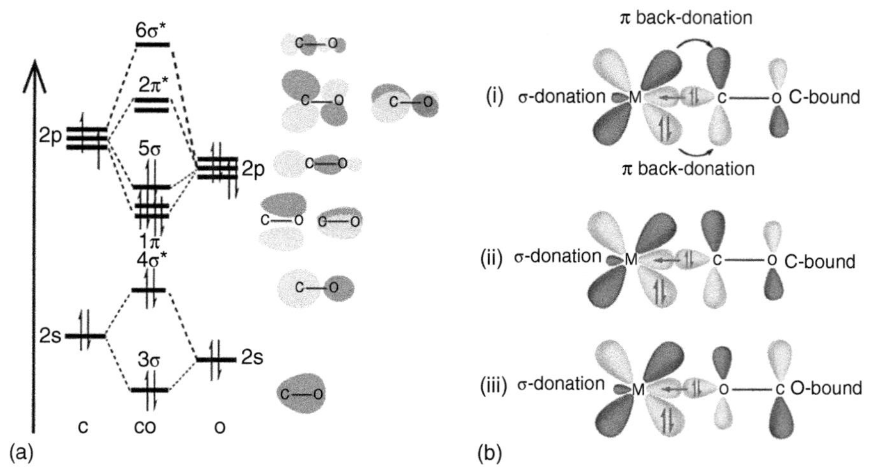

Figure 10.8 (a) Molecular orbital diagram for carbon monoxide (the vertical arrow points in the direction of increasing energy levels. Source: Emily V Eames. CC BY 4.0. (b) Bonding of carbon monoxide to a transition metal. (i) Classical bonding with σ-donation from C to M and π back-donation from M to CO. (ii) Non-classical C-bound carbonyl in which the π back-bonding component is absent or negligible. (iii) O-bonded, or isocarbonyl, bonding in which the carbonyl oxygen atom is bonded to the metal. Source: Maron et al. [111]. Reproduced with permission of American Chemical Society.

As a result, the classical Blyholder model of a combined CO-to-metal σ-donation and metal-to-CO π-back-donation, with the CO ligand bound to the metal through its C atom, is the most common description of carbon monoxide adsorption in metal carbonyls (Figure 10.8b-i). Given the antibonding character of the carbon monoxide LUMO, electron back-donation phenomena tend to weaken the C—O bond, and it is often central to the activation of carbon monoxide on metal centers. Even in instances where the π* component of the M—CO bond is lessened or eliminated, for instance on metal centers with a d^0 electron configuration and hence unable to engage in π bonding or highly electrophilic cations whose d electrons are too low in energy to interact strongly with the CO π* orbitals, the carbonyl ligand binds to the metal through molecular orbitals associated with the C atom (Figure 10.8b-ii). Finally, the O-bound coordination configuration of CO on metal centers, albeit energetically feasible as an isomeric form of the corresponding C-bound adducts on highly electropositive metal sites, is very rare (Figure 10.8b-iii) [111].

Carbon monoxide might be activated either associatively or dissociatively with the assistance of metal catalysts. Associative CO activation does not involve the cleavage of the C—O bond, and thus precedes the catalytic insertion of the CO molecule into M—R and M—H bonds, where R represents a general alkyl group and M the metal center. Migratory insertion of CO into these bonds is a primary reaction step in several reactions catalyzed by organometallic catalysts, such as methanol synthesis, the hydroformylation of olefins, or various types of carbonylation transformations. Without any known exception, this associative CO activation and insertion involves C-bound carbon monoxide metal complexes (Figure 10.8b-i,ii) [112]. Given the known reactivity of mononuclear organometallic catalysts in solution, chemical reactions involving such associative activation of CO have been the primary focus for the identification and optimization of supported SACs, as potential all-inorganic replacements for their molecular counterparts.

A different picture emerges for full hydrogenation reactions, such as CO methanation or the FTS of higher hydrocarbons, which require the dissociative activation of CO. In this case, the CO dissociation has been shown to require the involvement of active centers consisting of poly-atom ensembles on the surface of metal nanoparticles. On extended metal surfaces, a larger variety of CO adsorption modes are possible. In addition to C-bound carbonyls with atop (μ_1-η^1), bridged (μ_2-η^1), and threefold (μ_3-η^1) coordination on flat metal surfaces, tilted CO adsorption configurations (μ_5-η^2 and μ_6-η^2), where both the C and the O atom of the adsorbate establish interactions with surface metal atoms, have been proposed to prevail at metal surface defects such as step edges (Figure 10.9a).

For the latter, DFT calculations have shown that the CO 2π*-metal dπ hybrid band responsible for the electron back-donation is more delocalized on both C and O atoms, which is believed to enhance metal-to-CO back-bonding and facilitate CO cleavage [113]. Macroscopically, the CO methanation as well as FTS reactions on Co and Ru metallic catalysts obey a marked structure sensitivity, i.e. the CO hydrogenation TOF (normalized per unit surface metal atom) increases significantly with increasing the metal nanoparticle size up to a particle diameter of c. 8–10 nm [115, 116]. This phenomenon has indeed been related to the critical participation of

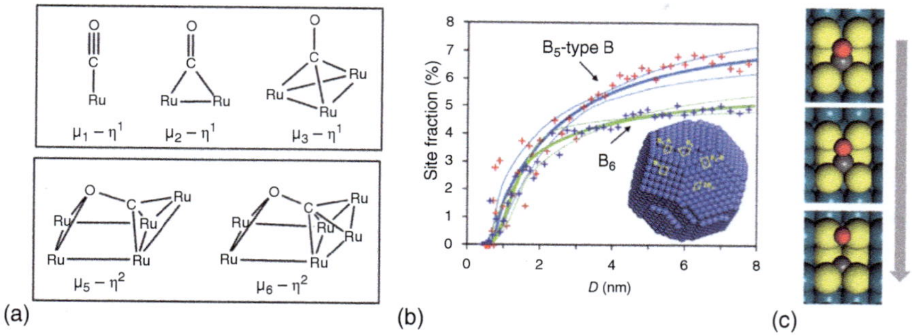

Figure 10.9 (a) Schematic representation of CO adsorption modes on a flat terrace (top) and at step edges (bottom) on extended Ru surfaces. Source: Foppa et al. [113]. Reproduced with permission of American Chemical Society. (b) Evolution of the abundance of B_5 (type B) and B_6 step-edge sites on the surface of cobalt *fcc* NP with the metal particle diameter (the inset shows a 3D model for a Co_{NP} with indication of different types of multi-atom surface sites). Source: Van Helden et al. [114]. Reproduced with permission of Elsevier B.V. (c) DFT-optimized structures for adsorbed CO (top), transition state (TS) for CO direct splitting (middle), and dissociated C^* and O^* atoms (bottom) on a B_6 on a Ru (100)B surface. The arrow points in the direction of the reaction coordinate. Source: Foppa et al. [113]. Reproduced with permission of American Chemical Society.

step-edge sites (e.g. B_5 and B_6 centers), whose abundance on the surface of the metal nanoparticles (Figure 10.9b), as well as intrinsic reactivity toward CO dissociation, both decrease sharply for NP with gradually smaller NP sizes [113, 114, 117, 118]. Unlike in chemical transformations involving non-dissociative CO insertion, the recognized requirement for poly-atom ensembles on metal nanoparticles of a minimum certain size for these reactions to proceed optimally is already an argument against the practicality of SACs.

10.4.2 Water–Gas-Shift Reaction

The WGSR is a central process for the production of hydrogen from hydrocarbon feedstocks, i.e. to enrich in hydrogen syngas mixtures obtained by steam reforming of methane and other light hydrocarbons in natural gas [119]. At present, more than half of the hydrogen obtained by this way is consumed in the production of ammonia, a base chemical with a global demand exceeding 170 million metric tons per year. The WGSR is also technologically important for the adjustment of the H_2/CO molar ratio in syngas streams produced via gasification of coal or biomass raw materials, prior to the Fischer–Tropsch conversion step, in coal-to-liquids (CTLs) and biomass-to-liquids (BTLs) processes. Additionally, it plays an important role in three-way catalytic converters [120]. It is long recognized that certain monoatomic molecular complexes are capable of steering this reaction under remarkably mild operation temperatures (<100 °C) [121]. Mechanistically, most of these monoatomic sites are believed to activate CO via routes analogous to the so-called Hieber base reaction, i.e. starting by the coordination of CO to the metal center, followed by the attack of the resulting metal carbonyl by a hydroxide species,

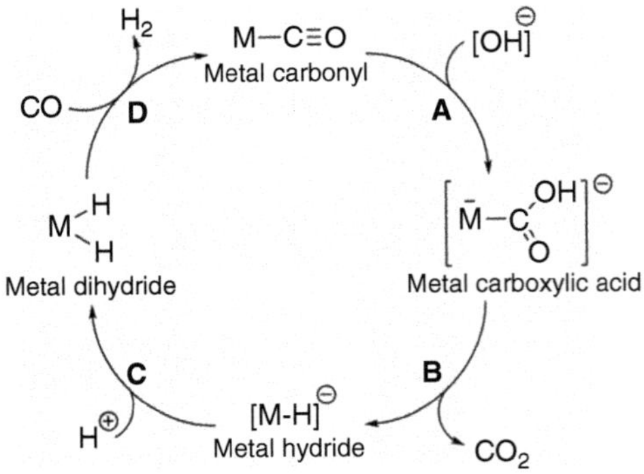

Figure 10.10 Catalytic cycle proposed for the water–gas-shift reaction (WGSR) on mononuclear metal centers. **A**: a metal-coordinated carbonyl ligand is attacked by hydroxide (Hieber base reaction), **B**: decarboxylation of carboxylic acid species to the corresponding metal hydride, **C**: formation of more stable dihydride in the presence of protons, and **D**: final hydrogen elimination in which the two hydrides are substituted by a carbonyl ligand, closing the catalytic cycle. Source: Schaper et al. [122]. Reproduced with permission of Wiley-VCH GmbH.

yielding a metal carboxylic acid, which undergoes degradation to the corresponding metal hydride with CO_2 liberation (Figure 10.10) [122].

The industrial operation of the WGSR is dominated by solid catalysts, which require significantly higher operation temperatures compared to their organometallic counterparts but are notably more stable and easier to handle. While all-oxide materials are utilized in so-called *high-temperature* WGSR processes (>350 °C) [123], oxide-supported metal nanoparticles are customarily applied in *low-temperature* WGSR operations (<350 °C) [124]. In the latter, catalysts based on copper as the major active metal are commonly applied industrially. However, there is significant interest in the development of alternative catalysts based on Pt-group precious metals owing to their higher stability. Therefore, obvious metal cost considerations have spurred intense research efforts, in an attempt to maximize activity and thus reduce the overall metal loading required to attain a given shift reaction rate, particularly in the intensive industry of automobile catalytic converters manufacture. This led to the proposal of isolated metal cations as active WGSR sites already decades ago [125]. It was several years later when Flytzani-Stephanopoulos and coworkers [126] demonstrated that few, isolated, and positively charged metal adatoms, which remained anchored on the oxide support by covalent M—O—$M_{support}$ bonds following the leaching of zerovalent metal species from nanoparticulate Pt- and Au-based catalysts via the formation of soluble cyanide complexes, were responsible for the overall WGSR catalytic activity. Their observation that the metal-specific reaction kinetics was ameliorated after cyanide leaching treatments, in spite of the removal of the majority of the metal from the catalyst surface, led these authors to conclude that those metallic aggregates existing on the surface of the starting catalysts were playing the role of mere spectators, with negligible contributions to the overall reaction rate. This landmark observation initiated a surge of research on SACs for the WGSR, particularly based on cationic Au_{SA} stabilized on different supports

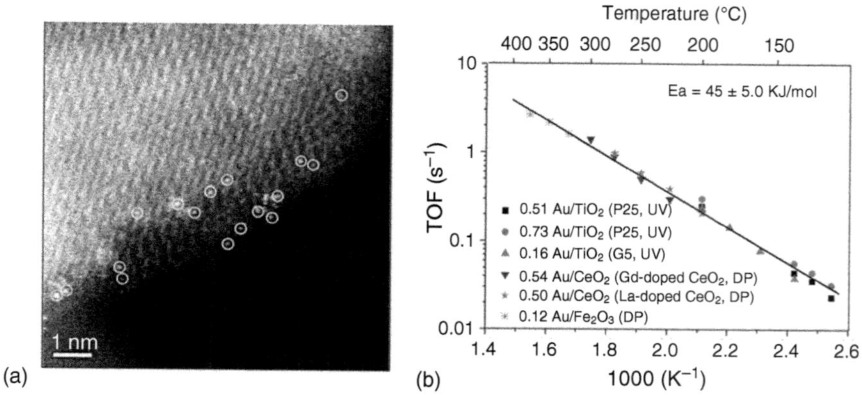

Figure 10.11 (a) HAADF–STEM micrograph of 1.16 wt% Au_{SA}/TiO_2 catalyst synthesized by ultraviolet (UV)-assisted deposition–precipitation, followed by air calcination, after catalytic use in the water–gas-shift reaction with a gas feed composition of 10% CO, 3% H_2O, balance He, at 100 °C for four hours and two treatments to leach metallic gold out of the catalyst surface with a 0.05 wt% NaCN aqueous solution. The circles on the micrograph indicate atom-sized high Z-contrast motifs interpreted as isolated gold atoms. (b) Arrhenius plot for the WGSR over cyanide-leached Au/CeO_2, Au/Fe_2O_3, and Au/TiO_2 catalysts with a feed gas composition simulating a reformate gas mixture (11% CO, 26% H_2O, 7% CO_2, 26% H_2, balance He). P25 designates commercial aeroxide TiO_2 P25 (Evonik) as catalyst support, while DP and UV mean deposition–precipitation and ultraviolet light-assisted deposition–precipitation as catalyst synthesis method, respectively. Source: Flytzani-Stephanopoulos et al. [127]. Adapted with permission of American Chemical Society.

(Figure 10.11). The nature of the oxide support was found to affect the catalytic performance [127]. However, in a follow-up study, Flytzani-Stephanopoulos and coworkers [128] observed that the major role of the support, as well as alkali promoters that had a positive effect on performance even at low loadings, is to ensure the stabilization of atomically dispersed, positively charged metal atoms, the latter being fully responsible for the activation of both CO and H_2O reactants. Later on, Zhang and coworkers [129] reported even higher WGSR activities with an Ir_{SA}/Fe_2O_3 SAC. In this study, however, a mechanism involving water dissociation on lattice oxygen centers of the Fe_2O_3 support and CO activation on the Ir_{SA} centers was proposed, hence involving the cooperation of the isolated 5d metal centers with adjacent sites on the partially reduced oxide support. More recently, a different and rather unusual reaction pathway was proposed by Rivero–Crespo et al. [130] to account for the very-low-temperature (50 °C) WGSR activity observed for Pt_{SA} centers stabilized by a regular first-coordination *shell* of water molecules within the channels of a metal–organic-framework scaffold. The proposed reaction is mediated by a dynamic behavior of the water cluster coordinated to the monoatomic metal center, with both oxygen atoms in the CO_2 product originating from the water reactant.

10.4.3 CO Oxidation

CO oxidation is a reaction of eminent significance in the context of the abatement of end-pipe emissions from internal combustion engines. A combination of low-temperature activity, resistance to CO poisoning as well as high-temperature stability is sought after to achieve sufficient performance under both cold-start and regime operation conditions in mobile applications [131]. In the presence of large excess of hydrogen, the preferential oxidation of CO (PROX) reaction [132] is pivotal for the purification of hydrogen streams to those levels required for their application as an energy vector in proton-exchange membranes (PEM) fuel cells, whose metal-based anode catalysts are highly susceptible to CO poisoning [133]. Under PROX conditions, the inhibition of competitive hydrogen dissociation and oxidation pathways is essential.

In part owing to the fact that CO oxidation is a technically simple reaction to operate, which does not pose major analytical challenges, CO oxidation is also very widely used as a showcase oxidation reaction in laboratory research. This also applies to the field of single-atom catalysis, where CO oxidation was the reaction of choice in the study, which first coined the term *single-atom catalyst* [134], and it continues to be routinely applied as a standard test reaction to assess the reactivity of atomically dispersed metals. This section discusses succinctly aspects of CO oxidation catalysis with SAC related to the adsorption and activation of CO, as well as active site dynamics triggered by gas atmosphere composition and operation temperature in this thermocatalytic process. Chapter 9 discusses mechanistic aspects for this and other oxidation reactions on SACs.

Three major mechanisms have been proposed for CO oxidation on metal catalysts. The Langmuir–Hinshelwood (LH) mechanism entails the co-adsorption of CO and O_2 reactants on a single site, and it typically proceeds via a peroxo-species ($^{\cdot}$OCOO) as intermediate [135]. In the Eley–Rideal (ER) mechanism, adsorption and activation of O_2 precedes the reaction with CO, or alternatively O_2 reacts directly to a surface metal carbonyl, typically leading to surface carbonate-like intermediates [136]. Finally, the MvK mechanism postulates that CO is oxidized by lattice oxygen, typically at the boundary between a supported metal and the oxide support, and O_2 dissociation replenishes those surface oxygen vacancies left behind upon CO_2 desorption [137].

In the former two mechanistic proposals, the adsorption and activation of CO on metal sites are expected to play a relevant role. Recently, Parkinson and coworkers [138] took a combined theoretical and experimental (surface science) approach to study the adsorption of CO on isolated adatoms of various transition metals (Cu, Ag, Au, Ni, Pd, Pt, Rh, and Ir) stabilized on an $Fe_3O_4(001)$ surface, and compared the resulting adsorption energetics to that of CO on single crystal surfaces of the same metals. While the former are well-defined models for SACs, the latter are more representative of conventional nanoparticulate catalysts, which expose comparatively extended metallic facets to gaseous reactants. They observed that, generally, the twofold binding of isolated metal atoms to the oxide support shifts the d-band center-of-mass toward the Fermi level, increasing the C—M bond strength with respect to CO adsorbed on the corresponding extended metal surfaces. Moreover, the magnitude of the differences in adsorption energy between SAs and metal facets was found to be metal dependent. These results with well-defined surface science systems illustrate that monoatomicity as well as the direct and multifold coordination of the isolated metal atoms to an oxide support in SACs are generally expected to strengthen CO binding. The CO binding strength is an important parameter to position different metals in virtue of their appropriateness as CO oxidation catalysts on a so-called Sabatier volcano plot (reactivity *vs.* reactant binding strength). Therefore, those changes identified in the CO binding strength for isolated metal atoms, compared to the corresponding extended metal surfaces, are expected to lead to displacements in the positions of different metals relative to the apex in such volcano plots when SAC, rather than nanoparticulate catalysts, are considered.

A large number of studies have addressed the performance of SACs for CO oxidation experimentally. However, no consensus has been achieved as to whether atomic dispersion may offer catalytic benefits compared to more conventional oxide-supported clusters or nanoparticles. Hence, while some previous studies underscored the high activity of SACs at low reaction temperatures (<100 °C) [134, 139], other works have reported only moderate performance and comparatively high reaction-onset temperatures [46, 140–142]. The origins for these apparent discrepancies remain unclear at present. They could be rooted in the different reaction environments applied in different studies, which range from lean [142–144] to stoichiometric [145] and rich conditions [146]. Working with $MgAl_2O_4$-supported Ir catalysts, Lu et al. [139] showed a notable difference in the reaction orders for O_2 and CO reactants on either Ir_{SA} or Ir_{NP}. While on Ir_{NP}, the experimentally determined reaction orders were -1.0 ± 0.1 and 0.9 ± 0.1 for CO and O_2, respectively, both reaction orders were positive and took values of 0.9 ± 0.1 and 0.1 ± 0.05, respectively, on Ir_{SA} under the same reaction conditions. These findings suggested that nanoparticles might be superior catalysts under lean reaction conditions, whereas atomically dispersed species could be more reactive under CO-rich conditions.

Next to kinetic considerations, structural changes have been proposed to perhaps underlie the disagreement as to the superiority of either oxide-supported SACs or supported metal aggregates as CO oxidation catalysts. Importantly, changes in the composition of the gas atmospheres to which the SACs are exposed during activation or catalysis have been observed to modify, in certain instances boost, their CO oxidation catalytic activity. Nie et al. [143] demonstrated that steaming a Pt_{SA}/CeO_2 SAC at 750 °C resulted in a significant decrease in the onset temperature for CO oxidation, and it was therefore critical to achieve activity at temperatures around 100 °C. The researchers ascribed the improved reactivity to a new type of $O_{lattice}[H]$ active site developed on CeO_2 in the direct vicinity of isolated Pt^{2+} sites. Hence, the involvement of redox-active sites from the oxide support was proposed to be essential to enhance the catalytic activity of the Pt-based SAC.

Dynamic structural changes in the local coordination structure and *confinement degree* of the monoatomic metal centers, in response to changes in temperature and atmosphere composition, have also been proposed on the basis of both experimental and computational evidences [142, 147]. In certain instances, variations in the oxidizing/reducing nature of the chemical environment have been found to drive clustering. These effects alter the starting monoatomicity of the metal centers and could thus, if unnoticed, lead to erroneous assignment of catalytic performance to single-atom centers. This was highlighted by Corma and coworkers [145] for a number of reactions, including selective hydrogenation of nitro-compounds, alkane dehydration, but also CO oxidation. They tracked the atomicity of Pt catalysts, atomically dispersed on various oxide supports (Al_2O_3, TiO_2, and CeO_2), by means of X-ray absorption and *in situ* FTIR spectroscopy using CO as the surface probe. Their results showed that metal clustering occurred in all cases during the exposure of the catalysts to CO oxidation conditions with a 1% O_2:2% CO:97% He feed stream, pointing to the nascent metal NPs as the sites responsible for the catalytic activity. Sarma et al. [142] studied CO oxidation reactivity trends in a series of 4d (Ru, Rh, Pd) and 5d (Ir, Pt) metals atomically dispersed on MgO using a lean gas feed (1% CO:20% O_2:79% He). The high energy of formation of oxygen vacancies on MgO ($>8 \times 10^2$ kJ/mol) makes this oxide particularly suitable to investigate the reactivity of the atomically dispersed metals without contribution from redox centers on the support. It was found that shortage of O_2 in the reaction atmosphere, even for short periods of time, i.e. a transient increase in the reduction potential of the gas environment, led to permanent enhancements in CO oxidation performance. Their scanning-transmission electron microscopy (STEM) and *in situ* CO-FTIR studies served to connect this improved catalytic activity to metal reductive agglomeration into small clusters, and thus suggested the higher reactivity of metal aggregates compared to the isolated metal centers. It stems from the aforementioned studies that tracking and understanding the atomic-level *fluxionality* of the monoatomic centers, in response to changes in their chemical environment, is essential to both rationalize their catalytic properties and to clarify whether isolated atoms or polynuclear structures derived therefrom dominate catalysis. In this regard, Maurer et al. [148] recently reported the observation of reversible clustering of initially single-atom Pt_{SA}/CeO_2 catalysts to be highly relevant for oxidation catalysis. Authors applied sophisticated spectroscopic methods, namely ultra-high-vacuum Fourier-transform infrared (UHV-FTIR) and high-energy resolution fluorescence-detection (HERFD) X-ray absorption spectroscopies to track the dynamics of Pt_{SA}/CeO_2 *at work*, i.e. during their action in the oxidation of CO, and light alkanes, respectively. The results suggest the cleavage of Pt—O—Ce bonds in the starting atomically dispersed catalyst, at temperatures >150 °C under the reaction atmosphere. This process leads to a fast assembly of the metal into polynuclear $Pt_x^{\delta+}$ clusters on the ceria surface (Figure 10.12).

The fact that this structural change was found to coincide with the onset of oxidation activity led the researchers to speculate that metal clusters derived from the single-atom precursor, rather than the isolated metal centers themselves, are the true (at least the dominant) active species. Interestingly, after a first catalysis light-off experiment, cooling the catalyst under O_2 led to the redispersion of these metal entities, and thus the reestablishment of the essentially inactive single Pt atom motifs on the catalyst surface. Their results not only underscore the importance of understanding the dynamic evolution of SACs during catalytic action, but in addition suggest that fast and fully reversible phenomena might be at play, which might be responsible for the *in situ* generation of polynuclear active sites even if only monoatomic species prevail both prior to and after catalysis.

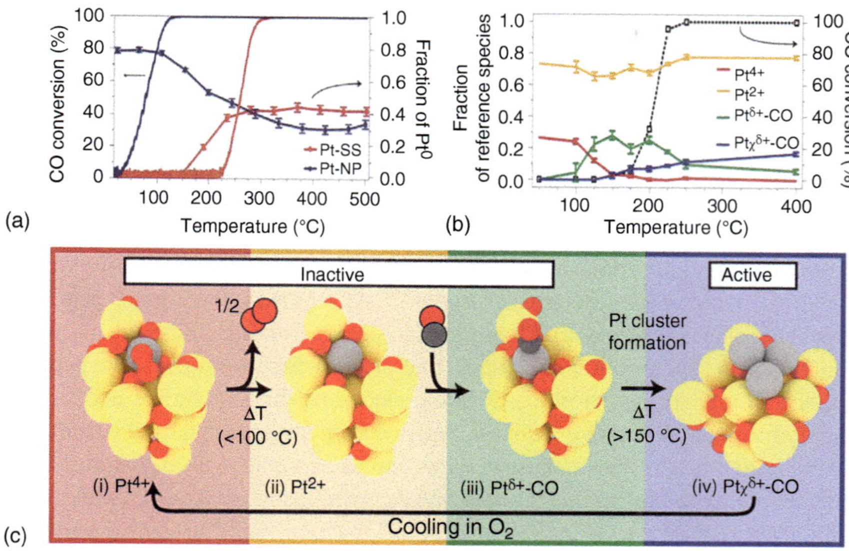

Figure 10.12 (a) CO conversion and linear combination fitting of transient X-ray absorption near-edge structure (XANES) spectra for nanoparticulate (Pt-NP) and atomically dispersed (Pt-SS) Pt/CeO_2 catalysts (0.94 wt% Pt) during heating (5 °C/min) in 1000 ppm CO and 10% O_2/He. (b) Catalytic activity (1000 ppm CO, 10% O_2 in He) during a temperature-resolved HERFD-XANES experiment, and fraction of Pt^{4+}, Pt^{2+}, $Pt^{\delta+}$ with adsorbed CO and $Pt_x^{\delta+}$ clusters, obtained by linear combination analysis (with standard error) using references obtained by a multivariate curve resolution alternating least square analysis of the HERFD-XANES spectra. (c) Schematic representation of the proposed reversible formation of the catalytically active $Pt_x^{\delta+}$ cluster based on *operando* HERFD-XANES analysis. Atom color codes: gray, Pt; yellow, Ce; red, O; dark gray, C. Source: Maurer et al. [148]. Reproduced with permission of Springer Nature.

10.4.4 Other CO Conversion Catalysis with SACs

SACs have also been reported to activate CO in important oxo-functionalization reactions such as the hydroformylation of olefins to aldehydes, a prototypical reaction of notable industrial significance, which is ordinarily catalyzed by monoatomic metal complexes in liquid phase [149]. Rhodium is particularly active for this reaction under mild conditions. Different strategies have been pursued to immobilize Rh organometallic complexes on various substrates to achieve the heterogenization of these catalysts and thus benefit from the technically uncomplicated recovery and recycling of solids [150]. Among these solid versions of hydroformylation catalysts, oxide-supported SACs are all-inorganic materials with higher thermal and chemical stabilities, which are features desired for catalyst synthesis and rejuvenation treatments. Zhang and coworkers [151] first reported the application of atomically dispersed Rh_{SA}/ZnO for the hydroformylation of aliphatic as well as aromatic olefins. With these solid catalysts, *TONs* up to 4×10^4 were achieved at 100 °C, higher than those attained with a conventional molecular $RhCl(PPh_3)_3$ catalyst under the same reaction settings. Zhang and coworkers exploited the activity of Rh_{SA}/CeO_2 for the water–gas-shift reaction under mild temperatures (120 °C) to catalyze a tandem WGSR–hydroformylation process using CO and styrene as reactants in water medium. The production of co-reactant hydrogen *in situ*, via the WGSR, was proposed to be essential for the enhanced selectivity observed to the linear (anti-Markovnikov) aldehyde product compared to the case where syngas was applied as the gas phase [152]. Single-atom Rh_{SA}/CoO were also found by Wang et al. [153] to be more effective than the corresponding cluster and nanoparticulate-based catalysts, as well as particularly regioselective to the linear (anti-Markovnikov) aldehyde product, in the hydroformylation of gaseous olefins such as propene in a gas–liquid–solid triphasic conversion process. In a complementary computational/experimental study, Amsler et al. [154] studied monoatomic rhodium carbonyl hydride complexes on different oxide surfaces as single-atom hydroformylation catalysts. The authors concluded that striking a compromise between site stability and activity, as determined by the degree of surface confinement of the monoatomic metal centers by the oxide support, is important to realize hydroformylation activities comparable to those of molecular catalysts. Rhodium centers loosely bonded to the oxide carrier might show higher hydroformylation *turnover* rates but are highly susceptible to leach into the liquid reaction medium as free carbonyl complexes. Oppositely, metal centers stabilized in higher coordination (even if near-surface) lattice sites of the oxide carrier are more stable but lack the coordination flexibility required to effectively co-activate the olefin and CO reactants.

10.5 Activation and Selective Conversion of Other Small Molecules with SACs

Aside from C_1 compounds, SACs have shown promise for the thermocatalytic selective activation and conversion of unsaturated hydrocarbons, alkenes, and alkynes, which are important building blocks in the current chemical industry. Selected examples shall be discussed in this section.

A prominent example is the hydrochlorination of acetylene, the lightest alkyne, to vinyl chloride. Vinyl chloride is the monomer for the manufacture of polyvinyl chloride (PVC), a polymer with a global annual demand exceeding 50 million ton. Alternative to ethylene oxychlorination, acetylene hydrochlorination represents a proven and high carbon-yield source of vinyl chloride from carbon feedstocks, and it is particularly attractive in areas with extensive exploitable coal reserves, such as Asia. Since the 1950s, $HgCl_2$ supported on activated carbon has been the benchmark catalyst in industrial acetylene hydrochlorination processes. However, obvious environmental and safety concerns are driving forces for the replacement of mercury-based catalysts [155]. Based on his observation of a scaling relationship between the reaction rate and the standard electrode potential of various carbon-supported metal chloride salts, Graham Hutchings, at that time employee at African Explosives and Chemical Industries (AECI), hypothesized that gold could be an alternative catalyst [156]. Only few years later, he and his team validated this proposal, and demonstrated that catalysts consisting of Au finely dispersed on carbon are active for this reaction [157, 158]. Ever since, significant research efforts were devoted to optimize catalyst performance and rationalize the nature of the active hydrochlorination sites [159]. It was early recognized that the synthesis route was determinant for the ultimate performance, with catalysts prepared by impregnation with strongly oxidizing *aqua regia* solutions of gold precursor salts leading to the best performances. Moreover, various *ex situ* characterization results pointed to the stabilization of cationic gold species on the catalyst surface to be necessary for activity. While the starting catalyst contained, in many cases, gold NPs, a combination of *operando* X-ray absorption spectroscopy (XAS) and post-reaction high-angle annular dark-field-scanning-transmission electron microscopy (HAADF–STEM) microscopy led Malta et al. [160] to propose atomically dispersed cationic Au adatoms, generated upon exposure of the gold catalysts to reaction conditions, as the true active centers. Similar to the case of supported gold catalysts for the WGSR, discussed previously in this chapter, these findings suggest that zerovalent gold aggregates do not contribute to the overall reactivity. Therefore, a sensible blueprint for the optimization of gold usage, and thus catalyst cost, is the achievement of a 100% metal dispersion. In 2015, Au/C catalysts entered industrial commercialization for acetylene hydrochlorination processes in China, representing the first example of a radical change in the catalyst formulation for an industrial process of manufacture of a major commodity chemical over the last half a century [161]. More recently, Pérez-Ramírez and coworkers have extended the concept to Pt and Pt–Au atomically dispersed catalysts, which showed improved stability under hydrochlorination reaction conditions [162, 163].

Olefins are very important base chemicals for the current chemical industry. The major use of light (C_2–C_4) olefins is as monomers in the production of polyolefins. Diene compounds, e.g. butadiene, are known to act as poisons for polymerization catalysts when present in even at minute concentrations (<10 ppm) in unpurified propylene feeds. Hence, diene removal by selective hemi-hydrogenation of 1,3-butadiene to butenes is an important process to deep-purify alkene feedstock prior to polymer manufacture processes. The reaction has been studied by Gates and coworkers [164] as a model reaction to understand the catalytic behavior of isolated metal atoms (and low-nuclearity metal nanoclusters) stabilized on the surface of oxide supports. This research group is one of the pioneers in the field of single-atom catalysis and provided, over the years, sustained and rigorous insights into the coordination flexibility, ligand effects, and dynamics of monoatomic metal centers during activation treatments as well as catalysis usage, features which they emphasize by referring to this catalytic entities as *supported molecular catalysts* [165]. Feeding parahydrogen as the reducing agent, Corma et al. [166] found that the hydrogenation of 1,3-butadiene to 1-butene (and butane) resulted in a pairwise hydrogen addition degree in the products, a prototypical feature of hydrogenation reactions proceeding on monoatomic molecular catalysts, which was at least an order of magnitude higher when catalyzed by a solid Au_{SA}/MWCNT SAC compared to supported Au_{NP}. This was taken as an evidence for the mimicry of the former with molecular catalysts. More recently, systematic studies by Babucci, Gates, Uzun, and coworkers [167, 168] demonstrated that the adjustment of the electronic density on the isolated metal sites in Ir_{SA}/Al_2O_3 SAC, via the tuning of the electron-donating/withdrawing character of the oxide support and/or the direct contact of the isolated metal centers with nm-thin ionic liquid overlays, provides an important handle on selectivity. Such studies unveil the ample possibilities existing to tune the catalytic performance of single-atom centers, beyond the mere selection of the metal identity, i.e. by additionally controlling the electron-donating/withdrawing character of their direct chemical environment. This strategy is expected to be more powerful with mononuclear sites, as those in SACs, compared to regular metal aggregates, e.g. nanoparticles. The higher metallic character of the latter and thus higher capacity to delocalize electronic charge render them less amenable to fine-tuning of the electronic density on all those metal centers exposed on their surface and thus available to contribute to the catalytic performance.

Other important olefin functionalization reactions, which are classical examples of reactions catalyzed by organometallic complexes in solution, have been demonstrated with SACs. The hydroformylation of olefins with syngas to produce aldehyde derivatives is a relevant example, which has already been discussed in Section 10.4.4 [151, 153, 154]. Additionally, olefin hydroamination, i.e. the formation of C—N bonds [169], as well as olefin hydrosilylation, i.e. the formation of C—Si bonds [170, 171], have been reported, with TOF comparable to those attained with molecular catalysts in solution. An interesting extension of single-atom catalysis is the integration of two different SACs to achieve sequential one-pot transformations. This was the concept adopted by Sarma et al. [171], which combined Rh_{SA}/CeO_2 and Ru_{SA}/CeO_2 catalysts in a tandem catalytic process integrating olefin isomerization and hydrosilylation reactions in a single pot. Owing to the high specificity observed for Ru and Rh isolated metal centers for olefin double-bond migration and the anti-Markovnikov hydrosylilation of terminal olefins, respectively, the integration of both catalysts enabled the transformation of industrial mixtures of terminal and internal aliphatic olefins into terminal organosilanes with a high regioselectivity, which was out of reach for either of the two catalysts individually. DFT calculations suggested that differences in the olefin binding strength onto CeO_2-stabilized Ru_{SA} and Rh_{SA} underlie the remarkable reaction specificity offered by these SACs, and provided an energetically plausible path for the tandem process wherein double-bond chain walking proceeds faster on the Ru-based catalyst, and terminal olefin intermediate products are transferred to the Rh-based catalyst, via the continuous liquid phase, where hydrosylilation is highly preferred over further double-bond isomerization (Figure 10.13).

The concept of tandem catalysis originated in the field of homogeneous catalysis with soluble molecular complexes [172]. It holds the promise for a multifold contribution to chemical process intensification: (i) by avoiding energy- and cost-intensive isolations of intermediate products, (ii) by improving safety and selectivity through the minimization of the residence time of highly reactive or unstable intermediate products in the reaction medium, or (iii) by driving reversible reactions to completion via the *in situ* processing of a primary reaction product in a subsequent irreversible conversion step. In the realm of tandem catalysis, oxide-supported SACs hold the potential to provide well-defined monoatomic sites, as well as their compartmentalization on non-contacting solid matrices. This is

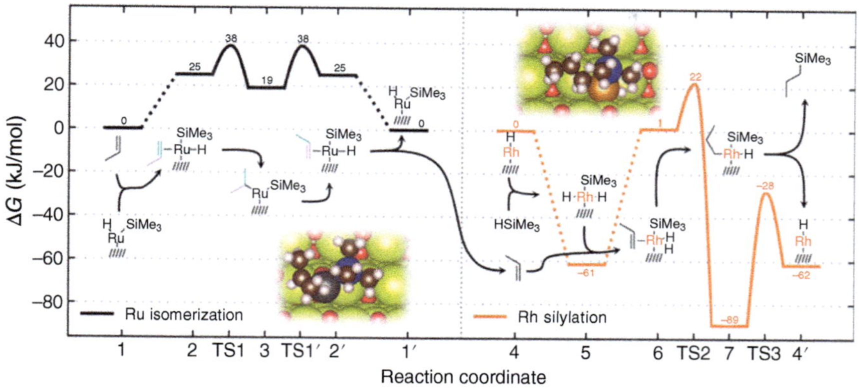

Figure 10.13 DFT-computed free-energy diagram of a tandem olefin isomerization/hydrosilylation process involving olefin isomerization catalyzed by a Ru_{SA}/CeO_2 SAC, and subsequent hydrosilylation with $HSiMe_3$ catalyzed by a Rh_{SA}/CeO_2 SAC. Reactants considered: propene as model olefin, $HSiMe_3$ as silylating agent. $T = 120\,°C$, $P = 10$ bar. Source: Sarma et al. [171]. John Wiley & Sons. CC BY 4.0.

reminiscent of the original compartmentalization strategies for soluble molecular catalysts in partially permeable vesicles, micelles, or dendrimer molecular structures [173, 174]; however, now with fully inorganic, thus easily recoverable, thermally and mechanically more stable materials. Hence, additional research efforts and further advances are to be expected in this area.

10.6 Concluding Remarks

Isolated metal atoms stabilized on oxide supports have shown to be promising catalysts for the activation of small molecules such as CO, CO_2, and CH_4 and light alkanes. Therefore, they currently deserve a significant deal of attention in the context of the valorization of unconventional C_1 carbon feedstock. SACs are interesting as mimics of mononuclear organometallic molecular catalysts, which are known to be particularly active for the activation of C—H bonds in methane and light alkanes, as well as the (r)WGSR, which interconverts both carbon oxides, thus potentially bridging waste CO_2 as reactant with the well-established conversion chemistry of CO.

For different catalytic reactions, atomically dispersed – thus partially charged – *single atoms* have been revealed to be responsible for catalytic performances previously ascribed to metal nanoclusters/nanoparticles, which coexist with the former on the surface of supported metal catalysts. This is the case, for instance of the WGSR and the hydrochlorination of acetylene, processes for which these discoveries suggest means to minimize the cost of precious-metal-based catalysts by achieving 100% metal dispersion on the support surface. On the contrary, SACs seem not to be a feasible solution for other thermocatalytic processes, which have proven to require poly-atom ensembles – on the surface of metal nanoparticles – as active centers, e.g. the dissociation and hydrogenation of CO in syngas conversion processes such as methanation and the FTS of hydrocarbons.

Progress has also been achieved in CH_4 conversion catalysis, with few examples of SACs already reported to deliver higher selectivities to high added-value partial oxygenation products, such as methanol and/or acetic acid, compared to their nanoparticulate counterparts. However, reaction rates and yields remain lower than the best performance levels demonstrated for processes catalyzed by molecular catalysts in solution, such as the so-called Periana-Catalytica process in concentrated sulfuric acid and oleum media. Further work is required to assess whether solid catalysts might provide a true alternative, effective in more benign reaction environments, should reaction throughputs be further improved to the magnitude required to stimulate technical interest.

Unlike in electrocatalytic processes, where the application of a voltage bias enables transformations to be activated at very mild temperatures, the application of SACs in thermocatalytic processes requires typically temperatures well above ambient to proceed. In gas–solid processes, this might trigger (reversible) metal clustering on the catalyst surface. In (gas-)liquid–solid processes, leaching of the metal into the solution in the form of soluble species – often following complexation with reactants and/or products – and possibly also metal agglomeration via Ostwald ripening phenomena, might take place under reaction conditions. It, therefore, behooves scientists to conclusively assess whether the catalytic activity ascribed to SACs is *bona fide* or, on the contrary, associated with molecular or colloidal species, which evolved *in situ* from the atomically dispersed "pre-catalyst."

Both the reactivity, intimately connected to the coordination flexibility around the isolated metal atom, and the stability against migration and clustering, desorption or leaching, depend on the degree of confinement experienced by the metal center from the oxide support. The latter might be significantly different for catalysts obtained via adsorption/impregnation methods or more severe oxidative redispersion of metal (oxide) crystallites at high temperatures (>600 °C). As exemplified in this chapter, the application of sensitive characterization techniques both prior to and after the catalytic action is important. Physicochemical methods, which provide a significant sampling, e.g. X-ray absorption spectroscopies, should always complement other techniques, which are limited to very local scrutiny, e.g. electron microscopy (see Chapter 5). In liquid-phase catalysis, hot-filtration tests, at sufficiently high reactant conversion levels, yet far from full conversion, are also highly recommended to check for the heterogeneity of the reaction. Yet, a dynamic *fluxionality* of the species under operation leaves room for other type of metal centers being responsible for the macroscopically observed performance, even if only monoatomic species are identified both before and after catalysis. Hence, the development and application of *operando* physicochemical methods to track the structure of the active centers during their catalytic action is expected to play a central role for the rational development of this thriving research area (see Chapter 6).

Acronym

AECI	African Explosives and Chemical Industries
BDE	Bond dissociation energy

BTL	Biomass-to-liquids
CT	Charge transfer
CTL	Coal-to-liquids
DFT	Density functional theory
DRIFT	Diffuse reflectance infrared Fourier transform
ER	Eley–Rideal
FDH	Formate dehydrogenase
FTIR	Fourier-transform infrared
GN	Graphene nanosheets
HAADF	High-angle annular dark field
HERFD	High-energy resolution fluorescence detection
HOMO	Highest occupied molecular orbital
LH	Langmuir–Hinshelwood
LUMO	Lowest unoccupied molecular orbital
MvK	Mars van Krevelen
PROX	Preferential oxidation of CO
PVC	Polyvinyl chloride
rWGSR	Reverse water–gas-shift reaction
SAC	Single-atom catalyst
STEM	Scanning-transmission electron microscopy
STY	Space–time yield
TEM	Transmission electron microscopy
TOF	Turnover frequency (number of reactant molecules converted per unit (metal) active center and time)
TON	Turnover number (cumulative number of reactant molecules converted per unit (metal) active center)
UHV	Ultra-high vacuum
UV	Ultraviolet
WGSR	Water–gas-shift reaction
XANES	X-ray absorption near-edge structure
XAS	X-ray absorption spectroscopy
ZSM	Zeolite Socony Mobil

References

1 Elvidge, C.D., Bazilian, M.D., Zhizhin, M. et al. (2018). The potential role of natural gas flaring in meeting greenhouse gas mitigation targets. *Energy Strategy Reviews* 20: 156–162.

2 Schwach, P., Pan, X., and Bao, X. (2017). Direct conversion of methane to value-added chemicals over heterogeneous catalysts: challenges and prospects. *Chemical Reviews* 117 (13): 8497–8520.

3 Tang, P., Zhu, Q., Wu, Z. et al. (2014). Methane activation: the past and future. *Energy & Environmental Science* 7 (8): 2580–2591.

4 Caballero, A. and Pérez, P.J. (2013). Methane as raw material in synthetic chemistry: the final frontier. *Chemical Society Reviews* 42 (23): 8809–8820.

5 Meng, X., Cui, X., Rajan, N.P. et al. (2019). Direct methane conversion under mild condition by thermo-, electro-, or photocatalysis. *Chem* 5 (9): 2296–2325.

6 Park, D. and Lee, J. (2013). Biological conversion of methane to methanol. *Korean Journal of Chemical Engineering* 30 (5): 977–987.

7 Li, C., Yan, W., and Xin, Q. (1994). Interaction of methane with surface of alumina studied by FT-IR spectroscopy. *Catalysis Letters* 24 (3–4): 249–256.

8 Yoshizawa, K. and Miyanishi, M. (2020). Orbital concept for methane activation. In: *Direct Hydroxylation of Methane: Interplay Between Theory and Experiment* (ed. K. Yoshizawa), 1–22. Springer Nature.

9 Sher Shah, M.S.A., Oh, C., Park, H. et al. (2020). Catalytic oxidation of methane to oxygenated products: recent advancements and prospects for electrocatalytic and photocatalytic conversion at low temperatures. *Advanced Science* 7 (23): 1–24.

10 Schwarz, H. (2011). Chemistry with methane: concepts rather than recipes. *Angewandte Chemie – International Edition* 50 (43): 10096–10115.

11 Roudesly, F., Oble, J., and Poli, G. (2017). Metal-catalyzed C–H activation/functionalization: the fundamentals. *Journal of Molecular Catalysis A: Chemical* 426: 275–296.

12 Ravi, M., Ranocchiari, M., and van Bokhoven, J.A. (2017). The direct catalytic oxidation of methane to methanol – a critical assessment. *Angewandte Chemie – International Edition* 56 (52): 16464–16483.

13 Shilov, A.E. and Shul'pin, G.B. (1997). Activation of C–H bonds by metal complexes. *Chemical Reviews* 97 (8): 2879–2932.

14 Labinger, J.A. and Bercaw, J.E. (2002). Understanding and exploiting C—H bond activation. *Nature* 417: 507–514.

15 Copéret, C. (2010). C—H bond activation and organometallic intermediates on isolated metal centers on oxide surfaces. *Chemical Reviews* 110 (2): 656–680.

16 Olivos-Suarez, A.I., Szécsényi, À., Hensen, E.J.M. et al. (2016). Strategies for the direct catalytic valorization of methane using heterogeneous catalysis: challenges and opportunities. *ACS Catalysis* 6 (5): 2965–2981.

17 Tang, Y., Li, Y., Fung, V. et al. (2018). Single rhodium atoms anchored in micropores for efficient transformation of methane under mild conditions. *Nature Communications* 9 (1): 1–11.

18 Bai, S., Liu, F., Huang, B. et al. (2020). High-efficiency direct methane conversion to oxygenates on a cerium dioxide nanowires supported rhodium single-atom catalyst. *Nature Communications* 11 (1): 1–9.

19 Kwon, Y., Kim, T.Y., Kwon, G. et al. (2017). Selective activation of methane on single-atom catalyst of rhodium dispersed on zirconia for direct conversion. *Journal of the American Chemical Society* 139 (48): 17694–17699.

20 Kokalj, A., Bonini, N., Sbraccia, C. et al. (2004). Engineering the reactivity of metal catalysts: a model study of methane dehydrogenation on Rh(111). *Journal of the American Chemical Society* 126 (51): 16732–16733.

21 Gandeepan, P., Müller, T., Zell, D. et al. (2019). 3D transition metals for C–H activation. *Chemical Reviews* 119 (4): 2192–2452.

22 Marcinkowski, M.D., Darby, M.T., Liu, J. et al. (2018). Pt/Cu single-atom alloys as coke-resistant catalysts for efficient C–H activation. *Nature Chemistry* 10 (3): 325–332.

23 Shan, J., Li, M., Allard, L.F. et al. (2017). Mild oxidation of methane to methanol or acetic acid on supported isolated rhodium catalysts. *Nature* 551 (7682): 605–608.

24 Huang, W., Zhang, S., Tang, Y. et al. (2016). Low-temperature transformation of methane to methanol on Pd_1O_4 single sites anchored on the internal surface of microporous silicate. *Angewandte Chemie* 128 (43): 13639–13643.

25 Sahoo, S., Suib, S.L., and Alpay, S.P. (2018). Graphene supported single atom transition metal catalysts for methane activation. *ChemCatChem* 10 (15): 3229–3235.

26 Cui, X., Li, H., Wang, Y. et al. (2018). Room-temperature methane conversion by graphene-confined single iron atoms. *Chem* 4 (8): 1902–1910.

27 Li, H., Zhang, H., Yan, X. et al. (2018). Carbon-supported metal single atom catalysts. *New Carbon Materials* 33 (1): 1–11.

28 Zhao, M., Feng, J., Yang, W. et al. (2021). Recent advances in graphitic carbon nitride supported single-atom catalysts for energy conversion. *ChemCatChem* 13 (5): 1250–1270.

29 Zimmermann, T., Bilke, M., Soorholtz, M. et al. (2018). Influence of catalyst concentration on activity and selectivity in selective methane oxidation with platinum compounds in sulfuric acid and oleum. *ACS Catalysis* 8 (10): 9262–9268.

30 Zimmermann, T. (2016). Selective Oxidation of Methane in Sulfuric Acid. PhD thesis. Ruhr Universität Bochum (Germany). URL: https://hss-opus.ub.ruhr-uni-bochum.de/opus4/frontdoor/index/index/docId/4816.

31 Bertau, M., Offermanns, H., Plass, L. et al. (2014). *Methanol: the Basic Chemical and Energy Feedstock of the Future*. Springer.

32 Sunley, G.J. and Watson, D.J. (2000). High productivity methanol carbonylation catalysis using iridium. *Catalysis Today* 58 (4): 293–307.

33 Lange, J.-P. (2001). Methanol synthesis: a short review of technology improvements. *Catalysis Today* 64 (1): 3–8.

34 Budiman, A.W., Nam, J.S., Park, J.H. et al. (2016). Review of acetic acid synthesis from various feedstocks through different catalytic processes. *Catalysis Surveys from Asia* 20 (3): 173–193.

35 Mironov, O.A., Bischof, S.M., Konnick, M.M. et al. (2013). Using reduced catalysts for oxidation reactions: mechanistic studies of the "Periana-Catalytica" system for CH_4 oxidation. *Journal of the American Chemical Society* 135 (39): 14644–14658.

36 Periana, R.A., Taube, D.J., Gamble, S. et al. (1998). Platinum catalysts for the high-yield oxidation of methane to a methanol derivative. *Science* 280 (5363): 560–564.

37 Zhang, Z.-G. (2019). Process, reactor and catalyst design: towards application of direct conversion of methane to aromatics under nonoxidative conditions. *Carbon Resources Conversion* 2 (3): 157–174.

38 Guo, X., Fang, G., Li, G. et al. (2014). Direct, nonoxidative conversion of methane to ethylene, aromatics, and hydrogen. *Science* 344 (6184): 616–619.

39 Liu, Y., Liu, J.C., Li, T.H. et al. (2020). Unravelling the enigma of nonoxidative conversion of methane on iron single-atom catalysts. *Angewandte Chemie – International Edition* 59 (42): 18586–18590.

40 Gao, Y., Neal, L., Ding, D. et al. (2019). Recent advances in intensified ethylene production – a review. *ACS Catalysis* 9 (9): 8592–8621.

41 Oh, S.C., Schulman, E., Zhang, J. et al. (2019). Direct non-oxidative methane conversion in a millisecond catalytic wall reactor. *Angewandte Chemie* 131 (21): 7157–7160.

42 Sakbodin, M., Wu, Y., Oh, S.C. et al. (2016). Hydrogen-permeable tubular membrane reactor: promoting conversion and product selectivity for non-oxidative activation of methane over an Fe©SiO_2 catalyst. *Angewandte Chemie* 128 (52): 16383–16386.

43 Postma, R.S. and Lefferts, L. (2021). Influence of axial temperature profiles on Fe/SiO_2 catalyzed non-oxidative coupling of methane. *ChemCatChem* 13 (4): 1157–1160.

44 Xie, P., Pu, T., Nie, A. et al. (2018). Nanoceria-supported single-atom platinum catalysts for direct methane conversion. *ACS Catalysis* 8 (5): 4044–4048.

45 Xiao, L. and Wang, L. (2007). Methane activation on Pt and Pt_4: a density functional theory study. *Journal of Physical Chemistry B* 111 (7): 1657–1663.

46 Jones, J., Xiong, H., DeLaRiva, A.T. et al. (2016). Thermally stable single-atom platinum-on-ceria catalysts via atom trapping. *Science* 353 (6295): 150–154.

47 Senftle, T.P., Van Duin, A.C.T., and Janik, M.J. (2017). Methane activation at the Pd/CeO_2 interface. *ACS Catalysis* 7 (1): 327–332.

48 Zuo, Z., Liu, S., Wang, Z. et al. (2018). Dry reforming of methane on single-site Ni/MgO catalysts: importance of site confinement. *ACS Catalysis* 8 (10): 9821–9835.

49 Tang, Y., Wei, Y., Wang, Z. et al. (2019). Synergy of single-atom Ni_1 and Ru_1 sites on CeO_2 for dry reforming of CH_4. *Journal of the American Chemical Society* 141 (18): 7283–7293.

50 Akri, M., Zhao, S., Li, X. et al. (2019). Atomically dispersed nickel as coke-resistant active sites for methane dry reforming. *Nature Communications* 10 (1): 5181.

51 Akri, M., El Kasmi, A., Batiot-Dupeyrat, C. et al. (2020). Highly active and carbon-resistant nickel single-atom catalysts for methane dry reforming. *Catalysts* 10 (6): 630.

52 Sarma, B.B., Agostini, G., Farpón, M.G. et al. (2021). Bottom-up assembly of bimetallic nanocluster catalysts from oxide-supported single-atom precursors. *Journal of Materials Chemistry A* 9 (13): 8401–8415.

53 Dokania, A., Ramirez, A., Bavykina, A. et al. (2019). Heterogeneous catalysis for the valorization of CO_2: role of bifunctional processes in the production of chemicals. *ACS Energy Letters* 4 (1): 167–176.

54 Prieto, G. (2017). Carbon dioxide hydrogenation into higher hydrocarbons and oxygenates: thermodynamic and kinetic bounds and progress with heterogeneous and homogeneous catalysis. *ChemSusChem* 10 (6): 1056–1070.
55 Chuanchai, A. and Ramaraj, R. (2018). Sustainability assessment of biogas production from buffalo grass and dung: biogas purification and bio-fertilizer. *3 Biotech* 8 (3): 151.
56 Schüth, F. (2011). Chemical compounds for energy storage. *Chemie Ingenieur Technik* 83 (11): 1984–1993.
57 Shao, X., Yang, X., Xu, J. et al. (2019). Iridium single-atom catalyst performing a quasi-homogeneous hydrogenation transformation of CO_2 to formate. *Chem* 5 (3): 693–705.
58 Millet, M.M., Algara-Siller, G., Wrabetz, S. et al. (2019). Ni single atom catalysts for CO_2 activation. *Journal of the American Chemical Society* 141 (6): 2451–2461.
59 Frei, M.S., Mondelli, C., García-Muelas, R. et al. (2019). Atomic-scale engineering of indium oxide promotion by palladium for methanol production via CO_2 hydrogenation. *Nature Communications* 10 (1): 1–11.
60 Vogt, C., Groeneveld, E., Kamsma, G. et al. (2018). Unravelling structure sensitivity in CO_2 hydrogenation over nickel. *Nature Catalysis* 1 (2): 127–134.
61 De, S., Dokania, A., Ramirez, A. et al. (2020). Advances in the design of heterogeneous catalysts and thermocatalytic processes for CO_2 utilization. *ACS Catalysis* 10 (23): 14147–14185.
62 Mondal, B., Song, J., Neese, F. et al. (2015). Bio-inspired mechanistic insights into CO_2 reduction. *Current Opinion in Chemical Biology* 25: 103–109.
63 Zhao, T., Wang, Q., and Jena, P. (2017). Rational design of super-alkalis and their role in CO_2 activation. *Nanoscale* 9 (15): 4891–4897.
64 Álvarez, A., Borges, M., Corral-Pérez, J.J. et al. (2017). CO_2 activation over catalytic surfaces. *ChemPhysChem* 18 (22): 3135–3141.
65 Aresta, M. and Angelini, A. (2015). The carbon dioxide molecule and the effects of its interaction with electrophiles and nucleophiles. *Carbon Dioxide and Organometallics*: 1–38.
66 Luther, G.W. (2004). Kinetics of the reactions of water, hydroxide ion and sulfide species with CO_2, OCS and CS_2: Frontier molecular orbital considerations. *Aquatic Geochemistry* 10 (1–2): 81–97.
67 Nakamura, S., Hatakeyama, M., Wang, Y. et al. (2015). A basic quantum chemical review on the activation of CO_2. *ACS Symposium Series* 1194: 123–134.
68 Mascetti, J. (2010). Carbon dioxide coordination chemistry and reactivity of coordinated CO_2. In: *Carbon Dioxide as Chemical Feedstock*, 55–88. Weinheim, Germany: Wiley-VCH Verlag GmbH & Co. KGaA.
69 Solymosi, F. (1991). The bonding, structure and reactions of CO_2 adsorbed on clean and promoted metal surfaces. *Journal of Molecular Catalysis* 65 (3): 337–358.
70 Freund, H.-J. and Roberts, M.W. (1996). Surface chemistry of carbon dioxide. *Surface Science Reports* 25 (8): 225–273.

71 Cheng, D., Negreiros, F.R., Aprà, E. et al. (2013). Computational approaches to the chemical conversion of carbon dioxide. *ChemSusChem* 6 (6): 944–965.

72 Taifan, W., Boily, J.F., and Baltrusaitis, J. (2016). Surface chemistry of carbon dioxide revisited. *Surface Science Reports* 71 (4): 595–671.

73 Liu, X., Sun, L., and Deng, W.-Q. (2018). Theoretical investigation of CO_2 adsorption and dissociation on low index surfaces of transition metals. *The Journal of Physical Chemistry C* 122 (15): 8306–8314.

74 Aresta, M., Dibenedetto, A., and Quaranta, E. (2016). *CO2 coordination to metal centres: modes of bonding and reactivity BT – Reaction Mechanisms in Carbon Dioxide Conversion* (eds. M. Aresta, A. Dibenedetto and E. Quaranta), 35–69. Berlin, Heidelberg: Springer Berlin Heidelberg.

75 Darensbourg, D.J. and Kudaroski, R.A. (1983). The activation of carbon dioxide by metai complexes. In: *Advances in Organometallic Chemistry*, vol. 22, 129–168.

76 Tanaka, R., Yamashita, M., and Nozaki, K. (2009). Catalytic hydrogenation of carbon dioxide using Ir(III)–pincer complexes. *Journal of the American Chemical Society* 131 (40): 14168–14169.

77 Hull, J.F., Himeda, Y., Wang, W.-H. et al. (2012). Reversible hydrogen storage using CO_2 and a proton-switchable iridium catalyst in aqueous media under mild temperatures and pressures. *Nature Chemistry* 4 (5): 383–388.

78 Modak, A., Bhanja, P., Dutta, S. et al. (2020). Catalytic reduction of CO_2 into fuels and fine chemicals. *Green Chemistry* 22 (13): 4002–4033.

79 Shen, X., Meng, Q., Dong, M. et al. (2019). Low-temperature reverse water–gas shift process and transformation of renewable carbon resources to value-added chemicals. *ChemSusChem* 12 (23): 5149–5156.

80 Sakaki, S., Kitaura, K., and Morokuma, K. (1982). Structure and coordinate bonding nature of nickel(0) and copper(I) carbon dioxide complexes. An ab initio molecular orbital study. *Inorganic Chemistry* 21 (2): 760–765.

81 Mori, K., Taga, T., and Yamashita, H. (2017). Isolated single-atomic Ru catalyst bound on a layered double hydroxide for hydrogenation of CO_2 to formic acid. *ACS Catalysis* 7 (5): 3147–3151.

82 Brookes, C., Bowker, M., and Wells, P.P. (2016). Catalysts for the selective oxidation of methanol. *Catalysts* 6 (7).

83 Olah, G.A. (2005). Beyond oil and gas: the methanol economy. *Angewandte Chemie International Edition* 44 (18): 2636–2639.

84 Arcoumanis, C., Bae, C., Crookes, R. et al. (2008). The potential of di-methyl ether (DME) as an alternative fuel for compression-ignition engines: a review. *Fuel* 87 (7): 1014–1030.

85 Kim, J., Sarma, B.B., Andrés, E. et al. (2019). Surface lewis acidity of periphery oxide species as a general kinetic descriptor for CO_2 hydrogenation to methanol on supported copper nanoparticles. *ACS Catalysis* 9 (11): 10409–10417.

86 Larmier, K., Liao, W.-C., Tada, S. et al. (2017). CO_2-to-methanol hydrogenation on zirconia-supported copper nanoparticles: reaction intermediates and the role

of the metal-support interface. *Angewandte Chemie International Edition* 56 (9): 2318–2323.
87 Kattel, S., Ramírez, P.J., Chen, J.G. et al. (2017). Active sites for CO_2 hydrogenation to methanol on Cu/ZnO catalysts. *Science* 355 (6331): 1296–1299.
88 Scharnagl, F.K., Hertrich, M.F., Neitzel, G. et al. (2019). Homogeneous catalytic hydrogenation of CO_2 to methanol – improvements with tailored ligands. *Advanced Synthesis and Catalysis* 361 (2): 374–379.
89 Wesselbaum, S., Moha, V., Meuresch, M. et al. (2015). Hydrogenation of carbon dioxide to methanol using a homogeneous ruthenium–Triphos catalyst: from mechanistic investigations to multiphase catalysis. *Chemical Science* 6 (1): 693–704.
90 Martin, O., Martín, A.J., Mondelli, C. et al. (2016). Indium oxide as a superior catalyst for methanol synthesis by CO_2 hydrogenation. *Angewandte Chemie International Edition* 55 (21): 6261–6265.
91 Frei, M.S., Mondelli, C., García-Muelas, R. et al. (2019). Atomic-scale engineering of indium oxide promotion by palladium for methanol production via CO_2 hydrogenation. *Nature Communications* 10 (1): 3377.
92 Ye, X., Yang, C., Pan, X. et al. (2020). Highly selective hydrogenation of CO_2 to ethanol via designed bifunctional Ir_1 – In_2O_3 single-atom catalyst. *Journal of the American Chemical Society* 142 (45): 19001–19005.
93 Qian, Q., Cui, M., He, Z. et al. (2015). Highly selective hydrogenation of CO_2 into C^{2+} alcohols by homogeneous catalysis. *Chemical Science* 6 (10): 5685–5689.
94 Moioli, E. and Züttel, A. (2020). A model-based comparison of Ru and Ni catalysts for the sabatier reaction. *Sustainable Energy & Fuels* 4 (3): 1396–1408.
95 Frontera, P., Macario, A., Ferraro, M. et al. (2017). Supported catalysts for CO_2 methanation: a review. *Catalysts* 7 (12): 59.
96 Kwak, J.H., Kovarik, L., and Szanyi, J. (2013). Heterogeneous catalysis on atomically dispersed supported metals: CO_2 reduction on multifunctional Pd catalysts. *ACS Catalysis* 3 (9): 2094–2100.
97 Kwak, J.H., Kovarik, L., and Szanyi, J. (2013). CO_2 reduction on supported Ru/Al_2O_3 catalysts: cluster size dependence of product selectivity. *ACS Catalysis* 3 (11): 2449–2455.
98 Aitbekova, A., Wu, L., Wrasman, C.J. et al. (2018). Low-temperature restructuring of CeO_2 – supported Ru nanoparticles determines selectivity in CO_2 catalytic reduction. *Journal of the American Chemical Society* 140 (42): 13736–13745.
99 Lu, B. and Kawamoto, K. (2014). Preparation of mesoporous CeO_2 and monodispersed NiO particles in CeO_2, and enhanced selectivity of NiO/CeO_2 for reverse water gas shift reaction. *Materials Research Bulletin* 53: 70–78.
100 Wu, H.C., Chang, Y.C., Wu, J.H. et al. (2015). Methanation of CO_2 and reverse water gas shift reactions on Ni/SiO_2 catalysts: the influence of particle size on selectivity and reaction pathway. *Catalysis Science & Technology* 5 (8): 4154–4163.

101 Wang, Y., Arandiyan, H., Scott, J. et al. (2018). Single atom and nanoclustered Pt catalysts for selective CO_2 reduction. *ACS Applied Energy Materials* 1 (12): 6781–6789.

102 Samantaray, M.K., D'Elia, V., Pump, E. et al. (2020). The comparison between single atom catalysis and surface organometallic catalysis. *Chemical Reviews* 120 (2): 734–813.

103 Zhao, D., Chen, Z., Yang, W. et al. (2019). MXene (Ti_3C_2) vacancy-confined single-atom catalyst for efficient functionalization of CO_2. *Journal of the American Chemical Society* 141 (9): 4086–4093.

104 Wang, C., Song, Q., Zhang, K. et al. (2019). Atomic zinc dispersed on graphene synthesized for active CO_2 fixation to cyclic carbonates. *Chemical Communications* 55 (9): 1299–1302.

105 Kung, H.H. (1980). Methanol synthesis. *Catalysis Reviews* 22 (2): 235–259.

106 Franke, R., Selent, D., and Börner, A. (2012). Applied hydroformylation. *Chemical Reviews* 112 (11): 5675–5732.

107 Dry, M.E. (2002). High quality diesel via the Fischer–Tropsch process – a review. *Journal of Chemical Technology & Biotechnology* 77 (1): 43–50.

108 Peng, J.-B., Geng, H.-Q., and Wu, X.-F. (2019). The chemistry of CO: carbonylation. *Chem* 5 (3): 526–552.

109 Weissermel, K. and Arpe, H. (2003). *Industrial Organic Chemistry*. Wiley.

110 Long, L.H. and Walsh, A.D. (1947). Remarks on the structure of carbon monoxide. *Transactions of the Faraday Society* 43: 342.

111 Maron, L., Perrin, L., Eisenstein, O. et al. (2002). Are the carbon monoxide complexes of Cp_2M(M = Ca, Eu, or Yb) carbon or oxygen bonded? An answer from DFT calculations. *Journal of the American Chemical Society* 124 (20): 5614–5615.

112 Berkefeld, A., Piers, W.E., Parvez, M. et al. (2012). Carbon monoxide activation via O-bound CO using decamethylscandocinium–hydridoborate ion pairs. *Journal of the American Chemical Society* 134 (26): 10843–10851.

113 Foppa, L., Copéret, C., and Comas-Vives, A. (2016). Increased back-bonding explains step-edge reactivity and particle size effect for CO activation on Ru nanoparticles. *Journal of the American Chemical Society* 138 (51): 16655–16668.

114 Van Helden, P., Ciobîcî, I.M., and Coetzer, R.L.J. (2016). The size-dependent site composition of FCC cobalt nanocrystals. *Catalysis Today* 261: 48–59.

115 Bezemer, G.L., Bitter, J.H., Kuipers, H.P.C.E. et al. (2006). Cobalt particle size effects in the Fischer–Tropsch reaction studied with carbon nanofiber supported catalysts. *Journal of the American Chemical Society* 128 (12): 3956–3964.

116 Carballo, J.M.G., Yang, J., Holmen, A. et al. (2011). Catalytic effects of ruthenium particle size on the Fischer–Tropsch Synthesis. *Journal of Catalysis* 284 (1): 102–108.

117 Pestman, R., Chen, W., and Hensen, E. (2019). Insight into the rate-determining step and active sites in the Fischer–Tropsch reaction over cobalt catalysts. *ACS Catalysis* 9 (5): 4189–4195.

118 den Breejen, J.P., Radstake, P.B., Bezemer, G.L. et al. (2009). On the origin of the cobalt particle size effects in Fischer–Tropsch catalysis. *Journal of the American Chemical Society* 131 (20): 7197–7203.

119 Chen, W.H. and Chen, C.Y. (2020). Water gas shift reaction for hydrogen production and carbon dioxide capture: a review. *Applied Energy* 258: 114078.

120 Su, E. (1986). Dynamic behavior of three-way catalysts. *Journal of Catalysis* 99 (2): 506–510.

121 Laine, R.M. and Crawford, E.J. (1988). Homogeneous catalysis of the water-gas shift reaction. *Journal of Molecular Catalysis* 44 (3): 357–387.

122 Schaper, L.A., Herrmann, W.A., and Kühn, F.E. (2012). Water–gas shift reaction. *Applied Homogeneous Catalysis with Organometallic Compounds: A Comprehensive Handbook in Three Volumes* 3 (Iii): 1689–1698.

123 Zhu, M. and Wachs, I.E. (2016). Iron-based catalysts for the high-temperature water–gas shift (HT-WGS) reaction: a review. *ACS Catalysis* 6 (2): 722–732.

124 Jacobs, G. and Davis, B.H. (2008). Low temperature water–gas shift catalysts. In: *Catalysis*, 122–285. Cambridge: Royal Society of Chemistry.

125 Murrell, L.L., Tauster, S.J., and Anderson, D.R. (1991). *Laser Raman Characterization of Surface Phase Precious Metal Oxides Formed on CeO_2*, vol. 71, 275–289. Elsevier.

126 Fu, Q., Saltsburg, H., and Flytzani-Stephanopoulos, M. (2003). Active non-metallic Au and Pt species on ceria-based water–gas shift catalysts. *Science* 301 (5635): 935–938.

127 Flytzani-Stephanopoulos, M. (2014). Gold atoms stabilized on various supports catalyze the water–gas shift reaction. *Accounts of Chemical Research* 47 (3): 783–792.

128 Zhai, Y., Pierre, D., Si, R. et al. (2010). Alkali-stabilized Pt-OH_x species catalyze low-temperature water–gas shift reactions. *Science* 329 (5999): 1633–1636.

129 Lin, J., Wang, A., Qiao, B. et al. (2013). Remarkable performance of Ir_1/FeO_x single-atom catalyst in water gas shift reaction. *Journal of the American Chemical Society* 135 (41): 15314–15317.

130 Rivero-Crespo, M.A., Mon, M., Ferrando-Soria, J. et al. (2018). Confined Pt_1^{1+} water clusters in a MOF catalyze the low-temperature water–gas shift reaction with both CO_2 oxygen atoms coming from water. *Angewandte Chemie International Edition* 57 (52): 17094–17099.

131 Singh, P. and Prasad, R. (2014). Catalytic abatement of cold-start vehicular CO emissions. *Catalysis in Industry* 6 (2): 122–127.

132 Bion, N., Epron, F., Moreno, M. et al. (2008). Preferential oxidation of carbon monoxide in the presence of hydrogen (PROX) over noble metals and transition metal oxides: advantages and drawbacks. *Topics in Catalysis* 51 (1–4): 76–88.

133 Rodrigues, A., Amphlett, J.C., Mann, R.F. et al. (1997). Carbon monoxide poisoning of proton-exchange membrane fuel cells. In: *IECEC-97 Proceedings of the Thirty-Second Intersociety Energy Conversion Engineering Conference (Cat. No.97CH6203)* 2, vol. 2, 768–773.

134 Qiao, B., Wang, A., Yang, X. et al. (2011). Single-atom catalysis of CO oxidation using Pt_1/FeO_x. *Nature Chemistry* 3 (8): 634–641.

135 Choi, K.I. and Vannice, M.A. (1991). CO oxidation over Pd and Cu catalysts IV. Prereduced Al_2O_3-supported copper. *Journal of Catalysis* 131 (1): 22–35.

136 Tang, D. and Zhang, J. (2013). Theoretical investigation on CO oxidation catalyzed by a copper nanocluster. *RSC Advances* 3 (35): 15225.

137 Kim, H.Y., Lee, H.M., and Henkelman, G. (2012). CO oxidation mechanism on CeO_2 supported Au nanoparticles. *Journal of the American Chemical Society* 134 (3): 1560–1570.

138 Hulva, J., Meier, M., Bliem, R. et al. (2021). Unraveling CO adsorption on model single-atom catalysts. *Science* 371 (6527): 375–379.

139 Lu, Y., Wang, J., Yu, L. et al. (2019). Identification of the active complex for CO oxidation over single-atom Ir-on-$MgAl_2O_4$ catalysts. *Nature Catalysis* 2 (2): 149–156.

140 Zhang, B., Asakura, H., and Yan, N. (2017). Atomically dispersed rhodium on self-assembled phosphotungstic acid: structural features and catalytic CO oxidation properties. *Industrial & Engineering Chemistry Research* 56 (13): 3578–3587.

141 Wang, H., Liu, J.-X., Allard, L.F. et al. (2019). Surpassing the single-atom catalytic activity limit through paired Pt–O–Pt ensemble built from isolated Pt_1 atoms. *Nature Communications* 10 (1): 3808.

142 Sarma, B.B., Plessow, P.N., Agostini, G. et al. (2020). Metal-specific reactivity in single-atom catalysts: CO oxidation on 4d and 5d transition metals atomically dispersed on MgO. *Journal of the American Chemical Society* 142 (35): 14890–14902.

143 Nie, L., Mei, D., Xiong, H. et al. (2017). Activation of surface lattice oxygen in single-atom Pt/CeO_2 for low-temperature CO oxidation. *Science* 358 (6369): 1419–1423.

144 Kang, L., Wang, B., Bing, Q. et al. (2020). Adsorption and activation of molecular oxygen over atomic copper(I/II) site on ceria. *Nature Communications* 11 (1): 4008.

145 Liu, L., Meira, D.M., Arenal, R. et al. (2019). Determination of the evolution of heterogeneous single metal atoms and nanoclusters under reaction conditions: which are the working catalytic sites? *ACS Catalysis* 9 (12): 10626–10639.

146 Pereira-Hernández, X.I., DeLaRiva, A., Muravev, V. et al. (2019). Tuning Pt-CeO_2 interactions by high-temperature vapor-phase synthesis for improved reducibility of lattice oxygen. *Nature Communications* 10 (1): 1358.

147 Tang, Y., Asokan, C., Xu, M. et al. (2019). Rh single atoms on TiO_2 dynamically respond to reaction conditions by adapting their site. *Nature Communications* 10 (1): 4488.

148 Maurer, F., Jelic, J., Wang, J. et al. (2020). Tracking the formation, fate and consequence for catalytic activity of Pt single sites on CeO_2. *Nature Catalysis* 3 (10): 824–833.

149 Börner, A. and Franke, R. (2016). *Hydroformylations: Fundamentals, Processes and Applications in Organic Synthesis*. Weinheim, Germany: Wiley-VCH Verlag GmbH & Co. KGaA.

150 Li, C., Wang, W., Yan, L. et al. (2018). A mini review on strategies for heterogenization of rhodium-based hydroformylation catalysts. *Frontiers of Chemical Science and Engineering* 12 (1): 113–123.

151 Lang, R., Li, T., Matsumura, D. et al. (2016). Hydroformylation of olefins by a rhodium single-atom catalyst with activity comparable to $RhCl(PPh_3)_3$. *Angewandte Chemie International Edition* 55 (52): 16054–16058.

152 Li, T., Chen, F., Lang, R. et al. (2020). Styrene hydroformylation with in situ hydrogen: regioselectivity control by coupling with the low-temperature water–gas shift reaction. *Angewandte Chemie International Edition* 59 (19): 7430–7434.

153 Wang, L., Zhang, W., Wang, S. et al. (2016). Atomic-level insights in optimizing reaction paths for hydroformylation reaction over Rh/CoO single-atom catalyst. *Nature Communications* 7 (1): 14036.

154 Amsler, J., Sarma, B.B., Agostini, G. et al. (2020). Prospects of heterogeneous hydroformylation with supported single atom catalysts. *Journal of the American Chemical Society* 142 (11): 5087–5096.

155 Liu, Y., Zhao, L., Zhang, Y. et al. (2020). Progress and challenges of mercury-free catalysis for acetylene hydrochlorination. *Catalysts* 10 (10).

156 Hutchings, G. (1985). Vapor phase hydrochlorination of acetylene: correlation of catalytic activity of supported metal chloride catalysts. *Journal of Catalysis* 96 (1): 292–295.

157 Nkosi, B., Coville, N.J., and Hutchings, G.J. (1988). Vapour phase hydrochlorination of acetylene with group VIII and IB metal chloride catalysts. *Applied Catalysis* 43 (1): 33–39.

158 Nkosi, B., Coville, N.J., and Hutchings, G.J. (1988). Reactivation of a supported gold catalyst for acetylene hydrochlorination. *Journal of the Chemical Society, Chemical Communications* 1: 71–72.

159 Malta, G., Freakley, S.J., Kondrat, S.A. et al. (2017). Acetylene hydrochlorination using Au/carbon: a journey towards single site catalysis. *Chem. Commun.* 53 (86): 11733–11746.

160 Malta, G., Kondrat, S.A., Freakley, S.J. et al. (2017). Identification of single-site gold catalysis in acetylene hydrochlorination. *Science* 355 (6332): 1399–1403.

161 Johnston, P., Carthey, N., and Hutchings, G.J. (2015). Discovery, development, and commercialization of gold catalysts for acetylene hydrochlorination. *Journal of the American Chemical Society* 137 (46): 14548–14557.

162 Kaiser, S.K., Fako, E., Manzocchi, G. et al. (2020). Nanostructuring unlocks high performance of platinum single-atom catalysts for stable vinyl chloride production. *Nature Catalysis* 3 (4): 376–385.

163 Kaiser, S.K., Clark, A.H., Cartocci, L. et al. (2021). Sustainable synthesis of bimetallic single atom gold-based catalysts with enhanced durability in acetylene hydrochlorination. *Small* 17 (16): 2004599.

164 Yardimci, D., Serna, P., and Gates, B.C. (2012). Tuning catalytic selectivity: zeolite- and magnesium oxide-supported molecular rhodium catalysts for hydrogenation of 1,3-butadiene. *ACS Catalysis* 2 (10): 2100–2113.

165 Cooper, C., Dooley, K.M., Fierro-Gonzalez, J.C. et al. (2020). Bruce gates: a career in catalysis. *ACS Catalysis* 10 (20): 11912–11935.

166 Corma, A., Salnikov, O.G., Barskiy, D.A. et al. (2015). Single-atom gold catalysis in the context of developments in parahydrogen-induced polarization. *Chemistry – A European Journal* 21 (19): 7012–7015.

167 Babucci, M., Fang, C.-Y., Hoffman, A.S. et al. (2017). Tuning the selectivity of single-site supported metal catalysts with ionic liquids. *ACS Catalysis* 7 (10): 6969–6972.

168 Babucci, M., Fang, C.-Y., Perez-Aguilar, J.E. et al. (2019). Controlling catalytic activity and selectivity for partial hydrogenation by tuning the environment around active sites in iridium complexes bonded to supports. *Chemical Science* 10 (9): 2623–2632.

169 Ma, X., An, Z., Song, H. et al. (2020). Atomic Pt-catalyzed heterogeneous Anti-Markovnikov C–N formation: Pt_1^{0} activating N–H for $Pt_1^{\delta+}$ – activated C=C attack. *Journal of the American Chemical Society* 142 (19): 9017–9027.

170 Cui, X., Junge, K., Dai, X. et al. (2017). Synthesis of single atom based heterogeneous platinum catalysts: high selectivity and activity for hydrosilylation reactions. *ACS Central Science* 3 (6): 580–585.

171 Sarma, B.B., Kim, J., Amsler, J. et al. (2020). One-pot cooperation of single-atom Rh and Ru solid catalysts for a selective tandem olefin isomerization-hydrosilylation process. *Angewandte Chemie* 132 (14): 5855–5864.

172 Fogg, D.E. and dos Santos, E.N. (2004). Tandem catalysis: a taxonomy and illustrative review. *Coordination Chemistry Reviews* 248 (21): 2365–2379.

173 Poe, S.L., Kobašlija, M., and McQuade, D.T. (2006). Microcapsule enabled multicatalyst system. *Journal of the American Chemical Society* 128 (49): 15586–15587.

174 Lu, J., Dimroth, J., and Weck, M. (2015). Compartmentalization of incompatible catalytic transformations for tandem catalysis. *Journal of the American Chemical Society* 137 (40): 12984–12989.

11

Supported Metal Single Atom Thermocatalysts for C—C, C—Si, and C—B Bond-Forming (Coupling) Reactions and Biomedical Applications

Rossella Greco, Marta Mon, and Antonio Leyva-Pérez

Universitat Politècnica de València-Consejo Superior de Investigaciones Científicas, Instituto de Tecnología Química (UPV-CSIC), Avda. de los Naranjos s/n, 46022 Valencia, Spain

11.1 Introduction

11.1.1 Chronology of Single-Atom Catalysts

In 1925, Taylor strongly claimed that "(catalyst) *activation consists in increasing the amount of the surface atoms, which are in a state of unsaturation relative to the main body of the catalyst material*" [1], after considering the Langmuir theories about the adsorption of species onto simple surfaces. Since then, and up to the 1970s, the importance of particle size in catalysis began to be highlighted, particularly by taking into account the number of metal atoms building up the active sites [2]. As a matter of fact, the decrease in the size of metal particles may improve the performance of catalysts from different points of view, which include a high number of exposed atoms and a low-coordination environment of the metal site. Figure 11.1 shows that small particles are also claimed to produce quantum size effects, where confinement of electrons brings to discrete energy level distributions and a defined highest occupied molecular orbital (HOMO)–lowest unoccupied molecular orbital (LUMO) gap [3, 4].

When downsizing macroscopic metal particles, a first geometry-electronic disruption is found for nanoparticles (NPs), commonly considered between 2 and 100 nm [5]. These particles still preserve the typical delocalized electronics of a macroscopic metal. However, they show plasmonic properties with discrete absorption and emission wavelengths. A representative example is a striking color of some metal NPs, for instance, the red NP of gold.

Going down deeper in size, clusters tickled scientists' interest. These species are described as monodispersed particles less than 2 nm in diameter with properties between bulk materials and single atoms [6]. Finally, single atoms can be considered the last stop in size for metal, with total unsaturation and complete availability for the reaction [4].

Going forward in time, during the 1990s, it was not common yet to speak about single atoms catalysts (SACs). Indeed, considering the techniques available in those

Supported Metal Single Atom Catalysis, First Edition. Edited by Philippe Serp and Doan Pham Minh.
© 2022 WILEY-VCH GmbH. Published 2022 by WILEY-VCH GmbH.

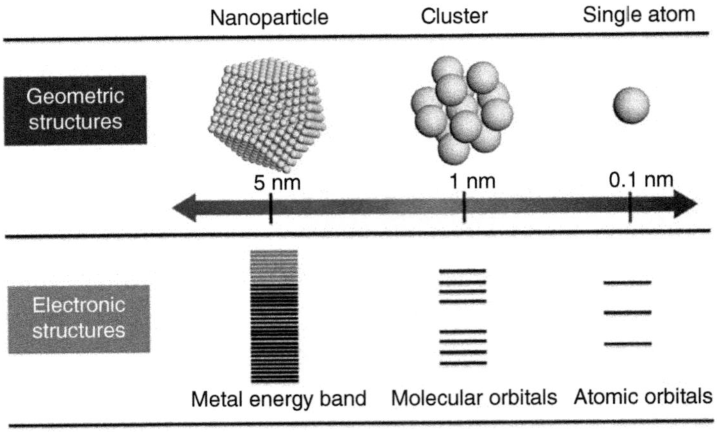

Figure 11.1 Influence of the size of the particles on the electronic distribution. Source: Liu and Corma [3]. Reproduced with permission of American Chemical Society.

times, the detection of these species would have been rather difficult. "Single-site heterogeneous catalysts" was the more frequent definition (see *Chapter 1*), where the isolated active sites (single site in this case) consisted of one or more atoms [7]. For instance, Figure 11.2 shows the epoxidation of cyclohexene catalyzed by $TiSi_3$–SBA15, and it can be seen that the yield of epoxide increases with the amount of Ti, but from the turnover number (TON) point of view, the behavior is the opposite: the TON increased with the decrease in Ti amount [8]. This trend suggested that the less amount of Ti we have, the more isolated the atoms are, and the more catalytically active they become.

The development of new techniques, such as aberration-corrected scanning transmission electron microscopy (AC-STEM), in combination with the improvement in other techniques such as X-ray absorption fine structure spectroscopy (EXAFS) and X-ray absorption near-edge spectroscopy (XANES) could definitively help to detect the "invisible" SAC (see *Chapter 5*), which pushed researchers to go deeper into their synthesis and applications [9]. Thus, in the 2000s, a single atom of palladium supported on MgO was finally spotted and soundly claimed to be catalytically active in the trimerization reaction of acetylene to benzene [10].

After 2010, the definition of single-atom catalysis started to be more common and less fancy. For example, Figure 11.3 shows the oxidation reaction of CO to CO_2 catalyzed by Pt_1/FeO_x, which was one of the first reported reactions to be expressly catalyzed by a SAC [11]. A brief description of the mechanism of this reaction based on density functional theory (DFT) calculations was presented, which confirmed what was already explicit from the catalytic experiments, *i.e.* that the adsorption of the CO molecule over one platinum atom has binding energy (1.27 eV in step iii) much lower than platinum clusters [11]. Since then and up to now, single-atom catalysis underwent a noteworthy development, and the applications of these small species moved to fairly challenging reactions. However, most of the catalytic applications of SACs are still devoted to the activation of carbon-free small

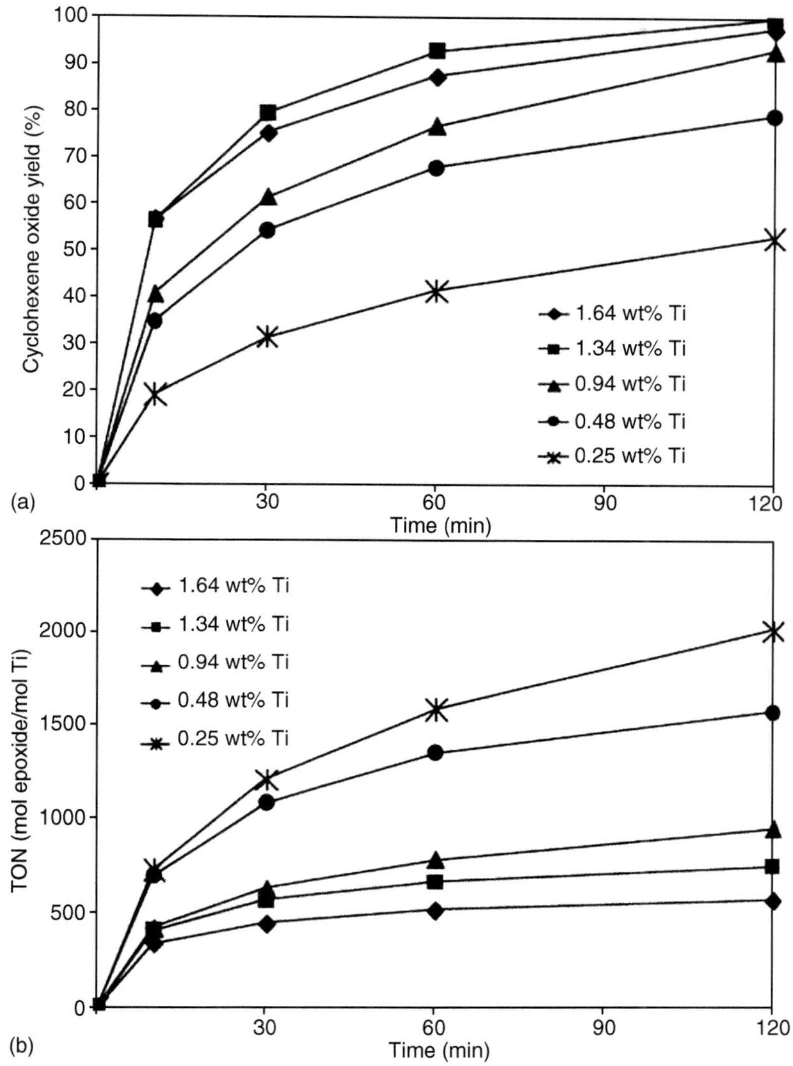

Figure 11.2 (a) Yield of cyclohexene oxide and (b) turnover number (TON) as a function of time during the cyclohexene epoxidation reaction. Source: Jarupatrakorn and Tilley [8]. Reproduced with permission of American Chemical Society.

molecules such as O_2, H_2, or even H_2O, often in electrochemical and photochemical processes. In comparison, much fewer studies have been reported in the catalytic performance of SACs in organic synthesis beyond hydrogenation (*Chapter 8*) and oxidation (*Chapter 9*) reactions [3]. Thus, the focus of this chapter will be on the recent catalytic applications of SACs concerning classical reactions in the organic synthesis where the activation of bonds containing carbon atoms occurs, such as carbon–carbon cross-coupling reactions, hydroboration, and hydrosilylation reactions. Some recent biological applications of SACs will also be commented.

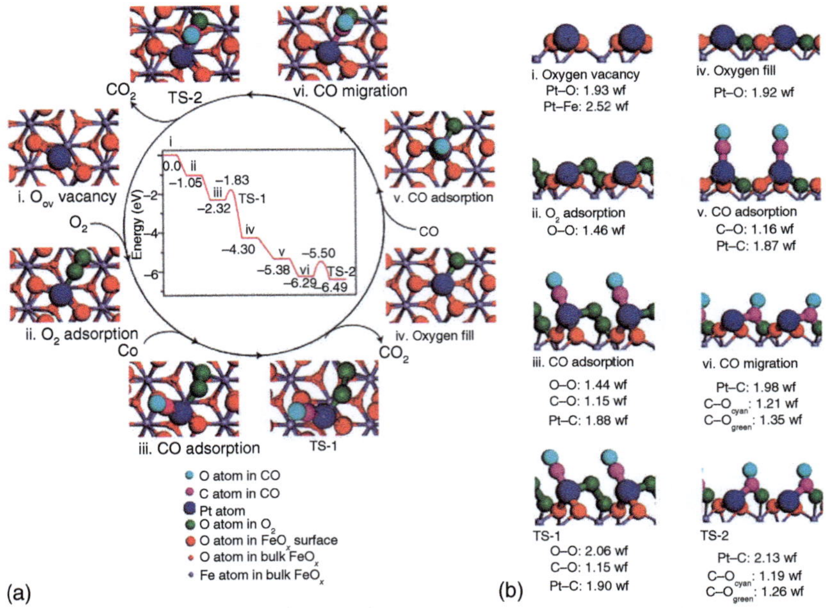

Figure 11.3 Schematic mechanism proposed for CO oxidation reaction on Pt_1/FeO_x. (a) Top view and (b) side view. Source: Qiao et al. [11]. Reproduced with permission of Springer Nature.

11.1.2 Use of SACs in Reactions of Interest for Organic Synthesis and Biomedical Applications

The recent development of SACs during the last decade has made sure that these tiny catalytic species are only starting to be studied in many fields, which include organic synthesis. Due to their structures, SACs could be used as heterogeneous catalysts; however, leaching effects could be dominant in many cases due to the somewhat barely adsorbed nature of the catalytic species [12–15]. Nevertheless, recent advances in catalyst design and solid recovery strategies after reaction, which includes control of metal-support interaction, solvent polarity, control of "boomerang" effects, and the use of milder reaction conditions, to name a few options, suggest that SACs could be of high interest for a plethora of organic reactions during the synthesis of complex organic molecules. Besides, the most studied SACs to date correspond to noble metals such as palladium and platinum, which are the catalysts of choice for many organic transformations. The fact that SACs may indeed be already present in classical solid metal catalysts such as palladium and platinum on charcoal [16, 17], used for many years in wet organic chemistry, strongly support that SACs could bear the occasional harsh reaction conditions of organic transformations and be efficiently used for organic synthesis, without leaching.

Metal-catalyzed carbon–carbon cross-coupling reactions are a paradigmatic example of useful reactions in organic synthesis [18]. The huge utility of

Table 11.1 Some examples of metal-catalyzed cross-couplings.

Entry	Coupling	Reaction scheme
1	Heck	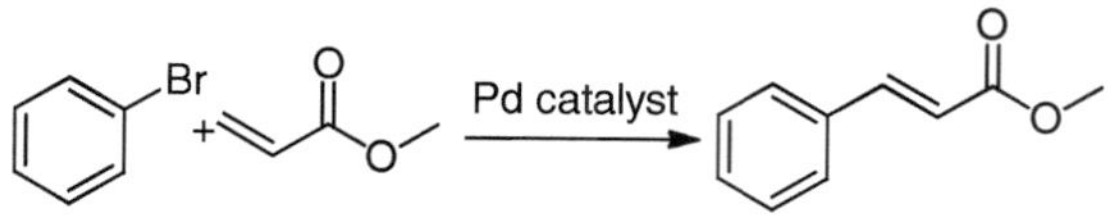
2	Suzuki	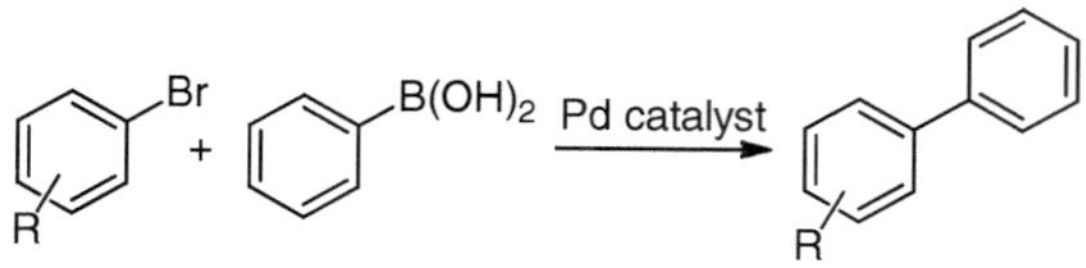
3	Sonogashira	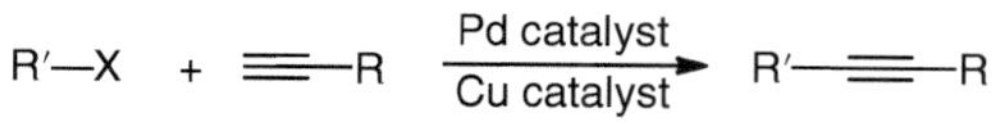 R′ = Aryl, Vinyl R = Aryl, Vinyl, Alkyl

metal-catalyzed carbon–carbon cross-coupling reactions is well-known to organic chemists since many decades [19]. Indeed, this approach allowed the synthesis of molecules that otherwise would be prepared with very arduous procedures. As shown in Table 11.1, the pioneering work of Heck, Suzuki, Sonogashira, and many others, made possible the application of metal complexes of palladium, nickel, silver, or copper to this limitless synthetic strategy [19–21].

Other classical organic transformations catalyzed by metals are the hydro-addition of X—H bonds (X = Si, B, O, N, halogen …) to unsaturated carbon–carbon bonds [22–24]. In particular, the hydrosilylation [25] and the hydroboration of alkenes and alkynes [26] are the methods of choice to obtain highly substituted alkyl silanes and boranes from the corresponding R_3Si–H and R_2BH precursors, respectively, typically catalyzed by platinum and palladium complexes. Not in vain, these reactions share with carbon–carbon cross-coupling reactions that are industrially catalyzed by noble metals, which make them ideal candidates to be studied with SACs.

In Sections 11.2 and 11.3, the utilization of SACs as catalysts for the above mentioned classical organic reactions will be reviewed. The high activity of SACs, sometimes higher than state-of-the-art homogeneous catalysts despite the few studies on the former, will be highlighted. Besides, the parallelism between the performance of SACs for these classical organic reactions, with relatively large molecules, with comparatively smaller molecule activation reactions such as the hydroformylation and the water gas-shift reaction (treated separately in *Chapter 10*), will be briefly commented.

Organic synthesis is traditionally related to biomedical applications. Of course, as mentioned above, SAC applications are in their childhood for organic synthesis, much more for biomedical applications. However, some pioneering studies have been reported during the last years, and this chapter will comment on them. Some of these studies constitute an evolution of the nanozyme field, where nanomaterials with intrinsic enzyme-like characteristics are modified with or treated as SACs for biocatalysis. But not only that, benchmark SAC can trigger or present by itself biological activity in, for instance, cancer treatment.

11.2 Carbon–Carbon Cross-Coupling Reactions

The use of SACs in carbon–carbon coupling reactions can overcome one of the main drawbacks of the traditional catalytic approach: the waste load. The most common coupling catalysts usually include metal complexes, where the metal is surrounded by ligands, *e.g.* tertiary phosphine or N–heterocycles [19–21], which represent a residue at the end of the reaction and a consequent cost for an eventual industrial application. Nowadays, the "green chase" is certainly calling for the use of cheaper and more eco-friendly catalysts. For this reason, during these years, the mechanism of palladium-catalyzed carbon–carbon cross-coupling reactions has been studied from different points of view, and some of these studies agreed that the lower the amount of active metal, the higher the TON [27]. It has been demonstrated, in fact, that $Pd(OAc)_2$ was more active at homeopathic concentrations, i.e. lower than 0.01 mol%, during the Heck coupling between PhBr and *n*–butyl acrylate. Figure 11.4 shows that this assertion can be explained considering the presence of very small metallic species that generate, stabilize, and act catalytically only in extremely diluted conditions, where palladium-isolated atoms and palladium clusters are in equilibrium with bigger species, such as palladium NPs and palladium "black" [28, 29]. Apart from palladium (II) complexes, the most widely used catalysts for cross-coupling reactions are supporting palladium NPs [30], which support that conclusion [31]. The smallest and more active palladium NPs tend to be easily deactivated by the aggregation to inactive NPs, which is another reason that has made the scientific community work on the development of more sustainable and efficient catalysts, such as SACs [16].

Evidence of the catalytic activity during the Heck coupling of ultra-small palladium species, including single atoms, was presented by an original study where the coupling was conducted with α-alumina-supported palladium SACs and nanoporous γ-alumina membrane, allowing to detect the diffusion of just catalytically active atomic palladium atoms. Figure 11.5 shows that on one side of

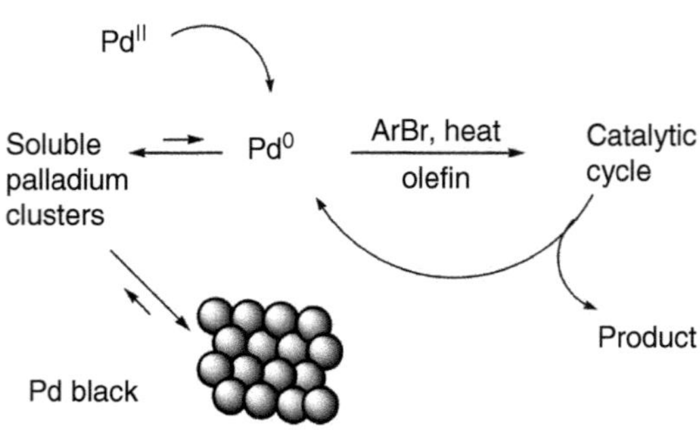

Figure 11.4 Equilibrium between palladium species in the Heck coupling of aryl bromides. Source: De Vries et al. [28]. Reproduced with permission of American Chemical Society.

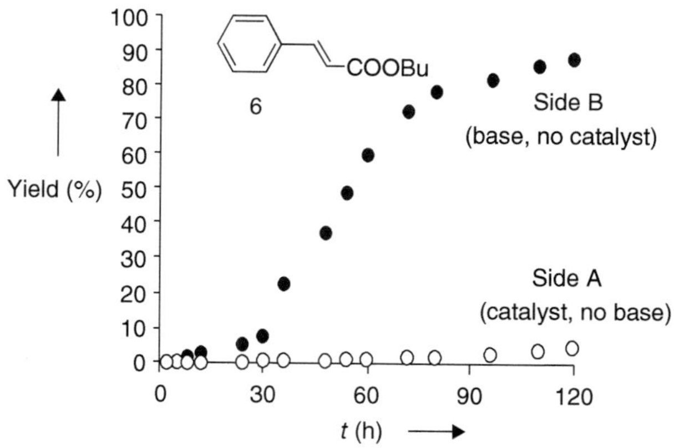

Figure 11.5 Time profile for the formation of Heck coupling product after diffusion of the catalyst through a membrane. Source: Thathagar et al. [32]. Reproduced with permission of John Wiley and Sons.

the membrane (A), there were palladium NPs in *N,N*-dimethyl formamide (DMF), and on the other side of the membrane (B), reagents and base, and that after an induction period of five hours, the product started to appear in the side B, confirming the catalytical activity of small species of palladium leached from the NPs [32].

The debate about the atomicity of the palladium catalyst during the Heck reaction is still not solved, and new players have come into the game since it has been recently reported that platinum, which is not usually active for this type of reaction, is very active when prepared as sub-nanometer clusters or single atoms [20]. This atom dynamics during the reaction is common for catalysts [33], and it is not surprising that different metal species (from SAC to supported NPs) can be claimed as the true catalyst. A paradigmatic reaction case in SACs is the low-temperature water–gas shift reaction ($CO + H_2O \rightarrow CO_2 + H_2$, WGSR), since it was early claimed that very diluted platinum and gold SACs on CeO_2 [34, 35], alumina [36], but also in zeolites [37], and metal-organic frameworks (MOFs) [38] showed very high activity toward the production of CO_2 and H_2. However, it was later claimed that the SACs were inactive, and only the *in situ* formation of Pt NPs under the WGSR conditions allowed the catalysis to proceed [39]. It was observed that species such as Pt–OH_x or Au–O–$(OH)_x$ are mainly stabilized by the support or by the presence of alkali atoms as Na or K [35, 37], and catalyze the WGSR. Assuming the mechanism as redox, the presence of an isolated single atom on a support that generates oxygen vacancies makes sense to be important for the dissociation of H_2O molecules [40].

Figure 11.6 shows diffuse reflectance infrared Fourier transform (DRIFT) studies on a Pd/FeO_x catalyst, where the original presence of both clusters and single atoms of palladium, with a bridged CO adsorption centered at 1909 cm^{-1}, can be clearly seen. It also shows with aberration-corrected high-angle annular dark-field scanning transmission electron microscopy (HAADF-STEM) images the presence of single atoms in the samples with lower loadings of palladium and their higher turnover frequency (TOF) compared to the more loaded ones [41]. From a kinetic point of view, the different activation energy (E_a) in the presence of Pt_{NP} and Pt_{SA} can also

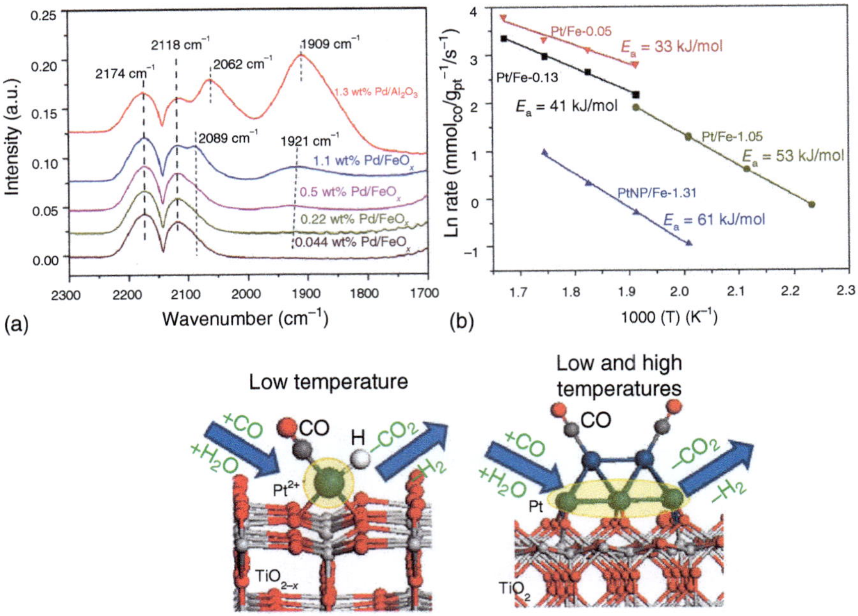

Figure 11.6 DRIFT spectra of the CO adsorption on different Pd/FeO$_x$ (a). Source: Sun et al. [41]. Reproduced with permission of John Wiley and Sons. (b) Activation energies of water gas-shift (WGS) reaction catalyzed by different Pt/FeO$_x$, with different loadings of platinum (b). Source: Chen et al. [42]. Reproduced with permission of American Chemical Society. Pt_{SA}/TiO_2 (a) and Pt-clusters/TiO_2 (b) and their chemical environment (bottom). Source: Ammal and Heyden [43]. Reproduced with permission of American Chemical Society.

underline the importance of isolated and low-coordinated metallic species. Actually, Figure 11.6 also shows that the E_a increases with the loading of platinum supported on FeO$_x$, giving the lowest value for the more diluted sample, which appears to be mainly composed by isolated single atoms, as shown by HAADF-STEM measurements [42]. Moreover, Figure 11.6 also shows DFT studies on Pt_{SA}/TiO_2 where the activity of single atoms is higher than NPs at low temperature due to their chemical environment. Indeed, one molecule of CO, one atom of H, and the support act as ligands for the Pt_{SA} and make more reactive the second molecule of CO, which will be more inclined to react with O_2 [43].

The parallelism between the Heck coupling and the WGSR during the rationalization of the metal active species, giving SACs as plausible platinum catalytic species, may be extended to studies on other reactions, thus opening new avenues for cross-fertilization between small molecule activation and classical organic reactions. This strategy will also be commented below for the case of oxidative aryl halide couplings with alcohols under CO atmosphere and hydroformylation reactions.

With regard to the Suzuki coupling, bromobenzene and phenylboronic acid pinacol ester couple, when catalyzed by palladium supported on an exfoliated high-surface form of graphitic carbon nitride (Pd/ECN), and the energy profile

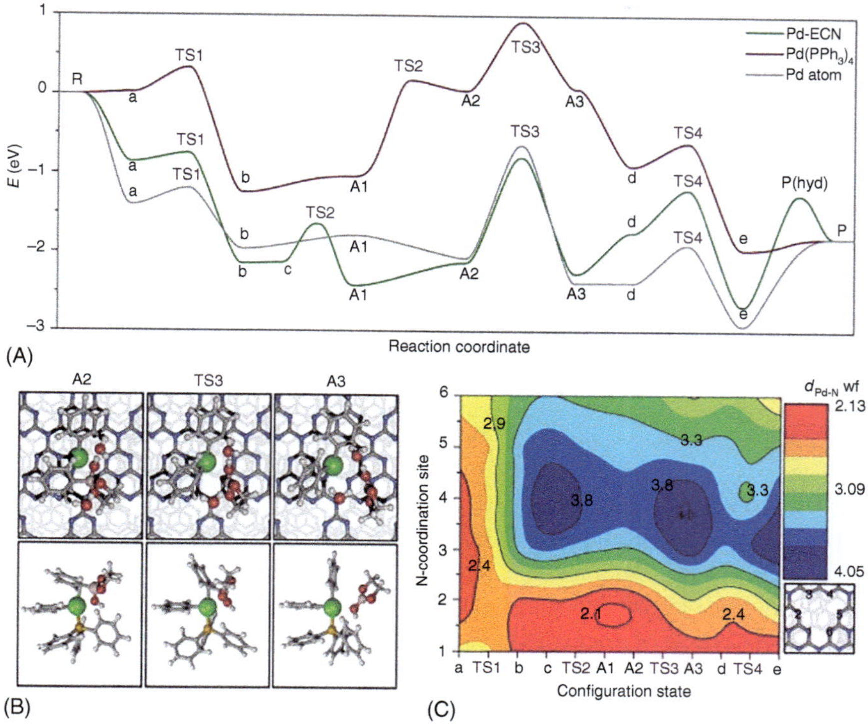

Figure 11.7 (A) Energy profiles of the Suzuki coupling of bromobenzene with phenylboronic acid pinacol ester over Pd–ECN, $Pd(PPh_3)_4$ and an isolated palladium atom. (B) The transmetallation step (A2 → A3) for the indicated configuration of Pd–ECN (top) and $Pd(PPh_3)_4$ (bottom). (C) Interatomic Pd–N distances (*d*) for the N-coordination site at each of the intermediate and transition states indicated in a). Source: Chen et al. [44]. Reproduced with permission of Springer Nature.

calculated by DFT confirmed the importance of isolated atoms. Figure 11.7 illustrates how the catalyst behaved as an isolated palladium atom and not as a classical complex. The latter clearly presents higher energy barriers due to the necessity of losing the ligands to have an open coordination sphere [44].

Pd single atoms supported on other solids catalyze different Suzuki couplings, such as on the related composite carbon nitride/reduced graphene oxide (C_3N_4/rGO) material [45] or engineered on a monolayered $Ti_{0.87}O_2$ nanosheet [46]. Remarkably, real-time fluorescence imaging of a heterogeneously-catalyzed Suzuki reaction using fluorescent substrates has revealed that solid-supported palladium catalysts leach out palladium during the reaction, or at least the catalytically active palladium atoms show long-distance mobility on the solid [47]. Thus, although these results cannot be extrapolated to all the solid-supported Pd single atoms reported, it seems that the leaching/re-deposition effect ("boomerang effect") can play a role in many of these solid catalytic systems [29].

The Suzuki coupling has also been reported with platinum single atoms on multi-walled carbon nanotubes (Pt_{SA}/MWCNT), catalyzing the coupling between

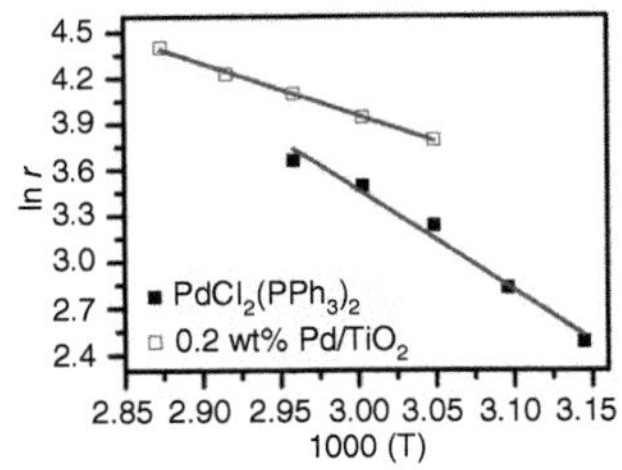

Figure 11.8 Arrhenius plots of heterogeneous catalyst Pd_{SA}/TiO_2 and homogeneous catalyst $Pd(PPh_3)_2Cl_2$. Source: Zhang et al. [49]. Adapted with permission of American Chemical Society.

4-iodoanisole and 4-methylbenzene boronic acid [48]. The use of thiol groups was necessary to stabilize the platinum single atoms and avoid their agglomeration to NPs, and the Pt(0) catalytic single site, confirmed by XANES, had a really high activity toward the reaction, even compared with the Pd(0) single sites prepared in the same way. Furthermore, the Pt_{SA} appeared to be very stable under the reaction conditions, as demonstrated by EXAFS pattern, and the catalyst was reusable up to 12 times.

The Sonogashira coupling between iodobenzene and phenylacetylene catalyzed by Pd_{SA}/TiO_2 has been studied. The kinetic studies confirmed the higher activity of pre-synthesized single atoms compared with the more conventional $[Pd(PPh_3)_2Cl_2]$ catalyst. Indeed, Figure 11.8 shows the remarkable difference among the two catalysts as evidenced by an Arrhenius plot, which shows activation energy (E_a) of 28.9 kJ/mol for the heterogeneous catalyst Pd_{SA}/TiO_2 and 51.7 kJ/mol for the homogeneous $[Pd(PPh_3)_2Cl_2]$ catalyst [49].

The *in situ* formation and catalytic activity of noble metal SACs in carbon–carbon bond-forming reactions such as cross-couplings is not surprising if one considers the performance of SACs for related reactions. For instance, the hydroformylation of propene to butyraldehyde proceeds over single atoms of rhodium supported on CoO (Rh_{SA}/CoO), with very high activity and regioselectivity (above 90%), and *in situ* DRIFT and X-ray photoelectron spectroscopy (XPS) experiments highlighted the presence of catalytically active single atoms of rhodium, reconstructed during the reaction by the adsorption of H_2 and CO. Figure 11.9 shows the correspondence between the two studies. Indeed, the spectra display how the interaction of the Rh_{SA} with the propene is much stronger in the presence of CO and H_2. On the basis of DFT calculations, it is possible to assert that the high selectivity of the process was due to the presence of single atoms, which can have just a few stable configurations in the presence of these reagents, leading to few reaction pathways and then fewer products [50].

Additionally, the catalyst Rh_{SA}/ZnO was demonstrated to be much more active than the classical homogeneous catalyst, *i.e.* $RhCl_3$ and the Wilkinson catalyst $[RhCl(PPh_3)_3]$ for the hydroformylation reaction. In this case, the rhodium single atoms exhibit a TON double than the more conventional catalysts and a higher selectivity toward the hydroformylation product. Actually, styrene hydroformylation can lead to secondary products coming from the hydrogenation of the double bond [51].

In accordance with the catalytic activity of SACs for the hydroformylation reaction, and as a logical extension to it, the alkoxycarbonylation of aryl iodides has been achieved with Pd_{SA}/CeO_2 as a catalyst under base-free and ligand-free

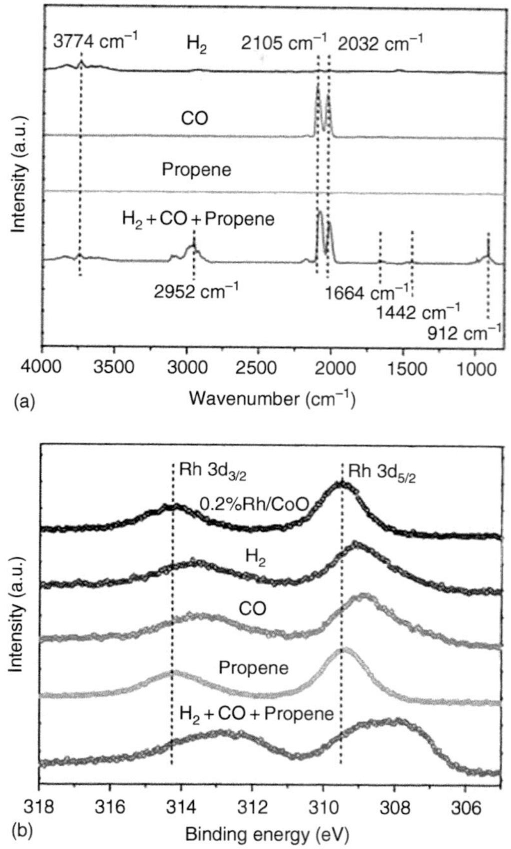

Figure 11.9 (a) DRIFT and (b) XPS *in-situ* spectra of the propene hydroformylation catalyzed by Rh_{SA}/CoO. Source: Wang et al. [50]. Springer Nature. CC BY 4.0.

conditions. This method allows the direct synthesis of carboxylic acid derivatives, commonly catalyzed by Pd–homogeneous complexes. Figure 11.10 shows that AC-HAADF-STEM images of the solid catalyst did not reveal any cluster or NPs, although the Pd_{SA} entities could not be spotted due to the presence of cerium, and that the solid catalyst could be reused four times without significant loss of reactivity [52].

Having a look into non-noble metal SACs, it is difficult to find examples of carbon–carbon bond-forming reactions with application in organic synthesis. However, the aerobic oxidative cross-coupling of secondary and primary alcohols can be certainly considered as a good example. Figure 11.11 shows that cobalt single atoms supported on graphite (Co–C–N/CMK-3) catalyzed the reaction and led to a plethora of α,β-unsaturated ketones, giving a high TON ($\approx 10^4$) despite the use of an untraditional non-noble metal. The activity of cobalt single atoms was confirmed by HAADF-STEM that detected the inaccessibility of cobalt NP deeply embedded in the graphitic walls [53].

Recently, many efforts have been focused on the search for supports that allow to isolate and anchor single atoms. As we have shown so far, the most common supports used for this type of reaction have been Al_2O_3 [32], TiO_2 [49], graphite [53],

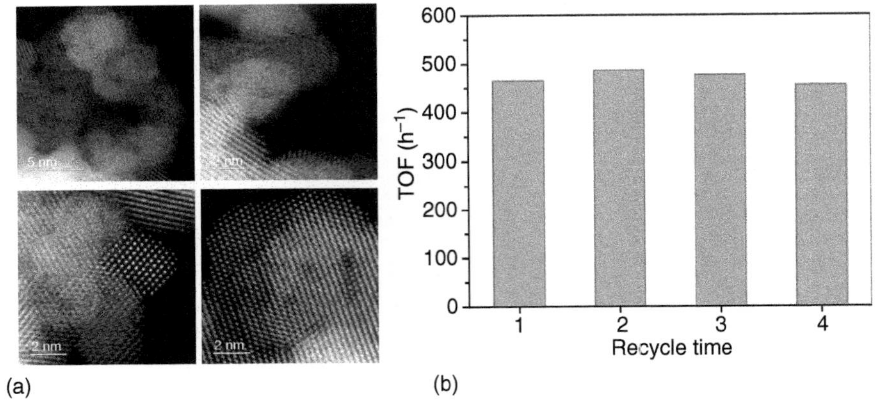

Figure 11.10 AC-HAADF-STEM images of Pd_{SA}/CeO_2 and reuses for the alkoxycarbonylation reaction of iodobenzene. Source: Chen et al. [52]. Reproduced with permission of Springer Nature.

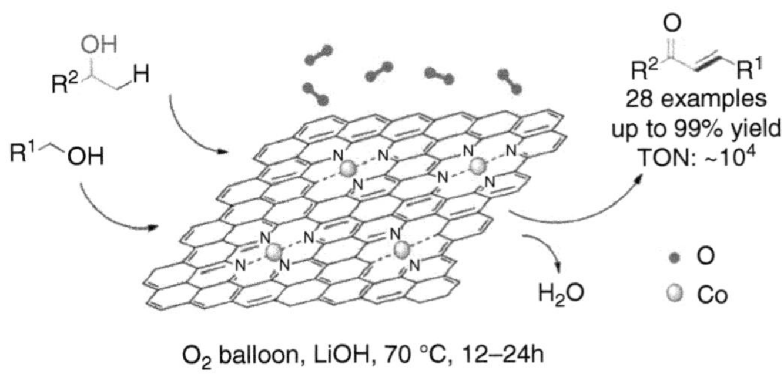

Figure 11.11 Reaction scheme of the aerobic oxidative cross-coupling of secondary and primary alcohols with Co-N-C/CMK-3. Source: Zhang et al. [53]. Reproduced with permission of American Chemical Society.

ECN [44], and ethylene vinyl alcohol (EVOH) [54]. However, unreactive metallic surfaces, such as Au(111), which has been doped with single atoms of palladium, have also been used. In this way, $Pd_{SA}Au$ single-atom alloys (SAAs) were obtained, which are capable of catalyzing carbon–carbon coupling reactions. In particular, the formation of ethane from deuterated methyl iodide using PdAu alloys with different palladium content has been experimentally and theoretically studied. Figure 11.12 shows that temperature-programmed desorption (TPD) experiments, in combination with other techniques, indicate that the alloys with isolated palladium atoms exhibited better catalytic activity and selectivity [55]. In addition, gold alloyed palladium single-atom ($Pd_{SA}Au$/resin) catalysts were found to be very active in the Ullman reaction of aryl chlorides in water. Figure 11.13 shows an exponential TON curve with respect to the palladium fraction present in the catalyst, thus confirming the importance of palladium isolation. Indeed, EXAFS and DRIFTS experiments support that palladium is isolated between gold atoms. Furthermore, this site was

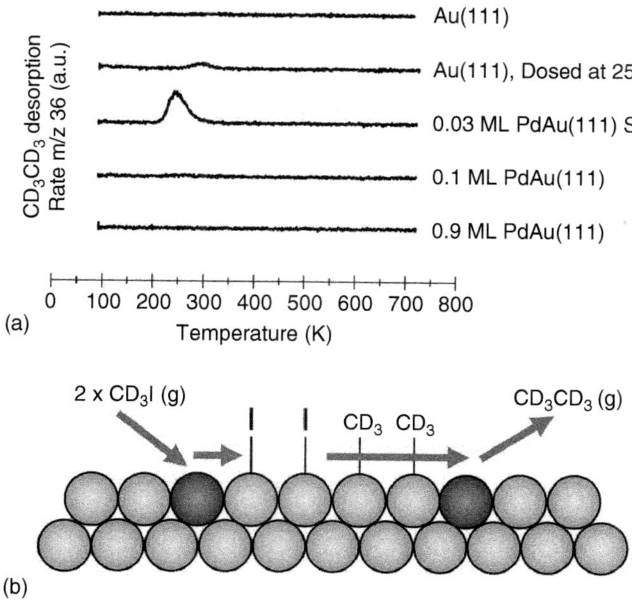

Figure 11.12 Ethane formation on various $Pd_{SA}Au(111)$ alloys and Au(111). TPD experiments (a) and schematic of the proposed reaction mechanism (b). Pd_{SA} are the dark spheres. Source: Réocreux et al. [55]. Reproduced with permission of Royal Society of Chemistry.

Figure 11.13 TON correlation with palladium fraction in the catalyst $Pd_{SA}Au$/resin for the Ullmann reaction. Source: Zhang et al. [56]. Reproduced with permission of American Chemical Society.

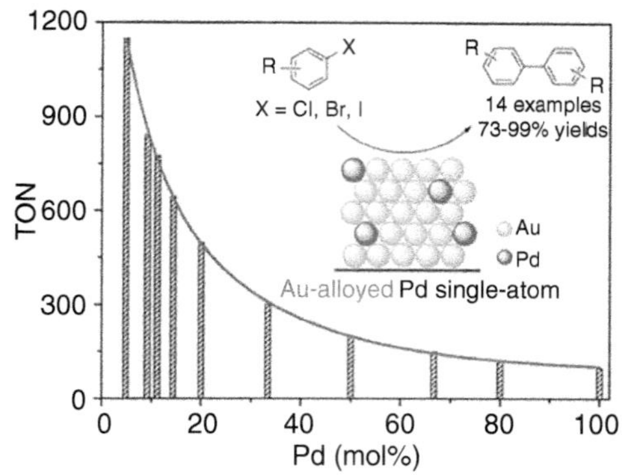

considered as the only active species in the catalyst since the catalyst did not work at all in its absence [56].

Other types of materials that have aroused great interest and have recently emerged as robust heterogeneous platforms are MOFs. Their structures are based on single metal ions or polynuclear metal clusters linked by a wide variety of organic ligands, through coordination bonds, to form infinite high-dimensional networks. MOFs offer multiple opportunities to create one or more active catalytic sites within the pores, and in addition, they have high porosity with tunable channels/cavities at the nanoscale level. Therefore, they can serve as templates for the preparation of single atoms [38, 57]. However, the use of SACs in MOFs for carbon–carbon coupling reactions has not yet been reported, unlike palladium complexes in MOFs [58] and smaller counterparts, the metal-organic polyhedra (MOP). The MOPs are discrete metal-organic molecular entities, which are also being used as supports to achieve higher catalytic efficiency. Specifically, a Zr-based MOP has been designed and subsequently modified, anchoring isolated palladium atoms, to carry out a Suzuki–Miyaura cross-coupling reaction in an aqueous medium. Table 11.2 shows the efficiency of this material [MOP–BPY(Pd)], which has been tested for the coupling reaction of bromophenyl derivatives and phenylboronic acid and compared with its analogues in the form of a molecular complex of palladium and a MOF with palladium atoms anchored. From this study, not only was it concluded that the designed MOP presents greater efficiency with respect to its analogues, but also its recyclability was demonstrated. Therefore, this is a clear example of both the great catalytic activity of the supported isolated atoms and the progress that has been made in the design of new heterogeneous catalysts for carbon–carbon coupling reactions [59]. However, more efforts in the material characterization and the reaction mechanism study are needed to have a complete picture of the SAC performance in this reaction.

Despite the great advances in obtaining SACs, where supports play a fundamental role, there is much work to be done, not only on their use for the reactions that

Table 11.2 Comparison of the efficiency of Zr-based MOP with its analogues to catalyze Suzuki-Miyaura reactions.

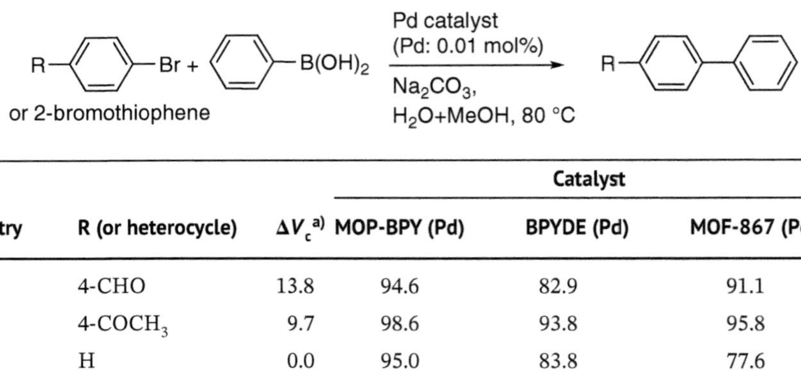

Entry	R (or heterocycle)	ΔV_c[a)]	Catalyst: MOP-BPY (Pd)	BPYDE (Pd)	MOF-867 (Pd)
1	4-CHO	13.8	94.6	82.9	91.1
2	4-$COCH_3$	9.7	98.6	93.8	95.8
3	H	0.0	95.0	83.8	77.6
4	4-OH	−2.0	93.1	93.0	74.6
5	4-CH_3	−3.3	86.2	41.6	65.8
6	4-OCH_3	−5.0	91.2	89.6	75.0
7	4-NH_2	−9.0	90.1	53.3	78.2
8	2-Bromothiophene	—	95.7	75.0	90.4

a) The degree of electron- donating and withdrawing effect for substitution groups of the substrates was quantified by ΔV_C calculated from the difference between molecular electrostatic potential at the nucleus of para carbon of the Br-substituted benzene and that of benzene. The larger positive ΔV_C value indicates the higher electron-withdrawing effect, while the larger negative ΔV_C value shows the higher electron-donating effect.

Source: Kim et al. [59]. Reproduced with permission of Springer Nature.

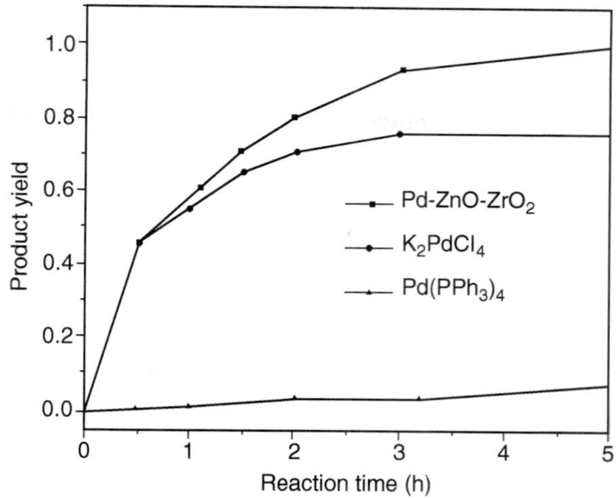

Figure 11.14 Kinetic studies of different palladium catalysts in Suzuki–Miyaura reactions. Source: Ding et al. [60]. Springer Nature. CC BY 4.0.

concern this chapter but also on the feasibility of synthetic methods for their large-scale preparation. A work reported last year, in which palladium atoms are homogeneously anchored on ZnO–ZrO_2, takes a step forward in these aspects since the catalyst can be produced on a multi-gram scale to catalyze Suzuki–Miyaura reactions under mild conditions [60]. Figure 11.14 shows kinetic studies for this reaction, which confirms the greater activity of the Pd/ZnO–ZrO_2 catalyst in comparison with conventional homogeneous catalysts. In addition, taking advantage of the development of characterization techniques, several of them were used to verify that the catalytic activity was due to the presence of Pd_{SA} and not to the formation of clusters or NPs.

11.3 Hydrosilylation and Hydroboration Reactions

11.3.1 Hydrosilylation Reactions

The hydrosilylation of alkenes is currently carried out in the industry for the production of silicones, using homogeneous platinum catalysts in part-per-million amounts, which remain in the product and arrive at the society. Despite the low amount of platinum remaining, the ubiquitous presence of these materials in daily objects makes this issue an environmental and toxicological concern [61, 62]. Thus, the search for sustainable and recoverable solid metal catalysts is of high interest [63], and in this sense, SACs have played an active role in the last years.

The first heterogeneous SAC for the hydrosilylation of alkenes was reported in 2017 [64], and consisted of alumina nanorods containing single platinum atoms. The resulting material catalyzed the hydrosilylation of a variety of alkenes, from linear and branched aliphatic olefins to vinylsiloxanes or a boron-containing alkene, and also included industrially relevant olefins and compounds with sensitive functional groups, with high TON ($\approx 10^5$) and significantly higher activity compared to related platinum NPs.

The hydrosilylation of alkenes has also been reported with a partially charged single platinum atom supported on anatase ($Pt_{SA}^{\delta+}/TiO_2$) [65]. The catalyst was prepared by electrostatic-induction ion exchange and catalyzed the hydrosilylation reaction with almost complete conversion and selectivity. The partial charge on platinum is essential for the catalytic activity, which is only achieved by the atomic dispersion on the support, and later studies showed that the generation of vacancies on TiO_2 helps to the catalytic activity of the Pt SAC [66]. Related to these studies, a N-doped graphene with isolated platinum single atomic sites (Pt_{SA}/N–G) was also reported for the hydrosilylation of alkenes. The solid catalyst was prepared by a Na_2CO_3 assisted one-pot pyrolysis strategy, with platinum loadings up to 5.3 wt% [67]. The platinum atoms are stabilized by the nitrogen in graphene through a Pt–N_4 structure, according to X-ray absorption fine structure (XAFS) and AC-STEM measurements, and show a fourfold higher catalytic activity than commercial Pt/C, with TOF of 180 h^{-1}. Following this, the use of platinum in MOF and zeolites was considered and reported [68]. Figure 11.15 shows that Pt_{SA} supported on zeolites and MOFs catalyzed the hydrosilylation of alkynes, alkenes, and alcohols with high efficiency in ppm amounts. Experimental and computational studies together with an *ad hoc* graphical method showed that the hydroaddition of alkynes proceeds through Pt–Si–H clusters of 3–5 atoms, what was called metal(oid) association, which decreased the energy of the transition state and directed the regioselectivity of the reaction. However, the hydroaddition of alkenes seemed to proceed through Pt SACs, based on the use of a Pt_{SA}-MOF [38].

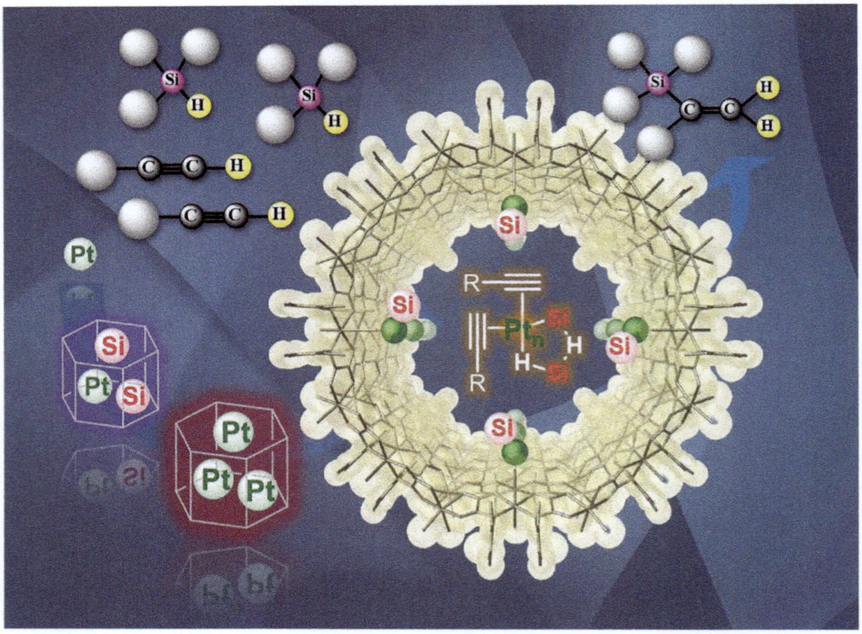

Figure 11.15 Hydrosilylation reaction of alkynes and alkenes catalyzed by platinum SAC and clusters confined in microporous solids, i.e. zeolites and MOFs. Source: Adapted from Rivero-Crespo et al. [68].

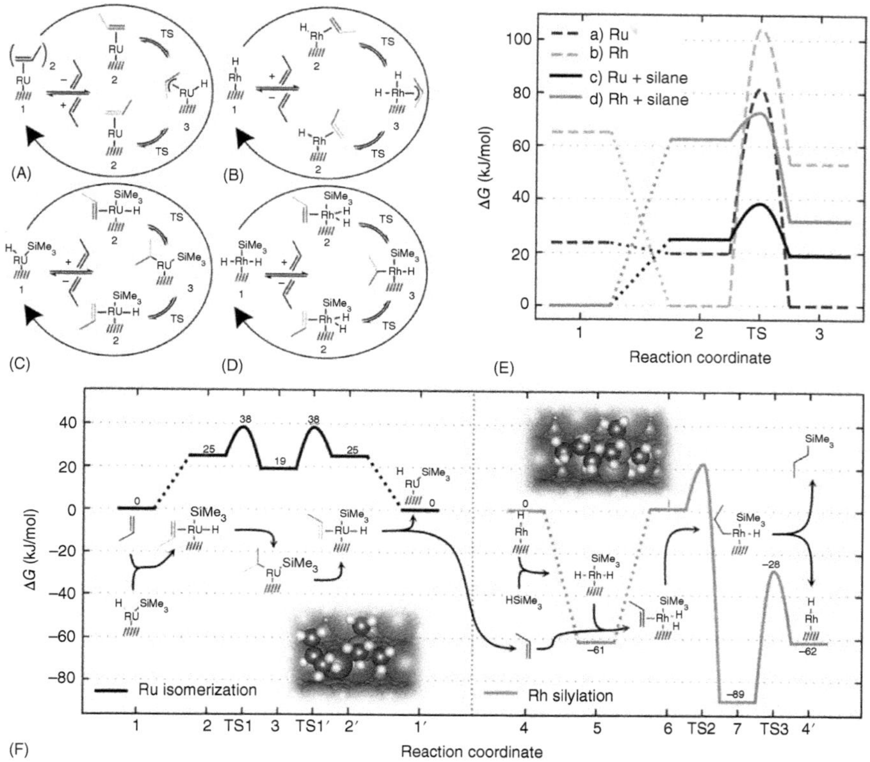

Figure 11.16 DFT calculations of the reaction mechanisms catalyzed by Ru_{SA}/CeO_2 and Rh_{SA}/CeO_2. Source: Sarma et al. [70]. John Wiley and Sons. CC BY 4.0.

The alkene hydrosilylation reaction has also been reported with a heterogeneous pseudo-single atom platinum catalyst in superparamagnetic Fe_3O_4–SiO_2 core–shell NPs, after applying a magnetic field. The solid material is functionalized with silane coupling agents, and the resulting Pt/vinyl/SiO_2/Fe_3O_4 catalyst can be reused four times without an appreciable decrease in activity [69].

Beyond platinum, it is difficult to find examples of SACs for the hydrosilylation of alkenes and alkynes. However, it has been recently reported that single metal atoms of ruthenium and rhodium on ceria (Ru_{SA}/CeO_2 and Rh_{SA}/CeO_2) catalyze the olefin isomerization-hydrosilylation tandem process [70]. Figure 11.16 shows that, according to DFT calculations, the binding strength of the olefin to the corresponding SAC controls the catalytic activity, and the SAC cooperation allows, in one pot, the synthesis of organosilane compounds with high regioselectivity and catalyst reuse.

In another example that points out to TiO_2 as a privileged support to stabilize single metal atoms for the hydrosilylation of alkenes, Co_{SA}/TiO_2 has also been reported as an active catalyst for this reaction [71]. A variety of alkenes were selectively converted to the corresponding alkylsilanes under solvent-free conditions with this catalyst, which showed high stability in air and high reusability. The formation of $CoTiO_3$ solid solution species was claimed to be at the origins of the catalytic activity of the material.

The deactivation of SACs is often promoted by sintering and NP formation. In a recent example, palladium/gold/platinum NPs were detached back into single atoms on oxide supports and nitrogen-doped carbon shells at high temperatures [72]. In particular, the so-produced palladium SAC catalyzes the alkene hydrosilylation reaction with high efficiency. In this way, deactivated metal NPs can be transformed to catalytically active metal clusters and SACs by thermal treatments. According to *in situ* transmission electron microscopy (TEM) images, the nitrogen defects of the carbon shells provide the diffusion sites for the atomic detachment. As commented above for carbon–carbon cross-coupling reactions, to prepare and stabilize SACs, metal alloys have been employed as alternative playgrounds to carbon-based and inorganic solid supports. Thus, Pd_{SA}-Au alloy catalysts were developed for the efficient hydrosilylation of α,β-unsaturated alkynes [73]. The Pd/Au molar ratio of the alloy affected the catalytic activity, and X-ray absorption spectroscopy (XAS) together with TEM and (XPS) measurements support the formation of isolated single palladium atoms in the low Pd/Au ratio alloy and the charge transfer from palladium to gold.

11.3.2 Hydroboration Reactions

The hydroboration of alkenes and alkynes, in contrast to the hydrosilylation reaction, is not necessarily catalyzed, and it can occur spontaneously for certain substrates. However, for most of the alkenes and alkynes studied, particularly for certain boranes, a catalyst is highly recommended to circumvent excessively long reaction times or partial selectivity. A typical metal catalyst for this reaction is copper, which implies a challenge from the SAC point of view, since, in contrast with carbon–carbon cross-coupling and hydrosilylation reactions, the metal of choice is not a relatively heavy atom but rather a first-raw transition metal atom. However, some examples have been published recently; in fact, all the examples correspond to the year 2020, which once again illustrates the very recent but rapid development of SACs in reactions for organic synthesis.

Cu_{SA}–CeO_2 has been reported as an effective catalyst for the hydroboration reaction [74]. Figure 11.17 shows that the Cu—O bond on the solid catalyst can be tuned from less to more ionic to control the catalytic activity, thus enabling the easier formation of a copper ethoxide intermediate and giving different vinylboronate products.

CeO_2 inherently presents vacancies; however, this is not the case for other conventional solid supports. As we have seen above, the presence of vacancies on the support is generally beneficial for the formation and stabilization of SACs, since these vacancies provide potential wells where the uncoordinated atoms can energetically be more comfortable. It has been recently reported a copper SAC on coordinatively unsaturated Al_2O_3 with an extraordinary high copper loading of 8.7%, for the 1,4-selective boration of the enals by Guo et al. [75]. Table 11.3 shows that this solid catalyst performs better than nano Cu/γ–Al_2O_3 and similarly to some homogeneous catalysts.

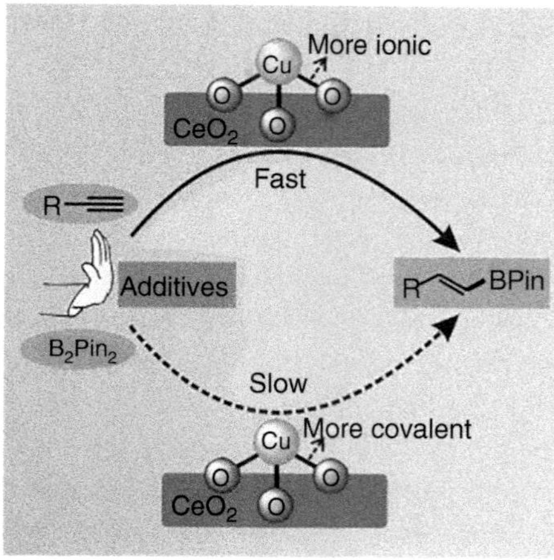

Figure 11.17 Scheme of the hydroboration reaction, according to the polarity of the Cu—O bond in the Cu_{SA}–CeO_2 catalyst. Source: Zhang et al. [74]. Reproduced with permission of Elsevier.

Table 11.3 The 1,4-selective boration of the enals using different catalysts.

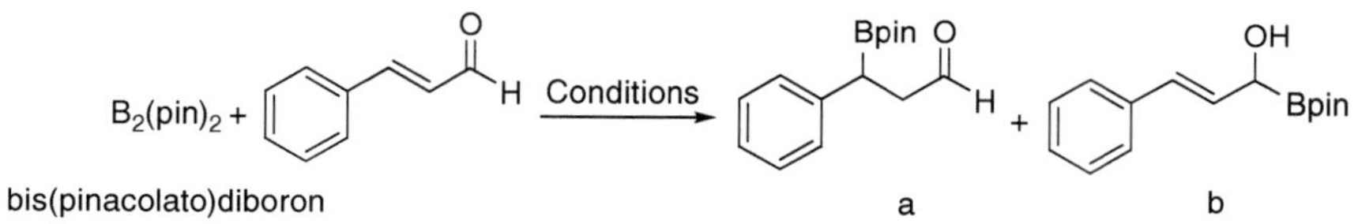

Entry	Catalyst	*T* (°C)	*t* (h)	Conversion (%)	Selectivity *a*/*b*
1	Cu_{SA}/rcu-Al_2O_3	65	0.5	99	95 : 5
2	nano Cu/γ-Al_2O_3	65	0.5	75	93 : 7
3	Pd complex	25	6	75	23 : 77
4	Cu complex	90	1	19	90 : 10
5	N-heterocyclic carbene (NHC)	70	1	95	95 : 5
6	Rh complex	22	1	99	—

Source: Guo et al. [75]. Reproduced with permission of John Wiley and Sons.

Platinum SACs, similar to that used in hydrosilylation reactions, have shown catalytic activity for the hydroboration reaction [76]. The study of the reaction of 1-octene with pinacolborane shows that the coordination structure of the platinum SAC definitively influences the catalytic activity for the hydroboration of alkenes, and the coordination of the Pt_{SA} with three O atoms in the support maximizes the catalytic activity, achieving a TON of 3288. DFT calculations support that the three O-coordinated platinum species show the lowest reaction energy during the limiting step of the reaction phase diagram.

The activation of boranes by SACs is not only limited to hydroboration reactions but also can occur during other reactions. For instance, Pt_{SA}/CeO_2 triggers the alcoholysis of ammonia borane to promote the hydrogenation of α-diazoesters and afford hydrazone esters with TOFs up to 566 h^{-1} [77].

11.4 Biomedical Applications

Not so long ago, NPs saw applications in medical science. Indeed, Fe_3O_4 NP mimics the horseradish peroxidase enzyme, catalyzing the oxidation of substrate 3,3,5,5-tetramethylbenzidine [78]. These metallic species, mimicking biological systems, were defined as nanozymes, to distinguish them from synzymes, functionalized polymers used as synthetic enzymes [79].

As in catalytic organic reactions, the active site isolation and downsizing started recently to be central points to increase the activity and selectivity of these enzyme-like materials. Consequently, single-atom nanozymes find their place in the story, considering that the presence of isolated active sites makes these species quite similar to bioenzymes [80]. Nowadays, despite the few applications in this area, they have been implied in cancer treatment, wound disinfection, and oxidative–stress cytoprotection [81].

Being enzyme-like, nanozymes show properties and abilities that are very similar to the natural ones, *e.g.* antioxidants. Indeed, the Fe–N/C SAC were able to oxidize 3,3′,5,5′-tetramethylbenzidine (TMB), 1,2-diaminobenzene (OPD) and dopamine (DA), and to decompose H_2O_2 to O_2 [82]. Moreover, the catalytic mechanism of this peroxidase process was confirmed by DFT calculations made on iron SAC, originated by the pyrolysis of zeolitic imidazolate frameworks (ZIFs), *i.e.* ZIF-8. The energy profile and reaction pathway of the peroxidase-like confined FeN_4 site confirmed that the oxidation of TMB by H_2O_2 is thermodynamically favorable [83].

What happens when the cell is in a pathologic situation? Usually, the natural enzymes cannot work properly and, for this reason, it is necessary to introduce external species that can operate instead of them. Taking into account what was elucidated above and that the cancer cells can be killed by an excess of oxygen species, nanozymes can be a good option for the treatment of tumors. It is of particular interest the use of iron, which is considered quite active in the production of OH ·radicals by the Fenton reaction. The radicals generated by this mechanism are considered to be killers of cancer cells [84, 85]. Figure 11.18 shows that single-atomic iron nanocatalysts (SAF NCs) with H_2O_2 presents high activity in the

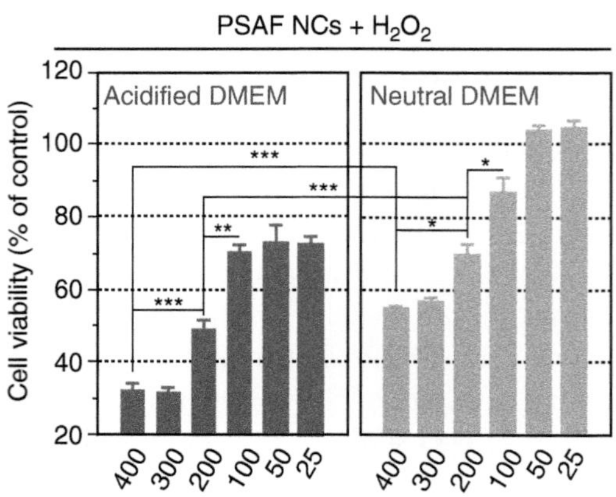

Figure 11.18 Cell viability of tumor cells in the presence of SAF NCs and H_2O_2. Source: Huo et al. [86]. Reproduced with permission of American Chemical Society.

treatment of tumor cells due to the high adsorption of H_2O_2 on the single atoms. The –O–O– is then weakened, and it undergoes homolytic breakage to generate OH· responsible for cell apoptosis [86].

In addition, porphyrin-based metal-organic frameworks (P-MOFs) were used as support for Fe_{SA} in cancer phototherapy. Indeed, the cavities of P-MOFs are formed by porphyrins that are quite known for their ability to generate complexes with iron. Then, these materials were applied in photothermal and photodynamic therapies of

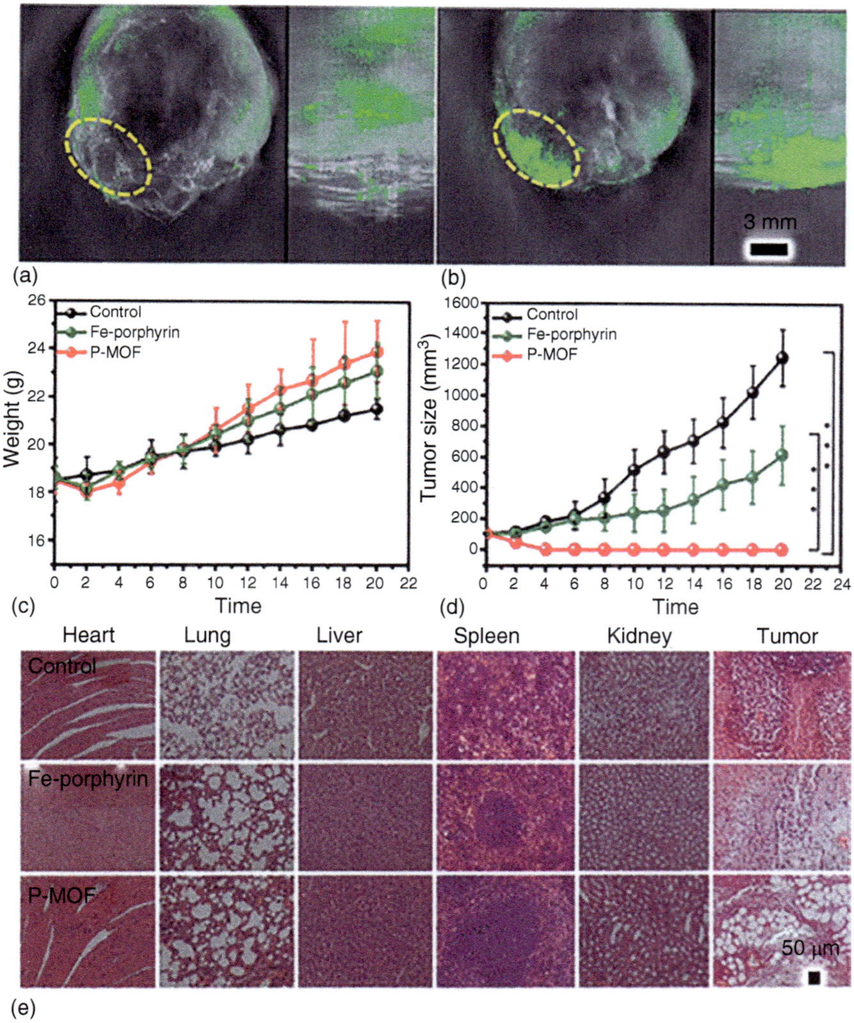

Figure 11.19 Photothermal and photodynamic therapeutic effects of P–MOF and irradiation. (a) 3D multispectral photoacoustic tomography (MSOT) image and views of tumor after the injection of phosphate buffer solution and (b) P–MOF. (c) Therapy effects on body weights of various mice groups after the therapy. (d) Tumor growth curves under different treatments. (e) H&E–stained main organs and tumor slices exposed to different therapies. Source: Wang et al. [87]. Reproduced with permission of American Chemical Society.

cancer cells at an irradiation wavelength of 808 nm. As shown in Figure 11.19, the comparison between the classical Fe-porphyrin and P-MOF with single atoms of iron demonstrates the faster answer to the therapy of the sample treated with P-MOF. Moreover, DFT calculations and mechanistic studies explain how this would not have been possible in the absence of Fe_{SA} and in the presence of bulk Fe_2O_3. The former, at 808 nm, can have an electron excited from the HOMO to the LUMO level, the first step to kill the tumor cell, and the latter cannot even undergo this first step [87].

In the same view, the application of Ru_{SA} to photodynamic treatment of cancer cells is quite novel. The self-assembled catalyst OxgeMCC-r single-atom enzyme (SAE) showed good biocompatibility thanks to the presence of polyvinylpyrrolidone (PVP), increasing the stability of the material in the physiological environment. This catalyst presented a high activity in the formation of O_2 coming from the breaking of H_2O_2. Furthermore, this high activity is due to the presence of single atoms, that in hypoxic conditions, can enhance the production of singlet $^1O_2$, which can induce cell apoptosis [88].

The ability of single atoms to generate radicals enhances their antibacterial properties, giving them also a place in the field of wound disinfection. An explicative example can be the material carbon nanospheres with a zinc-centered porphyrin-like structure (PMCS), where the isolated zinc atoms showed high activity against *Pseudomonas aeruginosa*. Indeed, Figure 11.20 shows that the presence of isolated zinc atoms made possible the homolysis of H_2O_2 and the

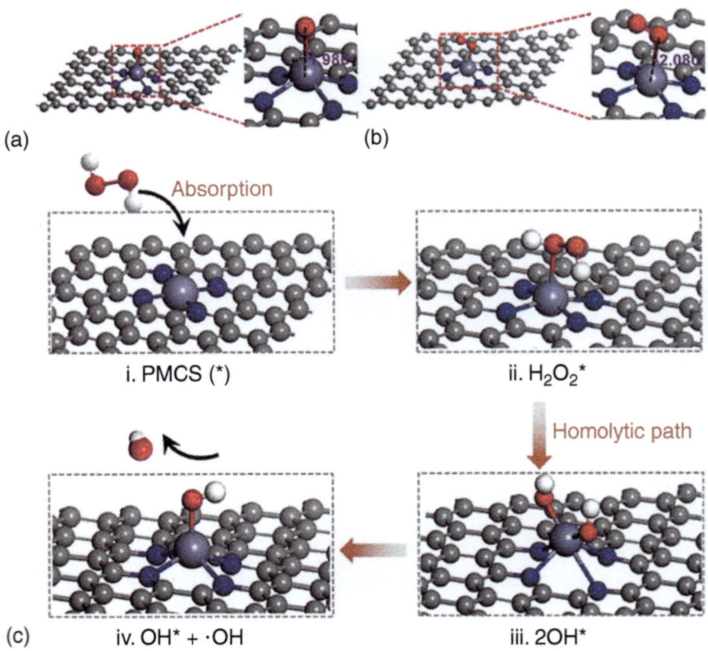

Figure 11.20 (a) and (b) Structures of O and O_2 species adsorbed on the surface of PMCS, respectively. (c) Proposed catalytic mechanism of the PMCS. Source: Xu et al. [89]. Reproduced with permission of John Wiley and Sons.

consequent generation of OH·, responsible for the bacteria death [89]. In addition, SAF NCs were also demonstrated to be active against *Staphylococcus aureus* and *Escherichia coli* [90].

11.5 Summary and Conclusions

We have seen in this chapter that different supported metal single-atom catalysts, from the first- to the third-transition raw, can be used for different reactions of interest in organic synthesis, with relatively complex substrates. Representative carbon–carbon cross-coupling reactions such as the Heck, Suzuki, and Sonogashira reactions have been accomplished with palladium and platinum SACs on different supports, including inorganic oxides, carbon materials (doped, graphene-type, etc.), and even metal alloys. The results are not so distinct to soluble organometallic complexes, which illustrates the high catalytic activity and the future feasibility of these and other SACs in cross-coupling reactions. The evolution of SAC shows parallelism with reactions involving small molecule activations, which assures cross-fertilization among the different reactivity realms. The hydrosilylation of alkenes and alkynes has also been accomplished with not only platinum but also other metal SACs, and the structure and catalytic activity of the former resembles the current industrial soluble catalyst, which paves the way for the production of more environmentally friendly and less toxic silicone materials. Following the hydrosilylation reaction, the hydroboration of alkenes can be carried out with similar platinum SACs and also with copper SACs. In most of the cases commented above, the catalytic activity can be controlled by the rational design of the active sites on the support. The choice of the support itself is important when preparing SACs since, for instance, the presence of vacancies and defects tends to help their formation and stabilization. This is not limited to vacant solids but also to conventional solid supports (Al_2O_3, TiO_2, carbon materials, etc.), for which stabilization sites can be easily generated by a variety of techniques.

In conclusion, the use of SACs in reactions of interest for organic synthesis is still in its infancy, and the next years promise a boost of studies in this regard. SACs are relatively stable to typically harsh organic reaction conditions (base or acid additives, polar solvents, heating, etc.) and seem not to leach under certain reaction conditions, thus allowing the recyclability of the solid material. Metal loadings in SACs are not anymore limited to highly diluted systems, and some SACs with metal loadings >5 wt% are already usual, which assures a high throughput for industrial applications. Not in vain, it has been proposed that classical solid catalysts for organic synthesis such as Pd/C and Pt/C indeed contain significant amounts of single metal atoms that could be responsible for some catalytic organic transformations. A revision of these processes may be seen in the last years. In any case, the richness in the design of SACs, devoted or not for a desired organic transformation, assures the success in different organic reactions during the next years and, sooner than later, their application in synthetic organic programs.

References

1 Taylor, H.S. (1925). A theory of the catalytic surface. *Proceedings of the Royal Society London A* 108 (745): 105–111.

2 Ozin, G.A. (1977). Metal atom matrix chemistry. Correlation of bonding with chemisorbed molecules. *Accounts of Chemical Research* 10 (1): 21–26. https://doi.org/10.1021/ar50109a004.

3 Liu, L. and Corma, A. (2018). Metal catalysts for heterogeneous catalysis: from single atoms to nanoclusters and nanoparticles. *Chemical Reviews* 118 (10): 4981–5079. https://doi.org/10.1021/acs.chemrev.7b00776.

4 Yang, X.–.F., Wang, A., Qiao, B. et al. (2013). Single-atom catalysts: a new frontier in heterogeneous catalysis. *Accounts of Chemical Research* 46 (8): 1740–1748. https://doi.org/10.1021/ar300361m.

5 Burda, C., Chen, X., Narayanan, R. et al. (2005). Chemistry and properties of nanocrystals of different shapes. *Chemical Reviews* 105 (4): 1025–1102. https://doi.org/10.1021/cr030063a.

6 Aiken, J.D. III, and Finke, R.G. (1999). A review of modern transition-metal nanoclusters: their synthesis, characterization, and applications in catalysis. *Journal of Molecular Catalysis A: Chemical* 145 (1-2): 1–44. https://doi.org/10.1016/S1381-1169(99)00098-9.

7 Thomas, J.M., Raja, R., and Lewis, D.W. (2005). Single-site heterogeneous catalysts. *Angewandte Chemie Internation Edition* 44 (40): 6456–6482. https://doi.org/10.1002/anie.200462473.

8 Jarupatrakorn, J. and Tilley, T.D. (2002). Silica-supported, single-site titanium catalysts for olefin epoxidation. A molecular precursor strategy for control of catalyst structure. *Journal of the American Chemical Society* 124 (28): 8380–8388. https://doi.org/10.1021/ja0202208.

9 Li, X., Yang, X., Zhang, J. et al. (2019). In situ/operando techniques for characterization of single-atom catalysts. *ACS Catalysis* 9 (3): 2521–2531. https://doi.org/10.1021/acscatal.8b04937.

10 Abbet, S., Sanchez, A., Heiz, U. et al. (2000). Acetylene cyclotrimerization on supported size-selected Pd_n clusters ($1 \leq n \leq 30$): one atom is enough! *Journal of the American Chemical Society* 122 (14): 3453–3457. https://doi.org/10.1021/ja9922476.

11 Qiao, B., Wang, A., Yang, X. et al. (2011). Single-atom catalysis of CO oxidation using Pt_1/FeO_x. *Nature Chemistry* 3 (8): 634–641. https://doi.org/10.1038/nchem.1095.

12 Djakovitch, L. and Koehler, K. (2001). Heck reaction catalyzed by Pd-modified zeolites. *Journal of the American Chemical Society* 123 (25): 5990–5999. https://doi.org/10.1021/ja001087r.

13 Dams, M., Drijkoningen, L., Pauwels, B. et al. (2002). Pd–zeolites as heterogeneous catalysts in heck chemistry. *Journal of Catalysis* 209 (1): 225–236. https://doi.org/10.1006/jcat.2002.3630.

14 Corma, A., García, H., Leyva, A. et al. (2003). Basic zeolites containing palladium as bifunctional heterogeneous catalysts for the Heck reaction.

Applied Catalysis A: General 247 (1): 41–49. https://doi.org/10.1016/S0926-860X(03)00060-7.

15 Corma, A., García, H., and Leyva, A. (2004). Controlling the softness–hardness of Pd by strong metal–zeolite interaction: cyclisation of diallylmalonate as a test reaction. *Journal of Catalysis* 225 (2): 350–358. https://doi.org/10.1016/j.jcat.2004.03.044.

16 Gawande, M.B., Fornasiero, P., and Zbořil, R. (2020). Carbon-based single-atom catalysts for advanced applications. *ACS Catalysis* 10 (3): 2231–2259. https://doi.org/10.1021/acscatal.9b04217.

17 Rivera-Cárcamo, C. and Serp, P. (2018). Single atom catalysts on carbon-based materials. *ChemCatChem* 10 (22): 5058–5091. https://doi.org/10.1002/cctc.201801174.

18 Wu, X.-F., Anbarasan, P., Neumann, H. et al. (2010). From noble metal to Nobel Prize: palladium-catalyzed coupling reactions as key methods in organic synthesis. *Angewandte Chemie Internation Edition* 49 (48): 9047–9050. https://doi.org/10.1002/anie.201006374.

19 Beletskaya, I.P. and Cheprakov, A.V. (2000). The Heck reaction as a sharpening stone of palladium catalysis. *Chemical Reviews* 100 (8): 3009–3066. https://doi.org/10.1021/cr9903048.

20 Brase, S. and de Meijere, A. (2004). Cross-coupling of organyl halides with alkenes: the Heck reaction. In: *Metal–Catalyzed Cross–Coupling Reactions*, vol. 2 (eds. A. de Meijere and F. Diederich), 217–315. Wiley-VCH Verlag GmbH & Co. KGaA.

21 Johansson Seechurn, C.C.C., Kitching, M.O., Colacot, T.J. et al. (2012). Palladium-catalyzed cross-coupling: a historical contextual perspective to the 2010 Nobel Prize. *Angewandte Chemie Internation Edition* 51 (21): 5062–5085. https://doi.org/10.1002/anie.201107017.

22 Alonso, F., Beletskaya, I.P., and Yus, M. (2004). Transition-metal-catalyzed addition of heteroatom–hydrogen bonds to alkynes. *Chemical Reviews* 104 (6): 3079–3160. https://doi.org/10.1021/cr0201068.

23 Corma, A., Leyva-Pérez, A., and Sabater, M.J. (2011). Gold-catalyzed carbon-heteroatom bond-forming reactions. *Chemical Reviews* 111 (3): 1657–1712. https://doi.org/10.1021/cr100414u.

24 Rubio-Marqués, P., Rivero-Crespo, M.A., Leyva-Pérez, A. et al. (2015). Well-defined noble metal single sites in zeolites as an alternative to catalysis by insoluble metal salts. *Journal of the American Chemical Society* 137 (36): 11832–11837. https://doi.org/10.1021/jacs.5b07304.

25 Rivero-Crespo, M.A., Leyva-Pérez, A., and Corma, A. (2016). A ligand-free Pt_3 cluster catalyzes the markovnikov hydrosilylation of alkynes with up to 106 turnover frequencies. *Chemistry – A European Journal* 23 (7): 1702–1708. https://doi.org/10.1002/chem.201605520.

26 Leyva, A., Zhang, X., and Corma, A. (2009). Chemoselective hydroboration of alkynesvs.alkenes over gold catalysts. *Chemical Communications* 33: 4947–4949. https://doi.org/10.1039/B901953G.

27 Reetz, M.T. and de Vries, J. (2004). Ligand-free Heck reactions using low Pd-loading. *Chemical Communications* 14: 1559–1563. https://doi.org/10.1039/b406719n.

28 de Vries, A.H.M., Mulders, J.M.C.A., Mommers, J.H.M. et al. (2003). Homeopathic ligand-free palladium as a catalyst in the heck reaction. A comparison with a palladacycle. *Organic Letters* 5 (18): 3285–3288. https://doi.org/10.1021/ol035184b.

29 Leyva-Pérez, A., Oliver-Meseguer, J., Rubio-Marqués, P. et al. (2013). Water-stabilized three- and four-atom palladium clusters as highly active catalytic species in ligand-free C-C cross-coupling reactions. *Angewandte Chemie Internation Edition* 52 (44): 11554–11559. https://doi.org/10.1002/anie.201303188.

30 Biffis, A., Centomo, P., Del Zotto, A. et al. (2018). Pd metal catalysts for cross-couplings and related reactions in the 21st century: a critical review. *Chemical Reviews* 118 (4): 2249–2295. https://doi.org/10.1021/acs.chemrev.7b00443.

31 Phan, N.T.S., Van Der Sluys, M., and Jones, C.W. (2006). On the nature of the active species in palladium catalyzed Mizoroki–Heck and Suzuki–Miyaura couplings – homogeneous or heterogeneous catalysis, a critical review. *Advanced Synthesis and Catalysis* 348 (6): 609–679. https://doi.org/10.1002/adsc.200505473.

32 Thathagar, M.B., ten Elshof, J.E., and Rothenberg, G. (2006). Pd nanoclusters in C–C coupling reactions: proof of leaching. *Angewandte Chemie Internation Edition* 45 (18): 2886–2890. https://doi.org/10.1002/anie.200504321.

33 Eremin, D.B. and Ananikov, V.P. (2017). Understanding active species in catalytic transformations: from molecular catalysis to nanoparticles, leaching, "Cocktails" of catalysts and dynamic systems. *Coordination Chemistry Reviews* 346: 2–19. https://doi.org/10.1016/j.ccr.2016.12.021.

34 Fu, Q., Saltsburg, H., and Flytzani-Stephanopoulos, M. (2003). Active nonmetallic Au and Pt species on ceria-based water–gas shift catalysts. *Science* 301 (5635): 935–938. https://doi.org/10.1126/science.1085721.

35 Zhai, Y., Pierre, D., Si, R. et al. (2010). Alkali-stabilized Pt-OH_x species catalyze low-temperature water–gas shift reactions. *Science* 329 (5999): 1633–1636. https://doi.org/10.1126/science.1192449.

36 Peterson, E.J., DeLaRiva, A.T., Lin, S. et al. (2014). Low-temperature carbon monoxide oxidation catalysed by regenerable atomically dispersed palladium on alumina. *Nature Communications* https://doi.org/10.1038/ncomms5885.

37 Yang, M., Li, S., Wang, Y. et al. (2014). Catalytically active Au-$O(OH)_x$-species stabilized by alkali ions on zeolites and mesoporous oxides. *Science* 346 (6216): 1498–1501. https://doi.org/10.1126/science.1260526.

38 Rivero-Crespo, M.A., Mon, M., Ferrando-Soria, J. et al. (2018). Confined Pt_1^{1+} water clusters in a MOF catalyze the low-temperature water–gas shift reaction with both CO_2 oxygen atoms coming from water. *Angewandte Chemie International Edition* 57 (52): 17094–17099. https://doi.org/10.1002/anie.201810251.

39 Ding, K., Gulec, A., Johnson, A.M. et al. (2015). Identification of active sites in CO oxidation and water-gas shift over supported Pt catalysts. *Science* 350 (6257): 189–192. https://doi.org/10.1126/science.aac6368.

40 Gokhale, A.A., Dumesic, J.A., and Mavrikakis, M. (2008). On the mechanism of low-temperature water gas shift reaction on copper. *Journal of the American Chemical Society* 130 (4): 1402–1414. https://doi.org/10.1021/ja0768237.

41 Sun, X., Lin, J., Zhou, Y. et al. (2017). FeO_x supported single-atom Pd bifunctional catalyst for water gas shift reaction. *AIChE Journal* 63 (9): 4022–4031. https://doi.org/10.1002/aic.15759.

42 Chen, Y., Lin, J., Li, L. et al. (2018). Identifying size effects of Pt as single atoms and nanoparticles supported on FeO_x for the water–gas shift reaction. *ACS Catalysis* 8 (2): 859–868. https://doi.org/10.1021/acscatal.7b02751.

43 Ammal, S.C. and Heyden, A. (2017). Water–gas shift activity of atomically dispersed cationic platinum versus metallic platinum clusters on titania supports. *ACS Catalysis* 7 (1): 301–309. https://doi.org/10.1021/acscatal.6b02764.

44 Chen, Z., Vorobyeva, E., Mitchell, S. et al. (2018). A heterogeneous single-atom palladium catalyst surpassing homogeneous systems for Suzuki coupling. *Nature Nanotechnology* 13 (8): 702–707. https://doi.org/10.1038/s41565-018-0167-2.

45 Fu, N., Liang, X., Li, Z. et al. (2020). Fabricating Pd isolated single atom sites on C_3N_4/rGO for heterogenization of homogeneous catalysis. *Nano Research* 13 (4): 947–951. https://doi.org/10.1007/s12274-020-2720-1.

46 Jin, Y., Lu, F., Yi, D. et al. (2020). Engineering electronic structure of single-atom Pd site on $Ti_{0.87}O_2$ nanosheet via charge transfer enables C–Br cleavage for room-temperature Suzuki coupling. *Chinese Chemical Society Chemistry* 2: 1453–1462. https://doi.org/10.31635/ccschem.020.202000388.

47 Costa, P., Sandrin, D., and Scaiano, J.C. (2020). Real-time fluorescence imaging of a heterogeneously catalysed Suzuki–Miyaura reaction. *Nature Catalysis* 3: 427–437. https://doi.org/10.1038/s41929-020-0442-0.

48 Lee, E.–.K., Park, S.–.A., Woo, H. et al. (2017). Platinum single atoms dispersed on carbon nanotubes as reusable catalyst for Suzuki coupling reaction. *Journal of Catalysis* 352: 388–393. https://doi.org/10.1016/j.jcat.2017.05.005.

49 Zhang, X., Sun, Z., Wang, B. et al. (2018). C–C coupling on single-atom-based heterogeneous catalyst. *Journal of the American Chemical Society* 140 (3): 95–962. https://doi.org/10.1021/jacs.7b09314.

50 Wang, L., Zhang, W., Wang, S. et al. (2016). Atomic-level insights in optimizing reaction paths for hydroformylation reaction over Rh/CoO single-atom catalyst. *Nature Communications* https://doi.org/10.1038/ncomms14036.

51 Lang, R., Li, T., Matsumura, D. et al. (2016). Hydroformylation of olefins by a rhodium single-atom catalyst with activity comparable to $RhCl(PPh_3)_3$. *Angewandte Chemie International Edition* 55 (52): 16054–16058. https://doi.org/10.1002/anie.201607885.

52 Chen, F., Li, T., and Pan, X. (2020). Pd_1/CeO_2 single-atom catalyst for alkoxycarbonylation of aryl iodides. *Science China Materials* 63: 959–964. https://doi.org/10.1007/s40843-019-1204-y.

53 Zhang, L., Wang, A., Wang, W. et al. (2015). Co–N–C catalyst for C–C coupling reactions: on the catalytic performance and active sites. *ACS Catalysis* 5 (11): 6563–6572. https://doi.org/10.1021/acscatal.5b01223.

54 Fernández, E., Rivero-Crespo, M.A., Domínguez, I. et al. (2019). Base-controlled heck, suzuki, and sonogashira reactions catalyzed by ligand-free platinum or palladium single atom and sub-nanometer clusters. *Journal of the American Chemical Society* 141 (5): 1928–1940. https://doi.org/10.1021/jacs.8b07884.

55 Réocreux, R., Uhlman, M., Thuening, T. et al. (2019). Efficient and selective carbon–carbon coupling on coke-resistant PdAu single-atom alloys. *Chemical Communications* 55 (100): 15085–15088. https://doi.org/10.1039/C9CC07932G.

56 Zhang, L., Wang, A., Miller, J.T. et al. (2014). Efficient and durable Au alloyed Pd single-atom catalyst for the ullmann reaction of aryl chlorides in water. *ACS Catalysis* 4 (5): 1546–1553. https://doi.org/10.1021/cs500071c.

57 Wei, Y.–.S., Zhang, M., Zou, R. et al. (2020). Metal–organic framework-based catalysts with single metal sites. *Chemical Reviews* 120 (21): 12089–12174. https://doi.org/10.1021/acs.chemrev.9b00757.

58 Adam, R., Mon, M., Greco, R. et al. (2019). Self-assembly of catalytically active supramolecular coordination compounds within metal–organic frameworks. *Journal of the American Chemical Society* 141 (26): 10350–10360. https://doi.org/10.1021/jacs.9b03914.

59 Kim, S., Jee, S., Choi, K.M. et al. (2021). Single-atom Pd catalyst anchored on Zr-based metal-organic polyhedra for Suzuki-Miyaura cross coupling reactions in aqueous media. *Nano Research* 14: 486–492. https://doi.org/10.1007/s12274-020-2885-7.

60 Ding, G., Hao, L., Xu, H. et al. (2020). Atomically dispersed palladium catalyses Suzuki–Miyaura reactions under phosphine-free conditions. *Communications Chemistry* 3: 43. https://doi.org/10.1038/s42004-020-0289-y.

61 Obligacion, J.V. and Chirik, P.J. (2018). Earth-abundant transition metal catalysts for alkene hydrosilylation and hydroboration. *Nature Reviews Chemistry* 2: 15–34. https://doi.org/10.1038/s41570-018-0001-2.

62 de Almeida, L.D., Wang, H., Junge, K. et al. (accepted for publication) (2021). Recent advances in catalytic hydrosilylations: developments beyond traditional platin catalysts. *Angewandte Chemie International Edition* 60 (2): 550–565.

63 Pagliaro, M., Ciriminna, R., Pandarus, V. et al. (2013). Platinum-based heterogeneously catalyzed hydrosilylation. *European Journal of Organic Chemistry* 2013: 6227–6235. https://doi.org/10.1002/ejoc.201300290.

64 Cui, X., Junge, K., Dai, X. et al. (2017). Synthesis of single atom based heterogeneous platinum catalysts: high selectivity and activity for hydrosilylation reactions. *ACS Central Science* 3 (6): 580–585. https://doi.org/10.1021/acscentsci.7b00105.

65 Chen, Y., Ji, S., Sun, W. et al. (2018). Discovering partially charged single-atom Pt for enhanced anti-markovnikov alkene hydrosilylation. *Journal of the American Chemical Society* 140 (24): 7407–7410. https://doi.org/10.1021/jacs.8b03121.

66 Zhou, X., Chen, L., Sterbinsky, G.E. et al. (2020). Pt-Ligand single-atom catalysts: tuning activity by oxide support defect density. *Catalysis Science & Technology* 10: 3353–3365. https://doi.org/10.1039/C9CY02594D.

67 Zhu, Y., Cao, T., Cao, C. et al. (2018). One-pot pyrolysis to N-doped graphene with high-density Pt single atomic sites as heterogeneous catalyst for alkene hydrosilylation. *ACS Catalysis* 8 (11): 10004–10011. https://doi.org/10.1021/acscatal.8b02624.

68 Miguel Rivero-Crespo, M.A., Oliver-Meseguer, J., Kapłónska, K. et al. (2020). Cyclic metal(oid) clusters control platinum-catalysed hydrosilylation reactions: from soluble to zeolite and MOF catalysts. *Chemical Science* 11: 8113–8124. https://doi.org/10.1039/D0SC02391D.

69 Zai, Y., Zhao, Y., Chen, S. et al. (2018). Heterogeneously supported pseudo-single atom Pt as sustainable hydrosilylation catalyst. *Nano Reserch* 11: 2544–2552. https://doi.org/10.1007/s12274-017-1879-6.

70 Sarma, B.B., Kim, J., Amsler, J. et al. (2020). One-pot cooperation of single-atom Rh and Ru solid catalysts for a selective tandem olefin isomerization-hydrosilylation process. *Angewandte Chemie International Edition* 59: 5806–5815. https://doi.org/10.1002/ange.201915255.

71 Mitsudome, T., Fujita, S., Sheng, M. et al. (2019). Air-stable and reusable cobalt ion-doped titanium oxide catalyst for alkene hydrosilylation. *Green Chemistry* 21: 4566–4570. https://doi.org/10.1039/C9GC01981B.

72 Zhou, H., Zhao, Y., Xu, J. et al. (2020). Recover the activity of sintered supported catalysts by nitrogen-doped carbon atomization. *Nature Communications* 11: 335. https://doi.org/10.1038/s41467-019-14223-w.

73 Miura, H., Endo, K., Ogawa, R. et al. (2017). Supported palladium–gold alloy catalysts for efficient and selective hydrosilylation under mild conditions with isolated single palladium atoms in alloy nanoparticles as the main active site. *ACS Catalysis* 7 (3): 1543–1553. https://doi.org/10.1021/acscatal.6b02767.

74 Zhang, J., Wang, Z., Chen, W. et al. (2020). Tuning polarity of Cu-O bond in heterogeneous Cu catalyst to promote additive-free hydroboration of alkynes. *Chem* 6 (3): 725–737. https://doi.org/10.1016/j.chempr.2019.12.021.

75 Guo, T., Tang, N., Lin, F. et al. (2020). High-loading single-atom copper catalyst supported on coordinatively unsaturated Al_2O_3 for selective synthesis of homoallylboronates. *ChemSusChem* 13 (12): 3115–3121. https://doi.org/10.1002/cssc.202000536.

76 Xu, Q., Guo, C., Tian, S. et al. (2020). Coordination structure dominated performance of single-atomic Pt catalyst for anti-Markovnikov hydroboration of alkenes. *Science China Materials* 63: 972–981. https://doi.org/10.1007/s40843-020-1334-6.

77 Liu, C., Chen, Z., Yan, H. et al. (2019). Expedient synthesis of E-hydrazone esters and 1H-indazole scaffolds through heterogeneous single-atom platinum catalysis. *Science Advances* 5 (12): eaay1537. https://doi.org/10.1126/sciadv.aay1537.

78 Gao, L., Zhuang, J., Nie, L. et al. (2007). Intrinsic peroxidase-like activity of ferromagnetic nanoparticles. *Nature Nanotechnology* 2 (9): 577–583. https://doi.org/10.1038/nnano.2007.260.

79 Manea, F., Houillon, F.B., Pasquato, L. et al. (2004). Nanozymes: gold-nanoparticle-based transphosphorylation catalysts. *Angewandte Chemie International Edition* 43 (45): 6165–6169. https://doi.org/10.1002/anie.200460649.

80 Zhang, H., Lu, X.F., Wu, Z.-P. et al. (2020). Emerging multifunctional single-atom catalysts/nanozymes. *ACS Central Science* 6 (8): 1288–1301. https://doi.org/10.1021/acscentsci.0c00512.

81 Xiang, H., Feng, W., and Chen, Y. (2020). Single-atom catalysts in catalytic biomedicine. *Advanced Materials* 32 (8): 1905994. https://doi.org/10.1002/adma.201905994.

82 Lu, M., Wang, C., Ding, Y. et al. (2019). Fe–N/C single-atom catalysts exhibiting multienzyme activity and ROS scavenging ability in cells. *Chemical Communications* 55 (96): 14534–14537. https://doi.org/10.1039/C9CC07408B.

83 Zhao, C., Xiong, C., Liu, X. et al. (2019). Unraveling the enzyme-like activity of heterogeneous single atom catalyst. *Chemical Communications* 55 (16): 2285–2288. https://doi.org/10.1039/C9CC00199A.

84 Dizdaroglu, M. and Jaruga, P. (2012). Mechanisms of free radical-induced damage to DNA. *Free Radical Research* 46 (4): 382–419. https://doi.org/10.3109/10715762.2011.653969.

85 Noh, J., Kwon, B., Han, E. et al. (2015). Amplification of oxidative stress by a dual stimuli-responsive hybrid drug enhances cancer cell death. *Nature Communications* 6: 6907. https://doi.org/10.1038/ncomms7907.

86 Huo, M., Wang, L., Wang, Y. et al. (2019). Nanocatalytic tumor therapy by single-atom catalysts. *ACS Nano* 13 (2): 2643–2653. https://doi.org/10.1021/acsnano.9b00457.

87 Wang, L., Qu, X., Zhao, Y. et al. (2019). Exploiting single atom iron centers in a porphyrin-like MOF for efficient cancer phototherapy. *ACS Applied Materials & Interfaces* 11 (38): 35228–35237. https://doi.org/10.1021/acsami.9b11238.

88 Wang, D., Wu, H., Phua, S.Z.F. et al. (2020). Self-assembled single-atom nanozyme for enhanced photodynamic therapy treatment of tumor. *Nature Communications* 11: 357. https://doi.org/10.1038/s41467-019-14199-7.

89 Xu, B., Wang, H., Wang, W. et al. (2019). A single-atom nanozyme for wound disinfection applications. *Angewandte Chemie International Edition* 58 (15): 4911–4916. https://doi.org/10.1002/anie.201813994.

90 Huo, M., Wang, L., Zhang, H. et al. (2019). Construction of single-iron-atom nanocatalysts for highly efficient catalytic antibiotics. *Small* 15 (31): 1901834. https://doi.org/10.1002/smll.201901834.

12

Supported Metal Single-Atom Thermo-Catalysts for Reforming Reactions

Xuan-Huynh Pham and Doan Pham Minh

Université de Toulouse, IMT Mines Albi, UMR CNRS 5302, Centre RAPSODEE, Campus Jarlard, F-81013 Albi cedex 09, France

12.1 Introduction

Syngas (a mixture of CO and H_2) has been getting more and more attention as it is an important industrial platform mixture with different applications such as energy (heat and electricity) generation, hydrogen or bio-methane production, liquid fuel synthesis, methanol and dimethyl ether synthesis, and ammonia production [1]. Nowadays, syngas can be produced from fossil fuels (petroleum, coal, and natural gas) or renewable sources (biomass, biogas, and organic wastes) (Figure 12.1). Reforming processes play an important role in determining the composition and quality of the produced syngas, which, in turn, determines the applications [1]. Although natural gas is still the main feedstock for hydrogen (and syngas) production, reforming of other renewable sources such as biogas, biomass, bio-alcohol or residues from fossil-fuel-refining processes is of great interest due to the related economic and environmental aspects.

Reforming is an endothermal process, i.e. it needs an external source of energy to maintain the process. In addition, an active and robust catalyst is generally needed to lower activation energy for the reforming reactions and to improve the catalytic selectivity toward desired products. To date, supported noble metals such as ruthenium have shown to be most active and stable for reforming reaction [2]. However, the high cost of noble metals is one of the major barriers for the development of this type of catalyst. Other transition metals, especially Ni, have been considered as alternative catalytic materials for reforming reactions [3, 4]. In addition, downsizing the active phase to maximize the metal utility is logically meaningful. Indeed, in reforming processes, metal particle size is considered as the most important factor determining catalytic performance, especially for coke limitation. The latter is mostly important for the applications of transition-metal-based catalysts. Accordingly, the smaller the metal particle size is, the higher the activity, and higher the coking resistance the catalyst reaches [5–7].

Within small metal particles [sub-nanoparticles or clusters, and nanoparticles (NPs)], metal atoms can be found on edge, terrace, or corner position. The share

Supported Metal Single Atom Catalysis, First Edition. Edited by Philippe Serp and Doan Pham Minh.
© 2022 WILEY-VCH GmbH. Published 2022 by WILEY-VCH GmbH.

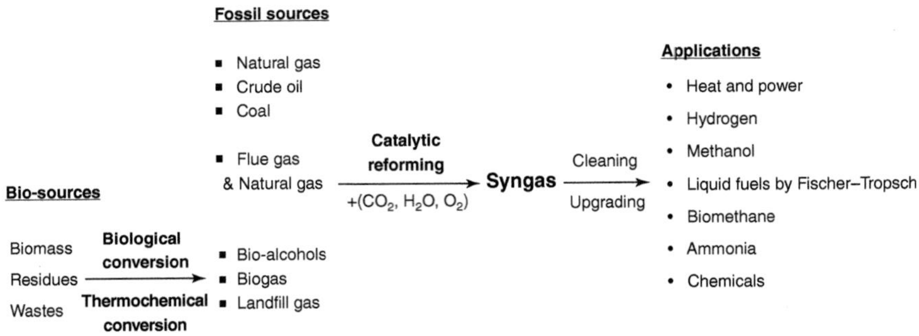

Figure 12.1 Simplified pathway for chemicals and energy production *via* reforming processes. Source: Xuan-Huynh Pham.

of these low coordinated surface atoms depends mostly on the particle size [7, 8]. According to Van Hardeveld and Hartog [8], for particles about 0.6–4 nm, the share of corner and edges atoms reaches its maximum. For larger particles (4–10 nm), the share of corner atoms decreases and that of terrace atoms increases. For particles larger than 10 nm, the share of terrace atoms is high and can reach up to 80%. The low coordinated surface atoms play an important role in reforming reaction as they can take part in binding toward the intermediates, lowering reaction barriers, changing the reaction pathway, and globally enhancing the catalytic activity [7].

In comparison with supported metal NP-based catalysts, supported metal single-atom catalysts (SACs) constitute a particular category of catalytic materials with specific properties as previously stated in Chapter 1, e.g. low metal contents, high metal dispersion, and possible homogeneity of active sites. Despite possessing a high performance and a potential low cost [9], SACs have high surface free energy [10], which is the main challenge for the stability of these catalysts (See Figure 1 in Chapter 1 of this book). This is more pronounced in reforming processes where high temperatures are usually needed (e.g. > 700 °C for methane steam reforming). Various synthesis approaches, including conventional and innovative methods, along with different materials used as support and/or promoters, have been investigated to overcome this issue. For example, flame spray pyrolysis was applied to stabilize Pt_{SA} on ZrO_2 support for DRM at 700 °C [11]. In addition, metal alloys [5, 12], support containing oxygen vacancies(O_v) [11], and promoters, i.e. cerium [13], have also shown their advantages in stabilization and dispersion of the metal atoms. Further details on appropriate support materials with different extents of metal–support interaction and innovative synthesis approaches have been discussed in Chapters 2 and 3 of this book.

Regarding the influence and advantages of SACs over cluster- and NP-based catalyst systems, fundamental features in catalysis science are generally discussed [14]:

- reaction mechanisms;
- binding energies of the reactants to the active sites;
- magnitude of the activation energy.

This chapter presents three main categories of reforming processes: methane reforming, hydrocarbon reforming, and alcohol reforming. We first provide a brief of fundamental aspects and main challenges and then highlight the recent advances in catalyst development related to SACs for each process. Experimental investigations along with some theoretical DFT calculations are analyzed to better understand the fundamental reaction mechanisms and the role of SACs in these processes.

12.2 Supported Metal Single Atoms for Methane Reforming

Methane is the main component of natural gas, biogas, landfill gas, and raw syngas from biomass gasification. Methane can be reformed into high value-added products, i.e. syngas or hydrogen, by using at least one of the three oxidants (CO_2, H_2O, and O_2) via various reactions [15] such as:

Steam reforming of methane (SRM):

$$H_2O + CH_4 \rightarrow CO + 3H_2 \quad \Delta H^0_{298} = +206.2\ \text{kJ/mol} \tag{12.1}$$

Dry reforming of methane (DRM):

$$CO_2 + CH_4 \rightarrow 2CO + 2H_2 \rightarrow \Delta H^0_{298} = +247.4\,\text{kJ/mol} \tag{12.2}$$

Partial oxidation of methane (POM):

$$CH_4 + 0.5O_2 \rightarrow CO + 2H_2 \quad \Delta H^0_{298} = -35.6\,\text{kJ/mol} \tag{12.3}$$

In addition to the main reactions above, side reactions along with the formation carbon deposits (coking) can also occur through a set of reactions:

Combustion of methane:

$$CH_4 + 2O_2 \rightarrow CO_2 + 2H_2O \qquad \Delta H^0_{298} = -802.2\,\text{kJ/mol} \tag{12.4}$$

Methane reforming with large excess of steam

$$CH_4 + 2H_2O \rightarrow CO_2 + 4H_2 \qquad \Delta H^0_{298} = +165.0\,\text{kJ/mol} \tag{12.5}$$

Methane decomposition or methane cracking:

$$CH_4 \rightarrow C + 2H_2 \qquad \Delta H^0_{298} = +74.9\,\text{kJ/mol} \tag{12.6}$$

Boudouard reaction:

$$2CO \rightarrow C + CO_2 \qquad \Delta H^0_{298} = -172.5\,\text{kJ/mol} \tag{12.7}$$

Water–gas shift reaction (WGSR):

$$CO + H_2O \rightarrow CO_2 + H_2 \qquad \Delta H^0_{298} = -41.2\,\text{kJ/mol} \tag{12.8}$$

Reverse WGSR:

$$CO_2 + H_2 \rightarrow CO + H_2O \qquad \Delta H^0_{298} = +41.2\,\text{kJ/mol} \tag{12.9}$$

Steam gasification of carbon:

$$C + H_2O \rightarrow CO + H_2 \qquad \Delta H^0_{298} = +131.3\ kJ/mol \tag{12.10}$$

Partial oxidation of carbon:

$$C + 0.5O_2 \rightarrow CO \qquad \Delta H^0_{298} = -110.5\ kJ/mol \tag{12.11}$$

Complete oxidation of carbon:

$$C + O_2 \rightarrow CO_2 \qquad \Delta H^0_{298} = -393.5\ kJ/mol \tag{12.12}$$

Along with the single reforming processes, *i.e.* SRM, DMR, and POM, other combined reforming processes can be also performed, such as CH_4 auto-thermal reforming (a combination of SRM and POM into a single reactor), bi-reforming (a combination of SRM and DRM), or tri-reforming (a combination of SRM, DRM, and POM). While the SMR, and POM at a lesser extent, are well developed for hydrogen production at industrial scale, the other processes are still at the R&D stage in several contexts, including CO_2 capture and conversion and biogas valorization [16].

Thermodynamically, high operating temperatures (>800 °C) are required to attain acceptable syngas yields [17]. However, at high temperatures, side reactions and carbon deposition are also promoted. Carbon deposits (whiskers, coke, and pyrolytic carbon) can encapsulate the active phase and inhibit accessibility of reactants to its surface, resulting in catalytic deactivation. In addition, at high temperatures, thermal deterioration such as active phase and support sintering can occur, leading to deactivation and uncontrollable side reactions. Regarding efforts of enhancing catalytic performance, downsizing of the active metal to single atoms dispersed on an appropriate support emerges as a solution. SACs can reduce activation energy and increase the selectivity into the desired products, along with inhibit catalyst deactivation by carbon deposition [3, 5, 18]. Since the last decade, SACs were successfully synthesized for methane but also hydrocarbon or alcohols reforming, using noble or other transition metals, even both of them. Here, we analyze relevant contributions from literature.

12.2.1 Noble-Metal Single-Atom Catalysts for Methane Reforming

As previously stated, one of the advantages of SACs is to maximize the utility of metal active phase; thus, it could compensate the high cost of noble-metal-based catalysts and can boost catalyst development. On the road to finding the most suitable catalyst for syngas production via methane-reforming processes, there has been extensive research on noble metal catalysts, including rhodium- [19–22], ruthenium- [23], platinum- [24, 25], and palladium-based [26] catalysts. However, to date, only Rh [18]- and Ru [27]-supported SACs have been investigated for methane reforming.

Rhodium single atoms (Rh_{SA}) supported on γ-Al_2O_3 or γ-Al_2O_3 modified with CeO_2 or CeO_2-Sm_2O_3 were successfully synthesized and evaluated for MSR at low temperature of 500 °C [18]. The catalysts were prepared by wet impregnation method. For the preparation of the supports, the authors used commercial γ-Al_2O_3, and solutions of $Sm(NO_3)_3 \cdot 6H_2O$ and/or $Ce(NO_3)_3 \cdot 6H_2O$ in ethanol.

After stirring at room temperature for 5 hours, the mixture was passed in a rotary evaporator at 60 °C to remove ethanol, then subsequently dried and calcined at 100 °C for 12 hours and 500 °C for 6 hours, respectively. The impregnation was then performed in a glovebox. A solution containing the appropriate concentration of rhodium acetate was mixed with the prepared support to obtain 0.5 wt% Rh in the final catalyst. As mentioned above, SAs have high surface-free energy and can potentially aggregate into clusters or NPs. In this catalyst system, CeO_2 was used to favor the formation and stabilization of Rh_{SA} (and of small NPs) due to a strong metal–support interaction created by the Rh—O—Ce bond [18]. In addition, the authors also used Sm_2O_3 to enhance the dispersion of Rh atoms on the CeO_2-Al_2O_3 support. The addition of Sm_2O_3 to ceria–alumina support allows the formation of more O_v, which in turn promotes the metal dispersion, as reported previously by the same research group for Rh- [28] and Pt-based catalysts [29]. In this study, singly dispersed Rh atoms were observed by scanning transmission electron microscopy (STEM) coupling with a high angle annular dark-field (HAADF) detector for the catalysts after calcination, reduction, and reaction (Figure 12.2). The relationship between Rh particle size and catalytic performance of the catalysts is summarized in Table 12.1.

As shown in Figure 12.2 and Table 12.1, the three catalysts calcined under air at 500 °C for 4 hours contained Rh_{SA}, with some NPs for Rh/Al_2O_3. However, after reduction under hydrogen at 600 °C for 1.5 hours, all Rh_{SA} in Rh/Al_2O_3 catalyst underwent sintering and evolved into Rh_{NP} with a particle size in the range of 0.5–1.7 nm. The reduced Rh/12 CeO_2-Al_2O_3 catalyst contained both Rh_{SA} and Rh_{NP} of 0.5–2.3 nm, while the reduced Rh/6 Sm_2O_3-6 CeO_2-Al_2O_3 catalyst mainly preserved Rh_{SA} with very few clusters of 0.3–1.3 nm. Therefore, these results mean that the addition of CeO_2 and Sm_2O_3 strongly enhanced the formation and stabilization of the atomically dispersed Rh atoms. Regarding the catalytic performance (reaction conditions: 5 mg of catalyst, MSR at 500 °C, flow rate of CH_4, H_2O(g), He = 5, 15,

Table 12.1 Particle size and catalytic performance of Rh based SAC under SRM conditions (reaction conditions: 5 mg of catalyst, flow rate of CH_4, H_2O(g), He = 5, 15, 520 mL/min).

	Rh_{NP} size (nm)			Specific activity ($mol_{CH4} site^{-1}/s$) under SRM at 500 °C		
	After reduction, H_2, 600 °C, 1.5 h	After SRM reaction, 500 °C, 48 h	Description	1 h	48 h	Deactivation (%)
Rh/Al_2O_3	0.5–1.7	0.9–1.9	Rh_{NP}	1.5	0.9	40
Rh/12CeO_2-Al_2O_3	0.5–2.3	0.5–2.3	Rh_{SA} and Rh_{NP}	2.3	1.9	17
Rh/6Sm_2O_3–6CeO_2–Al_2O_3	0.3–1.3	<1	Only Rh_{SA}	1.8	1.2	33

Source: Duarte et al. [18]. Reproduced with permission of American Chemical Society.

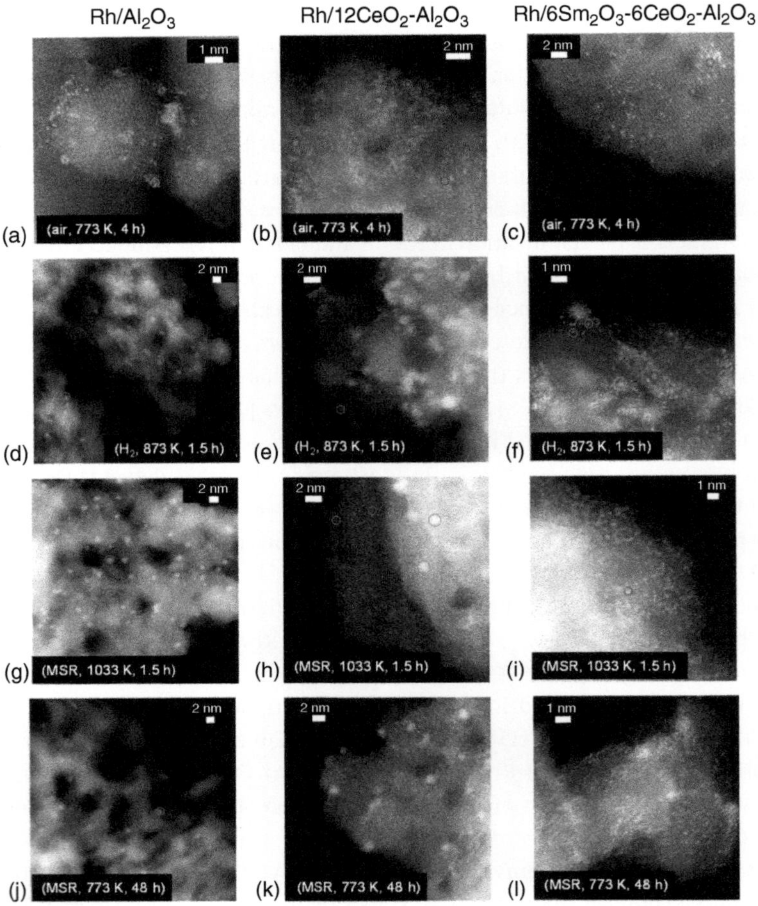

Figure 12.2 HAADF-STEM images of Rh/Al_2O_3, $Rh/12CeO_2-Al_2O_3$, and $Rh/6Sm_2O_3-6CeO_2-Al_2O_3$ catalysts after calcination at 500 °C (a–c), reduction at 600 °C (d–f), MSR for 1.5 hours at 760 °C (g–i), and MSR for 48 hours at 500 °C (j–l). (Reaction conditions: 5 mg of catalyst, flow rate of CH_4, $H_2O(g)$, He = 5, 15, 520 mL/min). Blue, green, and red circles mark the NP, clusters, and atomically dispersed rhodium, respectively. Source: Duarte et al. [18]. Reproduced with permission of American Chemical Society.

520 mL/min, TOS = 48 hours), it was reported that particle size strongly determines methane activation and CO formation, which in turn controls the kinetic of MSR reaction [18]. As observed in Table 12.1, the reduced $Rh/12CeO_2$-Al_2O_3 catalyst (containing both Rh_{SA} and Rh_{NP}) had the highest activity, followed by the reduced $Rh/6Sm_2O_3$-$6CeO_2$-Al_2O_3 catalyst (containing mainly Rh_{SA}), while the reduced Rh/Al_2O_3 catalyst (containing only Rh_{NP}) showed the lowest activity. In addition, catalyst deactivation was observed for all catalysts, which was mostly caused by carbon deposition. The reduced Rh/Al_2O_3 catalyst containing only Rh_{NP} favored carbon deposition and thus deactivated more than the other ones. The reduced $Rh/12CeO_2$-Al_2O_3 catalyst containing both Rh_{SA} and Rh_{NP} showed the highest activity and carbon resistance. According to the authors, Rh_{SA} favored methane

activation, but they also promoted carbon deposition [following Eq. (12.6)], while Rh_{NP} of few nanometers (*e.g.* those in the reduced Rh/12CeO_2-Al_2O_3 catalyst) promoted the partial oxidation of carbon through step-edge sites of NPs to form CO. The coexistence of both Rh_{SA} and Rh_{NP} provides a cooperative effect on the catalytic performance of the reduced Rh/12CeO_2-Al_2O_3 catalyst in SRM. On the other hand, the reduced Rh/6Sm_2O_3-6CeO_2-Al_2O_3 catalyst containing only Rh_{SA} was less performing than the reduced Rh/12CeO_2-Al_2O_3 catalyst because of the deactivation by carbon deposition. Later, the coexistence of Rh_{SA} and Rh_{NP} is revealed as an optimal design for a performing catalyst in DRM by Wu et al. [22]. Accordingly, CH_4 dissociation was prone on Rh_{SA}-containing sites but hard to continue due to carbon deposition. Coexistence of Rh_{SA} and Rh_{NP} plays an important role in maintaining high reactivity due to the electron transfer effect to accelerate the mobility of active oxygen O* species generated from CO_2 dissociation. The active oxygen O* species limit carbon deposition thanks to oxidation reactions.

The influence of particle size when downsizing from NPs to SAs was also highlighted by Zhang et al. [27], who investigated Ru SACs for DRM at high temperature (reaction conditions: 30 mg of catalyst, 850 °C, $CH_4/CO_2/N_2$ = 49.5/49.5/1, GHSV = 100 000–800 000 mL/h/g). A $MgAl_2O_4$-supported Ru_{SA} catalyst containing 0.07 wt% Ru was synthesized by physical vapor deposition (PVD) method, denoted as 0.07Ru/$MgAl_2O_4$-PVD. Ru clusters-supported catalyst containing 0.15 wt% Ru (denoted as 0.15Ru/$MgAl_2O_4$-PVD), and Ru_{NP} catalysts prepared by conventional incipient wetness impregnation (IWI) method, denoted as Ru/$MgAl_2O_4$-IWI (containing 0.14 or 1 wt% Ru), were used for comparison. The Ru particle size in the prepared catalysts was measured by aberration-corrected HAADF–STEM. The mean particle size, activation energy, and specific activity of the Ru-based catalysts for DRM are given in Table 12.2.

The catalysts prepared by IWI containing large Ru particles had a low performance and showed a quick catalyst deactivation by carbon deposition. In contrast,

Table 12.2 Correlation between the size of Ru species and activation energy for DRM on the prepared Ru-based catalysts.

Catalyst	**0.07Ru/ $MgAl_2O_4$-PVD**	**0.15Ru/ $MgAl_2O_4$-PVD**	**0.15Ru/ $MgAl_2O_4$-PVD-(TOS-600 h)**	**1 wt%Ru/ $MgAl_2O_4$-IWI**	**0.14 wt%Ru/ $MgAl_2O_4$-IWI**
Particle size (nm)	0.6	1.1	2.7	3.1	1.6 to hundreds of nanometers
Activation energy (kJ/mol)	163.7	135.4	129.7	209.0	207.1
Specific activity $mol_{CH4}/mol_{Ru}/s$	191.5	257.0	385.5	28.4	108.2

Reaction conditions: 30 mg of catalyst, 850 °C, $CH_4/CO_2/N_2$ = 49.5/49.5/1, GHSV = 100 000–800 000 mL/h·g).
Source: Adapted from Zhang et al. [27].

catalysts prepared by PVD containing Ru_{SA} and Ru clusters performed better [27]. Regarding the catalytic performance, the well-dispersed 0.07Ru/$MgAl_2O_4$-PVD catalyst containing Ru_{SA} showed a lower specific activity (191.5 $mol_{CH4}/mol_{Ru}/s$) than 0.15Ru/$MgAl_2O_4$-PVD (257.0 $mol_{CH4}/mol_{Ru}/s$), containing Ru clusters. In the temperature range of 700–850 °C, the activation energy was found at 209.0, 207.1, 163.7, and 135.4 kJ/mol, for 1 wt% Ru/$MgAl_2O_4$-IWI, 0.14 wt% Ru/$MgAl_2O_4$-IWI, 0.07Ru/$MgAl_2O_4$-PVD, and 0.15Ru/$MgAl_2O_4$-PVD, respectively (Table 12.2). In addition, the analysis of the used catalysts showed that after 600 hours' time-on-stream (TOS) at 850 °C, the thermal sintering of the active phase was significant for both catalysts prepared by PVD method, since the mean particle size increased from 1.1 to 2.7 nm for 0.15Ru/$MgAl_2O_4$-PVD, while Ru_{SA} disappeared for 0.07Ru/$MgAl_2O_4$-PVD [27]. However, this sintering didn't cause catalytic deactivation, but it increased the catalytic activity by 1.5 times for the case of the 0.15Ru/$MgAl_2O_4$-PVD-(600 hours TOS) (Table 12.2). By assuming the surface Ru atoms as active sites, the TOF values were calculated for the 0.15Ru/$MgAl_2O_4$-PVD and 0.15Ru/$MgAl_2O_4$-PVD-(TOS-600 hours) catalysts to be 34.6, 74.5, 159.6 s^{-1} and 113.0, 269.3, 517.8 s^{-1} during DRM at 700, 750, 800 °C, respectively. The authors explained this result by the fact that Ru sites on the faceted Ru surface are more active in DRM than unfaceted Ru clusters and Ru_{SA} [27]. In fact, by DFT calculation (Figure 12.3), the authors showed that the adsorption energies of methane onto Ru_{SA}, unfaceted cluster, and Ru(111) are −4.8, −56.0, and −20.3 kJ/mol, respectively. Then, the barriers for breaking the first C—H bond of methane over Ru_{SA}, unfaceted cluster, Ru(111) are 187.2, 106.1, and 95.5 kJ/mol.

The adsorption energies of CH_3* and H* adatoms onto the isolated Ru atom, unfaceted cluster, and Ru(111) were found to be −21.2, −38.6, and −4.9 kJ/mol. The calculated barriers were found to be consistent with experimental results. These results indicated that the DRM over the Ru/$MgAl_2O_4$-PVD catalyst is

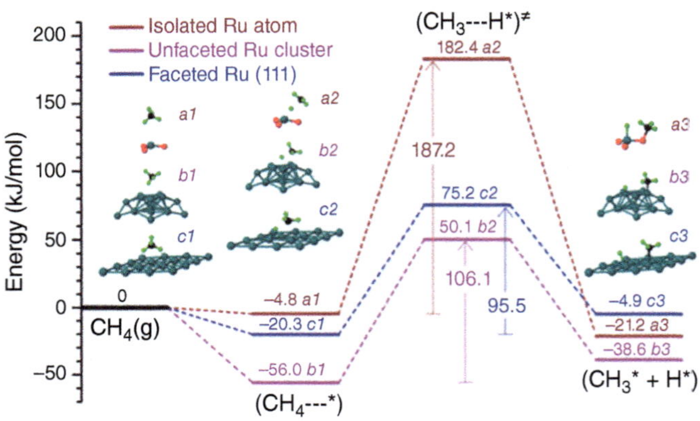

Figure 12.3 DFT calculated potential energy landscape, barriers of adsorption of CH_4, and abstraction of the first hydrogen atom from CH_4 on model structures of isolated Ru_{SA} (a1, a2, a3), unfaceted Ru cluster (b1, b2, b3), and faceted Ru(111) (c1, c2, c3). Color code: carbon is gray, hydrogen is green, oxygen is red, and ruthenium is dark cyan. Source: Zhang et al. [27]. Reproduced with permission of American Chemical Society.

structure-sensitive, and faceted Ru_{NP} are more active than the unfaceted clusters and Ru_{SA}. This experimental/theoretical study demonstrated the important role of the size and the associated structure of Ru-based catalysts in the activation of the C—H bond of methane, and therefore on the catalytic performance. Ru_{NP} of around 2 nm seem to be the optimal particle sizes for DRM. It is worth noting that this rule cannot simply be extrapolated to other metals in DRM. The nature of the active phase and the support must be considered case-per-case, as demonstrated below with Ni-based catalysts.

12.2.2 Ni-Based Single-Atom Catalysts for Methane Reforming

Apart from the noble-metal-supported SAC, Ni-based SAC were also successfully synthesized and evaluated for reforming of methane [3, 13, 30]. Compared to noble metals, transition metals are more abundant and have lower cost. However, it is worth noting that in nanoparticle-containing catalysts, carbon deposition is favored on the Ni-based catalysts compared to the noble-based catalysts [31].

Zou et al. [30] prepared a Ni/MgO catalyst derived from NiO–MgO solid solutions to perform the DRM at 800 °C. By using coprecipitation, synthesis with different metal loadings of 2.5, 5, and 10 wt%, Ni_{SA}- or Ni_{NP}-containing catalysts could be obtained. The 2.5 wt% Ni/MgO catalyst mainly contained Ni_{SA} with a size centered at 0.3 nm (1–2 Ni atoms), while the 5 wt% Ni/MgO and 10 wt% Ni/MgO catalysts contained more Ni clusters (3–4 Ni atoms with size of 0.6–0.8 nm) and some 1.1 nm Ni_{NP}. In DRM, experimental study showed that no carbon deposition was observed for all Ni/MgO catalysts (reaction conditions: 200 mg of catalyst, flow rate of CH_4, CO_2, N_2 = 15, 15, 15 mL/min, 800 °C). The SAC (2.5 wt% Ni/MgO) exhibited the lowest activity in DRM, in comparison with both the Ni cluster catalyst (5 wt% Ni/MgO) and the NP catalyst (10 wt% Ni/MgO). By DFT calculation, the authors estimated the adsorption energy of reactants and intermediates on the surface of the catalysts with different active sites and on the surface of MgO(100) as a reference (Table 12.3).

Table 12.3 DFT calculation results for binding energies of the reactants and intermediates involved in DRM on MgO(100), Ni_{SA}/MgO(100), and Ni_4/MgO(100).

	Adsorption energy, E_{ads} (kJ/mol)		
Adsorbate	**MgO(100)**	**Ni_{SA}/MgO(100)**	**Ni_4/MgO(100)**
$*CH_4$	−3.9	−3.9	−16.4
$*CH_3$	−11.6	−149.6	−250.9
*H	−43.4	−185.3	−288.5
*C	−266.3	—	−723.7
$*CO_2$	−40.5	−22.2	−95.5
*CO	−16.4	−70.4	−209.4
*O	−231.6	−289.5	−601.1

Source: Adapted from Zuo et al. [30].

The SAC Ni/MgO(100) is not active enough to adsorb the reactants and the intermediates and to perform C—O bond scission, C—H bond cleavage, and so the overall DRM reaction. In detail, the formation of SAs on Ni_{SA}/MgO(100) did not significantly change the binding energy of the reactants and intermediates in comparison with those on the bare MgO(100) support. For example, the adsorption energy of CH_4* remains at −3.9 kJ/mol. However, the single-site Ni_4/MgO catalyst is able to provide stronger bindings than Ni_{SA}/MgO, with the adsorption energy of CH_4* of −16.4 kJ/mol. These clusters of 4 Ni atoms are able to work cooperatively for the activation of both CH_4 and CO_2, enabling the production of CO, H_2 and H_2O, and completely eliminating carbon deposition.

Other important works on Ni-based supported SACs were performed by Akri et al. [13, 32]. The authors synthesized atomically dispersed Ni supported on hydroxyapatite (HAP) using strong electrostatic adsorption (SEA) method. By varying the Ni contents, they obtained catalysts containing either Ni_{SA} (at 0.5 wt%, 0.5Ni/HAP) or Ni_{NP} (at 10 wt%, 10Ni/HAP). The catalysts were then evaluated for DRM (reaction conditions: 750 °C, 1 atm, molar ratio of CH_4:CO_2:He = 10 : 10 : 30, total gas flow rate = 50 mL/min, GHSV = 60 000 mL/h/g). Under the same reaction conditions, the initial activity of 0.5Ni/HAP catalyst was only slightly lower than that of 10Ni/HAP catalyst. Then, both catalysts underwent rapid deactivations during DRM at 750 °C. The CO_2 conversion decreased from 84% to 49% and 90% to 73% for 0.5Ni/HAP and 10Ni/HAP catalysts after 7 hours TOS, respectively. Carbon deposition was reported as the main reason for the deactivation in the case of the 10Ni/HAP catalyst. However, carbon deposition was not observed in the case of 0.5Ni/HAP. The deactivation of 0.5Ni/HAP catalyst was due to thermal sintering of the active metal during high-temperature DRM. The latter problem was solved by adding Ce as promoter (5 wt% in the final catalyst) to stabilize Ni_{SA}. To this end, a mixture of calcium nitrate and cerium nitrate was precipitated with the phosphate precursor to obtain cerium-containing HAP support (denoted as HAP-Ce), which was then used to deposit Ni at 0.5, 1, 2, or 10 wt% by the same SEA method (Ni/HAP-Ce) [13]. XRD characterization revealed that Ce was not found as crystalline phase, but could be highly dispersed on the external HAP surface or even incorporated in the apatite structure. The main characteristics of the catalysts after Ce promotion are summarized in Figure 12.4. The 0.5Ni/HAP-Ce catalyst contained a majority of isolated Ni atoms (95%). The 2 Ni/HAP-Ce catalyst presented a minor amount of *c.* 1 nm clusters (30%) and a predominant fraction of Ni_{SA} (70%). In the case of the 10 Ni/HAP-Ce catalyst, Ni_{SA} were not observed, and Ni clusters of 1–2 nm (c. 70%) and Ni_{NP} of 2–17 nm (c. 30%) were formed. Under the same conditions in DRM, the support HAP-Ce showed negligible activity in DRM.

Among the different Ni/HAP-Ce catalysts, 10Ni/HAP-Ce catalyst containing only Ni_{NP} shows the lowest initial activity and a rapid deactivation. The three catalysts containing 0.5, 1, and 2 wt% Ni show similar initial activity, which was higher than that of the 10Ni/HAP-Ce catalyst. Consequently, the turnover frequency (TOF) calculated for both CH_4 and CO_2 strongly decreased when Ni content in Ni/HAP-Ce catalysts increased. These results completely differed from those observed for Ni/MgO [30], which suggests the important role of the support in DRM. About the

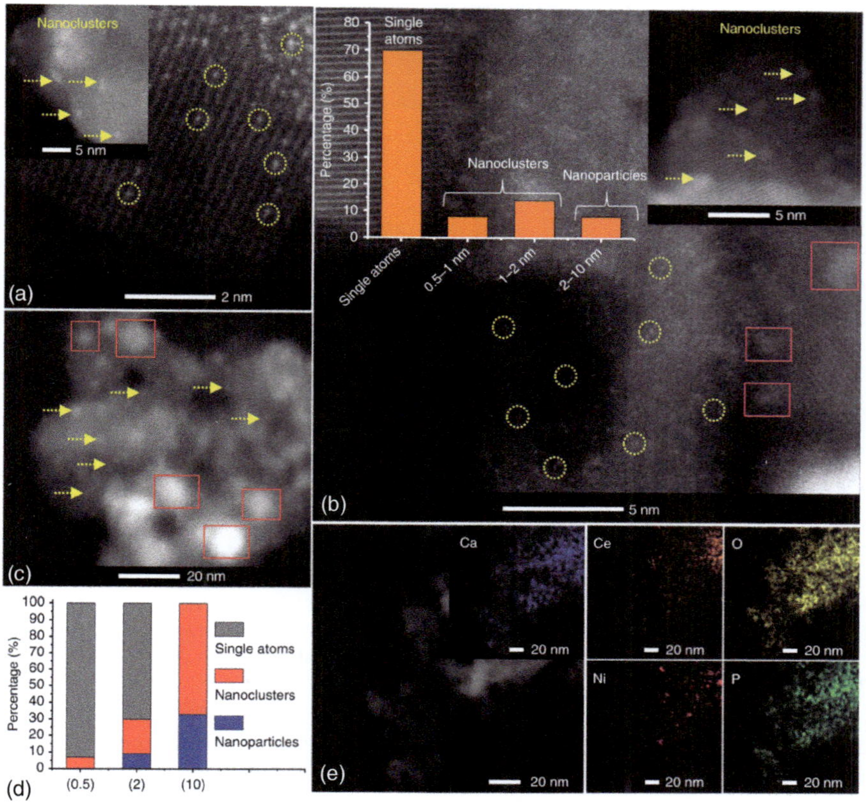

Figure 12.4 Electron microscopy images and size distribution of Ni/HAP-Ce samples. (a) HAADF–STEM images of 0.5Ni/HAP-Ce, (b) 2Ni/HAP-Ce, and (c) 10Ni/HAP-Ce samples after 500 °C H_2 reduction; yellow circles indicate atomically dispersed Ni and red squares indicate Ni metals nanoparticles. (d) Particle size distributions of (a–c). (e) EDX element maps of 0.5Ni/HAP-Ce. Source: Akri et al. [13]. Springer Nature/CC BY-SA 4.0.

catalytic stability, a slight deactivation was observed for 0.5Ni/HAP-Ce for 70 hours TOS, while no deactivation was observed for 1Ni/HAP-Ce and 2Ni/HAP-Ce catalysts. The latter showed the best activity and stability. The authors tried to explain the lower stability of 0.5Ni/HAP-Ce catalyst compared to those of 1Ni/HAP-Ce and 2Ni/HAP-Ce catalysts by the fact that the 0.5Ni/HAP-Ce catalyst contains more unstable Ni_{SA}, which were in lower coordination environments, e.g. on the edges of HAP crystal, and thus aggregated during DRM. Furthermore, the authors clarified the origin of coke resistance over the 2Ni/HAP-Ce and 10Ni/HAP-Ce catalysts by experimental study coupled with DFT calculations for only CH_4 decomposition reaction (equation (12.6)). The results are summarized in Figure 12.5 [13]. Experimental results indicated a significant amount of carbon formed on the used 10Ni/HAP-Ce catalyst after reaction, while almost no carbon was observed with the used 2Ni/HAP-Ce catalyst. The Ni_{SA} were reported to favor the partial CH_4 dehydrogenation and the C–C coupling reactions (for instance, to form C_2H_4) instead of the complete decomposition reaction leading to carbon formation.

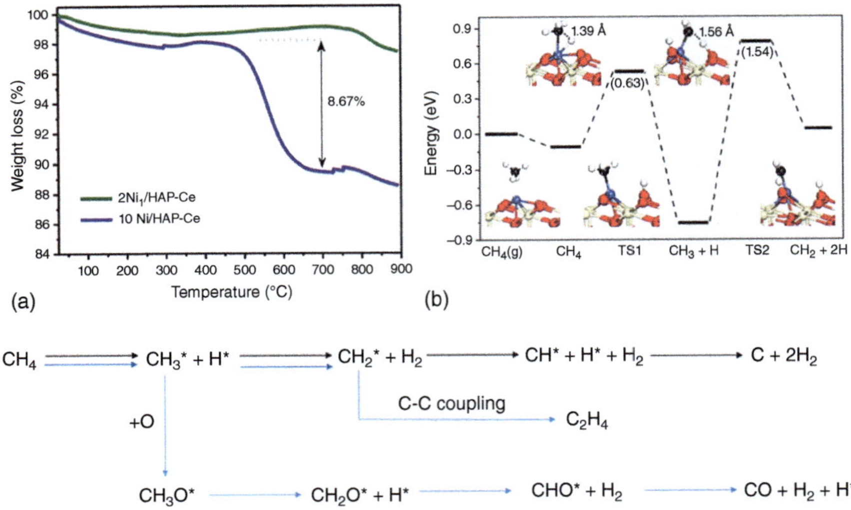

Figure 12.5 Investigation into origin of carbon resistance over the 2Ni/HAP-Ce and 10Ni/HAP-Ce catalysts during methane decomposition. Top panel: (a) TPO results for CH_4 decomposition over 2Ni/HAP-Ce and 10Ni/HAP-Ce catalysts; (b) DFT calculation for CH_4 decomposition over Ni_{SA}/CeO_2. Bottom panel: CH_4 decomposition pathway over 10Ni/HAP-Ce catalyst (black arrow) and 2Ni/HAP-Ce catalyst (blue arrow). Source: Akri et al. [13]. Springer Nature. CC BY 4.0.

In addition, the DFT calculations revealed that the Ni_{SA}/CeO_2 promotes the CH_4 reforming to CO reaction. Accordingly, CH_4 is first adsorbed on $Ni_{SA}/CeO_2(111)$ via physisorption ($E_a = -10.6$ kJ/mol). The first C—H bond cleavage is exothermic by 62.7 kJ/mol with an activation barrier of 60.8 kJ/mol, which is significantly lower than that over a Ni(111) surface (86.8 kJ/mol). Therefore, the activation of the first C—H bond in CH_4 is extremely efficient over $Ni_{SA}/CeO_2(111)$. The resulting CH_3^* species binds atop the Ni atom, while the H adatom moves to a neighboring O. The next dehydrogenation step, $CH_3^* \rightarrow CH_2^* + H^*$, is thermodynamically unfavorable (endothermic by 77.2 kJ/mol) with a high activation barrier of 148.6 kJ/mol. Alternatively, a new pathway for transforming CH_3^* to CO without carbon deposition was proposed. It involves the oxidation of CH_3^* to CH_3O^*, which is subsequently dehydrogenated to CHO^* and finally to CO. Thus, the presence of Ni_{SA} intrinsically inhibited the carbon formation by changing activation and decomposition pathways of methane.

Apart from this Ce-promoted SACs, the same research group reported that polyvinylpyrrolidone (PVP) is also a good promoter for HAP-supported Ni SACs [32]. The catalyst was prepared by co-precipitation with a low Ni loading of 0.5 wt%. An adequate amount of nickel and PVP was dissolved separately in deionized water and mixed together with the support of stoichiometric hydroxyapatite (Ca/P = 1.67). The obtained solid was filtered, rinsed carefully with deionized water, and finally dried at 80 °C overnight before being calcined at 500 °C for 4 hours. The formation of Ni_{SA} was confirmed by various techniques, including aberration-corrected scanning transmission electron microscopy (AC-STEM), *in situ* extended X-ray absorption

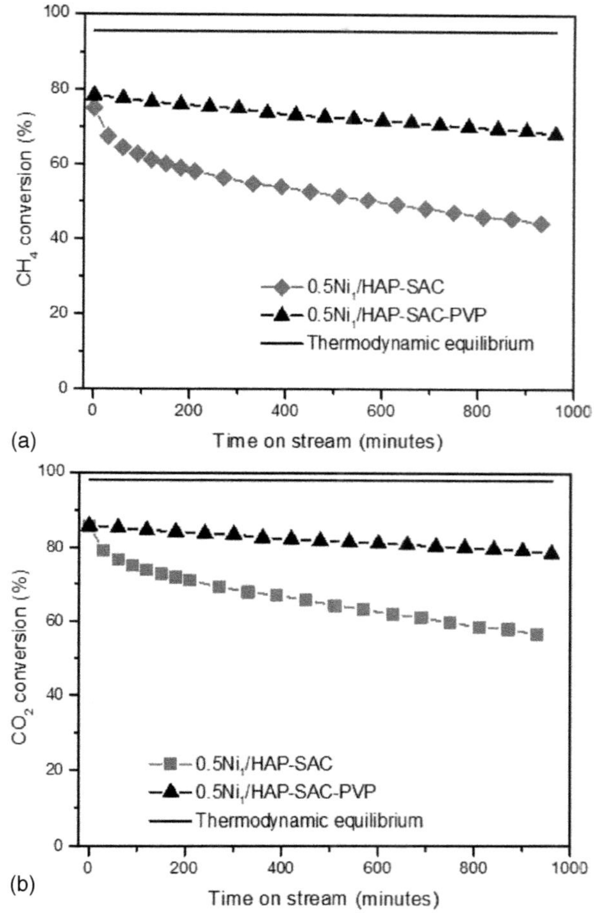

Figure 12.6 Comparison of Ni/HAP SAC with and without PVP promoter: (a) CH_4 conversion and (b) CO_2 conversion during DRM at 750 °C, GHSV = 12 × 10^6 mL/g_{Ni}/h, (CH_4:CO_2: He = 1 : 1 : 3). Source: Akri et al. [32]. MDPI. CC BY 4.0.

fine structure (EXAFS), *in situ* X-ray absorption near-edge spectroscopy (XANES), and *ex situ* XPS. The Ni/HAP SAC promoted by PVP was evaluated in DRM for 960 minutes and compared with a 0.5 Ni/HAP SAC prepared by impregnation without promoter as previously presented [13]. The results of catalytic tests are presented in Figure 12.6.

The addition of PVP limited the sintering of the Ni active phase. After 960 minutes' TOS, the CH_4 and CO_2 conversion over the Ni/HAP SAC promoted by PVP decreased from 78% and 83% to 68% and 78%, respectively, while the CH_4 and CO_2 conversion dropped to only 45% and 55% for the Ni/HAP SAC without PVP (Figure 12.6). This work presents a new insight in the preparation and utilization of Ni SACs for DRM.

12.2.3 Synergy Between Noble and Transition Metals in SACs

Catalyst systems containing two sets of singly dispersed active site (Ni and Ru) were also reported [3]. The authors prepared the catalyst with 2.5% of Ni and Ru supported on ceria to obtain a $Ce_{0.95}Ni_{0.025}Ru_{0.025}O_2$ catalyst for being evaluated in DRM at

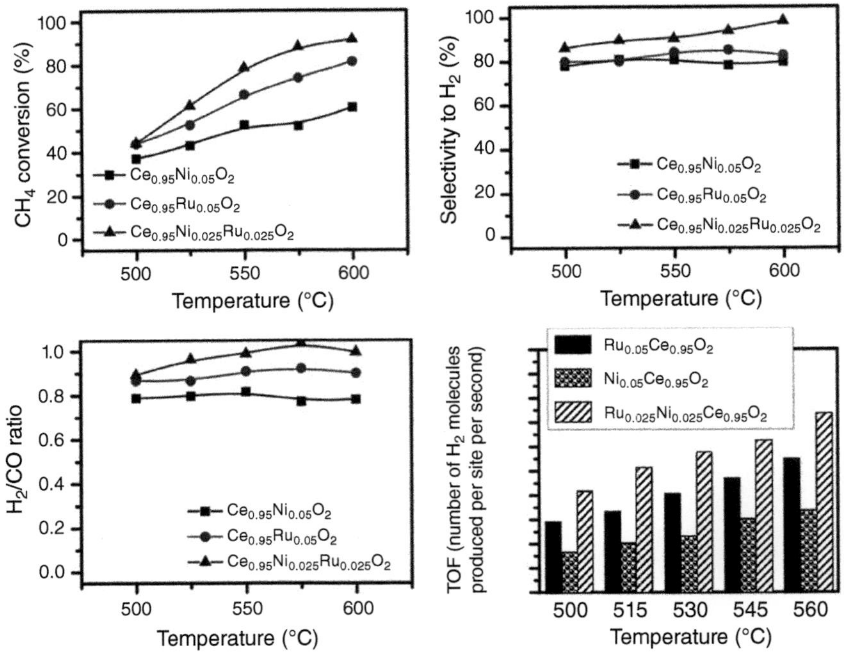

Figure 12.7 Catalytic performances of Ni-supported SAC $Ce_{0.95}Ni_{0.05}O_2$, Ru-supported SAC $Ce_{0.95}Ru_{0.05}O_2$, and Ni-Ru SAC $Ce_{0.95}Ni_{0.025}Ru_{0.025}O_2$ in DRM (reaction conditions: 50 mg catalyst, 40 mL/min of a mixture containing 1% CH_4 and 1% CO_2, Ar balance, $T = 500$–600 °C). Source: Tang et al. [3]. Reproduced with permission of American Chemical Society.

low temperature of 500–600 °C (reaction conditions: 50 mg catalyst, 40 mL/min of a mixture containing 1% CH_4 and 1% CO_2, Ar balance). For comparison, SAC with separated 5% metal Ni and Ru, $Ce_{0.95}Ni_{0.05}O_2$, and $Ce_{0.95}Ru_{0.05}O_2$ were used. Operando characterizations and DFT calculations were performed for mechanistic study. As can be seen in Figure 12.7, the $Ce_{0.95}Ru_{0.05}O_2$ catalyst showed a higher CH_4 conversion but a comparable selectivity toward hydrogen compared to the $Ce_{0.95}Ni_{0.05}O_2$ catalyst. The $Ce_{0.95}Ni_{0.025}Ru_{0.025}O_2$ catalyst showed the highest activity and selectivity within the temperature range investigated, indicating a synergistic effect of the bimetallic phase in $Ce_{0.95}Ni_{0.025}Ru_{0.025}O_2$ SAC.

Computational studies revealed that the molecular-level origins of synergistic effects between Ni_{SA} and Ru_{SA} sites are: (i) the complementary function of Ni_{SA} for CH_4 activation and of Ru_{SA} (coupled with O_v on the CeO_2 surface) for activation of CO_2, and (ii) the sequential role of first generating H atoms in activating CH_4 on a Ni_{SA} site and then coupling H atoms to form H_2 on a Ru_{SA} site.

Notably, in this study, *operando* analyses (AP-XPS and EXAFS) indicated that in the $Ce_{0.95}Ni_{0.025}Ru_{0.025}O_2$ SAC, the active sites for DRM are Ni and Ru atoms in cationic state instead of metallic atoms. The cationic state provides lower adsorption energy for methane compared to the metallic state, thus allowing performing the DRM reaction at low temperature.

Recently, by DFT-simulated X-ray photoelectron spectroscopy (XPS), Mohan et al. [12] revealed high thermodynamic stabilization of boron-promoted Ni (NiB) single-atom alloy (SAA) catalysts doped with Pt, Pd, Rh, Cu, and Mn. Among them, Mn-NiB SAA showed the highest capacity to reduce the activation barriers for CO_2 and CH_4. This catalyst provides a high endergonicity for CH_4 stepwise dehydrogenation as well as a low barrier for Boudouard reaction, thus reducing carbon deposition.

As partial conclusion, to date, only a few works have been devoted to methane reforming into syngas over SACs. For instance, studies on Rh-, Ru-, and Ni-based SACs in methane reforming were reported in the open literature. From these first researches, some conclusions can be drawn:

- Methane reforming needs a high activation energy. A robust catalyst should be designed to convert it into syngas by using various oxidative agents (CO_2, H_2O, O_2).
- Depending on the support and active metal, as well as operating conditions, the deactivation by carbon deposition and/or by thermal sintering of metal single atoms can take place. Rh_{SA} supported on alumina-based supports allowed thermal stabilization but suffered slight catalyst deactivation by carbon deposition in SRM at 500 °C [18]. Similar observation was found with Ni_{SA}/CeO_2-based composite [3]. On the other hand, Ni_{SA}/HAP allowed inhibition of carbon deposition, but this SAC underwent catalyst deactivation by thermal sintering in DRM at 750 °C [13]. To fix these problems, the utilization of a promoter, e.g. cerium or the combination of two active metals in the form of single atoms, can be envisaged.
- Particle size plays a pivotal role in methane-reforming processes. SACs reduce activation energy for methane activation and changes reaction pathway toward syngas. Thus, SACs are generally found to be more active in comparison with homologous NP catalysts. However, singly dispersed metal atoms (Rh, Ru, and Ni for instance) seem to favor carbon deposition as they accelerate the deep methane decomposition reaction. Catalysts containing both singly dispersed metal atoms and metal nanoclusters of c. 2 nm appear to be the most performing due to their cooperative effect. However, it is worth noting that along with the metal particle size of the active phase, other parameters must also be taken into consideration, such as the nature of the support and the strength of metal–support interaction. For instance, in contrast to Ni_{SA}/MgO, Ni_{SA}/HAP is highly resistant to carbon deposition.
- A first study has been devoted to the determination of the nature of the active phase in DRM [3]. For the $Ce_{0.95}Ni_{0.025}Ru_{0.025}O_2$ catalyst, cationic Ru and Ni species were found as active sites. This finding provides a new insight into the low-temperature operation of methane-reforming processes.

12.3 Supported Metal Single Atoms for Hydrocarbon Reforming

Apart from methane, raw natural gas containing C_2–C_5 alkanes and crude oil rich in naphtha are also valuable sources for syngas production via steam reforming [33, 34].

To start, let's consider the steam-reforming reactions of propane (as representative for aliphatic hydrocarbons) and of toluene (as representative for aromatic hydrocarbons) leading to the production of CO and H_2 as examples:

Propane steam reforming:

$$C_3H_8 + 3H_2O \rightarrow 3CO + 7H_2 \quad \Delta H^0_{298} = +498.5\ \text{kJ/mol} \tag{12.13}$$

Propane dry reforming:

$$C_3H_8 + 3CO_2 \rightarrow 6CO + 4H_2 \quad \Delta H^0_{298} = +759.2\ \text{kJ/mol} \tag{12.14}$$

Propane partial oxidation:

$$C_3H_8 + 1.5O_2 \rightarrow 3CO + 4H_2 \quad \Delta H^0_{298} = -226.9\ \text{kJ/mol} \tag{12.15}$$

Toluene steam reforming:

$$C_7H_8 + 7H_2O \rightarrow 7CO + 11H_2 \quad \Delta H^0_{298} = +906.7\ \text{kJ/mol} \tag{12.16}$$

Toluene dry reforming:

$$C_7H_8 + 7CO_2 \rightarrow 14CO + 4H_2 \quad \Delta H^0_{298} = +1194.8\ \text{kJ/mol} \tag{12.17}$$

Toluene partial oxidation:

$$C_7H_8 + 3.5O_2 \rightarrow 7CO + 4H_2 \quad \Delta H^0_{298} = -786.0\ \text{kJ/mol} \tag{12.18}$$

With aliphatic hydrocarbon, coke deposition can occur following the reaction:

$$C_3H_8 \rightarrow 3C + 4H_2 \quad \Delta H^0_{298} = +104.6\ \text{kJ/mol} \tag{12.19}$$

With aromatic hydrocarbon, the formation of benzene, methane, and coke can occur:

$$C_7H_8 + H_2O \rightarrow C_6H_6 + CO + 2H_2 \quad \Delta H^0_{298} = +168.2\ \text{kJ/mol} \tag{12.20}$$

$$C_7H_8 + H_2 \rightarrow C_6H_6 + CH_4 \quad \Delta H^0_{298} = -38.0\ \text{kJ/mol} \tag{12.21}$$

$$C_7H_8 + 10H_2 \rightarrow 7CH_4 \quad \Delta H^0_{298} = -536.7\ \text{kJ/mol} \tag{12.22}$$

$$C_7H_8 \rightarrow 7C + 4H_2 \quad \Delta H^0_{298} = -12.4\ \text{kJ/mol} \tag{12.23}$$

In addition, the homogeneous and heterogeneous reactions related to the light gas CO_2, CO, H_2O, O_2, CH_4, H_2, and C could also occur, as mentioned in Section 12.2 dealing with methane reforming.

Thermodynamically, all the reactions involved in the hydrocarbon steam reforming are strongly endothermic. In addition, from the stoichiometry of these reactions and according to Le Chatelier's principle, these reactions are favored at high temperatures and low pressures. Interestingly, compared to methane reforming, the reforming of higher hydrocarbons can be performed at a lower temperature. For instance, complete conversion of toluene can be achieved at c. 600 °C and 1 bar from an initial mixture of H_2O/toluene = 7 (Figure 12.8) [4]. At this high steam-to-carbon ratio, neither methane nor carbon is formed.

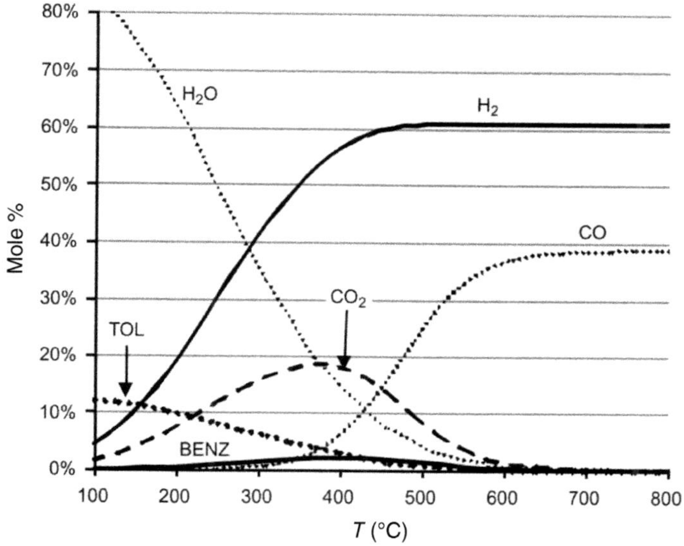

Figure 12.8 Equilibrium composition of the toluene steam reforming at 1 bar without carbon and methane formation (initial conditions H_2O/toluene = 7). Source: Bion et al. [4]. Reproduced with permission of Royal Society of Chemistry.

Working at high steam-to-carbon ratio allows limiting carbon deposition, but steam-added amount favors endothermic reactions, and thus globally results in high energy consumption. By reducing this ratio, carbon deposition remains a main challenge. Angeli et al. [31] compared carbon accumulation during steam reforming of methane, ethane, and propane over Ni_{NP} and Rh_{NP} catalysts at moderate temperatures of 400–550 °C (reaction conditions: steam-to-carbon ratio (S/C) = 3, 10 hours, WHSV = 70 000 h^{-1}). They found that carbon deposition is significant in the case of Ni_{NP} catalysts, while it is nearly absent in the case of Rh_{NP} catalysts. The accumulation of carbon species on the surface of Ni_{NP} in the cases of ethane and propane reforming was found much higher than that in the case of methane reforming (Table 12.4).

Seeburg et al. [35] investigated catalysts containing Rh (including isolated Rh atoms and Rh nanoparticles of 1–3 nm) for hydrocarbon reforming. A Rh/Al_2O_3

Table 12.4 Results of carbon depositions after steam reforming of methane, ethane, and propane over nanoparticle Ni-based and Rh-based catalysts (reaction conditions: at T = 400–550 °C, S/C = 3, 10 h, WHSV = 70 000 h^{-1}).

	Carbon accumulation on catalyst (wt%)			Conversion of inlet C to solid C (%)		
	Methane reforming	Ethane reforming	Propane reforming	Methane reforming	Ethane reforming	Propane reforming
Ni-based catalyst	0.010	0.243	0.326	1.99×10^{-4}	7.20×10^{-3}	9.17×10^{-3}
Rh-based catalyst	0.003	0.007	0.006	1.31×10^{-6}	1.78×10^{-4}	1.66×10^{-4}

Source: Angeli et al. [31]. Reproduced with permission of Elsevier.

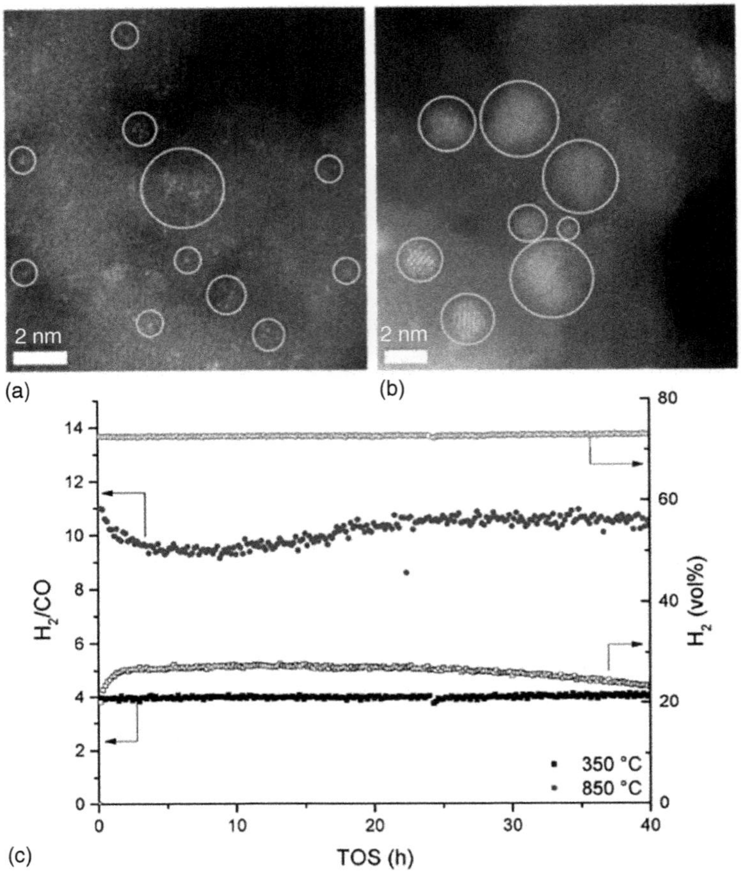

Figure 12.9 HAADF–STEM images of (a) fresh Rh_{SA}/Al_2O_3 catalyst; (b) spent Rh_{SA}/Al_2O_3 catalyst; and (c) catalyst performance during steam reforming of permeate natural gas (reaction conditions: 150 mg of catalyst, at S/C = 4, WHSV = 8000 h^{-1}, and temperatures of 350 and 850 °C). Source: Seeburg et al. [35]. MDPI/CC BY-SA 4.0.

SAC with Rh loading of 1.0 mol% was synthesized by wet impregnation and evaluated in steam reforming of pure C_{1-4} alkanes, mixtures of C_{2-4} with methane, and simulated natural gas at 200–850 °C. The characterization of the catalyst before and after catalysis and the catalytic performance are given in Figure 12.9. Accordingly, isolated Rh atoms were clearly observed in the fresh catalyst, while Rh_{NP} of 1–3 nm were more visible in the spent catalyst. However, this sintering phenomenon of the active phase seems to have negligible impact on the catalyst performance. At low reaction temperature of 350 °C, the hydrogen amount increased during the first hour before being relatively stable. At high reaction temperature of 850 °C, no deactivation was observed during 40 hours of TOS. The hydrogen yield reached 73 vol% and the molar ratio of H_2/CO remained at around 11. Notably, carbon deposition was not found in the spent catalyst composed mainly of Rh_{NP} of 1–3 nm (Figure 12.9). Previously, Duarte et al. [18] found that Rh_{SA}/Al_2O_3 prepared by the

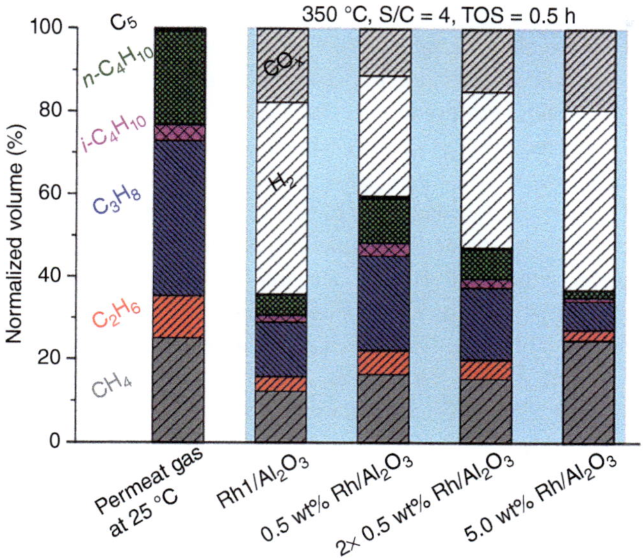

Figure 12.10 Steam reforming of permeate natural gas over the Rh_{SA}/Al_2O_3 catalyst and the commercial catalysts at 350 °C (reaction conditions: 150 mg of catalyst, at S/C = 4, WHSV = 8000 h^{-1}). Source: Seeburg et al. [35]. MDPI. CC BY 4.0.

same wet impregnation method suffered catalyst deactivation by coke formation. On the other hand, the evolution of Rh_{SA} into small Rh_{NP} during hydrocarbon reforming found by Seeburg et al. allows avoiding coke formation, and so maintaining the catalytic activity [35]. To rationalize the influence of the Rh_{SA} on the catalytic performance, the authors performed also a series of tests under similar conditions with two commercial NP-based catalysts (0.5% Rh/Al_2O_3 and 5% Rh/Al_2O_3) for comparison. The results show that the Rh_{SA}/Al_2O_3 catalyst showed higher activity toward reforming of permeate gas (composition of C_1, C_2, C_3, C_{n-4}, C_{i-4}, C_{n-5} were 25.2, 10.2, 37.5, 22.6, 3.9, 0.6 vol%, respectively) compared to those of the two commercial catalysts (Figure 12.10). This might be due to the smaller Rh particles (1–3 nm) of the synthesized catalyst in comparison with those of commercial ones [about 3.2 mm for the 0.5% Rh/Al_2O_3 (206172) from Sigma-Aldrich (Steinheim, Germany) and the 5% Rh/Al_2O_3 (C301099-5) from Alfa Aesar].

The global syngas yield produced from the reforming over the Rh_{SA}/Al_2O_3 catalyst was comparable to that over the 5% Rh/Al_2O_3 commercial catalyst, and was higher than that over 0.5% Rh/Al_2O_3 commercial catalyst, even if the amount of the latter was double. Among these different catalysts, methane conversion was highest with Rh_{SA}/Al_2O_3 catalyst, highlighting the beneficial effect of small Rh particle size in methane reforming.

The influence of the S/C ratio and the reaction temperature on the performance of the Rh_{SA}/Al_2O_3 catalyst in the reforming reaction of raw natural gas and permeate natural gas was also reported. The authors indicated that, when the S/C increased, the H_2/CO molar ratio increased and the complete conversion of methane and higher alkanes occurred at lower temperatures. At S/C = 1, the full conversion of higher alkanes and of methane took place at 450 and 850 °C, respectively. At S/C = 4, the conversion of higher alkanes and methane was complete at around 450 and 650 °C, respectively. High temperature (above 650 °C) was necessary to get full conversion, but this also impacted the H_2/CO ratio, which reached the highest value at around 750 °C. This behavior could be explained by the WGS and methanation reactions, which directly implies the participation of CO, H_2, and methane.

To sum up, despite requiring lower reaction temperatures than those in methane reforming reaction, carbon deposition is still the main challenge and even more pronounced in hydrocarbon-reforming processes when using conventional nanoparticle-based catalysts. This could be resolved by downsizing the metal particle to single atom as initiated by Seeburg et al. [35]. Further mechanistic and kinetic investigations coupling with computational modeling are needed to evidence the role of SAC in hydrocarbon reforming.

12.4 Supported Metal Single Atoms for Aqueous-Phase Reforming of Alcohols

Bio-alcohols have been considered as one of the most promising renewable energy resources since they present low toxicity and can be safely stored and easily transported [36–40]. Bio-alcohols can be produced by fermentation of biomass such as sugar cane, corn, or lignocellulose. Alcohols are reactive molecules whose decomposition over catalyst surfaces (or in gas phase) is generally faster than other hydrocarbons. Basically, let's consider the ethanol steam reforming as an example. Ethanol–steam mixtures can give rise to numerous reactions [4], the most important being:

The steam reforming leading to CO and H_2:

$$C_2H_5OH + H_2O \rightarrow 2CO + 4H_2 \qquad \Delta H^0_{298} = +297.8\ \text{kJ/mol} \tag{12.24}$$

The steam reforming leading to CO_2 and H_2:

$$C_2H_5OH + 3H_2O \rightarrow 2CO_2 + 6H_2 \qquad \Delta H^0_{298} = +215.5\ \text{kJ/mol} \tag{12.25}$$

The hydrogenolysis to methane

$$C_2H_5OH + 2H_2 \rightarrow 2CH_4 + H_2O \qquad \Delta H^0_{298} = -114.6\ \text{kJ/mol} \tag{12.26}$$

The ethanol dehydration to ethylene

$$C_2H_5OH \rightarrow C_2H_4 + H_2O \qquad \Delta H^0_{298} = +87.8\ \text{kJ/mol} \tag{12.27}$$

The dehydrogenation to acetaldehyde

$$C_2H_5OH \rightarrow CH_3CHO + H_2 \qquad \Delta H^0_{298} = +85.4\ \text{kJ/mol} \tag{12.28}$$

The cracking to methane, CO, and H_2

$$C_2H_5OH \rightarrow CH_4 + CO + H_2 \qquad \Delta H^0_{298} = +91.6\ \text{kJ/mol} \tag{12.29}$$

The cracking to methane and CO_2

$$C_2H_5OH \rightarrow 0.5CO_2 + 1.5CH_4 \qquad \Delta H^0_{298} = -32.1\,\text{kJ/mol} \tag{12.30}$$

The cracking to carbon, CO and H_2

$$C_2H_5OH \rightarrow C + CO + 3H_2 \qquad \Delta H^0_{298} = +166.5\,\text{kJ/mol} \tag{12.31}$$

The cracking to carbon, water, and H_2

$$C_2H_5OH \rightarrow 2C + H_2O + 2H_2 \qquad \Delta H^0_{298} = +35.2\,\text{kJ/mol} \tag{12.32}$$

The cracking to carbon, methane, and water

$$C_2H_5OH \rightarrow CH_4 + C + H_2O \qquad \Delta H^0_{298} = -39.7\,\text{kJ/mol} \tag{12.33}$$

Compared to hydrocarbon reforming, alcohol-reforming process is more complex due to the formation of these intermediates [4].

Thermodynamically, ethanol should be totally converted in the whole range of temperatures (Figure 12.11). At low temperatures (100–300 °C), the cracking of ethanol into methane and carbon dioxide is favored. As temperature increases, syngas is progressively produced. The production of methane is thus favored at low temperatures, while the production of syngas is favored at high temperatures [4]. For CH_4 production, carbon deposition seems to be the main challenge, while for syngas production, selectivity and stability of the catalyst are the major issues. In this field, catalysts are developed mostly based on noble metals and thus downsizing the active phase is crucial. Several investigations on SACs for alcohol reforming have been reported in the literature [41, 42].

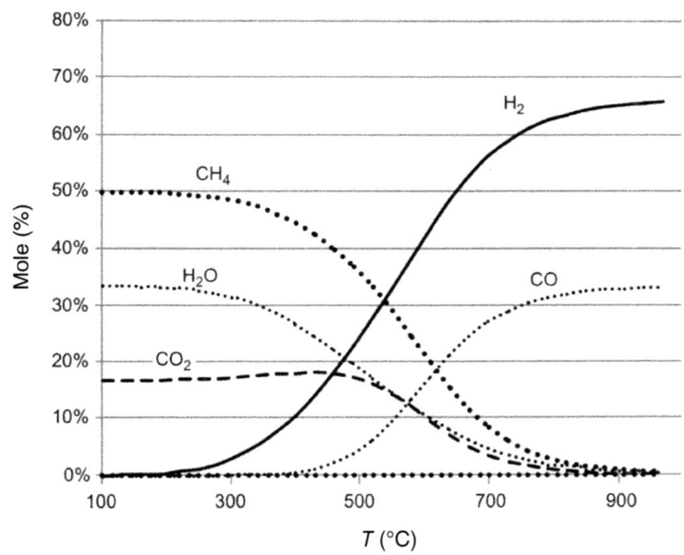

Figure 12.11 Equilibrium composition of gases in the reaction of ethanol with steam (with carbon formation). Initial state: H_2O/EtOH molar ratio = 1, P = 1 bar. Source: Bion et al. [4]. Reproduced with permission of Royal Society of Chemistry.

As the first contribution, Gu et al. [41] prepared catalysts containing Pt_{SA} or Au_{SA} (0.0125 wt%) supported on ZnO nanowires (NW) and evaluated them for methanol steam reforming (experiments performed at 390 °C with 50 mg of catalyst). Using HAADF–HRTEM characterization, the authors concluded that in the prepared catalyst system, Pt and Au atoms were highly dispersed and neither clusters nor NPs were found. Pt_{SA} and Au_{SA} were anchored to the surface of ZnO_{NW} by lattice oxygen stems. By DFT calculation, the authors confirmed that Pt and Au atoms exist in the cationic state, which act as active phase in methanol steam reforming. These active sites were reported to strongly bind toward the intermediates, favoring reaction energetics and kinetics, and changing the reaction pathways. Figure 12.12a shows the catalytic performance of three catalysts: Au_{SA}/ZnO_{NW}, Pt_{SA}/ZnO_{NW}, and ZnO_{NW}. All of the three catalysts were very stable during methanol steam-reforming reaction at 390 °C. Regarding the catalytic activity, Pt_{SA}/ZnO_{NW} catalyst showed the highest methanol conversion of about 43% at steady state, followed by Au_{SA}/ZnO_{NW} with c. 28% methanol conversion, and finally ZnO_{NW} with c. 10% methanol conversion. The TOF were calculated for the three catalysts: Au_{SA}/ZnO_{NW}, Pt_{SA}/ZnO_{NW}, and ZnO_{NW}, which reached 4.7, 18.9, and 1.8×10^{-2} s^{-1}, respectively. In all cases, the selectivity into CO_2 was very high reaching nearly 100% with both ZnO_{NW} and Au_{SA}/ZnO_{NW} catalyst, and c. 88% with Pt_{SA}/ZnO_{NW}. It is worth noting that in view of syngas production, this high CO_2 selectivity should be avoided.

The presence of surface-embedded Pt_{SA} or Au_{SA} dramatically changes the adsorption sites, their binding strength, and reaction energetics and barriers. Indeed, the enhanced binding of the reactants/intermediates and decreased O_v formation energy due to Pt_{SA} or Au_{SA} improve reaction energetics and the subsequent reactivity. The detailed calculated binding energies for the reactants and intermediates involved in methanol steam reforming at the most favorable site are shown in Figure 12.12b. The larger spatial extension of Pt_{SA} and Au_{SA} d-orbitals than that of Zn allows more extensive charge transfer and redistribution, forming a stronger chemical bond with the intermediates. Regarding mechanism of the reactivity enhancement, Figure 12.12c,d present potential energy surfaces and all the transition states (TS) for the dehydrogenation of CH_3OH^* and H_2O^* (toward CO_2 and H^*) on the surface of ZnO_{NW}, Au_{SA}/ZnO_{NW}, and Pt_{SA}/ZnO_{NW}. The formation of CO_2 from methanol reforming can follow two pathways: (i) association of formaldehyde CH_2O^* from methanol with hydroxyl OH^* from water; and (ii) decomposition of formaldehyde CH_2O^* to CO followed by WGSR. The authors indicated that on the ZnO_{NW} surface, the pathway (i) is kinetically and energetically favorable, and (ii) is likely favorable on the surfaces of Au_{SA}/ZnO_{NW}, and Pt_{SA}/ZnO_{NW}. The presence of Au_{SA} and Pt_{SA} enhanced the binding energies of the reactants and provided higher activation energies for the O—H bond scissions, thus energetically hindering the selectivity toward CO_2.

Xu et al. [42] investigated Pd_{SA}/ZnO catalyst for methanol steam reforming at 380 °C (reaction conditions: 50 mg of catalyst, 37 mL/min of a mixture containing 8 vol% CH_3OH + 12 vol% H_2O and He balance, GHSV = 44 400 mL/h/g). The ZnO_{NW} support was prepared by a one-step, non-catalytic, and template-free physical vapor deposition process. The Pd_{SA}/ZnO_{NW} (0.05 wt%/Pd.) catalyst was

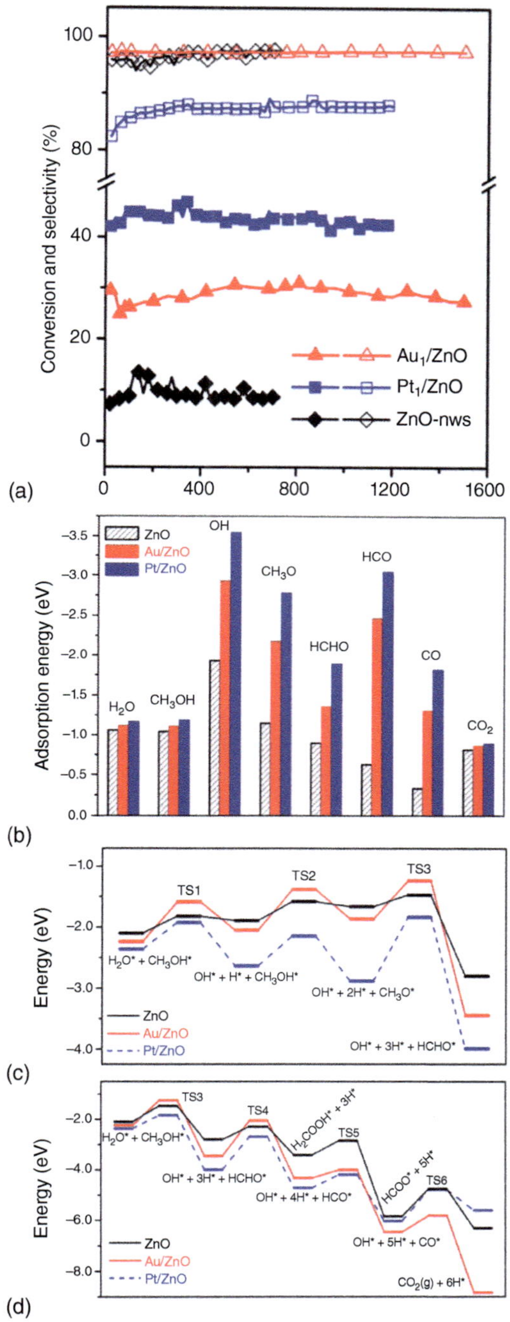

Figure 12.12 (a) Catalytic test results: CH_3OH conversion (solid symbol) and CO_2 selectivity (open symbol); (b) calculated binding energies of the reactants and intermediates involved in methanol steam reforming over ZnO{1010}(shadow), Au_{SA} (red solid), and Pt_{SA} (blue solid) embedded ZnO{1010}; and (c) and (d): calculated potential energy surfaces for $CH_3OH^* + H_2O^* \rightarrow CO_2 + 6H^*$ on ZnO_{NW}, Au_{SA}/ZnO_{NW}, and Pt_{SA}/ZnO_{NW}. Source: Gu et al. [41]. Reproduced with permission of American Chemical Society.

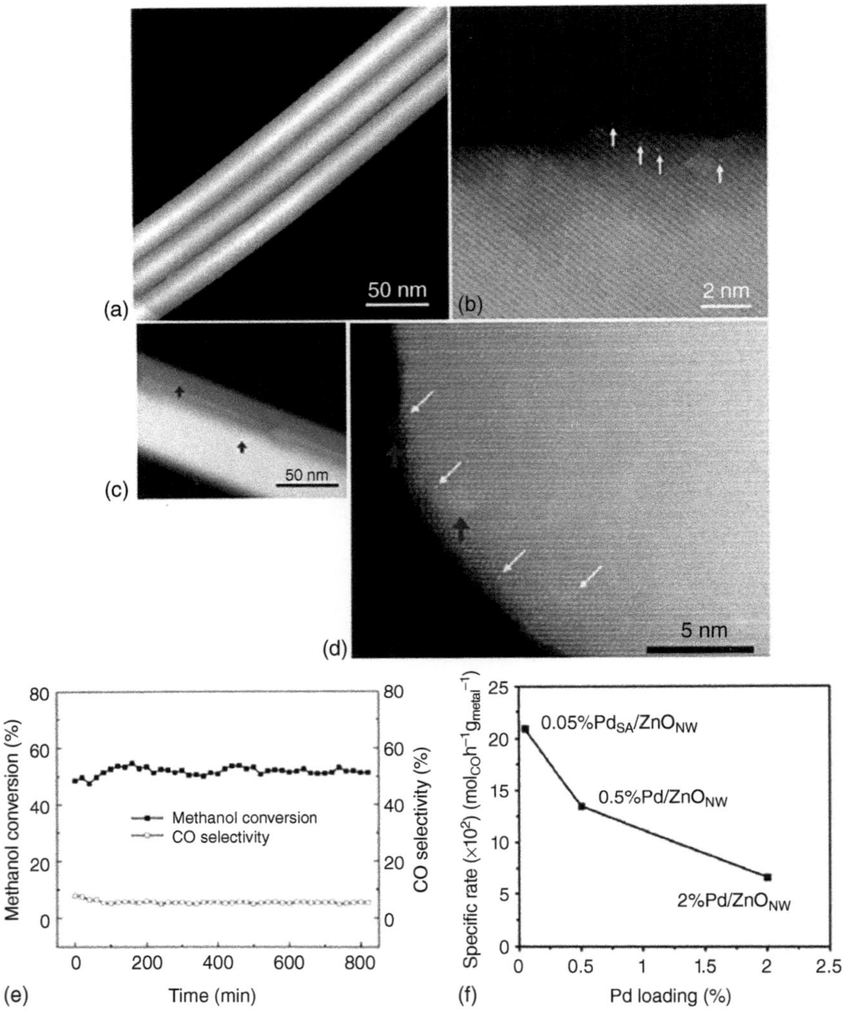

Figure 12.13 HAADF–HRTEM images of the Pd_{SA}/ZnO_{NW} SAC before (a, b) and after (c, d) methanol steam reforming; (e) catalytic activity and CO selectivity during methanol steam reforming over Pd_{SA}/ZnO_{NW} catalyst; and (f) specific rates of ZnO_{NW}-supported Pd_{SA}, Pd clusters, and Pd_{NP}. Pd single atoms are indicated by the bright arrows and Pd clusters are indicated by the dark arrows (reaction conditions: 50 mg catalyst, total inlet gas flow rate: 37 mL/min, inlet gas composition: 8 vol% CH_3OH + 12 vol% H_2O and He balance, 380 °C, GHSV = 44 400 mL/h·g, TOS = 800 minutes). Source: Xu et al. [42]. Reproduced with permission of Elsevier.

prepared by a modified adsorption method. This low Pd loading allowed avoiding the formation of clusters and NPs. Catalysts containing higher Pd loadings were also investigated resulting in Pd clusters (0.5 wt%) and Pd_{NP} (2 wt%). The catalysts were then characterized by various techniques, including SEM, HAADF–HRTEM, and AC–STEM. The results of characterization and catalytic tests are summarized in Figure 12.13 for Pd_{SA}/ZnO_{NW}.

Before catalysis, the Pd_{SA}/ZnO_{NW} catalyst only contained Pd_{SA}, while Pd clusters and NPs were not detected (Figure 12.13a,b). As clearly shown in Figure 12.13c,d, the Pd_{SA}/ZnO_{NW} SAC was very stable for the SRM reaction at 380 °C even after the 800 minutes' run. In accordance, catalytic deactivation was not observed. The methanol conversion remained stable at around 50% for 800 minutes TOS, but the CO selectivity was low (c. only 5%) (Figure 12.13e). Compared with other catalysts containing higher Pd loading, 0.5 and 2 wt%, the authors showed that the Pd_{SA}/ZnO_{NW} catalyst performed the highest CO yield (Figure 12.13f). The methanol conversion rate reached 68 and nearly 100% at 380 and 400 °C, respectively; this conversion being only 15% and 27% over pristine ZnO_{NW}, respectively. In addition, the TOF value of the Pd_{SA}/ZnO_{NW} catalyst reached $3.2\,s^{-1}$, while that of ZnO_{NW} was only $0.018\,s^{-1}$.

In this study, the author highlighted the role of Pd_{SA} on changing the reaction pathway. Accordingly, the ZnO-supported Pd NPs or clusters catalyzed methanol decomposition ($CH_3OH = CO + 2H_2$), while Pd_{SA} directed the reaction toward CO_2 production. Their DFT calculations suggested that for the SRM reaction, the reaction pathway toward CO_2 can be accomplished via two different routes: (i) association of formaldehyde (from methanol) with hydroxyl (from water), and (ii) decomposition of formaldehyde to CO followed by WGSR [42].

From the investigations mentioned above, we can highlight that:

- SAC provides active sites to bind the intermediates, lower reaction barriers, and change the reaction pathway in alcohol-reforming process.
- Cationic state of active phase is active for ethanol and methanol reforming. However, isolated noble metal atoms seemed to be not favorable for the formation of CO.

12.5 Conclusions and Outlook

Reforming is the key process for syngas production. Currently, steam reforming of natural gas is the commercialized process to produce dihydrogen, while other reforming processes are under R&D stage. Supported single-metal atoms catalysts are emerging as a solution to overcome eco-technical problems related to metal utilization and catalytic performance in reforming processes. The application of SACs in reforming processes could on the one hand help to optimize the use of noble metals and thus lower the cost, and on the other hand, to improve catalyst reactivity and selectivity toward the syngas product as well as prevent the catalyst from deactivation. In many reforming processes, carbon deposition is the main deactivation factor. The presence of singly dispersed atoms in SACs could change the conversion pathway, and thus inhibit the carbon deposition over catalyst surface. In addition, SACs could provide active sites to oxidize the carbon deposit during reaction. Another deactivation related to thermal deterioration such as active-phase sintering or support sintering could also be avoided or at least limited due to the advantage of using SACs at lower reaction temperature or with the presence of catalytic promoters. Thus, the development of supported single metal atoms catalyst is expected to ensure the supply for the renewable syngas and hydrogen in the future.

Acknowledgments

This work was supported by the Agence Nationale de la Recherche (France) via CARNOTM.I.N.E.S in the context of the project HyTREND, which is gratefully acknowledged.

References

1 Engvall, K., Kusar, H., Sjöström, K. et al. (2011). Upgrading of raw gas from biomass and waste gasification: challenges and opportunities. *Topic in Catalysis* 54: 949–959.

2 Aziz, M.A.A., Setiabudi, H.D., Teh, L.P. et al. (2019). A review of heterogeneous catalysts for syngas production via dry reforming. *Journal of the Taiwan Institute of Chemical Engineers* 101: 139–158.

3 Tang, Y., Wei, Y., Wang, Z. et al. (2019). Synergy of single-atom Ni_1 and Ru_1 sites on CeO_2 for dry reforming of CH_4. *Journal of the American Chemical Society* 141 (18): 7283–7293.

4 Bion, N., Epron, F., and Duprez, D. (2010). Bioethanol reforming for H_2 production. A comparison with hydrocarbon reforming. In: *Catalysis 22*, vol. 2 (eds. J.J. Spivey and K.M. Dooley), 1–55. Royal Society of Chemistry.

5 Zhang, Y., Yao, Y.F., Qiao, Y.Y. et al. (2021). First-principles theoretical study on dry reforming of methane over perfect and boron-vacancy-containing h-BN sheet-supported Ni catalysts. *Physical Chemistry Chemical Physics* 23: 617–627.

6 Wu, G., Zhang, C., Li, S. et al. (2013). Hydrogen production via glycerol steam reforming over Ni/Al_2O_3: influence of nickel precursors. *ACS Sustainable Chemistry & Engineering* 1: 1052–1062.

7 Słowik, G., Greluk, M., Rotko, M. et al. (2017). Evolution of the structure of unpromoted and potassium-promoted ceria-supported nickel catalysts in the steam reforming of ethanol. *Applied Catalysis B: Environmental* 221: 490–509.

8 Van Hardeveld, R. and Hartog, F. (1972). Influence of metal particle size in nickel-on-aerosil catalysts on surface site distribution, catalytic activity, and selectivity. *Advances in Catalysis* 22: 75–113.

9 Cheng, N., Zhang, L., Doyle-Davis, K. et al. (2019). Single-atom catalysts: from design to application. *Electrochemical Energy Reviews* 2: 539–573.

10 Yang, X.F., Wang, A., Qiao, B. et al. (2013). Single-atom catalysts: a new frontier in heterogeneous catalysis. *Accounts of Chemical Research* 46 (8): 1740–1748.

11 Ding, S., Chen, H.A., Mekasuwandumrong, O. et al. (2021). High-temperature flame spray pyrolysis induced stabilization of Pt single atom catalysts. *Applied Catalysis B: Environmental* 281: 119471.

12 Mohan, O., Xu, R., and Mushrif, S.H. (2021). Novel nickel-based single-atom alloy catalyst for CO_2 conversion reactions: computational screening and reaction mechanism analysis. *Journal of Physical Chemistry C* 125 (7): 4041–4055.

13 Akri, M., Zhao, S., Li, X. et al. (2019). Atomically dispersed nickel as coke-resistant active sites for methane dry reforming. *Nature Communications* 10 (5181): 1–10.

14 Thomas, J.M. (2008). How far is the concept of isolated active sites valid in solid catalysts? *Topics in Catalysis* 50: 98–105.

15 Song, C. and Pan, W. (2004). Tri-reforming of methane: a novel concept for catalytic production of industrially useful synthesis gas with desired H_2/CO ratios. *Catalysis Today* 98 (4): 463–484.

16 Minh, D.P., Siang, T.J., Vo, D.-V.N. et al. (2019). Hydrogen production from biogas reforming: an overview of steam reforming, dry reforming, dual reforming, and tri reforming of methane. In: *Hydrogen Supply Chains Design, Deployment and Operation* (ed. C. Azzaro-Pantel), 111–166. Amsterdam: Academic Press.

17 Zhang, Y., Zhang, S., Gossage, J.L. et al. (2014). Thermodynamic analyses of tri reforming reactions to produce syngas. *Energy Fuels* 28: 2717–2726.

18 Duarte, R.B., Krumeich, F., and van Bokhoven, J.A. (2014). Structure, activity, and stability of atomically dispersed Rh in methane steam reforming. *ACS Catalysis* 4 (5): 1279–1286.

19 Horn, R., Williams, K.A., Degenstein, N.J. et al. (2006). Syngas by catalytic partial oxidation of methane on rhodium: mechanistic conclusions from spatially resolved measurements and numerical simulations. *Journal of Catalysis* 242 (1): 92–102.

20 Donazzi, A., Maestri, M., Michael, B.C. et al. (2010). Microkinetic modeling of spatially resolved auto-thermal CH_4 catalytic partial oxidation experiments over Rh-coated foams. *Journal of Catalysis* 275 (2): 270–279.

21 Richardson, J.T. and Paripatyadar, S.A. (1990). Carbon dioxide reforming of methane with supported rhodium. *Applied Catalysis* 61 (1): 293–309.

22 Wu, J., Qiao, L.Y., Zhou, Z.F. et al. (2019). Revealing the synergistic effects of Rh and substituted $La_2B_2O_7$ (B = Zr or Ti) for preserving the reactivity of catalyst in dry reforming of methane. *ACS Catalysis* 9: 932–945.

23 Lanza, R., Järås, S.G., and Canu, P. (2008). Microemulsion-prepared ruthenium catalyst for syngas production via methane partial oxidation. *Applied Catalysis A: General* 337 (1): 10–18.

24 Souza, M.M.V.M., Macedo Neto, O.R., and Schmal, M. (2006). Synthesis gas production from natural gas on supported Pt catalysts. *Journal of Natural Gas Chemistry* 15 (1): 21–27.

25 Silva, F.d.A., Ruiz, J.A.C., de Souza, K.R. et al. (2009). Partial oxidation of methane on Pt catalysts: effect of the presence of ceria-zirconia mixed oxide and of metal content. *Applied Catalysis A: General* 364 (1–2): 122–129.

26 Ryu, J.H., Lee, K.Y., Kim, H.J. et al. (2008). Promotion of palladium-based catalysts on metal monolith for partial oxidation of methane to syngas. *Applied Catalysis B: Environmental* 80 (3–4): 306–312.

27 Zhang, J.C., Ge, B.H., Liu, T.F. et al. (2020). Robust ruthenium-saving catalyst for high-temperature carbon dioxide reforming of methane. *ACS Catalysis* 10 (1): 783–791.

28 Duarte, R.B., Safonova, O.V., Krumeicha, F. et al. (2014). Atomically dispersed rhodium on a support: the influence of a metal precursor and a support. *Physical Chemistry Chemical Physics* 16: 26553–26560.

29 Duarte, R.B., Damyanova, S., de Oliveira, D.C. et al. (2011). Study of Sm_2O_3-doped CeO_2–Al_2O_3-supported Pt catalysts for partial CH_4 oxidation. *Applied Catalysis A: General* 399 (1–2): 134–145.

30 Zuo, Z., Liu, S., Wang, Z. et al. (2018). Dry reforming of methane on single-site Ni/MgO catalysts: importance of site confinement. *ACS Catalysis* 8 (10): 9821–9835.

31 Angeli, S.D., Pilitsis, F.G., and Lemonidou, A.A. (2015). Methane steam reforming at low temperature: effect of light alkanes' presence on coke formation. *Catalysis Today* 242 (A): 119–128.

32 Akri, M., El Kasmi, A., Batiot-Dupeyrat, C. et al. (2020). Highly active and carbon-resistant nickel single-atom catalysts for methane dry reforming. *Catalysts* 10 (6): 630.

33 Rakib, M.A., Grace, J.R., Lim, C.J. et al. (2010). Steam reforming of propane in a fluidized bed membrane reactor for hydrogen production. *International Journal of Hydrogen Energy* 35 (12): 6276–6290.

34 Reyes, S.C., Sinfelt, J.H., and Feeley, J.S. (2003). Evolution of processes for synthesis gas production: recent developments in an old technology. *Industrial & Engineering Chemistry Research* 42 (8): 1588–1597.

35 Seeburg, D., Liu, D., Dragomirova, R. et al. (2018). Low-temperature steam reforming of natural gas after LPG-enrichment with MFI membranes. *Processes* 6: 263.

36 Palma, V., Ruocco, C., Cortese, M. et al. (2020). Bioalcohol reforming: an overview of the recent advances for the enhancement of catalyst stability. *Catalysts* 10: 665.

37 Ju Han, S., Bang, Y., Yoo, J. et al. (2013). Hydrogen production by steam reforming of ethanol over mesoporous Ni–Al_2O_3–ZrO_2 aerogel catalyst. *International Journal of Hydrogen Energy* 38 (35): 15119–15127.

38 Kim, S., Prajitno, H., Yoo, J. et al. (2021). Dispersion behavior of various single metals on carbonaceous coal supports and their reactivity in methanol steam reforming. *Journal of Industrial and Engineering Chemistry* 94 (25): 317–325.

39 Tian, H., Pei, C., Wu, Y. et al. (2021). Tunable metal-oxide interaction with balanced Ni^0/Ni^{2+} sites of $Ni_xMg_{1-x}O$ for ethanol steam reforming. *Applied Catalysis B: Environmental* 293 (15): 120178.

40 Zhou, P., Zhang, Q., Chao, Y. et al. (2021). Partially reduced Pd single atoms on CdS nanorods enable photocatalytic reforming of ethanol into high value-added multicarbon compound. *Chem Catalysis* 7 (4): 1033–1049.

41 Gu, X.K., Qiao, B., Huang, C.Q. et al. (2014). Supported single Pt_1/Au_1 atoms for methanol steam reforming. *ACS Catalysis* 4 (11): 3886–3890.

42 Xu, J., Song, Y., Wu, H. et al. (2017). Probing the catalytic behavior of ZnO nanowire supported Pd1 single-atom catalyst for selected reactions. *Chinese Journal of Catalysis* 38 (9): 1549–1557.

13

Electrocatalysis with Single-Metal Atom Sites in Doped Carbon Matrices

Tristan Asset[1], Frédéric Maillard[1], and Frédéric Jaouen[2]*

[1] *Institut Charles Gerhardt de Montpellier, Université de Montpellier, CNRS, ENSCM, 34090 Montpellier, France*
[2] *Université Grenoble Alpes, Université. Savoie Mont Blanc, CNRS, Grenoble INP, LEPMI, 38000 Grenoble, France*

13.1 Introduction

Electrochemical reactions take place at an electrode–electrolyte interface. If the electrode material accelerates the rate of the electrochemical reaction, the phenomenon is referred to as "electrocatalysis." Electrochemical potential (E) and current density (j), two easily accessible parameters, are used in electrocatalysis to assess the performance of catalytic materials with so-called polarization curves, $E = f(j)$. In addition to temperature and pressure effects encountered in heterogeneous catalysis, the electrochemical potential is a key parameter in electrocatalysis, which can be used to dramatically increase the reaction rate of thermodynamically favorable reactions (e.g. in fuel cells), or, to force the occurrence of thermodynamically unfavorable reactions (e.g. electrolysis). Very important fields of application of electrocatalysis are the conversion of electrical energy into chemical energy and of chemical energy into electrical energy. Major examples of the former are water electrolysis, and co-electrolysis of water and carbon dioxide (CO_2), or water and nitrogen (N_2). In these devices, the following reactions need to be catalyzed: hydrogen evolution reaction (HER) and oxygen evolution reaction (OER) in water electrolyzers, selective electroreduction of CO_2 (CO_2RR) into C_1 [formic acid (HCOOH), carbon monoxide (CO), and methane (CH_4)] and C_2 products (ethylene, ethanol) or the electroreduction of N_2 in NH_3 (N_2RR). For chemical to electric energy conversion in electrochemical devices, fuel cells fed with H_2 fuel are the most prominent example, and the reactions to be catalyzed are the hydrogen oxidation reaction (HOR) and the oxygen (O_2) reduction reaction (ORR), even if other fuels have been or are being investigated, such as methanol, ethanol, or hydrazine.

The development of fuel cells began in the 1950–1960s for manned space missions. Attention was focused on low-temperature fuel cells in the 1990s due

* Corresponding author email: frederic.jaouen@umontpellier.fr

Supported Metal Single Atom Catalysis, First Edition. Edited by Philippe Serp and Doan Pham Minh.
© 2022 WILEY-VCH GmbH. Published 2022 by WILEY-VCH GmbH.

to the growing need of carbon-free power sources. In alkaline fuel cells, many metals, including non-precious metals, are thermodynamically stable in their metallic, oxide, or hydroxide form, and electrochemical reactions kinetics are usually more facile than in acidic media, although with the exception of the HOR. In acidic fuel cells (i.e. PEMFC, proton-exchange membrane fuel cell), the most industrially mature so far, only few metals are stable and active, typically platinum (Pt) and platinum group metals (PGM, comprising platinum, iridium, osmium, palladium, rhodium, and ruthenium). Today, the state-of-art catalysts for the ORR and the HOR in PEMFCs are Pt-based nanoparticles (NPs) supported onto high surface area carbon (Pt/C). The Pt loading in PEMFCs has been decreased radically from the 1980s from ca $10\,mg_{Pt}/cm^2{}_{geo}$ to c. $0.2\,mg_{Pt}/cm^2{}_{geo}$ today, by adding Nafion® in catalytic layers to improve the electrochemical utilization, and by nanostructuring Pt. However, even with 2–3 nm Pt_{NP}, less than c. 50% of the atoms are located on the surface. Intuitively, further increasing the utilization of the rare and expensive Pt in PEMFC (especially for the ORR, since the HOR in acid medium requires only a low amount of Pt in PEMFC) may have been expected via the development of catalysts comprising single Pt atom sites. However, it is for the chlorine evolution reaction [1] and the HER [2–4] that promising single atom catalysts (SACs) involving PGMs have been developed. The main reason behind this is the high surface energy of metal single atoms (SAs), which requires strong interactions with the support to avoid migration and aggregation. Metal oxides are well suited to strongly bind PGM-based SAs, but most of them are insulating or, in the best case, semi-conducting, which is not desired for electrochemical applications targeting high *j*. Furthermore, the Gibbs–Thompson relation predicts that the redox potential of NPs is inversely proportional to the particle size and shifts toward negative potentials with respect to the bulk material. As an example, the Gibbs–Thompson equation (with $V_{m,Pt} = 9.1 \times 10^{-6}\ m^3/mol$ the volume of a Pt atom [5] and a surface tension of $\gamma_{Pt} = 2.370\,J/m^2$ [6]) predicts that Pt_{SA} are thermodynamically under corrosion situation at any ORR-relevant potential in acidic solutions. Finally, the binding strength of oxygen intermediates is strongly correlated to the Pt–Pt coordination number, e.g. Pt with low coordination number (CN = 1–2), and therefore closer to Pt SAC, exhibit a OH_{ads} adsorption strength, which is c. 1 eV higher than optimal [7].

In contrast to single PGM atoms embedded in a carbon-based matrix, SACs using non-precious metals embedded in the carbon structure have been successfully developed and demonstrated to catalyze the ORR. Early findings from Jasinski in 1964 that macrocyclic compounds based on 3d transition metal [such as chromium (Cr), manganese (Mn), iron (Fe), cobalt (Co), nickel (Ni), copper (Cu) or zinc (Zn)] and nitrogen (N) with a metal–N_4 configuration catalyze the ORR activity in both alkaline [8] and acidic [9] electrolyte initiated the development of SACs in electrocatalysis. Catalysts were first synthesized by adsorbing Fe and Co macrocycles with well-defined square-planar metal–N_4 configuration onto a carbon support. In a second step, mild annealing (500 °C) or higher temperature pyrolysis was shown to increase the stability of such catalysts, but questioned the nature of the active sites after the thermal treatment. The first breakthrough toward materials with metal SA sites embedded into a nitrogen-doped carbon matrix (metal–N–C) was achieved in 1989 by Gupta et al. who revealed that catalysts with ORR activity could be synthesized by pyrolyzing a mix of metal, nitrogen, and carbon precursors, with no metal–N_4 coordination pre-existing the pyrolysis [10]. A second breakthrough was made in 2009–2011: for the first time, ORR performance compatible with PEMFC application was demonstrated for Fe–N–C materials [11, 12]. The introduction of thermally decomposable metal–organic frameworks (MOFs) as a carbon and nitrogen source further improved the ORR activity compared to the existing state of art, and more importantly, improved the accessibility of the reactants to the active sites [13, 14]. This chapter focuses on metal–N–C SACs involving one or several 3d transition metal, since they are scientifically and technologically the most mature SACs for application in various electrochemical devices. For completeness and outlook into the future, we also shortly discuss latest reports on carbon-embedded SACs prepared with other metals and/or other light-element dopant (e.g. Pt–S–C and tin (Sn)-doped N–C). This review chapter is organized as follows. In Section 13.2, we describe how these materials can be synthesized for application in electrocatalysis. In Section 13.3, the focus is on the structure of the active sites for electrocatalytic reactions, and key physico-chemical techniques that have been used to characterize these materials. When available, we detail how these techniques can be used *in situ/operando*. Section 13.4 is devoted to the mechanisms of the ORR, the CO_2RR, and the N_2RR or the NO_3 reduction reaction (NO_3RR) on such materials. As research has recently moved to the next important step for application, namely stability and degradation mechanisms of metal–N–C catalysts, we cover these aspects in Section 13.5. Finally, Section 13.6 gives conclusions and an outlook into the future of carbon-embedded SACs.

13.2 Synthesis Methods

Before the advent and broad availability of characterization techniques that can unambiguously identify metal SA sites (e.g. X-ray absorption spectroscopy (XAS), high-resolution scanning transmission electron microscopy (HR-STEM), and cryostatic ^{57}Fe Mössbauer spectroscopy), the synthesis of metal–N–C SACs by pyrolysis was an intention rather than a demonstrated fact. The initial reports on the ORR activity of unpyrolyzed macrocycles [8, 15, 16] revealed, that Fe and Co were, by far, the most active ones among 3d transition metals for catalyzing the ORR, both in alkaline and acidic media. This important learning was found to be valid for metal–N–C SACs prepared via pyrolysis in the following decades. Hence, we focus mainly on the synthesis methods of Fe- and Co–N–C SACs in this section. While in those early days, the use of metal–N_4 macrocycles without pyrolysis secured the well-defined metal–N_4 active centers, the different approaches used to interface them with a conductive support impacted the local environment of metal cations and their accessibility. Phthalocyanines and porphyrins are poorly soluble in water, and different solvents or suspensions in concentrated sulfuric acid were used, removing the solvent by evaporation or precipitating the macrocycle on the support. The dispersion quality of metal macrocycles on supports was shown to depend strongly on the morphology, specific surface area (SSA) but also on acido-basic properties of carbon supports [17, 18]. For example, the characterization by ^{57}Fe Mössbauer spectroscopy, a technique that allows identifying the different coordinations and spin states of Fe (discussed in more detail in Section 13.3), of iron phthalocyanine (Fe–Pc) precipitated on a carbon powder revealed multiple coordinations and spin states for Fe, whereas a single coordination and electronic state was seen for the Fe–Pc monomer [17]. Controlled deposition or solvent removal however resulted in Fe–Pc/C composites with a single (or vast majority of one) ^{57}Fe Mössbauer spectroscopic signature [19]. For example, removal of the pyridine solvent for Fe–Pc by boiling it off at 420 °C resulted in an Fe–Pc/C composite with 95% of the spectral signal assigned to one specific doublet [19]. The latter can nowadays be assigned to an O_2–Fe(III)–N_4 coordination [20], implying high dispersion and accessibility to O_2 of the Fe–N_4 sites.

While some heterogeneities of metal coordination and site accessibility were exemplified above for SACs prepared via interfacing metal–N_4 macrocycles without high-temperature pyrolysis, the extent of heterogeneities was progressively increased, and the true nature of the active site blurred for several decades, with the introduction of high-temperature treatments of macrocycles in a first step, and the switch to the synthesis of metal–N–C SACs from separate metal, nitrogen, and carbon precursors at high temperature, in a second step. Mild (200–500 °C) and then more aggressive thermal treatment (>600 °C) in inert or reductive atmosphere was applied to metal–N_4 macrocycles supported on carbon (labeled henceforth as metal–N_4/C), starting in 1976 [15]. This was triggered in response to the quickly decaying ORR activity of non-heat-treated metal–N_4/C materials in acidic medium, in an attempt to improve their lifetime. Improved electrochemical durability in acid medium was always observed after pyrolysis, while the magnitude of increase in the ORR activity highly depended on the nature of the metal (Fe, Co) and of the macrocycle (phthalocyanines, porphyrins, tetraazaannulene, etc.) [21–26]. As a general rule, the improvement in ORR activity was however >20 times. The effect of the annealing temperature on a Co–Pc/C material was studied, e.g. with Raman spectroscopy, showing that the peaks characteristic for pristine Co–Pc disappeared after a pyrolysis at 700 °C [27]. The transformation was proposed to be the loss of peripheral functional groups first while the Co–N_4 core was retained, and then above 700 °C the complete disintegration of the phthalocyanine (Pc) structure. The fate of the metal ions and their coordination after a high-temperature treatment of metal–N_4/C materials has been the object of a long scientific debate that will only be shortly discussed here. The three main hypotheses on the nature of the active site resulting from pyrolysis were (i) metal–N_4 coordination after any pyrolysis temperature, similar to the core of macrocycles [28], (ii) a C–N_x–metal complex, with the metal adsorbing on C–N_x moieties *in situ* during electrochemistry [10], and (iii) the formation of specific metal–N_x–C_y site at high temperature [18]. While unoptimized syntheses initially led to the formation of metallic Co or Fe particles during pyrolysis (possibly along with SA sites) [28, 29], the application of specific techniques that can distinguish metal aggregates from SA sites, correlation between the metal coordination, ORR activity and synthetic approach, resulted in the early 2000s in the identification of the existence of SA sites of the type metal–N_x, even after pyrolysis [30–32]. These sites should however be regarded as structurally and electronically different from metal–N_4 sites in macrocycles, since the metal–N_x sites are covalently integrated in the carbon support (which we therefore refer henceforth to as a "carbon matrix," when the final material is discussed) after a pyrolysis at 700–1000 °C [32]. The covalent integration of metal–N_x sites in the carbon matrix generally explains the improved stability and electrocatalytic activity relative to unpyrolyzed metal–N_4/C materials. The active site in such SACs is no longer a molecule, but a moiety in a conductive material. The effect of the temperature and macrocycle loading on carbon supports was shown to impact the type of metal species present after pyrolysis [28, 33–35]. Excessive macrocycle loading leads to metal aggregation at high temperature and their carbo-reduction. In turn, Fe and Co metallic particles catalyze the graphitization of the carbon support, which decreases the nitrogen content, accelerating the transformation of metal–N_x sites into metallic aggregates. Optimized macrocycle loading and initial dispersion, pyrolysis temperature, and duration can however lead to metal–N–C materials with all metal atoms as metal–N_x sites [36]. The ORR performance of pyrolyzed metal-macrocycle SACs was greatly improved through the simultaneous removal of the carbon support and the addition of porogens. This approach was originally developed by the group of Tributsch and then continued by Kramm [37, 38]. It however necessitates the careful removal by acid wash of excessive metal after pyrolysis, a natural consequence of the high metal content in macrocycles (e.g. 10 wt% Fe in Fe–Pc with molecular formula $C_{32}H_{16}FeN_8$), well above the current limit of Fe as Fe–N_x sites in Fe–N–C materials (c. 2–3 wt% Fe). The addition of sulfur (S) before pyrolysis was shown to result in the formation of Fe or Co sulfide phases after a first pyrolysis that can be easily acid-washed, leaving behind only the acid-stable metal–N_x sites. In addition, Fe and Co sulfides do not catalyze graphitization at high temperature, also helping in retaining a high nitrogen content, important for high number of metal–N_x sites. Nowadays, the metal–N_4 macrocyclic approach to prepare metal–N–C SACs is still an active field, with original approaches such as the preparation of MOF structures integrating porphyrins as building blocks [39, 40], combining metal porphyrins with the silica hard-templating method [41], or forming aerogels of porphyrins [42] before pyrolysis.

While the clear experimental proof for the existence of ORR-active metal–N_x sites in metal–N–C materials derived from the pyrolysis of metal–N_4 macrocycles was perhaps available only in the early 2000s, the idea emerged long before the proof was obtained that perhaps all that is needed for driving the spontaneous formation of ORR-active sites is the simultaneous presence of metal (Fe, Co, Sn), carbon, and nitrogen at high temperature. The first pyrolytic synthesis of a metal–N–C material not resorting to a metal–N_4 macrocycle was probably reported by Yeager's group in 1989 [10]. Polyacrylonitrile, Fe, or Co acetate salt, and a carbon black powder were mixed by wet chemistry in dimethylformamide, then dried, and pyrolyzed in inert gas. The samples showed no ORR activity after annealing at low temperature and optimum ORR activity after annealing at 750 or 900 °C, depending on the nature of the metal. This work opened the door to metal–N–C synthesis from simple metal salts, nitrogen, and carbon precursors. Between 1989 and 2009, most pyrolytic syntheses of metal–N–C materials involved a pre-existing carbon support, adding metal precursors (most often, simply salts) and nitrogen precursors in optimized amounts [43–45]. The use of a pre-existing carbon support offered a reasonably good control of the morphology, graphitic ordering, porosity, and SSA of the final catalyst. The disadvantages were the lack of "mobility" of carbon atoms from such a support (especially for the more graphitic carbon supports [46]), as well as propensity for metal aggregation into particles in the case of carbon supports with a low SSA. The carbon mobility aspect is important to explain at this stage. The site formation at high temperature requires the rearrangement of carbon, nitrogen, and metal atoms. While the nitrogen-rich organic precursors and metal salts are highly reactive above 700 °C, this is much less the case for carbon atoms from the carbon support due to the high melting point of graphite (>3600 °C). It was revealed in 2004–2006 that the presence of a disordered carbon phase in the starting carbon support is critical to reach high ORR activity of the resulting Fe–N–C materials [47–49]. It was also shown that a negative correlation exists between the SSA and the content of disordered carbon in carbon blacks. Since both high SSA and high content of disordered carbon are desirable to prepare a metal–N–C catalyst precursor, the negative correlation mentioned above was a limiting factor for metal–N–C materials until 2009. This led to (i) the parallel formation of metal particles and metal–N_x sites, (ii) a low density of metal–N_x active sites, and an overall unsatisfactory ORR activity and performance of metal–N–C catalysts, with ORR activity of only 1–2 mA/cm^2_{geo} at 0.9 V cell voltage [47, 50].

Another breakthrough was achieved in 2009, with a novel synthesis leaning on the functionalization of a high SSA carbon black, namely Black Pearls 2000 (BP2000), with an organic molecule rich in nitrogen and carbon, namely 1,10-phenanthroline [11]. Functionalizing the micropores of BP2000 with optimized amounts of phenanthroline and ferrous acetate resulted in a catalyst precursor with high amount of disordered carbon and optimized amounts of Fe and nitrogen. The pyrolysis of such catalyst precursors resulted in Fe–N–C materials achieving a current density of c. 30 mA/cm^2_{geo} at 0.9 V in H_2/O_2 PEMFC (with cathode loading of 4 $mg_{Fe\text{–}N\text{–}C}/cm^2_{geo}$), similar to the value obtained with a state-of-art cathode with 0.4 mg_{Pt}/cm^2_{geo} [11]. In more detail, it was shown that the micropores of the high SSA support were filled with phenanthroline, and the pores re-opened during pyrolysis as a result of the transformation of phenanthroline and Fe salt into Fe–N_x sites, with c. two-third of the mass of phenanthroline leaving the reactor as volatile products. The concept of the preparation of catalyst precursors comprising "disordered carbon," nitrogen atoms, and metal atoms was found to be valid for other high surface area carbon supports, including carbide-derived carbons [51, 52], and for other organic molecules than phenanthroline [53]. One remaining drawback of the "functionalized carbon support approach" was the high relative amount of inactive carbon atoms in the catalyst. The idea therefore progressively emerged to design syntheses that do not lean on pre-existing carbon supports, but rather involve organic or metal–organic materials as carbon and nitrogen precursors. Beside the pore-forming agent approach already discussed earlier for preparing highly active metal–N–C materials from metal porphyrins, four other synthetic approaches are discussed below that have found success in the preparation of high-performance metal–N–C catalysts for ORR, leading ultimately to SACs.

13.2.1 Hard Templating with Silica

In this approach, a macrocycle, or separate metal, nitrogen, and carbon organic precursors, are mixed in optimized amounts with silica particles, and pyrolyzed (Figure 13.1a). The removal of silica particles by hydrofluoric acid and/or potassium hydroxide (KOH) leaves a metal–N–C catalyst with high amount of mesopores by hard templating effect, with some micropores also present in the carbon phase resulting from the pyrolysis of the organic precursors. This approach was originally developed for metal–N–C materials by Atanassov's group in 2008 [54], still resorting to metal–porphyrins. More active Fe–N–C materials were prepared later by resorting to Fe salts and N–C precursors such as aminoantipyrine and nicarbazin [55–58]. Under optimized conditions, the synthesis results in the majority or complete dispersion of Fe into Fe–N_x sites [56, 59]. For optimized ORR activity, re-annealing the catalyst after silica removal is however important. The silica templating synthesis of metal–N–C SACs has been scaled-up by the company Pajarito Powder. The scaled-up material however shows a lower relative amount of Fe–N_x sites than laboratory batches, with concurrent metallic Fe particles (compare the PAJ and UNM samples in Ref. [59]). This gives an example of the challenge of scaling-up SAC production [59]. Interestingly, although the site density (SD) of Fe–N_x sites was low in the scaled-up material, their turnover frequency (TOF, number of electrons transferred per site per second) was high, leading to an overall competitive ORR activity. The particle size, morphology, and SSA of the silica template were found to affect the resulting Fe–N–C catalysts [60]. An alternative use of silica was reported by Joo's group in 2016, by depositing a silica overlayer on top of carbon nanotubes that had been previously coated with a monolayer of Fe-porphyrin [61]. The work showed that the silica overlayer minimized the formation of Fe particles during pyrolysis, thereby increasing the final Fe–N_x SD and ORR activity.

13.2.2 Soft Templating with Metal–Organic Frameworks

The use of MOFs for preparing Fe–N–C materials via pyrolysis was first reported in 2011 by the groups of Dodelet and Liu [13, 14]. In Liu's conceptual approach, a Co-based zeolithic imidazolate framework (ZIF) was a promising catalyst precursor, featuring a high density of tetrahedral Co–N_4 coordination sites. Following pyrolysis at 750 °C, the material showed promising ORR activity. The amount of Co in their ZIF (c. 30 wt%) was however well above the limit that can be integrated as Co–N_x sites in Co–N–C. As a result, metallic Co NPs formed during pyrolysis. In Dodelet's approach, MOFs were of interest as a highly and well-defined microporous substrate (micropores in Fe–N–C had previously been showed to correlate with high ORR activity by his group [62]), while simultaneously containing nitrogen and carbon atoms. The specific nature of the metal in the MOF was regarded as less important, since an optimized (small) amount of Fe or Co can be added as a salt,

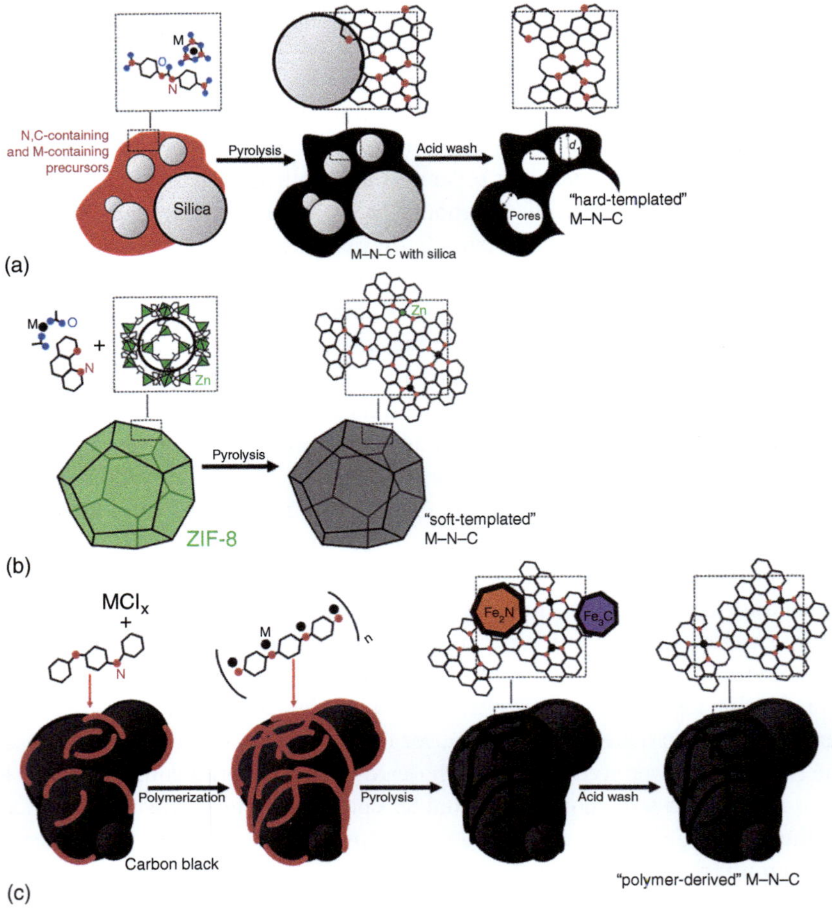

Figure 13.1 Main synthesis methods to achieve metal–nitrogen–carbon electrocatalysts: (a) hard-templating approach with silica as an example; (b) soft-templating approach with ZIF-8 as an example; (c) sacrificial polymer approach with polyaniline as an example – the red lines are representative of the aniline before and after polymerization. Source: Tristan Asset, Frédéric Maillard, Frédéric Jaouen.

by wet chemistry approach, before pyrolysis. The Zn-based ZIF-8 structure was commercially available already in 2011, and this MOF was used in replacement of BP2000, in an otherwise similar synthesis as that reported in 2009 [11]. The resulting optimized Fe–N–C material, whose synthesis is illustrated in Figure 13.1b, showed c. twice increased power performance in H_2/O_2 PEMFC compared to their report of 2009, reaching a peak power density of $0.9\,W/cm^2_{geo}$ [13]. This demonstrated improved accessibility of the Fe-based active sites with the sacrificial ZIF-8 approach. Scanning electron microscopy (SEM) images showed that the nitrogen-doped carbon matrix resulting from ZIF-8 pyrolysis had a defined geometry, with size and shape similar to that of the ZIF-8 crystals before pyrolysis, while TEM images showed that the catalytic particles contained interconnected pores

[13]. The availability of a commercial Zn-based ZIF-8 in 2011 was fortuitous, and was later proved to be (near)optimal, with the easy removal of Zn during pyrolysis, due to its low boiling point of 907 °C. These two works from 2011 spurred an entire line of research on metal–N–C and N–C materials derived from MOF pyrolysis (ZIF-8 in particular), as reviewed, e.g. in Refs. [63, 64]. Subsequent works showed the importance of nanostructuring the ZIF crystals, leading to their soft-templating into nanosized Fe–N–C or Co–N–C particles with the same dodecahedron shape [65–67]. One-pot syntheses of Fe-doped ZIF-8 were then reported [68, 69]. Other one-pot syntheses were developed, with ZIF-8 acting as a host, and selecting Fe precursors small enough to be encapsulated in the inner cavity of ZIF-8 or other MOFs. This was achieved, e.g. with the encapsulation of ferrocene in ZIF-8 [70] or of the Fe(phenanthroline)$_3$ complex in the cavity of three different Zn-based ZIFs [71]. In the latter study, the ZIF with the largest cavity size led to the Fe–N–C with the highest SSA and highest ORR activity. Similarly, a comprehensive study on the preparation of Fe–N–C materials from different Zn-based ZIFs revealed the importance of the cavity size to reach high ORR activity of Fe–N–C materials after pyrolysis [72]. Despite numerous ZIFs and MOFs investigated to date, ZIF-8 remains close to the optimum for the preparation of metal–N–C SACs, due to its large cavity size of 11.8 Å (Ångström). The synthesis of an Fe–N–C material from ZIF-8, featuring only atomically dispersed Fe–N$_x$ sites, was demonstrated for the first time in 2015 [73]. The synthesis leaned on the dry-mixing of Fe acetate, phenanthroline, and ZIF-8, avoiding any wet chemistry step that easily leads to Fe clustering by capillarity effects. Since then, other groups also achieved the complete atomic dispersion of the metal in Co- and Fe–N–C materials prepared from ZIF-8 under optimized conditions, with full atomic dispersion demonstrated using either ^{57}Fe Mössbauer spectroscopy, HR-STEM, or XAS techniques [66, 68, 74, 75]. Recently, two approaches relying on the soft-templating method were also investigated that take advantage of chemical vapor deposition (CVD). In the first study, CVD was used to synthesize an Fe-doped ZIF, by reacting the gaseous ligand (2-methyl-imidazole) at 350 °C with Fe-doped ZnO particles. This CVD-prepared catalyst precursor was then pyrolyzed at 1000 °C to form an Fe–N–C SAC [76]. In the second study, a Zn–N–C material with high density of Zn–N$_4$ sites was first prepared via pyrolysis of ZIF-8 and phenanthroline. Zn–N$_4$ sites were then used as platform for the formation of Fe–N$_4$ sites via CVD of $FeCl_{3(g)}$ at 750 °C, leading to the high-temperature transmetallation between Zn and Fe in the N$_4$ cavity, and to a highly ORR-active Fe–N–C SAC with Fe–N$_x$ sites located preferentially on the surface [77]. A low-temperature transmetallation had previously been proposed by Fellinger's group, with a magnesium(Mg)–N–C material comprising Mg–N$_x$ sites [78], but the resulting Fe–N–C had however moderate ORR activity.

13.2.3 Sacrificial Polymers

Nitrogen-doped polymers have been studied as a suitable precursors of nitrogen and carbon, with the nitrogen groups being also useful to bind metal cations and disperse them atomically before pyrolysis. Polyaniline (PANI) and polypyrrole are perhaps the two most prominent examples of polymers that have been investigated for preparing Fe–N–C or other metal–N–C SACs, as reviewed in Ref. [79]. Wood et al. reported in 2007 on a highly active Co–N–C catalyst prepared from the polymerization of nitroaniline monomers in the presence of Fe chloride, followed by pyrolysis (illustrated in Figure 13.1c). PANI can bind Fe cations either through the imine or amine groups, helping the Fe dispersion [80]. Since 2008, the Zelenay's group has been highly active in research on Fe–N–C materials prepared via the sacrificial polymer approach [12, 81–83]. PANI has often resulted in the best performing Fe–N–C catalysts in their work, but the addition of some carbon support or even titanium oxide support during the polymerization step was however found useful to optimize the pore size distribution and electrocatalytic performance of the resulting catalysts. The aniline polymerization step was accelerated by the addition of an oxidant, such as ammonium peroxydisulfate. In this case, the introduced sulfur can also react during pyrolysis with excess Fe that cannot be hosted as Fe–N_x sites. Iron sulfur is then easily leached away during a subsequent acid-wash step. In 2017, Zelenay's group reported on an optimized synthesis involving PANI, cyanamide, BP2000 as a carbon black support, and $FeCl_3$, leading to an Fe–N–C catalyst containing only or mostly Fe–N_x sites, as evidenced by HR-STEM [84]. Cyanamide acted mainly as a pore-forming agent, and it was shown that much lower SSA was achieved for a similar synthesis but without cyanamide. HR-STEM evidenced the presence of Fe–N_x single sites, with a particularly high concentration at or near the edges of graphene sheets. The optimized catalyst resulted in a high PEMFC performance of c. 0.9 and 0.39 W/cm^2_{geo} in H_2/O_2 and H_2/air conditions, respectively. A recent work showed that designing PANI-like polymers having a stronger interaction with Fe cations could lead to higher amount of Fe being adsorbed before pyrolysis, and higher density of Fe–N_x sites after pyrolysis [85]. Polymers with different nitrogen-bearing side chains were prepared to increase the basicity of nitrogen groups relative to PANI. With these polymers, the Fe content in the final catalysts could be increased by a factor of c. 4 compared to the Fe–N–C prepared from PANI, indicating that more Fe was converted to Fe–N_x sites. This increase was also reflected by increased ORR activity (c. x 2.5 at $E = 0.8$ V vs. the reversible hydrogen electrode, RHE) and performance in PEMFC (+30% peak power density). This shows that, generally, the basicity of the nitrogen functional group in the starting polymers must be optimized.

13.2.4 Electrospun Polymer/MOF Composites

Another type of synthesis approach combining the sacrificial polymer approach and the sacrificial MOF approach was reported by Liu's group in 2015 [86]. It leaned on the electrospinning technique to prepare mats of metal-doped polymer/nano-MOF composite fibers, followed by their pyrolysis [86]. In this approach, the Fe–N–C phase derived from the pyrolysis of the nano-MOFs is the highly active phase, and the integration of such nano-scaled domains in carbon fibers (derived from the pyrolysis of the polymer fibers) allows preparing hierarchically structured catalyst with high amount of micropores (for hosting Fe–N_x sites) and high volume of macropores (for high mass-transport properties). With such an approach, it is critical that the diameter of the MOF crystals be much smaller than the diameter of the polymer fibers formed during electrospinning; otherwise, the MOF domains cannot be integrated inside the composite fibers. The ORR activity and PEMFC performance obtained with such advanced synthesis has been high, but not significantly higher than those obtained with less complicated separate polymer or MOF approaches, thus far. The electrospinning approach can also provide a pathway for the preparation of self-standing Fe–N–C electrodes, when the pyrolysis of the Fe, nitrogen-doped polymer fiber mat retains mechanical integrity after pyrolysis. However, due to the void volume between fibers with electrospinning, the macropore volume in such self-standing electrodes is usually exceedingly high. This leads to overly thick and porous cathode layer for a given Fe–N–C loading, which is beneficial for O_2 transport in the gas phase, but detrimental for the transport of protons.

13.2.5 Synthesis of Metal–N–C SACs Beyond Fe and Co

Without or with only a minor modification in the synthesis, it is most often possible to prepare Mn-, Ni-, Cu-, and Zn–N–C SACs with similar metal–N_x coordination as found in Fe- and Co–N–C SACs [32, 87, 88]. Ni- and Cu–N–C SACs typically have low ORR activity (comparable to N–C) but have been shown to have high selectivity and activity for CO_2RR, as reported from 2015 in [89] (see Section 13.4.2). A Cr–N–C SAC was reported in 2019, with promising ORR activity in acid medium [90]. One important application issue is the catalysis of Fenton reactions between *in situ* formed hydrogen peroxide (H_2O_2) during ORR (especially, in acidic medium of PEMFCs) and 3d transition metals, leading to autocatalytic degradation or degradation of other components in the fuel cell, such as the membrane or the ionomer in the cathode. From this perspective, Fe is regarded (perhaps misleadingly) as the worst choice of 3d metal. The replacement of Fe by Co, Mn, or Cr in metal–N_x sites might therefore be appealing for improved *operando* stability at the cathode of PEMFC. However, all 3d transition metals have some Fenton reactivity, and among metal–N–C materials, Fe is the most active for ORR, due to the highest TOF of Fe–N_4 sites. In 2020, the first highly active metal–N–C material based on a p-block metal, namely Sn, was reported, with Sn–N_x sites having similar TOF for ORR in acid medium than Fe–N_x sites [91]. However, the material contained also SnO_x nanoclusters in addition to Sn–N_x sites, showing the challenge of selectively forming Sn–N_x sites. Interestingly, the latter have high catalytic activity and selectivity toward CO_2 reduction, leading to a different CO_2RR product than with Ni–N–C and Fe–N–C, as discussed in Section 13.4.2. Zn–N_x sites in Zn–N–C have generally no or low reactivity, but Zn–N–C SACs with high density of Zn–N_x sites can be utilized as a platform for preparing, e.g. Fe–N–C SACs with a high density of sites as discussed earlier. The synthesis of metal–N–C SACs involving PGM has also been reported recently with various synthetic routes but that are similar to those established previously for Fe–N–C, e.g. for Pt [92], rhodium [93]. Pt–N_x sites in Pt–N–C have been found to efficiently catalyze the HER, with the advantage of increased utilization of expensive PGM, compared to PGM NPs, where less than 100% of the metal is on the surface. In some studies, Pt_{SA}

were prepared on nitrogen-doped carbon supports by simple adsorption, without pyrolysis, and high HER activity was also observed [94, 95]. The nitrogen doping of the carbon was shown to be the key for retaining the atomic dispersion in *operando* conditions, with a reference Pt SAC on a nitrogen-free carbon leading to clusters and NP formation during HER [94]. Thus, depending on the strength of the metal—N bond (in turn possibly related to the radius of the metal cations, since the N_4 cavity in the carbon matrix must have a certain size, with limited flexibility) and the electrochemical operating conditions, subjecting the metal to a high-temperature treatment is not always necessary for imparting activity and stability to SAs. Higher HER activity after high-temperature treatment was however reported for Pt–N–C in Ref. [92]

13.2.6 Synthesis of Metal–S–C SACs

While nitrogen is perhaps the strongest ligand for metal cations, other light elements can be targeted to stabilize metal SAs in conductive carbon matrices, such as S, P, and O. Several recent reports have demonstrated the pyrolytic synthesis of metal–S–C materials, with metal–S_x sites. For example, Choi et al. reported in 2016 on the synthesis of a Pt–S–C material, comprising c. 5 wt% of Pt atoms, all being present as Pt–S_4 SA site [96]. A S-doped carbon was first prepared with a hard-template method, resorting to a zeolite and depositing carbon on it via a CVD approach in acetylene/hydrogen sulfide at 550 °C. The S–C material, featuring 17 wt% S, was then wet impregnated with a Pt salt, and reduced in H_2 at 250 °C (Pt–S–C). No Pt NsP was observed in Pt–S–C, but atomically dispersed Pt, coordinated by 4 S atoms according to extended X-ray absorption fine structure (EXAFS) analysis. The Pt–S–C material catalyzed the ORR to selectively form H_2O_2 (96%) in stark contrast to metallic Pt that mainly catalyzes the ORR to form water (see more discussion on two-electron ORR electrocatalysis by carbon-embedded SACs in Section 13.4.1). Another Pt–S–C material with Pt–S_4 sites was reported in 2019 [97], prepared via silica hard-templating method but in an otherwise similar approach than in Ref. [96]. Increasing the annealing temperature to 700 °C resulted in the conversion of Pt–S_4 sites into Pt clusters of c. 1.5 nm, and resulted in markedly different HER mass activity at an overpotential of 20 mV vs. RHE in 0.5 M H_2SO_4.

13.3 Characterization Methods and Structure

13.3.1 Structure of metal SA Sites

The structure of metal SA sites in the metal–N–C electrocatalysts bears similarities with that in the metal–N_4 macrocyclic complexes, as discussed in Section 13.2, and the main trends between the nature of the metal cation and the activity of metal–N–C SACs for key electrochemical reactions, such as the ORR (see Section 13.4), are also similar to those observed with unpyrolyzed metal macrocycles. The pyrolytic processes explored to synthesize the metal–N–Cs aimed to design electrocatalysts with higher stability and activity than their molecular counterparts, consisting of a graphitic structure doped with metal–N_x moieties and, usually, also doped with additional nitrogen atoms that are not involved in metal ligation. In this section, we review the structural properties of the metal–N–C electrocatalysts from atomic to macroscopic scale, and, specifically: (i) the nature of the different moieties, namely the metal–N or nitrogen functionalities observed within the carbon structure and (ii) the carbon structural disorder and its effect on the electron density near the atomically dispersed metal. The pyrolysis leading to the formation of the metal–N–C electrocatalysts incidentally generates a wide variety of different moieties, which can be divided into two main categories: (i) the metal-free moieties and (ii) the metal-containing moieties. Metallic or other metal-rich NPs are also observed in electrocatalysts synthesized with a high metal loading or with unoptimized synthesis approach. They encompass a wide range of carbides, nitrides, and metallic phases, which are often detrimental for the electrochemical performances and stability for the ORR [73, 81], especially in acidic environment [98, 99], or detrimental to the CO_2RR selectivity [100]. The metal-free moieties consist in diverse nitrogen-based structures, namely *N*-pyridinic, *N*-pyrrolic, *N*-graphitic, and N-protonated[101] whose exact contribution to the electrocatalytic activity of the metal–N–Cs remains highly debated, and probably depends on the pH (for ORR), and on the relative amount of N-moieties and metal–N_x moieties in each metal–N–C material.

It is widely accepted, however, that the metal-based moieties are often the main contributor to the metal–N–C's performances, in the case of optimized materials with relatively high amount of metal–N_x SA sites. As such, they are the main focus of this discussion. More specifically, the discussion is to focus on Fe–N_x moieties, owing to the fact that ^{57}Fe Mössbauer spectroscopy has been applied extensively to Fe–N–C materials *ex situ*, and even *in situ* or *operando* recently. This technique is particularly well suited to identify the oxidation state and spin state of atomically dispersed Fe, and combined with XAS or HR-STEM, makes the understanding of the Fe–N–C active site structure more advanced than for other metal–N–C SACs for which Mössbauer spectroscopy is typically not achievable or not easily accessible. However, other metal–N–C were studied and some of the structural conclusions discussed in this section can be extended to them (e.g. Co–N–C and Sn–N–C). For example, Co–N–C SACs have been studied instead with XAS, resorting to the calculation of X-ray absorption near-edge spectroscopy (XANES) spectra for a library of Co–N_x sites, and comparing to the experimental XANES spectrum collected on all Co atoms in the material [102].

13.3.1.1 Different Fe–N_xC_y Moieties

For Fe-N-C, the Fe–N_x moieties have been hypothesized to exist under various forms, e.g. (i) Fe–N_4 [103] (illustrated in Figure 13.1a), (ii) Fe–N_{2+2} [104], and (iii) N–Fe–N_{2+2} [104]. Furthermore, the Fe coordination is believed to change with its position in the carbon basal plane, leading to the presence of Fe–N_2 (on the carbon edge), Fe–N_3 [105], or generically Fe–N_xC_y sites (e.g. Fe coordinated to x nitrogen atoms in the first coordination sphere, and y carbon atoms in the second coordination sphere). N–Fe–N_{2+2} and Fe–N_{2+2} are often assumed to be found as "bridges" between two graphite crystallites defining micropores (pore size < 2 nm) [104, 106], whereas the Fe–N_4 exists in carbon basal planes and, thus, likely, in mesopores (2 nm < mesopores < 50 nm). Their different coordination with nitrogen directly impacts their electronic structure and the spin state of Fe. For example, it was discussed that Fe–N_x moieties with different orbital filling can present either an empty d_{z2}, a filled d_{z2}, or a half-filled d_{z2} orbital [106]. This impacts the binding strength and conformation of various reactive intermediates on their surface. With a filled d_{z2} orbital, the Fe cation in Fe–N_x cannot bind with oxygen in a non-bent configuration, hence implying a weaker binding than Fe–N_x site with emptied d_{z2} orbital, able to form a σ-bond with O_2. When addressing the Fe–N_x sites, their local coordination, and therefore the nature (pyridinic vs. pyrrolic) of the nitrogen atoms bound to Fe, was experimentally investigated through the comparison of calculated and measured XANES spectra, for an Fe–N–C material that was independently demonstrated by EXAFS and Mössbauer to contain no Fe clusters. The coordination was evidenced to be mainly a near-planar Fe–N_4–C_{12} on an Fe–N–C synthesized from ZIF-8, i.e. an Fe coordinated with four *N*-pyrrolic [73].

Here, it is important to note that the label of (N–)Fe–N_{2+2} moieties was previously proposed and used to reflect the fact that these sites are hosted in micropores, and therefore probably involving carbon atoms from two different graphitic crystallites. This is as opposed to in-plane Fe–N_4 site, with all carbon atoms belonging to the same crystallite. However, by introducing structural disorder in a graphene sheet (including five-carbon atom rings), it is also possible to produce a variety of Fe–N_x moieties, from the "perfect defect" corresponding to two-carbon defects (for creating space for the metal cation) in a perfect graphene sheet and the replacement of four carbon atoms by nitrogen atoms (labeled usually as FeN_4C_{10} site), to a highly defective graphene sheets in which an FeN_4C_{12} site can be created, with structure of its two first coordination spheres very similar to that existing in Fe porphyrin. We also highlight that the terminology of pyrrolic and pyridinic nitrogen in the field of metal–N–C is not free of ambiguity. Strictly speaking, this is derived from the chemical state of nitrogen in pyridine and in pyrrole molecules. In the field of metal–N–C, the terminology of pyridinic and pyrrolic is usually referred to on the basis of X-ray photoelectron spectroscopy (XPS) N1s deconvolution. Although, originally, the binding energies of said molecules were used for deconvolution of the metal–N–C N1s envelope, recent works have been highlighting the need to consider a binding energy shift when assessing the N-containing moieties by XPS [107, 108]. This "core-level-shift" is especially important when discussing the metal—N bond, and leads to shifts in binding energy >1 eV, but also impacts the binding energy of nitrogen depending on its position in the carbon basal plane, e.g. edge vs. plan. In the more recent XANES work applied to Fe–N–C, however, the terminology of pyrrolic or pyridinic is more related to the way the nitrogen atoms are embedded in the carbon matrix (as this is the information provided by XANES by Zitolo et al. [73]), and in particular, if it is involved in six- or five-membered rings of light elements. In this context, the definition of pyrrolic/pyridinic is therefore geometrical, and not chemical, by nature.

13.3.1.2 Macroscopic Structure

As a result of its embedded nature in the basal plane or on the edges of graphitic crystallites, the reactivity of Fe–N_x moieties is not only related to the local structure in the first 1–3 coordination spheres, but also highly dependent of the long-range carbon neighbors, including the graphitization degree of the carbon matrix, the presence of structural defects (e.g. missing atoms, nitrogen-containing moieties) but also its acido-basicity and electron delocalization. Some of these long-range parameters of a carbon material surface are intertwined (see Section 13.5.4). The overall morphology of a metal–N–C catalyst, i.e. size of graphitic crystallites, micro/mesoporous pore size distribution, functionalization with nitrogen and/or oxygen (oxygen is always present on surface, after exposure to air), depends on the synthesis procedure, as detailed in Section 13.2. For example, a metal–N–C synthesized from a hard-templated method, using silica spheres >20 nm, will certainly result in mesoporous or micro-mesoporous metal–N–C. In contrast, a metal–N–C synthesized from a MOF by sacrificial approach may show only/mostly micropores [59]. However, the materials that comprise only or mostly metal–N_x sites (no metal particles) are always carbon-rich and therefore share morphological properties similar to other carbon supports, i.e. several levels of porous networks (micro-, meso-, macroporous) that together lead to the formation of a hierarchical structure. The variations in porosities and their distribution can impact the proportion of different Fe–N_x actives sites, i.e. the soft-templated Fe–N–C materials (from ZIF-8) are believed to exhibit an enhanced presence of micropore-hosted sites vs. hard-templated Fe–N–C electrocatalysts. A clear interconnection between the type of Fe–N_x sites and the carbon matrix morphology is however not yet established. Developing methods that can differentiate a basal graphite surface from surfaces created by the edges in complex carbon materials could help progressing on this topic. In addition, N_2 sorption isotherms, while efficient at capturing the micro- and mesopores, are not suited to assess the macropores in catalyst powders and in catalytic layers. Macropores are often the result of the electrode preparation method, and ink-drying process, and key to gas-phase transport in catalytic layers.

It is believed that the oxidation degree and the out-of vs. in-plane configuration of the Fe–N_4 moieties impact their reactivity, whether it is for the CO_2RR to CH_4, with the Fe(II) → Fe(I) hypothetical transition at very low potential, or the ORR, with the *out-of-plane* Fe(II) → *in-plane* Fe(III) transition [109, 110]. The potential at which the Fe(III)/Fe(II) transition occurs for the ORR is correlated to how delocalized the π-electron are in the carbon basal plane, which in turn depends on the carbon structural disorder [111].

13.3.1.3 Importance of the Carbon Surface and π-Electron Delocalization

The structural order of the carbon matrix takes various aspects, which impact the π-electron delocalization on two levels (i) the crystallite size, i.e. smaller crystallites, and thus more disordered C, result in a disruption of the π-electron delocalization, similar to (ii) the presence of heteroatoms (e.g. nitrogen groups, Fe–N_x, or oxygen groups) and other structural defects (holes, edges sites, etc.), as illustrated in Figure 13.2a. A disruption of the π-electron delocalization increases

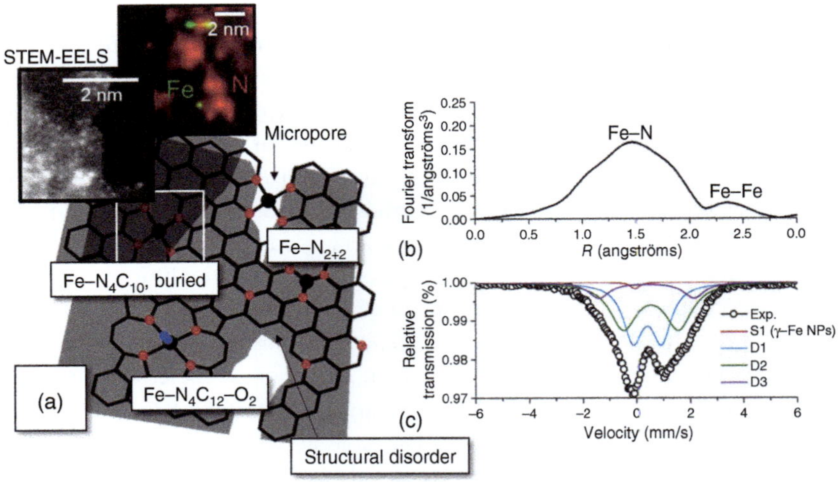

Figure 13.2 Illustration of the Fe-moieties local coordination and of the experimental tools used to assess their existence: (a) graphical representation of a disordered graphene plane, with an illustration of the various metal-containing moieties observed in metal–nitrogen–carbon electrocatalyst and a micrograph of iron atoms embedded in the carbon structure along with an elemental map distribution (the micrograph and the elemental map do not correspond to the same region); (b) Fourier transform of the extended X-ray absorption fine structure obtained for an Fe–N–C electrocatalyst, with the main features, and (c) Mössbauer experimental data and fitting with a singlet component (corresponding to γ-Fe particles) and three doublets corresponding to different Fe–N$_x$ SMA sites. Source: Tristan Asset, Frédéric Maillard, Frédéric Jaouen.

the electron-withdrawing nature of the graphene plane, thus affecting the electron density at Fe–N$_x$ sites and their TOF and selectivity. Ramaswamy et al. showed that the initial ORR activity of Fe–N–C catalysts prepared from the pyrolysis of an Fe-porphyrin and vastly different carbon black supports (from low to high surface area, and from highly ordered to highly disordered) could be interpreted in terms of modified π-electron delocalization around Fe–N$_x$ sites [111]. Choi et al. showed, with a single Fe–N–C catalyst, that introducing numerous oxygen functional groups on the carbon surface with different H_2O_2 treatments resulted in an ORR activity decrease that correlated with the increased acidity of the surface and increased work function, i.e. decreased π-electron delocalization (see also Section 13.5.4) [112].

13.3.2 Characterization Methods Dedicated to Metal–N–C SACs

One of the key challenges encountered when working with SACs is that traditional physico-chemical characterization techniques such as X-ray diffraction (XRD), scanning electron microscopy combined with energy dispersive X-ray spectroscopy (SEM/X-EDS), accessible in many laboratories, do not have sufficient precision/resolution to distinguish if atoms are dispersed in the atomic state or not. In fact, only a certain number of selected characterization techniques are suited to unveil the presence of metal SA sites. In what follows, we restrict ourselves to these must-have characterization techniques, describe their principle, the information they allow to reach and discuss representative papers, which used them. The most straightforward technique to examine SACs with a resolution of the order of the Ångström is scanning transmission electron microscopy (STEM). In this technique, an ultra-small electron probe is swept over the sample, and the transmitted electrons are collected by a high angle annular dark field detector (HAADF). Correction of spherical and chromatic aberrations (aberration-corrected scanning transmission electron microscopy, AC-STEM) significantly improves the resolution of traditional STEMs, providing sub-Ångström resolution well adapted to SACs. Combination with electron energy-loss spectroscopy (EELS) provides additional insights into chemical environment. For example, STEM-EELS allows visualizing how transition metal atoms are distributed in the carbon matrix (either as SAs or under the form or metal/metal carbides/metal oxide NPs), and distinguishing whether nitrogen atoms are located in their vicinity [113]. This is illustrated in Figure 13.2a, showing Fe_{SA} dispersed on a carbon matrix and demonstrating the presence of nitrogen atoms in their vicinity.

Complementary to AC-STEM, XAS is an element-specific characterization technique providing insights into the oxidation state of metal SAs and into their coordination [73, 84, 102, 114]. While the XAS technique is only accessible in synchrotron facilities (except for the recent bench-top equipment, but not suited to look at dilute samples such as metal–N–C), the developed EXAFS analysis is easily accessible and allows to routinely demonstrate if a SAC has been synthesized, or if a metal–N–C SAC comprises mix of SAs and particles, or mostly metal particles. Practically speaking, SAC powder or a pellet of powder is illuminated with a monochromatic beam of X-ray photons either *ex situ* [73, 115], *in situ* [116, 117], or *operando* [102, 118]. Absorption of X-rays by the transition metal atoms leads to the ejection of photoelectrons and formation of excited core-hole states. For X-ray photons energies near the absorption edge (E_0, referred to as the XANES region), backscattering of the excited photoelectrons by the surrounding atoms provides information on the oxidation state, the orbital occupancy, and the local coordination of the transition metal atoms. For energies well above the absorption edge (i.e. starting at $E_0 + 100\,eV$), the EXAFS region is a fingerprint of the nature and number of atoms present in the first coordination spheres (i.e. from 1 to 5–6 Å, see Figure 13.2b). Typically, Fourier transform extended X-ray absorption fine structure (FT-EXAFS) can identify a peak at c. 1.5 Å, related to metal–N coordination, while peaks due to Fe–Fe backscattering can be expected at c. 1.2 and 4.5 Å, if metallic Fe particles are present (similar atomic distances are assessed for different metal atoms such as Co, Sn, Mn. We detail here the case of Fe–N–C for the sake of clarity). Fitting of the FT-EXAFS of pure SACs can provide more information, such as exact distance, and the average coordination number of Fe with its first neighbors, etc. While in the case of SACs free of metal clusters, the FT-EXAFS analysis and output is relatively facile to understand, for SAsC containing even a small fraction of metal clusters, the output becomes quickly complex. This is because the EXAFS signal is averaged on all Fe atoms present, and in the case of mix $Fe–N_x$ sites and Fe particles for example, the coordination number found by fitting for Fe–N, for example (and same for Fe–Fe) loses its physical meaning at the single-site level. For example, a theoretical mix of 50% Fe as Fe–N_4 sites and 50% Fe as Fe clusters would result in an Fe–N apparent average coordination number by FT-EXAFS of 2.0, even though there is no single site Fe–N_2 present in the material. Therefore, detailed structural characterization of SA sites in SACs is preferably done in the absence of any metal clusters.

Complementary to experimental EXAFS and XANES as well as calculated EXAFS (leading to the knowledge of the type, number, and bond distance between the absorber (Fe) and surrounding atoms), the recent development of methods to calculate XANES spectra from optimized density functional theory (DFT) structures and to compare them to experimental spectra has provided further information on the site geometry of SAs, since XANES is more sensitive than EXAFS to the spatial distribution of surrounding atoms. The combined application of calculated XANES spectra for optimized hypothetical Fe–N_x structures and comparison to XANES experimental spectrum of a model Fe–N–C material free of Fe clusters leading to pioneering advances as the work of Zitolo et al. evidenced the porphyrinic planar architecture with an FeN_4C_{12} core and with O_2 adsorbed in the side-on or end-on configuration of metal-containing moieties observed in Fe–N–C, as discussed previously [73, 102, 108]. Complementary to XAS is X-ray emission spectroscopy (XES). This technique is based on the measurement of fluorescence photons emitted upon transfer of electrons from an outer shell to the electron hole, and renders changes in the oxidation state and spin state observable. Saveleva et al. reported recently that XES is capable of tracking *in situ* the changes of oxidation state and spin state of single Fe atoms in ORR conditions [119].

With some similarity to the XANES principle, ^{57}Fe Mössbauer spectroscopy is based on the recoilless absorption of γ-rays by ^{57}Fe atoms. It does not require any crystalline order of the Fe atoms and is therefore well suited for Fe–N–C SACs [120]. The technique is extremely sensitive to tiny changes in the energy levels of the Fe nucleus, implied by the local environment around each Fe atom, changing the oxidation state, spin state, and the electric field gradient around Fe. In addition, ^{57}Fe Mössbauer spectroscopy is powerful at distinguishing crystalline phases of Fe from Fe single-atom sites, since most Fe crystalline phases in Fe–N–C materials have magnetic properties, resulting in sextet signals in the Mössbauer spectra (e.g. Fe carbides, Fe oxides, and α-Fe). In contrast, Fe-based SA sites have no magnetic properties due to their dispersed nature, resulting in a doublet signature in Mössbauer spectra. This key difference allows an easy quantification of Fe_{SA} and Fe clusters, for most Fe–N–C materials, by fitting the Mössbauer spectra with a set of doublets and sextets. In addition, a singlet component for the γ-Fe metallic phase is sometimes necessary. This metastable phase of metallic Fe is often found in Fe–N–C catalysts comprising Fe clusters, due to the rapid quenching from high temperature (where γ-Fe is more stable than α-Fe) at the end of pyrolysis.

In the field of metal–N–C materials, Mössbauer spectroscopy has most often been used to study monometallic Fe–N–C catalysts [20, 38, 73, 120], as illustrated in Figure 13.2c, but has recently been used for Sn–N–C catalysts comprising Sn atomically dispersed in a nitrogen-doped carbon matrix. Mössbauer spectra are on the first hand used to assess the relative amounts of Fe crystalline phases and of SA sites [102]. From *ex situ* Mössbauer spectroscopy on many Fe–N–C SACs, two main doublets (most often labeled D1 and D2 in the literature) have been ubiquitously observed, differing mainly in their quadrupole splitting (QS, represented by the peak to peak "distance" of a given doublet). These doublets have initially been interpreted by comparison to doublets experimentally measured for unpyrolyzed FeN_4 macrocycles with known oxidation and spin states. Recently, DFT methods have been developed to calculate the QS and the isomer shift (IS) of different Fe–N_x hypothetical sites embedded in (disordered) graphene sheets, which rationalizes the interpretation of Mössbauer spectra of pyrolyzed Fe–N–C SACs [20, 121, 122]. The DFT results nevertheless showed the ambiguity of the spectroscopic response. For example, three different sites (namely Fe(II)N_4C_{12} site with spin state 0 or 1 but also Fe(III)N_4C_{12} with spin state 5/2 or Fe(III)N_4C_{10} with spin 5/2) all lead to QS-values around 1 mm/s, compatible with the experimental D1 [20]. In contrast, only the QS-values calculated for the Fe(II)N_4C_{10} (in spin state either 0 or 1) moiety matched the experimentally observed value of D2 (2.4–3.0 mm/s) [20]. Generally, Mössbauer spectroscopy must therefore be combined with other techniques probing the oxidation and/or spin state of Fe to safely conclude on the different doublets. While *ex situ* Mössbauer spectroscopy has been applied for a long time to Fe–N–C SACs, *in situ* or *post mortem* investigations have been almost inexistent until 2019, due to the high dilution of Fe (typically, less than 2 wt% Fe as atomic site) and the low percentage of the ^{57}Fe isotope (Mössbauer-active) in natural Fe, making such experiments challenging, even with ^{57}Fe enriched Fe–N–C.

In 2021, *in situ* Mössbauer spectroscopy in a PEMFC allowed revealing the potential-dependent nature of D1 (switching between Fe(III) and Fe(II) states) and potential-independent nature of D2 (remaining Fe(II)) [123]. Such measurements are however also challenged by the low stability of Fe–N–C materials in acidic medium, with a fraction of the D1 sites quickly converting to nanosized Fe oxides. Due to their size, such oxides are superparamagnetic, and contribute with a doublet signal that is very similar to the true D1 signal. The nanosized Fe oxides can be separated from the true D1 signal only at low temperature (typically below −253 °C, i.e. far away from electrochemical conditions), since at such low temperature they contribute with a sextet signal, as large Fe oxides do. This observation conducts us to mention the promising application of Mössbauer spectroscopy to characterize Fe–N–C electrodes after electrochemical operation in PEMFC or other devices, in which case the low-temperature measurement is critical to distinguish nano-Fe oxides from a true D1 signal related to Fe–N_x sites [123]. First reports have shown that the D2 signal is stable in operating PEMFC, while D1 quickly converts to nano-Fe oxides. In this frame, *post mortem* low-temperature Mössbauer spectroscopy proved much more efficient than XANES-EXAFS at quantifying the contribution of D2, D1, and nano-Fe-oxides [123]. The reason is that nano-Fe-oxides have a XANES spectrum very similar to that of Fe–N_x sites, and the Fe–Fe signal from amorphous Fe-oxides is poorly visible in EXAFS.

Complementary to Mössbauer spectroscopy, nuclear inelastic scattering (NIS) consists of measuring the energy spectrum of inelastic absorption of X-rays, from a synchrotron beam, by nuclei. A signal of elastic nuclear absorption arises, when the energies of incident X-rays and nuclear transition coincide. Nuclear absorption can also occur inelastically with creation or annihilation of lattice vibrations. The latter processes cause inelastic sidebands in the energy spectra around the elastic peak, which lead to the partial density of phonon states. NIS provides information of the energy spectra of the recoil, i.e. the frequency distribution of the lattice vibrations and allows the identification of modes which are not available otherwise. By combining Mössbauer spectroscopy, NIS and DFT calculations, Kneebone et al. identified the Fe sites that are ORR active in a PANI-based Fe–N–C catalyst [124]. Wagner et al. recently used NIS to detect Fe oxide species on the surface of an Fe–N–C material [125].

Except for *in situ* Mössbauer spectroscopy, which can be surface-specific, but is challenging and definitely not a routine method, and *in situ* EXAFS-XANES, other techniques have been developed to probe the electrochemically active sites in Fe–N–C and other metal–N–C SACs. By definition, both Mössbauer spectroscopy and XAS are probing the bulk of the material, due to the high energy (c. 7 keV for XAS at Fe *K* edge and 14.4 keV for ^{57}Fe Mössbauer). To tackle this shortcoming, cyclic voltammetry, a technique in which the electrode potential is swept linearly with time between two potential limits, is of prime interest. Kucernak's group [126] first reported an electrochemical method, based on a sodium nitrite probe molecule ($NaNO_2$) that can be strongly adsorbed on SAs and electrochemically removed, via electroreduction, from the electrocatalyst surface thus allowing estimation of the density of metal SAs located on the surface (ASD). Unfortunately, $NaNO_2$ molecules indifferently probe single-metal Fe atoms and Fe-oxides/Fe-carbide species [127]. An *ex situ* chemisorption/desorption-based method, using low-temperature (−100 °C) CO chemisorption [128], was recently reported by Strasser's group to determine the site density of SACs based on Fe or Mn atoms. Low temperature was in this case necessary, since CO does not adsorb strongly on Fe–N_x or Mn–N_x sites at room temperature (explaining why such sites are tolerant to CO in electrochemical conditions, in contrast to Pt). Another recent quantification method using a cyanide anion as a probe molecule was published by Choi's group [129]. The method proved valid to determine the active SD of various transition/noble metal SACs beyond Fe and Mn, and also in a broad pH range (1–13).

13.4 Applications in Electrocatalysis

13.4.1 Oxygen Reduction Reaction

The ORR has been the first electrochemical reaction investigated for the application of metal–N–C SAsC, and of Fe–N–C SACs in particular. The ORR can lead to the formation of water (four-electron ($4e^-$) ORR) or H_2O_2 (two-electron ($2e^-$) ORR) depending on the catalytic sites, electrode morphology, and experimental conditions. In fuel cells, water is the desired ORR product, not only to maximize the current density per flux of O_2 diffusing inside the electrode, but also to prevent undesired reactions between H_2O_2 and the catalyst, ionomer, or membrane

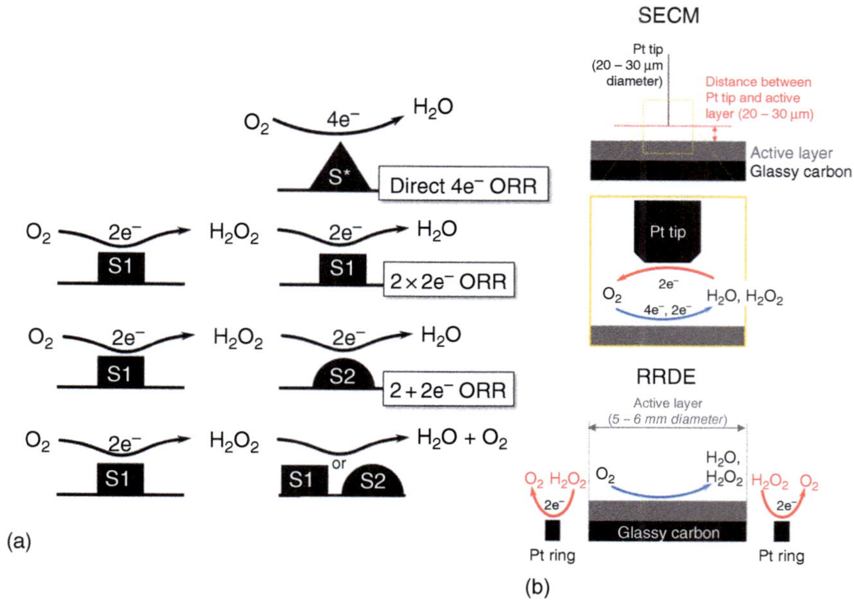

Figure 13.3 Illustration of the mechanistic and instrumental concepts involved in the study of the oxygen reduction reaction on metal–nitrogen–carbon electrocatalysts. (a) Direct and indirect ORR mechanisms and (b) scheme of SECM and RRDE techniques that can measure the selectivity of electrocatalysts during ORR. Source: Tristan Asset, Frédéric Maillard, Frédéric Jaouen.

materials. H_2O_2 is a strong oxidant, which can chemically react with transition metals (in particular, Fe), forming even more oxidizing reactive oxygen species, such as °OH˙ radicals. In a fuel cell cathode, the presence of multitudinous active sites often leads to the co-existence of different paths for the ORR, including direct $4e^-$ ORR, $2e^-$ ORR (releasing desorbed H_2O_2), and indirect $4e^-$ ORR, implying the formation of H_2O_2 on one active site (first site), its desorption, and re-adsorption on another (second) site where H_2O_2 is electrochemically reduced to water. When the nature of the second active site is the same as that of the first site, the ORR pathway is described as being $2e^- \times 2e^-$, while when the first and second sites differ in nature, the pathway is described as being $2e^- + 2e^-$ [109] (see Figure 13.3a for an illustration of the pathways).

In practice, identifying whether the ORR pathway on a given catalyst (always containing a high number of active sites) is $4e^-$, $2e^- + 2e^-$, or $2e^- \times 2e^-$ is very challenging. In addition, $2e^-$ ORR followed by the non-electrochemical H_2O_2 disproportionation into O_2 and H_2O, can, after several "cycles," lead to an apparent low production of H_2O_2 during ORR. Answering the question whether a given catalyst mainly catalyzes the ORR through $4e^-$, $2e^- + 2e^-$, $2e^- \times 2e^-$, or $2e^-$ + disproportionation reaction bears not only fundamental importance, but also practical ones. Even if a low apparent amount of peroxide is produced initially, the existence of desorbed "free" H_2O_2 inside the cathode with all indirect ORR mechanisms can lead to rapid degradation of performance over time. For example, Tylus et al. investigated the ORR pathway on three different Fe–N–C catalysts (with only SA Fe–N_x sites, or a mix of Fe–N_x sites and Fe particles) both in acid and alkaline media [109]. They concluded that, in alkaline medium, Fe–N_x sites can catalyze the direct 4e$^-$ ORR and the 2e$^-$ × 2e$^-$ indirect pathway, while in acid, the role of Fe particles (Fe oxides, or metallic Fe or Fe carbide encapsulated in a carbon shell) in catalyzing H_2O_2 electroreduction to water was invoked for explaining the rotating ring disk electrode (RRDE) experimental data. Other studies however reported low % peroxide during ORR in acid media for Fe–N–C catalysts free of Fe particles (as measured *ex situ* [130]). To experimentally differentiate direct 4e$^-$ ORR from indirect ORR mechanisms, it is important to keep in mind that the catalytic layer should be made as thin as possible, to quickly remove from the layer any peroxide that is produced, to prevent further reactions and to assess the true ORR mechanism. While this effect is generally valid for all ORR catalysts, including Pt/C catalysts [131], the effect was shown to be particularly strong for a particular Fe–N–C catalyst for ORR in acid medium in 2008 [132], with the percentage H_2O_2 measured by RRDE increasing from 5% to 95% when the loading was decreased from 800 to 50 $\mu g/cm^2_{geo}$.

The role of the nature of the metal cation in defining the ORR selectivity of metal–N_x sites has been regularly investigated, first on metal–N–C catalysts with undefined metal speciation, or comprising a mix of metal particles and metal–N_x sites. Recent works on metal–N–C materials comprising only or mostly metal–N_x sites confirm the previous reported trends, with generally high 4e$^-$ selectivity for Fe–N–C and high 2e$^-$-selectivity for Co–N–C. The low selectivity generally observed of Co–N_x sites for 4e$^-$ ORR is in line with the experience gained in ORR electrocatalysis by unpyrolyzed Co–N_4 macrocycles, where the adsorption of OOH on Co(II) is weak relative to that of OOH_{ads} intermediate on Fe(II). Beside the classical RRDE method, other methods have been developed to more locally probe tiny amounts of peroxide produced on surfaces or thin active films during ORR, such as the scanning electrochemical microscopy (SECM) [133, 134]. In this approach, the position and electrochemical potential of a Pt tip (diameter of typically 20–30 μm) situated directly above the ORR active surface is controlled (Figure 13.3b). This SECM setup allows a shorter distance (typically 20–30 μm as well) between the ORR active surface and the Pt tip, where H_2O_2 is electro-oxidized to water, than is achievable with RRDE setups [135].

Another way to investigate the ability or not of metal–N–C catalysts to indirectly electrocatalyze the ORR is to study their activity to electroreduce H_2O_2 to water, in the absence of ORR. This can be done in classical RDE setup, with a known concentration of H_2O_2 added into the electrolyte and bubbling with inert gas, to continuously remove O_2 from the system. An early study showed the low activity of Fe–N–C and other metal–N–C (with unknown metal speciation) for H_2O_2 electroreduction, when compared to Pt-based electrocatalysts [136]. In particular, the polarization curve for H_2O_2 electroreduction by all metal–N–C in acid medium shows a linear shape, uncanny for electrocatalysis where exponential behaviors are expected, at least at low current density and in the absence of mass-transport limitation. Fe–N–C was found to be the most active for H_2O_2 electroreduction, followed by Cu–N–C [136]. A recent study investigated the H_2O_2 electroreduction in acid of three

Fe–N–C catalysts with well-identified Fe speciation, involving an Fe–N–C material free of any Fe clusters, an Fe–N–C material with a small amount of metallic Fe (c. 8 % of the signal) along with Fe–N_x sites, and a third Fe–N–C with a large fraction of Fe present as metallic iron and iron carbide (47 rel%) [130]. The results suggested that metallic Fe particles directly exposed to the electrolyte or metal-free N–C surface do not significantly catalyze H_2O_2 electroreduction in acidic medium. In contrast, both Fe–N_x sites and metallic Fe particles wrapped in N–C protective shell seem to be able to catalyze both the direct 4e$^-$ ORR and the H_2O_2 electroreduction.

While 4e$^-$ ORR is desired in the context of fuel cell application, H_2O_2 is a useful chemical product and its selective synthesis via 2e$^-$ ORR is of potential interest to displace the current anthraquinone process used to produce it at scale. Until recently, platinum–mercury (Pt–Hg) and palladium–mercury (Pd–Hg) alloys or amalgams were the state-of-art electrocatalysts to that end, but suffer from high cost and environmental issues related to PGM and Hg, respectively [137, 138]. In a study on metal–N–C materials with different 3d transition metals (Mn, Fe, Co, Ni, and Cu) but with all catalysts comprising only metal–N_x sites, Sun et al. found that Co–N–C results in the optimal balance between high selectivity for 2e$^-$ ORR and low overpotential [118]. While Fe–N–C has a slightly lower overpotential than Co–N–C and other metal–N–C to form peroxide from O_2, it also has a much higher activity to further electroreduce it to water. The latter aspect becomes increasingly a drawback, when the peroxide concentration increases over time, which is desired from an application viewpoint. The initial selectivity of Co–N–C catalysts toward H_2O_2 via 2e$^-$ ORR was further improved by intentionally introducing oxygen functional groups on the top surface [139]. In this approach, the negative effect of H_2O_2 in PEM fuel cells (namely, reducing the TOF and 4e$^-$ ORR selectivity of Fe–N–C and Co–N–C, see Section 13.5) was brought to profit for modifying a Co–N–C catalyst aimed for H_2O_2 electrosynthesis. The high activity of Co–N–C for 2e$^-$ ORR combined with low activity for further electroreduction of H_2O_2 was explained at the atomic level by DFT, predicting an optimal OH-binding energy on Co–N_x sites, strong enough to bind O_2 and OOH but not overly strong, allowing the desorption of H_2O_2 [118] instead of its further electroreduction. These features identify Co–N–C and other doped versions (doped with oxygen, etc.) as promising and affordable catalysts for the electrosynthesis of peroxide [140].

This section is concluded with a short discussion on the deconvolution of the overall ORR activity of a given metal–N–C catalyst into its TOF and SD values, according to:

$$I_m = \mathrm{TOF} \times \mathrm{SD}_m \times F/N_\mathrm{A} \tag{13.1}$$

$$I_v = \mathrm{TOF} \times \mathrm{SD}_v \times F/N_\mathrm{A} \tag{13.2}$$

where I is the ORR activity (mass- or volume-normalized, m or v, respectively) at a given potential E, TOF the TOF at the same potential, SD the mass- or volume-normalized site density, N_A Avogadro's number, and F Faraday's constant [141]. ORR volumetric activity targets were defined for the first time in 2005 for PGM-free catalysts, and deconvoluted into possible combinations of SD and TOF values to reach the same targeted overall ORR activity. Since then, methods have been developed that can quantify the SD, and hence (in combination with ORR activity measurement), the average TOF value of Fe–N–C catalysts [126, 128, 129]. While increasing the number of active sites seems a generally good idea to increase the total activity of an electrocatalyst, one would *a priori* assume the TOF to be relatively independent of the SD value. However, there is recent growing evidence that the TOF value of metal–N_x sites embedded in a nitrogen-doped carbon matrix is intimately related to the physico-chemical properties of the latter. A recent benchmark study on four Fe–N–C catalysts revealed a negative correlation between TOF and SD values, i.e. decreasing TOF with increasing SD values of Fe–N_x sites [59]. The doping of nitrogen-doped carbon with many metal–N_x sites may induce electronic and/or surface changes that reduce the TOF. As a second example of TOF changes, the introduction of oxygen surface groups on top of Fe–N–C after its synthesis has also been shown to decrease the TOF [112] (discussed in detail in Section 13.5). The in-depth understanding of these inter-relations between TOF of metal–N_x sites and the physico-chemical properties of the nitrogen-doped carbon matrix (doped with other non-metal or metal elements) will be critical to combine high SD while retaining a high TOF in metal–N–C materials in the future.

13.4.2 CO_2 Reduction, N_2, and NO_3^- Reduction

The range of reactions that are catalyzed by single-atom metal–N–C electrocatalysts is not limited to the four- or two-electron ORR, but also extends, e.g. to the electrochemical reduction of CO_2 (CO_2RR). CO_2RR is of paramount importance for several reasons: (i) the Industrial Revolution and the resulting dramatic increase in CO_2 in the atmosphere (from 250 ppm in the early 1800s maximum to 390 ppm nowadays [142]) are primarily responsible of the greenhouse effect, thus making CO_2 recapture essential to hinder global warming; (ii) CO_2RR products are numerous, i.e. ranging from simple CO or formic acid C molecules to complex multi-carbon molecules (here referred as C_{2+}) obtained through direct electrochemical reduction or using cascade electrocatalysts (which consist in a combination of different catalysts each catalyzing parts of the CO_2RR that can be not only metallic or carbon-based, but also enzymatic, molecular, etc. [143]), each with their own added value. As such, the CO_2RR has been under growing attention throughout the past decade, and the metal–N–C electrocatalysts were investigated in detail. The CO_2RR is a complex reaction, which is in competition with the HER. Here arises the first advantage of the metal–N–C electrocatalysts: As evidenced by Bagger et al. the metal–N_4 sites are poorly active for the HER, as a result of (i) a different binding conformation (top vs. hollow) compared to extended metallic surfaces, and (ii) a different, less favorable mechanistic pathway (Volmer–Heyrovsky vs. Volmer–Tafel), which overall leads to a 0.3 eV energy penalty for the HER for single-atom sites vs. metallic surfaces [144]. Owing to this depreciated activity for the HER, the metal–N–C electrocatalysts are highly selective for the CO_2RR, but the nature of its products on said electrocatalysts needs to be discussed.

When addressing the CO_2RR on metallic extended surfaces, the CO_2RR products can be categorized in three groups, excluding the elements that are more selective for H_2 generation [145]: (i) CO; (ii) formic acid (if the CO_2 adsorbs through one of its oxygen atoms instead of the carbon atoms [146]) and, on Cu-based electrocatalysts; (iii) C_{2+} (including C_2H_4, ethanol, etc.), although with a low selectivity for a given product. Metal–N–Cs mainly belong to the first category, due to (i) the initial adsorption of CO_2 through its carbon atom and (ii) the fact that the adsorption of two adsorbed CO (CO_{ads}–CO_{ads}) on a single metal–N_4 moiety is highly unfavorable ($\Delta G_{\text{COads}\rightarrow\text{COads–COads}}$ = c. 0.7 eV [147]), thus greatly limiting the metal–N–Cs capability to form a C—C bond and thus generate C_{2+} species. Despite the simple nature of its main product, i.e. CO, performing the CO_2RR on metal–N–C SACs remains of interest: the CO can be combined with H_2 to generate syngas. The latter is a c. 1 : 2 combination of CO and H_2 and an intermediate in the synthesis of Fischer–Tropsch products [148]. It is currently mainly produced from fossil fuels, thus making the CO_2RR on metal–N–Cs electrocatalysts a viable alternative for syngas generation.

First, it is important to limit the scope of this discussion to metal–N–C electrocatalysts, i.e. pyrolyzed materials presenting atomically dispersed metallic atoms surrounded by nitrogen, embedded in the carbon structure, therefore excluding carbon-supported molecular catalysts, chalcogenides, and other non-pyrolyzed materials [149], although the latter also presented relevant performances for the CO_2RR (e.g. a faradaic efficiency >80% toward CO was recently reported by Pan et al. on Co phthalocyanine on nitrogen-doped porous carbon nanospheres [150]). However, they remain limited in terms of catalytic activity vs. the metal–N–Cs, and exhibit a lower stability due to the non-embedded nature of the active sites. It is also important to note that as a result of their synthesis process, the metal–N–Cs do not solely exhibit metal–N_4 sites as their catalytic sites. As discussed in Section 13.3, three potential types of actives sites can coexist in such materials, namely: (i) the metal-free, nitrogen-containing sites (e.g. *N*-pyridinic, *N*-pyrrolic, and *N*-graphitic), (ii) the metallic NPs and clusters, and (iii) the metal–N_x sites. Owing to the focus of this chapter, (i) and (ii) will only be briefly discussed, to mainly focus on metal SA sites. First, the role of metal-free N-moieties to the electrocatalytic activity of metal–N–C electrocatalysts is non-negligible: (i) an entire sub-field of the CO_2RR is exclusively dedicated to the study of nitrogen-doped carbon [151–153], which often exhibit great faradaic efficiency at low overpotentials (e.g. FE_{CO} = 80% on nitrogen-doped carbon nanotubes or graphene); (ii) it was also observed that, on Fe–N–Cs, the activity was increasing with the *N*-pyridinic abundancy, and that those moieties acted as preferential CO_2 chemisorption sites [147]. On the other hand, the presence of metallic NPs/cluster leads to a decrease of the faradaic efficiency toward CO, and catalyze the HER, as recently shown by Huan et al. [100].

The reactivity of the metal–N_x moieties in metal–N–C for the CO_2RR was explored through two main axes: (i) achieving an understanding of the reactivity of metal-containing moieties (metal–N_4, metal–N_{2+2}, etc. some of which are illustrated in Figure 13.4a) toward the CO_2RR through DFT, and (ii) experimental assessment of various metal–N–C electrocatalysts. For metal–N–Cs, the rate-determining step of the CO_2RR is the first electron transfer of the CO_2 reduction in $COOH_{ads}$ ($CO_2 \rightarrow CO_2^-$) [157, 158], and the role played by the nature of the metal element of a metal–N–C electrocatalyst, under its metal–N_4 form, on the free energy transition (ΔG) has been widely studied. Early 3d transition metals (Cr–N_4, Mn–N_4, Fe–N_4, Co–N_4) only require energies ranging from 0.25 to 0.60 eV to achieve the complete step ($CO_2 \rightarrow COOH$), whereas for Ni–N_4 and Cu–N_4 ΔG range up to 1.60 eV. The metal nature also impacts the CO_{ads} binding strength. The 3d early transition metals metal–N_4 moieties act as strong CO_{ads} binders, whereas Ni–N_4, Cu–N_4 and Zn–N_4 spontaneously desorb CO below 0 V vs. RHE [147]. These theoretical considerations predict fairly well the trends observed for the CO_2RR activity and selectivity of various metal–N–Cs electrocatalysts, as shown in Figure 13.4b. For example, Ni–N–C electrocatalysts present a high CO_2RR onset potential, along with up to 90–100% faradaic efficiency for CO generation, which arise from (i) the high free energy for hydrogen adsorption on Ni–N_4, (ii) the high ΔG for the first electron transfer in the CO_2RR, and (iii) the weak binding of the CO_{ads} [88, 154, 159]. On the other hand, Co–N–C shows lower selectivity (20–40%) for CO generation, as a result of its far lower ΔG for the $H^+ \rightarrow H_{ads}$ HER step [88]. The reactivity and

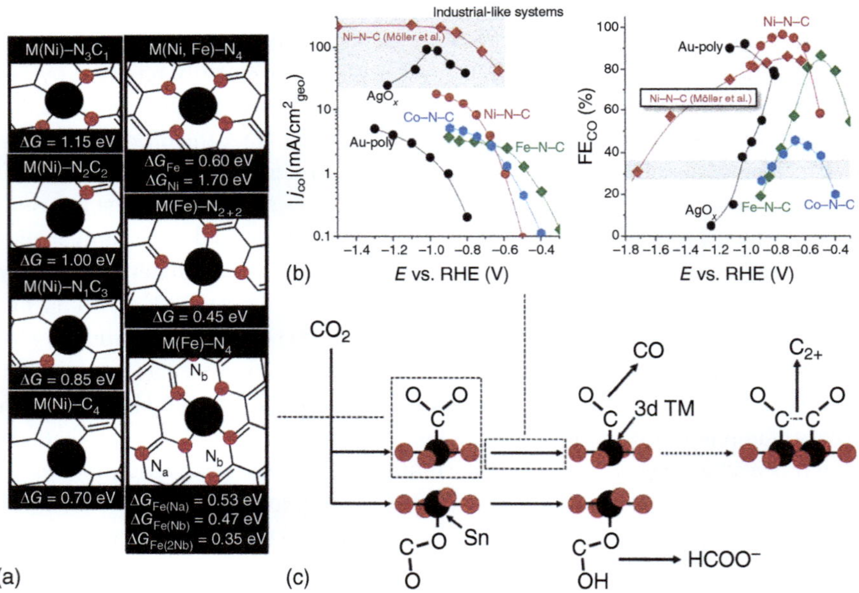

Figure 13.4 Nature of the active sites, activity, and reaction paths for the electrochemical reduction of CO_2 by metal–nitrogen–carbon catalysts. (a) Illustration of various metal–N_x moieties investigated by DFT and the free energy transition for the first proton + electron transfer ($CO_2 \rightarrow COOH_{ads}$); (b) partial current density for CO generation and faradaic efficiency for CO generation for various metal–N–C electrocatalysts originating from the work of Pan et al. [154] and Möller et al. [155], along with polycrystalline Au, Hori et al. [156] and AgO_x, Möller et al. [155] and (c) illustration of the mechanistic pathway on early transition 3d metal–N–C and Sn–N–C for the electrochemical reduction of CO_2, along with an hypothetical pathway in presence of dual-atom moieties. Source: Tristan Asset, Frédéric Maillard, Frédéric Jaouen.

selectivity of the metal-containing moieties also depends on their environment, coordination, and the oxidation degree of the metal. Removing nitrogen atoms from the metal–N_4 structure, or replacing carbon by nitrogen in the neighboring benzene ring was investigated by DFT and was shown to greatly impact the ΔG of the $CO_2 \rightarrow COOH_{ads}$ step [160]. For example, whereas a Ni–N_4 site exhibits a $\Delta G = 1.7\,eV$ for the first electrochemical step, a Ni–C_4 site should only exhibit a $\Delta G = 0.7\,eV$ [155]. The free energy variations were also impacted by other changes in the moiety conformation, e.g. switching toward metal–N_3 sites or transitioning toward bridges sites (metal–N_{2+2}) [161]. Those different structures are summarized in Figure 13.4a. Notably, these structures were studied through DFT. An experimental catalyst designed through the approaches discussed in Section 13.2 is unlikely to exhibit a single, optimal type of metal-containing moieties, i.e. usually several moieties are simultaneously present in a metal–N–C catalyst, such as the metal–N_4 and the metal–N_3 in the basal plane, and the metal–N_{2+2} as "bridge" over small pores (see Section 13.3.1.1). Beyond its impact on the ΔG, the conformation of these moieties modifies the binding strength of given intermediates, e.g. CO_{ads}. The relative abundance of the metal-containing moieties therefore impacts the ΔG for the first electrochemical step, along with the binding energy of the intermediates species and can be tuned by modifying the metal–N–C morphology, e.g. a more microporous electrocatalyst leads to a higher proportion of metal–N_{2+2}, whereas a mesoporous electrocatalyst exhibits a higher proportion of basal planes in the mesopores, and, thus, of metal–N_4. A practical example of the impact of the morphology onto the electrocatalyst selectivity is found in the recent work of Hursàn et al., which evidenced that the faradaic efficiency toward CO_2RR (vs. HER) on nitrogen-doped porous carbon was directly correlated to the pore size [162].

Both from a theoretical and experimental standpoint, metal–N–C electrocatalysts are extremely promising for syngas generation due to the limited number of products, i.e. almost exclusively H_2 and CO, and their high selectivity toward the latter, especially for Ni–N–C. This selectivity can be finely tuned by modifying the nature and chemistry of the main active sites, e.g. the metal–N_x sites, or by adding metallic NPs, along with the porous network of the electrocatalyst, hence allowing the production of syngas with various CO:H_2 ratios. However, such electrocatalysts should not be considered on the sole basis of their expected performances, or the CO_2RR performances obtained in liquid, static, electrolyte, but already considered as *to-be* commercial electrocatalysts and designed in that frame. Recently, notable efforts have been made to transfer this technology toward zero-gap electrolyzer [163] (i.e. systems where the electrocatalysts are in direct contact with the membrane, not separated by an electrolyte layer, static or circulating) or flow cells, leading to performances up to 200 $mA/cm^2{}_{geo}$ on Ni–N–C [155]. Transferring from a liquid to a solid electrolyte induces new design-related challenges, and further optimization of the electrocatalyst pore distribution, hydrophobicity, etc. Thankfully, the wide range of synthesis processes available to achieve metal–N–C electrocatalysts greatly facilitate the tuning of such parameters (see Section 13.2), e.g. use of different MOFs or silica spheres in the soft- and hard-templating methods, respectively; use of different carbon precursors, etc. An important question is now whether the selectivity of metal–N–C electrocatalysts can be modified toward other C_1 species, e.g. CH_4 or HCOOH. CH_4 generation is sometimes observed on metal–N–C, but remains a minor product (<1%), with an enhanced selectivity at $E < -1.1$ V vs. RHE on Fe–N–C, coincidental with a modification of the Fe–N_4 oxidation degree, from Fe(II)–N_4 to a hypothesized Fe(I)–N_x [164]. The formation of C_2 species, such as ethanol (up to 40% at −1.2 V vs. RHE), was reported on a Cu–N–C SAC [165]. However, this unexpected CO_2RR product for SA sites was shown to be related to the rapid *operando* formation of metallic Cu clusters from Cu–N_x sites during CO_2RR. Interestingly, more opportunities exist for HCOOH production with metal–N–C. On extended surfaces or NPs, Sn-based electrocatalysts reach 80% faradaic efficiency for formate generation [166, 167], making them natural candidates for investigating the CO_2RR reactivity of tin in Sn–N–C. Recent work from Zu et al. [168] and Paul et al. [169] focused on the design and utilization of Sn–N–C and reported faradaic efficiency toward formate generation up to 75%, owing to the stabilization of the $CO_2^{\cdot-}$ intermediates on the Sn–N_x moieties. Finally, achieving C_{2+} products on non-NP based electrocatalysts requires the adsorption of CO_{ads} onto neighboring sites to allow the formation of the C—C bond, and, therefore, the design of sites alloying such conformation, e.g. dual-atom sites (Fe_2–N_x, Fe–Co–N_x, etc.). They possibly already exist in minor proportions in metal–N–C but are extremely challenging to characterize, along to synthesize as the sole type of sites of a metal–N–C electrocatalyst. Those different points are illustrated in Figure 13.4c.

The attractiveness of metal–N–C electrocatalysts for electroreduction extends toward other complex reactions, e.g. the nitrogen reduction reaction (N_2RR). The N_2RR aims at the electrochemical production of ammonia (NH_3), at near-room temperature and near-ambient pressure. The latter challenges arise from (i) the energy stored in the N≡N bond (941 kJ/mol) and (ii) the large band gap (10.82 eV) between the highest unoccupied molecular orbital (HUMO) and the lower occupied molecular orbital (LOMO) [170]. This explains the extreme conditions under which it is usually performed, i.e. through the Haber–Bosch process at 300–500 °C and 200–300 bars on Fe-based catalysts. The metal–N–C (M = Fe, Mo, etc.) were investigated for the N_2RR, taking inspiration from both the Haber–Bosch electrocatalysts and the nitrogenase (an enzymatic complex existing in the nature, able to perform the $N_2 \rightarrow NH_3$ through the use of an Fe_4S_4-based complex and a molybdenum-iron protein [171]) and advantage of the low HER activity of metal–N–C electrocatalysts. The choice of those specific elements also arises from the work from Skúlason et al., which identified those elements, along with ruthenium and rhodium as the thermodynamically most promising for N_2RR [172]. This results from their near-optimal N_{ads} binding (leading to an N_2RR thermodynamically favorable at E vs. RHE < −0.5 V). However, the same work underlined the far more favorable H_{ads} adsorption vs. N_{ads} and the subsequent easier nature of the HER vs. N_2RR on a wide variety of extended surfaces. This gap cannot be closed by the 0.3 eV penalty brought by the use of SACs. Hence, despite promising faradaic efficiency for NH_3 generation reported in the literature on metal–N–C electrocatalysts (with a maximum of 55% on Fe–N–C) [173–175], it is inferred that the NH_3 produced by those materials does not originate from the N_2RR, but from other sources. Throughout the past years, a tremendous effort has been carried out to investigate the alternative sources of NH_3 when investigating the electrochemical NH_3 synthesis, not only for metal–N–C, but for all electrocatalysts families, to avoid false positives. In that frame, two principal sources were identified (i) NH_3 contamination from neighboring contaminations, e.g. electrolyte, gas, air, which is usually addressed by using $^{15}N_2$ gas as a reactant (the NH_3 produced by the N_2RR being then $^{15}NH_3$) and (ii) the presence of NO_3 impurity as a minor species in $^{14}N_2$ and $^{15}N_2$ gas sources [176–178]. If the NH_3 generation observed with metal–N–C arises from the NO_3 impurities reduction and not from N_2, this implies that they may exhibit a low activity for the N_2RR, but might be promising for the NO_3RR. This reaction recently encountered a renewed interest [179], as an alternative to the Haber–Bosch process as an NH_3 source and thus act as a natural direction for future investigation of the electrocatalytic properties of metal–N–C electrocatalysts.

13.5 Stability of Metal–N–C Electrocatalysts

As three chemical elements, metal, nitrogen, and carbon, constitute the active sites in metal–N–C electrocatalysts, it can thus be intuitively perceived that any change in the electronic/physical structure of one of these elements can modify the TOF of the active sites or lead to their complete destruction. At least five degradation mechanisms of metal–N–C catalysts have been identified: (i) demetallation, (ii) changes in the chemical and physical nature of the central metal ion, (iii) protonation of the N-groups, (iv) reversible or irreversible corrosion of the carbon matrix (carbon oxidation reaction, COR) and associated phenomena such as (v) migration and aggregation of the SAs (Figure 13.5). The predominant degradation mechanism of SACs in electrocatalysis strongly depends on the nature of the potential range, the temperature, and the nature of the atmosphere, and these experimental conditions will be detailed in what follows.

13.5.1 Demetallation in the Absence of Carbon Oxidation

Demetallation of SACs easily occurs upon corrosion of the carbon matrix [98, 115]. This degradation mechanism will be covered in the following sub-sections, and we focus here on the demetallation of SACs in the absence of COR. Pourbaix diagrams, which represent the domain of stability of different phases of an element as a function of pH, potential range (in Pourbaix diagrams, the potential is given with respect to the standard hydrogen electrode, SHE), are powerful tools to estimate the resistance to demetallation of metal–N–C SACs. In particular, Pourbaix diagrams reveal that acidic pH, high potential, and O_2 are very aggressive stressors for SAC. These conditions are typical of the operating conditions of a PEMFC cathode, in which metal cations (Fe(II)/Fe(III), Co(II), Mn(II) and Cr(III) cations) are the stable phases of Fe, Co, Mn, and Cr. There are two driving forces for demetallation: (i) competition between the metallic ion in the metal–N–C catalyst and protons for the

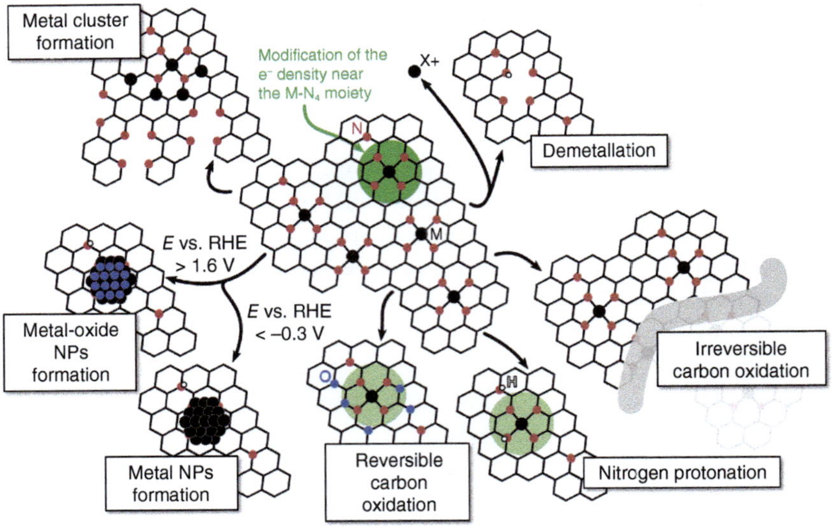

Figure 13.5 Illustration of the main degradation processes occurring during a metal–N–C electrocatalyst operation, i.e. demetallation, reversible or irreversible carbon oxidation reaction, nitrogen protonation, and metal cluster formation. The intensity of the color of the green circles is indicative of the electron density near the M–N_4 moiety. Source: Tristan Asset, Frédéric Maillard, Frédéric Jaouen.

□–N_x (i.e. nitrogen structures where the metal cation is removed), (ii) change in the oxidation state of the metal SA. In the frame of the first scenario, there is a risk of demetallation when the SACs are exposed to cations with higher ionic potential (the ratio of the charge density to the radius of the atom over which this charge is distributed) than the central metal cation. As the ionic radius of H^+ is 134 pm and that of Fe(II)/Fe(III) cations is 77/63 pm [180], the higher ionic potential of Fe cations provides thermodynamic stability to Fe–N–C SAC in acidic electrolyte. The same reasoning applies to Co(III) (68.5 pm), Mn(II) (80 pm) and Cr(III) (75.5 pm) cations. Hence, the demetallation observed at the beginning of life of some metal–N–C catalysts may simply be perceived as improper stabilization of SAs (incomplete coordination of metal cations to N_x atoms) or dissolution of some NPs (carbides, nitrides, sulfides, oxides depending on the synthesis conditions) present in the catalyst, which are more prone to dissolution than SAs [98, 115]. Demetallation is also prone to occur because the oxidation state of the central metal ion changes during electrocatalysis. In this frame, metal porphyrins and related derivatives such as metal-phthalocyanines (metal-Pc) have often been studied to gain insights into the electrocatalytic activity and the stability of metal–N–C catalysts. Baranton et al. reported drop in ORR activity on Fe–Pc after holding the potential at 0.5 V vs. RHE or upon cycling between 0.05 and 0.90 V vs. RHE in O_2-saturated 0.5 M H_2SO_4 [181]. The authors suggested that the electron transfer from the out-of-plane Fe(II) to the adsorbed O_2 molecule results in a transient Fe(III) state, destabilized in acid medium due to its lower ionic radius than Fe(II), and they backed up their findings with infrared (IR) spectroscopy measurements showing formation of the protonated, metal-free macrocycle (H_2–Pc). Similar observation and explanation have been earlier proposed by Meier et al. for polymeric Fe(II)–Pc [182]. However, no change of IR spectra was observed in Ar-saturated 0.5 M H_2SO_4, thus indicating that the lack of stability of Fe–Pc is not related to oxidation of Fe(II) in Fe(III) but to ORR electrocatalysis. A change from Fe–Pc to H_2–Pc was also reported by Chen et al. upon stepping to 0.4 V vs. RHE in aerated 0.1 M $HClO_4$, not in N_2-saturated electrolyte, using *operando* tip-enhanced Raman spectroscopy (EC-TERS) [183]. Similarly, Kumar et al. showed that the active sites of Fe–N–C and Co–N–C SACs remained unchanged after 10 000 potential steps between 0.6 and 1.0 V vs. RHE in Ar-saturated 0.1 M H_2SO_4 thermostated at 80 °C [115, 127]. In contrast, the same accelerated stress test (AST) performed in O_2-saturated 0.1 M H_2SO_4 leads to formation of Fe nano-oxides. The pivotal role of O_2 adsorption and dissociation on demetallation will be discussed in detail in Section 13.5.7.

13.5.2 Changes in the Chemical and Physical Nature of the Metal Ion: Metal–N–C as a Pre-catalyst

Although metal–N–C SACs are stable over a wide range of potentials in acidic and alkaline electrolytes, their atomic-scale structure and chemistry change significantly in highly reductive ($E < -0.3$ V vs. SHE) and highly oxidative ($E > 1.5$ V vs. SHE) conditions, such as those encountered during the HER/CO_2RR/N_2RR and the OER, respectively. Consequently, the question arises whether the active site identified *ex situ* remains identical under reaction conditions. Based on Pourbaix diagrams, the stable phase of many transition metals is higher oxidation state or metal oxide in acidic or alkaline OER conditions, respectively. Recently, Bai et al. reported that a Co–N–C SAC can be transformed into a Co–Fe double atom catalyst that is highly active toward the OER in alkaline electrolyte [184]. High OER activity nevertheless involved multiple preliminary steps: a Co–N–C catalyst was first synthesized, and then pre-treated ("activated") in Fe-containing 0.1 M KOH, leading to formation of Co—Fe bonds as revealed from EXAFS. The OER activity rapidly declined over time but could be restored upon reactivation in Fe-containing 0.1 M KOH, thus providing evidence for the key role of Fe in alkaline OER catalysis. Upon long-term electrolysis, the oxidation state of Co atoms changed from Co(II) to Co(IV) as revealed from EXAFS, which the authors ascribed to formation of Co(IV)=O bonds. No TEM image of the aged material was provided, thus raising doubts about the nature of the active sites. Indeed, by combining EXAFS and TEM, Hocking et al. reported that tetranuclear-Mn cluster $[Mn_4O_4L_6]^+$ (L = diarylphosphinate) transform in 5–10 nm Mn-oxide particles after polarization at 1.0 V vs. Ag/AgCl/3 M NaCl (potential of 0.2 V vs. the SHE at 25 °C) in 0.1 M Na_2SO_4 electrolyte [185]. These results must be also compared with those of Kanan and Kocera [186], who reported that extremely active OER catalysts of micrometer-size dimension form upon polarization of tin oxide electrode in phosphate-buffered water containing Co^{2+} ions. It may thus be proposed that metal–N–C do not electrocatalyze the OER but rather act as a "pre-catalyst" in OER conditions, providing metal ions for the formation of the catalytically active metal oxides. Pourbaix diagrams also indicate that 3d transition metals are prone to reduce to their metallic form at a modest cathodic potential, such as that employed to electrocatalyze the HER and the CO_2RR. Anxolabéhère-Mallart et al. [187] showed that metallic Co NPs form from Co bis glyoximato complexes during HER in aprotic solvents (acetonitrile + 0.1 M $HClO_4$ in this study), independently on the initial ligands [188, 189]. Similarly, Karapinar et al. evidenced that Cu(II) species are reduced into Cu(O) (metallic Cu) in the reductive conditions of the CO_2RR in close to neutral pH, as mentioned in Section 13.4.2 [165]. Using EXAFS, the authors showed that the change in oxidation state was reversible; however, Cu NPs, not Cu–N–C, were considered to be the active sites for the CO_2RR, in line with former conclusions of Weng et al. [190]. These findings underline the importance of *operando* measurements, and nicely illustrate that metal–N–C shape the formation of metallic or oxidized NPs in highly reductive or oxidative conditions.

13.5.3 Protonation of Nitrogen Atoms

Protonation of N-groups in acidic media is another well-established deactivation mechanism of metal–N–C catalysts, especially for catalysts pyrolyzed under NH_3 (NH_3 increases the basicity of the N–C matrix [73, 191]). The Popov group [192, 193] argued that *N*-pyridinic species are more prone to protonate in acidic media due to their lone pair of electrons. In contrast, catalysts featuring mainly *N*-graphitic perform initially worse toward the ORR but are more stable [192]. According to Herranz et al., these are the nitrogen groups located close to the metal–N_x sites (hence not directly participating to the ORR), which are protonated thus leading to indirect deactivation of the active sites [194]. The authors operated a PEMFC containing an acid-leached Fe-N–C catalyst at the cathode and reported a drop in performance that they related to protonation of the nitrogen groups quickly followed by anion adsorption. Importantly, such protonation proved to be partially or totally reversible, i.e. part of the catalyst's activity could be restored after an *ex situ* thermal or chemical treatment. Now, besides the deactivation of the catalytic sites, it should be mentioned that protonated nitrogen groups poorly transport H^+, due to weakened ionomer–catalyst interactions [195]. In conclusion, the protonation of the nitrogen groups partially accounts for the ORR activity decay of some metal–N–C catalysts but is not a real degradation mechanism *per se*.

13.5.4 Carbon Oxidation Reaction

The carbon matrix ensures electron conduction and transport of reactants to and products from the electrode, and is compatible with most of the ionomers required for efficient ion transport in electrochemical devices. Thermodynamically speaking, the carbon matrix can be oxidized in highly acidic conditions (pH 0–1) to CO_2 or CO at $E \geq 0.21$ V vs. SHE and $E \geq 0.52$ V vs. SHE, respectively. Kinetically speaking, the COR rate is negligible below 1.0 V vs. SHE [196, 197], but it increases with temperature (especially above 50 °C) [197–199] and at high relative humidity [200–202]. Mild COR leads first to the formation of oxygen-containing groups on the surface of the carbon matrix [203]. As the TOF of metal–N_x moieties strongly depends on their local environment (coordination number and electronegativity of the neighboring atoms [112, 204, 205]), these oxygen-containing surface groups lead to the depletion of the delocalized π electrons and result in a drop of the TOF of metal–N–C catalysts. For example, recent combined DFT calculations and experimental work by Jung et al. have shown that hydrogen- or oxygen-containing carbon surface groups change the binding strength of transition metal to OOH_{ads} species and control the ORR selectivity [139]. Using these findings, the authors selectively synthesized H_2O_2 on Co atoms dispersed on oxidized graphene catalysts. A consequence of the formation of oxygen-containing surface groups is that the carbon surface turns from hydrophobic to hydrophilic [206]. According to Zhang et al. [203] and Wang et al. [207], hydrophilic carbon surfaces are more prone to micropores' flooding, and beside demetallation and/or transformation of Fe–N_x sites into Fe-oxides, mild COR might also partially account for the rapid decay in performance observed on Fe–N–C catalysts, e.g. after a cell voltage hold at $U = 0.5$ V for 100 hours in a H_2/O_2 PEMFC operating at $T = 80\,°C$ [203]. To reduce the adverse effects of flooding, Yang et al. prepared a series of doubly pyrolyzed MOF-derived Fe–N–C catalysts, which differ by the pyrolysis temperature used [208]. The most robust catalyst was obtained after a first pyrolysis at $T = 1150\,°C$ (the highest pyrolysis temperature employed in this work) and featured the lowest concentration of nitrogen and oxygen atoms at its surface, i.e. it was more hydrophobic.

In contrast to mild COR, partially reversible, prolonged COR leads to CO_2 evolution, and as a result, to the destruction of metal–N–C site. Using identical-location transmission electron microscopy, Choi et al. reported thinning of carbon particles after potential sweeping between 1.2 and 1.5 V *vs.* RHE at $T = 50\,°C$ in Ar-saturated 0.1 M $HClO_4$ [98]. Online inductively coupled plasma mass spectrometry and differential electrochemical mass spectrometry (ICP-MS and DEMS, respectively) experiments supported the conclusions, and X-EDS analyses revealed pronounced demetallation under these experimental conditions [98, 115]. Ultimately, the porous structure of metal–N–C catalysts collapses, thus restricting the transport of gases to the active sites. For example, during the ORR, the cathodic catalytic layer becomes starved from O_2 leading to a drop in performance at high current density [209].

13.5.5 Effect of Hydrogen Peroxide

H_2O_2 is the only stable intermediate or by-product of the ORR in acid medium and particularly stresses metal–N–C catalysts [210]. Indeed, Fenton reactions between transition metal cations and H_2O_2 lead to the formation of reactive oxygen species (ROS). Fe(III) cations are known to produce ROS species; however, Cu(II) > Cr(III) ~ Co(II) are much more active toward ROS production at pH 7, underlining the importance of this degradation mechanism [211]. Nitrone spin-trap coupled to electron paramagnetic resonance (EPR) experiments of Choi et al. evidenced that ROS form during a 2 hours exposure of Fe–N–C to a 5 wt% H_2O_2 solution at open circuit potential [112]. Post-treatment XAS and Mössbauer analyses revealed intact Fe–N_x core structure, but a modification of the second coordination sphere of Fe. Moreover, XPS analyses suggested that the surface of the carbon matrix was chemically oxidized during the H_2O_2 treatment and the introduction of the oxygen groups resulted in an acidification of the surface, modified potential of zero charge, and increased work function. Oxidation of the carbon surface leads to the depletion from π-electrons and to a decrease in the TOF of the Fe–N_x sites. RRDE experiments of Choi et al. reported that 32% of O_2 molecules are transformed into H_2O_2 for a thin cathode (Fe–N–C catalyst loading of 100 μg/$cm^2{}_{geo}$) [130]. In contrast, only 6% H_2O_2 was detected for a thicker cathode (catalyst loading of 800 μg/$cm^2{}_{geo}$), most likely because the produced H_2O_2 molecules decomposed within the thick electrode and were not detected at the ring, as inferred in Section 13.4.1. Similar to the effect of bulk COR, chemical oxidation of the carbon matrix leads to (i) decreased TOF values [112, 212] and (ii) increased hydrophilic character, and associated flooding issues [203, 208, 213]. For example, Goellner et al. reported that the ORR activity of Fe–N–C and Co–N–C catalysts is divided by 6 and 3 after *ex situ* H_2O_2 treatment, respectively [212]. Increase of the H_2O_2 concentration [210], longer duration of exposure to H_2O_2 solution [29, 83], and higher temperature of the H_2O_2 solution ($T \geq 50\,°C$) [112] accelerate the extent of degradation of the metal–N–C catalysts. Choi et al. also reported that the ORR selectivity changed from 4e⁻ (O_2 molecules to water) to 2e⁻ ORR on H_2O_2-treated Fe–N–C catalysts in acidic medium, showing that the effect of H_2O_2 is autocatalytic. In contrast, exposing the same Fe–N–C catalyst to H_2O_2 in alkaline electrolyte does not change the ORR activity nor selectivity, as revealed by combined RRDE and spin-trap measurements. Importantly, as the effect of H_2O_2 mostly relates to formation of oxygen-containing surface groups onto the carbon surface, it is partially reversible. Indeed, both the ORR activity and the selectivity of the Fe–N–C catalyst could be recovered after electrochemical reduction of the oxygen-containing groups present on the surface of the carbon matrix [112].

13.5.6 Migration and Aggregation of metal SAs

Recently, Kumar et al. used Cs-corrected single-atom resolved STEM to investigate the mobility of Fe_{SA} in Fe–N–C materials [214]. The authors synthesized two Fe–N–C SACs: both were first pyrolyzed at $T = 975\,°C$ in N_2:H_2 atmosphere (93 : 7), and one encountered a second pyrolysis at $T = 950\,°C$ in N_2:NH_3 atmosphere (90 : 10). A secondary pyrolysis usually reveals to be beneficial to the initial ORR activity, more particularly when the material is pyrolyzed under NH_3 [103, 215]; however, this effect is known to be non-durable [11, 13, 216]. Kumar et al. reported that Fe atoms are mobile and cluster on the doubly pyrolyzed Fe–N–C material, not on the Fe–N–C material submitted to a single pyrolysis after square-wave potential cycling between 0.6 and 1.0 V vs. RHE in Ar-saturated 0.1 M H_2SO_4 thermostated at 80 °C. Raman spectroscopy results suggested that the carbon matrix of the doubly pyrolyzed material was more prone to COR than that of singly pyrolyzed material, thus facilitating the movement of the SAs in the oxidative potential range investigated in this study.

13.5.7 Combined Effects

Thanks to well-designed ASTs, degradation mechanisms and the associated effects on electrocatalytic reactions have been isolated. However, several stressors are present during the life of SACs in the operating conditions of an electrochemical device, and a combination of stressors may influence or even revert the predominant degradation mechanism. Perhaps the most striking example of this assertion was provided by Kumar et al. [127]. The authors assessed the long-term ORR activity of a MOF-derived Fe–N–C SAC using the same AST (10 000 square-wave potential steps between 0.6 and 1.0 V vs. RHE – 3 seconds at each potential – $T = 80\,°C$) in Ar-saturated or O_2-saturated 0.1 M H_2SO_4. Four times higher ORR activity loss were assessed when performing load cycling AST in O_2- vs. Ar-saturated electrolyte. Combined results from cyclic voltammetry, nitrite stripping, X-ray absorption, ^{57}Fe Mössbauer, Raman and X-EDS spectroscopies revealed that the more pronounced drop in ORR activity was due to ROS-induced carbon corrosion, and associated formation of nanometric Fe-oxide particles upon reprecipitation of Fe cations leached from Fe–N_x sites. Recently, Osmieri et al. also reported that small Fe clusters form after a 100 hour aging test in potentiostatic conditions ($U = 0.85$ V under air), and a loss of 70% of the initial ORR activity was monitored [217]. EELS analyses revealed that Fe_{SA} were not coordinated with nitrogen anymore; hence, they became mobile on the carbon support. In contrast, the same aging test performed under inert atmosphere had little to no effect. The pivotal role of O_2, in combination with electrochemical potential and acidic pH, was also underlined in the *operando* Mossbauer study of Li et al. [123]. Using two Fe–N–C catalysts, pyrolyzed either in Ar or NH_3 atmosphere, the authors provided evidence that SACs initially comprising two distinct Fe–N_x sites (S1 and S2) degrade via the transformation of S1 into Fe nano-oxides, while the structure and number of S2 were unmodified after 50 hours operation in a PEMFC cathode running on H_2/O_2. These findings show that the main reason for the loss of performance of Fe–N–C catalysts at a PEMFC cathode is a ROS-triggered C corrosion, yielding Fe–N_4 site demetallation and reprecipitation, in agreement with former results of Chenitz et al. [218].

13.6 Summary and Conclusions

Notable progress has been achieved from the early works of Jasinski highlighting the potential use of 3d transition metal macrocyclic complexes as fuel cell electrocatalysts, to the structurally diverse, highly active, metal–N–C electrocatalysts from today. The existence of metal–N_x sites covalently embedded in the carbon matrix for pyrolyzed metal–N–C materials is now well established and recognized in the material's field, with a first coordination structure bearing some analogy with that of metal–N_4 macrocycles. Such metal–N–C materials have however key advantages over non-pyrolyzed macrocycles, such as enhanced stability and activity, and enhanced electron conductivity. The methods to synthesize the metal–N–C are diverse, ranging from the functionalization of carbon black with nitrogen and metal-based precursors to approaches entirely foregoing the use of a pre-existing carbon support, instead combining carbon, nitrogen, and metal precursors with different approaches such as hard-template (e.g. with silica) and soft-template approaches (e.g. MOF and polymers). Although the doping with Fe and Co has hitherto been most often studied, it has also been recently shown that most of the 3d transition metals and also other elements (e.g. tin and platinum-group metals) can be integrated as SAs in nitrogen-doped carbon materials. Progresses of metal–N–carbon SACs are not limited to the optimization and further development of the syntheses processes, but also evolve around improved understanding of the chemistry, structure, and reactivity of those materials. A broad range of metal–N_x moieties can be observed, with the detailed structure, proportion, and site density driven by the synthesis process, including nitrogen-based groups and metal, nitrogen-containing sites. The latter are of critical importance. Their structure is now partially understood, owing to a wide range of analytical tools, including Mössbauer and X-ray absorption spectroscopy, electrochemistry, and microscopy, which provide insights on their direct environment and electronic structure.

The metal–N–C SACs exhibit interesting electrocatalytic properties for the ORR, the H_2O_2 generation, and the CO_2RR. In acidic environment, the Fe–N_4 sites are able to catalyze the $4e^-$ exchange required to transform O_2 in water. Their activity can be tuned by modifying the degree of graphitization and disorder (i.e. the functionalization, density of nitrogen, oxygen, and moieties) of the carbon basal plane the metallic atoms are embedded into. As for the CO_2RR, they do exhibit a wide range of selectivities for CO generation, ranging from c. 25% on Co–N–C to near 100% on Ni–N–C, thus allowing the direct production of syngas (a 1 : 2 – 1 : 3 CO:H_2 mixture) when using metal–N–C electrocatalysts. Other value-added products can be synthesized through CO_2RR, e.g. formate by using Sn–N–C or ethanol by using Cu–N–C. For the latter, it was shown that ethanol production resulted from the dynamic clustering of Cu–N_x sites. Structural changes during electrocatalysis are not uncommon, and metal–N–C catalysts can undergo various degradation processes when used as ORR cathode catalysts, including not only formation of clusters, but also (i) carbon corrosion, which incidentally leads to the destruction of the active sites; (ii) demetallation, i.e. the replacement of the metal cation by protons in the metal–N_4 moiety; iii) protonation of the nitrogen groups, and (iv) the partial (reversible) oxidation of the carbon basal plane, which impacts the electron density at metal–N_x sites and thus their TOF toward the ORR.

This chapter aimed to provide a comprehensive overview of the different aspects of metal–N–C electrocatalysts, including their synthetic processes, structures, activity for various electrochemical reactions, and durability; to highlight the notable progresses achieved throughout the past decade, but also to hint at the remaining challenges. Further investigation into the structural properties of metal–N–C electrocatalysts and, more specifically, of the still-debated exact structure and electronic state of the active sites, will require techniques that specifically probe such moieties, e.g. advanced microscopic techniques or other local techniques. Most approaches to assess the active site structure in metal–N–C SACs nowadays heavily rely on theoretical models to interpret and/or fit the experimental data, while experimental data are most often collected on a high number of SA sites (even for techniques regarded as probing a small surface area, such as XAS). Metal–N–C's recent venture in new reactions pastures was a success, as they showed promise as CO_2 to CO or CO_2 to formate electrocatalysts, but those paths can further be explored (i) in CO_2RR, by enabling the metal–N–C to synthesize value-added-products "beyond" CO, i.e. C_{2+} species either through the design of dual sites or cascade catalysts; and (ii) in other reactions, such as nitrate reduction, by investigating the potentiality of those materials, hinted by the recent results obtained for the nitrogen reduction reaction. Finally, the metal–N–C electrocatalysts were originally thought as an alternative for platinum-based electrocatalysts in proton-exchange membrane fuel cells. Hence, the most important challenge was, and remains, to close the activity gap that exists between those two families, along with going forward in implementing metal–N–C in actual devices to achieve, in the upcoming years, a successful implementation in noble-metal-free fuel cell vehicles.

References

1 Lim, T., Jung, G.Y., Kim, J.H. et al. (2020). Atomically dispersed Pt–N_4 sites as efficient and selective electrocatalysts for the chlorine evolution reaction. *Nature Communications* 11: 412.

2 Liu, D., Li, X., Chen, S. et al. (2019). Atomically dispersed platinum supported on curved carbon supports for efficient electrocatalytic hydrogen evolution. *Nature Energy* 4 (6): 512–518.

3 Wang, J., Tan, H.-Y., Kuo, T.-R. et al. (2021). In situ identifying the dynamic structure behind activity of atomically dispersed platinum catalyst toward hydrogen evolution reaction. *Small* 17 (16): 2005713.

4 Chao, T., Luo, X., Chen, W. et al. (2017). Atomically dispersed copper–platinum dual sites alloyed with palladium nanorings catalyze the hydrogen evolution reaction. *Angewandte Chemie International Edition* 56 (50): 16047–16051.

5 Rinaldo, S.G., Stumper, J., and Eikerling, M. (2010). Physical theory of platinum nanoparticle dissolution in polymer electrolyte fuel cells. *The Journal of Physical Chemistry C* 114 (13): 5773–5785.

6 Darling, R.M. and Meyers, J.P. (2003). Kinetic model of platinum dissolution in PEMFCs. *Journal of the Electrochemical Society* 150 (11): A1523.

7 Calle-Vallejo, F., Martínez, J.I., García-Lastra, J.M. et al. (2014). Fast prediction of adsorption properties for platinum nanocatalysts with generalized coordination numbers. *Angewandte Chemie International Edition* 53 (32): 8316–8319.

8 Jasinski, R. (1964). A new fuel cell cathode catalyst. *Nature* 201: 1212–1213.

9 Jasinski, R. (1965). Cobalt phthalocyanine as a fuel cell cathode. *Journal of the Electrochemical Society* 112 (5): 526–528.

10 Gupta, S., Tryk, D., Bae, I. et al. (1989). Heat-treated polyacrylonitrile-based catalysts for oxygen electroreduction. *Journal of Applied Electrochemistry* 19 (1): 19–27.

11 Lefèvre, M., Proietti, E., Jaouen, F. et al. (2009). Iron-based catalysts with improved oxygen reduction activity in polymer electrolyte fuel cells. *Science* 324 (5923): 71–74.

12 Wu, G., More, K.L., Johnston, C.M. et al. (2011). High-performance electrocatalysts for oxygen reduction derived from polyaniline, iron, and cobalt. *Science* 332 (6028): 443–447.

13 Proietti, E., Jaouen, F., Lefèvre, M. et al. (2011). Iron-based cathode catalyst with enhanced power density in polymer electrolyte membrane fuel cells. *Nature Communications* 2: 416.

14 Ma, S., Goenaga, G.A., Call, A.V. et al. (2011). Cobalt imidazolate framework as precursor for oxygen reduction reaction electrocatalysts. *Chemistry European Journal* 17 (7): 2063–2067.

15 Jahnke, H., Schönborn, M., and Zimmermann, G. (1976). Organic dyestuffs as catalysts for fuel cells. *Topics in Current Chemistry* 61: 133–181.

16 Bagotzky, V.S., Tarasevich, M.R., Radyushkina, K.A. et al. (1978). Electrocatalysis of the oxygen reduction process on metal chelates in acid elecvtrolyte. *Journal of Power Sources* 2: 233–240.

17 Blomquist, J., Moberg, L.C., Johansson, L.Y. et al. (1981). Mössbauer measurements on iron phthalocyanines. *Journal of Inorganic and Nuclear Chemistry* 43 (10): 2287–2292.

18 Dodelet, J.-P. (2006). Oxygen reduction in PEM fuel cell conditions: heat-treated non-precious metal-N_4 macrocycles and beyond. In: *N_4-Macrocyclic Metal Complexes* (eds. J. Zagal, F. Bedioui and J.-P. Dodelet), 83–147. New York: Springer Science.

19 Scherson, D.A., Yao, S.B., Yeager, E.B. et al. (1983). In situ and ex situ Moessbauer spectroscopy studies of iron phthalocyanine adsorbed on high surface area carbon. *The Journal of Physical Chemistry* 87 (6): 932–943.

20 Mineva, T., Matanovic, I., Atanassov, P. et al. (2019). Understanding active sites in pyrolyzed Fe-N-C catalysts for fuel cell cathodes by bridging density functional theory calculations and ^{57}Fe Mössbauer spectroscopy. *ACS Catalysis* 9 (10): 9359–9371.

21 van Veen, J.A.R. and Visser, C. (1979). Oxygen reduction on monomeric transition metal phthalocyanines in acid electrolyte. *Electrochimica Acta* 24 (9): 921–928.

22 Wiesener, K., Ohms, D., Neumann, V. et al. (1989). N_4 macrocycles as electrocatalysts for the cathodic reduction of oxygen. *Materials Chemistry and Physics* 22 (3): 457–475.

23 van Veen, J.A.R. and Colijn, H.A. (1981). Oxygen reduction on transition-metal porphyrins in acid electrolyte II. Stability. *Berichte der Bunsengesellschaft für Physikalische Chemie* 85 (9): 700–704.

24 Faubert, G., Lalande, G., Côté, R. et al. (1996). Heat-treated iron and cobalt tetraphenylporphyrins adsorbed on carbon black: physical characterization and catalytic properties of these materials for the reduction of oxygen in polymer electrolyte fuel cells. *Electrochimica Acta* 41: 1689–1701.

25 Bagotzky, V.S., Tarasevich, M.R., Radyushkina, K.A. et al. (1978). Electrocatalysis of the oxygen reduction process on metal chelates in acid electrolyte. *Journal of Power Sources* 2 (3): 233–240.

26 Gouérec, P., Biloul, A., Contamin, O. et al. (1997). Oxygen reduction in acid media catalyzed by heat treated cobalt tetraazaannulene supported on an active charcoal: correlations between the performances after longevity tests and the active site configuration as seen by XPS and ToF-SIMS. *Journal of Electroanalytical Chemistry* 422 (1): 61–75.

27 Vallejos-Burgos, F., Utsumi, S., Hattori, Y. et al. (2012). Pyrolyzed phthalocyanines as surrogate carbon catalysts: initial insights into oxygen-transfer mechanisms. *Fuel* 99: 106–117.

28 van Wingerden, B., van Veen, J.A.R., and Mensch, C.T.J. (1988). An extended X-ray absorption fine structure study of heat-treated cobalt porphyrin catalysts supported on active carbon. *Journal of the Chemical Society, Faraday Transactions 1: Physical Chemistry in Condensed Phases* 84 (1): 65–74.

29 Schulenburg, H., Stankov, S., Schünemann, V. et al. (2003). Catalysts for the oxygen reduction from heat-treated iron(III) tetramethoxyphenylporphyrin chloride: structure and stability and active sites. *The Journal of Physical Chemistry B* 107 (34): 9034–9041.

30 Lefèvre, M., Dodelet, J.P., and Bertrand, P. (2002). Molecular oxygen reduction in PEM fuel cells: evidence for the simultaneous presence of two active sites in Fe-based catalysts. *The Journal of Physical Chemistry B* 106 (34): 8705–8713.

31 Lefèvre, M., Dodelet, J.P., and Bertrand, P. (2005). Molecular oxygen reduction in PEM fuel cell conditions: ToF-SIMS analysis of Co-based electrocatalysts. *The Journal of Physical Chemistry B* 109 (35): 16718–16724.

32 Tributsch, H., Koslowski, U.I., and Dorbandt, I. (2008). Experimental and theoretical modeling of Fe-, Co-, Cu-, Mn-based electrocatalysts for oxygen reduction. *Electrochimica Acta* 53 (5): 2198–2209.

33 Chang, S.-T., Wang, C.-H., Du, H.-Y. et al. (2012). Vitalizing fuel cells with vitamins: pyrolyzed vitamin B12 as a non-precious catalyst for enhanced oxygen reduction reaction of polymer electrolyte fuel cells. *Energy and Environmental Science* 5 (1): 5305–5314.

34 Bouwkamp-Wijnoltz, A.L., Visscher, W., van Veen, J.A.R. et al. (2002). On active-site heterogeneity in pyrolyzed carbon-supported iron porphyrin catalysts for the electrochemical reduction of oxygen : an in situ Mössbauer study. *The Journal of Physical Chemistry B* 106 (50): 12993–13001.

35 Ziegelbauer, J.M., Olson, T.S., Pylypenko, S. et al. (2008). Direct spectroscopic observation of the structural origin of peroxide generation from Co-based pyrolyzed porphyrins for ORR applications. *The Journal of Physical Chemistry C* 112 (24): 8839–8849.

36 Kramm, U.I., Zana, A., Vosch, T. et al. (2016). On the structural composition and stability of Fe–N–C catalysts prepared by an intermediate acid leaching. *Journal of Solid State Electrochemistry* 20 (4): 969–981.

37 Herrmann, I., Kramm, U.I., Fiechter, S. et al. (2009). Oxalate supported pyrolysis of CoTMPP as electrocatalysts for the oxygen reduction reaction. *Electrochimica Acta* 54 (18): 4275–4287.

38 Kramm, U.I., Herrmann-Geppert, I., Behrends, J. et al. (2016). On an easy way to prepare metal-nitrogen doped carbon with exclusive presence of MeN_4-type sites active for the ORR. *Journal of the American Chemical Society* 138 (2): 635–640.

39 Liberman, I., Shimoni, R., Ifraemov, R. et al. (2020). Active-site modulation in an Fe-porphyrin-based metal–organic framework through ligand axial coordination: accelerating electrocatalysis and charge-transport kinetics. *Journal of the American Chemical Society* 142 (4): 1933–1940.

40 Lions, M., Tommasino, J.-B., Chattot, R. et al. (2017). Insights into the mechanism of electrocatalysis of the oxygen reduction reaction by a porphyrinic metal organic framework. *Chemical Communications* 53 (48): 6496–6499.

41 Cheon, J.Y., Kim, T., Choi, Y. et al. (2013). Ordered mesoporous porphyrinic carbons with very high electrocatalytic activity for the oxygen reduction reaction. *Scientific Reports* 3 (1): 2715.

42 Zion, N., Douglin, J.C., Cullen, D.A. et al. (2021). Porphyrin aerogel catalysts for oxygen reduction reaction in anion-exchange membrane fuel cells. *Advanced Functional Materials* 31 (24): 2100963.

43 Bron, M., Radnik, J., Fieber-Erdmann, M. et al. (2002). EXAFS, XPS and electrochemical studies on oxygen reduction catalysts obtained by heat treatment of iron phenanthroline complexes supported on high surface area carbon black. *Journal of Electroanalytical Chemistry* 535 (1): 113–119.

44 Jaouen, F., Marcotte, S., Dodelet, J.P. et al. (2003). Oxygen reduction catalysts for polymer electrolyte fuel cells from the pyrolysis of iron acetate adsorbed on various carbon supports. *The Journal of Physical Chemistry B* 107 (6): 1376–1386.

45 Villers, D., Jacques-Bédard, X., and Dodelet, J.-P. (2004). Fe-based catalysts for oxygen reduction in PEM fuel cells. *Journal of the Electrochemical Society* 151 (9): A1507.

46 Lefèvre, M. and Dodelet, J.P. (2008). Fe-based electrocatalysts made with microporous pristine carbon black supports for the reduction of oxygen in PEM fuel cells. *Electrochimica Acta* 53 (28): 8269–8276.

47 Jaouen, F., Charreteur, F., and Dodelet, J.P. (2006). Fe-based catalysts for oxygen reduction in PEMFCs. *Journal of the Electrochemical Society* 153 (4): A689.

48 Charreteur, F., Jaouen, F., Ruggeri, S. et al. (2008). Fe/N/C non-precious catalysts for PEM fuel cells: influence of the structural parameters of pristine commercial carbon blacks on their activity for oxygen reduction. *Electrochimica Acta* 53 (6): 2925–2938.

49 Charreteur, F., Ruggeri, S., Jaouen, F. et al. (2008). Increasing the activity of Fe/N/C catalysts in PEM fuel cell cathodes using carbon blacks with a high-disordered carbon content. *Electrochimica Acta* 53 (23): 6881–6889.

50 Jaouen, F. and Dodelet, J.-P. (2007). Average turn-over frequency of O_2 electro-reduction for Fe/N/C and Co/N/C catalysts in PEFCs. *Electrochimica Acta* 52 (19): 5975–5984.

51 Ratso, S., Ranjbar Sahraie, N., Sougrati, M.T. et al. (2018). Synthesis of highly-active Fe–N–C catalysts for PEMFC with carbide-derived carbons. *Journal of Materials Chemistry A* 6 (30): 14663–14674.

52 Ratso, S., Kruusenberg, I., Käärik, M. et al. (2018). Highly efficient transition metal and nitrogen co-doped carbide-derived carbon electrocatalysts for anion exchange membrane fuel cells. *Journal of Power Sources* 375: 233–243.

53 Lefèvre, M., Proietti, E., Jaouen, F. et al. (2009). Iron-based catalysts for oxygen reduction in PEM fuel cells: expanded study using the pore-filling method. *ECS Transactions* 25: 105–115.

54 Pylypenko, S., Mukherjee, S., Olson, T.S. et al. (2008). Non-platinum oxygen reduction electrocatalysts based on pyrolyzed transition metal macrocycles. *Electrochimica Acta* 53: 7875–7883.

55 Gokhale, R., Chen, Y., Serov, A. et al. (2017). Novel dual templating approach for preparation of highly active Fe-N-C electrocatalyst for oxygen reduction. *Electrochimica Acta* 224: 49–55.

56 Serov, A., Artyushkova, K., Niangar, E. et al. (2015). Nano-structured non-platinum catalysts for automotive fuel cell application. *Nano Energy* 16: 293–300.

57 Serov, A., Robson, M.H., Halevi, B. et al. (2012). Highly active and durable templated non-PGM cathode catalysts derived from iron and aminoantipyrine. *Electrochemistry Communications* 22: 53–56.

58 Liu, G., Li, X., Ganesan, P. et al. (2009). Development of non-precious metal oxygen-reduction catalysts for PEM fuel cells based on N-doped ordered porous carbon. *Applied Catalysis B: Environmental* 93 (1): 156–165.

59 Primbs, M., Sun, Y., Roy, A. et al. (2020). Establishing reactivity descriptors for platinum group metal (PGM)-free Fe–N–C catalysts for PEM fuel cells. *Energy and Environmental Science* 13 (8): 2480–2500.

60 Monteverde Videla, A.H.A., Osmieri, L., Armandi, M. et al. (2015). Varying the morphology of Fe-N-C electrocatalysts by templating iron phthalocyanine precursor with different porous SiO_2 to promote the oxygen reduction reaction. *Electrochimica Acta* 177: 43–50.

61 Sa, Y.J., Seo, D.-J., Woo, J. et al. (2016). A general approach to preferential formation of active Fe–N_x sites in Fe–N/C electrocatalysts for efficient oxygen reduction reaction. *Journal of the American Chemical Society* 138 (45): 15046–15056.

62 Jaouen, F., Lefèvre, M., Dodelet, J.P. et al. (2006). Heat-treated Fe/N/C catalysts for O_2 electroreduction: are active sites hosted in micropores? *The Journal of Physical Chemistry B* 110 (11): 5553–5558.

63 Barkholtz, H.M. and Liu, D.-J. (2017). Advancements in rationally designed PGM-free fuel cell catalysts derived from metal-organic frameworks. *Materials Horizons* 4 (1): 20–37.

64 Morozan, A. and Jaouen, F. (2012). Metal organic frameworks for electrochemical applications. *Energy and Environmental Science* 5 (11): 9269–9290.

65 Xia, W., Zhu, J., Guo, W. et al. (2014). Well-defined carbon polyhedrons prepared from nano metal–organic frameworks for oxygen reduction. *Journal of Materials Chemistry A* 2 (30): 11606–11613.

66 Zhang, H., Hwang, S., Wang, M. et al. (2017). Single atomic iron catalysts for oxygen reduction in acidic media: particle size control and thermal activation. *Journal of the American Chemical Society* 139 (40): 14143–14149.

67 Armel, V., Hannauer, J., and Jaouen, F. (2015). Effect of ZIF-8 crystal size on the O_2 electro-reduction performance of pyrolyzed Fe-N-C catalysts. *Catalysts* 5: 1333–1351.

68 Wang, X., Zhang, H., Lin, H. et al. (2016). Directly converting Fe-doped metal–organic frameworks into highly active and stable Fe-N-C catalysts for oxygen reduction in acid. *Nano Energy* 25: 110–119.

69 Zhao, D., Shui, J.-L., Grabstanowicz, L.R. et al. (2014). Highly efficient non-precious metal electrocatalysts prepared from one-pot synthesized zeolitic imidazolate frameworks. *Advanced Materials* 26: 1093–1097.

70 Wang, J., Han, G., Wang, L. et al. (2018). ZIF-8 with ferrocene encapsulated: a promising precursor to single-atom Fe embedded nitrogen-doped carbon as highly efficient catalyst for oxygen electroreduction. *Small* 14 (15): 1704282.

71 Palaniselvam, T., Biswal, B.P., Banerjee, R. et al. (2013). Zeolitic imidazolate framework (ZIF)-derived, hollow-core, nitrogen-doped carbon nanostructures for oxygen-reduction reactions in PEFCs. *Chemistry European Journal* 19 (28): 9335–9342.

72 Armel, V., Hindocha, S., Salles, F. et al. (2017). Structural descriptors of zeolitic-imidazolate frameworks are keys to the activity of Fe-N-C catalysts. *Journal of the American Chemical Society* 139 (1): 453–464.

73 Zitolo, A., Goellner, V., Armel, V. et al. (2015). Identification of catalytic sites for oxygen reduction in iron- and nitrogen-doped graphene materials. *Nature Materials* 14 (9): 937–942.

74 He, Y., Hwang, S., Cullen, D.A. et al. (2019). Highly active atomically dispersed CoN_4 fuel cell cathode catalysts derived from surfactant-assisted MOFs: carbon-shell confinement strategy. *Energy and Environmental Science* 12 (1): 250–260.

75 Yin, P., Yao, T., Wu, Y. et al. (2016). Single cobalt atoms with precise N-coordination as superior oxygen reduction reaction catalysts. *Angewandte Chemie International Edition* 55 (36): 10800–10805.

76 Liu, S., Wang, M., Yang, X. et al. (2020). Chemical vapor deposition for atomically dispersed and nitrogen coordinated single metal site catalysts. *Angewandte Chemie International Edition* 59 (48): 21698–21705.

77 Jiao, L., Li, J., Richard, L.L. et al. (2021). Chemical vapor deposition of Fe-N-C oxygen reduction catalysts with full utilization of dense Fe-N_4 sites. *Nature Materials*, https://doi.org/10.1038/s41563-021-01030--2 20: 1385–1391.

78 Mehmood, A., Pampel, J., Ali, G. et al. (2018). Facile metal coordination of active site imprinted nitrogen doped carbons for the conservative preparation of non-noble metal oxygen reduction electrocatalysts. *Advanced Energy Materials* 8 (9): 1701771.

79 Jiang, Z., Yu, J., Huang, T. et al. (2018). Recent advance on polyaniline or polypyrrole-derived electrocatalysts for oxygen reduction reaction. *Polymers* 10 (12): 1397.

80 Ghosh, D., Giri, S., Mandal, A. et al. (2013). H_+, Fe_{3+} codoped polyaniline/MWCNTs nanocomposite: superior electrode material for supercapacitor application. *Applied Surface Science* 276: 120–128.

81 Ferrandon, M., Kropf, A.J., Myers, D.J. et al. (2012). Multitechnique characterization of a polyaniline-iron-carbon oxygen reduction catalyst. *The Journal of Physical Chemistry C* 116 (30): 16001–16013.

82 Wu, G., Nelson, M.A., Mack, N.H. et al. (2010). Titanium dioxide-supported non-precious metal oxygen reduction electrocatalyst. *Chemical Communications* 46 (40): 7489–7491.

83 Wu, G., Artyushkova, K., Ferrandon, M. et al. (2009). Performance durability of polyaniline-derived non-precious cathode catalysts. *ECS Transactions* 25: 1299–1311.

84 Cullen, D.A., Chung, H.T., More, K.L. et al. (2017). Direct atomic-level insight into the active sites of a high-performance PGM-free ORR catalyst. *Science* 357 (6350): 479–484.

85 Yin, X., Chung, H.T., Martinez, U. et al. (2019). PGM-free ORR catalysts designed by templating PANI-type polymers containing functional groups with high affinity to iron. *Journal of the Electrochemical Society* 166 (7): F3240–F3245.

86 Shui, J., Chen, C., Grabstanowicz, L. et al. (2015). Highly efficient nonprecious metal catalyst prepared with metal–organic framework in a continuous carbon nanofibrous network. *Proceedings of the National Academy of Sciences of the United States of America* 112 (34): 10629–10634.

87 Li, J., Pršlja, P., Shinagawa, T. et al. (2019). Volcano trend in electrocatalytic CO_2 reduction activity over atomically dispersed metal sites on nitrogen-doped carbon. *ACS Catalysis* 9 (11): 10426–10439.

88 Ju, W., Bagger, A., Hao, G.P. et al. (2017). Understanding activity and selectivity of metal-nitrogen-doped carbon catalysts for electrochemical reduction of CO_2. *Nature Communications* 8: 1–9.

89 Varela, A.S., Ranjbar Sahraie, N., Steinberg, J. et al. (2015). Metal-doped nitrogenated carbon as an efficient catalyst for direct CO_2 electroreduction to CO and hydrocarbons. *Angewandte Chemie International Edition* 54 (37): 10758–10762.

90 Luo, E., Zhang, H., Wang, X. et al. (2019). Single-atom $Cr-N_4$ sites designed for durable oxygen reduction catalysis in acid media. *Angewandte Chemie International Edition* 58 (36): 12469–12475.

91 Luo, F., Roy, A., Silvioli, L. et al. (2020). *P*-block single-metal-site tin/nitrogen-doped carbon fuel cell cathode catalyst for oxygen reduction reaction. *Nature Materials* 19 (11): 1215–1223.

92 Chen, S., Lv, C., Liu, L. et al. (2021). High-temperature treatment to engineer the single-atom Pt coordination environment towards highly efficient hydrogen evolution. *Journal of Energy Chemistry* 59: 212–219.

93 Xiong, Y., Dong, J., Huang, Z.-Q. et al. (2020). Single-atom Rh/N-doped carbon electrocatalyst for formic acid oxidation. *Nature Nanotechnology* 15 (5): 390–397.

94 Zhang, Z., Chen, Y., Zhou, L. et al. (2019). The simplest construction of single-site catalysts by the synergism of micropore trapping and nitrogen anchoring. *Nature Communications* 10: 1657.

95 Ye, S., Luo, F., Zhang, Q. et al. (2019). Highly stable single Pt atomic sites anchored on aniline-stacked graphene for hydrogen evolution reaction. *Energy and Environmental Science* 12 (3): 1000–1007.

96 Choi, C.H., Kim, M., Kwon, H.C. et al. (2016). Tuning selectivity of electrochemical reactions by atomically dispersed platinum catalyst. *Nature Communications* 7: 10922.

97 Yan, Q.-Q., Wu, D.-X., Chu, S.-Q. et al. (2019). Reversing the charge transfer between platinum and sulfur-doped carbon support for electrocatalytic hydrogen evolution. *Nature Communications* 10: 4977.

98 Choi, C.H., Baldizzone, C., Grote, J.P. et al. (2015). Stability of Fe-N-C catalysts in acidic medium studied by operando spectroscopy. *Angewandte Chemie International Edition* 54 (43): 12753–12757.

99 Choi, C.H., Baldizzone, C., Polymeros, G. et al. (2016). Minimizing operando demetallation of Fe-N-C electrocatalysts in acidic medium. *ACS Catalysis* 6 (5): 3136–3146.

100 Huan, T.N., Ranjbar, N., Rousse, G. et al. (2017). Electrochemical reduction of CO_2 catalyzed by Fe-N-C materials: a structure-selectivity study. *ACS Catalysis* 7: 1520–1525.

101 Artyushkova, K., Serov, A., Rojas-Carbonell, S. et al. (2015). Chemistry of multitudinous active sites for oxygen reduction reaction in transition metal-nitrogen-carbon electrocatalysts. *The Journal of Physical Chemistry C* 119 (46): 25917–25928.

102 Zitolo, A., Ranjbar-Sahraie, N., Mineva, T. et al. (2017). Identification of catalytic sites in cobalt-nitrogen-carbon materials for the oxygen reduction reaction. *Nature Communications* 8: 957.

103 Kramm, U.I., Abs-Wurmbach, I., Herrmann-Geppert, I. et al. (2011). Influence of the electron-density of FeN_4-centers towards the catalytic activity of pyrolyzed FeTMPPCl-based ORR-electrocatalysts. *Journal of the Electrochemical Society* 158 (1): B69–B78.

104 Kramm, U.I., Lefèvre, M., Larouche, N. et al. (2014). Correlations between mass activity and physicochemical properties of Fe/N/C catalysts for the ORR in PEM fuel cell via ^{57}Fe Mössbauer spectroscopy and other techniques. *Journal of the American Chemical Society* 136 (3): 978–985.

105 Kabir, S., Artyushkova, K., Kiefer, B. et al. (2015). Computational and experimental evidence for a new TM-N_3/C moiety family in non-PGM electrocatalysts. *Physical Chemistry Chemical Physics* 17 (27): 17785–17789.

106 Kramm, U.I., Herranz, J., Larouche, N. et al. (2012). Structure of the catalytic sites in Fe/N/C-catalysts for O_2-reduction in PEM fuel cells. *Physical Chemistry Chemical Physics* 14 (33): 11673–11688.

107 Matanovic, I., Artyushkova, K., Strand, M.B. et al. (2016). Core level shifts of hydrogenated pyridinic and pyrrolic nitrogen in the nitrogen-containing graphene-based electrocatalysts: in-plane vs edge defects. *The Journal of Physical Chemistry C* 120 (51): 29225–29232.

108 Matanovic, I., Artyushkova, K., and Atanassov, P. (2018). Understanding PGM-free catalysts by linking density functional theory calculations and structural analysis: perspectives and challenges. *Current Opinion in Electrochemistry* 9: 137–144.

109 Tylus, U., Jia, Q., Strickland, K. et al. (2014). Elucidating oxygen reduction active sites in pyrolyzed metal-nitrogen coordinated non-precious-metal electrocatalyst systems. *The Journal of Physical Chemistry C* 118 (17): 8999–9008.

110 Li, J., Ghoshal, S., Liang, W. et al. (2016). Structural and mechanistic basis for the high activity of Fe-N-C catalysts toward oxygen reduction. *Energy and Environmental Science* 9 (7): 2418–2432.

111 Ramaswamy, N., Tylus, U., Jia, Q. et al. (2013). Activity descriptor identification for oxygen reduction on nonprecious electrocatalysts: linking surface science to coordination chemistry. *Journal of the American Chemical Society* 135 (41): 15443–15449.

112 Choi, C.H., Lim, H., Chon, G. et al. (2018). The Achilles' heel of iron-based catalysts during oxygen reduction in an acidic medium. *Energy and Environmental Science* 11 (11): 3176–3182.

113 Martinez, U., Holby, E.F., Babu, S.K. et al. (2019). Experimental and theoretical trends of PGM-free electrocatalysts for the oxygen reduction reaction with different transition metals. *Journal of the Electrochemical Society* 166 (7): F3136–F3142.

114 Wan, X., Liu, X., Li, Y. et al. (2019). Fe–N–C electrocatalyst with dense active sites and efficient mass transport for high-performance proton exchange membrane fuel cells. *Nature Catalysis* 2: 259–268.

115 Kumar, K., Gairola, P., Lions, M. et al. (2018). Physical and chemical considerations for improving catalytic activity and stability of non-precious-metal oxygen reduction reaction catalysts. *ACS Catalysis* 8 (12): 11264–11276.

116 Jia, Q., Ramaswamy, N., Tylus, U. et al. (2016). Spectroscopic insights into the nature of active sites in iron–nitrogen–carbon electrocatalysts for oxygen reduction in acid. *Nano Energy* 29: 65–82.

117 Osmieri, L. (2019). Transition metal–nitrogen–carbon (M–N–C) catalysts for oxygen reduction reaction. Insights on synthesis and performance in polymer electrolyte fuel cells. *ChemEngineering* 3 (1): 16.

118 Sun, Y., Silvioli, L., Sahraie, N.R. et al. (2019). Activity-selectivity trends in the electrochemical production of hydrogen peroxide over single-site metal-nitrogen-carbon catalysts. *Journal of the American Chemical Society* 141 (31): 12372–12381.

119 Saveleva, V.A., Ebner, K., Ni, L. et al. (2021). Potential-induced spin changes in Fe/N/C electrocatalysts assessed by in situ X-ray emission spectroscopy. *Angewandte Chemie International Edition* 60 (21): 11707–11712.

120 Kramm, U.I., Ni, L., and Wagner, S. (2019). ^{57}Fe Mössbauer spectroscopy characterization of electrocatalysts. *Advanced Materials* 31 (31): 1805623.

121 Gallenkamp, C., Kramm, U.I., Proppe, J. et al. (2021). Calibration of computational Mössbauer spectroscopy to unravel active sites in FeNC catalysts for the oxygen reduction reaction. *International Journal of Quantum Chemistry* 121 (3): 1–19.

122 Ni, L., Gallenkamp, C., Paul, S. et al. (2021). Active site identification in FeNC catalysts and their assignment to the oxygen reduction reaction pathway by in situ ^{57}Fe Mössbauer spectroscopy. *Advanced Energy and Sustainability Research* 2 (2): 2000064.

123 Li, J., Sougrati, M.T., Zitolo, A. et al. (2021). Identification of durable and non-durable FeNx sites in Fe–N–C materials for proton exchange membrane fuel cells. *Nature Catalysis* 4 (1): 10–19.

124 Kneebone, J.L., Daifuku, S.L., Kehl, J.A. et al. (2017). A combined probe-molecule, Mossbauer, nuclear resonance vibrational spectroscopy, and density functional theory approach for evaluation of potential iron active sites in an oxygen reduction reaction catalyst. *The Journal of Physical Chemistry C* 121 (30): 16283–16290.

125 Wagner, S., Auerbach, H., Tait, C.E. et al. (2019). Elucidating the structural composition of an Fe-N-C catalyst by nuclear- and electron-resonance techniques. *Angewandte Chemie International Edition* 58 (31): 10486–10492.

126 Malko, D., Kucernak, A., and Lopes, T. (2016). In situ electrochemical quantification of active sites in Fe–N/C non-precious metal catalysts. *Nature Communications* 7: 13285.

127 Kumar, K., Dubau, L., Mermoux, M. et al. (2020). On the influence of oxygen on the degradation of Fe-N-C catalysts. *Angewandte Chemie International Edition* 59 (8): 3235–3243.

128 Sahraie, N.R., Kramm, U.I., Steinberg, J. et al. (2015). Quantifying the density and utilization of active sites in non-precious metal oxygen electroreduction catalysts. *Nature Communications* 6: 8618.

129 Bae, G., Kim, H., Choi, H. et al. (2021). Quantification of active site density and turnover frequency: from single-atom metal to nanoparticle electrocatalysts. *JACS Au* 1 (5): 586–597.

130 Choi, C.H., Choi, W.S., Kasian, O. et al. (2017). Unraveling the nature of sites active toward hydrogen peroxide reduction in Fe-N-C catalysts. *Angewandte Chemie International Edition* 56 (30): 8809–8812.

131 Biddinger, E.J., von Deak, D., Singh, D. et al. (2011). Examination of catalyst loading effects on the selectivity of CN_x and Pt/VC ORR catalysts using RRDE. *Journal of the Electrochemical Society* 158 (4): B402.

132 Bonakdarpour, A., Lefevre, M., Yang, R. et al. (2008). Impact of loading in RRDE experiments on Fe-N-C catalysts: two- or four-electron oxygen reduction? *Electrochemical and Solid-State Letters* 11 (6): B105–B108.

133 Dobrzeniecka, A., Zeradjanin, A., Masa, J. et al. (2013). Application of SECM in tracing of hydrogen peroxide at multicomponent non-noble electrocatalyst films for the oxygen reduction reaction. *Catalysis Today* 202 (1): 55–62.

134 Sánchez-Sánchez, C.M., Rodríguez-López, J., and Bard, A.J. (2008). Scanning electrochemical microscopy. 60. Quantitative calibration of the SECM substrate generation/tip collection mode and its use for the study of the oxygen reduction mechanism. *Analytical Chemistry* 80 (9): 3254–3260.

135 Dobrzeniecka, A., Zeradjanin, A.R., Masa, J. et al. (2016). Evaluation of kinetic constants on porous, non-noble catalyst layers for oxygen reduction - a comparative study between SECM and hydrodynamic methods. *Catalysis Today* 262: 74–81.

136 Jaouen, F. and Dodelet, J.-P. (2009). O_2 reduction mechanism on non-noble metal catalysts for PEM fuel cells. Part II: a porous-electrode model to predict the quantity of H_2O_2 detected by rotating ring-diskelectrode. *The Journal of Physical Chemistry C* 113 (34): 15433–15443.

137 Verdaguer-casadevall, A., Deiana, D., Karamad, M. et al. (2014). Trends in the electrochemical synthesis of H_2O_2. *Nano Letters* 14: 1603–1608.

138 Siahrostami, S., Verdaguer-Casadevall, A., Karamad, M. et al. (2013). Enabling direct H_2O_2 production through rational electrocatalyst design. *Nature Materials* 12 (12): 1137–1143.

139 Jung, E., Shin, H., Lee, B.H. et al. (2020). Atomic-level tuning of Co–N–C catalyst for high-performance electrochemical H_2O_2 production. *Nature Materials* 19 (4): 436–442.

140 Gao, J. and Liu, B. (2020). Progress of electrochemical hydrogen peroxide synthesis over single atom catalysts. *ACS Materials Letters* 2 (8): 1008–1024.

141 Gasteiger, H.A., Kocha, S.S., Sompalli, B. et al. (2005). Activity benchmarks and requirements for Pt, Pt-alloy, and non-Pt oxygen reduction catalysts for PEMFCs. *Applied Catalysis B: Environmental* 56 (1–2): 9–35.

142 Lacis, A.A., Schmidt, G.A., Rind, D. et al. (2010). Atmospheric CO_2: principal control knob governing earth's temperature. *Science* 330 (6002): 356–359.

143 Guo, S., Asset, T., and Atanassov, P. (2021). Catalytic hybrid electrocatalytic/biocatalytic cascades for carbon dioxide reduction and valorization. *ACS Catalysis* 11: 5172–5188.

144 Bagger, A., Ju, W., Varela, A.S. et al. (2017). Single site porphyrine-like structures advantages over metals for selective electrochemical CO_2 reduction. *Catalysis Today* 288: 74–78.

145 Bagger, A., Ju, W., Varela, A.S. et al. (2017). Electrochemical CO_2 reduction: a classification problem. *ChemPhysChem* 18: 3266–3273.

146 Birdja, Y.Y., Pérez-Gallent, E., Figueiredo, M.C. et al. (2019). Advances and challenges in understanding the electrocatalytic conversion of carbon dioxide to fuels. *Nature Energy* 4: 732–745.

147 Asset, T., Garcia, S.T., Herrera, S. et al. (2019). Investigating the nature of the active sites for the CO_2 reduction reaction on carbon-based electrocatalysts. *ACS Catalysis* 9: 7668–7678.

148 Schulz, H. (1999). Short history and present trends of Fischer-Tropsch synthesis. *Applied Catalysis A: General* 186: 3–12.

149 Varela, A.S., Ju, W., and Strasser, P. (2018). Molecular nitrogen–carbon catalysts, solid metal organic framework catalysts, and solid metal/nitrogen-doped carbon (MNC) catalysts for the electrochemical CO_2 reduction. *Advanced Energy Materials* 8: 1–35.

150 Pan, Y., Lin, R., Chen, Y. et al. (2018). Design of single-atom Co-N_5 catalytic site: a robust electrocatalyst for CO_2 reduction with nearly 100% CO selectivity and remarkable stability. *Journal of the American Chemical Society* 140: 4218–4221.

151 Sharma, P.P., Wu, J., Yadav, R.M. et al. (2015). Nitrogen-doped carbon nanotube arrays for high-efficiency electrochemical reduction of CO_2: on the understanding of defects, defect density, and selectivity. *Angewandte Chemie International Edition* 54: 13701–13705.

152 Wu, J., Yadav, R.M., Liu, M. et al. (2015). Achieving highly efficient, selective, and stable CO_2 reduction on nitrogen-doped carbon nanotubes. *ACS Nano* 9: 5364–5371.

153 Wu, J., Liu, M., Sharma, P.P. et al. (2016). Incorporation of nitrogen defects for efficient reduction of CO_2 via two-electron pathway on three-dimensional graphene foam. *Nano Letters* 16: 466–470.

154 Pan, F., Deng, W., Justiniano, C. et al. (2018). Identification of champion transition metals centers in metal and nitrogen-codoped carbon catalysts for CO_2 reduction. *Applied Catalysis B: Environmental* 226: 463–472.

155 Möller, T., Ju, W., Bagger, A. et al. (2019). Efficient CO_2 to CO electrolysis on solid Ni-N-C catalysts at industrial current densities. *Energy and Environmental Science* 12: 640–647.

156 Hori, Y., Wakebe, H., Tsukamoto, T. et al. (1994). Electrocatalytic process of CO selectivity in electrochemical reduction of CO_2 at metal electrodes in aqueous media. *Electrochimica Acta* 39: 1833–1839.

157 Ju, W., Bagger, A., Wang, X. et al. (2019). Unraveling mechanistic reaction pathways of the electrochemical CO_2 reduction on Fe-N-C single-site catalysts. *ACS Energy Letters* 4: 1663–1671.

158 Varela, A.S., Kroschel, M., Leonard, N.D. et al. (2018). pH effects on the selectivity of the electrocatalytic CO_2 reduction on graphene-embedded Fe-N-C motifs: bridging concepts between molecular homogeneous and solid-state heterogeneous catalysis. *ACS Energy Letters* 3: 812–817.

159 Hu, X.M., Hval, H.H., Bjerglund, E.T. et al. (2018). Selective CO_2 reduction to CO in water using earth-abundant metal and nitrogen-doped carbon electrocatalysts. *ACS Catalysis* 8: 6255–6264.

160 Zhang, C., Yang, S., Wu, J. et al. (2018). Electrochemical CO_2 reduction with atomic iron-dispersed on nitrogen-doped graphene. *Advanced Energy Materials* 1703487: 1–9.

161 Pan, F., Zhang, H., Liu, K. et al. (2018). Unveiling active sites of CO_2 reduction on nitrogen-coordinated and atomically dispersed iron and cobalt catalysts. *ACS Catalysis* 8: 3116–3122.

162 Hursán, D., Samu, A.A., Janovák, L. et al. (2019). Morphological attributes govern carbon dioxide reduction on N-doped carbon electrodes. *Joule* 3: 1719–1733.

163 Endrődi, B., Kecsenovity, E., Samu, A. et al. (2019). Multilayer electrolyzer stack converts carbon dioxide to gas products at high pressure with high efficiency. *ACS Energy Letters* 4 (7): 1770–1777.

164 Leonard, N., Ju, W., Sinev, I. et al. (2018). The chemical identity, state and structure of catalytically active centers during the electrochemical CO_2 reduction on porous Fe-nitrogen-carbon (Fe-N-C) materials. *Chemical Science* 9: 5064–5073.

165 Karapinar, D., Huan, N.T., Ranjbar Sahraie, N. et al. (2019). Electroreduction of CO_2 on single-site copper-nitrogen-doped carbon material: selective formation of ethanol and reversible restructuration of the metal sites. *Angewandte Chemie International Edition* 58 (42): 15098–15103.

166 Zhao, Y., Liang, J., Wang, C. et al. (2018). Tunable and efficient tin modified nitrogen-doped carbon nanofibers for electrochemical reduction of aqueous carbon dioxide. *Advanced Energy Materials* 8 (10): 1–9.

167 Zhang, S., Kang, P., and Meyer, T.J. (2014). Nanostructured tin catalysts for selective electrochemical reduction of carbon dioxide to formate. *Journal of the American Chemical Society* 136: 1734–1737.

168 Zu, X., Li, X., Liu, W. et al. (2019). Efficient and robust carbon dioxide electroreduction enabled by atomically dispersed $Sn^{\delta+}$ sites. *Advanced Materials* 31 (15): 1–8.

169 Paul, S., Kao, Y.-L., Ni, L. et al. (2021). Influence of the metal center in M–N–C catalysts on the CO_2 reduction reaction on gas diffusion electrodes. *ACS Catalysis* 11 (9): 5850–5864.

170 Jia, H.P. and Quadrelli, E.A. (2014). Mechanistic aspects of dinitrogen cleavage and hydrogenation to produce ammonia in catalysis and organometallic chemistry: relevance of metal hydride bonds and dihydrogen. *Chemical Society Reviews* 43 (2): 547–564.

171 Seefeldt, L.C., Hoffman, B.M., and Dean, D.R. (2009). Mechanism of Mo-dependent nitrogenase. *Annual Review of Biochemistry* 78 (1): 701–722.

172 Skúlason, E., Bligaard, T., Gudmundsdóttir, S. et al. (2012). A theoretical evaluation of possible transition metal electro-catalysts for N_2 reduction. *Physical Chemistry Chemical Physics* 14 (3): 1235–1245.

173 Han, L., Liu, X., Chen, J. et al. (2019). Atomically dispersed molybdenum catalysts for efficient ambient nitrogen fixation. *Angewandte Chemie International Edition* 58 (8): 2321–2325.

174 Liu, W., Han, L., Wang, H.T. et al. (2020). FeMo sub-nanoclusters/single atoms for neutral ammonia electrosynthesis. *Nano Energy* 77: 105078.

175 Wang, M., Liu, S., Qian, T. et al. (2019). Over 56.55% Faradaic efficiency of ambient ammonia synthesis enabled by positively shifting the reaction potential. *Nature Communications* 10: 341.

176 Suryanto, B.H.R., Du, H.L., Wang, D. et al. (2019). Challenges and prospects in the catalysis of electroreduction of nitrogen to ammonia. *Nature Catalysis* 2 (4): 290–296.

177 Choi, J., Suryanto, B.H.R., Wang, D. et al. (2020). Identification and elimination of false positives in electrochemical nitrogen reduction studies. *Nature Communications* 11: 1–10.

178 Andersen, S.Z., Čolić, V., Yang, S. et al. (2019). A rigorous electrochemical ammonia synthesis protocol with quantitative isotope measurements. *Nature* 570 (7762): 504–508.

179 van Langevelde, P.H., Katsounaros, I., and Koper, M.T.M. (2021). Electrocatalytic nitrate reduction for sustainable ammonia production. *Joule* 5 (2): 290–294.

180 Shannon, R.D. (1976). Revised effective ionic radii and systematic studies of interatomic distances in halides and chalcogenides. *Acta Crystallographica Section A: Foundations* 32 (5): 751–767.

181 Baranton, S., Coutanceau, C., Roux, C. et al. (2005). Oxygen reduction reaction in acid medium at iron phthalocyanine dispersed on high surface area carbon substrate: tolerance to methanol, stability and kinetics. *Journal of Electroanalytical Chemistry* 577 (2): 223–234.

182 Meier, H., Tschirwitz, U., Zimmerhackl, E. et al. (1977). Application of radioisotope techniques for the study of phthalocyanine catalyzed electrochemical processes in fuel cells. *The Journal of Physical Chemistry* 81 (8): 712–718.

183 Chen, Z., Jiang, S., Kang, G. et al. (2019). Operando characterization of iron phthalocyanine deactivation during oxygen reduction reaction using electrochemical tip-enhanced Raman spectroscopy. *Journal of the American Chemical Society* 141 (39): 15684–15692.

184 Bai, L., Hsu, C.-S., Alexander, D.T.L. et al. (2019). A cobalt–iron double-atom catalyst for the oxygen evolution reaction. *Journal of the American Chemical Society* 141 (36): 14190–14199.

185 Hocking, R.K., Brimblecombe, R., Chang, L.-Y. et al. (2011). Water-oxidation catalysis by manganese in a geochemical-like cycle. *Nature Chemistry* 3 (6): 461–466.

186 Kanan, M.W. and Nocera, D.G. (2008). In situ formation of an oxygen-evolving catalyst in neutral water containing phosphate and Co^{2+}. *Science* 321 (5892): 1072–1075.

187 Anxolabéhère-Mallart, E., Costentin, C., Fournier, M. et al. (2012). Boron-capped tris(glyoximato) cobalt clathrochelate as a precursor for the electrodeposition of nanoparticles catalyzing H_2 evolution in water. *Journal of the American Chemical Society* 134 (14): 6104–6107.

188 El Ghachtouli, S., Guillot, R., Brisset, F. et al. (2013). Cobalt-based particles formed upon electrocatalytic hydrogen production by a cobalt pyridine oxime complex. *ChemSusChem* 6 (12): 2226–2230.

189 El Ghachtouli, S., Fournier, M., Cherdo, S. et al. (2013). Monometallic cobalt-trisglyoximato complexes as precatalysts for catalytic H_2 evolution in water. *The Journal of Physical Chemistry C* 117 (33): 17073–17077.

190 Weng, Z., Wu, Y., Wang, M. et al. (2018). Active sites of copper-complex catalytic materials for electrochemical carbon dioxide reduction. *Nature Communications* 9: 415.

191 Santori, P.G., Speck, F.D., Li, J. et al. (2019). Effect of pyrolysis atmosphere and electrolyte pH on the oxygen reduction activity, stability and spectroscopic signature of FeN_x moieties in Fe-N-C catalysts. *Journal of the Electrochemical Society* 166 (7): F3311–F3320.

192 Liu, G., Li, X., and Popov, B.N. (2009). Stability study of nitrogen-modified carbon composite catalysts for oxygen reduction reaction in polymer electrolyte membrane fuel cells. *ECS Transactions* 25: 1251–1259.

193 Liu, G., Li, X., Lee, J.-W. et al. (2011). A review of the development of nitrogen-modified carbon-based catalysts for oxygen reduction at USC. *Catalysis Science and Technology* 1 (2): 207.

194 Herranz, J., Jaouen, F., Lefèvre, M. et al. (2011). Unveiling N-protonation and anion-binding effects on Fe/N/C catalysts for O_2 reduction in proton-exchange-membrane fuel cells. *The Journal of Physical Chemistry C* 115 (32): 16087–16097.

195 Artyushkova, K., Workman, M.J., Matanovic, I. et al. (2017). Role of surface chemistry on catalyst/ionomer interactions for transition metal–nitrogen–carbon electrocatalysts. *ACS Applied Energy Materials* 1 (1): 68–77.

196 Binder, H., Kohling, A., Richter, K. et al. (1964). Über die anodische oxydation von aktivkohlen in wässrigen elektrolyten. *Electrochimica Acta* 9: 255.

197 Castanheira, L., Silva, W.O., Lima, F.H.B. et al. (2015). Carbon corrosion in proton-exchange membrane fuel cells: effect of the carbon structure, the degradation protocol, and the gas atmosphere. *ACS Catalysis* 5: 2184–2194.

198 Roen, L.M., Paik, C.H., and Jarvi, T.D. (2004). Electrocatalytic corrosion of carbon support in PEMFC cathodes. *Electrochemical and Solid-State Letters* 7 (1): A19–A22.

199 Stevens, D.A. and Dahn, J.R. (2005). Thermal degradation of the support in carbon-supported platinum electrocatalysts for PEM fuel cells. *Carbon* 43 (1): 179–188.

200 Appleby, A.J. (1987). Corrosion in low and high temperature fuel cells - an overview. *Corrosion* 43 (7): 398–408.

201 Stevens, D.A., Hicks, M.T., Haugen, G.M. et al. (2005). Ex situ and in situ stability studies of PEMFC catalysts. *Journal of the Electrochemical Society* 152 (12): A2309–A2315.

202 Maass, S., Finsterwalder, F., Frank, G. et al. (2008). Carbon support oxidation in PEM fuel cell cathodes. *Journal of Power Sources* 176 (2): 444–451.

203 Zhang, G., Chenitz, R., Lefèvre, M. et al. (2016). Is iron involved in the lack of stability of Fe/N/C electrocatalysts used to reduce oxygen at the cathode of PEM fuel cells? *Nano Energy* 29: 111–125.

204 Zhang, J. and Dai, L. (2015). Heteroatom-doped graphitic carbon catalysts for efficient electrocatalysis of oxygen reduction reaction. *ACS Catalysis* 5 (12): 7244–7253.

205 Zhao, S., Li, M., Han, M. et al. (2018). Defect-rich Ni_3FeN nanocrystals anchored on N-doped graphene for enhanced electrocatalytic oxygen evolution. *Advanced Functional Materials* 28 (18): 1706018.

206 Giordano, N., Antonucci, P.L., Passalacqua, E. et al. (1991). Relationship between physicochemical properties and electrooxidation behavior of carbon materials. *Electrochimica Acta* 36 (13): 1931–1935.

207 Wang, X.X., Prabhakaran, V., He, Y. et al. (2019). Iron-free cathode catalysts for proton-exchange-membrane fuel cells: cobalt catalysts and the peroxide mitigation approach. *Advanced Materials* 31 (31): 1805126.

208 Yang, L., Larouche, N., Chenitz, R. et al. (2015). Activity, performance, and durability for the reduction of oxygen in PEM fuel cells, of Fe/N/C electrocatalysts obtained from the pyrolysis of metal-organic-framework and iron porphyrin precursors. *Electrochimica Acta* 159: 184–197.

209 Reshetenko, T., Serov, A., Artyushkova, K. et al. (2016). Tolerance of non-platinum group metals cathodes proton exchange membrane fuel cells to air contaminants. *Journal of Power Sources* 324: 556–571.

210 Lefèvre, M. and Dodelet, J.P. (2003). Fe-based catalysts for the reduction of oxygen in polymer electrolyte membrane fuel cell conditions: determination of the amount of peroxide released during electroreduction and its influence on the stability of the catalysts. *Electrochimica Acta* 48 (19): 2749–2760.

211 Strlič, M., Kolar, J., Šelih, V.S. et al. (2003). A comparative study of several transition metals in Fenton-like reaction systems at circum-neutral pH. *Acta Chimica Slovenica* 50 (4): 619–632.

212 Goellner, V., Armel, V., Zitolo, A. et al. (2015). Degradation by hydrogen peroxide of metal-nitrogen-carbon catalysts for oxygen reduction. *Journal of the Electrochemical Society* 162 (6): H403–H414.

213 Banham, D., Kishimoto, T., Sato, T. et al. (2017). New insights into non-precious metal catalyst layer designs for proton exchange membrane fuel cells: improving performance and stability. *Journal of Power Sources* 344: 39–45.

214 Kumar, K., Asset, T., Li, X. et al. (2021). Fe-N-C electrocatalysts' durability: effects of single atoms' mobility and clustering. *ACS Catalysis* 11 (2): 484–494.

215 Domínguez, C., Peña, M.A., Rojas, S. et al. (2016). Effect of the pyrolysis atmosphere and nature of iron precursor on the structure and activity of Fe/N based electrocatalysts for the oxygen reduction reaction. *International Journal of Hydrogen Energy* 41 (47): 22560–22569.

216 Charreteur, F., Jaouen, F., and Dodelet, J.-P. (2009). Iron porphyrin-based cathode catalysts for PEM fuel cells: influence of pyrolysis gas on activity and stability. *Electrochimica Acta* 54 (26): 6622–6630.

217 Osmieri, L., Cullen, D.A., Chung, H.T. et al. (2020). Durability evaluation of a Fe–N–C catalyst in polymer electrolyte fuel cell environment via accelerated stress tests. *Nano Energy* 78: 105209.

218 Chenitz, R., Kramm, U.I., Lefèvre, M. et al. (2018). A specific demetalation of Fe-N_4 catalytic sites in the micropores of NC-Ar + NH_3 is at the origin of the initial activity loss of the highly active Fe/N/C catalyst used for the reduction of oxygen in PEM fuel cells. *Energy and Environmental Science* 11 (2): 365–382.

14

Supported Metal Single-Atom Photocatalysis

Bruno F. Machado[1], Lifeng Liu[2], Zhipeng Yu[2,3], and Joaquim L. Faria[3]

[1] *CoLAB Net4CO$_2$ – Network for a Sustainable CO$_2$ Economy, Rua Dr. Júlio de Matos 828-882, 4200-355 Porto, Portugal*
[2] *International Iberian Nanotechnology Laboratory (INL), Avenida Mestre José Veiga, 4715-330 Braga, Portugal*
[3] *Laboratory of Separation and Reaction Engineering – Laboratory of Catalysis and Materials (LSRE-LCM), Faculdade de Engenharia Universidade do Porto (FEUP), Rua Dr. Roberto Frias s/n, 4200-465 Porto, Portugal*

14.1 Introduction

Atomically efficient photocatalysis is at the heart of any future meaningful breakthroughs in the fascinating field of photochemical-assisted catalytic processes. Modern society evolves at paces never seen before, especially due to the uncertainty caused by healthiness, well-being, climate, and societal changes. Our future as individuals, society, and a global planetary community depends on how we tackle the challenges concerning health, energy, and the environment in the near future. By no means society will reduce its need for energy voluntarily, even in the face of obvious severe environmental impacts and loss of biological diversity. The scientific challenge for one and every researcher is to develop physical, chemical, and biological solutions to build a sustainable future with clean and renewable energy supply by diminishing the detrimental environmental impact associated with its generation. Semiconductor heterogeneous photocatalysis can effectively solve energy and environmental problems such as fuel generation and water treatment using solar radiation. As noted by Scaiano and Lanterna [1] in the green road map for heterogeneous photocatalysis, more than 100 years ago, the Italian researcher from the University of Bologna paved the way for the energy transition to renewable energy by establishing the principles behind the potential use of solar energy [2].

With single-atom photocatalysis, we move toward maximizing the atom-to-utilization efficiency, searching for enhanced performance arising from the unstrained environment congestion and unique electronic structures. By embedding single atoms (SAs) into optical semiconductors, we can engineer the energy band structure and electronic structure of the solid toward increased light absorption in the region of interest (visible range if solar is the target) and improved charge transfer dynamics. In addition, the surface structures of single-atom photocatalysts (SAPs) can be customized for enhancing the adsorption and affinity of

Supported Metal Single Atom Catalysis, First Edition. Edited by Philippe Serp and Doan Pham Minh.
© 2022 WILEY-VCH GmbH. Published 2022 by WILEY-VCH GmbH.

the photocatalyst toward reactants. Single-atom photocatalysis has been recently reviewed [3] and outlooked [4] in several instances owing to their compelling potential and tremendous active research in enhancing light-harvesting, charge transfer dynamics, and characterization of surface reactions in the photochemical semiconductor photo-assisted systems. In the present chapter, we propose a look at the potential applications of single-atom catalysts (SACs) in photocatalysis and photoelectrocatalysis (PEC) in a selection of ground-breaking domains, including solar fuels derived from water splitting, CO_2 reduction, environmental protection, and production of value-added chemicals.

Water splitting has been recently reviewed as the main route to produce energetic solar vectors like molecular hydrogen, with particular attention to various synthesis strategies for SACs [5]. Attention was drawn to the very recent and exciting support graphitic carbon nitride (g-C_3N_4) that follows its rediscovery as an optical semiconductor [6] allied to its extraordinary ability to be tailored by functionalization using mechanical, chemical, or thermal treatments [7]. Depending on the synthesis method, different nanostructures are obtained, corresponding photoactive sites with variable photocatalytic activity for the SAs loaded on g-C_3N_4 support. The most common methods for preparing SAs on g-C_3N_4 include the wet chemical route, thermal copolymerization, atomic layer deposition (ALD), and molecular metal complexes, and were recently reviewed [8].

The critical aspects in the synthesis of these materials concern the dynamics of anchoring and the identification of the SA active sites of the photocatalysts. The interactions between metal SAs and g-C_3N_4 support are determinant for the stability of the photocatalytic system, and the adequacy of the process to solar applications, owing to the interaction based on the electron transfer from the metal to the support and the consequences on the adsorption properties of reactive species, and therefore on the catalytic activity of active metal sites [9].

It was elegantly demonstrated that the cooperative interplay of atomic metal sites and adjacent environments significantly influences the overall photocatalytic performance in ways like what occurs with metalloenzymes and homogeneous catalysts, typically not considered in heterogeneous catalysts [10, 11].

Although catalysts based on single atoms are not a novelty, the continuous developments in the characterization techniques combined with recent advances in computational methods and materials manufacture enabled the atomic design of host materials and the exploration of trends at a scale not available before. Artificial intelligence and additive manufacture become essential tools in materials design, building unimaginable bridges between experiments and computations, especially in the field of PEC [12]. The extraordinary catalytic performance of SACs compared to conventional heterogeneous catalysts has excited many researchers. However, there is still too much to unveil, especially if we want to understand what is going on under operating conditions. Enormous leaps were made recently in the science of measurement. At the same time, we can use high-resolution imaging techniques to gain crystal-clear pictures of a molecular structure and capture real-time visuals of chemical reactions happening on surfaces in greater detail than ever before [13–15]. A long way has been traveled since the first report on the successful synthesis of

TiO_2 photocatalysts loaded with isolated noble metal atoms (Pt, Pd, Rh, or Ru) used in the photocatalytic generation of hydrogen [16]. In the following, we will illustrate using some crucial examples of general application, the current potential to obtain solar fuels, added-value chemicals, or develop sustainable processes based on SAP.

14.2 Synthesis and Characterization Methods

It is not the intention of the authors of this chapter to provide in this section a detailed description of all the available techniques used to prepare and characterize SACs. The methods commonly used to synthesize and characterize photocatalysts are inherently similar/identical to those described for other SACs and have already been thoroughly described in Chapters 2–7. Instead, we will provide only a brief description of the most important techniques in the photocatalysis toolbox and how these reveal important photocatalytic features for structure–activity correlations.

14.2.1 Synthesis

Despite its rather advanced equipment requirements, one of the most popular synthesis methods for preparing SAPs is ALD [17, 18]. This method allows for precise control of the deposition of SAs and clusters over supporting materials. One other standard method includes wet-chemical routes, where the precursor materials consist of single-atom species, and metal SAs can be dispersed on the supports through a chemical reaction [17, 19]. In addition to these methods, others based on photochemical processes have attracted gowning interest due to their potential for large-scale SAC production [20–23].

The photochemical approach is a mild and cost-effective preparation method based on ultraviolet (UV) irradiation, which prevents the aggregation of SAs to nanoparticles (NPs). One example has been reported by Zheng and coworkers [20]. The authors developed a room-temperature photochemical route to synthesize an atomically dispersed and stable Pd_{SA}/TiO_2 catalyst with a Pd loading as high as 1.5 wt%. To achieve this, they used ethylene glycolate (EG)-stabilized ultrathin (two-atom-thick) TiO_2 nanosheets to disperse the Pd atoms. The EG radicals generated by UV radiation (low-density provided by a Xe lamp) promoted the removal of the Cl^- ligands from the Pd precursor, playing a pivotal role in stabilizing the Pd single atoms.

Wei et al. developed an iced-photochemical reduction method for anchoring atomically dispersed Pt onto various supports, including mesoporous carbon, graphene, carbon nanotubes, titanium dioxide NPs, and zinc oxide nanowires [21]. During the SAC synthesis (Figure 14.1a), the H_2PtCl_6 solution was frozen in a container with liquid nitrogen, followed by irradiation with UV light to ensure the photochemical reduction of Pt(IV). Using such iced-photochemical reduction, the aggregation of atoms was significantly prevented, and Pt_{SA} were successfully stabilized. Moreover,

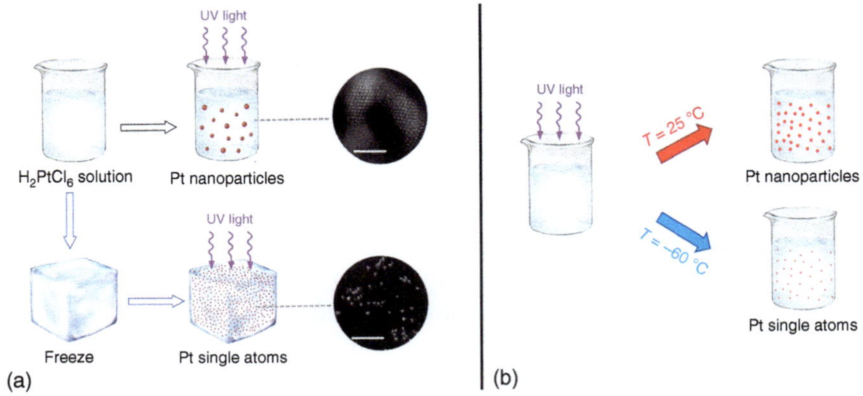

Figure 14.1 (a, b) Schematic illustration of the iced-photochemical process together with the conventional photochemical reduction of H_2PtCl_6 aqueous solution for Pt SAC preparation. Source: (a) Wei et al. [21]. Springer Nature / CC BY 4.0; (b) Wei et al. [22]. Adapted with permission of Royal Society of Chemistry.

the authors also showed that this method could be extended to other single atoms such as gold and silver.

In another work, the same authors also showed that this method could also be used to fabricate other Pt_{SA} catalysts (with approximately 0.1 nm in size), such as that depicted in Figure 14.1b, where the formation of Pt_{NP} can be effectively suppressed at temperatures of −60 °C [22]. Here, the authors confirmed that ultralow reaction temperature could efficiently suppress the nucleation and aggregation of isolated metal atoms in solution, generating atomically dispersed catalysts. However, they also observed that if the temperature was raised to 25 °C, the Pt_{SA} aggregated, forming Pt_{NP}. The authors explained this behavior using *ab initio* molecular dynamics (AIMD) simulations: the volume of ethanol liquid is negatively correlated with temperature, such that Pt_{SA} are more tightly surrounded by ethanol molecules when the temperature is lower. Thus, the nucleation process at 25 °C proceeded with an extremely high-rate constant compared to −60 °C in liquid ethanol.

A variation of these photochemical reduction methods can also be used to deposit SAs over supports such as carbons. These materials have been successfully incorporated into traditional photocatalytic systems and enhance the activity by minimizing the electron/hole recombination acting as an electron sink (role of electron acceptor) or through bandgap tuning/photosensitization effect (role of electron donor) [24].

A modified photochemical solid-phase reduction method was developed to deposit Pt_{SA} on a nitrogen-doped porous carbon (NPC) [23]. Using this synthesis strategy, $PtCl_6^{2-}$ ions were directly reduced by UV light irradiation and then deposited on the carbon surface, without further post-physical or chemical treatments. The authors observed that a Pt loading as high as 3.8 wt% could be attained, while avoiding the aggregation of Pt atoms, as evidenced by the characterization (scanning transmission electron microscopy, STEM; and X-ray absorption fine structure, XAFS, spectroscopy) results.

14.2.2 Characterization

Confirming the existence and corresponding distribution of metal SAs on a given surface is crucial for the development of SACs. To obtain this information at the atomic level requires advanced characterization techniques since traditional ones often present resolution limitations. Significant progress has been made in recent years, and the characterization of single atoms requires a toolbox of complementary techniques, including microscopy, spectroscopy, and theoretical studies [25–27]. See Chapters 5–7 in this book for additional details.

Overall, the most intuitive approach for the characterization of SAPs is the direct imaging of the metal single atoms, using, for example, scanning tunnelling microscopy (STM) (imaging material surfaces at the atomic level, where individual atoms can be imaged and manipulated by appropriate conducting tips), or high-angle annular dark-field scanning transmission electron microscopy (HAADF-STEM) (single imaging atoms with an atomic resolution with the help of energy dispersive X-ray analysis (EDS), or electron energy loss spectroscopy (EELS), to confirm the composition). Other standard techniques include X-ray absorption spectroscopy (X-ray absorption near-edge structure, XANES, and extended X-ray absorption fine structure, EXAFS) based on synchrotron radiation (enabling the evaluation of metal–support interactions and identifying the coordination and electronic structures of SAC), diffuse reflectance infrared Fourier transform spectroscopy (DRIFTS) (directly monitor the interaction between adsorbed probe molecules such as carbon monoxide, ammonia, pyridine, etc., and supported metal species) and X-ray photoelectron spectroscopy (XPS) (composition and valence state information on SAPs).

These techniques are among the most commonly used to characterize SAPs, as described in Chapters 5 and 6. Nevertheless, other techniques are of particular importance for the characterization of materials used in photocatalysis. These typically include photoluminescence (PL), providing information on the optical and photochemical properties of semiconductors and their electronic structure, and ultraviolet-visible (UV–vis) diffuse reflectance spectroscopy (DRS), which measures electronic transitions and can be used to determine the bandgap of the semiconductors.

However, there is still no direct evidence in the literature that these techniques currently possess the resolution necessary to help distinguish between metal single atoms and metal nanoparticles on the surface of semiconductor material.

14.2.3 Effects of Single Atoms in Photocatalysis

From a photocatalytic point of view, the presence of metal single atoms can significantly affect the three main steps typically associated with these systems:

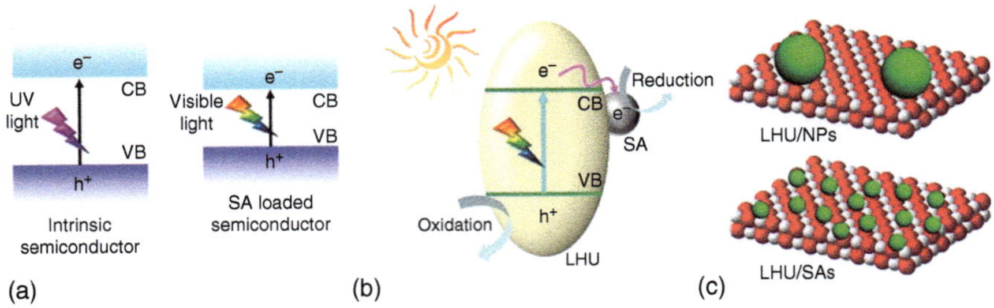

Figure 14.2 Schematic illustrations for the role of single-atoms in (a) light-harvesting, (b) charge separation and transfer, and (c) surface catalytic reaction in photocatalysis (CB: conduction band; VB: valence band; LHU: light-harvesting units; NPs: nanoparticles). Source: Gao et al. [3]. Reproduced with permission of American Chemical Society.

(i) light-harvesting, (ii) charge separation and transfer, and (iii) surface reaction [3] (Figure 14.2), compared to the common systems where the metal NPs are the catalytically active sites, and the support is responsible for the light-harvesting step.

The main limitations in photocatalysis include rapid recombination of photoexcited charge carriers (electrons/holes) and the limited number of active sites for the catalytic reactions, which result in poor experimental performances compared to those theoretically possible. By embedding metal SA_son light-harvesting supports, these increase the number of active sites due to a maximum atom-utilization efficiency and modify the supports' energy band and electronic structures, thus adjusting their light-absorption performance and charge transfer dynamics. Moreover, the surface structures of SAPs can also be tailored by adjusting metal–support interactions, resulting in an enhanced adsorption and activation potential of the photocatalyst due to their unique geometric and electronic features.

Light harvesting (Figure 14.2a) is the initial step in photocatalysis and determines the capacity of a photocatalytic system to absorb incident photons and generate sufficient electron–hole pairs for the corresponding catalytic reactions depending on the energy band structure of the semiconductor support. Achieving the absorption of visible or infrared radiation is the key to full utilization of sunlight (spectrum consists of 5% ultraviolet, 43% visible, and 52% infrared). When the single atoms are embedded in the semiconductor, the energy band structure of the support could be modified toward absorption of the incident light in the specific spectral region, especially visible/infrared light [3]. To understand the effect of metal SAs on light harvesting, the spectral range of light absorption, optical bandgap, conduction band (CB), and valence band (VB) positions for inorganic semiconductors (or highest occupied molecular orbital, HOMO, and lowest unoccupied molecular orbital, LUMO, energy levels for organic semiconductors) should be carefully characterized for both the naked and SA-containing support [26].

The charge transfer between metal SAs and light-harvesting support plays a critical role in improving the overall efficiency of the photocatalytic process (Figure 14.2b). Efficient separation and transfer of photogenerated electrons/holes at the metal–support interface is critically important to hinder carrier recombination and obtain high photocatalytic performances. Generally, an electronic equilibrium is established between the metal particles and semiconductor support. Since the electronic structure of the semiconductor support is usually constant, the electronic structures of metal species significantly influence the electronic structures of the metal-semiconductor junction, with the photogenerated electrons transferring onto the metal species through a Schottky barrier [3, 28]. This charge carrier separation could be potentially present on SAPs as well, meaning that the unique electronic structures of SAs could influence the metal–support interaction, thus further tailoring the charge transfer behavior between light-harvesting material and SAs.

Finally, surface reactions (Figure 14.2c) play a crucial role in determining both the activity and selectivity of a photocatalytic system. When metal species are excited by light, electrons in the ground state will be transferred to higher energy levels, which can be further transferred to molecules or participate in surface reaction directly. One important aspect related to metal particles is the effect of their size on optical properties. It is recognized that particle size and morphology significantly influence the plasmonic properties of metal particles, as the UV–vis absorption band will generally shift to lower wavelengths with decreasing particle size [26]. Hence, different photocatalytic properties are observed depending on the size of the metal species, with the latter acting as light sensitizers or cocatalysts.

The controllable surface structure of different materials provides distinct coordination environments for supported SAs. Thus, the geometric and electronic structure of SAPs can be tailored through metal–support interaction. This allows a modification of the species adsorption kinetics and ultimately modify the reaction pathway. Interestingly, the metal SAs have specific coordinatively unsaturated sites and unique electronic structures, providing large numbers of active sites for the catalytic reaction compared with their nanocluster, nanoparticle, or bulk equivalents [29].

14.3 SAC Performance in Photocatalysis

SAC, with 100% utilization efficiency of metal atoms, are emerging as a class of promising catalysts that are potentially cost efficient for many reactions [30]. Unlike metal cluster and nanoparticle catalysts, SACs consisted in isolated SAs on suitable support giving rise to coordinative unsaturation and strong metal–support interaction [31], which may eventually result in remarkable activity, improved stability, and high selectivity in the catalytic process [32].

14.3.1 Photocatalytic Water Splitting

Hydrogen is a promising environment-friendly energy carrier that can replace fossil fuels. Solar-driven photoelectrocatalytic hydrogen evolution reaction (HER) has been regarded as one of the most important catalytic processes because it offers an effective route for storing solar energy in chemical bonds in the form of dihydrogen.

The key technology for producing dihydrogen from electricity is water splitting. However, the big challenge in water splitting is that noble metals, such as platinum, are often needed as catalyst. The high activity is attributed to the isolated Pt atom serving as the leading active site for photocatalysis hydrogen evolution. Therefore, the performance and cost efficiency of catalysts are greatly improved by engineering atomically dispersed Pt on the support surface. Atom efficiency is undoubtedly important when energetic bills are to be considered.

Graphitic carbon nitride with a layered N/C-coordinating framework has recently emerged as a promising photocatalyst for hydrogen production, showing a favorable photo-response in the visible light region of the solar spectrum [33]. High solar-to-hydrogen (STH) conversion efficiency requires careful design to suppress the accumulation of photoexcited electrons and holes, driving the yield of HER close to quantum efficiencies. This can be achieved by carefully selecting the cocatalysts that can produce consecutive charge transfers with minimum charge recombination in the conceptual cocatalyst/photocatalyst system [34]. Guided by density functional theory (DFT) calculations, Pt_{SA}-doped carbon nitride (CN) was successfully synthesized by a high-temperature calcination process for use as the cocatalyst to the optical semiconductor CuS. The synthesized photocatalyst system (Pt_{SA}-CN@CuS) exhibited an enhanced photocatalytic performance for water splitting, producing an H_2 generation of 13.9, 25.4, 11.8, and 10.2 mmol h^{-1} under the illumination of LED lamps (450, 530, 550, and 610), with the corresponding apparent quantum yield values of 30.8%, 50.3%, 22.2%, and 19.2%, respectively. STH conversion efficiency is calculated to be 0.5% under AM 1.5 illumination. This is the very first report of Pt_{SA} as the cocatalyst, which decreases the overpotential of CN during the water splitting and lowers interfacial resistance of the catalyst/cocatalyst and cocatalyst/electrolyte [35]. The photocatalyst Pt_{SA}-CN@CuS exhibits a good electrocatalytic H_2 generation compared to the reference CN@CuS (Figure 14.3a), and the lowest overpotential was even comparable to a commercial 20 wt% Pt/C. For the CN@CuS and Pt_{SA}-CN@CuS samples, the charge transfer resistance (R_{CT}) values were fitted to be 130.7 and 40.8 Ω/cm^{-2} (Figure 14.3b), respectively, further confirming that the monatomic Pt doping significantly decreases the interface resistance of CN, and thus boost the carrier migration across the catalyst–solution interface. The unique electronic properties of Pt_{SA} are responsible for the moderate interaction between hydrogen atoms and Pt_{SA}. As shown in Figure 14.3c,d, the hydrogen atom binds with the Pt_{SA} and the neighboring C. Upon adsorption, the Pt_{SA} transfers c. 0.6e to the bonded C atom, making the proton combining with the electron easily both from the Pt and C atoms. The single sites of Pt–C are thus responsible for H_2 generation, as shown in Figure 14.3e.

The development of SAPs is a huge step in the direction of reducing noble metal (like Pt) particle sizes from clusters to dots and eventually to single atoms, which effectively decreases noble metal usage and maximizes atom-utilization efficiency. Another recent example of a SAP supported on CN (Pt_{SA}/C_3N_4) is processed by a simple etching-anchoring strategy [36]. The Pt_{SA}/C_3N_4 photocatalyst exhibits a superb efficiency with an H_2 evolution rate value ~94.0 times higher than that of C_3N_4 alone. Again, with the aid of DFT calculations, it is possible to extrapolate that

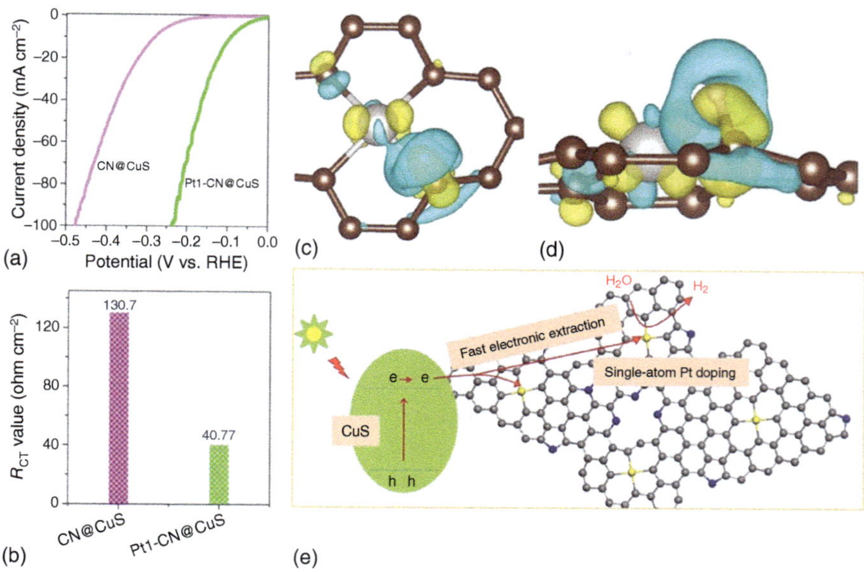

Figure 14.3 (a) Polarization curve, (b) the R_{CT} values of the sample measured in 0.5 M Na_2SO_4 electrolyte. (c) Top view and (d) side view of charge density difference map for the hydrogen adsorbed on the Pt_{SA}-$4C_0N$ site. Yellow and cyan represent positive and negative charge areas, respectively; the isosurface value of electron density is set to 0.005 e$Å^{-3}$. (e) Proposed photocatalytic H_2 generation of Pt_1-CN@CuS. Source: Yan et al. [35]. Reproduced with permission of John Wiley and Sons.

the introduction of Pt_{SA} onto C_3N_4 altered the band structure and created a bridging Pt—N bond for highly efficient electron transfer, thereby significantly enhancing the photocatalytic activity.

Using the benchmark TiO_2 as photocatalyst support for single-atom noble metals is also referred to as increasing the activity toward HER by water splitting. Pt_{SA}/TiO_2 photocatalysts can be well achieved by flame spray pyrolysis (FSP), controlling the loading of the noble metal at a very low level. The process offers an effective way to maximize the atom dispersion for surface catalysis. By this method, the FSP-made photocatalyst (0.1 Pt_{SA}/TiO_2 with 0.1% molar ratio of Pt to Ti) produced 108.5 times the amount of H_2 compared to the benchmark sample from commercial flame-made TiO_2 (P25) [37].

In a different approach, a set of photocatalysts was synthesized through an adsorption-limited wet impregnation process using bare and phosphate-modified TiO_2 as model supports and earth-abundant metals (Cu and Ni) with various loadings (0.008–5 wt%) as cocatalyst species. Based on the site-isolation strategy, increased turnover frequencies (TOF) enhancement factors, from 14 to 39 in Cu, were obtained. The PO_4/TiO_2 support was shown to further facilitate the morphological change toward smaller species, thus confirming the beneficial structural effect of surface modification on single-site stabilization. Overall, maximal TOF values of up to 657 and 1077 h^{-1} were achieved for Ni and Cu on bare TiO_2. In the case of the PO_4-modified support, the maximal TOF for $Cu/PO_4/TiO_2$ even surpassed maximal TOF for the analogous reference Au-based catalyst [38].

The uses of Ni-based cocatalysts were recently reviewed concerning the various concept schemes, including morphology engineering, interface engineering, nickel single atoms, alloys, solid solution, complexes, MOFs and dual cocatalysts, all of which effectively enhanced the photocatalytic performance. The photocatalytic H_2 production performance of some nickel alloys and complexes can compete and even exceed some of the noble metal Pt cocatalysts [39].

From the described works, it is evident that there is already a great deal of information, which urges to use rationally in further studies. At present, a stage has emerged where the trial, high-throughput experimentation, and kinetic studies need to be combined with computational techniques to predict and conceptualize the desired catalysts. Morphology and interface engineering need to be addressed by DFT and other computational techniques, profiting from the rapid development in artificial intelligence for materials development [40–44].

14.3.2 Photocatalytic CO_2 Reduction

The fast-growing consumption of fossil fuels and large carbon dioxide emissions have led to a severe energy crisis and global warming. Solar-driven CO_2 reduction reaction (CO_2RR) is considered a promising way to confront the predicaments.

Very recently, a cobalt single atom incorporated into Bi_3O_4Br was found to favor the charge transition, carrier separation, CO_2 adsorption, and activation. Using Co_{SA} and two-dimensional ultrathin Bi_3O_4Br atomic layers, light-driven CO_2 reduction was achieved with a selective CO formation rate, roughly 4 and 32 times higher than that of atomic layer Bi_3O_4Br and bulk Bi_3O_4Br, respectively [45]. By lowering the CO_2 activation energy barrier by stabilizing the electronic excited COOH* intermediates, the reaction path is tuned through the rate-limiting step from forming adsorbed intermediate COOH* to CO* desorption.

Surface strategies for catalytic CO_2 reduction have recently been reviewed, including several types of two-dimensional materials and single atoms. The first immediate advantage of 2D materials comes from a higher surface-to-volume ratio, which provides abundant surface area for solar harvesting (photocatalysis and PEC), surface active sites for CO_2 adsorption, and surface catalytic sites for initiating redox reactions. Some forms of carbon surfaces, such as carbon nanotubes, graphene, and graphene oxide, are excellent candidates as support for SAPs, mainly due to their large surface area and the ability to create active surface defects [46].

Among carbon materials, the already referred CN is also a promising visible-light-driven photocatalyst for CO_2 reduction. Besides being a photocatalyst on its own, it can also act as a photocatalyst support for dispersing metal single atoms.

On the other hand, rare-earth elements are gradually emerging in SAPs due to their unique electronic structure and optical properties. Using a strategy based on atom confinement and coordination is possible to decorate carbon nitride nanotubes (CN-NTs) with rare-earth single erbium (Er) atoms, with a tunable dispersion density. The photocatalyst Er_{SA}/CN-NT is highly efficient and robust, and exhibits outstanding CO_2RR performance in a pure-water system, with the crucial role of Er_{SA} being revealed by experimental results and DFT [47]. Using *in situ* synthesis and chemisorption is possible to manufacture Er_{SA} composite photocatalysts, coupling CN with Zn_2GeO_4 (a material widely used in photocatalysis, electrochemical sensors, and lithium-ion batteries due to the advantages of high chemical stability, thermal stability, and light stability). The photocatalyst Zn_2GeO_4:Er^{3+}/CN is more conducive to the tight junction of Zn_2GeO_4:Er^{3+} and CN and more favorable for CN to anchor rare-earth atoms. Under visible-light irradiation, the catalytic performance of Zn_2GeO_4:Er^{3+}/CN is more than five times higher than that of pure CN. There is evidence that Er^{3+} single atoms play a dual role in CO_2 molecular activation and 4f levels as energy transfer bridge [48].

Rare-earth La single atoms on CN were synthesized with La–N charge-transfer bridge acting as the active center for photocatalytic CO_2 reaction [49]. The electronic structure of the La–N bridge enables a high CO-yielding rate and CO selectivity superior to most of the CN-based photocatalytic CO_2 reductions referred to in the literature. The CO production rate remained nearly constant under light irradiation for five cycles of 20 hours. Experimental and DFT calculations indicate that the 4f and 5d orbitals of the La_{SA} and the p–d orbital hybridization of La–N atoms enabled the formation of the charge-transfer channel. The La–N charge bridges function as the key active center for CO_2 activation, rapid COOH* formation, and CO* desorption.

In addition to rare-earth elements, adding single atoms of selected transition metals to graphitic carbon nitrides allows the tailoring of the electronic and chemical properties of these 2D nanomaterials. A set of SAPs prepared with Ni^{2+}, Pt^{2+}, or Ru^{3+} by cation exchange, over CN was tested in several photocatalytic reactions as a proof of concept to demonstrate the impact of the surface charge and its polarization under illumination, such as H_2 evolution, CO_2 reduction, and photooxidation in general [50].

Photocatalysts containing gold single atoms on amino-group-modified graphitic carbon nitride (U-ACN) were synthesized through a mild and eco-friendly urea reduction method. In the photocatalytic CO_2 reduction, provided CO and CH_4 yields 1.97 and 4.15 times higher than the parent CN upon 150 minutes of visible-light irradiation [51]. The performance was due to the introduction of Au_{SA}, which lowered the energy barrier of CH_4 formation while narrowing the CN bandgap and preventing the recombination of charge carriers. In addition, the U-ACN photocatalysts displayed an enhanced CO_2 affinity owing to the amino groups in the photocatalysts introduced by the urea.

The activation of CO_2 on CN supported Co_{SA} sites was demonstrated by a remarkable yield of CO formation (94.9 $\mu mol\,g^{-1}\,h^{-1}$) when compared with that of pristine CN (0.25 $\mu mol\,g^{-1}\,h^{-1}$) [52]. Using DFT, the reaction pathways of CO_2 to CO on naked and Co_{SA}-modified CN were calculated and shown that single-atom Co sites (Co-CN) possess stronger CO_2 adsorption ability and lower the barrier for CO_2 hydrogenation activation. The C=O bonds on CO_2 were effectively activated owing to the strong interaction between Co 3d electrons and C 2p electrons of CO_2.

Following the increasing interest in SACs owing to the high atom utilization efficiency and high catalytic activity in the CO_2RR, recent reviews appeared to provide a detailed overview of the photocatalytic and photoelectrocatalytic applications to this reaction [53]. The exceptional features of CN, as a unique visible-light-responsive photocatalyst with excellent chemical stability and a tunable electronic structure, possesses a network of periodically separated N atoms that can act as anchoring sites for metal singleatoms. In this way, the exceptional features of the resulting SAPs containing Pd, Pt, Cu, Co, Ni, or Fe make them very efficient to the CO_2RR in solar applications as recently reviewed [54].

14.3.3 Photocatalytic Fixation of Nitrogen

Ammonia is undoubtedly one of the essential compounds for the chemical industry and agriculture. Currently, NH_3 is mainly produced by natural nitrogen fixation and industrial nitrogen fixation through the traditional Haber–Bosch process. While this process for ammonia production requires harsh reaction conditions (up to 25 MPa and 550 °C) and emits a large amount of greenhouse gases, the photocatalytic alternative normally works under milder conditions. The photocatalytic nitrogen reduction reaction (NRR) is, in principle, sustainable if sunlight is used as an energy source, and the dissolved nitrogen can come from natural or wasted water sources. Photocatalytic fixation of nitrogen to ammonia by a single Ru atom over TiO_2 nanosheets, rich in oxygen vacancies, has been reported to afford an NH_3 formation rate of 56.3 $\mu g\,h^{-1}\,g_{cat}{}^{-1}$ [55]. It was found that isolated Ru atoms likely weakened the hydrogen evolution by promoting the chemisorption of N_2. In addition, the presence of the noble atoms improved the charge carrier separation, which led to an improved photocatalytic reaction yield.

The importance of a rational design and development of atomic-scale catalysts for nitrogen fixation has been recently reviewed [56]. In that work, attention is drawn to the different mechanistic aspects of NRR under mild conditions. The importance of developing atomic defects to improve the NRR yield is evidenced in the synthetic approaches of generation SACs.

14.3.4 Photocatalytic Production of H_2O_2 with Environmental Significance

Hydrogen peroxide (H_2O_2) has been widely used as a liquid phase oxidant, being preferred in many industrial applications to other alternatives like gaseous molecular oxygen due to safety and transportation issues. Its use covers a wide range of applications from organic synthesis to environmental remediation, including energy applications, as an alternative to hydrogen as an energy vector. Au-modified MoS_2 nanosheets (Au@MoS_2) prepared by a deposition–reduction–immobilization process was reported in the photocatalytic production of H_2O_2 [57]. The Au modification increased the charge carrier separation, leading to longer-lived conductive electrons producing a more negative flat band potential for MoS_2. The Au@MoS_2 photocatalytic performance was remarkable for five cycles of reuse, in the absence of any added molecular oxygen (just normal aerobic conditions) or organic sacrificial reagent. The H_2O_2 productivity at pH 9 was further enhanced 7.4 times from that at pH 2. These results pave the way to novel variants of photo-Fenton-like applications.

In fact, there is a diversity of heterogeneous Fenton-like reactions that deal with the inherent limitations of the classic Fenton systems [58]. A collection of transition metal single-atom sites (SA-TM, TM = Cr, Mn, Fe, Co, Cu) coordinated onto pyrrolic N-rich g-C_3N_4 (PN-g-C_3N_4) scaffold was studied in relation with heterogeneous Fenton-like reactions [59]. The pyrrolic N-rich SA-TM catalytic sites exhibit excellent performances for the catalytic oxidation of bisphenol A, under visible light with outstanding cyclic stability and wide effective pH range (3.0–11.0), owing to accelerated production and separation of charge carriers as well as the cycling of $TM^{(n+1)+}/TM^{n+}$ couple. This study expands the family of the Fe-free Fenton-like systems and sets new strategies for the design of on-demand multifunctional SACs for advanced water remediation.

The emerging tendencies on synthetic and characterization strategies of SACs in advanced oxidation processes (AOP) were recently reviewed [60]. The removal of pollutants concomitant to the fast generation of reactive oxo-species ($SO_4^{\cdot -}$, $HO^{\cdot}$, $^1O_2$, and $O_2^{\cdot -}$) and high atomic efficiencies are appointed as the main reasons for future breakthroughs in single-atom photocatalytic AOP.

14.3.5 Photocatalytic Organic Synthesis

Non-oxygen coupling of benzylamine and several substituted derivatives to the corresponding imines in the presence of carbon nitride nanosheets (HCNS) with atomically dispersed $Cu_{SA}N_3$ moieties (Cu_{SA}@HCNS) was achieved with high performances under visible light activation ($\lambda > 420$ nm) [61]. The photocatalyst was prepared by a molecular assembly approach where the Cu_{SA} are embedded within the polymeric carbon nitride (PCN) nanosheets. The embedded Cu_{SA} acts as a modifier to effectively modulate the electron structure and promote interfacial charge transfer of PCN. The effect strongly depends on the location of Cu_{SA} in the PCN. The coordination method is a very effective strategy in terms of the improvement in photocatalysis of PCN owing to the intensified metal–PCN interaction.

The effect of the metal particle size on the support has been addressed in the artificial photocatalytic oxidation of the biomass-derived benzaldehyde with simultaneous proton reduction (to generate H_2) in a closed redox cycle driven by supported Pt_{SA} (0.2 nm), nanoclusters (1 nm), and nanoparticles (4 and 7 nm diameter), over g-C_3N_4 [62]. The reaction rate follows the dimension of the Pt particle size in the order of single atom > nanoclusters > NPs. Once again, it is crucial that the photo-induced charge carriers, more likely to be trapped by Pt_{SA}, warranting the efficiency of charge separation, and slowing down the recombination. The amount of H_2 was found to be halved to the amount of oxidation production (benzaldehyde) owing to the electron donation/consumption rate for proton reductions (two electrons for one H_2) with the rate for oxidation (one electron). For the observed reactivity enhancement for Pt_{SA} catalyst is crucial its preference for photoinduced hole delivery, leading to oxidation. Consequently, free electrons accumulate on the SAC for proton reduction, contributing to the H_2 production on the other half-reaction.

In pharmaceutical synthesis, β-ketosulfones, are an essential class of sulfone-containing organic molecules. Biomimetic photocatalysis using Fe_{SA} sites developed over carbon nitride coupled with hemin (CNH) are reported for the visible light-promoted sulfonation of alkenes to produce β-ketosulfones with up to 94% yield [63]. The experimental results show that efficient charge carrier separation is attained by the CN structural motif inserted in the CNH. The SAP can be irradiated under near-infrared light with a satisfactory yield and is also feasible for the sulfonation reactions of androstenones. Owing to its high catalytic performance, reuse ability, high turnover number, and excellent functional group tolerance, this biomimetic catalysis-based synthesis system can be further explored in other organic transformations to produce several other bioactive drugs.

In the case of the benchmark photocatalytic support TiO_2, using first-principles simulation calculation of the energy barrier of acetone dehydrogenation and conversion over a series of loaded noble metal SAs (Ru, Rh, Pd, Ag, Os, Ir, Pt, and Au) photocatalysts (M_{SA}/TiO_2), predictions show that a Pt single atom-loaded TiO_2 photocatalyst (Pt_{SA}/TiO_2) can enable efficient methyl activation, $CH_3COCH_2^{\cdot}$ radical formation, and hydrogen production most from acetone, as an example of C–C coupling [64]. The predictions were confirmed by the photocatalytic experiments, where Pt_{SA}/TiO_2 exhibited a photocatalytic activity of 3.87 mmol g^{-1} h^{-1} for the direct coproduction of 2,5-hexanedione and hydrogen from acetone with a selectivity of 93% – the results are well above those obtained for other M_{SA}/TiO_2 or when compared to catalysts containing Pt-loaded nanoparticle clusters.

The reactivity of a SAP Ni/TiO_2 was explored in the visible-light-induced site-selective sulfonation of enamides to give amidosulfones with significant yields up to 99% [65]. These reactions are a good example of site-selective sulfonation of enamides to construct α-amidosulfones and β-propionamidosulfones with the possibility of using solar activation. The synthesized Ni SAPs prove adequate to gram-scale product yield with a high turnover number and excellent recyclability.

14.4 SACs for Photoelectrocatalysis

Apart from the applications in photocatalysis, SACs have recently also been explored as cocatalysts to couple with semiconductor electrodes for PEC. Compared to photocatalysis, where redox reactions occur simultaneously on the semiconductor particle surface, in PEC, the reduction and oxidation reactions are spatially separated and take place at the photocathode and photoanode, respectively. In this way, the stringent requirements for semiconductor bandgap and band edge positions can be relaxed, and the recombination of photogenerated electrons and holes be suppressed to a large extent. By far, PEC has been extensively investigated for many reactions such as HER, oxygen evolution reaction (OER), CO_2RR, NRR, and organic compound oxidation [66–70]. Like photocatalysis, the cocatalysts play an important role in PEC and are almost indispensable for achieving a decent photo-to-fuel efficiency [71]. Although the commonly used nanoparticulate or thin-film cocatalysts in PEC can substantially expedite the reaction kinetics, they will also introduce unfavorable parasitic light absorption, compromising PEC performance. SACs, to this end, are better alternatives since they are nearly transparent to the incident light. Moreover, recent studies also demonstrated that the loading of SACs as cocatalysts could alter the electronic structure of the host material, leading to better and more efficient charge separation and providing suitable sites for adsorption of hydrogen and accumulation of photo-induced electrons [72]. In the following, recent advances in using SACs for photoelectrocatalytic HER, OER, CO_2RR, and NRR will be summarized and discussed.

14.4.1 Photoelectrocatalytic Hydrogen Evolution

Graphitic carbon nitride has proved to be a suitable support material able to offer a large number of coordinatively unsaturated or defective sites, which enable the immobilization of metal atoms as cocatalysts [73]. Bu and coworkers have reported Rh_{SA} cocatalysts loaded on porous g-C_3N_4 (denoted as Rh_{SA}/CN(H)) and investigated their photocatalytic and photoelectrocatalytic performance toward HER [74]. The layered, porous structure of g-C_3N_4 was capable of anchoring Rh_{SA} via strong interfacial interaction. Furthermore, the researchers further performed a phosphating treatment at elevated temperature considering that Rh-P may show better HER performance. They converted Rh_{SA}/CN(H) into rhodium phosphide single-site cocatalysts supported on the porous g-C_3N_4 (denoted as Rh-P/CN(H)).

Interestingly, even if the phosphating treatment was carried out at 300 °C for 2 hours, the as-obtained Rh-P was still dispersed atomically on g-C_3N_4 without noticeable aggregation (Figure 14.4a). Photoelectrocatalytic measurements revealed that the Rh-P/CN(H) shows an early onset for HER and a much higher PEC current density under simulated sunlight, than those of pure g-C_3N_4 (Figure 14.4b,c). This indicates that the Rh-P single-site can significantly reduce the surface energy barrier of C_3N_4 to accelerate the consumption of photogenerated holes and increase the density of photogenerated electrons for hydrogen evolution. This assumption is further supported by DFT calculations, which show that electrons become accumulated around the Rh-P single-site, rendering fast electron transfer for the proton reduction. Furthermore, compared to pure g-C_3N_4 and Rh_{SA}/CN(H), the Rh-P/CN(H) also exhibited a smaller standard Gibbs free energy (ΔG) for hydrogen adsorption, suggesting that Rh-P/CN(H) is more intrinsically active for HER.

Besides direct coupling with semiconductors, SAs can also be deposited first on a conductive support and then altogether loaded on a photocathode as cocatalysts for photoelectrocatalytic HER. For example, Ramalingam et al. recently demonstrated that Ru_{SA} could be dispersed on a two-dimensional (2D) titanium carbide ($Ti_3C_2T_x$) MXene support through the coordination interaction between Ru_{SA} and the nitrogen (N) and sulfur (S) heteroatom dopants [75]. The strong electronic coupling between Ru_{SA} and the $Ti_3C_2T_x$ MXene support via coordinative bonding

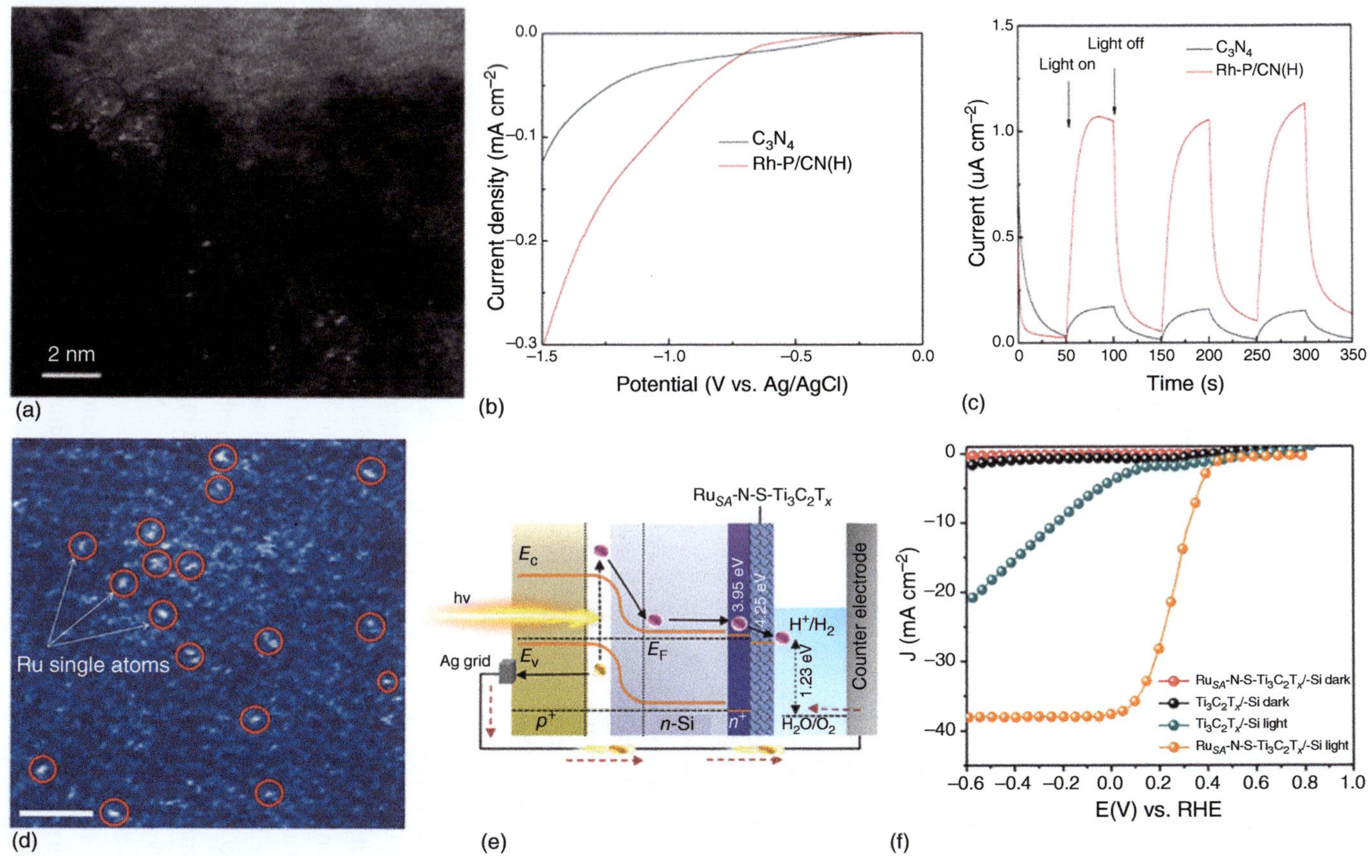

Figure 14.4 (a) HAADF-STEM image of Rh-P/CN(H). (b) Linear scan voltammograms of the Rh-P/CN(H)/FTO and g-C_3N_4/FTO were measured under simulated sunlight (100 mW cm^{-2}). Scan rate: 0.02 V s^{-1}. (c) Photo-response of the Rh-P/CN(H)/FTO and g-C_3N_4/FTO electrode measured at 0 V vs. Ag/AgCl. Electrolyte: 10 vol% triethanolamine solution. Source: Chen et al. [74]. Reproduced with permission of Elsevier. (d) Magnified HAADF-STEM image of Ru_{SA}-N-S-$Ti_3C_2T_x$ (bright dots marked with red circles indicate the Ru_{SA} on the $Ti_3C_2T_x$ MXene support). (e) Band structure diagram of the Ru_{SA}-N-S-$Ti_3C_2T_x$/n^+np^+-Si photocathode. (f) Current density–voltage (J–V) characteristic curves of the $Ti_3C_2T_x$ and Ru_{SA}-N-S-$Ti_3C_2T_x$ electrocatalysts integrated on the n^+np^+-Si photocathode (0.5 M H_2SO_4, scan rate: 20 mV s^{-1}, AM 1.5 illumination). Source: Ramalingam et al. [75]. Reproduced with permission of John Wiley and Sons.

with N and S atoms, as verified by XPS, XANES and EXAFS measurements, enables isolated Ru atoms to be anchored on the surface (Figure 14.4d) and thereby a high HER activity to be achieved in the as-prepared Ru_{SA}-N-S-$Ti_3C_2T_x$. When drop-casting the Ru_{SA}-N-S-$Ti_3C_2T_x$ on the n^+np$^+$-Si photocathode, a favorable band structure is formed since Ru_{SA}-N-S-$Ti_3C_2T_x$ has a higher work function than that of n^+np$^+$-Si, as illustrated in Figure 14.4e, which facilitates the continuous shuttle of photogenerated electrons to the active sites of Ru_{SA}-N-S-$Ti_3C_2T_x$ cocatalysts and prevents charge carrier recombination both at Si and at the Si/Ru_{SA}-N-S-$Ti_3C_2T_x$ interface. Consequently, a photocurrent density as large as 37.6 mA cm^{-2} has been achieved (Figure 14.4f), about 10 times higher than that of the $Ti_3C_2T_x$/n^+np$^+$-Si photocathode in the absence of Ru_{SA}.

14.4.2 Photoelectrocatalytic Oxygen Evolution

Transition metal oxides such as TiO_2, WO_3, α-Fe_2O_3, and $BiVO_4$ are the most commonly used semiconductor photoanode materials, given their outstanding chemical stability in aqueous electrolyte and their positive valence band position relative to the O_2/H_2O redox potential, which is large enough to enable the injection of photogenerated holes into the electrolyte to produce O_2 [76]. However, the large bandgap of metal oxide semiconductors usually only allows them to absorb a small portion of light in the solar spectrum, and their low electrical conductivity is unfavorable for charge transport. For these reasons, it is of particular importance to introduce cocatalysts with minimal parasitic light absorption. To this end, SACs meet the needs and have therefore recently emerged as effective cocatalysts for photoanodes.

Using a hydrothermal synthesis method with the presence of titanium butoxide and bismuth nitrate ($Bi(NO_3)_3$), Wang et al. lately reported the fabrication of TiO_2 nanorod arrays decorated with Bi_{SA} as cocatalysts. The best-performing Bi_{SA}/TiO_2 photoanode exhibited a high photocurrent density of 1.65 mA cm^{-2} at 1.23 V vs. reversible hydrogen electrode (RHE), being the highest value compared to the state-of-the-art TiO_2-based photoanode materials surveyed by the authors [77]. HAADF-STEM imaging confirmed the formation of isolated Bi_{SA} and tiny clusters on the TiO_2 nanorod surfaces, and XPS investigation suggested that Bi was present in a slightly oxidized state ($Bi^{\delta+}$) and formed a Bi—O—Ti bond on the surface. Comprehensive PEC tests revealed that besides the high photocurrent density, Bi_{SA}/TiO_2 showed a higher applied bias photon-to-current efficiency (ABPE), more negative saturation potential, higher incident photon to current conversion efficiency (IPCE), improved light absorption, and longer electron lifetime, concerning the pure TiO_2 photoanode, which can be attributed to the effective charge separation and transport as well as the localized surface plasmon effect arising from the Bi_{SA} cocatalyst.

While TiO_2 has long been studied as a photocatalyst and photoanode for PEC water splitting, it only absorbs the ultraviolet region of the solar spectrum, which inherently limits its PEC performance. Hence, α-Fe_2O_3, i.e. hematite with a suitable bandgap (1.9–2.2 eV) for visible light absorption, has emerged as a promising photoanode material and has been extensively investigated in the past decade

[78, 79]. Recently, Cui and coworkers reported a robust and atomically dispersed hybrid catalyst formed *in situ* on a hematite semiconductor photoanode during photoelectrocatalytic oxygen evolution [80]. This was realized by firstly depositing NiO_x sites onto the hematite surface (i.e. H-NiO_x) through a transient electroreduction of nickel cations, followed by *in situ* electrostatic absorption of negatively charged alkali-stable $[Ir(OH)_6]^{2-}$ dissolved in the NaOH electrolyte. Successful anchoring of Ir_{SA} on H-NiO_x was confirmed by HAADF-STEM imaging. Furthermore, the authors investigated in detail the influence of different MO_x (M = Fe, Cu, Co, Ni, and Ti) sublayers, the layer thickness and the concentrations of $[Ir(OH)_6]^{2-}$ on the Ir adsorption and the PEC performance of the hematite, and they obtained an optimal photoanode structure. The reaction mechanism was further comprehensively examined, and the NiO_x sites were found to possess the highest capacity of hole accumulation, compared to other MO_x sites investigated, and be able to efficiently transfer the photogenerated holes to the Ir sites for OER catalysis through the *in situ* formed weak "Fe/Ni—O—Ir" bonds. This rationally explains why H-NiO_x outperforms other H-MO_x photoanodes. Consequently, the H-NiO_x photoanode coupled with Ir single sites showed a marked activity enhancement relative to the bare hematite, with a TOF of 2.4–12.7 s^{-1} per Ir site at 1.23 V vs. RHE. Remarkably, even if the bonding of Ir_{SA} to NiO_x was supposed to be weak, the photoanode exhibited outstanding stability for more than 80 hours in 4.0 M NaOH under AM 1.5 illumination upon the application of a constant potential of 1.23 V vs. RHE.

Given that the OER involves four proton-coupled electron transfer steps, having more than one metal atom per active site is favorable to accelerate the reaction kinetics and achieve better selectivity, but this remains challenging. Recently, Zhao and coworkers designed a dinuclear heterogeneous catalyst (DHC) consisting of two Ir atoms per catalytic site bridged by oxygen (Ir–O–Ir) that are stably bound to the hematite photoanode [81]. To do so, they soaked a hematite support in the Ir homodimer solution, followed by photochemical treatment with a UV light for a certain time, which removed the organic ligands in the homodimer. The HAADF-STEM characterization unambiguously confirmed the dinuclear nature of the cocatalysts, and the *in situ* CO DRIFT spectroscopy and X-ray absorption spectroscopy (XAS) further elucidated the formation of the Ir—O—Ir, but not Ir—Ir bonding. DFT calculations suggested that the threefold hollow site on the OH-terminated α-Fe_2O_3 surface could stabilize the Ir–O–Ir structure and the atomic model thus-built showed excellent agreement with the experimental HAADF-STEM images. The hematite photoanode coupled with Ir DHC cocatalysts exhibited substantially enhanced photoelectrocatalytic performance compared to the photoanodes decorated with Ir_{SA} cocatalysts because the Ir atomic pairs can reduce the energy need for the O—O bond formation, as proved by DFT calculations. More impressively, the anchored Ir DHC cocatalysts showed outstanding stability against detachment or aggregation upon continuous PEC water oxidation for at least 10 hours.

In addition to the direct usage of SACs as cocatalysts on semiconductor photoanodes, researchers also managed to load SAs on a doped carbon support, then used as cocatalysts for PEC water oxidation. For example, Hou and coworkers recently reported the synthesis of a composite nanocarbon electrocatalyst composed of atomically dispersed S|NiN$_x$ species embedded in porous carbon nanosheets (S|NiN$_x$-PC), by pyrolysis of ternary dicyandiamide–thiophene–nickel salt nanocomposites grown on electrochemically exfoliated graphene foil, followed by an acid leaching treatment [82]. Comprehensive experimental and computational studies revealed that the atomically isolated Ni atoms are coordinated with three N atoms and one S atom in the carbon matrix, which can create sufficient localized reactive sites by modifying the local charge distribution on the carbon surface and reducing the potential barriers of the elementary reactions, boosting its OER kinetics. Consequently, when loading the S|NiN$_x$-PC on the Fe_2O_3 nanorod array (Fe_2O_3-NA) for PEC water splitting, the photoanode showed a remarkable cathodic shift in the onset potential, high charge transport and charge transfer efficiency, leading to a photocurrent density of 1.58 mA cm^{-2} at 1.23 V vs. RHE and an ABPE of 0.24% at 0.92 V vs. RHE. Similarly, Yang et al. demonstrated the chemical interfacing of carbon nanosheets embedded with Ni single active sites to the α-Fe_2O_3 nanorod photoanode [83], which achieved a higher photocurrent density of 1.85 mA cm^{-2} at 1.23 V vs. RHE and good stability for over 15 hours. Besides, Song et al. recently reported a unique composite photoanode consisting of Co_{SA} anchored between the lamina g-C_3N_4 (l-g-C_3N_4) and porous carbon spheres (PCS) [84]. The as-prepared l-g-C_3N_4@Co/PCS photoanode exhibited higher PEC performance with respect to l-g-C_3N_4 and Co/PCS references. Both experimental studies and DFT calculations revealed that Co_{SA} act as electron mediators, accelerating interfacial electron transfer between l-g-C_3N_4 and PCS. The l-g-C_3N_4 works as an electron collector preventing the recombination of photogenerated carriers. As a result, a photocurrent density of 5.08 mA cm^{-2} was achieved at 1.23 V vs. RHE. Furthermore, the N site on the N–Co–C "bridges" interface was proposed to be the main active center of the composite photoanodes.

It is worth mentioning that photoelectrocatalytic OER is generally taking place in strongly oxidative conditions where the carbon support is very likely to be corroded over a long period of operation, which would deactivate the SAC. However, the long-term stability of these carbon-supported SAC cocatalysts for PEC OER has not been thoroughly assessed so far.

14.4.3 Photoelectrocatalytic Carbon Dioxide Reduction and Nitrogen Reduction

CO_2 reduction would be a key enabler to close the "carbon cycle," contributing to reaching global carbon neutrality. While intensive research effort has been dedicated to electrochemical CO_2RR in the past decade, PEC–CO_2RR has recently received considerable attention [68, 85, 86]. However, using SACs as cocatalysts for PEC–CO_2RR has been rarely investigated so far. Zhang and coworkers recently proposed an artificial photosynthetic (APS) cell [87], where different semiconductors including N-doped rutile TiO_2 (N-TiO_2), Sn-doped $BiVO_4$ (Sn-$BiVO_4$) and Sn-doped Fe_2O_3 (Sn-Fe_2O_3) were used as the photoanode, and sulfur and nitrogen codoped graphene with atomically dispersed Ni active sites (Ni-SNG) was used as CO_2RR-catalyst cathode (Figure 14.5a). Such APS cells had a total

Faradaic efficiency of unity toward the conversion of CO_2 to syngas ($CO + H_2$) under simulated solar irradiation ($100\,mW\,cm^{-2}$), and the gaseous product yield showed strong dependence on the photoanode material (Figure 14.5b) with the N-TiO_2 APS cell being the best-performing one.

Interestingly, the H_2/CO ratio in the resulting syngas could be tuned over a wide range between 5 : 1 and 1 : 5 by changing the work voltage of the cell (Figure 14.5c). In particular, in the N-TiO_2 APS cell, a maximal CO evolution rate of $154.9\,mmol\,g^{-1}\,h^{-1}$ was achieved, with a solar-to-syngas energy conversion efficiency as high as 13.6% and a turnover frequency of $529.5\,h^{-1}$ for CO conversion. Through scanning electrochemical microscopy–atomic force microscopy (SECM–AFM) observations, the researchers confirmed that the Ni_{SA} sites function as catalytic active centers for the multielectron reduction of CO_2 to CO. Furthermore, they demonstrated that the CO_2-to-syngas conversion could be driven by commercial PV cells. Overall quantum efficiency of 6.29% was achieved if four PV cells were

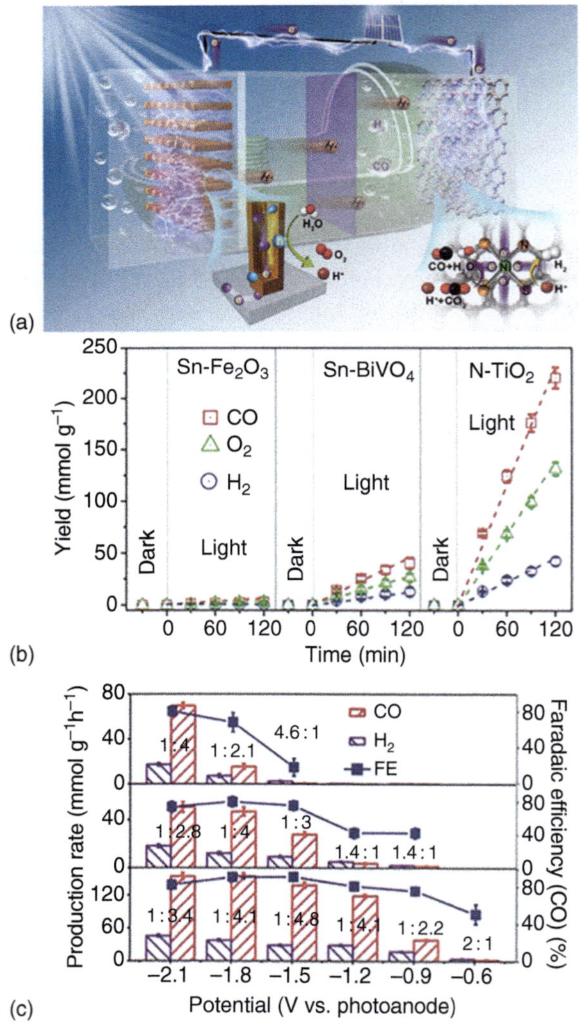

Figure 14.5 (a) Model and mechanism of the photovoltaic-coupled two-compartment artificial photosynthetic (APS) cell for syngas production from CO_2 and H_2O. (b) Time course for the formation of gaseous products in the N-TiO_2, Sn-$BiVO_4$, and Sn-Fe_2O_3 APS cells at 1.5 vs. the photoanode. (c) Gas-production rates and Faradaic efficiency in the N-TiO_2, Sn-$BiVO_4$, and Sn-Fe_2O_3 APS cells as a function of the work voltage. Source: Zhang et al. [87]. Reproduced with permission of John Wiley and Sons.

used in tandem to power the APS cell for syngas production, suggesting that the PV–APS cells are viable solutions for the chemical industry to produce syngas in a sustainable way.

Using a similar photoanode + electrocathode configuration, Chen and coworkers have reported efficient photoelectrocatalytic conversion of CO_2 into methanol [88]. They used anodized TiO_2 nanotube arrays as the photoanode. They designed a hybrid cathode catalyst comprising atomically dispersed metal-N_x species anchored on a porous carbon (NC) matrix with embedded CuCo bimetallic NPs (CuCo/NC). Through XANES characterization and DFT calculations, the researchers confirmed that electrons transfer from Co to Cu species, leading to preferable adsorption of the CO* intermediates on the surface of CuCo/NC to be further reduced to the CHO* intermediates, which accelerates the formation of methanol. Furthermore, the 3D macroporous carbon support with a coordinated N component allows numerous effective catalytically active sites to interact with CO_2. Consequently, a maximal Faradaic efficiency of 60.2% and a maximal carbon atom conversion rate of 4098 nmol h^{-1} cm^{-2} can be achieved during the photoelectrocatalytic conversion of CO_2 to methanol, substantially higher than those of control samples composed of monometallic components (i.e. Cu/NC and Co/NC).

Aside from CO_2RR, solar-driven photoelectrocatalytic reduction of nitrogen to ammonia has also drawn considerable attention in recent years [67], showing promise as a sustainable process for replacing the energy-intensive Haber–Bosch process that contributes significantly to the global CO_2 emissions (c. 2%) [89]. While a variety of semiconductor materials have been explored for PEC–NRR [90–92], there are so far very few reports on the use of SACs to catalyze the NRR in a PEC cell. Wang et al. recently reported the conversion of N_2 into NH_3 with the aforementioned photoanode + electrocathode configuration, where the photoanode is composed of cobalt phosphate (CoPi) modified Ti-doped Fe_2O_3 (CoPi/Ti-Fe_2O_3) nanoarrays, and the cathode comprises Co_{SA} [93]. They found that the Ti-doping and introducing a CoPi layer can promote the charge conductivity and water oxidation efficiency of the Fe_2O_3 photoanode, respectively. At the same time, the Co SAC shows high electrocatalytic activity toward the NRR. As a result, a large NH_3 yield rate of 1021.5 μg $mg_{Co}{}^{-1}$ h^{-1} (12.26 μg $mg_{cat.}{}^{-1}$ h^{-1}) and high Faradaic efficiency of 11.9% can be achieved at an applied potential of 1.2 V vs. RHE on the photoanode under AM 1.5 G simulated solar irradiation (100 mW cm^{-2}). Notably, the NRR performance closely depends on the PEC performance of the photoanode. When the same Co-SAC cathode was coupled with the bare Fe_2O_3 or Ti-Fe_2O_3 nanoarray photoanode and tested under the same conditions, the NH_3 yield rate and Faradaic efficiency substantially decreased, highlighting the importance of semiconductor photoelectrodes in boosting the NRR performance.

14.5 Summary and Outlook

The Holy Grail of any photocatalytic or photoelectrocatalytic process resides on the performance of charge separation of the photogenerated carriers. Ideally, noncompeting channels for the oxidative and reductive pathways must coexist for maximum performance. The contribution of the single atoms to effective separation of the photogenerated charge carriers was put in evidence in most of the examples highlighted in this chapter. Most of the effort must be directed toward any strategies leading to an effective dispersion and use of the metal single atoms within this objective of maximizing charge separation. The paths to fulfil such an objective include new conceptual synthesis with the development of entirely new methods or adopting techniques from other fields of materials fabrication. In this aspect, computer-aided additive manufacture will undoubtedly come to play a role soon [94]. Additive manufacturing can have an enormous potential to intensify catalytic processes to make them more efficient and sustainable, producing advanced designs of photocatalytic structures with increased adaptation and superior performance.

Equally important to the enhancement of the manufacturing techniques are the characterization methods. In parallel, these two aspects essential to precisely control the location and coordination environments of the metal single atoms are determinant to the SAP performance [53, 60, 95]. The characterization will encompass in the future great developments, especially in the field of time-resolved imaging for following surface reactions [13, 63, 65, 95, 96]. Using *in situ* time-resolved spectroscopic techniques [97] or hard X-ray imaging techniques [98], the details of the dynamics of the surfaces can be followed in what concerns not only its structure but also the evolution of the adsorbed species with real access to the key features of reaction mechanisms. Kinetic analysis is, therefore, on the verge of a significant leap aided by these techniques.

Despite high atomic efficiency owing to the almost complete dispersion of metal atoms over surfaces, the SAPs need to address issues of cost, robustness, and efficient solar light activation to become industrially relevant. Robustness is related chiefly to stability and reusability, therefore closely linked to the nature of the support and interactions of the metal SA with the support. Exploiting these interactions can only be achieved in a case-to-case analysis; therefore, a universal SAP is still far from set [60, 64]. However, the tendency toward adaptable support, pristine metal-free, light-sensitive, and capable of stabilizing the metal atoms with the tailoring of the electronic and chemical properties of the resulting photocatalyst is finding resonance in the recently adopted 2D structures of graphitic carbon nitrides [8, 38, 50, 53, 54]. The advantage of this support is that it combines visible light activation, chemical stability, modulated electronic structure. At the same time, its unique 2D architecture contains periodically separated N atoms that can effectively anchor the metal single atoms.

Grubbing these very recent issues and challenges will undoubtedly establish the SAPs as the new highways of technological development.

Acknowledgments

This work was financially supported by Project 2DMAT4FUEL (POCI-01-0145-FEDER-029600-COMPETE2020 – FCT/MCTES – PIDDAC, Portugal) funded by ERDF through COMPETE2020 – Programa Operacional Competitividade e

Internacionalização (POCI) – and by national funds through FCT – Fundação para a Ciência e a Tecnologia; Base-UIDB/50020/2020 and Programmatic-UIDP/50020/2020 Funding of Associate Laboratory LSRE-LCM by national funds through FCT/MCTES (PIDDAC). BFM acknowledges funding from NORTE-06-3559-FSE-000069 – Highly Qualified Human Resources for CoLAB NET_4CO_2 – Network for a Sustainable CO_2 Economy – funded by the European Social Fund (FSE).

References

1 Scaiano, J.C. and Lanterna, A.E. (2020). A green road map for heterogeneous photocatalysis. *Pure and Applied Chemistry* 92 (1): 63–73.

2 Ciamician, G. (1912). The photochemistry of the future. *Science* 36 (926): 385–394.

3 Gao, C., Low, J., Long, R. et al. (2020). Heterogeneous single-atom photocatalysts: fundamentals and applications. *Chemical Reviews* 120 (21): 12175–12216.

4 Xia, B., Zhang, Y., Ran, J. et al. (2021). Single-atom photocatalysts for emerging reactions. *ACS Central Science* 7 (1): 39–54.

5 Zhang, Q. and Guan, J. (2020). Recent progress in single-atom catalysts for photocatalytic water splitting. *Solar RRL* 4 (9): 2000283.

6 Wang, X., Maeda, K., Thomas, A. et al. (2009). A metal-free polymeric photocatalyst for hydrogen production from water under visible light. *Nature Materials* 8 (1): 76–80.

7 Lima, M.J., Silva, A.M.T., Silva, C.G. et al. (2017). Graphitic carbon nitride modified by thermal, chemical and mechanical processes as metal-free photocatalyst for the selective synthesis of benzaldehyde from benzyl alcohol. *Journal of Catalysis* 353 (1): 44–53.

8 Fu, J., Wang, S., Wang, Z. et al. (2020). Graphitic carbon nitride based single-atom photocatalysts. *Frontiers of Physics* 15 (3): 33201.

9 Millet, M.-M., Algara-Siller, G., Wrabetz, S. et al. (2019). Ni single atom catalysts for CO_2 activation. *Journal of the American Chemical Society* 141 (6): 2451–2461.

10 Lee, B.-H., Park, S., Kim, M. et al. (2019). Reversible and cooperative photoactivation of single-atom Cu/TiO_2 photocatalysts. *Nature Materials* 18 (6): 620–626.

11 Pan, Y., Qian, Y., Zheng, X. et al. (2020). Precise fabrication of single-atom alloy co-catalyst with optimal charge state for enhanced photocatalysis. *National Science Review* 8 (1): 1–8.

12 Zhang, X., Chen, A., Chen, L. et al. (2021). 2D materials bridging experiments and computations for electro/photocatalysis. *Advanced Energy Materials* https://doi.org/10.1002/aenm.202003841.

13 Filez, M., Vesely, M., Garcia-Torregrosa, I. et al. (2021). Chemical imaging of hierarchical porosity formation within a zeolite crystal visualized by small-angle X-ray ccattering and in-situ fluorescence microscopy. *Angewandte Chemie International Edition* 60 (25): 13803–13806.

14 Pei, G.X., Dzade, N.Y., Zhang, Y. et al. (2021). Identification of photoexcited electron relaxation in a cobalt phosphide modified carbon nitride photocatalyst. *ChemPhotoChem* 5 (4): 330–334.

15 Stanciakova, K. and Weckhuysen, B.M. (2021). Water-active site interactions in zeolites and their relevance in catalysis. *Trends in Chemistry* 3 (6): 456–468.

16 Xing, J., Chen, J.F., Li, Y.H. et al. (2014). Stable isolated metal atoms as active sites for photocatalytic hydrogen evolution. *Chemistry – A European Journal* 20 (8): 2138–2144.

17 Wang, Q., Zhang, D., Chen, Y. et al. (2019). Single-atom catalysts for photocatalytic reactions. *ACS Sustainable Chemistry & Engineering* 7 (7): 6430–6443.

18 Eswar, N.K.R., Singh, S.A., and Heo, J. (2019). Atomic layer deposited photocatalysts: comprehensive review on viable fabrication routes and reactor design approaches for photo-mediated redox reactions. *Journal of Materials Chemistry A* 7 (30): 17703–17734.

19 Liu, J., Bunes, B.R., Zang, L. et al. (2018). Supported single-atom catalysts: synthesis, characterization, properties, and applications. *Environmental Chemistry Letters* 16 (2): 477–505.

20 Liu, P., Zhao, Y., Qin, R. et al. (2016). Photochemical route for synthesizing atomically dispersed palladium catalysts. *Science* 352 (6287): 797–800.

21 Wei, H., Huang, K., Wang, D. et al. (2017). Iced photochemical reduction to synthesize atomically dispersed metals by suppressing nanocrystal growth. *Nature Communications* 8 (1): 1490.

22 Wei, H., Wu, H., Huang, K. et al. (2019). Ultralow-temperature photochemical synthesis of atomically dispersed Pt catalysts for the hydrogen evolution reaction. *Chemical Science* 10 (9): 2830–2836.

23 Li, T., Liu, J., Song, Y. et al. (2018). Photochemical solid-phase synthesis of platinum single atoms on nitrogen-doped carbon with high loading as bifunctional catalysts for hydrogen evolution and oxygen reduction reactions. *ACS Catalysis* 8 (9): 8450–8458.

24 Machado, B.F. and Serp, P. (2015). Chapter 8: Photocatalysis on nanostructured carbon supported catalysts. In: *Nanostructured Carbon Materials for Catalysis* (eds. B.F. Machado and P. Serp), 412–444. UK: The Royal Society of Chemistry.

25 Cheng, N., Zhang, L., Doyle-Davis, K. et al. (2019). Single-atom catalysts: from design to application. *Electrochemical Energy Reviews* 2 (4): 539–573.

26 Kaiser, S.K., Chen, Z., Faust Akl, D. et al. (2020). Single-atom catalysts across the periodic table. *Chemical Reviews* 120 (21): 11703–11809.

27 Li, X., Yang, X., Zhang, J. et al. (2019). In situ/operando techniques for characterization of single-atom catalysts. *ACS Catalysis* 9 (3): 2521–2531.

28 Bai, S., Jiang, J., Zhang, Q. et al. (2015). Steering charge kinetics in photocatalysis: intersection of materials syntheses, characterization techniques and theoretical simulations. *Chemical Society Reviews* 44 (10): 2893–2939.

29 Gao, C., Chen, S., Wang, Y. et al. (2018). Heterogeneous single-atom catalyst for visible-light-driven high-turnover CO_2 reduction: the role of electron transfer. *Advanced Materials* 30 (13): 1704624.

30 Wang, A., Li, J., and Zhang, T. (2018). Heterogeneous single-atom catalysis. *Nature Reviews Chemistry* 2 (6): 65–81.
31 Yang, X.-F., Wang, A., Qiao, B. et al. (2013). Single-atom catalysts: a new Frontier in heterogeneous catalysis. *Accounts of Chemical Research* 46 (8): 1740–1748.
32 Zhu, Y., Sun, W., Luo, J. et al. (2018). A cocoon silk chemistry strategy to ultrathin N-doped carbon nanosheet with metal single-site catalysts. *Nature Communications* 9 (1): 3861.
33 Volokh, M., Peng, G., Barrio, J. et al. (2019). Carbon nitride materials for water splitting photoelectrochemical cells. *Angewandte Chemie International Edition* 58 (19): 6138–6151.
34 Takata, T., Jiang, J., Sakata, Y. et al. (2020). Photocatalytic water splitting with a quantum efficiency of almost unity. *Nature* 581 (1): 411–414.
35 Yan, J., Ji, Y., Batmunkh, M. et al. (2021). Breaking platinum nanoparticles to single-atomic Pt-C_4 Co-catalysts for enhanced solar-to-hydrogen conversion. *Angewandte Chemie International Edition* 60 (5): 2541–2547.
36 Hu, Y., Qu, Y., Zhou, Y. et al. (2021). Single Pt atom-anchored C_3N_4: a bridging Pt—N bond boosted electron transfer for highly efficient photocatalytic H_2 generation. *Chemical Engineering Journal* 412 (1): 128749.
37 Gao, F., Xu, Z., and Zhao, H. (2021). Flame spray pyrolysis made Pt/TiO_2 photocatalysts with ultralow platinum loading and high hydrogen production activity. *Proceedings of the Combustion Institute* 38 (4): 6503–6511.
38 Ayala, P., Giesriegl, A., Nandan, S.P. et al. (2021). Isolation strategy towards earth-abundant single-site Co-catalysts for photocatalytic hydrogen evolution reaction. *Catalysts* 11 (4): 417.
39 Wang, Z., Fan, J., Cheng, B. et al. (2020). Nickel-based cocatalysts for photocatalysis: hydrogen evolution, overall water splitting and CO_2 reduction. *Materials Today Physics* 15 (1): 100279.
40 Ball, P. (2019). Using artificial intelligence to accelerate materials development. *MRS Bulletin* 44 (5): 335–344.
41 Dimiduk, D.M., Holm, E.A., and Niezgoda, S.R. (2018). Perspectives on the impact of machine learning, deep learning, and artificial intelligence on materials, processes, and structures engineering. *Integrating Materials and Manufacturing Innovation* 7 (3): 157–172.
42 Hardian, R., Liang, Z., Zhang, X. et al. (2020). Artificial intelligence: the silver bullet for sustainable materials development. *Green Chemistry* 22 (21): 7521–7528.
43 Huang, J.S., Liew, J.X., Ademiloye, A.S. et al. (2021). Artificial intelligence in materials modeling and design. *Archives of Computational Methods in Engineering* 28 (1): 3399–3413.
44 Sha, W., Guo, Y., Yuan, Q. et al. (2020). Artificial intelligence to power the future of materials science and engineering. *Advanced Intelligent Systems* 2 (4): 1900143.
45 Di, J., Chen, C., Yang, S.Z. et al. (2019). Isolated single atom cobalt in Bi_3O_4Br atomic layers to trigger efficient CO_2 photoreduction. *Nature Communications* 10 (1): 2840.

46 Wang, L., Chen, W., Zhang, D. et al. (2019). Surface strategies for catalytic CO_2 reduction: from two-dimensional materials to nanoclusters to single atoms. *Chemical Society Reviews* 48 (21): 5310–5349.

47 Ji, S., Qu, Y., Wang, T. et al. (2020). Rare-earth single erbium atoms for enhanced photocatalytic CO_2 reduction. *Angewandte Chemie International Edition* 59 (26): 10651–10657.

48 Han, Z., Zhao, Y., Gao, G. et al. (2021). Erbium single atom composite photocatalysts for reduction of CO_2 under visible light: CO_2 molecular activation and 4f levels as an electron transport bridge. *Small* 17 (26): 2102089.

49 Chen, P., Lei, B., Dong, X. et al. (2020). Rare-earth single-atom La–N charge-transfer bridge on carbon nitride for highly efficient and selective photocatalytic CO_2 reduction. *ACS Nano* 14 (11): 15841–15852.

50 Colombari, F.M., Da Silva, M.A.R., Homsi, M.S. et al. (2021). Graphitic carbon nitrides as platforms for single-atom photocatalysis. *Faraday Discussions* 227 (1): 306–320.

51 Yang, Y., Li, F., Chen, J. et al. (2020). Single Au atoms anchored on amino-group-enriched graphitic carbon nitride for photocatalytic CO_2 reduction. *ChemSusChem* 13 (8): 1979–1985.

52 Fu, J., Zhu, L., Jiang, K. et al. (2021). Activation of CO_2 on graphitic carbon nitride supported single-atom cobalt sites. *Chemical Engineering Journal* 415 (1): 128982.

53 Wei, Z.X., Zhu, Y.T., Liu, J.Y. et al. (2021). Recent advance in single-atom catalysis. *Rare Metals* 40 (4): 767–789.

54 Xiao, X., Zhang, L., Meng, H. et al. (2021). Single metal atom decorated carbon nitride for efficient photocatalysis: synthesis, structure, and applications. *Solar RRL* 5 (6): 2000609.

55 Liu, S., Wang, Y., Wang, S. et al. (2019). Photocatalytic fixation of nitrogen to ammonia by single Ru atom decorated TiO_2 nanosheets. *ACS Sustainable Chemistry & Engineering* 7 (7): 6813–6820.

56 Yan, X., Liu, D., Cao, H. et al. (2019). Nitrogen reduction to ammonia on atomic-scale active sites under mild conditions. *Small Methods* 3 (9): 1800501.

57 Song, H., Wei, L., Chen, C. et al. (2019). Photocatalytic production of H_2O_2 and its in situ utilization over atomic-scale Au modified MoS_2 nanosheets. *Journal of Catalysis* 376 (1): 198–208.

58 Hamd, W.S. and Dutta, J. (2020). Chapter 11: Heterogeneous photo-Fenton reaction and its enhancement upon addition of chelating agents. In: *Nanomaterials for the Detection and Removal of Wastewater Pollutants* (eds. B. Bonelli, F.S. Freyria, I. Rossetti and R. Sethi), 303–330. Elsevier.

59 Chen, F., Wu, X.L., Shi, C. et al. (2021). Molecular engineering toward pyrrolic N-rich M-N4 (M = Cr, Mn, Fe, Co, Cu) single-atom sites for enhanced heterogeneous Fenton-like reaction. *Advanced Functional Materials* 31 (13): 2007877.

60 Huang, B., Wu, Z., Zhou, H. et al. (2021). Recent advances in single-atom catalysts for advanced oxidation processes in water purification. *Journal of Hazardous Materials* 412 (1): 125253.

61 Wang, G., Zhang, T., Yu, W. et al. (2020). Modulating location of single copper atoms in polymeric carbon nitride for enhanced photoredox catalysis. *ACS Catalysis* 10 (10): 5715–5722.

62 Wang, L., Tang, R., Kheradmand, A. et al. (2021). Enhanced solar-driven benzaldehyde oxidation with simultaneous hydrogen production on Pt single-atom catalyst. *Applied Catalysis B: Environmental* 284 (1): 119759.

63 Wen, J., Yang, X., Sun, Z. et al. (2020). Biomimetic photocatalytic sulfonation of alkenes to access β-ketosulfones with single-atom iron site. *Green Chemistry* 22 (1): 230–237.

64 Zhou, P., Chao, Y., Lv, F. et al. (2020). Metal single atom strategy greatly boosts photocatalytic methyl activation and C–C coupling for the coproduction of high-value-added multicarbon compounds and hydrogen. *ACS Catalysis* 10 (16): 9109–9114.

65 Yang, J., Sun, Z., Yan, K. et al. (2021). Single-atom-nickel photocatalytic site-selective sulfonation of enamides to access amidosulfones. *Green Chemistry* 23 (7): 2756–2762.

66 Yao, T., An, X., Han, H. et al. (2018). Photoelectrocatalytic materials for solar water splitting. *Advanced Energy Materials* 8 (21): 1800210.

67 Yan, Z., Ji, M., Xia, J. et al. (2020). Recent advanced materials for electrochemical and photoelectrochemical synthesis of ammonia from dinitrogen: one step closer to a sustainable energy future. *Advanced Energy Materials* 10 (11): 1902020.

68 Chang, X., Wang, T., Yang, P. et al. (2019). The development of cocatalysts for photoelectrochemical CO_2 reduction. *Advanced Materials* 31 (31): 1804710.

69 Walter, M.G., Warren, E.L., McKone, J.R. et al. (2010). Solar water splitting cells. *Chemical Reviews* 110 (11): 6446–6473.

70 Li, T., Kasahara, T., He, J. et al. (2017). Photoelectrochemical oxidation of organic substrates in organic media. *Nature Communications* 8 (1): 390.

71 Thalluri, S.M., Bai, L.C., Lv, C.C. et al. (2020). Strategies for semiconductor/electrocatalyst coupling toward solar-driven water splitting. *Advancement of Science* 7 (6): 33.

72 Liu, L.P., Wu, X., Wang, L. et al. (2019). Atomic palladium on graphitic carbon nitride as a hydrogen evolution catalyst under visible light irradiation. *Communications Chemistry* 2 (1): 18.

73 Chen, Z., Mitchell, S., Vorobyeva, E. et al. (2017). Stabilization of single metal atoms on graphitic carbon nitride. *Advanced Functional Materials* 27 (8): 1605785.

74 Chen, Z., Bu, Y., Wang, L. et al. (2020). Single-sites Rh-phosphide modified carbon nitride photocatalyst for boosting hydrogen evolution under visible light. *Applied Catalysis B: Environmental* 274 (1): 119117.

75 Ramalingam, V., Varadhan, P., Fu, H.-C. et al. (2019). Heteroatom-mediated interactions between ruthenium single atoms and an MXene support for efficient hydrogen evolution. *Advanced Materials* 31 (48): 1903841.

76 Xia, C.K., Wang, H., Kim, J.K. et al. (2021). Rational design of metal oxide-based heterostructure for efficient photocatalytic and photoelectrochemical systems. *Advanced Functional Materials* 31 (12): 2008247.

77 Pang, Y., Zang, W., Kou, Z. et al. (2020). Assembling of Bi atoms on TiO_2 nanorods boosts photoelectrochemical water splitting of semiconductors. *Nanoscale* 12 (7): 4302–4308.

78 Sivula, K., Le Formal, F., and Grätzel, M. (2011). Solar water splitting: progress using hematite (α-Fe_2O_3) photoelectrodes. *ChemSusChem* 4 (4): 432–449.

79 Lin, Y., Yuan, G., Sheehan, S. et al. (2011). Hematite-based solar water splitting: challenges and opportunities. *Energy & Environmental Science* 4 (12): 4862–4869.

80 Cui, C., Heggen, M., Zabka, W.-D. et al. (2017). Atomically dispersed hybrid nickel-iridium sites for photoelectrocatalysis. *Nature Communications* 8 (1): 1341.

81 Zhao, Y., Yang, K.R., Wang, Z. et al. (2018). Stable iridium dinuclear heterogeneous catalysts supported on metal-oxide substrate for solar water oxidation. *Proceedings of the National Academy of Sciences* 115 (12): 2902–2907.

82 Hou, Y., Qiu, M., Kim, M.G. et al. (2019). Atomically dispersed nickel–nitrogen–sulfur species anchored on porous carbon nanosheets for efficient water oxidation. *Nature Communications* 10 (1): 1392.

83 Yang, G., Li, Y., Lin, H. et al. (2020). Constructing chemical interaction between hematite and carbon nanosheets with single active sites for efficient photo-electrochemical water oxidation. *Small Methods* 4 (12): 2000577.

84 Song, Q., Li, J., Wang, L. et al. (2019). Stable single-atom cobalt as a strong coupling bridge to promote electron transfer and separation in photoelectrocatalysis. *Journal of Catalysis* 370 (1): 176–185.

85 White, J.L., Baruch, M.F., Pander, J.E. et al. (2015). Light-driven heterogeneous reduction of carbon dioxide: photocatalysts and photoelectrodes. *Chemical Reviews* 115 (23): 12888–12935.

86 Xie, S., Zhang, Q., Liu, G. et al. (2016). Photocatalytic and photoelectrocatalytic reduction of CO_2 using heterogeneous catalysts with controlled nanostructures. *Chemical Communications* 52 (1): 35–59.

87 Zhang, H.W., Ming, J.T., Zhao, J.W. et al. (2019). High-rate, tunable syngas production with artificial photosynthetic cells. *Angewandte Chemie International Edition* 58 (23): 7718–7722.

88 Cheng, J., Yang, X., Xuan, X. et al. (2020). Efficient conversion of carbon dioxide on atomically dispersed metal–N_x species-anchored porous carbon with embedded Cu_xCo_y nanoparticles by accelerating electron separation. *ACS Sustainable Chemistry & Engineering* 8 (15): 5994–6002.

89 Comer, B.M., Lenk, M.H., Rajanala, A.P. et al. (2021). Computational study of transition-metal substitutions in rutile TiO_2 (110) for photoelectrocatalytic ammonia synthesis. *Catalysis Letters* 151 (1): 1142–1154.

90 Ithisuphalap, K., Zhang, H.G., Guo, L. et al. (2019). Photocatalysis and photoelectrocatalysis methods of nitrogen reduction for sustainable ammonia synthesis. *Small Methods* 3 (6): 1800352.

91 Zheng, J.Y., Lyu, Y.H., Qiao, M. et al. (2019). Photoelectrochemical synthesis of ammonia on the aerophilic–hydrophilic heterostructure with 37.8% efficiency. *Chem* 5 (3): 617–633.

92 Bai, Y.J., Bai, H.Y., Qu, K.G. et al. (2019). In-situ approach to fabricate BiOI photocathode with oxygen vacancies: understanding the N_2 reduced behavior in photoelectrochemical system. *Chemical Engineering Journal* 362 (1): 349–356.

93 Wang, W., Zhang, S., Liu, Y. et al. (2021). Integration of Fe_2O_3-based photoanode and atomically dispersed cobalt cathode for efficient photoelectrochemical NH_3 synthesis. *Chinese Chemical Letters* 32 (2): 805–810.

94 Laguna, O.H., Lietor, P.F., Godino, F.J.I. et al. (2021). A review on additive manufacturing and materials for catalytic applications: milestones, key concepts, advances and perspectives. *Materials & Design* 208 (1): 109927.

95 Zhou, X. (2021). TiO_2-supported single-atom catalysts for photocatalytic reactions. *Wuli Huaxue Xuebao/ Acta Physico - Chimica Sinica* 37 (6): 2008064.

96 Wang, Z., Hu, X., Zou, G. et al. (2019). Advances in constructing polymeric carbon-nitride-based nanocomposites and their applications in energy chemistry. *Sustainable Energy & Fuels* 3 (3): 611–655.

97 An, H., Wu, L., Mandemaker, L.D.B. et al. (2021). Sub-second time-resolved surface-enhanced Raman spectroscopy reveals dynamic CO intermediates during electrochemical CO_2 reduction on copper. *Angewandte Chemie International Edition* 60 (30): 16576–16584.

98 Veselý, M., Valadian, R., Merten Lohse, L. et al. (2021). 3-D X-ray nanotomography reveals different carbon deposition mechanisms in a single catalyst particle. *ChemCatChem* 13 (10): 2494–2507.

15

Supported Double and Triple Metal Atom Catalysts

Zhiwen Chen[1,2], *Chandra V. Singh*[2,3], *and Qing Jiang*[1]

[1] *Jilin University, Key Laboratory of Automobile Materials, Ministry of Education, and School of Materials Science and Engineering, Changchun 130022, China*
[2] *University of Toronto, Department of Materials Science and Engineering, 184 College Street, Suite 140, Toronto, ON M5S 3E4, Canada*
[3] *University of Toronto, Department of Mechanical and Industrial Engineering, 5 King's College Road, Toronto, ON M5S 3G8, Canada*

15.1 Introduction

Single-atom catalysts (SACs) have emerged as one of the most popular research topics in the field of catalysis owing to their high catalytic activity, high selectivity, and nearly 100% utilization of active metal atoms [1, 2]. Double atom catalysts (DACs) and triple atom catalysts (TACs) are atomic catalysts with active sites constituted of two and three bonded metal atoms, respectively. While the development of DACs and TACs is relatively new, they could offer a comprehensive and distinct understanding of atomic catalysts. The concurrent challenges restricting their practical utilization include: (i) limited and expensive experimental techniques for reliable synthesis, and (ii) their characterization [3]. With the rapid development of science technology, accurate regulation of atomic-level structure and morphology has been achieved through rational synthesis strategies, and the use of advanced characterization techniques, benefiting the research of atomic catalysts [4, 5]. DACs and TACs are developing extensively hand in hand with SAC in the field of catalysis.

The rapid development of DACs and TACs shortens the gap between experimental results and theoretical calculations due to their simple geometric structures. Experimentally, two Fe metal atoms that were covalently bonded within the graphene lattice were first prepared through manipulating a focused electron beam (80 kV) on graphene containing an intentionally deposited Fe precursor reservoir [6]. By density functional theory (DFT) calculations, Li et al. further extended this system to other metal dimers supported on graphene with adjacent single vacancies and used them as catalysts for CO_2 reduction reaction [7]. These works indicate that experimental methods and theoretical calculations should complement each other in the study of DACs and TACs. Such catalysts are beginning to catch the attention of many researchers, and more and more DAC and TAC have been investigated and

Supported Metal Single Atom Catalysis, First Edition. Edited by Philippe Serp and Doan Pham Minh.
© 2022 WILEY-VCH GmbH. Published 2022 by WILEY-VCH GmbH.

applied for various catalytic reactions [8, 9]. The research on DACs and TACs is very meaningful for both practical applications and theoretical exploration in catalysis.

Firstly, their emergence makes full use of the advantages of homogeneous catalysts in heterogeneous catalysis due to the very similar active centers, where all the active atoms of DACs and TACs interact with reactants or products. The 100% active atom utilization results in high catalytic efficiency, just like that in homogeneous catalysis [10]. The simplification, uniformity, and legibility of active sites in DACs and TACs can avoid the complexity of chemical reactions, and thereby show high selectivity, which is often a difficulty for traditional heterogeneous catalysis. The supports for DACs and TACs can be viewed as ligands to active centers. Especially, the supports are always more stable than the ligands of homogeneous catalysts. Moreover, the nature of heterogeneous catalysis makes it easier to reuse DACs and TACs, which is a serious issue in homogeneous catalysis. It means that DACs and TACs combine the advantages of homogeneous and heterogeneous catalysts while spurning the corresponding shortcomings of homogeneous and heterogeneous catalysts.

Secondly, SACs, DACs, and TACs have the smallest active site units of top, bridge, and hollow sites, respectively. The simple, explicit, and uniform active sites are ideal models for exploring catalytic mechanisms. In fact, the practically synthesized traditional catalysts are always inhomogeneous and usually consist of mixtures of nanoparticles or subnanometer clusters in a wide range of size distributions, which means that the corresponding active sites are various, such as vertex, ridge, and surface defects, resulting in complex reaction mechanisms [11]. As it is well-known, the theoretical calculation is a powerful approach to reveal the catalytic reaction mechanism, where the theoretical models usually abstract experimental samples to further study reaction mechanisms. Owing to the special active sites of DACs and TACs, the gap between theoretical models and experimental samples becomes smaller. Consequently, theoretical calculations can explain experimental phenomena reasonably well, and conversely, experimental results can verify theoretical predictions accurately. Thus, DACs and TACs are ideal models for exploring new catalytic mechanisms and further rationally design catalytic materials, which will greatly accelerate developments in the field of catalysis.

In this chapter, the synthesis routes of DACs and TACs are analyzed first because they are not only the foundation for the development of DACs and TACs but also the greatest challenge for atomic catalysts. In addition to synthetic methods, characterization techniques are also essential for further recognizing DACs and TACs. Therefore, we summarize various characterization techniques for DACs and TACs, which include high-angle annular dark-field scanning transmission electron microscopy (HAADF-STEM), X-ray absorption near-edge spectroscopy (XANES), extended X-ray absorption fine structure (EXAFS), pressure-dependent diffuse reflectance infrared Fourier transform (DRIFT) measurement, wavelet transform (WT), and so on. Then, the applications of DACs and TACs in thermocatalysis, electrocatalysis, and photocatalysis are summarized. Finally, we discuss, in-depth, the great opportunities and challenges for the development of this type of catalyst.

15.2 Synthesis Routes

The biggest challenge of DACs and TACs is their successful synthesis with high loading of active atoms. The high surface energy of active atoms makes it difficult to stabilize them, especially for DACs and TACs. Therefore, some novel and effective strategies have been developed for promoting the formation of DACs and TACs, including increasing metal atom loading, grafting single atom (SA) to SA, preselecting precursors, and preselecting supports, as discussed in detail below.

15.2.1 High Metal Atom Loading

The loading of active atoms in SACs is generally low (<5 wt%), preventing the sintering of active atoms through short diffusion distances [12, 13]. There are very few double or triple bonded metal atom clusters present for the too low loading of active atoms, while too high loading inevitably results in metal clustering. Thus, increasing the appropriate loading amount could be a strategy to get DACs and TACs. Zeng et al. examined this strategy by increasing the Pt mass loading on MoS_2 [14]. When the loading of Pt atoms on MoS_2 is 0.2%, all active sites are Pt_{SA}, the double atom active sites appear gradually with the increase of the loading. Up to 7.5%, the most double-atom active sites are reached with 65.5% of all Pt atoms, which exhibits higher catalytic activity and lower activation energy than Pt_{SA} in CO_2 hydrogenation into methanol. Unfortunately, there are still 23.7% of Pt clusters and 10.8% of Pt_{SA} present in the system of 7.5% Pt/MoS_2. Thus, DACs or TACs could be obtained by increasing the loading of active atoms by a suitable amount, which determines the homogeneity of active sites.

15.2.2 Further SAC Grafting

Another potential strategy is grafting SA to SA, which means that the first synthesized SA active sites are used as nucleation sites for another atom to form double-atom active sites and even triple-atom active sites. At first, the precursors with different active atoms should be chosen during the successive synthesis processes. Moreover, the secondary atom is thermodynamically more likely to be adsorbed on the previously fabricated SA active sites due to the stronger adsorption capacity of active SAs with a low coordination number. Lu et al. applied this strategy to synthesize Pt_2 active sites for hydrolytic dehydrogenation of ammonia borane through a bottom-up approach from atomic layer deposition [15], where graphene was used as support for Pt_{SA} deposition, and then the preliminary Pt_{SA} selectively adsorbed a secondary Pt atom to form Pt_{DA} active sites, as shown in Figure 15.1a. This work points out an avenue for bottom-up fabrication of supported atomically precise DACs for practical applications.

15.2.3 Preselected Precursors for Double or Triple Atom Active Centers

The precursors of atomic catalysts are critical for precise synthesis with high uniformity and dispersion, especially for double- and triple-atom active sites. For

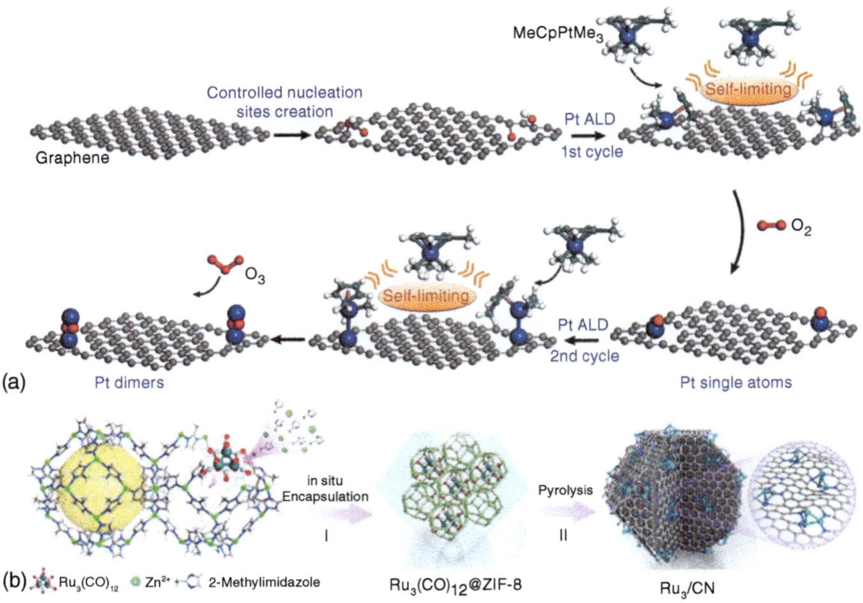

Figure 15.1 (a) Schematic illustration of bottom-up synthesis of dimeric Pt_2/graphene catalysts. Controlled creation of isolated anchoring sites on pristine graphene; one cycle of Pt ALD on the anchoring sites for Pt_{SA} formation by alternately exposing [$MeCpPtMe_3$] and molecular O_2 at 250 °C; second cycle of Pt ALD on the Pt_{SA}/graphene to selectively deposit the secondary Pt atoms on the preliminary ones for Pt_{DA} dimers formation at 150 °C. (b) Illustration of the Ru_{TA}/CN preparation process. (a): Source: Yan et al. [15]. Springer Nature. CC BY 4.0; (b): Source: Ji et al. [16]. Reproduced with permission of American Chemical Society.

avoiding the existence of SA or other metal clusters as the active sites in DACs and TACs, some molecular precursors with double or triple atom centers protected by some ligands have been preselected during the synthesis of double-atom and triple-atom active sites. What is certain is that SA active sites will be absent because of the structural characteristic of the preselected precursors. Moreover, the steric hindrance of their ligands presents the potential to prevent the aggregation of metal atoms. For instance, the bis(dicarbonyl-cyclopentadienyl-iron) precursor, $[(\eta^5\text{-}C_5H_5)Fe(CO)_2]_2$, was preselected to successfully synthesize highly dispersed Fe_2 clusters supported on mesoporous C_3N_4 [17]. Furthermore, this group synthesized atomically dispersed uniform catalysts of Ru_{TA}/CN through the $[Ru_3(CO)_{12}]$ precursor and suitable molecular-scale cages of zeolitic imidazolate frameworks (ZIFs), as shown in Figure 15.1b [16]. The steric hindrance provided by the ligands of $[(\eta^5\text{-}C_5H_5)Fe(CO)_2]_2$ and $[Ru_3(CO)_{12}]$ prevents the agglomeration of Fe_2 and Ru_3 clusters and facilitates the Fe_2 and Ru_3 clusters to be uniformly distributed on the C_3N_4 and CN supports, respectively. The general applicability indicates that this "precursor-preselection" strategy may be the most promising approach, which can be applied to produce other DACs and TACs, establishing a substantial foundation for further studies of the structure–activity relationship in catalysis.

15.2.4 Preselected Supports for Supporting DACs or TACs

In addition to the precursors, preselecting supports are more complex, diverse, and important for the successful synthesis of DACs and TACs. The requirements of the support vary with different chemical reactions and reaction environments. For instance, photocatalysis requires the support to possess an appropriate optical gap for specific reactions, electrocatalysis needs highly conductive supports, and thermocatalysis hopes catalysts with high thermal stability. Moreover, the interaction between the support and active atoms not only determines the stability of atomic catalysts but also affects the catalytic performance for chemical reactions. Therefore, it is undeniable that an optimal support is a prerequisite for successful DACs or TACs. In terms of this issue, this section introduces the selection and role of support in DACs and TACs from the following types of support:

15.2.4.1 Metallic Supports

Metallic supports can be used for supporting DACs and TACs. It is a process of surface alloying. Metallic alloys and metallic core–shell structures provide opportunities to alter electronic structures and chemical properties of the active sites. Through DFT calculations, Chen et al. investigated TACs of Pt_3 cluster-decorated core–shell structures, where the core consisted of cobalt-palladium or nickel-palladium, and the shell was Pt [18]. This strategy could significantly reduce the cost of catalysts due to the non-noble metal core. Moreover, the catalytic activity of Pt_3 active sites is further improved by tuning the shell thickness. Interestingly, this system had been synthesized by the same group in the next year [9]. They applied this TAC of a Pt_3 cluster decorated cobalt-palladium core–shell structure with a low Pt loading of only 2.4 wt% as fuel cell cathodes in an alkaline electrolyte through a developed wet chemical reduction method. This TAC shows a specific mass activity that is enhanced by a factor of 30.6 relative to a commercial platinum catalyst, which is attributed to the unique charge localization induced by Pt_3 decoration. Another advantage of metallic support is high electrical conductivity, which denotes a potential in the field of electrocatalysis. These findings about metal supports for DACs and TACs are expected to be useful for surface engineering and the design of advanced DACs and TACs.

15.2.4.2 Oxide Supports

Oxides are usually selected as supports for atomic catalysts mainly due to their high thermal stability [19]. Moreover, the surface oxygen coordination could fix the active metal atoms on the surface, resulting in the high stability of DACs and TACs. Wang et al. synthesized Ir dinuclear heterogeneous catalysts on α-Fe_2O_3 support through a simple method [20]. Firstly, the Fe_2O_3 support is soaked in an Ir homodimer solution and then thoroughly rinsed with deionized water to form an Ir intermediate. Secondly, photochemical treatments are performed using a UV-ozone (UVO) cleaner system equipped with UV light. In a typical procedure, the Ir intermediate product is placed in the middle of the UVO chamber. The process lasts for 28 minutes to finally obtain Ir_2 DAC, which exhibits outstanding stability and catalytic activity for solar water oxidation.

Both theoretical and experimental works were devoted to oxide supports for TACs. Liu et al. applied first-principle theoretical study and microkinetic analysis to study an anchored Fe_3 cluster on the θ-Al_2O_3(010) surface as a heterogeneous catalyst for ammonia synthesis [21]. The whole catalytic mechanism of conversion of N_2 to NH_3 on Fe_{TA}/θ-Al_2O_3(010) was studied, which follows an associative mechanism, in which the adsorbed N_2 was first hydrogenated to NNH, with high turnover frequency. This was attributed to the large spin polarization, low oxidation state, and multistep redox capability of Fe_3 clusters. This theoretical work shows that TACs on oxides are promising candidates as heterogeneous catalysts with high activity and selectivity. However, in the actual synthesis process, there is a huge challenge to synthesize TACs with high uniformity and dispersion. Vajda et al. synthesized a narrow size distribution of Ag_n ($n = 2$–4) clusters with a dominant trimer contribution on Al_2O_3 at 2.2% atomic monolayer coverage using a quadrupole mass filter and deflector assembly [22]. The preparation process is very complex and challenging, as shown in Figure 15.2a. Firstly, a Nd:YAG laser is used to generate

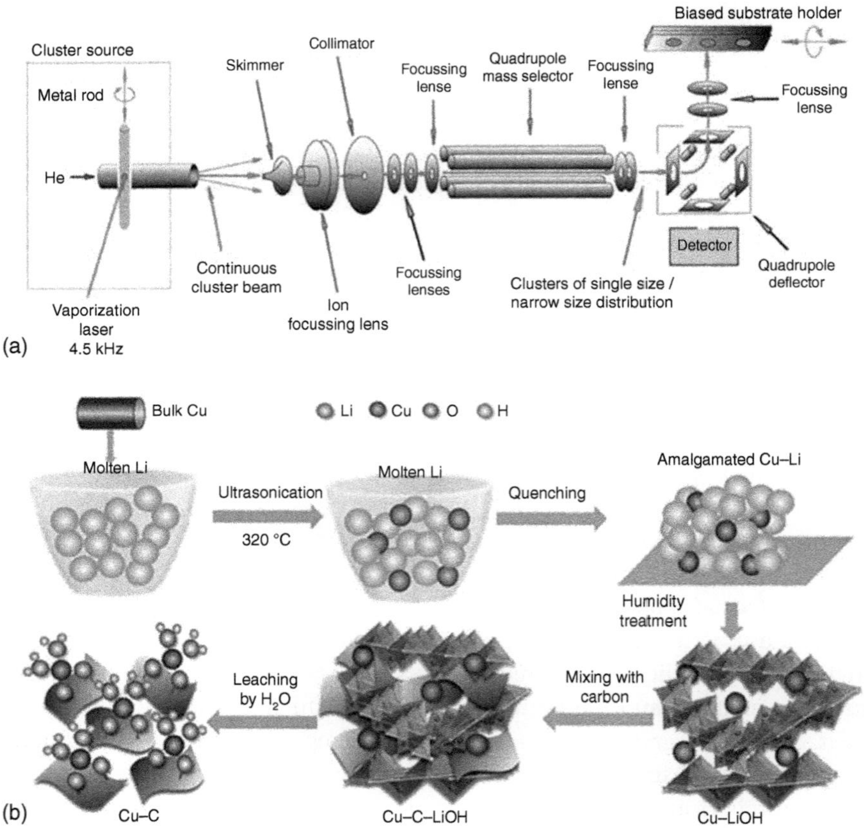

Figure 15.2 (a) Schematic representation of the cluster deposition apparatus. (b) Step-by-step preparation of the carbon-supported Cu_{SA} catalyst using an amalgamated Cu–Li method. (a): Source: Lei et al. [22]. Reproduced with permission of American Association for the Advancement of Science; (b): Source: Xu et al. [23]. Reproduced with permission of Nature Publishing Group.

a continuous beam of Ag clusters in a new high-flux laser vaporization cluster source at 4.5 kHz. Secondly, the generated beam of neutral and charged clusters passes through a biased skimmer into the ion guide of the second differentially pumped vacuum stage and then into the third stage of the collimator. And then, during the focusing of lenses, the clusters with a positive charge are guided and gone into a quadrupole mass spectrometer for mass analysis. Finally, narrow cluster distributions in the size range Ag_n^+ ($n = 1, \ldots, 10$) with three dominant sizes can be produced by optimizing the temperature of the cluster source, the pressure of the helium carrier gas, and the potential settings on the individual ion optics elements. SACs and DACs could also be prepared through this advanced synthesis technology.

15.2.4.3 2D Material Supports

2D materials have great potential for supporting active atoms in atomic catalysts due to the distinct structural feature of a large surface/volume ratio, which is beneficial to achieve the high loading of active atoms. Graphene, as the most popular 2D material, has been particularly investigated as support for DACs and TACs [8]. He et al. observed through an atomic-resolution structural study that two Fe metal atoms were covalently bonded within the graphene lattice, and investigated their interaction with each other [6]. They found four stable structures: two variants of a Fe dimer in a graphene trivacancy, a Fe dimer embedded in two adjacent monovacancies, and a Fe dimer trapped by a quadvacancy, which indicated four different types of DACs. The corresponding catalytic performance is worthy of further exploration. Another interesting work is the precise synthesis of Pt_2 supported on graphene through a bottom-up approach using atomic layer deposition (ALD) on the created nucleation sites [15], where the pretreated graphene nanosheets were used as the nucleation sites (or support) and trimethyl(methylcyclopentadienyl)-platinum(IV) [$MeCpPtMe_3$] was taken as the precursor. Graphene was also used as support for supporting the Cu_3 cluster through an amalgamated Cu–Li method, as shown in Figure 15.2b [23]. The bulk Cu is added into molten lithium until it is fully dissolved by ultrasonication. After quenching of lithium melting and humidifying air treatment, the atomically isolated Cu atoms form a mixture with LiOH (Cu–LiOH). Then, this compound is mixed with a carbon black support (XC-72). After leaching by water, the LiOH is eliminated and the Cu atoms are transferred over to the carbon surface at ambient temperature. The isolated Cu atoms are transformed into metallic Cu_3 under a reducing potential of −1.0 V, indicating that Cu_{TA}/C is formed, which provides Cu_3 active centers for CO_2-to-ethanol conversion. This work suggests a new strategy for synthesizing TAC with graphene support.

N-doping makes graphene even more attractive, especially in the field of catalysis. Xiao et al. designed a novel binuclear active site structure Co_{DA}/N_xC_y and successfully fabricated this DAC by precisely controlling the atomic-scale structure of bimetal-organic frameworks before pyrolysis [24]. Combined with DFT calculation and experimental characterization techniques, the active sites Co_2N_5 were identified and confirmed for the first time. In addition to homometallic DACs, some heterobimetallic DACs were also synthesized on N-doped carbon supports, such as Ni–Fe [25], Fe–Co [26], Pt–Fe [27], or Co–Pt [28]. All of them show excellent catalytic activity and selectivity, indicating the great potential of N-doped graphene as support for DACs in the field of catalysis. For the more challenging synthesis of TACs, Li et al. reported a novel approach to achieve uniform Ru_3 clusters stabilized by nitrogen species of carbon through a cage-separated precursor preselection and pyrolysis strategy [16], which included two steps: (i) synthesis of $Ru_3(CO)_{12}$@ZIF-8; and (ii) synthesis of Ru_{TA}/CN. The key for a successful synthesis resides in the choice of optimal support, N-doped carbon, but also an appropriate precursor, $[Ru_3(CO)_{12}]$.

Coming up behind graphene, MoS_2 can also be taken as support of atomic catalysts [29, 30]. Li et al. [29] applied the strategy of increasing the loading of Pt atoms to synthesize Pt atomic catalysts on MoS_2 with DA active sites. Therefore, the internal Pt atoms are not the active atoms while the surface S atoms bonded to Pt atoms become the active centers because of the redistribution of electrons after the substitution of Pt atoms. Importantly, this DAC has high stability due to the structural feature of the substitution sites resulted in the difficult diffusion of Pt atoms in the inner layer of MoS_2 nanosheets. This work opens a new perspective of DAC in the field of catalysis.

15.2.4.4 Highly Porous and Specific Supports

Currently, the accurate synthesis of DACs and TACs on supports of metals, oxides, or 2D materials is still very difficult. Furthermore, the agglomeration of active atoms on the supports seems inevitable during the catalytic process. Compared with traditional supports, highly porous materials present a confinement effect due to their porous morphology, which may inhibit the diffusion of active atoms, providing high stability to DACs or TACs.

Metal-organic frameworks (MOFs) have been successfully used as support for SAC [31, 32]. Mon et al. reported a DAC of Pt_2 clusters homogenously distributed and densely packed within the functional channels of a MOF decorated with thioether recognition sites [33]. The synthesis method includes the following three steps: first, a robust 3D MOF with high stability in water and featuring pores decorated with thioalkyl chains is prepared as support. Then, the metal salt $K_2[PtCl_4]$ is anchored by S atoms on the coordination network in 3D MOF for introducing Pt_2^+ cations. Finally, the Pt_2 units are obtained by reduction with $NaBH_4$. This synthesized Pt_2 DAC has a high loading of 8 wt% and high stability under low temperature.

C_2N, with uniform pore distribution, consists of benzene rings bridged by pyrazine rings and has first been synthesized by Baek et al. in 2015 [34]. The abundant pyridine-like sp^2-bonded nitrogen atoms present in the uniform porous structure of C_2N provide coordination sites to bond tightly with metal atoms due to their electron lone pairs and inhibit the diffusion and aggregation of metal atoms [35]. Chen et al. investigated a series of DAC supported on C_2N for electrocatalytic nitrogen reduction reaction (NRR) by DFT calculations [36]. Through the analyses of thermodynamics and kinetics, all DACs on C_2N showed high stability. Similarly, another carbon nitride C_3N_4 also has a uniformly distributed porous structure. Li et al. used C_3N_4 as support for supporting Fe_2 clusters through a "precursor-preselection" wet chemistry strategy [17]. This strategy could be extended to other DACs on C_3N_4, such as Pd_{DA}/C_3N_4 and Ir_{DA}/C_3N_4, demonstrating the general applicability of the approach for synthesizing DAC on C_3N_4.

Relevant works about TACs on highly porous supports are scarce, especially experimental works. By DFT calculations, Chen et al. investigated porous heterostructure support consisting of graphdiyne and graphene (GDY-Gra) for supporting atomic catalysts, including SACs, DACs, and TACs [37]. This support could satisfy the following four criteria for electrocatalytic NRR: (i) large surface/volume ratio; (ii) high stability; (iii) fast electron transport; and (iv) facile synthesis. The GDY-Gra support could stand ultrahigh loading of triple atoms with a theoretical loading of ~30 wt%. A possible synthesis route for Fe_{TA}/GDY-Gra was provided for achieving the high loading of active atoms through the preselected precursor $[Fe_3(CO)_{12}]$. This catalyst design is worth exploring a corresponding experimental synthesis process.

15.3 Characterization Techniques

In addition to the decisive role of synthesis routes, major advances in DACs and TACs are also relying on the development of advanced characterization techniques. The catalytic performance and reaction mechanism are determined by the active sites of DACs and TACs. Therefore, it is necessary to characterize the bonding characteristics, oxidation state, and coordination environments of active sites in DACs and TACs. Advanced characterization techniques enable us to identify active sites of DACs and TACs, which is crucial for exploring the catalytic reaction mechanism, constructing new catalytic theories, and further designing optimal catalysts. For DACs and TACs, an integrated electron microscopy approach, together with spectroscopy techniques and theoretical calculations, could provide detailed and reliable information for describing electronic and geometric structures at the atomic scale.

The most direct means to detect double-atom and triple-atom active sites is a high-resolution scanning transmission electron microscope (HR-STEM) with spherical aberration correction due to its subatomic structure resolution. Based on HAADF-STEM, the atomic-scale resolution images of double-atom and triple-atom could be obtained because the brightness of atoms is proportional to the square of their atomic number [5]. Wang et al. detected Fe–Co double-atom active sites on carbon nanotubes by magnified aberration-corrected HAADF-STEM, as shown in Figure 15.3a [38]. The homogeneously distributed bright dual dots show the existence of Fe–Co dual sites because the Fe and Co atoms are heavier than C and N atoms, which is confirmed by electron energy-loss spectroscopy (EELS), where the atomic coordination among Fe, Co, and N atoms was found at the atomic level (Figure 15.3b). This consistent conclusion of the coexistence of Fe and Co in Fe–Co dual sites was further verified by the homogeneous elemental distribution of Co, Fe, C, and N on the (Fe, Co)/carbon nanotubes (CNT), as shown in the EELS mapping images (Figure 15.3c).

The electron microscope for investigating DACs and TACs is ideal to image and observe the coordination environments of active centers. However, some physical and chemical properties of active sites on DACs and TACs are related to

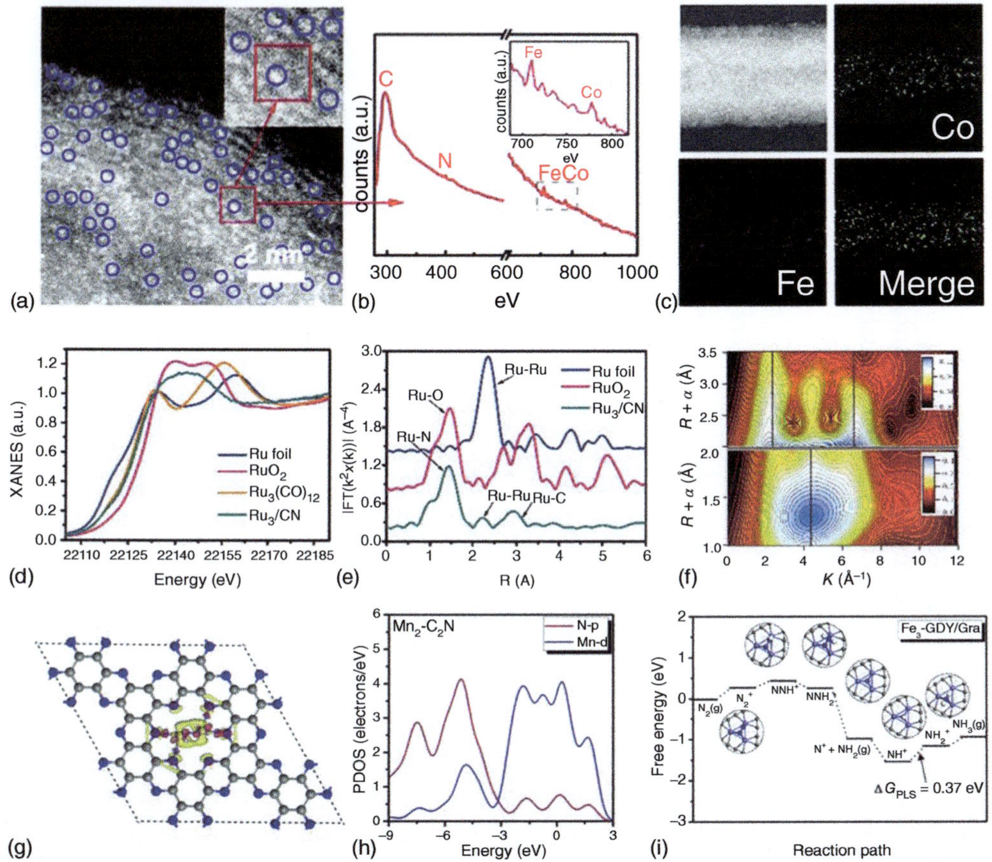

Figure 15.3 (a) Magnified HAADF-STEM images of the (Fe, Co)/CNT, showing that Fe–Co dual sites are dominant in the sample. (b) The EEL spectrum was taken on the red rectangle in (a) and shows Fe, Co, N, and C indicating that Fe, Co is coordinated with N at the atomic scale. (c) Corresponding EELS mapping images of Co, Fe, and N in (Fe, Co)/CNT. (d) Ru K-edge XANES spectra. (e) EXAFS Fourier transformed (FT) k^2-weighted $\chi(k)$ function spectra of Ru_{TA}/N–C and reference. (f) Wavelet transforms for the k^2-weighted Ru K-edge EXAFS signals for the high-coordination shells in Ru_{TA}/N-C. The colors in the contour figures indicate the moduli of the Morlet wavelet transform. (g) Geometric structure of Mn_{DA}/C_2N with the difference in charge density. (h) Partial density of states (PDOS) of Mn_{DA}/C_2N, including the N-p and Mn-d orbitals. (i) Reaction paths of nitrogen reduction reaction and their corresponding free energies on Fe_{TA}/GDY-Gra. Source: (a–c) Wang et al. [38]. Reproduced with permission of Royal Society of Chemistry; (d–f) Tian et al. [17]. Reproduced with permission of American Chemical Society; (g–h) Chen et al. [36]. Reproduced with permission of Wiley-VCH; (i) Chen et al. [37]. Reproduced with permission of Royal Society of Chemistry.

the electronic structures, which are difficult to identify only with imaging methods. Spectroscopy and spectral analysis could provide the statistical average information, electronic structures, chemical valence, and coordination environment of active atoms on DACs and TACs, particularly X-ray synchrotron radiation absorption spectroscopy, which mainly includes XANES and EXAFS. To identify the structure of Ru_3 clusters on N-doped carbon (Ru_{TA}/N–C), Ji et al. carried out XANES, EXAFS, and a WT, as shown in Figure 15.3d–f [16]. From the Ru K-edge XANES curves in Figure 15.3d, a middle energy absorption threshold value of Ru_{TA}/N–C is found compared with Ru foil and RuO_2 systems, indicating some electron transferring from Ru_3 to the support. Therefore, the Ru_3 cluster carries a positive charge, which is smaller than that of RuO_2. The corresponding EXAFS Fourier transformed spectra (Figure 15.3e) exhibited the main peak at about 1.4 Å, denoting the contribution of the first shell of Ru–N scattering. The other two smaller-shell peaks at distances of 2.3 and 3.0 Å in Ru_{TA}/N–C were ascribed to Ru–Ru scattering form Ru_3 active sites and Ru–C scattering, respectively. The electron states and coordination environments of Ru_3 active sites were further confirmed by the WT analysis of the Ru K-edge EXAFS oscillations (Figure 15.3f). Moreover, quantitative EXAFS fitting results displayed that the coordination number for the first shell of Ru–N (Ru–Ru) is 2.2 ± 0.7 (1.6 ± 0.5), and the corresponding mean bond length is 1.99 ± 0.02 Å (2.53 ± 0.02 Å).

Although few other experimental characterization techniques, such as Fourier transform infrared spectroscopy (FTIR), X-ray photoelectron spectroscopy (XPS), solid-state magic-angle spinning-nuclear magnetic resonance (MAS-NMR), electron paramagnetic resonance (EPR), have been applied for studying DACs and TACs, they have great potential to offer useful information for identifying the electronic configurations and coordination types of DACs and TACs from different aspects. These advanced techniques regarding DACs and TACs are currently ongoing nowadays, which are highly essential to be more aware of atomic catalysts and deep understanding of structure-performance relationships for guiding a more rational design of DACs and TACs.

In addition to these experimental characterization techniques, theoretical calculations could not only identify invaluable information about atomic and electronic structures of the active sites of DACs and TACs but also provide detailed reaction mechanisms for various chemical reactions [39, 40]. For the study of DACs and TACs, theoretical calculations are currently indispensable for getting insights into the structure–performance relationships during the catalytic process by some calculation results, such as geometric optimization structures, charge distribution, electronic structures, thermodynamic data (adsorption energy, binding energy, reaction energy, etc.), and dynamic data (reaction barrier, diffusion coefficient, mean square displacement, etc.). For example, the geometric optimization structure of Mn_{DA}/C_2N is shown in Figure 15.3g along with the charge density difference [36]. The Mn_2 cluster is fixed into the pores of the C_2N support and bonds with the N atoms, where the pronounced charge density redistributions on the Mn—N bonds demonstrate strong chemical Mn—N bonds. The binding energy of Mn_2 on C_2N is 7.66 eV with the average Mn—N bond length of 2.02 Å, indicating the high stability of Mn_{DA}/C_2N. The strong interaction between Mn_2 and C_2N can be further verified by PDOS of Mn-d orbitals and N-p orbitals in the system of Mn_2-C_2N, as shown in Figure 15.3h, where the orbital hybridization between Mn-d orbitals and N-p orbitals is observed to be strong near the Fermi level. Another interesting theoretical work studied a new TAC consisting of a Fe_3 cluster supported on the heterostructured support graphdiyne and graphene (Fe_{TA}/GDY-Gra) for NRR [37]. The reaction path of NRR on FeTA/GDY-Gra was explored, as depicted in Figure 15.3i. The potential limiting step is the formation of NH_2* with a reaction-free energy of 0.37 eV, indicating the high catalytic performance of Fe_{TA}/GDY-Gra for electrocatalytic NRR. Currently, with the blossoming development of computational efficiency and computational accuracy, the perspective of theoretical calculations could be divided into three aspects for DACs and TACs: (i) explaining experimental phenomena, (ii) predicting experimental results, and (iii) guiding experimental investigation.

15.4 Applications

Both advanced synthesis routes and accurate characterization techniques serve for the potential applications of DACs and TACs. Although the studies on DACs and TACs are emerging fields and in their infancy, their applications have covered thermocatalysis, electrocatalysis, and photocatalysis. In this section, several cases are described to illustrate the application of DACs and TACs in the above catalysis fields. Through these situations, we hope to gain a deeper understanding of DACs and TACs and further elucidate their advantages. Note that although the insight comes from these applications, it does not mean that DACs and TACs are limited to these applications.

15.4.1 Thermocatalysis

Thermocatalysis is the most traditional catalytic process with the driving force of high temperature or high pressure. Therefore, the catalysts in thermocatalysis should satisfy high stability under reaction conditions, usually at high temperatures or pressures. Unfortunately, the atomic catalysts of DACs and TACs often have worrying stability due to the susceptibility to sintering caused by the large surface energy of active atoms. Thus, the works on DACs and TACs in thermocatalysis have a tremendous challenge, even though they have great potential in the field of catalysis.

15.4.1.1 CO Oxidation

A large amount of carbon monoxide emission from human life and industrial processes endangers human's health and causes serious pollution to the environment. One effective strategy in limiting CO emission is CO oxidation. Moreover, CO oxidation is an efficient way for hydrogen purification for fuel cells, and as a benchmark catalytic reaction, is usually used for evaluating the catalytic activity of specific catalysts. Chen et al. studied the stability and electronic properties of copper dimers supported on a porous C_2N monolayer (Cu_{DA}/C_2N) for CO oxidation by means of systematic DFT computations [41]. Compared with its monometallic counterpart

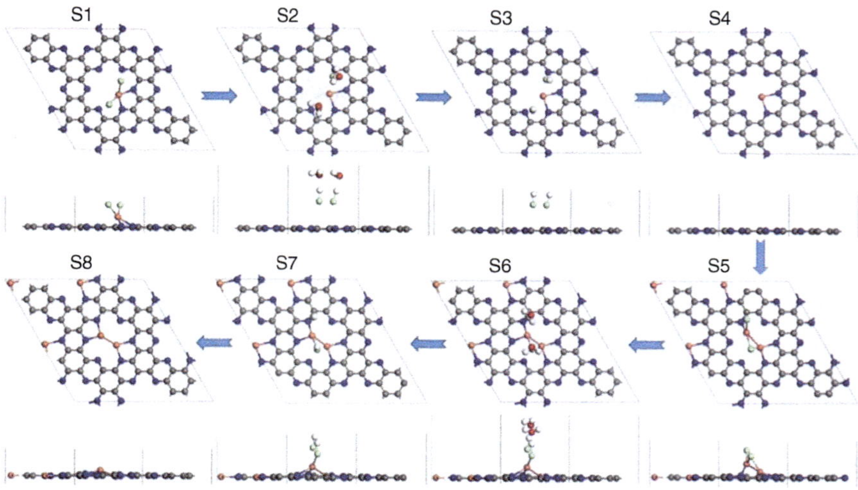

Figure 15.4 Top and side views of the designed synthetic route for Cu_{DA}/C_2N. Color scheme: Cu, orange; C, gray; N, blue; O, red; Cl, cyan. Source: Li and Chen [41]. Reproduced with permission of Royal Society of Chemistry.

(Cu_{SA}/C_2N), Cu_{DA}/C_2N exhibits higher stability and better catalytic performance toward CO oxidation. For confirming the experimental production of Cu_{DA}/C_2N, they proposed a synthetic route, as shown in Figure 15.4. All the reaction steps could easily occur since they are either spontaneous (barrierless) or only slightly endothermic, which is further verified by first-principles molecular dynamics at 350 K, where $CuCl_2$ is adsorbed on C_2N, then Cl^- ions are desorbed, and finally, the Cu_{DA}/C_2N is formed at 0.5 ps of the first-principles molecular dynamics. Furthermore, the heteronuclear $Fe_{SA}Cu_{SA}/C_2N$ was investigated in comparison with homonuclear counterparts Fe_{DA}/C_2N and Cu_{DA}/C_2N for CO oxidation [42]. The results show that the heteronuclear species $Fe_{SA}Cu_{SA}/C_2N$ possesses high stability and it should feasible to prepare it experimentally. Moreover, the heteronuclear $Fe_{SA}Cu_{SA}/C_2N$ catalyst has even better catalytic activity toward CO oxidation than its homonuclear counterparts, especially, without suffering the CO-poisoning problem. Considering the myriad of unexplored heteronuclear dimers that could be potentially anchored at appropriate supports, DACs and TACs with heteronuclear atoms open a broader avenue and provide a useful guideline for further developing atomic catalysts.

15.4.1.2 Ammonia Synthesis

Since World War II, ammonia synthesis, based on the Haber–Bosch process ($N_2 + H_2 \rightarrow NH_3$), becomes one of the most important industrial catalytic reactions due to its significant role in the growth of the human population [43]. In recent years, ammonia is also considered to have great potential as an energy storage material due to its high energy density, and environmental friendliness of its products of decomposition (N_2 and H_2) [44]. Therefore, ammonia synthesis from abundant nitrogen in the Earth's atmosphere is crucial for the sustainable development of the economy and society. However, it remains a significant challenge to activate dinitrogen because of its stability and chemical inertness (bond energy: 9.75 eV) [45]. Nowadays, industrial-scale NH_3 production still requires harsh reaction conditions (high temperature: ~700 K, high pressure: ~100 bar), consumes approximately 2% global fossil energy, and releases large amounts of greenhouse gases. Thus, the exploration of highly efficient catalysts for ammonia synthesis is a pressing issue.

Liu et al. applied first-principles theoretical study and microkinetic analysis to propose a new strategy to anchor Fe_3 clusters on the θ-Al_2O_3 (010) surface as a heterogeneous catalyst for ammonia synthesis [21]. Inspired by nitrogenase, in which the iron-molybdenum cofactor (FeMo–co) is responsible for N_2 activation and ammonia synthesis due to the highly efficient redox cycle between Fe(II) and Fe(III) of the Mo–Fe–S–C cluster of FeMo–co, the Fe_3 clusters supported by θ-Al_2O_3 (010) surface should be able of efficient N_2 activation because of the even more reduced states for metal–metal bonded Fe_3 clusters on the inert support. Moreover, the theoretical model of this Fe_3 cluster on alumina surface may be experimentally prepared by the soft-landing cluster method or thermal treatment of ligated tri-iron cluster, such as [$Fe_3(CO)_{12}$]. As expected, the designed Fe_3/θ-Al_2O_3 (010) has a large magnetic moment of 10 μB on Fe_3 clusters, which serves as an electron reservoir for regulating the charge variation of the whole process. More importantly, the classical dissociation mechanism on the metal surface, such as the B5 site of Ru(0001) step surface and the C7 site of Fe(111) or Fe(211) surfaces [46], changes to associative hydrogenation mechanism on Fe_{TA}/θ-Al_2O_3 (010), where N_2 is first activated on Fe_3 active sites followed by attacking by dissociated H atom, which differs from the procedure in the electrochemical condition where a proton attacks the adsorbed N_2 in solution [47]. The associative process is believed to occur in the homogenous catalysis and enzymatic mechanism, which is the advantage of atomic catalysts [3].

15.4.1.3 CO_2 Reduction Reaction

The CO_2 reduction reaction has long been of concern not only because of its environmental implications, but also its close relation to energy. Currently, the state-of-the-art catalysts for CO_2 reduction reaction are Ni-, Ru-, and Cu-based catalysts [48]. Methane is the major product that forms during CO_2 hydrogenation over Ni and Ru catalysts, while methanol and CO are mainly produced on Cu catalysts. Therefore, atomic catalysts with high selectivity should play an important role in catalyzing CO_2 reduction reaction [49, 50].

Li et al. designed a Pt_{DA}/MoS_2 catalyst by increasing Pt mass loading up to 7.5% [14]. The interaction between two neighboring Pt_{SA} significantly raises the catalytic performance and deepens the mechanistic understanding of heterogeneous catalysis. The catalytic performance of the designed Pt_{DA}/MoS_2 was evaluated for CO_2 reduction reaction in 30 ml of DMF under 32 bar (CO_2:H_2 = 1 : 3). Compared with Pt SAC, Pt DAC exhibits dramatic enhancement in activity: the yields of formic acid and methanol were 0.2 and 0.8 mmol, respectively, at 150 °C after three hours, while at 210 °C the yields were 2.3 and 7.3 mmol, respectively. The turnover frequency number of Pt DAC was 162.5 h^{-1}, which is 14.8 times that of Pt SAC at 150 °C, while the turnover frequency number of Pt DAC was 1.9 times that of Pt SAC at 210 °C (Figure 15.5a). The calculated activation energy from the

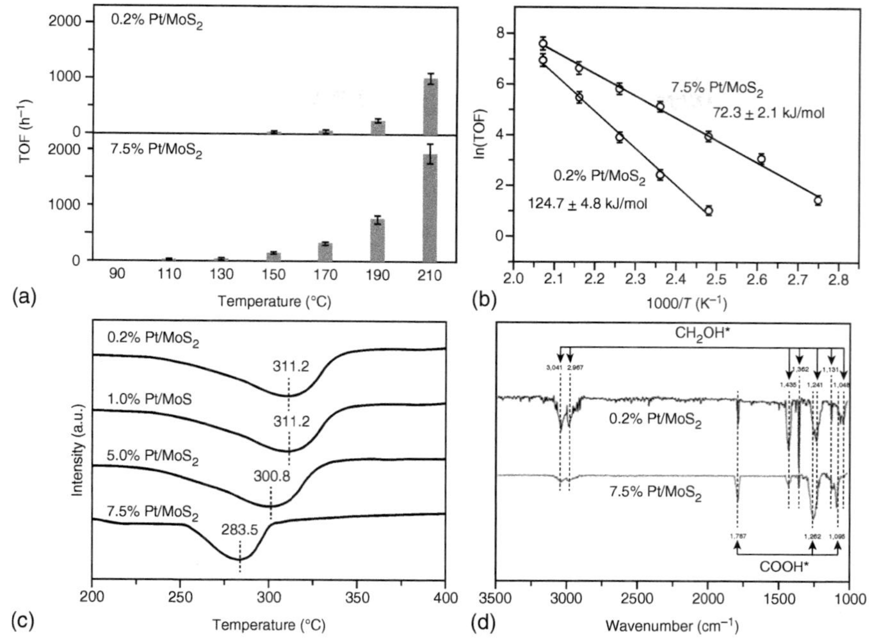

Figure 15.5 (a) Comparison of turnover frequency with 0.2% Pt/MoS_2 (SAC) and 7.5% Pt/MoS_2 (DAC) at different temperatures. (b) Arrhenius plots of 0.2% Pt/MoS_2 and 7.5% Pt/MoS_2. (c, d) Optimized reaction paths in CO_2 hydrogenation for Pt_{SA}/MoS_2 and Pt_{DA}/MoS_2, respectively. Source: Li et al. [14]. Reproduced with permission of Nature Publishing Group.

Arrhenius plots was 72.3 kJ/mol for the Pt DAC and 124.7 kJ/mol for the Pt SAC, as shown in Figure 15.5b. The enhanced catalytic performance is mainly originated from the synergetic interaction between neighboring monomers in DACs, where the neighboring Pt atoms not only work in synergy to vary the reaction barrier but also undergo distinct reaction paths compared with isolated monomers, as shown in Figure 15.5c,d. The discovery of this DAC is expected to get insight into the mechanistic understanding of heterogeneous catalysis.

15.4.1.4 Other Chemical Reactions

The applications of DACs and TACs are not limited to the above three important chemical reactions. Their applications in the field of thermocatalysis included hydrogenation of acetylene [51], oxidation of various molecules [16, 17, 22, 52], hydrolytic dehydrogenation of ammonia borane [15], ring-opening polymerization (ROP) of propylene oxide (PO) [53], and NO decomposition [54]. All of them demonstrate the excellent catalytic activity of DACs and TACs. Moreover, through the study of dimeric Cu active sites in Cu-ZSM-5 catalyst, Sajith et al. discovered that the presence of protons played an important role in the production of N_2O from two NO molecules [54]. In the proton-free mechanism, this process requires a large activation barrier of 56.3 and 55.3 kcal/mol for two dicopper model systems (where the Cu(I) atoms are separated by one and two SiO_4 tetrahedra, respectively), while the inclusion of protons reduces it to 31.4 and 17.3 kcal/mol. This work denotes that the study on DACs and TACs is helpful to identify new reaction mechanisms. In another interesting study, 8 wt% Pt_2 dimers homogenously distributed in the channels of a MOF were produced through multi-gram-scale chemical synthesis for technically easier, cheaper, and dramatically less dangerous industrial reactions [33]. The unusual Pt_2 species have a powerful catalytic behavior, allowing the synthesis of NH_4CN at room temperature, the methanation of CO_2 at low temperature (<140 °C), and the hydrogenation of alkenes at 60 °C, as shown in Figure 15.6. This work points out the universality of DACs in the field of thermocatalysis and the great potential for mass production in industrial catalysis.

15.4.2 Electrocatalysis

15.4.2.1 Hydrogen Evolution Reaction (HER)

Hydrogen with the highest energy density of 142 MJ/kg has been considered as the most promising energy carrier for the exploitation of clean and renewable energy sources [55]. Hydrogen could be obtained by hydrogen evolution reaction (HER) during electrochemical water splitting, which needs low cost, highly stable and highly active catalysts to minimize the overpotential. The state-of-the-art catalysts for HER are Pt and Pt-based materials. However, their high cost and scarcity severely inhibit their large-scale applications [56]. In that context, DACs and TACs have an advantage in cost due to the 100% utilization of active atoms.

Among nonprecious metal catalysts, complexes of transition metals, nitrogen, and carbon (M–N–C) are considered to be one type of most promising candidates to replace Pt and Pt-based catalysts for HER [57, 58]. However, some highly active M–N–C catalysts usually suffer a quick current decay under the electrochemical reaction condition [59]. Zeng et al. designed a DAC through a precise "single-atom to single-atom" grafting of Pt atom onto the Fe center via a bridging oxygen link, creating a new active moiety of Pt_{SA}-O_2-Fe_{SA}-N_4 [27]. The $Pt_{SA}Fe_{SA}$/N–C DAC shows a high hydrogen evolution reaction activity with an onset potential of 15 mV, which is smaller than other reference samples (Pt_{SA}/C and Fe–N–C). Tafel plots based on linear sweep voltammetry (LSV) curves were studied for the HER kinetics of the catalysts, where the $Pt_{SA}Fe_{SA}$/N–C follows the electrochemical desorption of Heyrovsky reaction with a Tafel slope of 42 mV/dec. This outstanding catalytic performance of $Pt_{SA}Fe_{SA}$/N–C can be attributed to the following three reasons: (i) the structural features (large surface area and abundant micro/mesopores) of $Pt_{SA}Fe_{SA}$/N–C certainly supply sufficient active sites and facilitate the mass transport for HER related species; (ii) the active sites of Pt_{SA}-O_2-Fe_{SA} favor the cleavage of H—OH bond and improve the kinetics of HER; (iii) the grafting of Pt_{SA}-O_2-would induce steric and electronic effects on Fe atoms in Fe–N_4 moieties that could also improve the water dissociation. This work provides a deeper understanding of the advantages of DACs for electrocatalysis.

15.4.2.2 Oxygen Evolution Reaction (OER)

Oxygen evolution reaction (OER) plays a critical role in many energy conversion and storage processes, for instance water splitting ($H_2O \rightarrow H_2 + 1/2O_2$) and rechargeable

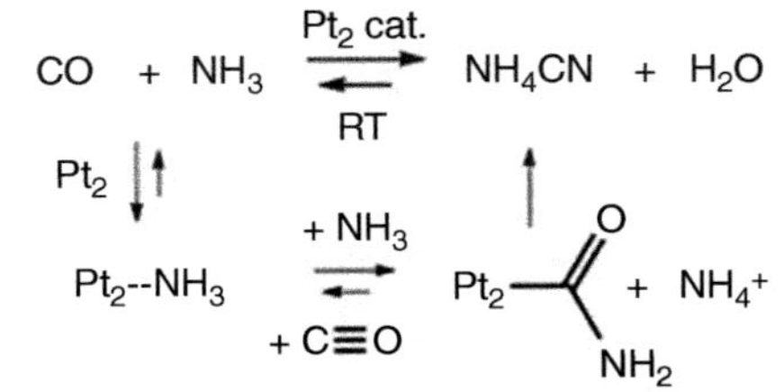

Entry	Catalyst	TON	TOF$_0$(s^{-1})
1	—	—	—
2	$PtCl_4$	<1	—
3	$PtCl_2$	—	—
4	$Pt(acac)_2$	<1	—
5	PtO_2	—	—
6	$Na_2PtCl_6.xH_2O$	<1	—
7	$H_2PtCl_6.xH_2O$	—	—
8	$Pt(NH_3)_4(NO_3)_2$	—	—
9	$Pt(NH_3)_4(OH)_2$	—	—
10	$Pt(cod)Me_2$	—	—
11	$PtMe_3Cp$	—	—
12	$Pt(NH_3)_4Cl_2.xH_2O$	—	—
13	$Pt(PhCN)_2Cl_2$	—	—
14	$Pt(cod)Cl_2$	<1	—
15	Pt/C	—	—
16	Pt^{2+}/NaY	—	—
17	Pt/NaY	—	—
18	**1**	—	—
19	**2**	25	0.17
20	**3**	53	0.35

(a)

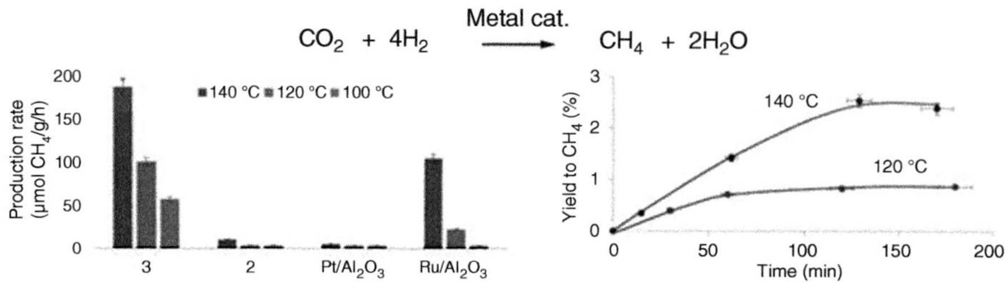

Catalyst	Temperature (°C)	Reaction pressure (bar)	Activation energy (kJ/mol)
Ni 58%	275–325	7–17	61
Ru/Al_2O_3 10 wt%	200–300	1	66
Ru/Al_2O_3 0.5 wt%	200–400	1	68
3	**100–140**	**4**	**35**

(b)

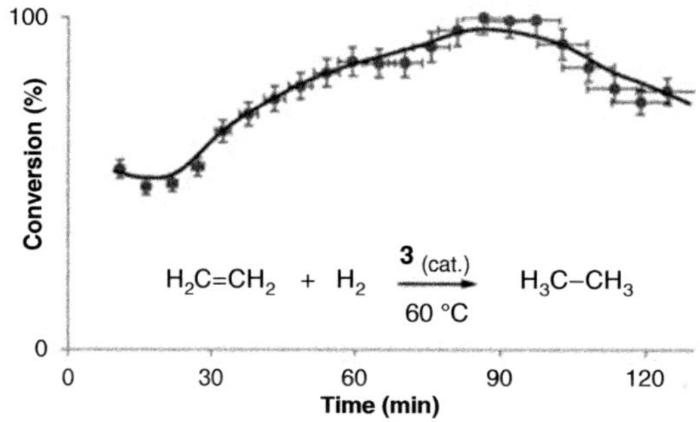

Catalyst	Temperature (°C)	Reaction pressure (bar)	Activation energy (kJ/mol)
Pd/SiO_2	80–120	1	36
Pd–Ag/SiO_2	80–120	1	32
3	40–80	5	32

(c)

Figure 15.6 (a) Pt-catalyzed synthesis of HCN at room temperature. The equation shows the reaction conditions: CO (4 bar, 1.5 mmol), NH_3 (2 bar, 0.75 mmol), and 0.0075 mmol Pt, and a plausible reaction mechanism. The table includes the set of Pt catalysts tested under these reaction conditions. (b) Pt-catalyzed methanation of CO_2: the equation shows the reaction conditions: 7 ml CO_2 (1 atm, 0.28 mmol), 7 ml N_2 (internal standard, 1 atm, 0.28 mmol), 7 ml H_2 (4 atm, 1.12 mmol), MOF catalyst 2 and 3 (8 wt%, 20 mg, 0.008 mmol metal) or M/Al_2O_3 (5 wt%, 32 mg, 0.008 mmol metal), 100–140 °C, six hours. (c) Pt-catalyzed hydrogenation of ethylene. The equation shows the reaction conditions: 2 ml/min C_2H_4, 6 ml/min H_2, atmospheric pressure, 60 °C, 50 mg **3** (0.010 mmol Pt). The table includes reaction values for reference literature catalysts and catalyst 3. Source: Mon et al. [33]. Reproduced with permission of Wiley-VCH.

metal-air batteries ($M_xO_2 \rightarrow M_x + O_2$), mainly due to the sluggish kinetics of OER. Therefore, the rational design of low-cost, highly active, and highly stable catalysts for this process is a critical issue in the societal pursuit of sustainable energy [60].

For understanding the thermal aggregation behaviors of metallic atoms during the precise synthesis of supported DACs and TACs, Wei et al. fabricated some DACs and TACs supported on a well-defined N-doped carbon layer by using some isostructural trinuclear-complex precursors (Fe_3/Fe_2Zn/Fe_2Co) [8]. Among these catalysts within the N-doped carbon layers, the active sites Fe_2 and Fe_2/Co_1 show excellent catalytic activity for OER with the overpotential of 355 and 350 mV at 10 mA/cm^2, which are lower than those of Fe_3 (447 mV) and Fe_1/Co_1 (480 mV) as well as IrO_2 (420 mV). Moreover, compared with other related materials, the active sites of Fe_2 and Fe_2/Co_1 have the highest turnover frequency, as shown in Figure 15.7a,b. The catalytic reaction was performed under different current densities (10–100 mA/cm^2) for over 200 hours for investigating the reaction durability. The high cycling stability (Figure 15.7c) strongly states the excellent robustness of the Fe–Fe site in the coordinating matrix under the operation condition of electrocatalytic OER. Moreover, the heterogeneous Co–Fe DAC, which is synthesized through the electrochemical activation of Co species atomically dispersed on N-doped carbon in Fe-containing alkaline electrolyte (as plotted in Figure 15.7d), exhibited higher catalytic performance [61]. For the Co–Fe DAC, the overpotential at 10 mA/cm^2 decreases to 309 mV and the Tafel slope reaches 37 mV/dec. This work introduces an easily produced, molecularly defined, and earth-abundant DACs for OER.

15.4.2.3 Oxygen Reduction Reaction (ORR)

As previously discussed, DFT calculation dealing with DACs and TACs is advantageous due to their simple theoretical models. Therefore, some DACs and TACs were designed and investigated for oxygen reduction reaction (ORR) through DFT calculation [9, 62]. Li et al. used DFT calculation to design DACs supported on 2D crystal C_2N and found that Co–Co on C_2N presented less than half the value of the reaction barrier of bulk Pt catalysts in ORR rate-limiting steps [62]. The adsorption energy of O_2 increases with the increase in transition metal (TM)-positive polarization charges following the order of Cu < Ni < Co < Cu–Cu < Ni–Ni < Co–Co. More interestingly, the O_2 dissociation barrier decreases almost linearly with the increase in negative polarization charge accumulated in O_2. It means that the catalytic performance of several TM atoms could be predicted by evaluating the induced polarization charges. Moreover, Zhuang et al. designed TAC of small Pt_3 clusters decorated core–shell structures (Co or Ni as the core, Pd as the shell) through DFT calculation [18]. The energy barriers of O–O dissociation on the Pt–Pt, Pt–Pd, and Pd–Pd sites of $Pt_{hcp\text{-}TA}$/1L-Pd/5L-Co and $Pt_{hcp\text{-}TA}$/1L-Pd/5L-Ni are 0.74, 0.79, 0.87, 0.59, 0.64, and 0.72 eV, respectively; all of which are smaller than that on Pt(111) (1.07 eV). This strategy not only reduces the catalyst cost but also tunes the catalytic performance by their adsorption sites and Pd thickness.

Fortunately, the TAC of platinum-trimer decorated cobalt-palladium core–shell nanocatalyst with a low platinum loading of only 2.4 wt% was synthesized via a developed wet chemical reduction method, following a controlled sequence of Co

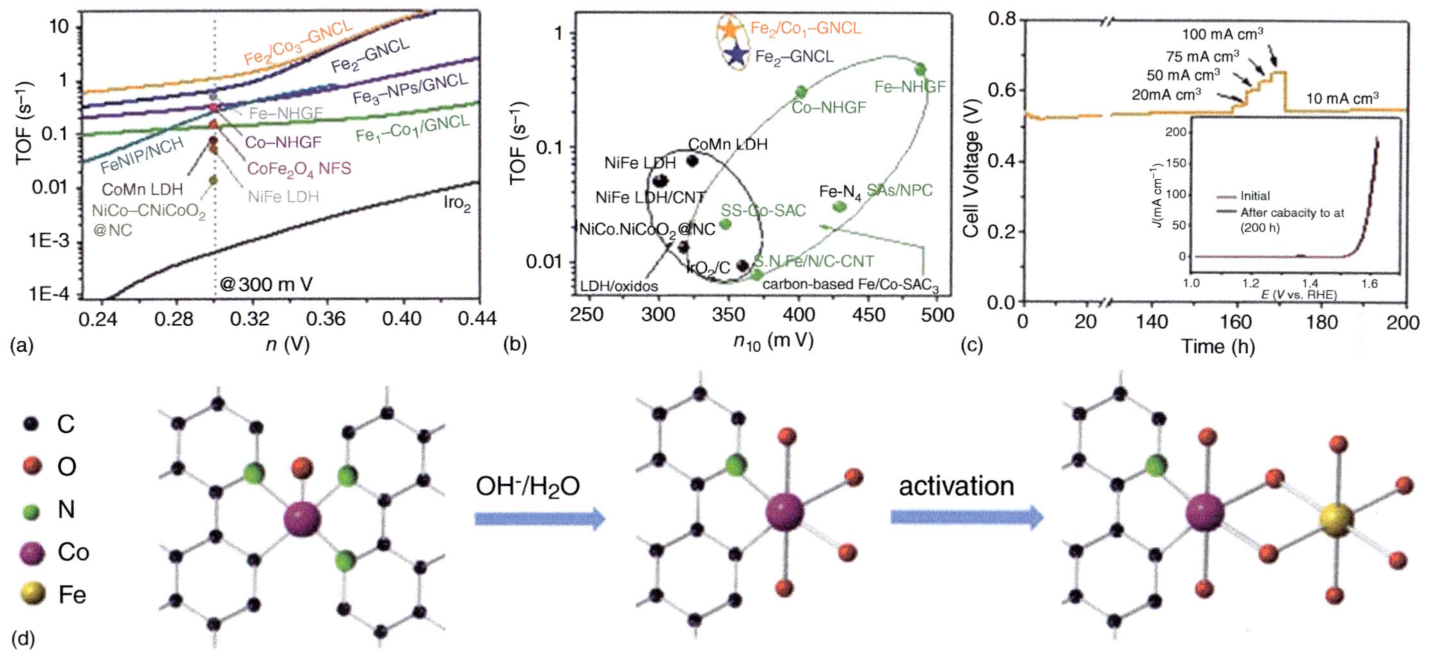

Figure 15.7 (a) Turnover frequency of the $Fe_{DA}Co_{SA}$/graphitized nitrogen-doped carbon layer (GNCL) catalyst and other recently reported OER Fe/Co-based catalysts. (b) Turnover frequency values of the $Fe_{DA}Co_{SA}$/GNCL catalyst with reported SACs (highlighted in green circle) and Fe/Co-based layered double hydroxide (LDH)/oxides (highlighted in black circle). (c) The time-dependent potential curve for $Fe_{DA}Co_{SA}$/GNCL at a constant current density of 10 mA/cm^2. Insets: LSV curves before and after the stability test. (d) Proposed model for the formation of Co–Fe DAC. After the precatalyst is immersed in the electrolyte solution, either two N or one N and one C were replaced by O. For simplicity, only the model where two N were replaced is shown. (a–c): Source: Wei et al. [8]. Reproduced with permission of Wiley-VCH; (d) Bai et al. [61]. Reproduced with permission of American Chemical Society.

nanoparticle formation, Pd shell growth, and Pt decoration [9]. This TAC shows a specific mass activity that is enhanced by a factor of 30.6 relative to a commercial platinum catalyst, which is attributed to the unique charge localization induced by platinum-trimer decoration, and is consistent with previous calculations [18]. Nitrogen doping makes graphene promising as support for supporting atomic catalysts [58, 63, 64]. A series of DACs supported on N-doped graphene have been synthesized for ORR, such as Co–Pt [28], Co_2 [24], Fe–Co [26], and Fe–Cu [65] DACs. All of them exhibit excellent catalytic performance for ORR, which are comparable to commercial Pt/C with lower cost. Moreover, carbon nanotubes were used for substituting graphene as support for Fe–Co DAC, which also indicated state-of-the-art ORR performance with an onset potential of 1.15 V and half-wave potential of 0.954 V [38]. More interestingly, when this DAC is employed as a cathode in a Zn–air battery, it exhibits high voltages of 1.31 and 1.23 V at discharge current densities of 20 and 50 mA/cm^2, respectively. In addition, the power density and the specific energy density reach 260 mW/cm^2 and 870 Wh/kg_{Zn}. The excellent performance of Fe–Co DAC is mainly due to the Fe–Co dual sites embedded in N-doped carbon nanotubes, which are beneficial for the activation of oxygen by weakening the O=O bonds.

15.4.2.4 CO_2 Reduction Reaction (CO_2RR)

Xu et al. reported a carbon-supported Cu atomic catalyst, mainly Cu_3 and Cu_4 active clusters identified by *operando* X-ray absorption spectroscopy, for direct electrochemical conversion of CO_2 to ethanol [23]. This catalyst, synthesized by an amalgamated Cu–Li method, achieves a single-product Faradaic efficiency (FE) of 91% at −0.7 V vs. the reversible hydrogen electrode (RHE) and onset potential as low as −0.4 V vs. RHE for electrocatalytic CO_2-to-ethanol conversion. Note that the FE of ethanol is highly sensitive to the initial dispersion of Cu atoms and decreased significantly when CuO and large Cu clusters become predominant species. In addition to the Cu_3 with high activity, the surface hydroxyl group plays a crucial role in modulating the interaction between the Cu_3 or Cu_4 and the carbon support, which determines the catalytic mechanism.

In addition to ethanol, the synthesis of other CO_2 electroreduction products (such as CO [66], CH_4 [67], and CH_3OH [14]) have also been reported on DACs. To figure out the distribution of the reaction products of CO_2 reduction on DACs, Li et al. screened a series of DACs supported on graphene by large-scale DFT calculation and microkinetics modeling [7]. As shown in the screening of CO_2 electroreduction intermediate (CO, COOH, CHO, H, O, and OH) adsorption with respect to CO and O species (Figure 15.8a,b), the COOH, CHO, and H intermediates exhibit a nonlinear adsorption strength correlation with that of CO on DACs, which is different from the traditional catalysts. This allows for the possibility of reducing the overpotentials of CO_2 protonation on most of the DACs. Interestingly, the adsorption energies of O and OH are linearly dependent on each other. According to this screening, some DACs have the potential to become catalyst candidates: Cu_2 DAC is the only catalyst with the product of CO, while CO on other DACs is more likely to undergo further reduction. Furthermore, Ni_2, MnCu, FeCu, and CoCu DACs might be active toward further CO reduction because OH or COOH adsorption is too strong, blocking the

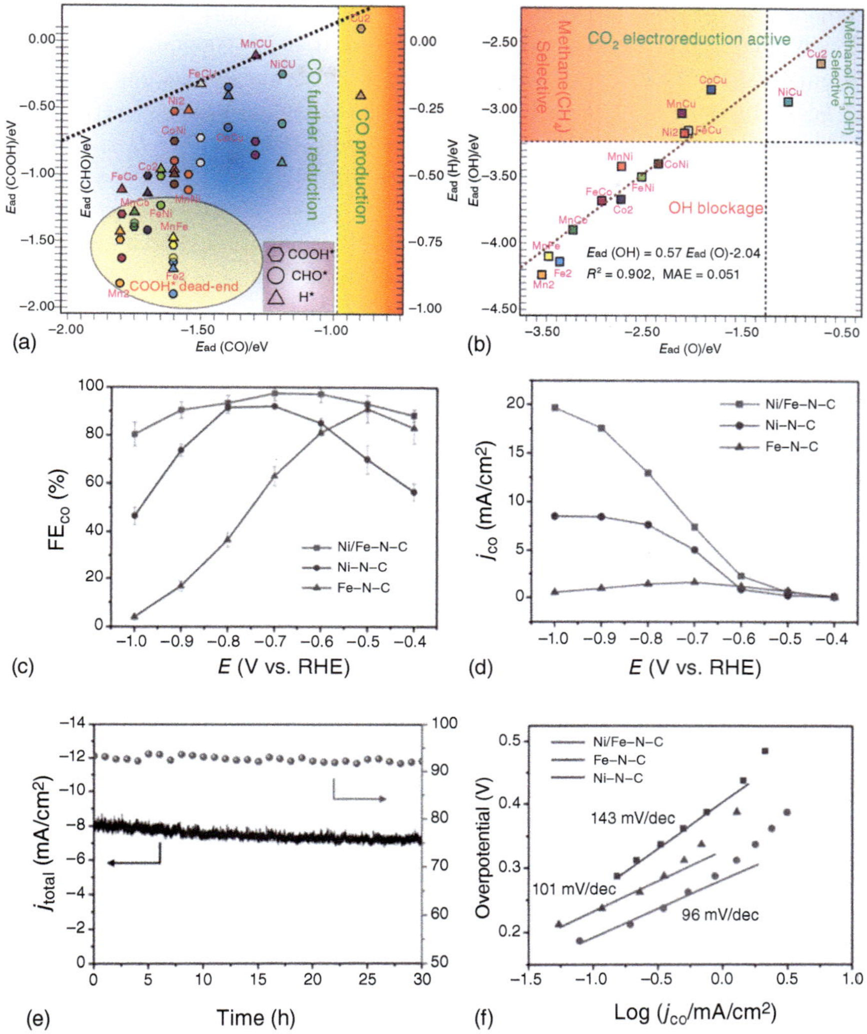

Figure 15.8 (a, b) Screening of CO_2 electroreduction intermediate (CO, COOH, CHO, H, O, and OH) adsorption with respect to (a) CO and (b) O species on different MN@2SV. The triangles, circles, and hexagons denote the adsorptions of H, CHO, and COOH, respectively, with respect to CO. (c) FE of CO (FE_{CO}) and (d) current density (j_{CO}) of Ni/Fe–N–C, Ni–N–C, and Fe–N–C at various applied potentials obtained in CO_2-saturated 0.5 m $KHCO_3$ solution. (e) Stability test for Ni/Fe–N–C at −0.7 V. (f) Tafel plots of the samples. Source: (a–b): Li et al. [7]. Reproduced with permission of American Chemical Society; (c–f) Ren et al. [25]. Reproduced with permission of Wiley-VCH.

active sites and becoming dead-end species for the whole reduction process. From the experimental point of view, Ren et al. synthesized isolated diatomic Ni–Fe sites as an efficient electrocatalyst for CO_2 reduction, with CO FE above 90% over a wide potential range from −0.5 to −0.9 V (98% at −0.7 V) [25]. The difference is that the support changes from graphene with double vacancies to N-doping graphene, which means the support plays a crucial role in the selectivity of CO_2 reduction reaction.

Compared with SACs of Fe–N–C and Ni–N–C, Ni/Fe–N–C shows a wider potential window and a higher FE, as shown in Figure 15.8c. At −0.7 V, Ni/Fe–N–C achieves a current density of 7.4 mA/cm^2, which is 1.5 and 4.6 times higher than that of Ni–N–C and Fe–N–C, respectively. With the increased applied potential, the performance gap becomes larger and larger (Figure 15.8d). The stability test for Ni/Fe–N–C at −0.7 V is shown in Figure 15.8e, where the outstanding durability for CO_2RR, maintaining 99% of the initial FE for CO production after 30 hours of continuous electrolysis, is found. The Tafel plots (Figure 15.8f) show that with the introduction of the second metal atoms into this DAC, the significantly lowered Tafel slope of Ni/Fe–N–C points out that the kinetics of this step is greatly enhanced. All these works about DACs and TAC indicate that they have great potential for electrocatalytic CO_2RR with high activity and selectivity.

15.4.2.5 Nitrogen Reduction Reaction (NRR)

Electrocatalytic NRR makes it possible to synthesize ammonia from abundant and inert nitrogen under ambient conditions [47, 68]. The key for this process is choosing the optimal catalyst with high activity. SACs have shown excellent catalytic activity and selectivity for electrocatalytic NRR [69–71]. With the advent of DACs and TACs, some researchers found that the catalytic activity of SACs is inferior to that of the corresponding DACs or TACs for NRR [36, 37, 72]. Chen et al. considered that DACs have the potential to replace SACs as an altar of atomic catalysts for NRR due to the stronger electron-donating ability of DACs [36]. Moreover, among TM_2–C_2N (M = Cr, Mn, Fe, Co, and Ni), the Mn_2–C_2N endows the highest catalytic activity with the lowest potential of −0.23 V vs. RHE, which is better than that most of the catalysts under ambient conditions. For exploring the intrinsic activity trend of DACs, Deng et al. systematically investigated NRR reactivity of homogeneous DACs formed by 20 metals supported on N-doped graphene via three reaction pathways [73]. A good correlation between the d-band center and the limiting potential of 20 DACs was found in the well-established volcano plot (Figure 15.9a), and the best candidate was Ru_2–N_6@G with limited potential of −0.36 V. The integral crystal orbital Hamilton population (ICOHP) by calculating the energy integral up to the highest occupied orbital (band) (below Fermi level), is shown in Figure 15.9b and has a linear correlation with the d-band center. More importantly, the selectivity of DACs for NRR compared with HER is shown in Figure 15.9c, where almost a half of the designed DACs have excellent NRR performance and inhibition of the competing HER, which is not the case for SACs and pure metal bulk surface catalysts. This important improvement opens another possibility of improving the FE of NRR. Guo et al. extended DACs from homogeneous to heterogeneous with a larger composition space for electrocatalytic NRR [74]. This work not only broadens the possibility of discovering more efficient DACs toward N_2 fixation but also provides a feasible strategy for exploring NRR electrocatalysts according to the descriptor-based design principle. For atomic catalysts, especially DACs and TACs, the key challenge is the relatively low loading of active materials, resulting in the low mass or volume activity of catalysts, which hinders their rapid development in the field of catalysis. Chen et al. designed a series of NRR TACs supported on the heterostructured supports

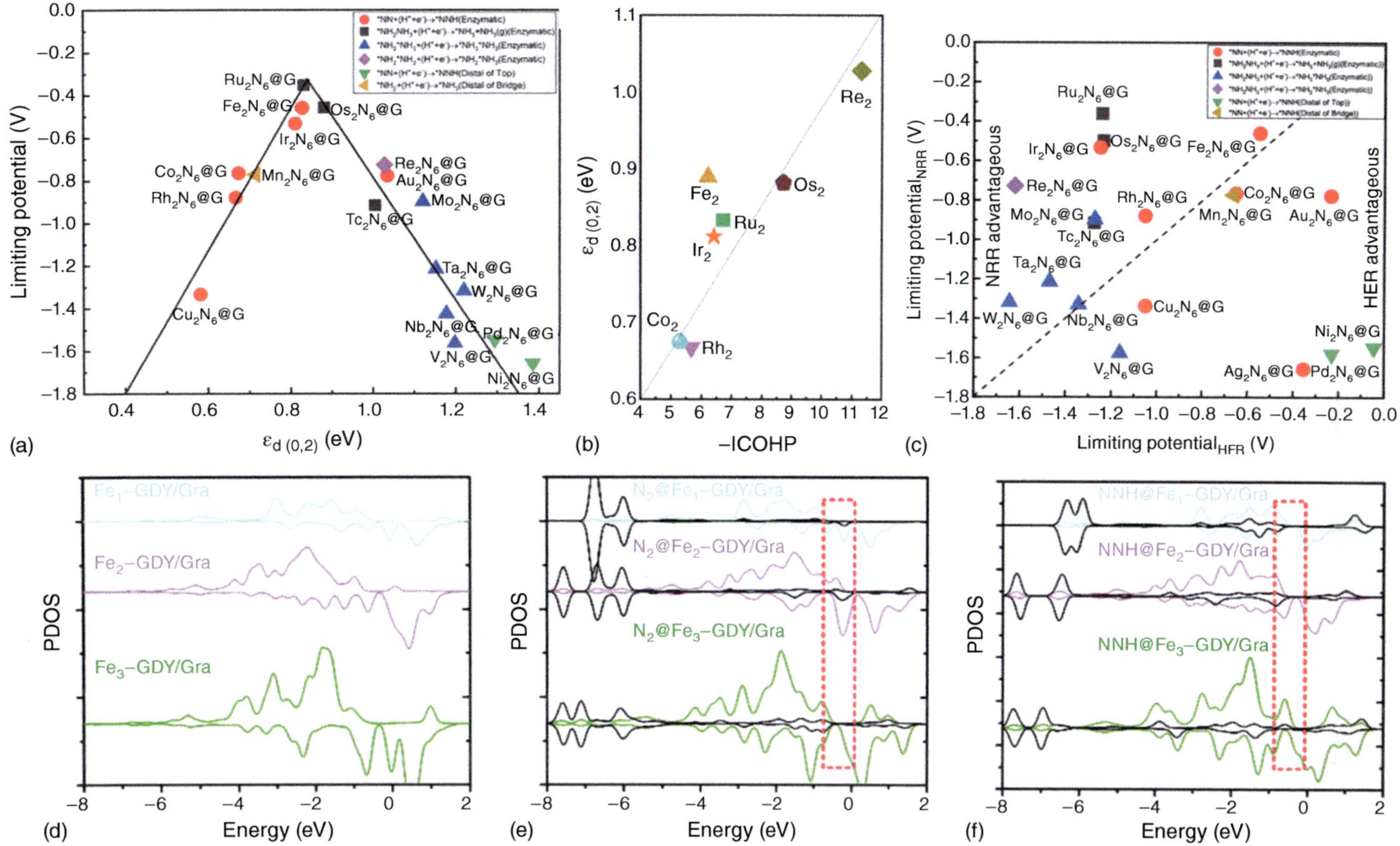

Figure 15.9 (a) Volcanic diagram of the limiting potential and d-band center of 0–2 eV $\varepsilon_{d(0,2)}$ above the Fermi level. (b) Illustration of the correlation between the integrated crystal orbital Hamilton population (ICOHP) and the d-band center $\varepsilon_{d(0,2)}$. Blue indicates bonding contributions, while cyan indicates antibonding contributions. (c) Comparation of the limiting potentials of 20 TM DACs for NRR and HER. NRR is dominant in the top left region, while HER is dominant in the bottom right region. (d–f) Partial density of states of N-p orbitals and Fe-d orbitals for the systems of Fe_x/GDY-Gra, N_2@Fe_x-GDY/Gra, and NNH@Fe_x-GDY/Gra ($x = 1$, 2, and 3), respectively. Azury, pink, green, and black lines indicate the Fe-d orbitals for the systems of Fe_{SA}/GDY-Gra, Fe_{DA}/GDY-Gra, Fe_{TA}/GDY-Gra, and N-p orbitals, respectively. The Fermi level is set to zero. Source: (a–c) Deng et al. [73]. Reproduced with permission of American Chemical Society; (d–f) Cheng et al. [37]. Reproduced with permission of Royal Society of Chemistry.

graphdiyne and graphene (GDY-Gra) with a high theoretical mass loading of active metal atoms [37]. The designed TACs show better catalytic activity for NRR than SACs and DACs due to the unique properties of M_3 (M = Mn, Fe, Co, and Ni) active sites, which provide more electrons for activating N_2, but also have a weak adsorption to more easily release the products. This was further verified by the analysis of PDOS, as shown in Figure 15.9d,f. The weak hybridization between the adsorbed N_2 and the Fe_3 near the Fermi level is found, denoting the weak adsorption energy of N_2 (−0.31 eV). However, after the first protonation of N_2, the d orbitals of Fe_3 have a strong hybridization with the p orbitals of the N atom. These two phenomena demonstrate that the active site of Fe_3 has strong adsorption of NNH* but weak adsorption of N_2*, both of which benefit for the first protonation of N_2*, which is usually the potential limiting step for NRR. Although experimental synthesis of DACs and TACs for NRR still presents great challenges, further study is needed due to their great potential based on these theoretical results.

15.4.3 Photocatalysis

Photocatalysis is considered as one of the most promising energy conversion methods as it allows converting an infinite amount of solar energy into chemical energy. However, the development of DAC and TAC photocatalysts is still in the burgeoning stage. Therefore, only limited examples of DACs and TACs have been explored for photocatalytic reactions.

Zhao et al. reported a facile photochemical method that produces catalytic centers consisting of two Ir metal cations, bridged by O and stably bound to a support for solar water oxidation [20]. Both DFT calculations and experimental data indicate that Ir_2 clusters are stabilized by support O, H_2O, and OH ligands, where the two Ir atoms in this DAC are bonded by five surface O atoms, and each Ir atom occupies a threefold hollow site on an OH-terminated α-Fe_2O_3 surface. The per-atom turnover frequency of Ir_{DA}/α-Fe_2O_3 at 1.23 V vs. RHE, was 2.6 and 5 times higher than the corresponding Ir SACs and Ir nanoparticles, respectively. Ir_{DA}/α-Fe_2O_3 shows a high catalytic activity, which attests to its functionality as an effective water-oxidation catalyst despite the fact that the organic ligands of the molecular precursor have been removed. Negligible degradation of the chronoamperometry data for the first 10 hours was observed from the stability characterization of Ir_{DA}/α-Fe_2O_3, implying that Ir_{DA}/α-Fe_2O_3 is stable under photoelectrochemical conditions. The O_3-binding site offered by the α-Fe_2O_3 support has been again proven critical for the formation and stability of Ir DACs.

Other important chemical reactions including photocatalytic CO_2RR, NRR, or even other energy-related chemical reactions have not been studied through using DACs or TACs. DACs and TACs have great potentials as photocatalysts for various chemical reactions and the diversity of supports for DACs and TACs provides more options for obtaining the most efficient light absorption and conversion, and further improving the photocatalytic efficiency. More importantly, the active centers of DACs and TACs are simple and definite, which will facilitate the exploration of catalytic mechanisms in photocatalysis.

15.5 Current Challenges and Future Outlook

Both experimental and theoretical works indicate that DACs and TACs endow great potential in the field of catalysis. However, opportunities are always accompanied by challenges. The challenges for DACs and TACs include the following four aspects: (i) high homogeneity of active centers. The active sites of DACs (TACs) often include not only double-atom (triple-atom) but also single atom or larger clusters, which affect the catalytic performance and puzzles the study of the catalytic mechanism. It means that homogeneous active centers of DACs or TACs are very important. Nevertheless, the cooperative effect of DACs and TACs with single atoms and nanoparticles should not be negligible. (ii) High loading of active atoms. The loading of active atoms for atomic catalysts is relatively low, especially for DACs and TACs, resulting in the specific low mass or volume activity of catalysts. The pursuit of high loading of active atoms has always been the more important challenge for atomic catalysts. (iii) Industrial production of DACs and TACs, which is inevitable for practical application. (iv) Rational design of DACs and TACs for specific catalytic reactions. In addition to the synthesis of DACs and TACs, rational design of DAC and TACs is also a challenge, which would avoid the traditional trial-and-error methods and help to develop the optimal catalyst for specific catalytic reactions.

To address these challenges for DACs and TACs, the research of DAC and TACs can be focused on the following three aspects: (i) Synthetic strategy. More appropriate supports and precursors for DAC and TACs should be exploited along with more advanced synthesis technology for achieving mass production of DACs and TACs with homogeneous active centers and high loading of active atoms. (ii) Structure–activity relationship. The catalytic activity of DACs and TACs is not only determined by the active centers but also affected by the support. A valid descriptor should be presented to reveal the interaction between active centers and the supports, and further get insights into the structure–activity relationship for DACs and TACs. A clear structure–activity relationship is a prerequisite for the rational design of such catalysts. (iii) New catalytic mechanisms. Due to the simple active centers of DACs and TACs, they are considered the most potential model catalysts, which provide new opportunities to explore new catalytic mechanisms. It means that the rapid development of DACs and TACs will further promote the study of catalysis.

15.6 Summary and Conclusions

In this chapter, published works on DACs and TACs are systematically summarized, which include synthesis routes from four aspects (increasing metal atom loading, grafting single-atom to single-atom, preselecting precursors, and preselecting supports), characterization techniques (microscopy, spectroscopy techniques, theoretical calculations, etc.), and applications in thermo, electro, and photocatalysis. Although the development of DACs and TACs is still in its early stages, both DACs and TACs have already shown great potentials in the field of catalysis.

DACs and TACs endow all advantages of SACs, such as high catalytic activity, high selectivity, and high utilization of active atoms, which build a bridge between homogeneous and heterogeneous catalysis. Moreover, DACs and TACs represent bridge and hollow active sites, respectively, as supplements to the top active site of SACs, which are more beneficial as model catalysts to explore the relationship between the catalytic performance and the structure of active sites. More importantly, heterogeneous DACs and TACs provide more designed spaces compared with SACs for the optimal catalysts. Note that some DACs and TACs have already been demonstrated to carry out better catalytic performance than SACs. Thus, DACs and TACs offer more opportunities for the development of the catalytic field. The rapid development of DACs and TACs not only raises the catalytic efficiency to a higher level, greatly promoting the development of a modern economy, but also provides a deep understanding of the catalytic mechanism, accelerating the formation of new theory in catalysis.

Acknowledgments

We wish to thank the National Natural Science Foundation of China (No. 51631004) and the JLU Science and Technology Innovative Research Team. CVS acknowledges financial support from the Nature Science and Engineer Research Council of Canada (NSERC), Hart Professorship, and the University of Toronto.

References

1 Yang, X.F., Wang, A., Qiao, B. et al. (2013). Single-atom catalysts: a new frontier in heterogeneous catalysis. *Accounts of Chemical Research* 46 (8): 1740–1748.
2 Cui, X., Li, W., Ryabchuk, P. et al. (2018). Bridging homogeneous and heterogeneous catalysis by heterogeneous single-metal-site catalysts. *Nature Catalysis* 1 (6): 385–397.
3 Chen, Z.W., Chen, L.X., Yang, C.C. et al. (2019). Atomic (single, double, and triple atoms) catalysis: frontiers, opportunities, and challenges. *Journal of Materials Chemistry A* 7 (8): 3492–3515.
4 Jiang, Y., Chen, Z., Han, Y. et al. (2018). Electron ptychography of 2D materials to deep sub-angstrom resolution. *Nature* 559 (7714): 343–349.
5 Zhang, J., Huang, Q., Wang, J. et al. (2020). Supported dual-atom catalysts: preparation, characterization, and potential applications. *Chinese Journal of Catalysis* 41 (5): 783–798.
6 He, Z., He, K., Robertson, A.W. et al. (2014). Atomic structure and dynamics of metal dopant pairs in graphene. *Nano Letters* 14 (7): 3766–3772.
7 Li, Y., Su, H., Chan, S.H. et al. (2015). CO_2 electroreduction performance of transition metal dimers supported on graphene: a theoretical study. *ACS Catalysis* 5 (11): 6658–6664.

8 Wei, Y.S., Sun, L., Wang, M. et al. (2020). Fabricating dual-atom iron catalysts for efficient oxygen evolution reaction: a heteroatom modulator approach. *Angewandte Chemie International Edition* 132 (37): 16147–16156.

9 Dai, S., Chou, J.P., Wang, K.W. et al. (2019). Platinum-trimer decorated cobalt-palladium core–shell nanocatalyst with promising performance for oxygen reduction reaction. *Nature Communications* 10: 440.

10 Chen, F., Jiang, X., Zhang, L. et al. (2018). Single-atom catalysis: bridging the homo- and heterogeneous catalysis. *Chinese Journal of Catalysis* 39 (5): 893–898.

11 Wang, Z., Liu, P., Han, J. et al. (2017). Engineering the internal surfaces of three-dimensional nanoporous catalysts by surfactant-modified dealloying. *Nature Communications* 8: 1066.

12 Cheng, N., Stambula, S., Wang, D. et al. (2016). Platinum single-atom and cluster catalysis of the hydrogen evolution reaction. *Nature Communications* 7: 13638.

13 Geng, Z., Liu, Y., Kong, X. et al. (2018). Achieving a record-high yield rate of 120.9 $\mu g_{NH3} mg_{cat.}^{-1} h^{-1}$ for N_2 electrochemical reduction over Ru single-atom catalysts. *Advanced Materials* 30 (40): 1803498.

14 Li, H., Wang, L., Dai, Y. et al. (2018). Synergetic interaction between neighbouring platinum monomers in CO_2 hydrogenation. *Nature Nanotechnology* 13 (5): 411–417.

15 Yan, H., Lin, Y., Wu, H. et al. (2017). Bottom-up precise synthesis of stable platinum dimers on graphene. *Nature Communications* 8: 1070.

16 Ji, S., Chen, Y., Fu, Q. et al. (2017). Confined pyrolysis within metal-organic frameworks to form uniform Ru_3 clusters for efficient oxidation of alcohols. *Journal of the American Chemical Society* 139 (29): 9795–9798.

17 Tian, S., Fu, Q., Chen, W. et al. (2018). Carbon nitride supported Fe_2 cluster catalysts with superior performance for alkene epoxidation. *Nature Communications* 9: 2353.

18 Zhuang, Y., Chou, J.P., Liu, P.Y. et al. (2018). Pt_3 clusters-decorated Co@Pd and Ni@Pd model core–shell catalyst design for the oxygen reduction reaction: a DFT study. *Journal of Materials Chemistry A* 6 (46): 23326–23335.

19 Nie, L., Mei, D., Xiong, H. et al. (2017). Activation of surface lattice oxygen in single-atom Pt/CeO_2 for low-temperature CO oxidation. *Science* 358 (6369): 1419–1423.

20 Zhao, Y., Yang, K.R., Wang, Z. et al. (2018). Stable iridium dinuclear heterogeneous catalysts supported on metal-oxide substrate for solar water oxidation. *Proceedings of National Academy of Sciences* 115 (12): 2902–2907.

21 Liu, J.C., Ma, X.L., Li, Y. et al. (2018). Heterogeneous Fe_3 single-cluster catalyst for ammonia synthesis via an associative mechanism. *Nature Communications* 9: 1610.

22 Lei, Y., Mehmood, F., Lee, S. et al. (2010). Increased silver activity for direct propylene epoxidation via subnanometer size effects. *Science* 328 (5975): 224–228.

23 Xu, H., Rebollar, D., He, H. et al. (2020). Highly selective electrocatalytic CO_2 reduction to ethanol by metallic clusters dynamically formed from atomically dispersed copper. *Nature Energy* 5: 623–632.

24 Xiao, M., Zhang, H., Chen, Y. et al. (2018). Identification of binuclear Co_2N_5 active sites for oxygen reduction reaction with more than one magnitude higher activity than single atom CoN_4 site. *Nano Energy* 46: 396–403.

25 Ren, W., Tan, X., Yang, W. et al. (2019). Isolated diatomic Ni-Fe metal-nitrogen sites for synergistic electroreduction of CO_2. *Angewandte Chemie International Edition* 58 (21): 6972–6976.

26 Wang, J., Huang, Z., Liu, W. et al. (2017). Design of N-coordinated dual-metal sites: a stable and active Pt-free catalyst for acidic oxygen reduction reaction. *Journal of the American Chemical Society* 139 (48): 17281–17284.

27 Zeng, X., Shui, J., Liu, X. et al. (2018). Single-atom to single-atom grafting of Pt_1 onto Fe-N_4 Center: Pt_1@Fe-N-C multifunctional electrocatalyst with significantly enhanced properties. *Advanced Energy Materials* 8 (1): 1701345.

28 Zhang, L., Fischer, J., Jia, Y. et al. (2018). Coordination of atomic Co-Pt coupling species at carbon defects as active sites for oxygen reduction reaction. *Journal of the American Chemical Society* 140 (34): 10757–10763.

29 Deng, J., Li, H., Xiao, J. et al. (2015). Triggering the electrocatalytic hydrogen evolution activity of the inert two-dimensional MoS_2 surface via single-atom metal doping. *Energy and Environmental Science* 8 (5): 1594–1601.

30 Liu, G., Robertson, A.W., Li, M.M. et al. (2017). MoS_2 monolayer catalyst doped with isolated Co atoms for the hydrodeoxygenation reaction. *Nature Chemistry* 9 (8): 810–816.

31 Fang, X., Shang, Q., Wang, Y. et al. (2018). Single Pt atoms confined into a metal-organic framework for efficient photocatalysis. *Advanced Materials* 30 (7): 1705112.

32 Zhao, C., Dai, X., Yao, T. et al. (2017). Ionic exchange of metal-organic frameworks to access single nickel sites for efficient electroreduction of CO_2. *Journal of the American Chemical Society* 139 (24): 8078–8081.

33 Mon, M., Rivero-Crespo, M.A., Ferrando-Soria, J. et al. (2018). Synthesis of densely packaged ultra-small $Pt(0)_2$ clusters within a thioether-functionalized MOF: catalytic activity in industrial reactions at low temperature. *Angewandte Chemie International Edition* 57 (21): 6186–6191.

34 Mahmood, J., Lee, E.K., Jung, M. et al. (2015). Nitrogenated holey two-dimensional structures. *Nature Communications* 6: 6486.

35 Mahmood, J., Li, F., Jung, S.M. et al. (2017). An efficient and pH-universal ruthenium-based catalyst for the hydrogen evolution reaction. *Nature Nanotechnology* 12 (5): 441–446.

36 Chen, Z.W., Yan, J.M., and Jiang, Q. (2019). Single or double: which is the altar of atomic catalysts for nitrogen reduction reaction? *Small Methods* 3 (6): 1800291.

37 Chen, Z.W., Chen, L., Jiang, M. et al. (2020). Triple atom catalyst with ultrahigh loading potential for nitrogen electrochemical reduction. *Journal of Materials Chemistry A* 8 (30): 15086–15093.

38 Wang, J., Liu, W., Luo, G. et al. (2018). Synergistic effect of well-defined dual sites boosting the oxygen reduction reaction. *Energy and Environmental Science* 11 (12): 3375–3379.

39 Chen, Z.W., Chen, L.X., Wen, Z. et al. (2019). Understanding electro-catalysis by using density functional theory. *Physical Chemistry Chemical Physics* 21 (43): 23782–23802.

40 Hou, C.C., Wang, H.F., Li, C. et al. (2020). From metal-organic frameworks to single/dual-atom and cluster metal catalysts for energy applications. *Energy & Environmental Science* 13 (10): 1658–1693.

41 Li, F. and Chen, Z. (2018). Cu dimer anchored on C_2N monolayer: low-cost and efficient Bi-atom catalyst for CO oxidation. *Nanoscale* 10 (33): 15696–15705.

42 Li, F., Liu, X., and Chen, Z. (2019). $1 + 1' > 2$: heteronuclear biatom catalyst outperforms its homonuclear counterparts for CO oxidation. *Small Methods* 3 (9): 1800480.

43 Ertl, G. (2008). Reactions at surfaces: from atoms to complexity (Nobel Lecture). *Angewandte Chemie International Edition* 47 (19): 3524–3535.

44 Lamb, K.E., Dolan, M.D., and Kennedy, D.F. (2019). Ammonia for hydrogen storage; A review of catalytic ammonia decomposition and hydrogen separation and purification. *International Journal of Hydrogen Energy* 44 (7): 3580–3593.

45 Chen, Z.W., Lang, X., and Jiang, Q. (2018). Discovery of cobweb-like MoC_6 and its application for nitrogen fixation. *Journal of Materials Chemistry A* 6 (20): 9623–9628.

46 Honkala, K., Hellman, A., Remediakis, I.N. et al. (2005). Ammonia synthesis from first-principles calculations. *Science* 307 (5709): 555–558.

47 Wang, Y., Shi, M.M., Bao, D. et al. (2019). Generating defect-rich bismuth for enhancing the rate of nitrogen electroreduction to ammonia. *Angewandte Chemie International Edition* 58 (28): 9464–9469.

48 Jalama, K. (2017). Carbon dioxide hydrogenation over nickel-, ruthenium-, and copper-based catalysts: review of kinetics and mechanism. *Catalysis Reviews* 59 (2): 95–164.

49 Han, L., Song, S., Liu, M. et al. (2020). Stable and efficient single-atom Zn catalyst for CO_2 reduction to CH_4. *Journal of the American Chemical Society* 142 (29): 12563–12567.

50 Back, S., Lim, J., Kim, N.Y. et al. (2017). Single-atom catalysts for CO_2 electroreduction with significant activity and selectivity improvements. *Chemical Science* 8 (2): 1090–1096.

51 Xing, D.H., Xu, C.Q., Wang, Y.G. et al. (2019). Heterogeneous single cluster catalysts for selective semi-hydrogenation of acetylene with graphdiyne-supported triatomic clusters. *The Journal of Physical Chemistry C* 123 (16): 10494–10500.

52 Sementa, L., Barcaro, G., and Fortunelli, A. (2015). Analogy between homogeneous and heterogeneous catalysis by subnanometer metal clusters: ethylene oxidation on Ag trimers supported on MgO (1 0 0). *Inorganica Chimica Acta* 431: 150–155.

53 Tran, C.H., Pham, L.T.T., Lee, Y. et al. (2019). Mechanistic insights on Zn(II)–Co(III) double metal cyanide-catalyzed ring-opening polymerization of epoxides. *Journal of Catalysis* 372: 86–102.

54 Sajith, P.K., Shiota, Y., and Yoshizawa, K. (2014). Role of acidic proton in the decomposition of NO over dimeric Cu(I) active sites in Cu-ZSM-5 catalyst: a QM/MM study. *ACS Catalysis* 4 (6): 2075–2085.

55 Chen, L.X., Chen, Z.W., Zhang, Y. et al. (2019). Insight into the excellent catalytic activity of (CoMo)S_2/graphene for hydrogen evolution reaction. *Applied Catalysis B: Environmental* 258: 118012.

56 Chen, L.X., Chen, Z.W., Wang, Y. et al. (2018). Design of dual-modified MoS_2 with nanoporous Ni and graphene as efficient catalysts for the hydrogen evolution reaction. *ACS Catalysis* 8 (9): 8107–8114.

57 Chen, W., Pei, J., He, C.T. et al. (2017). Rational design of single molybdenum atoms anchored on N-doped carbon for effective hydrogen evolution reaction. *Angewandte Chemie International Edition* 56 (50): 16086–16090.

58 Chen, W., Pei, J., He, C.T. et al. (2018). Single tungsten atoms supported on MOF-derived N-doped carbon for robust electrochemical hydrogen evolution. *Advanced Materials* 30 (30): 1800396.

59 Zhao, D., Shui, J.L., Grabstanowicz, L.R. et al. (2014). Highly efficient non-precious metal electrocatalysts prepared from one-pot synthesized zeolitic imidazolate frameworks. *Advanced Materials* 26 (7): 1093–1097.

60 Suntivich, J., May, K.J., Gasteiger, H.A. et al. (2011). A perovskite oxide optimized for oxygen evolution catalysis from molecular orbital principles. *Science* 334 (6061): 1383–1385.

61 Bai, L., Hsu, C.S., Alexander, D.T.L. et al. (2019). A cobalt-iron double-atom catalyst for the oxygen evolution reaction. *Journal of the American Chemical Society* 141 (36): 14190–14199.

62 Li, X., Zhong, W., Cui, P. et al. (2016). Design of efficient catalysts with double transition metal atoms on C_2N layer. *The Journal of Physical Chemistry Letters* 7 (9): 1750–1755.

63 Zhu, C., Shi, Q., Xu, B.Z. et al. (2018). Hierarchically porous M-N-C (M = Co and Fe) single-atom electrocatalysts with robust MN_x active moieties enable enhanced ORR performance. *Advanced Energy Materials* 8 (29): 1801956.

64 Li, X., Huang, X., Xi, S. et al. (2018). Single cobalt atoms anchored on porous N-doped graphene with dual reaction sites for efficient fenton-like catalysis. *Journal of the American Chemical Society* 140 (39): 12469–12475.

65 Du, C., Gao, Y., Chen, H. et al. (2020). Cu and Fe dual-atom nanozyme mimicking cytochrome c oxidase to boost the oxygen reduction reaction. *Journal of Materials Chemistry A* 8 (33): 16994–17001.

66 Li, Y., Chen, C., Cao, R. et al. (2020). Dual-atom Ag_2/graphene catalyst for efficient electroreduction of CO_2 to CO. *Applied Catalysis B: Environmental* 268: 118747.

67 Chen, S., Yuan, H., Morozov, S.I. et al. (2020). Design of a graphene nitrene two-dimensional catalyst heterostructure providing a well-defined site accommodating one to three metals, with application to CO_2 reduction electrocatalysis for the two-metal case. *The Journal of Physical Chemistry Letters* 11 (7): 2541–2549.

68 Bao, D., Zhang, Q., Meng, F.L. et al. (2016). Electrochemical reduction of N_2 under ambient conditions for artificial N_2 fixation and renewable energy storage using N_2/NH_3 cycle. *Advanced Materials* 29 (3): 1604799.

69 Wang, M., Liu, S., Qian, T. et al. (2019). Over 56.55% Faradaic efficiency of ambient ammonia synthesis enabled by positively shifting the reaction potential. *Nature Communications* 10: 341.

70 Liu, X., Jiao, Y., Zheng, Y. et al. (2019). Building up a picture of the electrocatalytic nitrogen reduction activity of transition metal single atom catalysts. *Journal of the American Chemical Society* 141 (24): 9664–9672.

71 Tao, H., Choi, C., Ding, L.X. et al. (2018). Nitrogen fixation by Ru single-atom electrocatalytic reduction. *Chem* 5 (1): 204–214.

72 Zhang, X., Chen, A., Zhang, Z. et al. (2018). Double-atom catalysts: transition metal dimer anchored C_2N monolayers as N_2 fixation electrocatalysts. *Journal of Materials Chemistry A* 6 (38): 18599–18604.

73 Deng, T., Cen, C., Shen, H. et al. (2020). Atom-pair catalysts supported by N-doped graphene for the nitrogen reduction reaction: d-band center-based descriptor. *The Journal of Physical Chemistry Letters* 11 (15): 6320–6329.

74 Guo, X., Gu, J., Lin, S. et al. (2020). Tackling the activity and selectivity challenges of electrocatalysts toward the nitrogen reduction reaction via atomically dispersed biatom catalysts. *Journal of the American Chemical Society* 142 (12): 5709–5721.

Index

Supported Metal Single Atom Catalysis, First Edition. Edited by Philippe Serp and Doan Pham Minh.
© 2022 WILEY-VCH GmbH. Published 2022 by WILEY-VCH GmbH.

i

t

z